Pot-Pollen in Stingless Bee Melittology

Patricia Vit • Silvia R.M. Pedro
David W. Roubik

Editors

Pot-Pollen in Stingless Bee Melittology

 Springer

Editors

Patricia Vit
Food Science Department
Faculty of Pharmacy and Bioanalysis
Universidad de Los Andes
Merida, Venezuela

Biomedical Sciences
School of Medical Sciences
The University of Sydney
Lidcombe, NSW, Australia

David W. Roubik
Smithsonian Tropical Research Institute
Balboa, Panama

Silvia R.M. Pedro
Biology Department
School of Philosophy, Sciences
and Literature
Universidade de São Paulo
Ribeirao Preto, Brazil

Additional material to this book can be downloaded from http://extras.springer.com

ISBN 978-3-319-61838-8   ISBN 978-3-319-61839-5   (eBook)
DOI 10.1007/978-3-319-61839-5

Library of Congress Control Number: 2017956133

Printed on acid-free paper

This Springer imprint is published by the registered company Springer International Publishing AG part of Springer Nature.
The registered company address is: Gewerbestrasse 11, 6330 Cham, Switzerland

To the evolution of pollen and stingless bees on planet Earth

Foreword

The world of stingless bees is fascinating. Do you ever wonder why these tiny creatures do not have stingers that most bee species have? How do they defend their colonies? Why are they so diverse and successful in nature?

By reading this book, you will gain more insights on the bees' complex behavior and the dynamics of plant-pollinator interaction. Based on previous studies, stingless bees use a wide range of floral resources. The palynological focus of the book contributes to the understanding of biodiversity, and the foraging and nesting behavior reported serves to guide management for pollination and the production of valuable hive products such as pot-honey, pot-pollen, cerumen, and propolis. Aware of the honey bee problems in some parts of the world, we need to nurture non-*Apis* species like stingless bees for biodiversity maintenance.

The authors' vast experience in stingless bee research makes this book a valuable source of information for students, researchers, scientists, beekeepers, and policy makers. Moreover, all contributors to this book have a demonstrated track record in publishing articles on stingless bee melittology based on original works that narrow the knowledge gaps in biology and management of the species. My personal interaction with one of the lead authors, Patricia Vit, made me a better stingless bee biologist.

Finally, I hope that this book will serve as a catalyst for valuing stingless bee research and development and formulating harmonized standards for its propagation and products.

Cleofas R. Cervancia

Cleofas R. Cervancia
Professor Emeritus, University of the Philippines Los Baños College Laguna, and President, Apimondia Regional Commission for Asia

Foreword

Stingless bee keeping is booming across their native range. The often docile, hard-working, cute, and fascinating bees are riding a wave of popularity. Perhaps social media and public education have sparked awareness and interest. Or maybe the time is just right for them. That said, now more than ever a good and general textbook is needed to provide a systematic overview of the basic natural history, characteristics of such bees, and the pot-pollen they process. Because of their wide distribution, from South and Central America to Africa, India, and across Southeast Asia to Australia, much local knowledge about the bees and their keeping requires communication to other parts of the world. Language barriers, the lack of publication outlets, and other factors often hinder that effort.

This book consolidates major research disciplines, from insect-plant mutualisms to general biodiversity, including landmarks in the world of little known associates of stingless honey bees. Further studies illuminate a millennium-old role in human culture. A thorough examination of the properties found in pot-pollen leads to a discussion of the marketing process for stingless bee products. In a single tome, this book guarantees the consumer the best and most authentic stingless bee experience, viewing with scientific rigor but in layman's terms some factors currently holding back the flow of stingless honey bee bounty to the world. The editors have invited and led an international group of researchers, students, and enthusiasts to produce the best introduction to the bees. No doubt readers will gather much new information from this book, but more importantly, they will be inspired to pursue future studies, thus keeping the pace of research and knowledge up to the demands of a well-deserved popularity enjoyed by the bees.

Claus Rasmussen, PhD
Assistant Professor, Aarhus University, Denmark

Acknowledgments

To the stingless bees and the stingless bee keepers of the world and for the pot-pollen and meliponiculture that have evolved.

Paula Calaça and Howard Junca kindly reviewed the list of microorganisms associated with bees. All botanical scientific names were checked and family names updated by Jorge Enrique Moreno Patiño in the lists of plants, according to the Missouri Botanical Garden (Tropics) database. Various authors updated plants listed in their chapters. The chapter reviewers provided timely and detailed comments and criticisms: Maria Lúcia Absy, Ingrid Aguilar, Rogerio Alves, Francisco de Assis R Santos, Ricardo Ayala, Ortrud Monika Barth, Fabio Borello, Peter Brook, Stephen Buchmann, Cleofas Cervancia, Carlos Echazarreta, Miguel Angel Fernandez Muiño, Tereza Giannini, Miroslava Kačániová, Vania Gonçalves Esteves, Víctor González, Tim Heard, Cynthia Maria de Lyra Neves, Camila Maia-Silva, Carmelina Flavia Massaro, Virginia Meléndez, Cristiano Menezes, Paula Morais, Patricia Nunes Silva, Sandra María Osés Gómez, Silvia RM Pedro, Mauro Ramalho, Elia Ramírez-Arriaga, Claus Rasmussen, Giancarlo Ricciardelli D'Albore, David W Roubik, Maria Teresa Sancho Ortiz, Cerilene Santiago Machado, Deborah Smith, Alfredo Usubillaga, Carlos Vergara, Rogel Villanueva Gutiérrez, Patricia Vit, Favio Vossler, Carmen Lucía Yurrita Obiols, and Julia Angélica Zavala Olalde. We also acknowledge our institutions and authorities for the academic support.

Contents

Contributors

Maria L. Absy Instituto Nacional de Pesquisas da Amazônia, Laboratório de Palinologia, Manaus, AM, Brazil

Natalia Acosta Quijano Posgraduate in Regional Rural Development, Universidad Autonoma Chapingo, Texcoco, CP, Mexico

Technical Coordination, Istaku Spinini A.C., Veracruz, CP, Mexico

Taofic Alabi University of Liege – Gembloux Agro-Bio Tech. Functional and Evolutionary Entomology, Gembloux, Belgium

Carlos Álvarez Asomoza Centro de Estudios Mayas, Instituto de Investigaciones Filológicas, Universidad Nacional Autónoma de México Ciudad Universitaria, México, DF, Mexico

Rogério Marcos de Oliveira Alves Instituto Federal de Educação, Ciência e Tecnologia Baiano, Catu, BA, Brazil

Yasmine Antonini Departamento de Biodiversidade Evolução e Meio Ambiente, ICEB, Universidade Federal de Ouro Preto, Ouro Preto, MG, Brazil

Ricardo Ayala Estación de Biología Chamela, Instituto de Biología, Universidad Nacional Autónoma de México (UNAM), San Patricio, Jalisco, Mexico

Wahizatul Afzan Azmi School of Marine and Environmental Sciences, Universiti Malaysia Terengganu, Terengganu, Malaysia

Ortrud Monika Barth Instituto Oswaldo Cruz, Fiocruz, Rio de Janeiro, Brazil

Laboratory of Palynology, Department of Geology, Institute of Geosciences, Federal University of Rio de Janeiro, Rio de Janeiro, Brazil

Esther Margarida Bastos Serviço de Recursos Vegetais e Opoterápicos, Diretoria de Pesquisa e Desenvolvimento, Fundação Ezequiel Dias (Funed), Rua Conde Pereira Carneiro, Belo Horizonte, MG, Brazil

Diego César Blettler Laboratory of Actuopalynology, CICyTTP-CONICET/ FCyT-UADER, Dr. Materi y España, Diamante, Entre Ríos, Argentina

Michael Burgett Department of Horticulture, Oregon State University, Corvallis, OR, USA

Paula Calaça Serviço de Recursos Vegetais e Opoterápicos, Diretoria de Pesquisa e Desenvolvimento, Fundação Ezequiel Dias (Funed), Rua Conde Pereira Carneiro, Belo Horizonte, MG, Brazil

Departamento de Botânica, ICB, CP 486, Universidade Federal de Minas Gerais, Belo Horizonte, MG, Brazil

Carlos Alfredo Lopes Carvalho Universidade Federal do Recôncavo da Bahia, Cruz das Almas, BA, Brazil

Bajaree Chuttong Science and Technology Research Institute, Chiang Mai University, Chiang Mai, Thailand

Wilberto Colli-Ucán El Colegio de la Frontera Sur, Unidad ChetumalAv. Centenario km 5.5, Chetumal, Quintana Roo, Mexico

Adriana Contreras-Oliva Colegio de Postgraduados Campus Córdoba, Carretera Córdoba-Veracruz km 348, Congregación Manuel León, municipio de Amatlán de los Reyes, Veracruz, CP, Mexico

María Magdalena Crosby-Galván Colegio de Postgraduados Campus Montecillo, Carretera México-Texcoco km 36.5, Montecillo, municipio de Texcoco, State of Mexico, Montecillo, CP, Mexico

Milagros Dalmazzo Entomología, Facultad de Humanidades y Ciencias, Universidad Nacional del Litoral, Santa Fe, Argentina

Hugo Delfín González Departamento de Zoología, Campus de Ciencias Biológicas y Agropecuarias, Universidad Autónoma de Yucatán, Mérida, Yucatan, Mexico

Juan Carlos Di Trani Instituto de Investigaciones Científicas y Servicios de Alta Tecnología (INDICASAT), Condominio Don Oscar, Ciudad de Panamá, República de Panamá

Edgard Cédric Fabre Anguilet University of Liege - Gembloux Agro-Bio Tech, Functional and Evolutionary Entomology, Gembloux, Belgium

Centre National de la Recherche Scientifique et Technologique (CENAREST), Institut de Recherches Agronomiques et Forestières (IRAF), Libreville, Gabon

Guillermina Andrea Fagúndez Laboratory of Actuopalynology, CICyTTP-CONICET / FCyT-UADER, Diamante, Entre Ríos, Argentina

Marcos G. Ferreira Instituto Nacional de Pesquisas da Amazônia, Laboratório de Palinologia, Manaus, AM, Brazil

Judith Figueroa-Ramírez Research Group AYNI, Bee Science and Technology, Veterinary Microbiology, Faculty of Veterinary Medicine and Zootechnics, Universidad Nacional de Colombia, Bogotá, DC, Colombia

Frédéric Francis University of Liege – Gembloux Agro-Bio Tech. Functional and Evolutionary Entomology, Gembloux, Belgium

Alex da Silva de Freitas Universidade Federal Fluminense, Niterói, Brazil

Viviana Frisone Museo di Archeologia e Scienze Naturali 'G. Zannato', Vicenza, Italy

Roziah Ghazi Agropolis Unisza, Universiti Sultan Zainal Abidin Kampus Besut, Besut, Terengganu, Malaysia

Luis M. Godínez-García Universidad Politécnica Mesoamericana, Departamento de Investigación y Desarrollo Sustentable, Tenosique, Tabasco, Mexico

Fernando Carlos Gómez-Merino Colegio de Postgraduados Campus Córdoba, Carretera Córdoba-Veracruz km 348, Congregación Manuel León, municipio de Amatlán de los Reyes, Veracruz, CP, Mexico

Éric Haubruge University of Liege – Gembloux Agro-Bio Tech, Functional and Evolutionary Entomology, Gembloux, Belgium

Caroline Hauxwell School of Earth, Environmental and Biological Sciences, Science and Engineering Faculty, Queensland University of Technology, Brisbane, Australia

Norma Ines Hilgert Instituto de Biología Subtropical (IBS), CONICET, Universidad Nacional de Misiones (UNaM), Facultad de Ciencias Forestales (UNaM), Puerto Iguazú, Misiones, Argentina

Michael Hrncir Departamento de Ciências Animais, Universidade Federal Rural do Semi-Árido, Mossoró, RN, Brazil

Vera Lucia Imperatriz-Fonseca Departamento de Ciências Animais, Universidade Federal Rural do Semi-Árido, Mossoró, RN, Brazil

Instituto Tecnológico Vale, Belém, PA, Brazil

Howard Junca RG Microbial Ecology, Division Ecogenomics and Holobionts, Microbiomas Foundation, Chia, Colombia

Robert Kajobe National Agricultural Research Organisation (NARO), Rwebitaba Zonal Agricultural Research and Development Institute (Rwebitaba ZARDI), Fort Portal, Uganda

Amanda Aparecida Castro Limão Departamento de Ciências Animais, Universidade Federal Rural do Semi-Árido, Mossoró, RN, Brazil

Luz Anel López-Garay Instituto Tecnológico Superior de Zongolica Campus Tequila, Carretera a la Compañía km 4, Tepetlitlanapa, municipio de Zongolica, Veracruz, CP, Mexico

Camila Maia-Silva Departamento de Ciências Animais, Universidade Federal Rural do Semi-Árido, Mossoró, RN, Brazil

Sol Martínez-Fortún Department of Natural Science, Universidad Técnica Particular de Loja, Loja, Ecuador

Carmelina Flavia Massaro School of Earth, Environmental and Biological Sciences, Science and Engineering Faculty, Queensland University of Technology, Brisbane, Australia

Virginia Meléndez Ramírez Departamento de Zoología, Campus de Ciencias Biológicas y Agropecuarias, Universidad Autónoma de Yucatán, Mérida, Yucatan, Mexico

Cristiano Menezes Embrapa Amazônia Oriental, Belém, PA, Brazil

Yolanda B. Moguel-Ordoñez Campo Experimental Mocochá, Instituto Nacional de Investigaciones Forestales, Agrícolas y Pecuarias, Km 25, antigua carretera Mérida-Motul, Mocochá, Yucatán, CP, Mexico

Paola Monserrate Research Group AYNI, Bee Science and Technology, Veterinary Microbiology, Faculty of Veterinary Medicine and Zootechnics, Universidad Nacional de Colombia, Bogotá, DC, Colombia

Jorge Enrique Moreno Patiño Smithsonian Tropical Research Institute, Calle Portobelo, Balboa, Ancon, Republic of Panama

Toussaint Ndong Bengone Centre National de la Recherche Scientifique et Technologique (CENAREST), Institut de Recherches Agronomiques et Forestières (IRAF), Libreville, Gabon

Bach Kim Nguyen University of Liege – Gembloux Agro-Bio Tech, Functional and Evolutionary Entomology, Gembloux, Belgium

Karina G. Pacheco-Palomo Universidad Autónoma de Campeche, Facultad de Ciencias Químico Biológicas, San Francisco de Campeche, Campeche, Mexico

Camila Raquel Paludo Faculdade de Ciências Farmacêuticas de Ribeirão Preto, Universidade de São Paulo, Ribeirão Preto, SP, Brazil

Silvia R.M. Pedro Departamento de Biologia, Faculdade de Filosofia, Ciências e Letras de Ribeirão Preto - FFCLRP, Universidade de São Paulo - USP, Av. Bandeirantes, Ribeirão Preto, SP, Brazil

María Peña-Vera Laboratory of Biotechnological and Molecular Analysis, Faculty of Pharmacy and Bioanalysis, Universidad de Los Andes, Mérida, Venezuela

Jaciara da Silva Pereira Departamento de Ciências Animais, Universidade Federal Rural do Semi-Árido, Mossoró, RN, Brazil

Elizabeth Pérez-Pérez Laboratory of Biotechnological and Molecular Analysis, Faculty of Pharmacy and Bioanalysis, Universidad de Los Andes, Mérida, Venezuela

Juan Antonio Pérez-Sato Colegio de Postgraduados Campus Córdoba, Carretera Córdoba-Veracruz km 348, Congregación Manuel León, municipio de Amatlán de los Reyes, Veracruz, Mexico

Rewat Phongphisutthinant Science and Technology Research Institute, Chiang Mai University, Chiang Mai, Thailand

Carla Portillo Research Group AYNI, Bee Science & Technology, Veterinary Microbiology, Faculty of Veterinary Medicine and Zootechnics, Universidad Nacional de Colombia, Bogotá, DC, Colombia

Mônica Tallarico Pupo Faculdade de Ciências Farmacêuticas de Ribeirão Preto, Universidade de São Paulo, Ribeirão Preto, SP, Brazil

Elia Ramírez-Arriaga Laboratorio de Paleopalinología, Departamento de Paleontología, Instituto de Geología, Universidad Nacional Autónoma de México, Ciudad Universitaria, Coyoacán, CP, Mexico

André R. Rech Universidade Federal dos Vales do Jequitinhonha e Mucuri, Curso de Licenciatura em Educação do Campo, Diamantina, MG, Brazil

Giancarlo Ricciardelli D'Albore Universitá degli Studi, Perugia, Italy

Carlos Augusto Rosa Departamento de Microbiologia, ICB, CP 486, Universidade Federal de Minas Gerais, Belo Horizonte, MG, Brazil

David W. Roubik Smithsonian Tropical Research Institute, Calle Portobelo, Balboa, Ancon, Republic of Panama

Carlos Ruíz Department of Natural Science, Universidad Técnica Particular de Loja, Loja, Ecuador
Animal Biology Department, Veterinary Department, Campus de Espinardo, Universidad de Murcia, Murcia, Spain

Bertha Santiago Apitherapy and Bioactivity, Food Science Department, Faculty of Pharmacy and Bioanalysis, Universidad de Los Andes, Mérida, Venezuela

Oswaldo Andrés Sánchez Research Group AYNI, Bee Science & Technology, Veterinary Microbiology, Faculty of Veterinary Medicine and Zootechnics, Universidad Nacional de Colombia, Bogotá, DC, Colombia

Cláudia Simeão Serviço de Recursos Vegetais e Opoterápicos, Diretoria de Pesquisa e Desenvolvimento, Fundação Ezequiel Dias (Funed), Rua Conde Pereira Carneiro, Belo Horizonte, MG, Brazil

Geni da Silva Sodré Universidade Federal do Recôncavo da Bahia, Cruz das Almas, BA, Brazil

Laura Elena Sotelo Santos Centro de Estudios Mayas, Instituto de Investigaciones Filológicas, Universidad Nacional Autónoma de México Ciudad Universitaria, México, Mexico

Korawan Sringarm Department of Animal and Aquatic Science, Faculty of Agriculture, Chiang Mai University, Chiang Mai, Thailand

Miguel Sulbarán-Mora Laboratory of Biotechnological and Molecular Analysis, Faculty of Pharmacy and Bioanalysis, Universidad de Los Andes, Mérida, Venezuela

Víctor Tibatá Research Group AYNI, Bee Science & Technology, Veterinary Microbiology, Faculty of Veterinary Medicine and Zootechnics, Universidad Nacional de Colombia, Bogotá, DC, Colombia

Libia Iris Trejo-Téllez Colegio de Postgraduados Campus Montecillo, Carretera México-Texcoco km 36.5, Montecillo, municipio de Texcoco, State of Mexico, Montecrillo, CP, Mexico

Margarito Tuz-Novelo El Colegio de la Frontera Sur, Unidad Chetumal, Chetumal, Quintana Roo, Mexico

Bart Vanderborgth Associação de Meliponicultores do Rio de Janeiro – AME-RIO, Rio de Janeiro, Brazil

Tommaso Francesco Villa School of Earth, Environmental and Biological Sciences, Science and Engineering Faculty, Queensland University of Technology, Brisbane, Australia

Rogel Villanueva-Gutiérrez El Colegio de la Frontera Sur, Unidad Chetumal, Chetumal, Quintana Roo, Mexico

Marcela Villegas-Plazas RG Microbial Ecology, Division Ecogenomics & Holobionts, Microbiomas Foundationl, Chia, Colombia

Patricia Vit Apitherapy and Bioactivity, Food Science Department, Faculty of Pharmacy and Bioanalysis, Universidad de Los Andes, Mérida, Venezuela

Cancer Research Group, Discipline of Biomedical Science, Cumberland Campus C42, The University of Sydney, Lidcombe, NSW, Australia

Favio Gerardo Vossler Laboratory of Actuopalynology, CICyTTP-CONICET/FCyT-UADER, Diamante, Entre Ríos, Argentina

Fernando Zamudio Instituto Multidisciplinario de Biología Vegetal (IMBIV), CONICET, UNC, Córdoba, Argentina

Raquel Zepeda García Moreno INANA, A.C. Calle Tajín 33. Colonia Campo Viejo, Coatepec, Veracruz, Mexico

Nur Syuhadah Zulqurnain School of Marine and Environmental Sciences, Universiti Malaysia Terengganu, Terengganu, Malaysia

Introduction

Our chosen term "melittology" means the biology of bees. We felt it is fitting, after the success of *pot-honey*, to highlight another mainstay of stingless bee biology, *pot-pollen*. Pollen, of course, is what the worker bees often bring on their hind leg corbicula, the "pollen basket," back to their colony. The corbicula's sole purpose is to assure efficient transport of material from the foraging environment to the nest. And pollen is the protein source, from which a developing bee larva draws part, but not all, of its sustenance. However, current studies show it is much more than a vegetable source of protein and the materia prima of the bee world. We explain why this is so and highlight the multiple applications and opportunities such information offers.

The behavior and interactions of bees in their environment guide their collection and elaboration of pollen as "bee bread." That surprisingly complex food can only be elaborated within the bee nest, where pot-pollen combines with microbes and nesting conditions. Pot-pollen has heretofore been a neglected topic. This book draws together the expanding knowledge and guideposts for keeping both the bees and ourselves healthy. With its diverse possibilities, not only for study, and biodiversity appreciation, but also for knowing how to run a small-scale business keeping native bees, we hope this is an example of applied science, at its best.

We observe that the influence of the massive volume of *Apis* or honeybees work in the literature seems to have a pernicious influence on the quality of our insight into the far larger and more varied group that comprise the stingless honey bees. We seek to remedy that problem with this book. Darwinian fitness—survival and reproduction—is brought into the forefront with articles from various research teams. In fundamental contrast to *Apis* or honey bees, meliponines place both honey and pollen in the new nest prior to colony reproduction (i.e., swarming). And stored food carries some of the microbial community established in the mother nest. That "inoculum", we maintain, is essential.

Why are nest microbes so important? As revealed in several chapters of this book, the organisms that live in pollen, and in honey, modify and improve the bee resource base. The food is given additional value in protein, vitamin, nutrient, and caloric content per unit volume. Yeasts (a fungus) in their stored food secrete enzymes that increase its nutritional value. Bacteria have additional roles, such as eliminating competitors that would destroy the food, or the bees themselves. Such economy and adaptation can only be described as outstanding examples of natural selection and Darwinian fitness. Honey made by

Meliponini is now well known to be more watery than the honey of *Apis*, and it even ferments, to a limited extent, when in storage. Yet, the microbes that cause spoilage or harm are eliminated by the mutualistic microbes that breed in the nests of Meliponini. At the same time, chemical processes modify the pH and increase acidity. Thus, pollen mixed with nectar, within pot-pollen, is "ripened," its food value increases, and the organic processes of digestion and conversion of the basic foodstuff used by coadapted microbial communities play an important role.

Meliponini reproduce very rarely, although floral resources are available during most or all of the year. Their new queens are routinely killed or rejected by the colony. But when a suitable new nesting site is situated nearby, and the food and worker population allow it, the colony can prepare the new nest site, populate it with a swarm and virgin queen from the mother colony, and enjoy a high probability of queen mating and colony survival.

Where does the "materia prima" of food and sustenance of Meliponini actually come from? We find that the bees are actively selective and constantly adjusting to the particular flowering plants in their environment. Without rigorously replicated studies on the pollen species used by bees, there is no conceivable conservation knowledge that will ultimately serve them, or their human shepherds. Further, without experiments on how the materia prima influence the performance of mutualist colony microbes, there is limited value in compiling taxonomic lists of pollen and nectar sources, or the microbes themselves. There is no evidence that a single microbial community suits all the different stingless bee species. We now recognize that meliponines are composed of several hundred species. We predict that the numbers of microbes, many still being taxonomically described and characterized by next-generation (NextGen) sequencing and other biological studies, will far surpass their number of host species. They are mutualists, as are the plants that provide food to meliponines and, often, receive their mutualistic benefit via pollination and plant propagation.

An introductory chapter explains how palynology is now being brought up to speed, by facing some of the shortcomings of past applications—such as describing "monofloral" honey from flowers incapable of providing nectar. *Solanum, Piper, Senna, Mimosa,* and *Acacia*—the list goes on—are frequently in honey, but only as added ingredients within the nest. They are incapable of providing sugary nectar, but their chemical contribution to honey cannot be denied. In addition, many tropical forest trees whose flowers are visited by the Meliponini are unisexual. Female individuals do not have pollen in their flowers; thus, bees that gather pollen from male flowers cannot be assumed to do any pollinating unless they also visit flowers of the female variety. Fully one-third of the tree genera known from the entire watershed of the Panama Canal are of this group. Pollen identified in a bee's nest, from any one of them, does not therefore constitute proof that the bee is a pollinator. The devil is in the details—only now being worked out by pot-pollen biologists, whose work we are pleased to share.

The Editors and Authors, April 2017

Part I

Pollen and the Evolution of Mutualism

Pot-Pollen as a Discipline: What Does It Include?

David W. Roubik and Jorge Enrique Moreno Patiño

1.1 Pot-Pollen and Palynology from an Ecological Point of View

Pollen mass is a key feature of a bee diet. Regardless of "functional ecology," in which names are meaningless, the plant species and variety in the bee nest provide fundamental information for biological study or analysis. The relative amounts of pollen can also indicate, in some cases, the type of resource, e.g., nectar, pollen, plant oil, or resin (Roubik 1989; Villanueva-Gutiérrez and Roubik 2016). Grain volumes are normally considered, irrespective of exine thickness or characteristics, or nonfood content of the pollen. Nonetheless, pollen importance may be approached by estimating pollen grain volume at a species or morphotype level. Using calculus, volume can be found for 3-D objects with an outline of a sphere, an ellipsoid, or a triangle (O'Rourke and Buchmann 1991). The simplest application is that of calculating sphere volume as approximately 4.2 × the cubed radius. Thus, a grain 7 microns in diameter has a volume one-thousandth that of a grain 70 microns in diameter. Alternatively, sample volume, irrespective of grain size or shape, can be measured and then, by using an internal spore count from *Lycopodium* (a fern), sold commercially in tablet form (Stockmarr 1971; O'Rourke and Buchmann 1991; Roubik and Moreno 2000, 2009, 2013); the relative volume of each pollen type can be found. Volumetric or mass estimates are much better, in any kind of analysis, than counting pollen grains that differ much in size. The use of a pollen trap on a hive, to remove and count or measure data from pollen pellets from returning foragers, is another technique applied in such quantitative study but primarily with honey bees (Richardson et al. 2015; Galimberti et al. 2014; Roubik et al. 1984; Roubik 1988, 1991; Villanueva-Gutiérrez and Roubik 2004; Roubik and Moreno 2013).

1.2 A Modern Synthesis of Bee Pollen and Pot-Pollen Study

Many authors discuss pollen grain number or volume in bee nests, with different terms and applications. However, all research predicated on the discovery of pollen importance to a given bee should consider the type or species of pollen and its relative abundance, if not nutritional value (Roulston and Cane 2000; Nicolson and Human 2013; Ziska et al. 2016) and interaction with

D.W. Roubik (✉)
Smithsonian Tropical Research Institute, Calle Portobelo, Balboa, Ancon, Republic of Panama
e-mail: roubikd@si.edu

J.E. Moreno Patiño
Smithsonian Tropical Research Institute, Calle Portobelo, Balboa, Ancon, Republic of Panama

© Springer International Publishing AG, part of Springer Nature 2018
P. Vit et al. (eds.), *Pot-Pollen in Stingless Bee Melittology*, DOI 10.1007/978-3-319-61839-5_1

microbes while in storage (e.g., Gilliam et al. 1985, 1990). There is also a prominent temporal aspect. Pollen is harvested and stored by meliponines in a relatively short time period (Roubik et al. 1986), but that pollen may exist in the nest and be used to feed brood and young adults for several months. We need to know, most of all, pollen species in bee nests by volume or weight—just as sugar concentration is measured in the floral nectar that bees imbibe (e.g., Kearns and Inouye 1993). Nectar can consist of roughly 10–50% sugar, and thus not all nectar is the same. This is obvious, and the concept also relates to pollen, as mentioned above. Further, to consider the pollen "type" a surrogate for pollen genus or species may be imprecise but helps to define plant importance to bees, in order to permit detailed ecological knowledge. In fact, for plants, bees, and other organisms, particularly in tropical regions, many species have no scientific species name—among arthropods, perhaps in 80% of all species (Hanson and Nishida 2016). Considering flowering plants, particularly those of the tropical areas shared by meliponine bees, Pitman and Jørgensen (2002) and Paton et al. (2008) give the impression that around 50% of plant species are either undescribed or endemic to a particular "hotspot" area—while at the global level, nearly half of all named species are in reality synonyms. That is, 41.6% of the plants we think are valid species in reality are not. Joppa et al. (2010) conclude that among the monocots, roughly 15% of species, mostly tropical, remain with no name. If the impetus for taxonomic and field study permits, we might someday know which plant species are present in the tropics, which are used by bees, and which are dependent on each other. In the meantime, a genus name is a "handle" and a practical method for trying to organize and understand natural diversity. A pollen type is a similar device. Herein, a concise introduction to pollen biology and palynology is provided. We eschew offering an esoteric term such as "meliponi-melittopalynology" or hyperspecialized jargon, because there is already abundant specialized terminology in our discipline.

Pollen taxonomy goes well beyond general morphological characteristics. We found >20% of plant species from Barro Colorado Island, in over 1269 species, 683 genera, and 90 plant families combined, had pollen of the "tricolporate–reticulate" type (Roubik and Moreno 1991). Further examination of acetolyzed pollen, nonetheless, permitted robust and accurate species or genus diagnosis from grain size, morphology, and shape—by including surface sculpture, exine, intine, and aperture morphology. We are thus able to identify pollen to species using light microscope techniques but sometimes cannot do this if we lack data on plant phenology, i.e., annual occurrence of flowering and bee foraging, considering a pollen type. There are too many possible pollen identities—just as in a sediment or other general, accumulated samples—without also knowing when and where bees find the pollen (Joosten and De Klerk 2002; De Klerk and Joosten 2007). However, as the predominant pollen types found in bee nests are identified to genus, further study using a combination of microscope or molecular techniques should establish which individual plant species are being used by a particular bee.

We suggest it is advisable to quantify or semi-quantify a pollen species in bee nests by morphology (Roubik and Moreno 1991; Colinvaux et al. 1999; and other pollen morphology guides) and, when possible, with molecular methods as well (Galimberti et al. 2014; Hawkins et al. 2015; Richardson et al. 2015), also considering the time and place of flowering and bee foraging. Some publications mention that it is costly to train palynologists and that molecular methods provide more rapid results. We agree but would like to point out that molecular labs are costly and an adequate reference collection of a flora, using "barcodes" or other gene markers, is an expensive undertaking. More importantly, in certain large plant families, like legumes, composites, or orchids, molecular approaches to identify species, or even genera, sometimes fail. Some of that failure is due to the new or uncatalogued species in any local tropical flora, as discussed above.

The stunning variety of natural forms found in tropical pollen on Barro Colorado Island, Panama, makes it a subject of renewed interest (Mander 2016), but practical application challenges remain. How can we come to know the resources used by pollinators, both seasonal and specialized non-colonial species, and among perennial tropical honey bees, including stingless honey bees and the honey bees, *Apis*, and their relative importance? In the following sections, we describe major tools for pollen identification, in order to illuminate the mainstays and applications of melittopalynology techniques.

1.3 Plant Reproduction

Reproduction involves investment in new individuals or offspring. Plant sexual reproduction mostly concerns *vascular* plants, Pteridophyta and Spermatophyta. The life cycle and reproduction of the pteridophytes are characterized by the complete absence of conspicuous flowers and fruit. In contrast, they make minute spores, often in copious quantities, and alternate generations between spore-producing plants (sporophytes) and gamete-producing plants (gametophytes). Among the vascular plants (Gymnospermae and Angiospermae), sexual reproduction involves two fundamental processes. Each parent contributes only half its genetic material (half the 2 N or diploid chromosome number) to a diploid offspring. Meiosis reduces the number of chromosomes by one-half, and fertilization restores the chromosomes to diploid number. Meiosis rearranges some genes by "crossover" which allows some DNA of each parent to occupy the same chromosome. Although the science of *palynology* of the vascular plants also includes the study of *spores* made by ferns, this book focuses on *pollen grains*. They are the male contribution to future plants and called *male gametophytes*. And pollen is the main source of protein used by many social and solitary arthropods, mainly bees, as well as vertebrates such as bats and birds.

Pollen grains are the fertilizing elements of flowering plants and in essence complete eukaryotic cells, specialized for transferring haploid male genetic material from the anther of one flower to the stigma of another (*cross-pollination*) or from the anther of a flower to the stigma of the same plant (*self-pollination*).

Each pollen grain contains a *vegetative* (non-reproductive) organelle—only a single organelle in most flowering plants—and a *generative* or reproductive organelle. After meiosis, within mother cells in the tapetum, an asymmetric *mitotic* division occurs (which produces two cells with the full 2 N chromosome number) resulting in a vegetative cell and a diminutive, generative cell (Edlund et al. 2004; also see Appendix). Subsequently, the generative cell undergoes a second mitosis to form the second sperm cell, required for *double fertilization*. The *tricellular* pollen completes that division before it is released from the anther, whereas *bicellular* pollen undergoes division later, within the elongating pollen tube.

1.4 Pollination

Wind, water, and animals are means by which pollen grains are transported from one flower to another, but insects as pollinators probably affect >80%, on a species-by-species basis (Richards 1997; Willmer 2011; Free 1993). The pollen from insect-pollinated plants is often sticky, to adhere to the bodies and hairs of insects, and flowering plants have developed a variety of strategies which involve floral morphology, traps, attractive odors, colors, and other stimuli to make pollinators visit flowers. The stigma or female receptive element is that which binds pollen and mediates tube migration into the style in order to reach and potentially fertilize an ovary.

Most pollen grains are metabolically quiescent and desiccated but still contain 15–35% water when released from anthers (Heslop-Harrison 1979). Water immediately surrounds grains that land on a wet stigma, but for those landing on a dry stigma, mobilization of the pollen coat occurs, leading to the mixing of lipids and proteins to form a point of contact on the stigma surface. Water, nutrients, and other small molecules are transported rapidly into the grain from the stigma exudate, and enzymes secreted by pollen may aid pollen tube entry into the stigma (Edlund et al. 2004; Heslop-Harrison 1977, 1979). Hydration transforms a pollen grain from a nonpolar cell to a highly polarized cell. Once the cell has established its internal polarity the pollen tube must breach the exine wall to pollinate an embryo. After crossing the exine, pollen tubes can only enter the style after transiting the stigmatic tube, and that tube constitutes a further barrier. When inappropriate pollen grains (e.g., genetic mismatches) reach this stage, further access is blocked by inhibitors of tube growth which originate in the stigma.

In principle, each vegetative cell of a pollen grain can develop a tube after reaching a receptive stigma. The much smaller generative cell or mitotic products—the two accessory haploid sperm cells—are enclosed inside the vegetative compartment and migrate forward along with the growing tube tip. All three (or two) travel through the style to an ovule inside the ovarium. Each *ovule* contains one *embryo sac*—a small *female gametophyte*—linked to one egg cell and one central cell. When the pollen tube reaches the embryo sac it bursts open and releases the two sperm cells, the actual sex cells. Double fertilization then may occur. One sperm cell fertilizes the egg cell, so that a *diploid zygote* arises, from which an embryo develops. The other sperm cell fertilizes the large central cell in the middle of the embryo sac and causes a series of divisions in the *endosperm*, which functions as a storage organ of nutrients for the developing seed. The dry ovule, with the mature embryo and the endosperm, composes the seed. The *fruit* is the ripened ovary that may hold one to many seeds. As a result of this process, a new diploid individual is formed, and the plant reproductive cycle is completed.

1.5 Pollen Biology and Palynology

Pollen exhibits great diversity, and each plant species produces its own kind of pollen (Punt et al. 2007). As already mentioned, female plants can detect the slight chemical differences between pollen from conspecific (of the same species) individuals, while they reject those of other species. The pollen–stigma interface can differ from species to species, probably the result of variability in the morphology and content of stigma exudates, exine layers, and pollen coats. At maturity, the pollen surface consists of an outer exine wall, itself multilayered, composed of a chemically resistant polymer of high molecular weight and containing fatty acids, known as *sporopollenin*. The inner intine of the pollen wall, also sometimes multilayered, is made primarily of cellulose and hemicellulose, as well as callose, which is always present. The exine surface often holds the *pollen kit*, considered the third stratum of a pollen grain. Pollen kit is composed of lipids, proteins, pigments, and aromatic compounds; it fills the sculptured cavities of the outer pollen exine. Sporopollenin protects the living vegetative and generative cell in the pollen grain from mechanical damage, chemical breakdown, or rapid desiccation and provides a shield against ultraviolet radiation.

The term *palynology*, proposed by Hyde and Williams (1944, 1945), refers to the study of pollen and spores, both structural and functional. Linnaeus provided, in *Systema Naturae* (1735), the first pollen descriptions. The discipline was established after that momentous event, taking advantage of all pollen features, which eventually led to "palyno-taxonomy," pollen databases, pollen reference collections, and applications in many scientific disciplines. Today, palynology includes both living and extinct organisms composed by materials highly resistant to degradation (sporopollenin, chitin, and related compounds), as well as microscopic organic and inorganic particles. All are recognized under the microscope and known as *palynomorphs*: mainly pollen, fern and fungal spores, diatoms, algae, foraminiferans, dinoflagellates, phytoliths, starch grains, acritarchs, chitinozoans, scolecodonts, charcoal, and arthropod cuticle (Traverse 2007; Fig. 1.1). Most of them occur in ancient sediments, and their assemblages are correlated with geological and biological past events, now principally applied to oil prospecting. Although

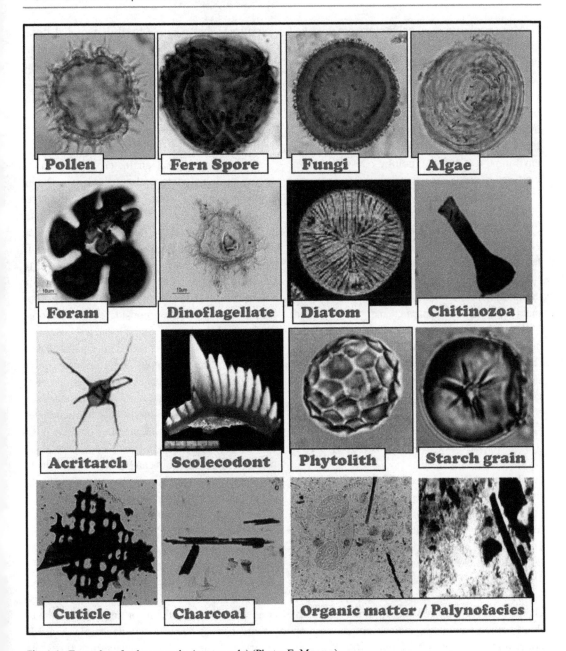

Fig. 1.1   Examples of palynomorphs (not to scale) (Photo: E. Moreno)

palynology was mainly developed by geologists to produce a chronology for different sediments, modern palynology continues to expand. Today, many specialized disciplines incorporate palynology. Pollen serves many different fields and can aid biological interpretation of complex natural phenomena.

Nowadays, ecologists and paleoecologists, paleontologists, archaeologists, taxonomic botanists, forensic technicians, medical allergists, pharmaceutic therapists, as well as practitioners of silviculture and agriculture, oil and mining industries, glaciology, and of course beekeeping and beekeepers use pollen analysis in research and exploration.

Many applications of palynology are thus well established. Paleoecology—studies using pollen and spores for reconstruction of past vegetation and paleoenvironments—remains central among palynological subdisciplines. In geology, palynology is widely used for oil prospecting, and in archeology/anthropology as an aid for interpretation of human activities and their effect on the environment. Copropalynology considers pollen content in feces. Aeropalynology involves the study of the pollen content in the atmosphere, its dispersal and distribution, and its sedimentation, e.g., in lakes or rivers. Iatropalynology considers effects of pollen (pollinosis/allergies) on human health. Pharmacopalynology emphasizes the use of pollen and spores in medicine and related disciplines. Two sub-categories that concern our main subject are separate and well defined. Melittopalynology and melissopalynology both aim at the study of pollen brought in the nest by returning bee foragers and in brood cells or storage pots (pot-pollen, as well as pot-honey, which contains some pollen).

1.6 Applied Pollen Taxonomy

Each pollen type has a "fingerprint," often unique, that can be recognized under the microscope. Several pollen morphology classics (Barth and Melhem 1988; Erdtman 1952; Faegri and Iversen 1989; Halbritter et al. 2008; Hesse et al. 2009; Kapps 2007; Kremp 1965; Moore et al. 1991; Pearsall 2000; Traverse 2007; Tschudy and Scott 1969) note the taxonomic utility of pollen characteristics. Electron microscope and light fluorescence techniques, versus optical microscopy, allow recognition of fine ultra-details in pollen grains, although their use is rather limited. Such different approaches have led to some confusion in pollen description and interpretation. The nomenclature used by palynologists has been, throughout its history, a theme of discussion. Through different glossaries of terms, authors explain their pollen descriptions. The current trend is the acceptance of a simple terminology, written in a language accessible to the palynologist community. The few published pol-

len atlases that compile local or regional floras are useful but have their own terms. Due to this fact, and the use of web searches—now routine for many fields of inquiry—coupled with the increase of online pollen databases, a consistent nomenclature is sought. Here, Punt et al.'s (2007) terms and definitions are used, which are followed in most palynological laboratories.

The morphological characteristics of pollen grains guide palynological nomenclature. Pollen grains are grouped mainly according to the presence/absence of apertures and their position, plus grain external ornamentation, size, and exine structure. Thus, because a pollen grain is a 3-D object, three known points can be plotted to describe the shape. Current taxonomy is based on this approach, contrary to those proposed groups where >3 elements are considered (e.g., Moore et al. 1991).

Pollen grain arrangement is another aspect of 3-D pollen morphology. The first descriptive level indicates individual or grouped grains, thus monad, tetrad, or polyad types. The second level, apertures or thinner regions of exine, includes ecto- or endoapertures, depending on their position in the different layers of exine, as explained below. The pollen grain can be inaperturate (lacking apertures) or aperturate, where two sides of cavities can be differentiated. Those cavities are of a pore type—more or less rounded hollows—or a colpus type, an elongate linear form. Their presence, or a combination of both types, as well as their position and number, allows further diagnosis. Exceptions occur when elongate apertures surround the grains and are spiral shaped, or are connected between them, or end at polar areas. As mentioned, because a pollen grain is a 3-D object, not more than three aperture types are considered. Therefore a grain with one circular aperture is named either monoporate or monocolpate. If two apertures are present the grain is diporate, dicolpate, or dicolporate, according to the observed combination. With three apertures, the procedure is the same, but if that number is exceeded, two general categories are used. Periaperturate applies if apertures are distributed at random, and stephano-aperturate, if concentrated at the grain's equatorial region. The terms stephano-porate, stephano-colpate, or stephano-colpo-rate then apply. The procedure is the same with peri-aperturate grains. Hypothetical combinations suggest definition of about 28 groups (Punt et al. 2007).

The next level concerns the sporoderm or cell wall. Divided into two units, it comprises exine and intine (Potonie 1934; Wodehouse, 1959) or ektexine and endexine (Faegri and Iversen 1950, 1989). Erdtman (1952) proposed division into sexine and nexine, both with a subdivision themselves (Colinvaux et al. 1999; Punt et al. 2007). Sexine consists of a closed lower layer (foot layer), generally covered by a highly variable species-specific stratification, found in an intermediate layer of columns (*baculae* or *columella*). Columns are generally connected at the outer side, so that they may form a roof (*tectum*) that contains linear or reticulate patterns or a surface bearing cavities (*lumina*). The presence or absence of a tectum means grains are tectate or intectate.

Sexine, nexine, or the entire exine generally exhibit breaks that give rise to colpi or pores. These are termed ectoaperturate if in sexine, endoaperturate if in nexine, or complete aperturate if incorporating all the exine. The pollen tube emerges from apertures. Exceptions—the inaperturate grains—require compatible (in the female tissue) enzymatic reactions to allow pollen tube growth. Many other features are associated with apertures, depending on outer or inner margin thickening or exine thickness. The margo, annulus, costae, operculum, and aperture form, including vestibulum, atrium, and drop shaped, among others, are features considered secondary but needed for identification.

When the tectum is smooth, pollen grains are *psilate*. The condition is found mainly in grains dispersed to the stigma by wind (anemophily). The majority of entomophilous plants have pollen ornamentation projecting beyond the tectum, called "positives," and/or toward the interior of the tectum, called "negatives" (Colinvaux et al. 1999). This trait is a good diagnostic. Grains with positive structures include spines (*echinate*), rounded elements (*verrucate*, *gemmate*), granules (*scabrate*), small posts (*baculate*, *clavate*),

and subtle variations thereof. Negatives derive from breaks, regular or irregular, that affect the tectum surface. The grains possess small, regular, or dispersed perforations (*punctate*), irregular and scattered perforation (*foveolate*), and irregular elongation (*fossulate*, *rugulate*, *striate*) or display a defined pattern (*reticulate*) or combinations thereof (details in Punt et al. 2007).

Aperture position and orientation, mainly colpus length, determine grain symmetry and polarity. The additional parameters also help to characterize pollen grains. Inaperturate grains lack an orientation. They are *apolar*. In aperturate grains, considering colpus length as a reference, it is possible to trace an imaginary line or polar axis that connects the two tips of the grain (*polar axis*). Likewise, an imaginary equatorial line that divides symmetrical grains into two hemispheres is called an *equatorial axis*. Thus, grains with similar hemispheres are *isopolar*, and if different, *heteropolar*. In the same way, when apertures are equidistant and the equator shows the same elements, grains are called *symmetric*, and if not, *asymmetric*. Special grains exhibit two different forms when the axis of rotation moves 90°, and then are considered *bilateral*. The relation between the polar (p) and equatorial (e) axes, in an idealized spheroidal pollen grain, has a value of 1. Grains exhibiting values >1 are *prolate* and <1 *oblate*. Shapes are given nine classes (Erdtman 1952).

Minor diagnostic features include aperture position, grain outline in polar and equatorial views, the polar index, and grain size. If the apertures are apical, in polar view, they are *angulo-aperturate*. If they are on the grain equator, they are *plan-aperturate*. The visual outlines of particular grains, both in polar and equatorial views, are variable, and such variation depends on grain hydration. Modern or fresh pollen grains are hygroscopic, and although the basic pattern tends to be constant, variation can make final description depend on analytical criteria. Basically, variation includes circular, angular, triangular, lobate, or square form for a particular taxon or species (Faegri and Iversen 1989).

Pollen grains vary in size from about 7 to >200 micrometers (Fig. 1.2) and may be classi-

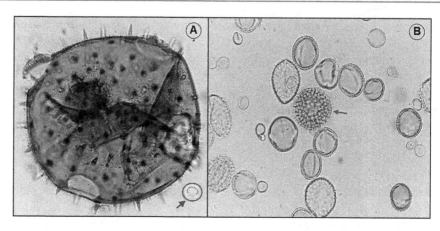

Fig. 1.2 Pollen grain size range and calibration spore among pollen. (**a**) *Cucurbita* (200 μm) versus *Piper* (7 μm, inside the *circle* indicated by *red arrow*). (**b**) Internal standard reference spore of fern, *Lycopodium* (indicated by *red arrow*), in general nest pollen sample from *Melipona panamica* (Photo: E. Moreno)

fied as very small (<10 μm), small (10–25 μm), medium (25–50 μm), large (50–100 μm), and very large (100 to >200 μm) (Barth and Melhem 1988; Erdtman 1952). Most pollen grains are 20–35 μm, mean 28 μm.

Certain mathematical relationships aid intra-specific diagnoses. Criteria are open to new proposals that allow separation of populations, although some do not serve this purpose, e.g., axis/pore ratio (Holst et al. 2007). The above-mentioned shape class (P/E index) is commonly applied (Kapps 2007). To calculate the number of pores in periporate pollen types, for example, among Malvaceae and Amaranthaceae, the pore length-to-grain diameter ratio is used (Kapps 2000). The *apocolpium index* or "polar area index" concerns the ratio of the distance between the apices of two colpi and the equatorial diameter of the grain; thus, if the index is low, then colpi are long, and vice versa.

There are many drawbacks to using simple, non-morphological, characteristics to identify a pollen type or species. Some of the pitfalls of molecular barcoding were already mentioned. The natural color of pollen grains is mostly white, cream, yellow, brown, or orange and used by beekeepers (sometimes erroneously) to recognize a pollen source. Pellets carried by the bees in their corbiculae can be caught with traps. However, once the pollen is washed its color changes. In addition, fresh pollen grains are not

easily recognized under the microscope due to their cellular content or pollen kit, which obscure morphological features. The cleaning and drying of pollen grains is essential, to permit seeing the pollen traits.

Pollen grains require cleaning and preparation with chemical and physical methods. Without them, light microscopy, especially in species-rich botanical areas, has limited value. Pollen is extremely resistant to strong acids (hydrofluoric acid [HF], hydrochloric acid [HCL], sulfuric acid [H_2SO_4], acetic acid [CH_3COOH]) but sensitive to oxidizing agents (nitric acid [HNO_3], potassium hydroxide [KOH], sodium hydroxide [NaOH], and *acetolysis* [a mixture of 1:9 of sulfuric acid and acetic anhydride]). Processing with such chemicals is applied mainly for recovering the pollen fraction of fossil samples. On the other hand, chemical treatment of modern pollen has been widely discussed by laboratory technicians who recommend avoiding its use as much as possible. Fortunately, pollen coming directly from flowers, honey, or bee nests needs minimal treatment (Brown 1960; Kearns and Inouye 1993). One washing with water, then application of acetic acid as a dehydrating agent, gentle application of acetolysis, and drying with alcohol are enough to obtain adequate preparations. We note that a number of tropical families, e.g., Zingiberaceae, Musaceae, Marantaceae, and Heliconiaceae,

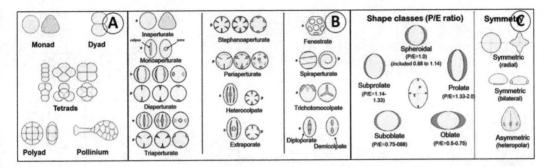

Fig. 1.3   Schematic representation of main pollen unit and grain types (after Punt et al. 2007). (**a**) pollen unit types. (**b**) pollen polarity, symmetry and shape classes. (**c**) grain type according to presence, number, and position of apertures

have large, thinly protected grains that are destroyed by a "gentle" acetolysis.

The addition of "markers" allows the calculation of the proportion of pollen species, their concentration, and ecological and floral characterization of honey or pollen loads (Roubik and Moreno 2013). Polystyrene microspheres, *Eucalyptus globulus* (Myrtaceae), and *Kochia scoparia* (Amaranthaceae) are some examples of pollen sample reference markers. The last two risk introducing contaminants mistaken for local flora. Tablets of fern spores (Fig. 1.2), *Lycopodium clavatum* (Lycopodiaceae), prove an adequate marker for pollen analysis (Stockmarr, 1971).

Although many palynologists recognize botanical taxa by identifying pollen grains, bibliographical resources (pollen atlases, worldwide web databases), formal pollen reference collections, and published descriptions (e.g., Roubik and Moreno 1991) continue to be critical. Some institutions maintain small and local collections that can be consulted (Moreno et al. 2014). Regional or local palynological floras, for routine analysis of pollen coming from bees, are scarce. Printed atlases published by Jones et al. (1995)

and Lewis et al. (1983) are useful for pollen of temperate regions. There are few works that emphasize the flora used by bees for tropical areas, where the number of plants greatly exceeds that of other regions of the world. Barth (1989), Silva et al. (2010), Giraldo et al. (2011), Martinez-Hernandez et al. (1993), Palacios-Chavez et al. (1991), Roubik and Moreno (1991), and Vit (2005) are examples of contributions, but none of them exceed 1200 species.

We offer an introductory glimpse of different things relevant to pot-pollen (Punt et al. 2007; Figs. 1.3, 1.4, 1.5 and 1.6). Readers are invited to consult the literature and explore further into practice and theory.

Appendix

http://www.yourarticlelibrary.com/biology/stamen-male-reproductive-organ-in-flowering-plants/11816/; http://www.vcbio.science.ru.nl/en/virtuallessons /cellcycle/postmeio/; https://www.boundless.com/biology/textbooks/boundless-biology-textbook/seed-plants-26/angiosperms-160/the-life-cycle-of-an-angiosperm-626-11847.

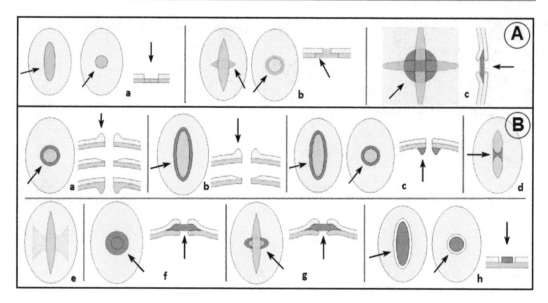

Fig. 1.4 Pollen aperture types and associated exine modifications, after Punt et al. (2007). (**a**) aperture types; endoaperture (*a*), ectoaperture (*b*), mesoaperture (*c*).(**b**) Schematic representation of exine modifications associated with apertures *a* annulus, *b* margo, *c* costae, *d* colpus constriction, *e* "H"-form colpus, *f* vestibulum, *g* fatigium, *h* operculum

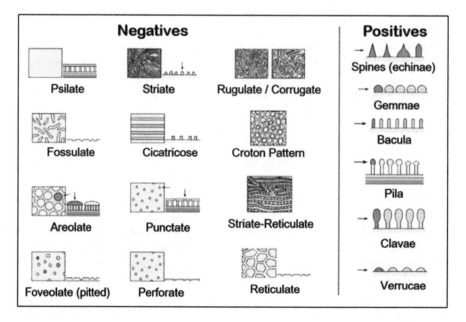

Fig. 1.5 Pollen exine ornamentation (after Punt et al. 2007, see also Colinvaux et al. 1999). Schematic representation of exine components and stratification. (**a**) exine components (**b**) exine stratification (after Punt et al. 2007)

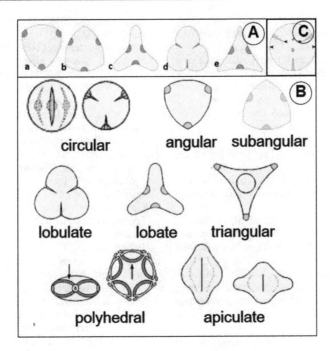

Fig. 1.6 Schematic representation of main pollen unit and grain types (after Punt et al. 2007). (**a**) pollen unit types. (**b**) pollen polarity, symmetry and shape classes. (**c**) grain type according to presence, number, and position of apertures *a* apolar, *e* equatorial, *p* polar views

References

Barth OM, Melhem TS. 1988. Glosario Ilustrado de palinología. Editora da UNICAMP; São Paulo, Brazil. 75 pp.

Barth OM. 1989. O polen no mel brasileiro. Grafica Luxor; Rio de Janeiro, Brazil. 151 pp.

Brown CA 1960. Palynological Techniques. [Published Privately]. Baton Rouge, USA. 188 pp.

Colinvaux P, de Oliveira PE, Moreno-Patiño JE. 1999. Amazon pollen manual and atlas –Manual e atlas palinologico da Amazônia. Harwood Academic Publishers; Amsterdam, Netherlands. 332 pp.

De Klerk P, Joosten, H. (2007). The difference between pollen types and plant taxa: a plea for clarity and scientific freedom. Eiszeitalter und Gegenwart, Quaternary Science Journal 56(3): 162–171 Doi 10.3285/eg.56.3.02

Edlund AF, Swanson R, Preuss D. 2004. Pollen and stigma structure and function: the role of diversity in pollination. The Plant Cell, Supplement 16: S84–S97.

Erdtman G. 1952. Pollen morphology and plant taxonomy. Angiosperms: an introduction to Palynology I. Almqvist and Wiksell: Stockholm, Sweden. 539 pp.

Erdtman G. 1957. Pollen and spore morphology/Plant taxonomy II. Gymnospermae, Pteridophyta, Bryophyta (Illustrations). Almqvist and Wiksell: Stockholm, Sweden. 151 pp.

Faegri K, Iversen J. 1950. Textbook of Modern Pollen Analysis. Munksgaard; Copenhagen, Denmark. 168 pp.

Faegri K, Iversen J. 1989. Textbook of pollen analysis. 4th Edn. Wiley; New York, USA. 328 pp.

Free JB. 1993. Insect Pollination of Crops. 2nd Ed. Academic Press; London, UK. 684 pp.

Galimberti A, De Mattia F, Ilaria B, Scaccabarozzi D, Sandionigi A, Barbuto M, Casiraghi M, Labra M. 2014. A DNA barcoding approach to characterize pollen collected by honeybees. PLoS ONE 9(10):1–13. e109363 doi: 10.1371/journal.pone.0109363

Gilliam M, Buchmann SL, Lorenz BJ, Roubik DW. 1985. Microbiology of the larval provisions of the stingless bee *Trigona hypogea*, an obligate necrophage. Biotropica 17: 28–325.

Gilliam M, Roubik DW, Lorenz BJ. 1990. Microorganisms associated with pollen, honey and brood provisions in the nest of a stingless bee, *Melipona fasciata*. Apidologie 21: 89–97.

Giraldo C, Rodriguez A, Chamorro FJ, Obregon D, Montoya P, Ramirez N, Solarte V, Nates-Parra G. 2011. Guia ilustrada de polen y plantas nativas visitadas por abejas. Fac. Ciencias, Univ. Nal. Colombia, Bogota, 230 pp.

Halbritter H, Weber M, Zetter R, Frosch-Radivo A, Buchner R, Hesse M. 2008. PalDat – Illustrated handbook on pollen terminology. Vienna, Austria. 61 pp.

Hanson PE, Nishida K. 2016. Insects and Other Arthropods of Tropical America. Cornell University Press, Zona Tropical; Ithaca, USA. 375 pp.

Hawkins J, De Vere N, Griffith A, Ford R, Allainguillaume J, Hegarty MJ, Baillie L, Adams-Groom B. 2015. Using DNA Metabarcoding to identify the floral composition of honey: a new tool for investigating honey bee foraging preferences. PLoS One. DOI: 10.1371/journal.pone.0134735

Heslop-Harrison J. 1979. An interpretation of the hydrodynamics of pollen. American Joiurnal of Botany 66: 737–743.Hesse

Heslop-Harrison Y. 1977. The pollen-stigma interaction: pollen tube penetration in crocus. Annals of Botany 41: 913–922.

Hesse M, Halbritter H, Zetter R, Weber M, Buchner R, Frosch-Radivo A, Ulrich S. 2009. Pollen terminology, an illustrated handbook. Springer, Wien, Austria. 261 pp.

Holst I, Moreno JE, Piperno D. 2007. Identification of teosinte, maize, and *Tripsacum* in Mesoamerica by using pollen, starch grains, and phytoliths. Proceedings of the National Academy of Science (USA) 104: 17608–17613.

Hyde HA, Williams DA. 1944. The right word. Pollen Anal. Circ. 8:6.

Hyde HA, Williams DA. 1945. Palynology. Nature 155: 264.

Jones GD, Bryant Jr. VM, Lieux MH, Jones SD, Lingren PD. 1995. Pollen of the Southeastern United States: with emphasis on melissopalynology and entomopalynology. American Association of Stratigraphic Palynologists (AASP), Contribuion Series No. 30, 104 pp.

Joosten H, De Klerk P. 2002. What's in a name? Some thoughts on pollen classification, identification, and nomenclature in Quaternary palynology. Rev. Palaeobot. Palynol. 122: 29–45.

Joppa LN, Roberts DL, Pimm SL. 2010. How many species of flowering plants are there?. Proceedings of the Royal Society of London B: Biological Sciences doi:10.1098/rspb.2010.1004

Kapps RO. 2007. Pollen and Spores. 2nd Edn. American Association of Stratigraphic Palynologists (AASP); USA, 279 pp.

Kearns CA, Inouye DW. 1993. Techniques for pollination biologists. University of Colorado Press; Niwot, USA. 583 pp.

Kremp GOW. 1965. Morphologic encyclopedia of palynology. Contribution No. 100, University of Arizona Press; Tucson, USA. 263 pp.

Lewis WH, Vinay P, Zenger PE. 1983. Airborne and Allergenic Pollen of North America. Johns Hopkins University Press; Baltimore, USA. 254 pp.

Linnaeus C. 1735. Systema Naturae, sive, Regna Tria Naturae systematice proposita per classes, ordines, genera, and species. Roterdamm Editore Theodorum Haak.

Mander L. 2016. A combinatorial approach to angiosperm pollen morphology. Proceedings of the Royal

Society B 283: 20162033. http://dx.doi.org/10.1098/rspb.2016.2033.

Martinez-Hernandez E, Cuadriello-Aguilar JI, Tellez-Valdez O, Ramirez-Arriaga E, Sosa-Najera MS, Melchor-Sanchez JEM, Medina-Camacho M, Lozano-Garcia MS. 1993 Atlas de las plantas y el polen utilizados por las cinco especies principales de abejas productoras de miel en la región del Tacana, Chiapas, Mexico. Inst. Geologia, Univ. Nal. Autonoma Mexico: Mexico DF, Mexico. 105 pp.

Moore PD, Webb JA, Collinson ME. 1991. Pollen Analysis. 2nd. Edn. Blackwell Scientific Publications; Oxford, UK. 217 pp.

Moreno JE, Vergara D, Jaramillo C. 2014. Las colecciones palinológicas del Instituto Smithsonian de Investigaciones Tropicales (STRI), Panamá. Boletin Asociación Latinoamericana de Paleobotanica y Palinologia 14: 207–222.

Nicolson SW, Human H. 2013. Chemical composition of the 'low quality' pollen of sunflower (*Helianthus annuus*, Asteraceae) Apidologie 44:144–152. DOI: 10.1007/s13592-012-0166-5

O'Rourke MK, Buchmann SL. 1991. Standardized analytical techniques for bee–collected pollen. Environmental Entomology 20: 507–513.

Palacios-Chavez R, Ludlow-Wiechers B, Villanueva R. 1991. Flora Palinológica de la Reserva de la Biosfera de Sian Ka'an, Quintana Roo, Mexico. Centro de Investigaciones de Quintana Roo (CIQRO); Chetumal, Mexico. 321 pp.

Paton AJ, Brummitt N, Govaerts R, Harman K, Hinchcliffe S, Allkin B, Lughadha EN. 2008. Towards Target 1 of the Global Strategy for Plant Conservation: a working list of all known plant species—progress and prospects. Taxon 57(2): 602–125.

Pearsall DM. 2000. Paleoethnobotany. A Handbook of Procedures. 2nd. Edn. Academic Press; Chicago, USA. 700 pp.

Pitman NC, Jørgensen PM. 2002. Estimating the size of the world's threatened flora. Science 298: 989.

Potonie R. 1934. Zur morphologie der fossilen pollen und spores. Arbeiten Ints. Paläobotanik Petrographie Brennsteine, 4: 5–24.

Punt W, Hoen, PP, Blackmore S, Nilsson S, Le Thomas A. 2007. Glossary of pollen and spore terminology. Review of Palaeobotany and Palynology 143: 1–825.

Richards AJ. 1997. Plant Breeding Systems. 2nd Edn. Chapman and Hall; London, UK. 529 pp.

Richardson RT, Chia-Hua L, Sponsler DB, Quijia JO, Goodell K, Johnson RM. 2015. Application of ITS2 metabarcoding to determine the provenance of pollen collected by honey bees in an agroecosystem. Applications in Plant Sciences 3: 1400066.; http://www.bioone.org/loi/apps.

Roubik DW. 1988. An overview of Africanized honey bee populations: reproduction, diet and competition. pp. 45–54. In: Needham GR, Page RE, Delfinado-Baker M, eds. Proccedings of International Conference on Africanized honey bees and bee mites: E. Horwood Ltd.; Chichester, UK. 100 pp.

Roubik DW. 1989. Ecology and natural history of tropical bees. Cambridge University Press; New York, USA. 514 pp.

Roubik DW. 1991. Aspects of Africanized honey bee ecology in tropical America. pp. 147-158 In: Spivak M, Breed MD, Fletcher DJC, eds. The African honey bee. Westview Press; Boulder, Colorado. 435 pp.

Roubik DW. 2009. Ecological impact on native bees by the invasive Africanized honey bee. Acta Biologica Colombiana 14: 115–124.

Roubik DW, Moreno JE. 1991. Pollen and Spores of Barro Colorado Island. Monographs in Systematic Botany from the Missouri Botanical Garden. St. Louis, USA. No. 36. 269 pp.

Roubik DW, Moreno JE. 2000. Generalization and specialization by stingless bees. pp. 112–118. In: Proceedings of the sixth international bee research conference on tropical bees. International Bee Research Association; Cardiff, UK. 190 pp.

Roubik DW, Moreno JE. 2009. *Trigona corvina*: An ecological study based on unusual nest structure and pollen analysis. Psyche. doi:10.1155/2009/268756

Roubik DW, Moreno JE. 2013. How to be a bee-botanist using pollen spectra. pp. 295–314. In Vit P, Pedro SRM, Roubik DW, eds. Pot–honey: a Legacy of Stingless Bees. Springer; New York, USA. 654 pp.

Roubik DW, Schmalzel RJ, Moreno JE. 1984. Estudio apibotanico de Panama: Cosecha y fuentes de polen y nectar usados por *Apis mellifera* y sus patrones estacionales y anuales. Technical Bulletin 24, OIRSA; San Salvador, El Salvador. 74 pp.

Roubik DW, Moreno JE, Vergara C, Wittmann D. 1986. Sporadic food competition with the African honey bee: projected impact on Neotropical social bees. Journal of Tropical Ecology 2: 97–111.

Roulston TH, Cane JH. 2000. Pollen nutritional content and digestibility for animals. Plant Systematics and Evolution 222: 187–209.

Silva, CI, Ortiz PL, Arista M, Bauermann SG, Evaldt ACP, Oliveira PE. 2010. Catalogo polinico: Palinologia aplicada em estudos de conservação de abelhas do gênero *Xylocopa* no Triângulo Mineiro. Editora da Universidade Federal de Uberlândia; Minas Gerais, Brazil. 153 pp.

Stockmarr J. 1971. Tablets with spores used in absolute pollen analysis. Pollen et Spores 13: 615–621.

Traverse A. 2007. Paleopalynology. 2nd Edn. Springer, New York, Vol. 28, 790 pp.

Tschudy RH, Scott RA. eds. 1969. Aspects of Palynology. Wiley-Interscience; New York, USA. 509 pp.

Villanueva-Gutiérrez R, Roubik DW. 2004. Why are African honey bees and not European bees invasive? Pollen diet determination in community experiments. Apidologie 35: 550–560.

Villanueva-Gutiérrez R, Roubik DW. 2016. More than protein? Bee-flower interactions and effects of disturbance regimes revealed by rare pollen in bee nests. Arthropod-Plant Interactions DOI 10.1007/s11829-015-9413-9

Vit P. 2005. Melissopalynology Venezuela. APIBA-CDCHT, Universidad de los Andes, Merida, Venezuela, 205 pp.

Willmer P. 2011. Pollination and floral ecology. Princeton University Press; Princeton, New Jersey, USA. 788 pp.

Wodehouse RP. 1959. Pollen Grains: their Structure, Identification, and Significance in Science and Medicine. Hafner Publishing Co.; New York, USA. 574 pp.

Ziska LH, Pettis JS, Edwards J, Hancock JE, Tomecek MB, Clark A, Dukes JS, Loladze I, Polley HW. 2016. Rising atmospheric CO_2 is reducing the protein content of a floral pollen source essential for North American bees. Proceedings of the Royal Society B 283: 20160414. http://dx.doi.org/10.1098/rspb.2016.0414

Are Stingless Bees a Broadly Polylectic Group? An Empirical Study of the Adjustments Required for an Improved Assessment of Pollen Diet in Bees

2

Favio Gerardo Vossler

2.1 Introduction

Bees—group Apiformes or Anthophila sensu Michener (2013)—are the richest and most widely spread group of pollinators around the world, and pollen and nectar are the most foraged food resources (Roubik 1989; Michener 2007, 2013). Evolutionary specialization to a nectar host has not been documented in bees, although morphological adaptations to extract nectar exist (Minckley and Roulston 2006). On the other hand, evolutionary specialization for pollen hosts in bees is well known and occurs at different levels, from monolecty to broad polylecty (Cane and Sipes 2006), and evolutionary shifts from specialist to generalist, and reversals, have been reported within a bee lineage (Larkin et al. 2008; Müller and Kuhlmann 2008; Michez et al. 2008, 2010). For these reasons, specialization in bees refers to pollen but not nectar resources, as even extreme specialists can take nectar from many floral species but forage pollen on a reduced number of host plants (Robertson 1925, 1926; Cane and Sipes 2006; Minckley and Roulston

2006; Müller and Kuhlmann 2008; Vossler 2013a, 2014). Pollen analysis, a helpful tool extensively used for studying bee diets (Cane and Sipes 2006; Müller and Kuhlmann 2008), was here applied to assess pollen specialization by Meliponini.

The tribe Meliponini is an eusocial bee group assumed to be extremely generalized in pollen foraging (i.e., *broadly polylectic*), as the numerous individuals of their perennial colonies with high reproductive rate are actively foraging on a large diversity of floral resources throughout the year (Roubik 1982, 1989; Ramalho et al. 2007; Michener 2007). However, *Melipona* species seem to be more specialized than the remaining genera according to palynological data (Kleinert-Giovannini and Imperatriz-Fonseca 1987; Ramalho et al. 1989, 2007; Vossler 2013b).

For the present entomopalynological study, the lexicon on pollen specialization of Cane and Sipes (2006) was applied to three species of the tribe Meliponini, a bee group with a well-known diet spectrum (*broad polylecty*). However, when comparing the relation between percentage values of foraged pollen types and the total number of available plant taxa, the three Meliponini were not assigned to *broad polylecty*. For this reason, the aims of the present study were to adjust these numerical values (steps are shown in Fig. 2.1). Because *broad polylecty* was rarely reached, an alternative classification was proposed for the generalist bees. Pollen diet studies including the

F.G. Vossler (✉)
Laboratorio de Actuopalinología, CICyTTP-CONICET/FCyT-UADER, Dr. Materi y España, E3105BWA, Diamante, Entre Ríos, Argentina
e-mail: favossler@yahoo.com.ar

© Springer International Publishing AG, part of Springer Nature 2018
P. Vit et al. (eds.), *Pot-Pollen in Stingless Bee Melittology*, DOI 10.1007/978-3-319-61839-5_2

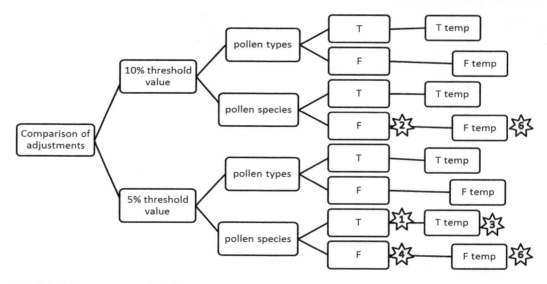

Fig. 2.1 Adjustments on calculations performed for foraged and available items. Stars under adjustment show the number of cases of *broad polylecty*. Only one case of *broad polylecty* was recorded for individual nests (Mo 4), three for nests grouped per season and two for all nests of a bee species. *Geotrigona* did not show any case

analysis of specialization for pollen plants are helpful to improve the understanding of natural history of bees and their use as pollinators of particular crops. The level of dispersion on floral resources during foraging activity (*polylecty*, *broad polylecty*, *degrees of polylecty*) could be considered of interest when selecting particular bee species for pollination (Vossler et al., Chap. 12).

2.2 Pollen Specialization Categories in Bees

The revised lexicon on pollen specialization by bees proposed by Cane and Sipes (2006) recognizes seven categories from the most specialized (*monolectic*) to the most generalized (*broad polylectic*) bees. These categories are assigned according to the number of pollen plant species, genera, and families foraged by bees for provisioning their brood cells and storing pollen. According to Cane and Sipes (2006), the term *monolecty* is used for bees relying on a single pollen plant species. In addition, the term is *narrow oligolecty* for bees collecting pollen from species belonging only to a genus or a small host clade, *oligolecty* when pollen collection is restricted to a few genera that belong to a same

clade (either tribe, subfamily, or family), *eclectic oligolecty* when it is restricted to few genera but from taxonomically disparate clades (different families), *mesolecty* when bee species gather pollen from more than four genera in two to three different clades (families or tribes of large families), and *polylecty* and *broad polylecty* when many species and genera from more than three families are foraged for pollen. Orders have not been considered as significantly important for analyzing pollen specialization in bees by Cane and Sipes (2006). However, the order is also useful for defining degree of generalization and resources used by a bee (Villanueva-Gutiérrez et al., Chap. 5). To distinguish between the categories *polylecty* and *broad polylecty*, both the number of pollen species foraged and the number of plant species available in the environment must be considered, as the former applies when <25% of available plant families or <10% of melittophilous species are foraged and the latter when these percentages are higher (Cane and Sipes 2006).

On the other hand, the lexicon proposed by Müller and Kuhlmann (2008) establishes seven categories that are in part different from those of Cane and Sipes (2006). They clarify the term *monolecty* by distinguishing it from a special case or narrow oligolecty, replace the term *oligolecty* by *broad oligolecty*, and add the concept of *polylecty* with *strong preference*, not considered by Cane and Sipes (2006). Because Müller and Kuhlmann (2008) did not differentiate the term *polylecty* from *broad polylecty*, their lexicon was not included in the present analysis.

Pollen analysis had successfully been used to identify specialization for pollen hosts in bees such as *Eremapis parvula* Ogloblin (Apidae, Exomalopsini) and *Calliopsis (Ceroliopoeum) laeta* (Vachal) (Andrenidae: Panurginae: Calliopsini) two solitary bees *narrowly oligolectic* on *Prosopis* (Vossler 2013a, 2014), *Melissodes* (Apidae: Eucerini) and *Megachile* (Megachilidae) species with different degrees of *oligolecty* (Cilla et al. 2012; Torretta et al. 2012, 2014), *Augochlora amphitrite* (Schrottky) (Halictidae: Augochlorini) a *broadly polylectic* species (Dalmazzo and Vossler 2015a, b), 25 North American *Diadasia* (Apidae: Emphorini), 60 palaearctic *Colletes* (Colletidae); and 28 nearctic *Andrena* (Andrenidae) from *narrow oligolecty* to *polylecty* with *strong preference* and *polylecty* s.s. (Sipes and Tepedino 2005; Müller and Kuhlmann 2008; Larkin et al. 2008).

The specialization categories proposed by Cane and Sipes (2006) and Müller and Kuhlmann (2008) are a powerful theoretical framework to identify the diverse degree of affinity of bees to their botanical counterparts. However, few studies have strictly analyzed bee diets following their recommendations. In most studies sampling is limited to a few sites, and therefore oligolecty is dubiously assigned. Moreover, both abundant and minor pollen types are generally considered at a same level of importance.

Palynological studies use the concept of "pollen type," which is the morphological entity equivalent to "plant taxon" in taxonomy (De Klerk and Joosten 2007). Based on exine morphology and palynologist knowledge, a pollen type can be attributed either to species, genus, family, part of a genus, etc., but it is not always possible to assign a pollen type to a plant taxon. For this reason, the classification in pollen types may not address actual specialization for certain pollen species, and therefore bee diet categorization can remain imprecise.

2.3 Pollen Analysis of Samples

For the present survey, nest pollen analysis was performed. Pollen type mean values were computed from all pot-pollen samples from each nest. A total of 171 slides were studied from 86 pot-pollen samples (i.e., pollen mass from 86 cerumen pots) from ten nests of *Tetragonisca fiebrigi*, ten pot-pollen samples from nine nests of *Melipona orbignyi*, and 75 pot-pollen samples from six nests of *Geotrigona argentina* from two types of xeric forests (Palosantal and Quebrachal) for the first species, while only Palosantal for the others. These types of forests are located no more than 250 km away from each other in the Chaco region of northern Argentina, both belonging to the western Chaco forests, sensu Prado (1993). The localities El Sauzalito (24°24′ S, 61°40′ W) and El Espinillo (25° 24′ S, 60° 27 W) are in Palosantal forest, while Miraflores (25°29′ S, 61°01′ W), Juan José Castelli (25°56′ S, 60°37′ W), and Villa Río Bermejito (25°37′ S, 60°15′ W) are in Quebrachal forest.

Pot-pollen samples were hydrated for up to 24 h and stirred with a magnetic stirrer for 10–15 min. A representative mixture of 5–10 mL was obtained and centrifuged at 472 × g for 5 min. Processing included acetolysis (Erdtman 1960). Pollen grain identification was carried out comparing nest pollen slides with pollen samples prepared on reference slides (see below). Under a Nikon Eclipse E200 light microscope, a total of 300–500 pollen grains per slide were counted.

As different pollen type assemblages are specific for each vegetation community, the number of pollen species ascribed to a particular pollen type varies regionally. For example, following the recommendations of Joosten and De Klerk (2002) and De Klerk and Joosten (2007) for pollen type nomenclature, in the Dry Chaco, the pollen type *Prosopis* was considered as the seven *Prosopis* species which have morphologically similar pollen grains: *Parkinsonia* included *Parkinsonia aculeata* and *P. praecox*, type *Schinopsis* as three species of two genera (*Schinopsis balansae*, *Schinopsis lorentzii*, and *Schinus fasciculata* var. *arenicola*), type *Maytenus* included *Maytenus vitis-idaea* and

Maytenus spinosa, *Ziziphus mistol* included only this species, type *Acacia aroma* included *A. aroma* and *A. curvifructa*—both having polyads composed of 16 colporate grains, and type *Acacia praecox* includes *A. praecox*, *A. furcatispina*, and *A. bonariensis*, which have polyads composed of 16 porate grains (see Table 2.1).

The reference pollen collection was made from flower buds of plant species collected in various localities from the Chaco province of Argentina. These plant specimens were pressed, dried, and identified by the author and deposited in the herbarium of the Museo de La Plata (LP); the herbarium of Museo Argentino de Ciencias Naturales "Bernardino Rivadavia" (BA), Buenos Aires; and the Herbarium Lorentz (DTE) of Diamante, Entre Ríos, Argentina. Plant taxa nomenclature follows Zuloaga et al. (2008). Bee specimens were collected from nests, identified by Arturo Roig-Alsina, and deposited in the Entomology Collection of the Museo Argentino de Ciencias Naturales "Bernardino Rivadavia", Buenos Aires, Argentina.

2.4　Adjustment Calculations to Assess Pollen Specialization Categories in Stingless Bees

To exceed the threshold of 10% for species or 25% for families established by Cane and Sipes (2006), the numerator value of the equation (the foraged items) must increase and/or the denominator value (the available taxa) decrease.

2.4.1　Modifying the Number of Foraged Resource Items: Threshold Values and Pollen Type Versus Pollen Species

To classify the three bee species into pollen specialization categories, only abundant pollen types were counted. To distinguish them from minor or potential contaminants, threshold values of at least 10% and 5% total counts were used.

Considering the number of foraged items was counted as the number of pollen types, i.e., palynological data as it is usually done (*italics* in Table 2.2a, for Tables 2.2b, c see supplemental material), 16 cases out of 38 here studied incorporated new pollen resources at 5% when comparing with 10% threshold value, and seven cases were assigned to a higher specialization category (e.g., nest Mo 9 was classified as *monolectic* at a 10% threshold value while as *eclectic oligolectic* at 5% (a higher specialization category), and nest Tf 5 as *eclectic oligolectic* at 10% while *polylectic* at 5%) (Table 2.2a, for Tables 2.2b, c see supplemental material). However, it had no impact for *Geotrigona* as no new resources were found at 5% (Table 2.2c, see supplemental material).

When *broad polylecty* was not detected using the number of pollen *types*, their conversion to pollen *species* (Table 2.1) was performed and a total of six cases was recorded both at 10% and 5% threshold values (Fig. 2.1). At 10%, 35 cases incorporated new pollen resources, but only 5 were assigned to a higher specialization category, while there were 36 and 5 at 5%, respectively (Table 2.2a, for Table 2.2b, c see supplemental material).

When working with pollen types, the 5% slightly improved the assessment of pollen specialization in some cases, but when working with pollen species, the two threshold values yield similar results.

2.4.2　Modifying the Number of Available Resources: Spatial and Temporal Adjustments

Cane and Sipes (2006) classified the bee tribe Meliponini as *broadly polylectic* and established that they utilize more than 10% of the pollen host species of the entire melittophilous flora at a site or more than 25% of the available plant families. For the present survey, these threshold values are 25 species and 21 families (the total number of plants available in the area (T) was 250 for species and 83 for families). However, when considering only plants available in the forest (F) the threshold

Table 2.1 Plant species ascribed to the pollen types occurring in >5% of representation in pot-pollen from nine nests of *Melipona orbignyi*, ten of *Tetragonisca fiebrigi*, and six of *Geotrigona argentina* in the Dry Chaco forest

Plant family	Plant species	Pollen type
Achatocarpaceae	*Achatocarpus praecox* Griseb.	*Achatocarpus praecox*
Anacardiaceae	*Schinopsis balansae* Engl.	Type *Schinopsis*
	Schinopsis lorentzii (Griseb.) Engl.	Type *Schinopsis*
	Schinus fasciculata (Griseb.) I.M. Johnst.	Type *Schinopsis*
Arecaceae	*Trithrinax schizophylla* Drude	*Trithrinax schizophylla*
Asteraceae	*Parthenium hysterophorus* L.	*Parthenium hysterophorus*
Bignoniaceae	*Handroanthus impetiginosus* (Mart. ex DC.) Mattos	Type *Tabebuia*
	Tabebuia nodosa (Griseb.) Griseb.	Type *Tabebuia*
Cannabaceae	*Celtis* spp.	*Celtis,*
Capparaceae	*Anisocapparis speciosa* (Griseb.) Cornejo & Iltis	*Capparicordis/Anisocapparis*, or *A. speciosa*
	Capparicordis tweediana (Eichler) Iltis & Cornejo	*Capparicordis/Sarcotoxicum* or *Capparicordis / Anisocapparis*
	Cynophalla retusa (Griseb.) Cornejo & Iltis	*Cynophalla retusa*
	Sarcotoxicum salicifolium (Griseb.) Cornejo & Iltis	*Capparicordis/Sarcotoxicum* or *S. salicifolium*
Celastraceae	*Maytenus vitis-idaea* Griseb.	Type *Maytenus*
	Maytenus spinosa Griseb.	Type *Maytenus*
Fabaceae, Caesalpinioideae	*Gleditsia amorphoides* (Griseb.) Taub.	*Gleditsia amorphoides*
	Parkinsonia aculeata L.	*Parkinsonia*
	Parkinsonia praecox (Ruiz & Pav. ex Hook.) Hawkins	*Parkinsonia*
Fabaceae, Mimosoideae	*Acacia bonariensis* Gillies ex Hook. & Arn.	Type *Acacia praecox*
	Acacia furcatispina Burkart	Type *Acacia praecox*
	Acacia praecox Griseb.	Type *Acacia praecox*
	Albizia inundata (Mart.) Barneby & J.W. Grimes	*Albizia inundata*
	Prosopis alba Griseb.	*Prosopis*
	Prosopis elata (Burkart) Burkart	*Prosopis*
	Prosopis kuntzei Harms	*Prosopis*
	Prosopis nigra (Griseb.) Hieron.	*Prosopis*
	Prosopis ruscifolia Griseb.	*Prosopis*
	Prosopis vinalillo Stuck.	*Prosopis*
	Prosopis (hybrids)	*Prosopis*
Malpighiaceae	*Mascagnia brevifolia* Griseb.	*Mascagnia brevifolia*
Rhamnaceae	*Ziziphus mistol* Griseb.	*Ziziphus mistol*
Sapotaceae	*Sideroxylon obtusifolium* (Roem. & Schult.) T.D. Penn.	*Sideroxylon obtusifolium*
Simaroubaceae	*Castela coccinea* Griseb.	*Castela coccinea*
Solanaceae	*Solanum* spp.	*Solanum*
Ximeniaceae	*Ximenia americana* L.	*Ximenia americana*
Zygophyllaceae	*Bulnesia sarmientoi* Lorentz ex Griseb.	*Bulnesia sarmientoi*

Table 2.2a *Melipona orbignyi*

Nests	Mo 1	Mo 4	Mo 5
Important pollen hosts (threshold value of 10%)	Fabaceae (*Prosopis*)	Fabaceae (*Prosopis*)	Fabaceae (*Prosopis*)
	Solanaceae (*Solanum*)	Solanaceae (*Solanum*)	Sapotaceae (*Sideroxylon obtusifolium*)
		Capparaceae (*Cynophalla retusa*)	
		Ximeniaceae (*Ximenia americana*)	
		Anacardiaceae (Type *Schinopsis*)	
Number of families (and pollen types) and category assigned			
Using the number of pollen types:	2 (2)	5 (5)	2 (2)
	Eclectic oligolecty	Polylecty [T: 6% (2%)]	Eclectic oligolecty
		Polylecty [T temp: 7.2% (2.6%)]	
		Polylecty [F: 7.5% (2.8%)]	
		Polylecty [F temp: 9.4% (3.6%)]	
Using the number of pollen species:	2 (11)	5 (16)	2 (8)
	Eclectic oligolecty	Polylecty [T: 6% (6.4%)]	Eclectic oligolecty
		Polylecty [T temp: 7.2% (8.3%)]	
		Polylecty [F: 7.5% (8.9%)]	
		Broad polylecty [F temp: 9.4% (11.6%)]	
Important pollen hosts (threshold value of 5%) (Only the additional resources are mentioned)	No modifications at 5%	No modifications at 5%	Fabaceae (*Albizia inundata*)
Number of families (and pollen types) and category assigned			
Using the number of pollen types:			2 (3)
			Eclectic oligolecty
Using the number of pollen species:			2 (9)
			Eclectic oligolecty

Mo 6	Mo 7	Mo 8	Mo 9	Mo 10	Mo 11
Fabaceae	Capparaceae	Ximeniaceae	Solanaceae	Anacardiaceae	Fabaceae
(Prosopis)	(Cynophalla retusa)	(Ximenia americana)	(Solanum)	(Type Schinopsis)	(Type Acacia praecox; Prosopis)
Capparaceae		Capparaceae			
(Cynophalla retusa)		(Anisocapparis speciosa)			
2 (2)	1 (1)	2 (2)	1 (1)	1 (1)	1 (2)
Eclectic oligolecty	Monolecty	Eclectic oligolecty	Monolecty	Monolecty	Oligolecty
2 (8)	No modifications	No modifications	1 (4)	1 (3)	1 (10)
Eclectic oligolecty			Narrow oligolecty	Oligolecty	Oligolecty
No modifications at 5%	Fabaceae	No modifications at 5%	Capparaceae	No modifications at 5%	No modifications at 5%
	(Prosopis)		(Cynophalla retusa)		
	Zygophyllaceae				
	(Bulnesia sarmientoi)				
	3 (3)		2 (2)		
	Eclectic oligolecty		Eclectic oligolecty		
	3 (9)		2 (5)		
	Eclectic oligolecty		Eclectic oligolecty		

Winter (nest Mo 11).	Spring (nests Mo 4, 5, 6, 7, 8)	Summer (nests Mo 1, 9, 10)	Bee species (all nests)
Fabaceae	Fabaceae	Solanaceae	Fabaceae
(Type Acacia praecox; Prosopis)	(Prosopis)	(Solanum)	(Prosopis; type Acacia praecox)
	Capparaceae	Anacardiaceae	Capparaceae
	(Cynophalla retusa; Anisocapparis speciosa)	(Type Schinopsis)	(Cynophalla retusa; Anisocapparis speciosa)
	Ximeniaceae	Fabaceae	Ximeniaceae
	(Ximenia americana)	(Prosopis)	(Ximenia americana)
	Sapotaceae		Sapotaceae
	(Sideroxylon obtusifolium)		(Sideroxylon obtusifolium)
	Solanaceae		Solanaceae
	(Solanum)		(Solanum)
	Anacardiaceae		Anacardiaceae
	(Type Schinopsis)		(Type Schinopsis)
1 (2)	6 (7)	3 (3)	6 (8)

(continued)

Table 2.2a (continued)

Oligolecty	*Polylecty [T: 7.2% (2.8%)]*	*Eclectic oligolecty*	*Polylecty [T: 7.2% (3.2%)]*
	Polylecty [T temp: 8.8% (3.7%)]		*Polylecty [T temp: 7.9% (3.6%)]*
	Polylecty [F: 9% (3.9%)]		*Polylecty [F: 9% (4.4%)]*
	Polylecty [F temp: 11.1% (5%)]		*Polylecty [F temp: 9.5% (5%)]*
1 (10)	6 (18)	3 (14)	6 (21)
Oligolecty	Polylecty [T: 7.2% (7.2%)]	Eclectic oligolecty	Polylecty [T: 7.2% (8.4%)]
	Polylecty [T temp: 8.8% (9.4%)]		Polylecty [T temp: 7.9% (9.5%)]
	Polylecty [F: 9% (10%)]		Broad polylecty [F: 9% (11.7%)]
	Broad polylecty [F temp: 11.1% (12.9%)]		Broad polylecty [F temp: 9.5% (13%)]
No modifications at 5%	Fabaceae	Capparaceae	Fabaceae
	(*Albizia inundata*)	(*Cynophalla retusa*)	(*Albizia inundata*)
	Zygophyllaceae		Zygophyllaceae
	(*Bulnesia sarmientoi*)		(*Bulnesia sarmientoi*)
	7 (9)	4 (4)	7 (10)
	Polylecty [T: 8.4% (3.6%)]	*Polylecty [T: 4.8% (1.6%)]*	*Polylecty [T: 8.4% (4%)]*
	Polylecty [T temp: 10.3% (4.7%)]	*Polylecty [T temp: 5.1% (1.8%)]*	*Polylecty [T temp: 9.2% (4.5%)]*
	Polylecty [F: 10.4% (5%)]	*Polylecty [F: 6% (2.2%)]*	*Polylecty [F: 10.4% (5.6%)]*
	Polylecty [F temp: 13% (6.5%)]	*Polylecty [F temp: 6.5% (2.5%)]*	*Polylecty [F temp: 11.1% (6.2%)]*
	7 (20)	4 (15)	7 (23)
	Polylecty [T: 8.4% (8%)]	Polylecty [T: 4.8% (6%)]	Polylecty [T: 8.4% (9.2%)]
	Broad polylecty [T temp: 10.3% (10.5%)]	Broad polylecty [T temp: 5.1% (6.9%)]	Broad polylecty [T temp: 9.2% (10.5%)]
	Broad polylecty [F: 10.4% (11.1%)]	Polylecty [F: 6% (8.3%)]	Broad polylecty [F: 10.4% (12.8%)]
	Broad polylecty [F temp: 13% (14.4%)]	Polylecty [F temp: 6.5% (9.4%)]	Broad polylecty [F temp: 11.1% (14.3%)]

Number of plant families and pollen types in pot-pollen of stingless bees found at 10% and 5% threshold values of representation per individual nest, nests grouped per season, and the total number of nests of a particular bee species. Pollen specialization categories were assigned to each of these cases. The same was applied for the number of pollen species potentially ascribed to each pollen type. For cases of *polylecty* and *broad polylecty*, percentages of families (without brackets) and species (in brackets) foraged by bees from the total number of available (T) and from only forest taxa (F) are shown separately. Temporal adjustment (temp) was also calculated for both total and forest

values decreased to 18 species and 17 families (180 species and 67 families recorded). For bees that forage for pollen on more than three families, only these values allow us to distinguish between the highest categories proposed by Cane and Sipes (2006): *polylecty* versus *broad polylecty*.

In the present study, when considering plant species and families available in the whole sampling area (*T*), the threshold values were only exceeded once (*T. fiebrigi* at 5% threshold value, when pollen types were converted to pollen species and when the total number of nests was considered) [*T, 13.3% (10.8%)*] (Fig. 2.1; Table 2.2a, for Tables 2.2b, c see supplemental material).

For this reason, adjustments in the number of available items were performed (i.e., by decreasing the denominator values of this ratio). For instance, the number of plant species available in the forest (*F*) was also estimated. Furthermore, temporal adjustment (*temp*) was calculated on flowerings available during the last 4 months before nest sampling date (a total of 5 months were therefore considered, in order to reduce the number of species available). Temporal adjustment was applied to *T* and *F* (Table 2.3, see supplemental material).

The plant species from the forest (*F*) included all woody species of the area and the herbs typical of forest environments. This spatial adjustment (*F*) was performed to reduce the number of resources considered available to stingless bee colonies, since most nests were well within the forest. These numbers were applied for nests sampled either in Palosantal or Quebrachal, as both forest types are composed of species that differ in their relative abundance (Prado 1993).

2.5 The Importance of an Appropriate Assessment of Pollen Specialization in Bees: Factors Causing Low Number of Foraged Items

After adjustments on the number of foraged and available items were applied to the calculations, a total of six cases of *broad polylecty* were recorded

(three for *M. orbignyi* and three for *T. fiebrigi*) (Fig. 2.1). This fact suggests that a combination of adjustments is needed for an appropriate assessment of pollen diet in bees.

Among the possible causes of low numerator values, the following factors either intrinsic to the Meliponini or extrinsic were postulated.

2.5.1 Abundant Versus Minor Pollen Types

In the present survey, pollen types foraged at lower percentages than the 10% or 5% threshold values were considered as contaminants and therefore not taken into account to classify these bees into pollen specialization categories. It has been suggested that their high richness and low abundance in pot-pollen of stingless bees are due to the quality as food, communicated by Meliponini scouts during their exploration of flowers (Kleinert-Giovannini and Imperatriz-Fonseca 1987), or coming from a nonprotein source (i.e., either nectar or oil), among others (Villanueva-Gutiérrez and Roubik 2016). Although Cane and Sipes (2006) propose that contaminant pollen was excluded when quantifying oligolecty, their definition of *broad polylecty* considers the whole number of plant species foraged and no pollen types (page 115 in Cane and Sipes 2006). It is suggested here that minor or contaminant pollen are also excluded when studying generalist species and working with "pollen types."

2.5.2 Recruitment Behavior

Few pollen types (from one to five at 10% threshold value or six at 5%) were considered as important for each nest studied. This foraging pattern was not caused by innate specialization on particular pollen hosts (i.e., a kind of oligolecty) but by recruitment behavior. Members of a stingless bee colony may temporarily gather pollen from only one or a few plant species in the presence of other attractive melittophilous resources. This was observed by the exclusive use of *Cynophalla retusa* in spring and *Solanum* in summer by *M. orbignyi* (nests 7 and 9), of type *Schinopsis* in winter by *T. fiebrigi* (nest 2), and of *Prosopis* in winter and *Cynophalla retusa* in spring by *G. argentina* (nests 3 and 6) (Table 2.2a, for Table 2.2b, c see supplemental material). Such selective but flexible foraging behavior is intrinsic to stingless bees and honey bees (Ramalho et al. 1990, 2007; Hrncir et al. 2000; Jarau et al. 2003) and would explain the reduced number of foraged items found in nests.

2.5.3 Intra-nest Pollen Analysis

Another cause for reduced number of foraged items is the averaging of the pollen composition of many pots or groups of pots to only one spectrum per nest (*nest pollen analysis*). This averaging does not identify many important pollen types such as those stored in some pots or cells which are identified by the *intra-nest pollen analysis* and therefore may profitably be included to assess pollen specialization in bees. For example, for the broadly polylectic *Augochlora amphitrite*, only five pollen types were identified as important (>10%) when *nest pollen analysis* was performed, and nine pollen types of importance were found with *intra-nest pollen analysis* (Dalmazzo and Vossler 2015a). The results of *intra-nest pollen analysis* performed for *T. fiebrigi* and *G. argentina* (Vossler, unpublished data) showed a higher number of important resources than those detected by the *nest pollen analysis* performed in the present study.

2.6 Factors Causing High Number of Available Items

A total of 250 species in 83 families of melittophilous vegetation was counted for the present survey, although many more species (mainly herbs) were available in the area. If these currently unidentified species were added, the number of available items would be higher, and the threshold levels of *polylecty* and *broad polylecty* become less likely.

Among stingless bees, *Tetragonisca* has a maximum foraging radius of approximately 600–950 m, *Geotrigona* between 1100 and 1700, and

Melipona more than 2000 m (Roubik and Aluja 1983; Araújo et al. 2004). They can forage only on the melittophilous flowers that are temporarily available in this area near the nests. Therefore, to assess pollen diets in bees, the total number of available melittophilous species might best be spatially and temporarily reduced, as performed for the present study.

2.7 Polylecty, Broad Polylecty, or Simply *Degrees of Polylecty*?

After the calculation adjustments were performed, two of the three Meliponini were identified as *broadly polylectic*, but *Geotrigona argentina* was not so identified. This ground-nesting species of low colony population only foraged on one or a few resources per nest at high percentages (more than 10%) as it efficiently recruits and apparently provisions pots during a short period of time (before summer–autumn floods), which could be interpreted as strategies for colony survival (Vossler et al. 2010).

The threshold value of 10% for species was not reached by *G. argentina*. The same held for the 25% suggested for families also not reached by *T. fiebrigi* (highest value of 19% for families after all adjustments).

Considering that the threshold values were proposed for all the foraged resources including contaminants, it is advisable to reduce values when only important resources are taken into account. For example, in the present survey, the maximum percentages recorded for *G. argentina* could have been used as thresholds. However, such low values may be due to the low number of samples analyzed (six nests) and/or the localized sampling to only two sites 250 km apart.

When a reduced number of samples is taken from a few or nearby sites or during a short period of the year, a false conclusion on the category of specialization can be reached, as can be seen in stingless bees when comparing individual nests versus several different nests (Table 2.2a, for Table 2.2b, c see supplemental material). Thus, a polylectic species that temporarily or locally specialized on a few resources (the only ones available at a period of time or a site) can be erroneously identified as oligolectic or other low categories of relative generalization (as shown in Table 2.2a, for Table 2.2b, c see supplemental material). This could be the case for some solitary bees with temporarily reduced foraging activity (Minckley et al. 2000; Cane and Sipes 2006) such as andrenid bees of the genus *Calliopsis* (Vossler 2014).

The threshold values of 10% for species and 25% for families are too high to reach *broad polylecty* when working with pollen data. The multispecies or multigenus pollen types further confuse the issue of exactly which flower species are visited and their relative importance to the bee species in question. For this reason, no direct comparisons between pollen types and pollen species can be made when working with certain palynological data.

As alternative, simple values are proposed to assess pollen diet of generalist bees: the maximum number of foraged items per nest (contaminants not considered) and the maximum percentage value of foraged versus available items (adjusted). These belong to a wider range of values that can be recognized as *degrees of polylecty*, allowing for a more precise identification than *polylecty* versus *broad polylecty*. For instance, Mo = five foraged types or 2% (*F temp*) (nest 4) and ten types for the nine nests studied together or 5%, Tf = 6% or 4.2% (nest 7) and 15% or 9.3% (all nests), and Ga = 3% or 2.3% (nest 5) and 7% or 4.3% (all nests), respectively. Thus, *T. fiebrigi* is considered more polylectic than *M. orbignyi* and both more polylectic than *G. argentina*. As can be seen, these values are useful to show degrees of generalization/specialization among species inside a bee group with similar foraging behavior such as Meliponini.

Acknowledgments I am especially thankful to Patricia Vit for her kind invitation to participate in this book and Alicia Basilio for recommending me, Nora Brea for her help in English language, David Roubik and Nora Brea for providing suggestions and critical comments on the manuscript, and Arturo Roig-Alsina for the identification of bees. This study was supported by CONICET (Consejo Nacional de Investigaciones Científicas y Técnicas).

References

Araújo ED, Costa M, Chaud-Netto J, Fowler HG. 2004. Body size and flight distance in stingless bees (Hymenoptera: Meliponini): inference of flight range and possible ecological implications. Brazilian Journal of Biology 64: 563–568.

Cane JH, Sipes S. 2006. Characterizing floral specialization by bees: analytical methods and a revised lexicon for oligolecty. pp 99–122. In Waser NM, Ollerton J, eds. Plant-Pollinator Interactions. From specialization to generalization. The University of Chicago Press; Chicago, USA. 488 pp.

Cilla G, Caccavari M, Bartoloni NJ, Roig-Alsina A. 2012. The foraging preferences of two species of *Melissodes* Latreille (Hymenoptera, Apidae, Eucerini) in farmed sunflower in Argentina. Grana 51: 63–75.

Dalmazzo M, Vossler FG. 2015a. Assessment of the pollen diet in a wood-dwelling augochlorine bee (Halictidae) using different approaches. Apidologie 46: 478–488.

Dalmazzo M, Vossler FG. 2015b. Pollen host selection by a broadly polylectic halictid bee in relation to resource availability. Arthropod-Plant Interactions 9: 253–262.

De Klerk P, Joosten H. 2007. The difference between pollen types and plant taxa: a plea for clarity and scientific freedom. Eiszeitalter und Gegenwart / Quaternary Science Journal 56: 162–171.

Erdtman G. 1960. The acetolysis method, a revised description. Svensk Botanisk Tidskrift 54: 561–564.

Hrncir M, Jarau S, Zucchi R, Barth FG. 2000. Recruitment behavior in stingless bees, *Melipona scutellaris* and *Melipona quadrifasciata*. II. Possible mechanisms of communication. Apidologie 31: 93–113.

Jarau S, Hrncir M, Schmidt VM, Zucchi R, Barth FG. 2003. Effectiveness of recruitment behavior in stingless bees (Apidae, Meliponini). Insectes Sociaux 50: 365–374.

Joosten H, De Klerk P. 2002. What´s in a name? Some thougths on pollen classification, identification, and nomenclature in Quaternary palynology. Review of Palaeobotany and Palynology 122: 29–45.

Kleinert-Giovannini A, Imperatriz-Fonseca VL. 1987. Aspects of the trophic niche of *Melipona marginata marginata* Lepeletier (Apidae, Meliponinae). Apidologie 18: 69–100.

Larkin LL, Neff JL, Simpson BB. 2008. The evolution of a pollen diet: Host choice and diet breadth of *Andrena* bees (Hymenoptera: Andrenidae). Apidologie 39: 133–145.

Michener CD. 2007. The bees of the world, 2 edn. The Johns Hopkins University Press; Baltimore, USA. 953 pp.

Michener CD. 2013. The Meliponini. pp 3–17. In Vit P, Pedro SRM, Roubik DW, eds. Pot honey: A legacy of stingless bees. Springer; New York, USA. 175 pp.

Michez D, Patiny S, Rasmont P, Timmermann K, Vereecken NJ. 2008. Phylogeny and host-plant evolution in Melittidae *s.l.* (Hymenoptera: Apoidea). Apidologie 39: 146–162.

Michez D, Eardley CD, Timmermann K, Danforth BN. 2010. Unexpected polylecty in the bee genus *Meganomia* (Hymenoptera: Apoidea: Melittidae). Journal of the Kansas Entomological Society 83: 221–230.

Minckley RL, Cane JH, Kervin L. 2000. Origins and ecological consequences of pollen specialization among desert bees. Proceedings of the Royal Society of London B 267: 265–271.

Minckley RL, Roulston TH. 2006. Incidental mutualisms and pollen specialization among bees. pp 69–98. In Waser NM, Ollerton J, eds. Plant-Pollinator Interactions. From specialization to generalization. The University of Chicago Press; Chicago, USA. 488 pp.

Müller A, Kuhlmann M. 2008. Pollen hosts of western palaearctic bees of the genus *Colletes* (Hymenoptera: Colletidae): the Asteraceae paradox. Biological Journal of the Linnean Society 95: 719–733.

Prado DE. 1993. What is the Gran Chaco vegetation in South America? I. A review. Contribution to the study of flora and vegetation of the Chaco. V. Candollea 48: 145–172.

Ramalho M, Kleinert-Giovannini A, Imperatriz-Fonseca VL. 1989. Utilization of floral resources by species of *Melipona* (Apidae, Meliponinae): Floral preferences. Apidologie 20: 185–195.

Ramalho M, Kleinert-Giovannini A, Imperatriz-Fonseca VL. 1990. Important bee plants for stingless bees (*Melipona* and *Trigona*) and Africanized honeybees (*Apis mellifera*) in Neotropical habitats: a review. Apidologie 21: 469–488.

Ramalho M, Silva MD, Carvalho CAL. 2007. Dinâmica de uso de fontes de pólen por *Melipona scutellaris* Latreille (Hymenoptera: Apidae): uma análise comparativa com *Apis mellifera* L. (Hymenoptera: Apidae), no Domínio Tropical Atlântico. Neotropical Entomology 36: 38–45.

Robertson CH. 1925. Heterotropic bees. Ecology 6: 412–436.

Robertson CH. 1926. Revised list of oligolectic bees. Ecology 7: 378–380.

Roubik DW. 1982. Seasonality in colony food storage, brood production and adult survivorship: studies of *Melipona* in tropical forest (Hymenoptera: Apidae). Journal of the Kansas Entomological Society 55: 789–800.

Roubik DW. 1989. Ecology and natural history of tropical bees. Cambridge University Press; New York, USA. 514 pp.

Roubik DW, Aluja M. 1983. Flight ranges of *Melipona* and *Trigona* in tropical forest. Journal of the Kansas Entomological Society 56: 217–222.

Sipes SD, Tepedino VJ. 2005. Pollen-host specificity and evolutionary patterns of host switching in a clade of specialist bees (Apoidea: *Diadasia*). Biological Journal of the Linnean Society 86: 487–505.

Torretta JP, Durante SP, Colombo MG, Basilio AM. 2012. Nesting biology of the leafcutting bee *Megachile (Pseudocentron) gomphrenoides* (Hymenoptera: Megachilidae) in an agro-ecosystem. Apidologie 43: 624–633.

Torretta JP, Durante SP, Basilio AM. 2014. Nesting ecology of *Megachile (Chrysosarus) catamarcensis* Schrottky (Hymenoptera: Megachilidae), a *Prosopis*-specialist bee. Journal of Apicultural Research 53: 590–598.

Villanueva-Gutiérrez R, Roubik DW. 2016. More than protein? Bee-flower interactions and effects of disturbance regimes revealed by rare pollen in bee nests. Arthropod-Plant Interactions 10: 9–20.

Vossler FG. 2013a. The oligolecty status of a specialist bee of South American *Prosopis* (Fabaceae) supported by pollen analysis and floral visitation methods. Organisms Diversity and Evolution 13: 513–519.

Vossler FG. 2013b. Estudio palinológico de las reservas alimentarias (miel y masas de polen) de "abejas nativas sin aguijón" (Hymenoptera, Apidae, Meliponini): un aporte al conocimiento de la interacción abeja-planta en el Chaco Seco de Argentina. Doctoral Thesis. Universidad Nacional de La Plata; La Plata, Argentina. 152 pp.

Vossler FG. 2014. A tight relationship between the solitary bee *Calliopsis (Ceroliopoeum) laeta* (Andrenidae, Panurginae) and *Prosopis* pollen hosts (Fabaceae, Mimosoideae) in xeric South American woodlands. Journal of Pollination Ecology 14: 270–277.

Vossler FG, Tellería MC, Cunningham M. 2010. Floral resources foraged by *Geotrigona argentina* (Apidae, Meliponini) in the Argentine Dry Chaco forest. Grana 49: 142–153.

Zuloaga FO, Morrone O, Belgrano MJ. 2008. Catálogo de las plantas vasculares del cono sur (Argentina, Sur de Brasil, Chile, Paraguay y Uruguay). Volumes 1–3. Monographs in Systematic Botany from the Missouri Botanical Garden 107: 1–983, 985–2286, 2287–3348.

Pollen Collected by Stingless Bees: A Contribution to Understanding Amazonian Biodiversity

Maria L. Absy, André R. Rech,
and Marcos G. Ferreira

3.1 Introduction

Every inch of land in the Amazon region presents a fascinating amount of biodiversity. If an observer is interested in plants, he will find up to 300 species per hectare in some areas of the Amazon (ter Steege et al. 2000). As for insects, it is possible to identify more than 480 species in 0.16 square kilometers (Wilkie et al. 2010). If interactions attract the eyes of the observer, one can find up to 35 species of bees pollinating the flowers of a single liana species in the Amazon region (Rech et al. 2011). Since the first trips of naturalists into the Amazon, its enormous biodiversity has raised questions about processes behind the observed patterns. Different lines of research have offered hypotheses to explain how Amazon biodiversity has evolved to reach the patterns that are currently observed. By looking at the interactions between bees and flowers, one may be led to ponder some tentative explanations for the observed diversity. In this chapter, we intend to review the studies previously conducted in the Brazilian Amazon that address the interactions between bees and flowers. We aim to give the reader a broad idea about what is presently known about bees and flowers in the Brazilian Amazon, which is primarily recorded using palynological tools, i.e., pollen grains for taxon identification.

3.1.1 Origin and Evolution of Plant-Bee Interactions

The earliest bees possibly appeared in the xeric interior of the paleo-continent Gondwana, which was presumably also the area of origin for flowering plants (Raven and Axelrod 1974). The morphological and behavioral diversity found in bees may be one of the evolutionary drivers toward the simultaneous expansion of angiosperms and bees during the mid-Cretaceous (Grimaldi 1999; Danforth and Poinar 2011).

Recently, Cardinal and Danforth (2013), supported by molecular and morphological evidence, have suggested that the origin of bees occurred approximately 123 Mya (113–132 Mya). This hypothesis is contemporary to an incremental expansion and abundance within the Eudicot group, a clade mostly dependent upon bees for reproduction. According to Ollerton et al. (2011), globally, approximately 85% of the angiosperms are animal pollinated, and the local percentage

M.L. Absy (✉) • M.G. Ferreira
Instituto Nacional de Pesquisas da Amazônia,
Laboratório de Palinologia,
Av. André Araujo, 2936 – Petrópolis, CEP,
69067-375 Manaus, AM, Brazil
e-mail: lucia.absy@gmail.com

A.R. Rech
Universidade Federal dos Vales do Jequitinhonha e
Mucuri, Curso de Licenciatura em Educação do
Campo, Alto da Jacuba 5000, CEP,
39100-000 Diamantina, MG, Brazil

© Springer International Publishing AG, part of Springer Nature 2018
P. Vit et al. (eds.), *Pot-Pollen in Stingless Bee Melittology*, DOI 10.1007/978-3-319-61839-5_3

varies from 78% to 94% depending on latitude, with the tropics being a location where animals are more important as pollinators at the community level (Rech et al. 2016).

Stingless bees (Meliponini) are species of social bees that produce honey. Nests are often inside hollow trees and constructed of wax that is secreted from dorsal metasomal glands, which bees combine with resin or propolis collected from plants (Engel and Michener 2013). Due to their permanent nests with large populations, most Meliponini are foragers. In the Amazon region, these bees are the main visitors of numerous plant species and, although they seem to pollinate a large number of them, there is a need for quantifying the importance of such apparent pollination (Roubik 1989). Considering that most of these plants are bisexual or obligate outcrossers, they usually need an animal pollinator to carry pollen from one flower to another (Bawa et al. 1985; Bawa 1990; Roubik 1989; Ollerton et al. 2011; Rech et al. 2016). Thus, the bees are essential for the maintenance of plants in the same way that plants are essential for bee survival, as pollen is the principal food of bee larvae. Social bees present a pantropical distribution (Indo-Australia, the Neotropics, and Africa-Madagascar) composed of continental disjunctions and showing a complex history of vicariance of great antiquity (Camargo 2013; Martins et al. 2014).

Bee diversification was likely rapid, with the earliest roots and stem members of the principal families appearing during and after the mid-Cretaceous and radiating through the latter part of the period (Engel 2001, 2004; Ohl and Engel 2007).

The earliest evidence of stingless bees comes from the latest stage of the Cretaceous (Engel and Michener 2013). The Cretaceous period lasted for approximately 80 million years. According to Michener and Grimaldi (1988), *Cretotrigona prisca* (Michener and Grimaldi) comprises the only Mesozoic record of the Meliponini, and it is the only definitive apid from the great middle age of Earth. This species showed a significant superficial similarity to the modern species of *Trigona* s. str. (Michener and Grimaldi 1988), while a detailed examination

has suggested that its phylogenetic affinities are more closely related to some Old World lineages (Engel 2000). Hence, the Meliponini are not only morphologically and biologically diverse but also ancient (Michener 2013).

According to Engel and Michener (2013), most of the ideas about stingless bee evolution have been grounded in our knowledge of the hundreds of living stingless bee species. At each moment during the evolution of Meliponini, there must have been numerous, and later hundreds, of species. The two authors mentioned above have emphasized that the nearly 550 extant species are a mere fraction of the total historical diversity of stingless bees (Rasmussen and Gonzalez 2013).

Looking for living species of bees and many other tropical organisms, many naturalists have visited the Amazon region since the early days after the European invasion of the Americas. Many species were described from this region by Adolfo Duke and other traveling naturalists (Hemming 2015). Nevertheless, it was only during the twentieth century when organized collection trips that focused on the study of bees were organized by Drs. João Maria Franco Camargo and Warwick Estevam Kerr. These two researchers and a group of collaborators traveled through the main rivers of the Amazon basin, collecting samples at several locations along the rivers not only of bees (mostly Meliponini) but also samples of their nest materials. These materials were very helpful for the further identification of the pollen that comprises the diet of most bees (Absy et al. 1984; Rech and Absy 2011a, b). It is in the context of these trips that these first authors became engaged in the discussion of the trophic ecology of bees in the Amazon region.

3.2 The Use of Pollen Analysis in the Study of Bees in the Amazon Rainforest

A number of studies have evaluated the pollen collected by stingless bees in the central Amazon region. Those studies have involved the use of classical protocols for the collection of pollen foraged by bees, such as from pollen pots (Absy

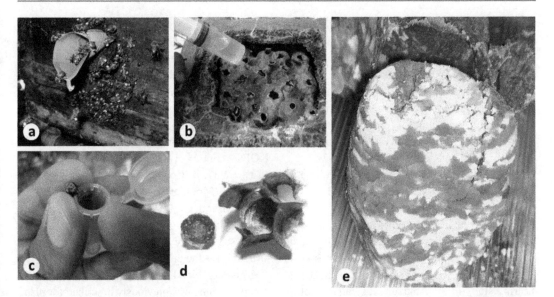

Fig. 3.1 Methodologies used in previous studies on pollen grains collected by stingless bees in the Central Amazon: (**a**) corbicular pollen in the nest entrance, (**b**) honey sam- ples in storage pots, (**c**) nectar regurgitated by workers in the nest entrance, (**d**) postemergence residue, and (**e**) pollen stored inside the nest (Photo MG Ferreira, and AQ Lima)

et al. 1984; Rech and Absy 2011a, b), from nectar and honey samples (Absy et al. 1980), or more commonly, from the pollen loads of worker bees (Absy and Kerr 1977; Marques-Souza et al. 1995, 1996, 2002; Marques-Souza 1996; Oliveira et al. 2009; Ferreira and Absy 2015). Recently, Ferreira and Absy (2013) established a new collection protocol for the postemergence residue that avoids the acetolysis processes without losing relevant information about the texture and orna- mentation of the pollen grain (see Fig. 3.1).

As noted in the introduction, the Amazon region encompasses a huge diversity of plants, and a large portion of them are used as food sources by bees (Absy et al. 1984; Rech and Absy 2011a, b). This diversity brings a great challenge to researchers interested in studying the diets of bees using pollen analysis. Therefore, having a large pollen library may not be suffi- cient to obtain satisfactory pollen identification. It is also necessary to have good samples and well-prepared slides to observe enough detail in the pollen grains to achieve good results in the identification of pollen types. This is the main reason that we support the use of the Erdtman's acetolysis method to prepare pollen samples from the Amazon. Following this technique, the

Palynology Laboratory of the National Institute of Amazon Research (INPA) has been conduct- ing research on pollen and bees since 1977 (Absy and Kerr 1977). These studies are divided primarily into two different groups. The first group of papers addressed large nest samples collected during taxonomic expeditions (Absy et al. 1984; Rech and Absy 2011a, b). The sec- ond group addressed annual studies, considering a single or few species of bees and the possible pollen sources used throughout the year (Absy et al. 1980; Marques-Souza et al. 1995, 1996, 2002, 2007; Marques-Souza 1996, 2010; Oliveira et al. 2009; Ferreira and Absy 2015).

3.3 Diversity of Plants, Stingless Bees, and Their Interactions in Central Amazon

Studies of pollen identification have revealed the increasing importance of botanical diversity with regard to bees. However, it is an indirect source of evidence; therefore pollen is a proxy for the poten- tially mutualistic pollination interaction and should not be used to deduce the interaction conse- quences. Also, considering the huge available diversity of plants and the similarities within some groups, there is a need for nomenclature and technique standardization with the aim of avoiding misinterpretation of pollen features. Therefore, not only within entomopalynology but also in the field of palynology, there is growing support to use the terminology "type" when referring to a taxon or a morphologically conspicuous group (Joosten and de Klerk 2002; de Klerk and Joosten 2007).

From a study of pollen collected by Meliponini in the Central Amazon region (Table 3.1), it was possible to recognize a wide variety of pollen types collected by 48 species of stingless bees and their interactions with the main botanical families that supply trophic resources (see Fig. 3.2). Using a large-scale approach, a pioneering study in the Amazon was carried out by Absy et al. (1984). For this study, pollen material was obtained from the nests of 24 stingless species distributed over the Baixo Tapajos, Trombetas, Medio Amazonas, and Baixo Uatumã rivers. From all analyzed nests, the authors found 122 pollen types, with a Myrtaceae type visited by 14 species of stingless bees, followed by *Attalea maripa* (Arecaceae) and *Tapirira guianensis* (Anacardiaceae), both visited by 13 species. Moreover, these authors emphasize that 18 bee species are generalists which collect the pollen of ten or more botanical species.

Following a similar approach, Rech and Absy (2011a) studied 10 different species of bees. Inside the nests of those bees, the authors found 78 pollen types, encompassing 70 different genera and 42 plant families. The same authors, in another study considering different sets of 14 bee species (from the genera *Partamona*, *Scaura*, and *Trigona*), found 78 pollen types from 36 plant families stored in the bee nests. After quantification of pollen abundance inside the pollen pots, Rech and Absy (2011b) defined 37 plants as attractive to bees (representation >10%) and 16 pollen types as a result of temporary specialization of the bees (representation >90% of a pollen pot).

The data from Absy et al. (1984) and Rech and Absy (2011a, b) were obtained from standard methods and covered a large terrestrial area. Using the species list from these studies, Sfair

and Rech (unpublished data) pooled together a large Amazonian network and checked for "modularity" and "nestedness." Considering the 74 stingless bees and 334 plant species (when a taxon was identified only to the genus level in different places, each was considered a different species), the "global connectance" was 0.03 and the network was considered nested (N total = 3.55, NODF (Er) = 1.83, p (Er) = 0.00; NODF (Ce) = 2.37, p (Ce) = 0.00), encompassing nine modules (Modularity = 0.56, Mrand = 0.466, sigma Mrand = 0.004). From the nine modules, two modules had only two species each, and the other seven were nested, suggesting a fractal structure for the entire network.

This combined structure, defined as nested compartments, was previously described for plant-herbivore interactions (Lewinsohn et al. 2006) and, as far as we know, is first shown here for mutualistic interaction. This structure probably emerges from the differences in abundance and attractiveness of the different pollen sources (Lewinsohn et al. 2006; Rech and Absy 2011a, b). Corroborating abundance as a possible driver of the nested compartments, as proposed for plant-herbivores, pollen from mass flowering plants was already found to represent >90% of the annual income of a given colony of stingless bee (Hrncir and Maia-Silva 2013). Hence, theoretical models suggest that nested networks tend to reduce competition and allow a greater number of species to coexist (Bastolla et al. 2009). In the same way, modularity is supposed to improve community stability (Fortuna et al. 2010). It may also be hypothesized that the general structure of plant-Meliponini interactions reflects its position halfway in the continuum of mutualism/antagonism, since bees sometimes are pollinators and sometimes mere herbivores—an open avenue for future research.

Contrasting with a large spatial scale, temporal studies usually covering 1 year have focused mainly on economically important stingless bee species. These studies have provided important data on the annual distribution of trophic resources for bees and their behavioral responses to fluctuations in food availability. In the Amazon, these types of studies began in 1977 with a collaboration

Table 3.1 Stingless bees referenced in pollen studies conducted in the Amazon between 1977 and 2015

Pollen studies

Stingless bees	Absy and Kerr (1977)	Absy et al. (1980)	Absy et al. (1984)	Kerr et al. (1986)	Marques-Souza et al. (1995)	Marques-Souza (1996)	Marques-Souza et al. (1996)	Marques-Souza et al. (2002)	Marques-Souza et al. (2007)	Oliveira et al. (2009)	Marques-Souza (2010)	Rech and Absy (2011a)	Rech and Absy (2011b)	Ferreira and Absy (2013)	Novais and Absy (2013)	Silva et al. (2013)	Ferreira and Absy (2015)	Novais and Absy (2015)	Novais et al. (2015)
Apararigona impunctata Ducke, 1916												PP							
Cephalotrigona femorata Smith, 1854										PL		PP							
Frieseomelitta silvestrii faceta [nom. nud.]			PP																
Frieseomelitta varia Lepeletier, 1836					PL						PL								
Frieseomelitta sp.								PL											
Melipona (Eomelipona) tumupasae Schwarz, 1932			PP																
Melipona (Melikerria) compressipes Fabricius, 1804						PL		PL											
Melipona (Melikerria) fasciculata Smith, 1854				PL															
Melipona (Melikerria) interrupta Latreille, 1811			PP											PER			PL		
Melipona (Michmelia) seminigra seminigra Friese, 1903								PL											

(continued)

Table 3.1 (continued)

Pollen studies

Stingless bees	Absy and Kerr (1977)	Absy et al. (1980)	Absy et al. (1984)	Kerr et al. (1986)	Marques-Souza et al. (1995)	Marques-Souza (1996)	Marques-Souza et al. (1996)	Marques-Souza et al. (2002)	Marques-Souza et al. (2007)	Oliveira et al. (2009)	Marques-Souza (2010)	Rech and Absy (2011a)	Rech and Absy (2011b)	Ferreira and Absy (2013)	Novais and Absy (2013)	Silva et al. (2013)	Ferreira and Absy (2015)	Novais and Absy (2015)	Novais et al. (2015)
Melipona (Michmelia) fulva Lepeletier, 1836			PP							PL									
Melipona (Michmelia) paraensis Ducke, 1916		N	PP		PL														
Melipona (Michmelia) rufiventris Lepeletier, 1836			PP																
Melipona (Michmelia) seminigra merrillae Friese, 1903	PL	N						PL		PL						H	PL		
Melipona (Michmelia) seminigra pernigra Moure and Kerr 1950			PP																
Nannotrigona (Scaptotrigona) postica flavisetis [n. nud.]			PP																
Nannotrigona minuta (Lepeletier, 1836)			PP																
Nogueirapis buteli Friese, 1900												PP							
Oxytrigona flaveola Friese, 1900												PP							
Oxytrigona tataira (Smith, 1863)			PP																
Partamona ailyae Camargo, 1980													PP						

Species															
Partamona epiphytophila Pedro and Camargo, 2007							PP								
Partamona ferreirai Pedro and Camargo, 2003							PP								
Partamona pearsoni (Schwarz, 1938)							PP								
Partamona vicina Camargo, 1980		PP					PP								
Partamona sp.1		PP													
Partamona sp.2		PP													
Plebeia minima Gribodo, 1893						PP									
Priotrigona lurida (Smith, 1854)		PP				PP									
Scaptotrigona fulvicutis Moure, 1964				PL											
Scaptotrigona polysticta Moure, 1950		PP													
Scaptotrigona sp.					PL	PP									
Scaura tenuis (Ducke, 1916)							PP								
Scaura latitarsis (Friese, 1900)							PP								
Schwarzula coccidophila Camargo and Pedro, 2002						PP									
Tetragona goettei (Friese, 1900)		PP													

(continued)

Table 3.1 (continued)

Stingless bees	Pollen studies																		
	Absy and Kerr (1977)	Absy et al. (1980)	Absy et al. (1984)	Kerr et al. (1986)	Marques-Souza et al. (1995)	Marques-Souza (1996)	Marques-Souza et al. (1996)	Marques-Souza et al. (2002)	Marques-Souza et al. (2007)	Oliveira et al. (2009)	Marques-Souza (2010)	Rech and Absy (2011a)	Rech and Absy (2011b)	Ferreira and Absy (2013)	Novais and Absy (2013)	Silva et al. (2013)	Ferreira and Absy (2015)	Novais and Absy (2015)	Novais et al. (2015)
Tetragonisca angustula (Latreille, 1811)												PP			PP			H	H
Trigona amalthea (Olivier, 1789)			PP																
Trigona branneri Cockerell, 1912			PP										PP						
Trigona chanchamayoensis Schwarz, 1948																			
Trigona cilipes (Fabricius, 1804)			PP										PP						
Trigona dalatorreana Friese, 1900													PP						
Trigona fulviventris Guérin, 1844										PL									
Trigona fuscipennis Friese, 1900			PP																
Trigona pallens (Fabricius, 1798)			PP																
Trigona williana Friese, 1900							PL						PP						
Trigona recursa Smith, 1863													PP						

PP pot-pollen, *PL* pollen loads, *H* honey, *N* nectar, *PER* postemergence residue

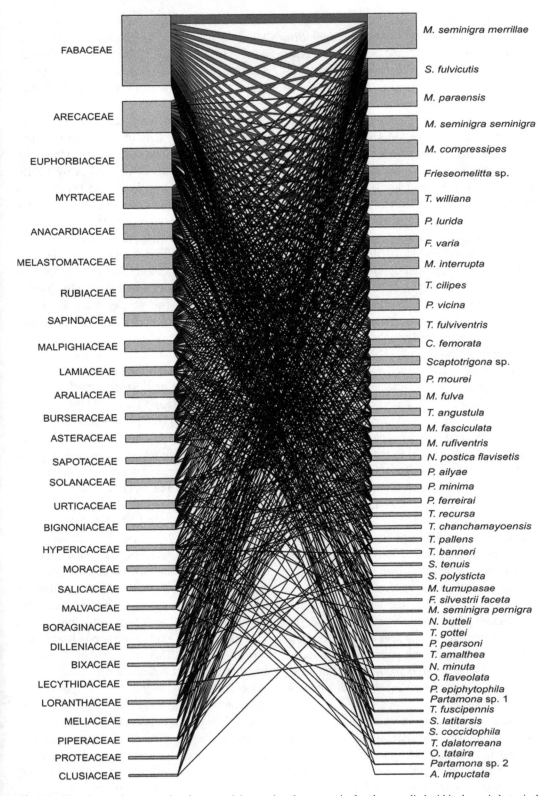

Fig. 3.2 Bipartite graph representing the network interactions between stingless bees studied within the main botanical families identified and represented by numbers of pollen types

between Drs. Absy and Kerr (1977). These authors found *Inga edulis* (Fabaceae/Mimosoideae), *Bixa orellana* (Bixaceae), and *Miconia* types (Melastomataceae) were three very important pollen sources for bees. Absy et al. (1980) found another 60 pollen types used by two species of bees. Later studies have confirmed the importance of Myrtaceae, Fabaceae, Melastomataceae, and Arecaceae as pollen sources for Meliponini. Hence, Marques-Souza et al. (1995) found *Miconia* type (Melastomataceae), *Myrcia* type, *Myrcia amazonica* (Myrtaceae), and *Leucaena* type (Fabaceae/ Caesalpinioideae) were important food sources for two species of bees. Moreover, Marques-Souza (1996) found 30 pollen types (22 genera, 19 families) used by *Melipona compressipes*, with *Cassia* type Fabaceae (Caesalpinioideae), *Miconia* type (Melastomataceae), and *Solanum* type (Solanaceae) being the most important pollen sources. Studying the bee *Trigona williana*, Marques-Souza et al. (1996) found *Cocos nucifera* (Arecaceae), *Attalea* sp. (Arecaceae), *Cassia* type (Fabaceae/Caesalpinioideae), *Carica papaya* (Caricaceae), *Bellucia grossularioides* (Melastomataceae), *Artocarpus altilis* (Moraceae), and *Stachytarpheta cayennensis* (Verbenaceae) being used by these stingless bees.

More recent studies that included new species of bees such as *Scaptotrigona fulvicutis* have noted some well-known important pollen source plant families, such as Fabaceae (Mimosoideae) and Myrtaceae while including others that have also been reported in previous studies, such as Sapindaceae, which was not as important for this particular stingless bee species (Marques-Souza et al. 2007). Moreover, the latter authors also find another 97 pollen types collected by *S. fulvicutis*, representing 73 genera and 36 plant families. The most important pollen types collected by these bees were *Stryphnodendron guianense* (Fabaceae/Mimosoideae) and *Schefflera morototoni* (Araliaceae).

Using a temporal approach and also considering different stingless bee species (*Melipona seminigra merrillae*, *Melipona fulva*, *Trigona fulviventris*, and *Cephalotrigona femorata*), Oliveira et al. (2009) recorded 90 pollen types from 31 plant families and 67 different genera. In this study, the most frequent pollen types were *Miconia myriantha* (Melastomataceae), *Leucaena leucocephala* (Fabaceae/Mimosoideae), *Tapirira guianensis* (Anacardiaceae), *Eugenia stipitata* (Myrtaceae), *Protium heptaphyllum* (Burseraceae), and *Vismia guianensis* (Hypericaceae). When considering different synchronopatric bee species, it is important to take into account the potential for food competition among the studied species. Ferreira and Absy (2015) have analyzed the trophic overlap between two common bee species that are managed for honey in the Central Amazon region. In their work, the authors show variation in the trophic overlap over time and by resource seasonality. Again, the most important plant families in which the proportion of shared pollen types raised the trophic overlap between the two bee species were Fabaceae, Melastomataceae, Myrtaceae, and Anacardiaceae.

3.4 Amazonian Bee Diet, Biology, and Suggested Interactions Potentially Leading to Pollination

The ecological relevance of social bees as pollinators comes from their dependence upon large amounts of pollen and nectar as food sources for constant brood production throughout the year (Simpson and Neff 1981; Roubik 1989; Corbet et al. 1991; Free 1993; Ramalho et al. 2007). By visiting flowers to collect trophic resources, these bees establish a complex network of interactions that may go from mutualism, when pollination really occurs, to antagonism, when resources are collected without any fruit set (Junker and Blüthgen 2010; Santamaría and Rodríguez-Gironés 2015). Therefore, although pollen evidence is a valid indication that an interaction had occurred, it is no guarantee of pollination, and more studies in this area are needed. This is especially true in the Amazon, where the canopy is high above the eyes of observers; therefore recording flower visitation with observation is not always an easy task. In this context, a powerful proxy such as pollen on the body of pollinators or in their nest is a welcome alternative to know the plants visited by each bee species. Once plants are known further studies will reveal whether interaction is mutualistic (resulting in pollination) or only trophic (resulting only in resource consumption).

Not only in the study of interactions that occur among tall native trees but also in the forest-agriculture landscape, pollen evidence of interactions may be of great value. From the 38 most common native plant species commercialized in the open market in Manaus-Amazonas (Rabelo 2012), 23 were recorded in pollen studies as collected by bees. Among the most important plant families found in the markets and in the pollen analysis, we may emphasize the following: Arecaceae (*Bactris gasipaes* Kunth, *Astrocaryum aculeatum* Mayer, *Mauritia flexuosa* L.f. and *Euterpe* ssp.), Anacardiaceae (*Spondias mombin* L.), and Myrtaceae (*Eugenia stipitata* McVaugh and *Myrciaria dubia* McVaugh).

Although evidence of bee visitation has been noted for many important plants in the Amazon, few have confirmed its importance in the process of pollination. However, many species of stingless bees seem to be promising for pollination in agroforestry systems, especially those of the genus *Melipona*. Hence, Roubik (1979) suggests that 84% of plants visited by *meliponine* bees are potentially benefiting from its pollination services. This may be associated with the bees' ability to extract pollen from plants with poricidal and non-poricidal anthers (Buchmann 1983; Proença 1992), especially several species of Myrtaceae, Fabaceae, Melastomataceae, and Solanaceae families (Roubik 1989; Endress 1996). The release of pollen is attained by the bees through the vibration of their thorax using the flight muscles. The whole process is called pollination by vibration or "buzz pollination" (Buchmann and Hurley 1978).

Furthermore, in other regions of Brazil, there is a growing body of evidence corroborating the great value of stingless bees as potential pollinators of economic importance (Malagodi-Braga and Kleinert 2004; Cruz et al. 2004; Del Sarto et al. 2005). The main interest in stingless bees as pollinators derives from the simple requirements to permit their management. Moreover, different species of stingless bees may be managed within the same area, adding conservation value to this activity. From the observation of different species interactions, it is possible to understand complex and significant ecological interactions (Rech et al. 2013), highlighting meliponiculture as a low impact and enjoyable practice.

Although social bees are true generalists that are able to temporarily specialize on profitable food sources, as noted above they are not always good pollinators. Rech and Absy (2011a, b) found that a large proportion of the species used as pollen sources by bees are not necessarily pollinated by them. Several authors note that social bees are very efficient at collecting pollen, sometimes destroying floral parts and behaving as thieves or robbers, and this may reduce pollen transfer to conspecific stigmas (Renner 1983). Considering this possibility, the authors recommended the association of different sources of evidence when the objective is to look at the mutualistic nature of this interaction. Otherwise, it is possible to confuse a potentially mutualistic network with another one that is purely trophic. Both cases are very interesting from an ecological point of view, but the different interpretations may lead to very different conclusions.

Comparing data from observations with pollen evidence, Rech and D'Apolito (unpublished data) studied plant-bee interaction networks in a Campina area (open scrubland surrounded by Amazonian forest) near Manaus-Amazonas. In total, the authors found 19 bee species interacting with nine plants. When data from pollen were considered together with data from observations, the network was considered nested (see Fig. 3.3). When considering only one source of evidence (observation or pollen), the network structure was not nested, and the interaction numbers were halved. The same study found an intense movement of the bees between open areas and the surrounding forest. The plant *Pradosia schomburgkiana* and the bee *Trigona fulviventris* were the most connected points in the network (see Fig. 3.3). Although bees were observed visiting many flowers, when only pollen was taken into account, there was more pollen from the forest on the bodies of the bees than from the plants in the

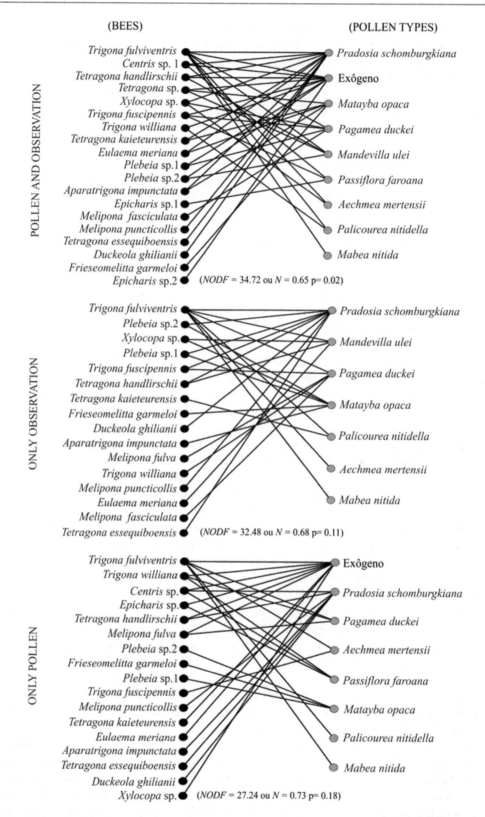

Fig. 3.3 Different bee-plant interaction networks from areas of Campina (open scrubland areas inside the Amazon forest) produced using evidence from pollen and observation (**a**) only pollen (**b**) and only observation (**c**)

open area. This is probably because bees had preferred larger amounts of pollen offered by trees instead of scattered portions of flowers from shrubs or herbs. Considering the bee perspective, many plants that are actually important as food sources will not be found if only pollen evidence is considered, especially when the plants are visited exclusively for nectar (Villanueva-Gutiérrez and Roubik 2016). In contrast, from the plant perspective, some of those bees are working as antagonists and are just acquiring resources without the outcome of pollination.

3.5 How to Improve Meliponiculture for Sustainable Development in the Amazon

Knowledge about the plants used as trophic or nesting resources by bees is a basis for the conservation and maintenance of promising species to produce honey in the Amazon region. The accumulated data on plants used by bees in the Amazon show that regardless of their huge generalist potential, stingless bees tend to rely upon a few continuous sources of pollen throughout the year (Roubik and Moreno 1990, 2000, 2013 and in this volume). These plants become factors that determine the maintenance of honey production over time. Studies on *Melipona* (*Michmelia*) *seminigra merrillae* and *Melipona* (*Melikerria*) *interrupta* (see Fig. 3.4), which are the two main species used in the beekeeping industry within the Central Amazon region, reveal a predominance of pollen

from Fabaceae, Melastomataceae, Myrtaceae, and Anacardiaceae in the bee diet from várzea (floodplains) and riverside areas (see Fig. 3.5). Moreover, ecological indexes show a large degree of seasonality and a high overlap in the pollen resources used, with a high proportion of a few shared species or pollen types (Ferreira and Absy 2015).

Starting with the many studies developed in the Amazon (Absy and Kerr 1977; Absy et al. 1980; Marques-Souza 1996; Oliveira et al. 2009; Ferreira and Absy 2013, 2015), the two main stingless species used for honey in the region have spread beyond the Central Amazon region. The main species managed in the Central Amazon region, *M. seminigra merrillae* and *M. interrupta*, have differing reasons for their popularity (Absy et al. 2013). For the first species, characteristic high honey production and easy adaptation to managed conditions are foremost. The second species, *M. interrupta*, is becoming popular due to the different types and amazing tastes of the honey it produces within the várzea riverside areas. The understanding of the plants mainly used by these managed species has shown the potential to justify forest preservation and increase fruit production due to better pollination, promoted by *meliponine* bees (Roubik 1995).

The largest human population of the Amazon is located in the várzea riverside region, and it is where most of the honey from stingless bees is produced. However, the várzea vegetation is less diverse than is that of the adjacent terra firme forest. In this region, both flora and fauna are adapted to the seasonally flooded conditions of the river (Kalliola et al. 1993; Peixoto et al.

Fig. 3.4 Stingless bees that are more frequently managed in beekeeping in the Central Amazon: (**a**) *Melipona interrupta* and (**b**) *Melipona seminigra merrillae* (Photo MG Ferreira)

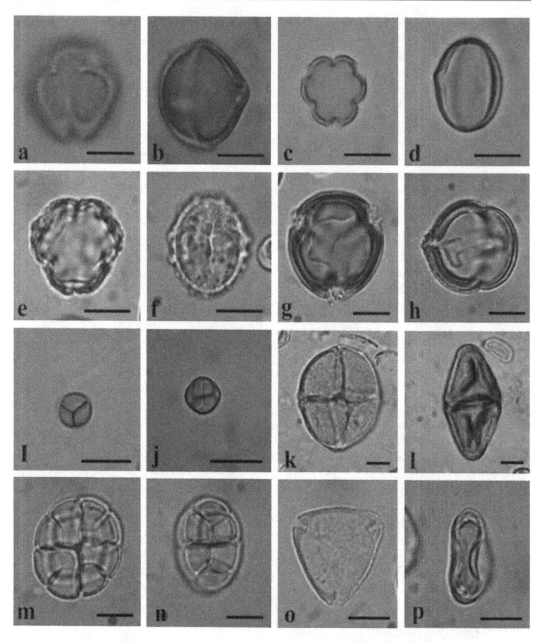

Fig. 3.5 Most of the representative pollen types shared by *Melipona seminigra merrillae* and *Melipona interrupta*. Anacardiaceae – *Tapirira guianensis* (**a**, **b**); Melastomataceae – *Miconia* type (**c**, **d**) and *Bellucia* type (**e**, **f**); Fabaceae/Faboideae – *Swartzia* type (**g**, **h**); Fabaceae/Mimosoideae, *Mimosa pudica* (**i**, **j**), *Mimosa sensitiva* (**k**, **l**), and *Mimosa guilandinae* (**m**, **n**); and Myrtaceae – *Plinia cauliflora* (**o**, **p**) (Adapted from Ferreira and Absy 2015)

2009). According to Moure and Kerr (1950), stingless bees maintain a very close relationship with floodplain areas in the Amazon, where most of the species of the genus *Melipona* are present, displaying a highly concentrated species diversity. Moreover, Dr. Warwick Estevam Kerr (personal communication) suggests that várzea areas promote the occurrence of tree hollows which provide the natural nest sites for many species of stingless bees.

Some species that are often used by bees to produce honey (both species *M. seminigra merrillae* and *M. interrupta* – Ferreira 2014) and feed larvae (Ferreira and Absy 2013), such as *Triplaris weigeltiana*, are endemic in the várzea riverside region and deserve special attention when sustainable strategies such as beekeeping are considered as an economic alternative for local communities (Wittmann et al. 2013).

3.6 Conclusions

Palynology has contributed to our knowledge about plant-bee interactions. The evidence produced using pollen grains has helped to shed light on the ecology of this potentially mutualistic interaction and also suggests ways to address questions regarding animal behavior and adaptive processes between plants and pollinators. Such knowledge can and must be applied in conservation and wildlife management programs, especially given the economic importance of bees as pollinators and the widespread perception of their decline in recent years.

Although most of the published studies concerning the plant-bee interaction networks are based on focal observations of flower visitation, recent studies also incorporate evidence from pollen analyses. As we have shown here, depending on the ecological context, this innovation may represent a significant difference in methods used to assess evidence for a particular interaction. Moreover, the rich diversity of plants already described in the Amazon region and the limitation to consistently observing all interactions in detail reinforce the need for different sources of evidence. In this context, palynology, especially as supported by Erdtman's method, emerges as a powerful tool to access evidence that is more closely related to what animals and plants are actually doing.

Finally, it was with the help of palynological tools that the trophic requirements of many bee species were elucidated, and this knowledge has promoted and improved meliponiculture in the Amazon region. This promising partnership between two areas of knowledge must be emphasized once meliponiculture becomes a sustainable economic activity, promoting not only bee multiplication and preservation but also the improvement of fruit production and the development of agro-ecological activities. It may thus come to encompass social, economic, and ecological dimensions. Meliponiculture is therefore likely to promote a real interchange of popular and scientific knowledge and is one of the ways to maintain forests and humans in this region of the planet, as it was for centuries before colonization.

Acknowledgments We thank Patricia Vit for inviting this contribution and the reviewers for all the corrections. We also thank the Laboratory of Palynology of the Instituto Nacional de Pesquisas da Amazônia (INPA) for sample preparation and analysis, Fundação de Amparo a Pesquisa do Estado do Amazonas (FAPEAM) for the scholarship awarded to the third author and for funding (Proc. 062.01180/2015), and Conselho Nacional de Desenvolvimento Científico e Tecnológico (CNPq) for the scholarship awarded to the first author and for funding (Proc. 477127/2011-8).

References

Absy ML, Bezerra EB, Kerr WE. 1980. Plantas nectaríferas utilizadas por duas espécies de *Melipona* da Amazônia. Acta Amazonica 10: 271-281.

Absy ML, Camargo JMF, Kerr WE, Miranda IPA. 1984. Espécies de plantas visitadas por Meliponinae (Hymenoptera: Apoidea), para coleta de pólen na região do médio Amazonas. Revista Brasileira de Biologia 44: 227-237.

Absy ML, Ferreira MG, Marques-Souza AC. 2013. Recursos tróficos obtidos por abelhas sem ferrão na Amazônia Central e sua contribuição a meliponicultura regional. PP. 147-158. In Bermúdez EGC, Teles BR, Rocha RA, eds. Entomologia na Amazônia Brasileira. Editora INPA; Manaus, Brasil. 234 pp.

Absy ML, Kerr WE. 1977. Algumas plantas visitadas para obtenção de pólen por operárias de *Melipona seminigra merrillae* em Manaus. Acta Amazonica 7: 309-315.

Bastolla U, Fortuna MA, Pascual-Garcia A, Ferrera A, Luque B, Bascompte J. 2009. The architecture of mutualistic networks minimizes competition and increases biodiversity. Nature 458: 1018-1020.

Bawa KS. 1990. Plant-Pollinator Interactions in Tropical Rain Forests. Annual Review of Ecology and Systematics 21: 399-422.

Bawa KS, Bulloch SH, Perry DR, Coville RE, Grayum MH. 1985. Reproduction biology of tropical lowland rain forest tree. II. Pollination system. American Journal of Botany 72: 346-356.

Buchmann SL. 1983. Buzz pollination in angiosperms. pp. 73-113. In Jones CE, Little RJ, eds. Handbook of experimental pollination biology. Van Nostrand Reinhold; New York, USA. 558 pp.

Buchmann SL, Hurley JP. 1978. A biophysical model for buzz pollination in angiosperms. Journal of Theoretical Biology 72: 639-657.

Camargo JMF. 2013. Historical biogeography of the Meliponini (Hymnoptera, Apidae, Apinae) of the Neotropical region. pp. 19-34. In Vit P, Pedro SRM, Roubik DW, eds. Pot-Honey: a legacy of stingless bees. Springer, New York. 653 pp.

Cardinal S, Danforth BN. 2013. Bees diversified in the age of eudicots. Proceedings of the Royal Society 280: 20122686.

Corbet SA, Williams IH, Osborne JL. 1991. Bees and the pollination of crops and wild flowers in the European Community. Bee World 72: 47-59.

Cruz DO, Freitas BM, Silva LA, Silva SEM, Bomfim IGA. 2004. Use of the stingless bee *Melipona subnitida* to pollinate sweet pepper (*Capsicum annuum* L.) flowers in greenhouse. In Proceedings of the 8th IBRA International Conference on Tropical Bees and VI Encontro sobre Abelhas, Ribeirão Preto, Brasil. 661 pp.

Danforth BN, Poinar GO. 2011. *Melittosphex burmensis* (Apoidea: Melittosphecidae): detailed description of the morphology, classification, antiquity, and implications for bee evolution. Journal Paleontology 85: 882-891.

de Klerk P, Joosten H. 2007. The difference between pollen types and plant taxa: a plea for clarity and scientific freedom. Eiszeitalt und Gegenwary Quaternary Science Journal 56: 162-171.

Del Sarto MCL, Peruquetti RC, Campos LA. 2005. Evaluation of the neotropical stingless bee *Melipona quadrifasciata* (Hymenoptera: Apidae) as pollinator of greenhouse tomatoes. Journal of Economic Entomology 98: 260-266.

Endress PK. 1996. Diversity and evolutionary biology of tropical flowers. Cambridge University Press. Cambridge, New York, USA. 528 pp.

Engel MS. 2000. A new interpretation of the oldest fossil bee (Hymenoptera: Apidae). American Museum Novitates 3296: 1-11.

Engel MS. 2001. A monograph of the Baltic amber bees and evolution of the Apoidea (Hymenoptera). Bulletin of the American Museum of Natural History 259: 1-192.

Engel MS. 2004. Geological history of the bees (Hymenoptera: Apoidea). Revista de Tecnologia e Ambiente 10: 9-33.

Engel MS, Michener CD. 2013. Geological history of the stingless bees (Apidae: Meliponini). pp. 1-7. In Vit P, Roubik DW, eds. Stingless bees process honey and pollen in cerumen pots. Facultad de Farmacia y Bioanálisis, Universidad de Los Andes; Mérida, Venezuela. xii+170 pp.

Ferreira MG. 2014. Exploração de recursos tróficos por *Melipona (Michmelia) seminigra merrillae* Cockerell, 1919 e *Melipona (Melikerria) interrupta* Latreille,

1811 (Apidae: Meliponini) criadas em meliponários na Amazônia Central. Tese de Doutorado, Instituto Nacional de Pesquisas da Amazônia, Amazonas, Brasil. 164 pp.

Ferreira MG, Absy ML. 2013. Pollen analysis of the post-emergence residue of *Melipona* (*Melikerria*) *interrupta* Latreille (Hymenoptera: Apidae), created rationally in the Central Amazon. Acta Botanica Brasilica 27: 709-713.

Ferreira MG, Absy ML. 2015. Pollen niche and trophic interactions between colonies of *Melipona* (*Michmelia*) *seminigra-merrillae* and *Melipona* (*Melikerria*) *interrupta* (Apidae:Meliponini) reared in floodplains in the Central Amazon. Arthropod-Plant Interactions 9: 263-279.

Fortuna MA, Stouffer DB, Olesen JM, Jordano P, Mouillot D, Krasnov BR., Poulin R, Bascompte, J. 2010. Nestedness versus modularity in ecological networks: two sides of the same coin? Journal of Animal Ecology 79: 1-7.

Free JB. 1993. *Insect pollination of crops*. 2. ed. London: Academic Press. 684 pp.

Grimaldi DA. 1999. The co-radiations of pollinating insects and angiosperms in the Cretaceous. Annals of the Missouri Botanical Garden 86: 373-406.

Hemming J. 2015. Naturalists in Paradise: Wallace, Bates and Spruce in the Amazon. Thames & Hudson, New York, USA. 368 pp.

Hrncir M, Maia-Silva C. 2013. The fast versus the furious - On competition, morphological foraging traits, and foraging strategies in stingless bees. pp. 1-13. In Vit P, Roubik DW, eds. Stingless bees process honey and pollen in cerumen pots. Facultad de Farmacia y Bioanálisis, Universidad de Los Andes; Mérida, Venezuela. xii+170 pp.

Joosten H, de Klerk P. 2002. What's in a name? Some thoughts on pollen classification, identification, and nomenclature in Quaternary palynology. Review of Palaeobotany and Palynology 122: 29-45.

Junker R, Blüthgen N. 2010. Floral scents repel facultative flower visitors, but attract obligate ones. Annals of Botany 105: 777-782.

Kalliola R, Puhakka M, Danjoy W. 1993. Amazonia peruana: vegetación húmeda tropical en el llano sudandino. Gummerus Printing; Turku, Finland. 265 pp.

Kerr WE, Absy ML, Marques-Souza AC. 1986. Espécies nectaríferas e poliníferas utilizadas pela abelha *Melipona compressipes fasciculata* (Meliponinae, Apidae), no Maranhão. Acta Amazonica 16: 145-156.

Lewinsohn TM, Prado PI, Jordano P, Bascompte J, Olesen JM. 2006. Structure in plant–animal interaction assemblages. Oikos 113: 174-184.

Malagodi-Braga KS, Kleinert AMP. 2004. Could *Tetragonisca angustula* Latreille (Apinae, Meliponini) be used as strawberry pollinator in greenhouses? Australian Journal of Agricultural Research 55: 771-773.

Marques-Souza AC. 1996. Fontes de pólen exploradas por *Melipona compressipes manaosensis* (Apidae, Meliponinae), abelha da Amazônia Central. Acta Amazonica 26: 77-86.

Marques-Souza AC. 2010. Ocorrência do pólen de *Podocarpus* sp. (Podocarpaceae) nas coletas de *Frieseomelitta varia* Lepeletier 1836 (Apidae: Meliponinae) em uma área urbana de Manaus, AM, Brasil. Acta Botanica Brasilica 24: 558-566.

Marques-Souza AC, Absy ML, Kerr W.E. 2007. Pollen harvest features of the Central Amazonian bee *Scaptotrigona fuvicutis* Moure 1964 (Apidae: Meliponinae), in Brazil. Acta Botanica Brasilica 211: 11-20.

Marques-Souza AC, Absy ML, Kerr WE, Aguilera-Peralta FJ. 1995. Pólen coletado por duas espécies de Meliponineos (Hymenoptera: Apidae) da Amazônia. Revista Brasileira de Biologia 55: 855-864.

Marques-Souza AC, Miranda IPA, Moura CO, Rabelo A, Barbosa EM. 2002. Características morfológicas e bioquímicas do pólen coletado por cinco espécies de meliponineos da Amazônia Central. Acta Amazônica 32: 217-229.

Marques-Souza AC, Moura CO, Nelson BW. 1996. Pollen collected by *Trigona williana* (Hymenoptera, Apidae) in Central Amazonia. Revista de Biologia Tropical 44: 567-573.

Martins AC, Melo GAR, Renner SS. 2014. The corbiculate bees arose from New World oil-collecting bees: Implications for the origin of pollen baskets. Molecular Phylogenetics and Evolution 80: 88-94.

Michener CD. 2013. The Meliponini. pp. 3-17. In Vit P, Pedro SRM, Roubik DW, eds. Pot honey: A legacy of stingless bees. Springer Verlag; Berlin, Germany. xxviii+654 pp.

Michener CD, Grimaldi DA. 1988. The oldest fossil bee: Apoid history, evolutionary stasis, and antiquity of social behavior. USA. Proceedings of the National Academyof Sciences 85: 6424-6426.

Moure JS, Kerr WE. 1950. Sugestões para a modificação da sistemática do gênero *Melipona*. Dusenia 1: 105-29.

Novais JS, Absy ML. 2013. Palynological examination of the pollen pots of native stingless bees from the Lower Amazon region in Pará, Brazil. Palynology 37: 1-13.

Novais JS, Absy ML. 2015. Melissopalynological records of honeys from Tetragonisca angustula (Latreille, 1811) in the Lower Amazon, Brazil: pollen spectra and concentration. Journal of Apicultural Research 54: 1-19.

Novais JS, Garcêz ACA, Absy ML, Santos FAR. 2015. Comparative pollen spectra of *Tetragonisca angustula* (Apidae, Meliponini) from the Lower Amazon (N Brazil) and caatinga (NE Brazil). Apidologie 46: 417-431.

Ohl M, Engel MS. 2007. Die Fossilgeschichte der Bienen und ihrer nächsten Verwandten (Hymenoptera: Apoidea). Denisia 20: 687-700.

Oliveira FPM, Absy ML, Miranda IS. 2009. Recurso polínico coletado por abelhas sem ferrão (Apidae, Meliponinae) em um fragmento de floresta na região de Manaus-Amazonas. Acta Amazônica 39: 505-518.

Ollerton J, Winfree R, Tarrant S. 2011. How many flowering plants are pollinated by animals? Oikos 120:321-326.

Peixoto JMA, Nelson BW, Wittmann F. 2009. Spatial and temporal dynamics of river channel migration and vegetation in central Amazonian white-water floodplains by remote-sensing techniques. Remote Sensing of Environment 113: 2258-2266.

Proença CEB. 1992. Buzz-pollination - older and more widespread than we think? Journal of Tropical Ecology 8: 115-120.

Rabelo A. 2012. Frutos nativos da Amazônia: comercializados nas feiras de Manaus-AM. Manaus, Brasil. 390 pp.

Ramalho M, Silva MD, Carvalho CAL. 2007. Dinâmica de uso de fontes de pólen por *Melipona scutellaris* Latreille (Hymenoptera: Apidae): uma análise comparativa com *Apis mellifera* L. (Hymenoptera: Apidae), no Domínio Tropical Atlântico. Neotropical Entomology 36: 38-45.

Rasmussen C, Gonzalez VH. 2013. Prologue. pp. v-iv. In Vit P, Roubik DW, eds. Stingless bees process honey and pollen in cerumen pots. Facultad de Farmacia y Bioanálisis, Universidad de Los Andes; Mérida, Venezuela. http://www.saber.ula.ve/handle/123456789/35292.

Raven PH, Axelrod DI. 1974. Angiosperm biogeography and past continental movements. Annals of the Missouri Botanical Garden 61: 539-673.

Rech AR, Absy ML. 2011a. Pollen sources used by species of Meliponini (Hymenoptera: Apidae) along the Rio Negro channel in Amazonas, Brazil. Grana 50: 150-161.

Rech AR, Absy ML. 2011b. Pollen storages in nests of bees of the genera *Partamona*, *Scaura* and *Trigona* (Hymenoptera: Apidae) along the Rio Negro channel in Amazonas, Brazil. Revista Brasileira de Entomologia 55: 361-372.

Rech AR, Dalsgaard B, Sonne J, Svenning JC, Holmes N, Ollerton J. 2016. The macroecology of animal versus wind pollination: the influence of plant richness, vegetation structure, topography, insularity and climate. Plant Ecology & Diversity 9: 253-262.

Rech AR, Manente-Balestieri FCL, Absy ML. 2011. Reproductive biology of *Davilla kunthii* A. St-Hil. (Dilleniaceae) in Central Amazonia. Acta Botanica Brasilica 25: 487-496.

Rech AR, Schwade MA, Schwade MRM. 2013. Abelhas-sem-ferrão amazônicas defendem meliponários contra saques de outras abelhas. Acta Amazonica 43: 389-393.

Renner S. 1983. The widespread occurrence of anther destruction by *Trigona* bees in Melastomataceae. Biotropica 15: 251-256.

Roubik DW. 1979. Africanized honeybees, stingless bees and the structure of tropical plant-pollinator communities. pp 403-417. In Caron D, ed., Proceedings Vth International Symposium on Pollination. Maryland Agricultural Experimental Station Miscellaneous Publication No. 1 College Park, Maryland. 190 pp.

Roubik DW. 1989. Ecology and natural history of tropical bees. Cambridge University Press, Cambridge; New York, USA. 514 pp.

Roubik DW. ed. 1995. Pollination of Cultivated Plants in the Tropics. Food and Agriculture Organization of the United Nations. Agricultural Services Bulletin 18. Rome, Italy. 196 pp.

Roubik DW, Moreno JE. 1990. Social bees and palm trees: what do pollen diets tell us? pp. 427-428. In Veeresh GK, Mallik B, Viraktamath CA, eds. Social insects and the environment: proceedings of the 11th international congress of IUSSI, Bangalore, India. Oxford and IBH Publishing Co. Pvt. Ltd; New Delhi, India. 765 pp.

Roubik DW, Moreno JE. 2000. Generalization and specialization by stingless bees. pp. 112-118. In Proceedings of the 6th International Bee Research Conference on Tropical Bees (IBRA), Cardiff, UK. 289 pp.

Roubik DW, Moreno-P JE. 2013. How to be a bee-botanist using pollen spectra. pp. 295-314. In Vit P, Pedro SRM, Roubik, DW, eds. Pot-honey: a legacy of stingless bees. Springer, New York. 654 pp.

Santamaría L, Rodríguez-Gironés MA. 2015. Are flowers red in teeth and claw? Exploitation barriers and the antagonist nature of mutualisms. Evolutionay Ecology 29: 311-322.

Silva IAA, Silva TMS, Camara CA, Queiroz N, Magnani M, Novais JS, Soledade LEB, Lima EO, Souza AL, Souza AG. 2013. Phenolic profile, antioxidant activity and palynological analysis of stingless bee honey from Amazonas, Northern Brazil. Food Chemistry 141: 3552-3558.

Simpson BB, Neff JL. 1981. Floral rewards: alternatives to pollen and nectar. Annals of the Missouri Botanical Garden 68: 301-322

ter Steege H, Sabatier D, Castellanos H, Van Andel T, Duivenvoorden J, de Oliveira AA, de Ek R, Lilwah R, Maas P, Mori S. 2000: An analysis of the floristic composition and diversity of Amazonian forests including those of the Guiana shield. Journal of Tropical Ecology 16: 801-828.

Villanueva-Gutiérrez R, Roubik DW. 2016. More than protein? Bee-flower interactions and effects of disturbance regimes revealed by rare pollen in bee nests. Arthropod Plant Interactions. doi:10.1007/s11829-015-9413-9.

Wilkie RKT, Mertl AL, Traniello JFA. 2010. Species diversity and distribution patterns of the ants of Amazonian Ecuador. PLoS ONE 5: e13146.

Wittmann F, Householder E, Piedade MTF, Assis RL, Schöngart J, Parolin P, Junk WJ. 2013. Habitat specificity, endemism and the neotropical distribution of Amazonian white-water floodplain trees. Ecography 36: 690-707.

The Stingless Honey Bees (Apidae, Apinae: Meliponini) in Panama and Pollination Ecology from Pollen Analysis

4

David W. Roubik and Jorge Enrique Moreno Patiño

4.1 An Introduction to the Stingless Honey bees and Pot-Pollen, in Panama

Stingless bees that make honey (the stingless honey bees, meliponines—after Michener 1974) are not defenseless, nor are they as uniform in general biology as the stinging honey bees, genus *Apis*, given an almost trademark name, "the honey bee," which is employed here. The former sometimes take animal flesh instead of protein; pick up pollen from leaves and petals instead of collecting it at anthers (also done by Africanized *A. mellifera*, DWR pers. obs.); chew holes in anthers, or floral corollas, to remove pollen or nectar; "buzz" or vibrate anthers to receive pollen; use "bugs" such as membracids or *Cryptostigma*, for wax or carbohydrate food; share nesting sites with termites or ants; have mutualistic fungi living in pollen; and chew into the sap layer in living plant stems, to remove carbohydrate resources. Meliponines include both parasites and relatively inoffensive inquilines (Roubik 1989, 2006; Michener 2013; Camargo 2013). Honey bees do some of those same things but can regulate brood nest temperature with incomparable precision during cold conditions and are occasionally parasites that rob food or usurp nests of conspecifics. However, they must not have evolved the morphology, biochemistry, or mutualisms that underlie many meliponine biological traits. Nor do they possess the antiquity or range, since meliponines are at least twice as old as honey bees, occupy (natural distribution) twice as much territory, and include >50 times more species (Michener 2007; Roubik 1989, 2006, 2013). The meliponines "win" in their niche range and adaptability—by age, distribution, and number.

The human processes causing tropical deforestation, however, may allow *Apis mellifera* to become more abundant, as the abovementioned biological factors are simplified, many likely causing meliponine populations or local species richness to decline (e.g., Roubik 1988, 1991, 2009; Roubik and Villanueva-Gutiérrez 2009; Vit et al. 2013; Vit and Roubik 2013). Meliponine colonies cannot move from where they live until they build and provision (with food) a new nest, to which a queen (almost always an unmated queen) and group of workers fly (Roubik 2006). They are an integral part of intact forest but unfortunately are often regarded simply as

D.W. Roubik (✉)
Smithsonian Tropical Research Institute, Balboa, Ancon, Republic of Panama
e-mail: roubikd@si.edu

J.E. Moreno Patiño
Smithsonian Tropical Research Institute, Calle Portobelo, Balboa, Ancon, Republic of Panama

© Springer International Publishing AG, part of Springer Nature 2018
P. Vit et al. (eds.), *Pot-Pollen in Stingless Bee Melittology*, DOI 10.1007/978-3-319-61839-5_4

generalists in flower visitation, without further analysis (but see, e.g., Sommeijer et al. 1983; Cane and Sipes 2006). In reality, they may have intricate yet scarcely appreciated *organized* ecological interactions within their foraging range, involving the local flora, landscape elements, and many organisms. Here we review meliponine biology in large forest areas, and apply melittopalynology, in our Panama fieldwork.

The analysis of total nest pollen was always a goal for our field studies in Panama. After a colony was taken from its tree or other natural nesting sites, using an axe, shovel, machete, or chainsaw, and phorid flies (*Pseudohypocera*) often destroyed it shortly thereafter, the pollen and honey, and some specimens for collections, were almost all that remained of the field effort or the bees (Fig. 4.1). Scientific study of pollen requires attention, curation, and appropriate techniques to provide insight that justifies removal of the bee colony from its environment. Because a comprehensive work on the *Flora of Barro Colorado Island, Panama*, had just been completed (Croat 1978), that island's flora helped us establish a goal in collecting pollen from identified plants and building a pollen reference collection for taxonomy. Barro Colorado Island contains >1369 vascular plants growing in a seasonal moist forest (9°09′N, 79°59′W) having approximately 2600 mm annual rainfall. Given the paucity of data on food importance to perennial bee populations in any of the broad classes of tropical environment (dry forest, moist forest, or rain forest), such studies, in natural habitat, should yield considerable insight. A reference collection of >683 genera and 1270 species and key to pollen were made for Barro Colorado Island, in the Panama Canal, where approximately 500 genera and 1000 species are considered representative of old, stable forest systems (Roubik and Moreno 1991). Further botanical survey has been conducted across the Isthmus and Canal Area (D'Arcy and Correa 1985; Pérez 2008), which contains 2000 angiosperms and >200 tree species alone.

The correctly identified plants with pollen material yielded pollen to mount on microscope slides and preserve for comparative study. Due to the location of Barro Colorado Island in low-

Fig. 4.1 Honey and pollen pots of the largest meliponine, *Melipona fallax* (Camargo and Pedro 2008), taken from the nest of a natural mixed colony, with *M. panamica*, in the same tree cavity (Roubik 1981) (Photo by D. W. Roubik)

lands at the center of the Isthmus of Panama, its flora contains both drier, seasonal forest elements of the Pacific corridor and those of the wetter Caribbean forests. Annual rainfall usually ranges from approximately 1400 to 3500 mm along the 76 km forest transect of the Panama Canal area in central Panama, between the Pacific Ocean and the Caribbean Sea. With over 500 angiosperm genera in our reference collection, we could begin to understand the details of plant-bee interactions, using the comprehensive data provided within the bee nests themselves. Since that time, many new data on Meliponini and the plants they use have become available through our hundreds of field studies in the Santa Rita Ridge and El Llano-Cartí areas, and elsewhere in the greater Chagres National Forest region, and continue today. Over a dozen new meliponine species and one new genus were discovered in Panama since Schwarz and Michener inspected this rich eusocial tropical bee biota (Table 4.1). Here we summarize existing knowledge, present new information, and analyze stingless bees and their pot-pollen ecology in Panama. We use the named bee species groups as taxonomic genera, as formulated from Michener (2007) and Camargo and Pedro (2007).

We intensively and extensively surveyed moist to wet lowland forest of Chagres, Portobelo, and Soberanía National Parks. Their 200,000 ha (conventional horizontal area, not topographic area) support the Panama Canal watershed. In Panama's 70,000 km², 63 meliponine species and 21 genera have been found (Table 4.1). Within the Panama Canal watershed, perennial bee colonies dominate the forest and interact with many plants but at unknown rates and among largely undocumented, coexisting mutualists, natural enemies, or commensals. The nest and colony characteristics, and discovery of obligate necrophagy—complete lack of pollen collection by a "free-living" (nonparasitic) bee—among meliponines, and mixed natural colonies, were described in the 1980s (Roubik 1981, 1982, 1983). Panamanian meliponines had already received some attention (Cockerell 1913; Schwarz 1932, 1934, 1948, 1951; Michener 1954). New species and a genus from Panama were later described by Camargo and Roubik (1991), Roubik et al. (1997), Pedro and Camargo (1997), Camargo and Pedro (2008), and Roubik and Camargo (2012), bringing to 24 the number of "valid" (non-synonymous) meliponine species first described from there (Table 4.1). A rather comprehensive regional meliponine faunal list is given for neighboring Costa Rica (Griswold et al. 1995; Aguilar et al. 2013), which shares the most meliponine taxa with Panama. Schwarz (1934) listed 26 stingless honey bees thought to be found on Barro Colorado Island. On BCI, there are ≥28 meliponine species (Wolda and Roubik 1986), but 2 of those mentioned in Schwarz (1934), *Melipona favosa phenax* and *Melipona beecheii* Bennett, 1831, do not exist there, as Schwarz (1934) and Michener (1954) recognized for the latter. *Trigona necrophaga* (Camargo and Roubik 1991) and *T.*

amalthea Olivier (1789) live in nearby mainland within Parque Soberanía but not on BCI (Wolda and Roubik 1986; DWR, pers. obs.). Across the central isthmian area, the drier Pacific side contains 22 meliponines, the moist forest near Barro Colorado Island in Soberanía National Park has 30, and the wetter Caribbean lowland forest contains 47 Meliponini (Roubik 1992a, 1993, and unpublished). Whereas Michener (1954) lists 46 Panamanian meliponines, nearly 45% more ≥63 species—some awaiting description and comparative study—have now been collected within Panama's borders (Table 4.1). At least one species, *Ptilotrigona occidentalis* (Schulz, 1904), is found in eastern Costa Rica and the Darién, but evidently nowhere in between, in all of Panama (Roubik and Camargo 2012), indicating biogeographic complexity within the region.

4.2 Pollen Niche, Relative Specialization, and Pollen Spectrum

At the International Bee Research Association meeting "Sixth International Conference on Apiculture in Tropical Climates" in Costa Rica in 1996, we examined the pollen spectrum from colonies of *Scaura latitarsis* (now *S. argyrea*), *Cephalotrigona capitata zexmeniae* (now *C. zexmeniae*), and *Tetragona dorsalis ziegleri* (now *T. ziegleri*—Table 4.2). We included bee colonies in the greater forested Chagres/BCI (Barro Colorado Island)—Soberanía Park area—and there were four, five (one sampled for current pollen stores and accumulated pollen feces), and three colonies considered, respectively.

4.2.1 Qualitative and Quantitative Analyses

Further examination of pollen data from our work with *Scaura*, *Cephalotrigona*, and *Tetragona* provides information on pollination per se (see Fig. 4.2 and below). Different approaches to pollen *quantification* or simple *qualification* (identification of pollen species, types, or genera) are employed in such results and conclusions. Further examples, mainly

Table 4.1 Stingless honey bees (Meliponini) found in Panama

Scientific name #	Previous name* or synonym applied	Biological notes
**Aparatrigona isopterophila (Schwarz, 1934)	Trigona (Paratrigona) isopterophila	Living termitarium nesting
Cephalotrigona zexmeniae (Cockerell, 1912)	Trigona (Cephalotrigona) capitata zexmeniae*	Dimorphic coloration
Dolichotrigona schulthessi (Friese, 1900)	Trigona (Trigonisca) schulthessi	
**Frieseomelitta paupera (Provancher, 1888)	Trigona (Tetragona) nigra paupera, T. (T.) nigra doederleini*	Subspecies status?
**Geotrigona kraussi (Schwarz, 1951)	Trigona (Geotrigona) acapulconis kraussi*	Ground nesting
**Geotrigona chiriquiensis (Schwarz, 1951)	Trigona (Tetragona) leucogastra chiriquiensis	Ground nesting
Lestrimelitta danuncia Oliveira and Marchi, 2005	Lestrimelitta limao*	Cleptoparasite
Melipona costaricensis Cockerell, 1919	Melipona fasciata panamica, M. costaricensis melanopleura*	W. Panama
**Melipona fallax Camargo and Pedro, 2008	M. fuliginosa*, M. flavipennis*	
**Melipona insularis Roubik and Camargo, 2012		Coiba Island endemic
**Melipona micheneri Schwarz, 1951	M. marginata micheneri*	
Melipona melanopleura Cockerell, 1919	Melipona fasciata melanopleura*, M. costaricensis*	
**Melipona panamica Cockerell, 1912	M. costaricensis*, M. fasciata paraensis*	
Melipona aff. phenax		Perlas Islands, Pacific
**Melipona phenax Cockerell, 1928	Melipona favosa phenax*	
**Melipona triplaridis Cockerell, 1925	Melipona interrupta triplaridis*, M. beecheii*	
Melipona aff. crinita	Melipona crinita*	Eastern Pacific
Meliwillea bivea Roubik, Lobo and Camargo, 1997		Far W. Highlands
Nannotrigona perilampoides (Cresson, 1878)	Trigona (Nannotrigona) testaceicornis perilampoides*	
**Nannotrigona mellaria (Smith, 1862)		
Nogueirapis mirandula (Cockerell, 1917)	Plebeia (Nogueirapis) mirandula	Ground or tree nesting
Oxytrigona daemoniaca Camargo, 1984		
**Oxytrigona isthmina Gonzalez and Roubik, 2008		
Oxytrigona mellicolor (Packard, 1869)	Trigona (Oxytrigona) tataira mellicolor*	
**Paratrigona lophocoryphe Moure, 1963		Exposed nesting
**Paratrigona opaca (Cockerell, 1917)	Trigona (Paratrigona) opaca pacifica*	Living ant nest
**Paratrigona ornaticeps (Schwarz, 1938)		
Partamona aequatoriana Camargo, 1980		Exposed nesting
Partamona grandipennis (Schwarz, 1951)		Presumed exposed nesting
Partamona musarum (Cockerell, 1917)		Partially exposed nesting
Partamona peckolti (Friese, 1901)	Trigona (Partamona) cupira*	Exposed nesting or partially exposed nesting, termite nest
Partamona orizabaensis (Strand, 1919)	P. cupira*	Partially exposed nesting

(continued)

Table 4.1 (continued)

Scientific name #	Previous name* or synonym applied	Biological notes
**Partamona xanthogastra* Pedro and Camargo, 1997		Partially exposed nesting
Plebeia minima (Gribodo, 1893)	*Trigona (Plebeia) minima*	Probably undescribed sp. exposed nesting on *Bactris*
Plebeia jatiformis (Cockerell, 1912)	*Trigona (Plebeia) jatiformis*	
Plebeia frontalis (Friese, 1911)	*Trigona (Plebeia) frontalis*	
Plebeia franki (Friese, 1900)	*Trigona (Plebeia) domiciliorum*	
Plebeia sp.		Caribbean lowland
Ptilotrigona occidentalis (Schulz, 1904)	*Trigona (Ptilotrigona) occidentalis, T. (P.) lurida occidentalis*	Darién
Scaptotrigona sp.		Pacific lowland and islands
**Scaptotrigona barrocoloradensis* (Schwarz, 1951)		
Scaptotrigona luteipennis (Friese, 1902)	*Trigona (Scaptotrigona) pachysoma*	
Scaptotrigona pectoralis (Dalla Torre, 1896)	*Trigona (S.) pectoralis panamensis*	
**Scaptotrigona subobscuripennis* (Schwarz, 1951)	*Trigona (S.) mexicana subobscuripennis*	Ground nesting, W. Panama
Scaura argyrea (Cockerell, 1912)	*Scaura latitarsis*	Living termitarium nesting
Scaura sp.		Caribbean lowland, aff. *S. longula*
**Tetragona perangulata* (Cockerell, 1917)	*Trigona (Tetragona) clavipes perangulata**	
**Tetragona ziegleri* (Friese, 1900)	*Trigona (Tetragona) dorsalis**	
Tetragonisca buchwaldi (Friese, 1925)		Ground nesting
Tetragonisca angustula (Latreille, 1811)	*Trigona (Tetragona) jaty*	Ground and tree nesting
Trigona amalthea (Olivier, 1789)	*Trigona silvestriana*, T. trinidadensis**	Exposed scutellum nesting
Trigona cilipes (Fabricius, 1804)	*T. compressa**	
Trigona corvina Cockerell, 1913		Exposed scutellum nesting
**Trigona ferricauda* (Cockerell, 1917)		Living termitarium nesting
Trigona fulviventris Guérin, 1844		Ground nesting
**Trigona fuscipennis* Friese, 1900	*Trigona amalthea**	Living termitarium nesting
Trigona muzoensis Schwarz, 1948	*Trigona pallida, Trigona pallens**	Living termitarium nesting
**Trigona necrophaga* Camargo and Roubik, 1991	*Trigona hypogea**	Obligate necrophage, Roubik 1982
Trigona silvestriana Vachal, 1908		Far W. Panama
Trigona aff. *silvestriana*		Darién
Trigonisca atomaria (Cockerell, 1917)	*Trigona (Trigonisca) duckei atomaria**	
Trigonisca sp. 1	*Trigonisca buyssoni**	n. sp. see Ayala (1999)
Trigonisca sp. 2		Darién
**Trigonisca roubiki* Albuquerque and Camargo, 2007		Eastern Panama

Nomenclature follows Camargo and Pedro (2008), Camargo (2013), and DWR, unpublished, uploaded (2013) to IABIN (Inter-American Biodiversity Information Network), and nesting biology from Roubik (1983, 1992a, b, 2006) and Michener (1974). Unless otherwise noted, colonies nest in hollow woody stems of trees and/or lianas. #Taxonomic author names and dates can almost always be found on the worldwide web at sites dedicated to taxonomy (e.g., ITIS, GBIF, "Moure Bee Catalog"); *previously misidentified, poorly known; see Schwarz (1934), Michener (1954), Yurrita et al. (2016), Roubik (1992a, b), and this volume; **primary type (holotype, etc.) from Panama

Table 4.2 Pollen identified and slide-mounted pollen counts in multiple nests of five meliponine species in Panama

Pollen	H	*	Carti Rd.				Santa Rita				Pipeline Rd.		BCI		Cerro Jefe	Por	For	Cur
Family/Genus/Species			Cz	Tz	Sa	Mp	Cz	Tz	Sa	Mp	Tz	Mp	Sa	Mp	Sa	Mp	Mp	Tc
Acanthaceae Mendoncia gracilis				p														
Acanthaceae Thunbergia sp. (exotic)												p					p	
Amaranthaceae Undetermined (two types)			p				p											
Anacardiaceae Spondias spp. (two types)	t		p		12		16	5	29	4		26	p	p	5	15	1	895
Annonaceae Annona spp. (two types)				1			4	4	2	1		p			p			
Apiaceae Eryngium sp.				p						1								
Apocynaceae Prestonia sp.								p										
prob. Apocynaceae Undetermined.								1										
Aquifoliaceae Ilex sp.			6				17			p							p	
Araceae Anthurium sp.	e	■			31			641	163	66					34	p		1
Araliaceae Dendropanax arboreus					97			1287										
Araliaceae Schefflera morototoni	t	■	85	8		481	65	40		48				229				317
Arecaceae aff. Bactris sp.				p				1	5									
Arecaceae Chamaedorea sp.	t	■			330	262	159	57	6									
Arecaceae Cryosophila warscewiczii	t	■	21		1275		93											13
Arecaceae Cocos nucifera (exotic)					128				8									74
Arecaceae Elaeis oleifera	t	■	11			106	41	11	273	717		25			p	83		509
Arecaceae Geonoma sp.										48								
Arecaceae Iriartea deltoidea	t	■	607	49	133	518	1155	877	155	169					p			
Arecaceae Oenocarpus mapora				3				22		10	p							
Arecacea Phytelephas macrocarpa							3		12	247						1		4
Arecaceae Attalea rostrata	t	■				3672	26	p	11	812						153	371	6519
Arecaceae Socratea exorrhiza	t	■	12	20	9	51	65	612	78	67	p	1				3		
Asteraceae Undetermined 1	h	■	23	4			276	29										
Asteraceae Undetermined 2	h	■	1055	423		27	1697	958	35	8				p	p		p	56
Asteraceae Vernonia sp.			77	p	p		134	11		3								1
aff. Begoniaceae Begonia sp.							36											

(continued)

Table 4.2 (continued)

Pollen	H	*	Carti Rd.				Santa Rita				Pipeline Rd.		BCI		Cerro Jefe	Por	For	Cur
Family/Genus/Species			Cz	Tz	Sa	Mp	Cz	Tz	Sa	Mp	Tz	Mp	Sa	Mp	Sa	Mp	Mp	Tc
Bignoniaceae *Arrabidaea* sp.	I	■	633		2	79	834	1	178	822		45	3	3		129	76	20
Bignoniaceae aff. *Mansoa* sp.										1								
Bignoniaceae *Martinella obovata*							p	p										
Bignoniaceae *Amphilophium* sp.							1											5
Bignoniaceae *Handroanthus guayacan*				p			3	1	72	12								2
Bixaceae *Bixa orellana*										17							26	
Boraginaceae *Cordia* sp.			1				1		1	11	1					p	19	22
Boraginaceae *Heliotropium* sp.															50			
Boraginaceae *Tournefortia* sp.										2								
aff. Bromeliaceae *Billbergia* sp.					p			p	9						15			
Bromeliaceae *Catopsis sessiliflora*																		5
aff. Bromeliaceae Undetermined			p				4		15									
Burseraceae *Bursera simaruba*	t					p			2	50								566
Burseraceae *Protium* sp.	t	■			58	9	52		56	49		16	2978	11		6	14	
aff. Cactaceae *Epiphyllum* sp.							p											
Calophyllaceae *Marila laxiflora*			3	p			p	11				1						
Cannabaceae *Celtis* sp.	I	■						544			1044							5
aff. Celastraceae *Maytenus schippii*	t	■	371				420											
Chrysobalanaceae *Hirtella racemosa*								p	2									
Cleomaceae *Cleome parviflora*										12							26	
Clusiaceae *Clusia odorata*								51										
Clusiaceae *Symphonia globulifera*			1															
Bixaceae *Cochlospermum vitifolium*			p				888											
Combretaceae *Combretum* sp.					p					35							4	5
Commelinaceae *Commelina* sp.				1														
Connaraceae *Connarus* sp.	I	■	205			180	52		50	260							3	
Convolvulaceae *Aniseia* sp.										1								
Convolvulaceae *Evolvulus* sp.																		1
Convolvulaceae *Iseia luxurians*					6					p								

(continued)

Table 4.2 (continued)

Pollen	H	*	Locality															
			Carti Rd.				Santa Rita				Pipeline Rd.		BCI		Cerro Jefe	Por	For	Cur
Family/Genus/Species			Cz	Tz	Sa	Mp	Cz	Tz	Sa	Mp	Tz	Mp	Sa	Mp	Sa	Mp	Mp	Tc
Convolvulaceae *Maripa panamensis*					39		13	p		1		1						2
Convolvulaceae *Merremia* sp.				p			2	1										
Cucurbitaceae *Cayaponia* sp.			p	3			1	41	p	4					p	5		
Cucurbitaceae *Fevillea cordifolia*				1			1											
Cucurbitaceae *Melothria* sp.										214				742	20			7
Cucurbitaceae *Momordica* sp. (exotic)							1	p		2								
Cucurbitaceae *Posadaea* sp.				1			3								p			
Cucurbitaceae *Sechium edule*				2			4											
Cyclanthaceae *Carludovica palmata*					47		2											34
Cyperaceae *Cyperus* sp.				88				564							1			
Dilleniaceae *Doliocarpus* sp.	l	■	2		19	476	177	1152	63	131			13					66
Dilleniaceae *Tetracera* sp.										5								
Muntingiaceae *Muntingia calabura*			105				821		9									
Ericaceae Unknown					1										p		1	
Erythroxylaceae *Erythroxylum* sp.										7			185					
Euphorbiaceae *Acalypha* spp.					371			p	1746	2			1111					
Euphorbiaceae *Alchornea latifolia*	t	■	76			p	218		p	85	1381							
Euphorbiaceae *Chamaesyce* sp.	h							626										2872
Euphorbiaceae *Croton* sp.			21		8		92		12			p		1				30
Euphorbiaceae *Dalechampia dioscoreifolia*															2			
Euphorbiaceae *Euphorbia* sp.																		1
Euphorbiaceae *Hura crepitans*				1	16				3									
Euphorbiaceae *Manihot esculenta*			4			2	3			p								1
Euphorbiaceae *Sapium* sp.			41				7	p	p						4			
Fabaceae-Caesalpinioideae *Bauhinia reflexa*	l	■	79				280			12							9	45
Fabaceae-Caesalpinioideae *Bauhinia ungulata*				1			p	1										
Fabaceae-Caesalpinioideae *Caesalpinia pulcherrima*									2	p								
Fabaceae-Caesalpinioideae *Cassia* sp.							7	1		8								

(continued)

Table 4.2 (continued)

Pollen	H	*	Locality															
			Carti Rd.				Santa Rita				Pipeline Rd.		BCI		Cerro Jefe	Por	For	Cur
Family/Genus/Species			Cz	Tz	Sa	Mp	Cz	Tz	Sa	Mp	Tz	Mp	Sa	Mp	Sa	Mp	Mp	Tc
Fabaceae-Caesalpinioideae Macrolobium sp.				p	2		28		p	3								
Fabaceae-Caesalpinioideae Peltophorum?			9				356		11									
Fabaceae-Faboideae Swartzia panamensis										p								
Fabaceae-Caesalpinioideae Tachigali versicolor							22			568								
Fabaceae-Faboideae Calopogonium mucunoides										139					52			
Fabaceae-Faboideae Clitoria sp.								1										
Fabaceae-Faboideae Dalbergia aff. brownei										p		1598						
Fabaceae-Faboideae Desmodium sp.					p			2		22		35	1					
Fabaceae-Faboideae Dioclea wilsonii							p			107								
Fabaceae-Faboideae Erythrina sp.							p	5										117
Fabaceae-Faboideae Machaerium sp.	l	■	1042	133			248	122		24		202	p	1780			1	253
Fabaceae-Faboideae Pterocarpus sp.	t	■	33			137	182		331	121				171				1
Fabaceae-Mimosoideae Acacia sp.			p	1			3	112		1							p	
Fabaceae-Mimosoideae Calliandra pittieri										p								
Fabaceae-Mimosoideae Enterolobium cyclocarpum							p			14		2					3	
Fabaceae-Mimosoideae Inga sp.			2	2		1	3	1	2	2				1		p	p	1
Fabaceae-Mimosoideae Mimosa sp. (*invisa/pigra*)	s	■				1120				26					2	1	14	
Fabaceae-Mimosoideae Mimosa sp. (*pudica/casta*)	s					61	1		2	191		p			34	1	17	323
Salicaceae Casearia sp.							11						p					1
Gentianaceae Xestaea lisianthoides							p					p						
Humiriaceae Humiriastrum sp.								16	p									4
Hypericaceae Vismia sp.							4					27						
Lamiaceae Hyptis sp.							2	p										
Lecythidaceae Gustavia sp.	t	■				932				301	p							
Lentibulariaceae. Utricularia sp.													p					
Asparagaceae Cordyline fruticosa																		1
Loganiaceae Spigelia scabra							p											
Loranthaceae Struthanthus sp.			3				26	1		1							p	

(continued)

Table 4.2 (continued)

Locality headers span: **Carti Rd.** (Cz, Tz, Sa, Mp); **Santa Rita** (Cz, Tz, Sa, Mp); **Pipeline Rd.** (Tz, Mp); **BCI** (Sa, Mp); **Cerro Jefe** (Sa); **Por** (Mp); **For** (Mp); **Cur** (Tc).

Pollen Family/Genus/Species	H	*	Carti Rd.				Santa Rita				Pipeline Rd.		BCI		Cerro Jefe	Por	For	Cur
			Cz	Tz	Sa	Mp	Cz	Tz	Sa	Mp	Tz	Mp	Sa	Mp	Sa	Mp	Mp	Tc
Loranthaceae aff. *Psittacanthus* sp.										1								
Lythraceae *Lagerstroemia* (exotic)							2		p	6						60		
Malpighiaceae *Hiraea* sp.			6	4			30	20										
Malpighiaceae *Mascagnia hippocrateoides*			p				p	p					20					
Malpighiaceae *Stigmaphyllon* sp.			66	51			10	69	6						18			
Malpighiaceae *Tetrapterys goudotiana*								1	1								3	
Malvaceae-Bombacoideae *Pachira quinata*					3		p			p				1				
Malvaceae-Bombacoideae *Pachira sessilis*								1		p						p		
Malvaceae-Bombacoideae *Cavanillesia platanifolia*																		66
Malvaceae-Bombacoideae *Ceiba pentandra*			1	6	1		p							1				
Malvaceae-Bombacoideae *Ochroma pyramidale*								12										
Malvaceae-Bombacoideae *Pachira aquatica*				2			5	2										6
Malvaceae-Bombacoideae *Pseudobombax septenatum*			p			492	1	1	14			p		4			p	467
Malvaceae-Bombacoideae *Quararibea asterolepis*			p															
Malvaceae-Bombacoideae *Quararibea magnifica*							p	p	1									
Malvaceae-Bombacoideae *Quararibea pterocalyx*										1								
Malvaceae-Bombacoideae Undetermined							3											4
Malvaceae-Grewioideae *Apeiba* sp.												17					43	
Malvaceae-Grewioideae *Heliocarpus* sp.	t	■	80			2	1688	39	95	23		1	1193					5
Malvaceae-Grewioideae *Luehea seemannii*								5	10									
Malvaceae-Malvoideae *Abutilon* sp.					p			1	p	p							1	
Malvaceae-Malvoideae *Hibiscus* sp.			1				15		3									
Malvaceae-Malvoideae *Sida* sp.			1				15	3										
Malvaceae-Byttnerioideae *Melochia* sp.													177					
Malvaceae-Tilioideae *Mortoniodendron* sp.			61			277	2		28	4								
Melastomataceae. *Miconia* sp.	s	■			5	2025	4		3053	33658	p	1259	1	867		4602	159	25
Meliaceae. *Cedrela odorata*							p	18	1									

(continued)

Table 4.2 (continued)

Pollen	H	*	Locality															
			Carti Rd.				Santa Rita				Pipeline Rd.		BCI		Cerro Jefe	Por	For	Cur
Family/Genus/Species			Cz	Tz	Sa	Mp	Cz	Tz	Sa	Mp	Tz	Mp	Sa	Mp	Sa	Mp	Mp	Tc
Meliaceae. Guarea sp.										1						23		
Meliaceae. Trichilia sp.			25	p		3	77	p		3	10						p	22
Menispermaceae. aff. Cissampelos sp.							5			p								
Moraceae. Brosimun sp.	t	■	79	351	285	35	710	4086	2486	1014		1	473		3069	63	12	
Myristicaceae. Virola sp.			100				492											
Primulaceae Ardisia sp.																		1
Myrtaceae. Eugenia/Psidium	t	■	48			51	115	2	45	1246		6	11	47		176	12	38
Myrtaceae. Syzygium jambos																		38
Onagraceae. Ludwigia octovalvis																		
Passifloraceae. Passiflora aff. auriculata				11			74	65	p	7				p		4		
Passifloraceae. Passiflora aff. nitida				8				p										
Pedaliaceae. Sesamum indicum (exotic)										p								
Phyllanthaceae. Hieronyma alchorneoides							252							80				
Pinaceae. Pinus sp. (exotic)							2											1
Piperaceae. Piper sp.	s	■	174	5800	3203	1431	5726	31320	7501	27		3			242		2	
Poaceae. Paspalum/Panicum/Bambusa<75µ	g	■	18	145	587		312	3172	307	12						p	27	1
Poaceae. Zea mays (exotic)				5			2	3	72	p								
Polygalaceae. Polygala sp.			9				p			p								
Proteaceae. Roupala montana			3			47	57	3	1	21				p		p		
Rubiaceae. Spermacoce alata			21	p		p	2	58		475						12		
Rubiaceae. Genipa americana						43				1								54
Rubiaceae. Macrocnemum roseum							118										4	
Rubiaceae. Posoqueria latifolia								38										
Rubiaceae. Psychotria sp.			p	1		1	30	2	1	2	236			8		1	1	2
Rubiaceae. Uncaria tomentosa												102						
Rutaceae. Citrus sp. (exotic)							10	p		23							99	
Rutaceae. Zanthoxylum sp.	t	■	55				213		5	4		12	164					
Salicaceae. Banara guianensis					p					53	768							

(continued)

Table 4.2 (continued)

Pollen / Family/Genus/Species	H	*	Carti Rd. Cz	Tz	Sa	Mp	Santa Rita Cz	Tz	Sa	Mp	Pipeline Rd. Tz	Mp	BCI Sa	Mp	Cerro Jefe Sa	Por Mp	For Mp	Cur Tc
Sapindaceae. Cupania sp.																		29
Sapindaceae. Paullinia sp.			34	p			21			7		p	1			2	p	1
Sapindaceae. Serjania sp.			46				107		p	7		9		5				10
Sapotaceae. Pouteria sp.			37		p		100		p	1			p		p			
Simaroubaceae. Simarouba amara							p											
Smilacaceae. Smilax sp.				p			1											
Solanaceae. Cestrum sp.											517							
Solanaceae. Brugmansia sp.																		1
Solanaceae. Solanum sp.	s	■			68	4	p	p	16	1369	p					137	713	
Urticaceae. Cecropia sp.	t	■	1		595		p	p	273	138			1660	1	4			4948
Verbenaceae. Aegiphila aff. elata			p	2					7	1								
Vochysiaceae. Vochysia ferruginea			p									p						28

p = presence of a few grains in total slide preparation. *Cz* = *Cephalotrigona zexmeniae* (nest "scutellum" pollen feces and total nest pollen), *Tz* = *Tetragona ziegleri*, *Sa* = *Scaura argyrea*, *Mp* = *Melipona panamica*, *Tc* = *Trigona corvina* (nest scutellum). *Hab* = plant habit, *t* = tree, *l* = liana, *s* = shrub, *e* = epiphyte, *g* = grass. Black bars indicate predominant pollen in either or both regions: eastern Panama lowland wet or moist forest (Santa Rita Ridge or El Llano-Carti Road) or central and western Panama—see text. Pipeline Road is located in Soberanía National Park; *BCI* = Barro Colorado Island, in Barro Colorado Island Nature Monument; *CJef* = Cerro Jefe area, Chagres National Park: *Por* = Portobelo area, Portobelo National Park: *For* = Fortuna Hydrological Forest Reserve, Chiriquí: *Cur* = Curundu, Panama City, Metropolitan Park area

of selectivity at the colony level, are given for Africanized honey bees (feral *Apis mellifera scutellata* x other subspecies) *Melipona*, *Nogueirapis*, *Trigona*, *Partamona*, and *Tetragonisca* in natural, forested foraging environments.

We presented a table on the three most important pollen genera, in order of descending importance, for the three species (Roubik and Moreno 2000). For *Tetragona ziegleri* they were *Paspalum* (Poaceae), *Trophis* (Moraceae), and *Piper* (Piperaceae); for *Cephalotrigona zexmeniae* they were *Iriartea* (Arecaceae), *Pterocarpus* (Fabaceae), and *Piper* (Piperaceae); and for *Scaura argyrea*, they were *Luehea* (Malvaceae-Grewoideae), *Alchornea* (Euphorbiaceae), and *Trophis* (Moraceae). Notwithstanding this assessment, the species (or "pollen types"—even at a family level) included >22 species. That subset of the total species was "relatively important"

when each kind constituted >5% total pollen biomass harvested by the bees, subsampled from the total pollen stores.

We presented a second table (Roubik and Moreno 2000) with a Sorensen index of similarity comparing all colony pollen stores, across all three bee species. Those data show a similarity between nests of a given bee species in mean average values: 0.32, 0.42, and 0.49, *Scaura*, *Cephalotrigona*, *Tetragona*, respectively, for species >5% total grain counts in samples, but further quantitative information was not applied. Using quantitative importance (*Lycopodium* spore internal standards providing volume estimates) from the same pollen data, however, the Sorensen index produces 0.69, 0.43, and 0.56 mean similarity, respectively. Clearly, *Scaura* shifted from relatively inconsistent to the most consistent of the three bees in its pollen use.

Fig. 4.2 Total pollen-type volume, calibrated from *Lycopodium* reference spore counts, determined in multiple nests of three meliponine species in Panama (see text)

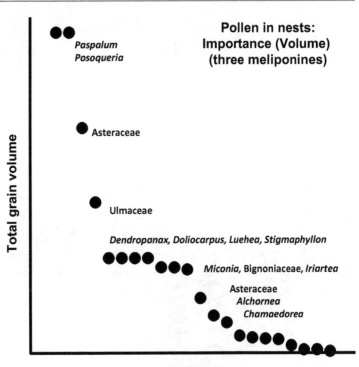

However, the data were compared by computing means, as may be inappropriate with such an index, when the variance is large and the sample size small (Henderson and Southwood 2016).

By using the quantitative data computed as summed pollen volume, rather than means (with such small sample size), the three bee genera discussed by Roubik and Moreno (2000, and above) were more similar to each other than each colony was to others of its own kind. As with small counts or samples in general, species abundance surveys normally add all data together (Wolda and Roubik 1986; Roubik and Wolda 2000). *Cephalotrigona/Scaura* averaged 0.62 (range 0.07–0.91), *Cephalotrigona/Tetragona* 0.61 (range 0.31–1.00), and *Tetragona/Scaura* 0.64 (range 0.24–0.96) in the Sorensen index values. Noting such large ranges, there is little to support the notion that the species are fundamentally different, but our global analysis suggests otherwise. The three genera were distinctive, but their pollen counts alone, without correction for a reference spore number, gave the impression that variation was high and specificity quite low and

could actually produce a significantly different portrayal of the second feature (specialization).

The three study species are not only different genera but differ much in their biology. *Scaura* is a rather small (5 mm) "pollen gleaner" (Laroca and Lauer 1973). That trait is demonstrated by its specialized morphology in having a very enlarged hind tibial basitarsus (see present book chapter on Yasuní, Ecuador), which it uses to collect fallen pollen on petals and leaves (after other foragers dislodge it), as though using a rake. That behavior can occur primarily at flowers that are "buzzed," such as *Cassia, Solanum, Bixa*, or melastomes. Further, *Cephalotrigona* is among the largest non-*Melipona* Neotropical stingless bees (9 mm), whereas *Tetragona* is of an intermediate size. In addition, the three bee genera together harvested a large amount of pollen that, at the colony or species level, *never* appeared particularly important. There were idiosyncrasies, at the colony level, which produced such an unexpected result. Almost one-sixth of the pollen types appeared in only one colony (Fig. 4.3, see also Table 4.2). The group of plants may be rare

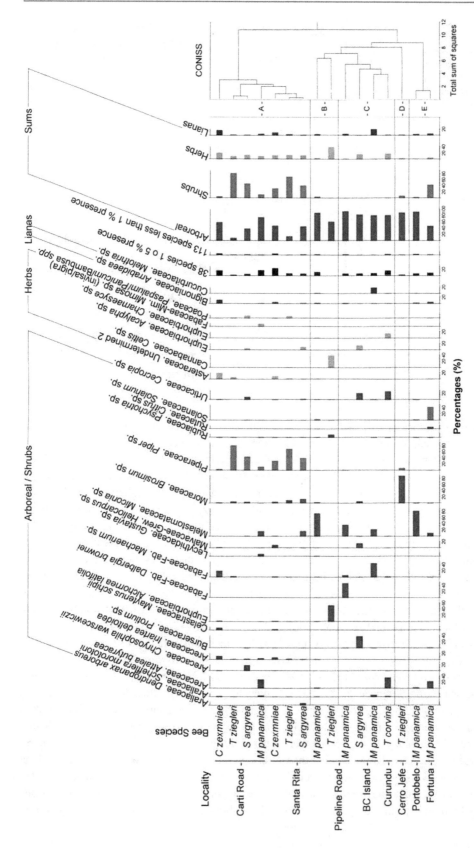

Fig. 4.3 Pollen diagram (Tilia software, 2006) of total pollen-type percentages, by plant habit, and similarities, among nests of five meliponine species in Panama

in space or brief in flower or those not "preferred" by the bees, but without extensive botanical and behavior surveys, the minor pollen types in bee diets will remain poorly understood (see Villanueva-Gutiérrez and Roubik 2016).

Further depiction of meliponine-plant interactions, with both long-term and extensive samples, and in different areas in forested Panama, is given in the pollen diagram and table of plant species (Fig. 4.3 and Table 4.2). Most of the pollen harvested was of trees and shrubs. Nonetheless, epiphytes, lianas, herbs, and many small trees also were prominent in bee diets. There was a moderate degree of similarity between colonies, both intraspecifically and interspecifically. The "mini-assemblage" was quite selective in which species were most used, despite having approximately 180 total native plant genera in the recorded diet. Many plants, perhaps few even sporadically important, surely remain unrecorded for their populations. Nonetheless, only 33 genera from a flora that has >600 vascular plant genera (Roubik and Moreno 1991) constituted the important pollen sources, further discussed below.

Melipona used many *Miconia* (shrubs), the cucurbit vine *Melothria*, and the tree *Gustavia* in particular; *Scaura* used the euphorb *Acalypha* heavily, as did *Cephalotrigona* use Asteraceae and *Maytenus*, while *Tetragona* heavily used *Celtis* (Table 4.2). *Trigona corvina*, in second growth habitat near a large park, concentrated on a palm (*Attalea*), secondary forest trees (*Bursera*, *Cecropia*), and *Chamaesyce*, an herb, among others (Roubik and Moreno 2009).

What is especially noteworthy in this bee pollen (pot-pollen) collection is that the largest number of pollen types (75–79 in single nests of *Melipona panamica*), and the most varied pollen gathered by individual colonies, occurs in the "advancing front" of human settlement, where some original forest persists (Santa Rita and El Llano-Cartí Rd). However, the flora is relatively richer and the rainfall higher in that area. Regardless of locality, the >30 plant genera that were prominent in the totals from five Meliponini included about 20 trees, 1 epiphyte, a few herbs, several lianas, and some grasses (Table 4.2).

Another plant habit or category, "shrub," is difficult to compare to the others. It includes *Miconia*, *Solanum*, *Mimosa*, and *Piper*, some with many species—virtually none (one *Miconia*) producing nectar—present in sunny conditions along forest edges and clearings but never in the canopy (unlike the trees and lianas). Furthermore, unless they outnumber all the other pollen types by several thousand times, they do not, in reality, constitute important pollen for the bees. Their grain size is <1/4 the length and hundreds of times less in volume that of an average species in the Panama flora (See Chap. 1). It is therefore not useful to compare such genera or pollen types to large grains like those of palms or cucurbits, by counts alone. Also unlike trees among the pollen types—normally only one species per type or genus—the noteworthy shrubs have several species, so that each genus presents flowers during most of the year. Therefore, the amount of pollen from a "pollen type" has vague ecological significance, at a more specific, seasonal level.

There are two basic forest types covered in the bee nest analysis in Table 4.2. Santa Rita Ridge and El Llano-Carti Road are wet forests (>3000 mm annual rainfall) and situated in eastern Panama, within a mosaic of old forest and small but ever-expanding subsistence "slash-and-burn" agriculture (Candanedo and Samudio 2005). Many palm species and a more South American forest plant assemblage live in the area (see, e.g., Croat 1978). There is moist or seasonally dry forest (ca. 2600 mm annual precipitation, BCI and Pipeline Road) and protected forest, or a secondary forest habitat (Curundu) in the other sites, and two highland areas—extensive wet, protected forest at Fortuna and moist forest at Cerro Jefe in the large, protected Chagres National Park, both near 1000 m elevation. As indicated, the wetter and more preserved forests correlated positively with the number of plant species harvested by the bees, and diet breadth is both greater and more evenly distributed among species or pollen types (Table 4.2).

4.2.2 Field Bee Short-Term Resource Selection

Field studies of colonies actively foraging—ten meliponine species and the Africanized honey bee—during 2 weeks of the wet season, May–June of 1983 (Roubik et al. 1986), in extensive forest of Soberanía National Park, Panama, are revisited here. The similarities and differences in total pollen harvest among the bees, quantified by capturing returning foragers and removing their pollen loads throughout the day, demonstrated the nearly universal heavy use of palms (*Socratea*, *Elaeis*) and also *Spondias* (Anacardiaceae), *Pouteria* (Sapotaceae), and *Cecropia* (Urticaceae)—see also Villanueva-Gutiérrez and Roubik (2004) and Roubik and Villanueva-Gutiérrez (2009). Whereas *Apis mellifera* used 33 pollen taxa during this period, the various meliponines used 4–16 plants, but the total samples taken were heterogeneous and clearly depended on colony size, foraging range, and general activity, all of which determine the flux and kind of pollen carried on returning foragers (all colonies were present throughout the study). Forty-seven pollen types, of 47 genera, were recorded. Significantly, for those stingless honey bees, most of their foraged pollen or nectar (near 50%) was taken to the nest in <5% of total foraging time. Therefore, their recruitment systems sustain them in a significant way, and a shorter time in monitoring would miss this important trait.

4.2.3 Pollen of Popular Meliponines, Africanized Honey bees, and Lesser Known Species

Pollen from nests within and outside of Panama composed 176 species used appreciably by *Tetragonisca angustula*, the bee probably second only to *Apis mellifera* in utilization by humans in the Americas (Roubik and Moreno 2013). A honey analysis was performed to make a list of pollen species used by a colony of *T. angustula* in Panama, both as pollen and nectar sources. Forty-three genera and 30 families were recorded (data not shown). The outstanding use of the leguminous liana *Machaerium*, and two trees of the Anacardiaceae, *Anacardium* and *Spondias*, among others, was noted in samples from several localities, including South America.

Palms, grasses, and even Malpighiaceae (flowers containing only pollen and oil, not nec-

tar) repeatedly were important, among a wide variety of Neotropical stingless bees and the naturalized Africanized honey bee. When there is an outstanding abundance of a few pollen types, that is perhaps sufficient quantitative information to signal specialization, in some time frame. Roubik and Moreno (1990) report that one nest of *Nogueirapis mirandula* and one of *Melipona panamica* contained almost all palm pollen, despite presence of 22 other pollen species. Two other bee nests, of *Trigona nigerrima* and *Oxytrigona mellicolor*, also contained 5–18% palm pollen species, but most of their pollen was Malpighiaceae (*Tetrapterys*) and Lecythidaceae (*Gustavia*), respectively. *Melipona fallax* and *M. titania*, the largest meliponines on earth, and also *Trigona fulviventris*, probably the most abundant Central American forest bee, as well as the western Amazonian *Melipona nebulosa*, had 5–17% total pollen species from palms. As for *Apis mellifera*, now invasive throughout tropical forests of America, it intensively uses pollen of the unisexual *Elaeis* oil palm flower, which it does not pollinate. Both the honey bees and *Tetragona ziegleri* are frequent visitors of grass flowers in early morning and store large amounts of grass pollen in their nests (see Table 4.2).

4.2.4 Pollination Ecology and Population Biology

Both field floral resource and bee nest quantification (e.g., flower frequency and abundance and bee storage and utilization of pollen—to survive and reproduce) are challenges equal to or surpassing the taxonomic challenge of identifying pollen species. Pollination ecology presents similar hurdles.

Plant breeding systems are routinely ignored in descriptive melittopalynology. Over 152 tree genera are known from the well-studied Panama Canal watershed (Pérez 2008). About 33% of those tree species have individuals that present only male (pollen-bearing) or female (fruit-bearing) flowers. We note two facts of further interest. No legumes (family Fabaceae, ABT-IV) are among the unisexuals. Of the total from 57 families, if we had bee nest pollen data, we would nonetheless know little about pollination of the 53 genera listed below, adding to the aforementioned list with breeding system information provided by Allen and Allen (1977), Croat (1978), and Ibáñez (2011). That is because if the same individual bee does not visit both male and female flowers (unless pollen is transferred between individuals within its nest, Roubik 2002), we do not know whether the pollen indicates a predominantly commensal or parasitic flower visitor or an outcrossing pollinator (Roubik 1989, 1995 and in press). The plant genera, here mostly trees, that are unisexual (either dioecious or monoecious) and which contain unisexual species in the Panama Canal watershed include *Alchornea, Alibertia, Amaioua, Amanoa, Astronium, Brosimum, Bursera, Carapa, Cecropia, Cedrela, Coccoloba, Croton, Cupania, Garcinia, Guarea, Hedyosmum, Heliocarpus, Hieronyma, Hippomane, Hura, Iriartea, Maclura, Maquira, Matayba, Morella, Trophis, Pera, Podocarpus, Poulsenia, Pourouma, Pouteria, Protium, Randia, Sapindus, Sapium, Simarouba, Siparuna, Sorocea, Sterculia, Swietenia, Ternstroemia, Tetragastris, Tovomita, Tovomitopsis, Trattinnickia, Trichilia, Trichospermum, Trema, Triplaris, Trophis, Virola, Xylosma*, and *Zanthoxylum* (Croat 1978; Pérez 2008; Allen and Allen 1977). More field studies, which consider not only flower visitors but also ascertain their contribution to fruit maturation, or even follow those with seed germination studies, are sorely needed.

4.2.5 Conclusions and Ecological Perspective

We found that although shrubs, epiphytes, herbs, and lianas are often important, trees provided most items in the diet of social bee colonies. Bee diet breadth, or the number of species included, and plant relative stature (i.e., large tree versus small herb) are often not a good indicator of resource importance. Although grain size may vary greatly among tropical species,

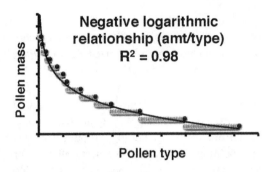

Fig. 4.4 Negative logarithmic relationship between species or pollen type and amount of pollen harvested by five stingless bee species (see Table 4.2)

the importance of pollen from palms is the only group we established as both of primary importance and consisting in many genera of large grains, well beyond the mean of the other taxa, so that their actual importance is underestimated when only semiquantitative (grain counts) are used. At the same time, certain frequent pollen grains, such as *Piper* or *Miconia*, being <0.1% the volume of an average palm, are relatively unimportant, unless many thousands of times more abundant than those of average or large-grained species.

The temporal specialization seen by individual colonies might arise from their keen recruitment to resources in bloom or the abundance of the flower within foraging range. However, over time and among multiple localities, resource selectivity by meliponines and honey bees seems fairly consistent. The bees are predictable. Pollen utilization is skewed toward a relatively small proportion of plant species, over time. The negative logarithmic relationship is well supported, between species and amount harvested, whether pollen volume or pollen grain counts are tabulated (Fig. 4.4). When we attempt to visualize any particular meliponine mini-assemblage and the similarity between bee species in resource utilization, we can appreciate how some data do not particularly represent pollen use, pollination, or specialization (Fig. 4.4). The subject of temporal specialization is ripe for investigation. We suggest its proper appreciation will not be an easy task to accomplish, and more scientific studies, and theories based on careful quantitative pollen analyses, are needed.

As summarized from data available considering bee generalists (Roubik 1992b), a large proportion of resources used by Africanized *Apis mellifera* are not especially important to them, and data coming now from the temperate zone provide a similar impression (Hawkins et al. 2015; Richardson et al. 2015). With quantitative pollen data from melissopalynology (honey samples) and melittopalynology (whole-nest pollen or foraging-day samples), we conclude that stingless honey bees and stinging honey bees alike are selective generalists that often specialize. Whether each meliponine specializes much on different things, in mature or regenerating forests (Roubik and Moreno 2000), remains to be seen.

References

Aguilar I, Herrera E, Zamora G. 2013. Stingless bees of Costa Rica. pp. 113–124 In: Vit P, Pedro SRM, Roubik DW, eds. Pot–honey: A Legacy of Stingless Bees. Springer; New York. 654 pp.

Allen PH, Allen DO. 1977. The Rain Forest of Golfo Dulce (2nd Edn.). Stanford University Press; Stanford. 417 pp.

Ayala R. 1999. Revisión de las abejas sin aguijón de México (Hymenoptera: Apidae: Meliponini). Folia Entomológica Mexicana 106: 1–123.

Camargo JMF. 2013. Historical biogeography of the Meliponini (Hymenoptera, Apidae, Apinae) of the Neotropical region. pp. 19–34. In: Vit P, Pedro SRM, Roubik DW, eds. Pot-Honey: A Legacy of Stingless Bees. Springer; New York. 654 pp.

Camargo JMF, Pedro SRM. 2008. Revisão das espécies de *Melipona* do grupo *fuliginosa* (Hymenoptera, Apoidea, Apidae, Meliponini). Revista Brasileira de Entomologia 52 (3). dx.doi.org/10.1590/S0085-56262008000300014

Camargo JMF, Pedro SRM. 2007. Meliponini Lepeletier, 1836. In: Moure JS, Urban D, Melo, GAR, eds. Catalogue of bees (Hymenoptera, Apoidea) in the Neotropical region – online version. Available at: www.moure.cria.org.br/catalogue.

Camargo JMF, Roubik DW. 1991. Systematics and bionomics of the apoid obligate necrophages: the *Trigona hypogea* group (Hymenoptera: Apidae; Meliponinae). Biological Journal of the Linnean Society 44: 13–39.

Candanedo I, Samudio R. 2005. Construyendo un mecanismo para medir el éxito de la conservación en el alto Chagres; Dilia Santamaría Espinosa, Panama. 80 pp.

Cane JH, Sipes S. 2006. Characterizing floral specialization by bees: analytical methods and a revised lexicon for oligolecty. pp 99–122 In: Waser NM, Ollerton J, eds. Plant pollinator interactions: from specialization

to generalization. University of Chicago Press, USA. 445 pp.

Cockerell TDA. 1913. Meliponine bees from Central America. Psyche 20: 10–14.

Croat TC. 1978. The flora of Barro Colorado Island. Missouri Botanical Garden, St. Louis, MO, USA. 958 pp.

D'Arcy WG, Correa, MDA, eds. 1985. The botany and natural history of Panama (La botánica e historia natural de Panamá). Monographs in Systematic Botany from the Missouri Botanical Garden. Vol. 10. 455 pp.

Gonzalez VH, Roubik DW. 2008. Especies nuevas y filogenia de abejas de fuego *Oxytrigona* (Hymenoptera, Apidae, Meliponini). Acta Zoologica Mexicana 24: 43–71.

Griswold T, Parker FD, Hanson PE. 1995. The bees (Apidae). pp. 650–691 In: Hanson PE, Gauld ID, eds. The Hymenoptera of Costa Rica. Oxford, UK. 893 pp.

Hawkins J, De Vere N, Griffith A, Ford R, Allainguillaume J, Hegarty MJ, Baillie L, Adams-Groom B. 2015. Using DNA Metabarcoding to identify the floral composition of honey: a new tool for investigating honey bee foraging preferences. PLoS One. doi: https://doi.org/10.1371/journal.pone.0134735.

Henderson PA, Southwood TRE. 2016. Ecological methods. 4th Edn. John Wiley and Sons, UK. 643 pp.

Ibáñez A. 2011. Guía botánica del parque nacional Coiba. International Cooperative Biodiversity Groups; Panama. 399 pp.

Laroca S, Lauer S. 1973. Adaptação comportamental de *Scaura laitarsis* para a coleta de pólen (Hymenoptera, Apoidea). Acta Biológica Paranaense 2: 147–152.

Michener CD. 1954. Bees of Panamá. Bulletin of the American Museum of Natural History 104: 5–175.

Michener CD. 1974. The Social Behavior of the Bees. A Comparative Study. Belknap Press of Harvard University Press, New York. 404 pp.

Michener CD. 2007. Bees of the World, 2nd Edn. Johns Hopkins University Press: Baltimore. 953 pp.

Michener CD. 2013. The Meliponini. pp. 3–17 In: Vit P, Pedro SRM, Roubik DW, eds. Pot-Honey: A Legacy of Stingless Bees. Springer; New York. 654 pp.

Pedro SRM, Camargo JMF. 1997. A new species of *Partamona* (Hymenoptera: Apidae) endemic to eastern Panama and notes on *P. grandipennis*. Revista de Biologia Tropical 44(3) –45(1): 199–208.

Pérez-M RA. 2008. Árboles de los bosques del canal de Panamá. Instituto Smithsonian de Investigaciones Tropicales; Panama. 465 pp.

Richardson RT, Chia-Hua L, Sponsler DB, Quijia JO,Goodell K, Johnson RM. 2015. Application of ITS2 metabarcoding to determine the provenance of pollen collected by honey bees in an agroecosystem. Applications in Plant Sciences 3: 1400066.; http://www.bioone.org/loi/apps.

Roubik DW. 1981. A natural mixed colony of *Melipona*. Journal of the Kansas Entomological Society 54: 263–268.

Roubik DW. 1982. Obligate necrophagy in a social bee. Science 217: 1059–1060.

Roubik DW. 1983. Nest and colony characteristics of stingless bees from Panama. Journal of the Kansas Entomological Society 56: 327–355.

Roubik DW. 1988. An overview of Africanized honey bee populations: reproduction, diet and competition. pp. 45–54. In: Needham G, Page R, Delfinado-Baker M, eds. Proc. Intl. Conf. on Africanized honey bees and bee mites. E. Horwood Ltd., Chichester, UK. 572 pp.

Roubik DW. 1989. Ecology and natural history of tropical bees. Cambridge University Press, New York. 514 pp.

Roubik DW. 1991. Aspects of Africanized honey bee ecology in tropical America. pp. 147–158 In: Spivak M, Breed MD, Fletcher DJC, eds. The "African" honey bee. Westview Press, Boulder, Colorado. 435 pp.

Roubik DW. 1992a. Stingless bees (Apidae: Meliponinae): a guide to Panamanian and Mesoamerican species and their nests. pp. 495–524. In: Quintero D, Aiello A, eds. Insects in Panama and Mesoamerica: selected studies; Oxford, UK. 692 pp.

Roubik DW. 1992b. Loose niches in tropical communities: Why are there so many trees and so few bees? pp. 327–354 In: Hunter MD, Ohgushi T, Price PW, eds. Resource Distribution and Animal-Plant Interactions. Academic Press, New York. 505 pp.

Roubik DW. 1993. Direct costs of forest reproduction, bee-cycling and the efficiency of pollination modes. Journal of Bioscience 18: 537–552.

Roubik DW. 2002. Tropical bee colonies, pollen dispersal and reproductive gene flow in forest trees. pp. 30–40. In: Degen B, Loveless M, Kremer A, eds. Proceedings of the symposium "Modelling and experimental research on genetic processes in tropical and temperate forests". Documentos de Embrapa Amazonia Oriental, Belém, Brazil. 168 pp.

Roubik DW. 2006. Stingless bee nesting biology. Apidologie 37: 124–143.

Roubik DW. 2009. Ecological impact on native bees by the invasive Africanized honey bee. Acta Biologica Colombiana 14: 115–124.

Roubik DW. 2013. Why they keep changing the names of our stingless bees (Hymenoptera: Apidae; Meliponini). A little history and guide to taxonomy. pp. 1–7. In Vit P, Roubik, DW, eds. Stingless bees process honey and pollen in cerumen pots. SABER-ULA, Universidad de Los Andes; Mérida, Venezuela. www.saber.ula.ve/handle/123456789/35292

Roubik DW. ed. 1995. Pollination of Cultivated Plants in the Tropics. Technical Bulletin 118; Food and Agriculture Organization of the United Nations, Rome. 269 pp.

Roubik DW. ed. In press. The Pollination of Cultivated Plants. A Compendium for Practitioners. Food and Agriculture Organization of the United Nations, Rome.

Roubik DW, Camargo JMF. 2012. The Panama microplate, island studies and relictual species of *Melipona (Melikerria)* (Hymenoptera: Apidae: Meliponini). Systematic Entomology 37: 189–199.

Roubik DW, Lobo JA, Camargo JMF. 1997. New endemic stingless bees genus from Central American cloudforests (Hymenoptera: Apidae; Meliponini). Systematic Entomology 22: 67–80.

Roubik DW, Moreno JE 1991. Pollen and spores of Barro Colorado Island. Monographs in Systematic Botany from the Missouri Botanical Garden. No. 36. 269 pp.

Roubik DW, Moreno JE, Vergara C, Wittmann D. 1986. Sporadic food competition with the African honey bee: projected impact on Neotropical social bees. Journal of Tropical Ecology 2: 97–111.

Roubik DW, Moreno JE. 1990. Social bees and palm trees: what do pollen diets tell us? pp. 427–428. In: Veeresh GK, Mallik B., Viraktamath CA, eds. Social Insects and the Environment; Oxford and IBH Publishing Co., New Delhi. 765 pp.

Roubik DW, Moreno JE. 2000. Generalization and specialization by stingless bees. pp. 112–118. In: Proceedings of the Sixth International Bee Research Conference on Tropical Bees. International Bee Research Association, Cardiff, UK. 212 pp.

Roubik DW, Moreno JE. 2009. *Trigona corvina*: An ecological study based on unusual nest structure and pollen analysis. Psyche. doi:10.1155/2009/268756.

Roubik DW, Moreno JE. 2013. How to be a bee-botanist using pollen spectra. pp. 295–314. In: Vit P, Pedro SRM, Roubik DW, eds. Pot–honey: a legacy of stingless bees. Springer; New York. 654 pp.

Roubik DW, Villanueva-Gutiérrez, R. 2009. Invasive Africanized honey bee impact on native solitary bees: a pollen resource and trap nest analysis. Biological Journal of the Linnean Society 98: 152–160.

Roubik DW, Wolda H. 2000. Male and female bee dynamics in a lowland tropical forest. pp. 167–174. In: Proceedings of the Sixth International Bee Research Conference on Tropical Bees: International Bee Research Association, Cardiff, UK. 226 pp.

Schwarz HF. 1932. Stingless bees in combat: observations on *Trigona pallida* Latreille on Barro Colorado Island. Natural History, New York. 32: 552–553.

Schwarz HF. 1934. The social bees (Meliponidae) of Barro Colorado Island, Canal Zone. American Museum Novitates. No. 731. pp. 1–23.

Schwarz HF. 1948. Stingless bees of the Western Hemisphere. *Lestrimelitta* and the following subgenera of *Trigona*: *Trigona, Paratrigona, Schwarziana, Parapartamona, Cephalotrigona, Oxytrigona, Scaura*, and *Mourella*. Bulletin of the American Museum of Natural History, Vol. 90: 546 pp. + xvii.

Schwarz HF. 1951. New stingless bees (Meliponidae) from Panama and the Canal Zone. Nuevas abejas jicotes (Meliponidae) de Panamá y la Zona del Canal. American Museum Novitates. 1505: 1–16.

Sommeijer MJ, de Rooy GA, Punt W, de Bruijn LLM. 1983. A comparative study of foraging behavior and pollen resources of various stingless bees (Hym., Meliponinae) and honey bees (Hym., Apinae) in Trinidad, West-Indies. Apidologie 14: 205–224.

Villanueva-Gutiérrez R, Roubik DW. 2004. Why are African honey bees and not European bees invasive? Pollen diet diversity in community experiments. Apidologie 35: 481–491.

Villanueva-Gutiérrez R, Roubik DW. 2016. More than protein? Bee-flower interactions and effects of disturbance regimes revealed by rare pollen in bee nests. Arthropod-Plant Interactions doi:https://doi.org/10.1007/s11829-015-9413-9.

Vit P, Pedro SRM, Roubik DW, eds. 2013. Pot-Honey: A Legacy of Stingless Bees. Springer; New York. 647 pp.

Vit P, Roubik DW, eds. 2013. Stingless Bees Process Honey and Pollen in Cerumen Pots. SABER-ULA, Universidad de Los Andes; Mérida, Venezuela. www.saber.ula.ve/handle/123456789/35292

Wolda H, Roubik DW. 1986. Nocturnal bee abundance and seasonal bee activity in a Panamanian forest. Ecology 67: 426–433.

Yurrita CA, Ortega-Huerta MA, Ayala R. 2016. Distributional analysis of *Melipona* stingless bees (Apidae: Meliponini) in Central America and Mexico: setting baseline information for their conservation. Apidologie DOI: 10.1007/s13592-016-0469-z.

The Value of Plants for the Mayan Stingless Honey Bee *Melipona beecheii* (Apidae: Meliponini): A Pollen-Based Study in the Yucatán Peninsula, Mexico

5

Rogel Villanueva-Gutiérrez, David W. Roubik,
Wilberto Colli-Ucán, and Margarito Tuz-Novelo

5.1 Understanding the Ecology of a Mayan Resource and Cultural Icon

Before the Spanish conquest, ancient Mayans of the Yucatán Peninsula used the products of one particular social bee colony, *Melipona beecheii* —"xunan kab" in Mayan language—for its unique waxy nesting material (cerumen, made of beeswax and plant resin), and the honey, to pay cultural tributes and in trade. The honey was used as not only a sweetener. It reportedly had medicinal value and ritualistic application among the ancient Maya (Labougle-Rentería and Zozaya-Rubio 1986). At present the cultivation or husbandry of the xunan kab in Mayan communities is quite limited. Some people continue with the tradition from their ancestors' times and venerate the bees that they still consider sacred. Now the bee cerumen and honey are taken without harm, and the plants with flowers often frequented by the bees are carefully conserved.

An extensive knowledge of the local flora is of great value for determining which species are primary food resources for *Melipona beecheii* or other bees and can best be studied by applied melittopalynology—bee pollen identification. We prepared a pollen reference collection with identified vouchers and newly collected material, deposited in herbaria, and we made a pollen atlas of the region (Palacios-Chavez et al. 1991). To this we can add knowledge of climate and other factors. According to the classification of Koeppen (1936), the type of climate that exists in the Yucatán Peninsula is Aw, which is defined as hot subhumid, with a mean annual temperature over 22°C and an annual precipitation between 700 and 1500 mm, with rainfall during the summer season.

Until now there have been no studies of the specific pollen plants most utilized by *Melipona* in Quintana Roo. Nor has there been reference to plant importance through multiple years or seasons. As noted by other authors (see Roubik and Moreno (2013), this volume; Villanueva-Gutiérrez and Roubik 2004), many tropical trees do not flower each year, while some vegetation has more predictable blooming and flowers during much of the year. Such details are often lacking in literature on bees and their reproductive schedules or resource utilization (see Roubik et al. 2005, e.g., from tropical Asia).

R. Villanueva-Gutiérrez (✉) • W. Colli-Ucán
M. Tuz-Novelo
El Colegio de la Frontera Sur. Unidad Chetumal. Av.
Centenario km 5.5, Chetumal, Quintana Roo,
Mexico, C. P. 77014
e-mail: rvillanu@ecosur.mx

D.W. Roubik
Smithsonian Tropical Research Institute, Calle
Portobelo, Balboa, Ancon, Republic of Panama

© Springer International Publishing AG, part of Springer Nature 2018
P. Vit et al. (eds.), *Pot-Pollen in Stingless Bee Melittology*, DOI 10.1007/978-3-319-61839-5_5

Much of the original vegetation in the Yucatán Peninsula has been disturbed by human activities, hurricanes, and fire, but some vegetation types regenerate after passing through different successional stages for 30 or more years (Sánchez-Sánchez et al. 2015; Rada et al. 2015). As the result of disturbances and natural low forest stature, there is extensive "edge habitat," with many pioneer trees, herbs, and shrubs. In one study, we found that *Cecropia* (a nectarless, dioecious pioneer tree) became a principal pollen resource for Africanized honey bees after hurricane disturbance (Villanueva-Gutiérrez and Roubik 2004; Roubik 2009). Such opportunism, related to habitat and light environment, may also apply to native highly eusocial bees, as they seek the most abundant flowers and recruit their nest mates. We carried out our work at four sites within medium-stature and low-stature deciduous forest, defined as vegetation with a canopy height of 15–25 meters. Our goal was to determine which kinds of pollen were used by *Melipona beecheii* in a north-south transect through Quintana Roo state (Fig. 5.1). This research was pursued at four sites, in which we analyzed the pollen stored by the bees in their nests.

5.2   Baseline Studies of Invasive Honey bees and Native Neotropical Bees

In the Neotropics, the most complete studies on flower use by the genus *Melipona* have been published from studies in Brazil, for example, by Kleirnet-Giovannini and Imperatriz-Fonseca (1987) who sampled nectar and pollen from two colonies of *Melipona marginata* during 1 year. Similarly, Ramalho et al. (1990) studied *Apis mellifera* and *Melipona,* finding that *M. quadrifasciata* visited 288 plant species, *M. marginata* 126, and Africanized honey bees a total of 125. Ramalho (1992) made a further study to quantify resource use. Villanueva-Gutiérrez (1984) also quantified the resources used by *Apis mellifera* from the provisions in its brood cells in Plan del Río, Veracruz. Later, Villanueva-Gutiérrez (2002) made a similar study of foraging and pollen

harvest by *Apis mellifera* in the Sian Ka'an Biosphere Reserve of Quintana Roo. At this time, the kind of honey bee was no longer that from Europe; it was a tropical honey bee hybrid from Africa. This was followed by Villanueva-Gutiérrez and Roubik (2004) who made a detailed quantitative analysis of pollen collected by Africanized and the formerly European variety of *Apis mellifera* in the same environment. Finally, Roubik and Villanueva-Gutiérrez (2009) analyzed the kind and proportion of different pollen species used by native solitary bees, *Megachile* and *Centris*, before and after the arrival of the colonizing competitor Africanized honey bees in the Yucatán Peninsula and compared those pollen types to species utilized by the honey bee that now has become a resident, not just a managed bee in apiaries. In other related work, Vossler et al. (2010) characterized, by pollen identification, the floral resources foraged by *Geotrigona argentina* (Apidae, Meliponini) in the Argentine Dry Chaco forest. In addition, a study on selectivity of pollen and nectar sources of *Melipona beecheii* was made in Matanzas, Cuba, by Fonte et al. (2012). Another study considering potential competition and food resources used by *Apis mellifera* and *Melipona beecheii* was made by Leal-Ramos and León-Sánchez (2013). Martins et al. (2011) performed a pollen spectrum analysis of honey from *Melipona fasciculata* in the municipality of Palmeirândia, Brazil, and Villanueva-Gutiérrez et al. (2015) analyzed floral phenology and potential bee competition in Yucatán.

5.3   Fieldwork

The present study was done from 1996 to 2016. Four sites were chosen: the first one (Site 1) was the Botanical Garden "Alfredo Barrera Marín" which is located 1 km south of Puerto Morelos (20° 50´ N. latitude and 86° 53´ W. longitude) and has an area of 66 ha. The second one (Site 2) was the biological station of Santa Teresa, located within the Sian Ka'an Biosphere Reserve, 30 km at the road from Felipe Carrillo Puerto to Vigía Chico (19° 43´ 20″ N., 87° 48´ 43″ W.). The third one (Site 3) is the Ranch of Palmas, located 36 km

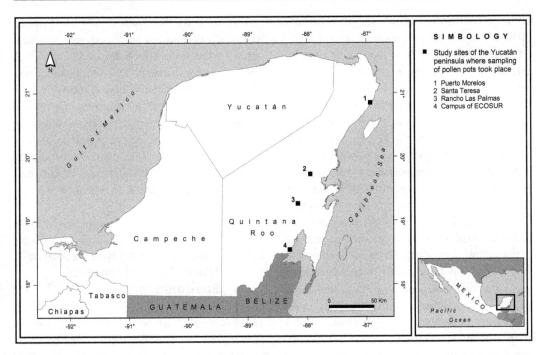

Fig. 5.1 Map of the Yucatán Peninsula with the four study sites where pollen samples were collected from *Melipona* colonies

south of Felipe Carrillo Puerto (19° 15′ 36″ N., 88° 7′ 57″ W.). The fourth one (Site 4) is within the campus of El Colegio de la Frontera Sur (ECOSUR), in Chetumal (18° 32′36″ N., 88° 15′44″W.). (See Fig. 5.1.) The botanical garden and Santa Teresa are located within the natural protected areas. Palmas and the campus at ECOSUR are part of the buffer zone of the Sian Ka'an Biosphere Reserve World Heritage area and the Manati Sanctuary, respectively.

A total of 34 pollen samples from 24 *Melipona* colonies, housed in both traditional log hives and wooden hive boxes, were taken; 11 samples at Site 1 (Botanical Garden "Alfredo Barrera Marín"), in Puerto Morelos; 8 samples at Site 2 (Santa Teresa); 6 samples at Site 3 (Palmas); and 9 samples at Site 4 (ECOSUR campus) (see Fig. 5.1). The samples were obtained from the entire stored pollen in cerumen pots within *Melipona* nests in order to determine its pollen sources. The work is the first to both characterize and quantify the relative importance of individual plant species and families for this bee, using spore reference internal standards in the labora-

tory and analytical innovations. The bee colonies were never fed artificially, neither with pollen nor sugar-water syrup.

Samples were collected during the whole year and thus included the wet season (June to October) and the dry season (November to May). During the course of the entire study, from 1996 to 2016, six colonies were lost after depredations of the "Nenem" fly (*Pseudohypocera kertezi*), the Tayra or "San hool" (*Eira barbara*), and the "Xulab" army ant (*Eciton burchellii*).

5.4 Pollen Analysis from Pot-Pollen Samples

Each pollen sample was obtained from all the pollen stored in the pots of a given colony. A metal spatula was inserted and rotated in a pot, touching the pot base (storage pots were approximately 3–4 cm wide and 4–6 cm in height). The collected pollen samples were dried at 45 °C for 24–48 h until reaching constant weight, and the final weight was recorded. A subsample of 10 g pollen was taken from each sample, soaked in 20 ml distilled water, and stirred magnetically for 1 h. The pollen grains from the sample were further desegregated using a sonicator or "cell disrupter" (O'Rourke and Buchmann method, O'Rourke and Buchmann 1991). Pollen samples were sonicated for 5 min at 24 kHz using a probe "ultrasonic disintegrator" (M.S.E. SONIPREP), at a medium power setting. *Lycopodium clavatum* spore tablets were added (ca. 13,000 spores per tablet) to serve as an internal calibration standard (Stockmarr 1971) to estimate the volume of pollen grains of different types. The number of spores counted in the reference slide corresponds to the sample size or volume registered by pollen counts made on a study slide. When considering more than one slide sample, pollen counts alone do not contain information on absolute pollen importance or number. This is because each slide from a pollen preparation contains a subsample that has a different pollen density per total volume. During pollen identification, we tabulate *Lycopodium* spores. This identification protocol (see below) allows the individual pollen species counts to be adjusted to a uniform density, consistent with that of the internal spore standard. Thus, if the spore count differs by a factor of two, when comparing two different reference slides of 600 counted and identified pollen grains, the relative number of each pollen grain species is either halved or doubled. This method functions well when the total quantity or volume of the acetolyzed mixture is kept at a standard amount for each sample that produces a pollen sample slide. In that way, the relative volume of each pollen type, irrespective of grain size, density, or shape, is obtained (Roubik and Moreno 2013). The pollen from the stored pollen samples was acetolyzed using the Erdtman technique (1943) and mounted on slides with glycerine jelly.

A palynological reference collection of the area, with more than 500 pollen species, was used to aid identification of the pollen grains from pollen pots. There are unknown species, and these were assignable to genus. There are over 850 angiosperm species listed for Sian Ka'an Biosphere Reserve, where our work is centered (Fig. 5.1), and pollen obtained from the herbarium at the Research Centre of Quintana Roo (CIQRO) allowed more complete representation of the flora than Palacios-Chavez et al. (1991) presented. For pollen identification, pollen grains were randomly counted (600 grains for each sample) in order to determine the relative frequency of the different pollen types or species. In this way, each sample composition could be analyzed in terms of (a) pollen frequency, (b) predominance (summed counts over all samples), and (c) pollen concentration summed over all samples.

5.5 Understanding Bee Resource Use in Dynamic Natural Environments

The documented use of pollen or nectar sources of *Melipona beecheii*, like that of other bees, has remained mostly unquantified or even anecdotal (see, e.g., Villanueva-Gutiérrez et al. 2015). Similarly, the preference for or use of a particular species lies beyond the scope of most studies, because no population is truly sampled nor have all the resources that flower throughout the tropical year been considered. Our work is an effort to remedy such universal shortcomings and challenges in determining the value of particular resources to a native bee population.

In our study area, within a transect containing varied vegetation over an expanse of nearly 350 km of lowlands (Fig. 5.1), we noted that *Melipona* uses a small portion of the flower species potentially available—68 of 850. Furthermore, only 32 species were included and only 5 plant families, among the 26 families and 47 genera found in its pollen stores (Fig. 5.2, Table 5.1). The Africanized honey bee in that area showed similar qualities. The pollen diversity of *Apis mellifera* is greater than that of *Melipona beecheii*, with a Jaccard coefficient of similarity of 0.52.

A remarkably few truly important resource species are abundant in the diet of these two colonial, perennially active bee colonies (Villanueva-Gutiérrez 2002, Villanueva-Gutiérrez and Roubik 2004, and present data) (Figs. 5.3 and 5.4, Table 5.1). As pointed out in general discussion of bee resource

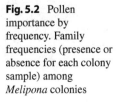

Fig. 5.2 Pollen importance by frequency. Family frequencies (presence or absence for each colony sample) among *Melipona* colonies

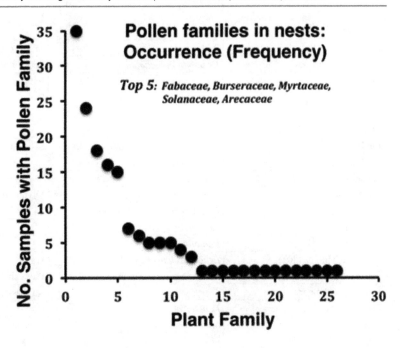

Table 5.1 Pollen taxa to level of family, genus, or species, and Maya common name, found in the pollen pots of *Melipona* at study sites (Fig. 5.1)

Family	Pollen type	Maya/local common name
Amaranthaceae	*Alternanthera ramosissima*	Sak mul, Sak pool tees
Picramniaceae	*Alvaradoa amorphoides*	Beel siinik, Xbesiinik che´
Asteraceae	Asteraceae	
Primulaceae	*Bonellia macrocarpa*	Chak si' ik, lengua de gallo
Burseraceae	*Bursera simaruba*	Chac chacaj, Chacaj
Burseraceae	*Bursera* sp.	
Malpighiaceae	*Byrsonima bucidifolia*	Sak paj, nance silvestre
Fabaceae-Caesalpinioideae	*Caesalpinia* sp.	
Salicaceae	*Casearia emarginata*	Amche, naranja che
Solanaceae	*Cestrum nocturnum*	Huele de noche, dama de noche
Fabaceae-Caesalpinioideae	*Chamaecrista nictitans*	Tamarindo xiu
Arecaceae	*Chamaedorea seifrizii*	Xiat, xiatil, yuyat
Rutaceae	*Citrus sinensis*	Chujuk pakal, China, naranja dulce
Polygonaceae	*Coccoloba spicata*	Boob, boob che
Polygonaceae	*Coccoloba* sp. 1	
Polygonaceae	*Coccoloba* sp. 2	
Amaranthaceae	*Dysphania ambrosioides*	Epasote, Apasote
Bixaceae	*Cochlospermum vitifolium*	Chun choy, chuun
Malvaceae-Grewioideae	*Corchorus siliquosus*	Chichi bej, nich iyuk, sak chichi bej
Euphorbiaceae	*Croton* sp.	
Sapotaceae	*Chrysophyllum cainito*	Caimito, nihkeh, cayumito
Sapotaceae	*Chrysophyllum mexicanum*	Chikeej, caimito
Cucurbitaceae	*Cucurbita argyrosperma*	Ka, xtop, calabaza gruesa
Sapindaceae	*Cupania glabra*	Sak poon

(continued)

Table 5.1 (continued)

Family	Pollen type	Maya/local common name
Fabaceae-Faboideae	*Desmodium incanum*	Pega pega, pak´ umpak
Myrtaceae	*Eugenia axillaris*	Ich juuj, Kiis yuuk
Myrtaceae	*Eugenia buxifolia*	Pichi che, sak loob
Myrtaceae	*Eugenia* sp. 1	
Myrtaceae	*Eugenia* sp. 2	
Euphorbiaceae	*Euphorbia schlechtendalii*	Sak chaca
Euphorbiaceae	*Euphorbia* sp.	
Fabaceae-Faboideae	*Gliricidia sepium*	Sak ya' ab, Balche keej, cocoite
Euphorbiaceae	*Gymnanthes lucida*	Yuyte, Yuytik
Polygonaceae	*Gymnopodium floribundum*	Tsi'tsi'lche
Acanthaceae	*Justicia campechiana*	
Verbenaceae	*Lantana velutina*	
Fabaceae-Mimosoideae	*Leucaena leucocephala*	Waxin
Fabaceae-Faboideae	*Lonchocarpus punctatus*	Ba'al che', ba'al che'
Fabaceae-Faboideae	*Lonchocarpus rugosus*	Ka' nasin, Kan' sin
Fabaceae-Faboideae	*Lonchocarpus* sp.	
Anacardiaceae	*Metopium brownei*	Chechen, box Chechen
Fabaceae-Mimosoideae	*Mimosa bahamensis*	Sak katsin, Katsin blanco
Muntingiaceae	*Muntingia calabura*	Capulin, capulincillo
Myrtaceae	*Myrcianthes fragrans*	Guallabillo, x-oko cha'an
Passifloraceae	*Passiflora foetida*	Poch'il, poch' k'aak', tu' tok
Solanaceae	*Physalis pubescens*	Yooch ik bach
Fabaceae-Faboideae	*Piscidia piscipula*	Ja' bin, U tsab ja' bin
Fabaceae-Mimosoideae	*Havardia albicans*	Chimay, chukun, sak chukun
Sapotaceae	*Pouteria* sp.	
Sapotaceae	*Pouteria reticulata*	Chi' iich' ya', ts'um ya'
Malvaceae-Bombacoideae	*Pseudobombax ellipticum*	Amapola, K'uxche
Myrtaceae	*Psidium sartorianum*	X-Pichi' che, guayabillo
Plantaginaceae	*Russelia campechiana*	
Fabaceae-Caesalpinioideae	*Senna pallida*	Okenkab
Fabaceae-Caesalpinioideae	*Senna racemosa*	X-k' anlol che, Ya' ax jabin
Fabaceae-Caesalpinioideae	*Senna* sp. 1	
Fabaceae-Caesalpinioideae	*Senna* sp. 2	
Fabaceae-Caesalpinioideae	*Senna* sp. 3	
Sapindaceae	*Serjania racemosa*	
Sapindaceae	*Serjania yucatanensis*	Ch'em pe'ek', X-kansep ak'
Sapotaceae	*Sideroxylon obtusifolium*	Ja' astoch, xcapoch, xpe' et kitan
Solanaceae	*Solanum americanum*	Pool kuts, hierba mora
Solanaceae	*Solanum lanceifolium*	Siclimuch
Solanaceae	*Solanum* sp. 1	
Solanaceae	*Solanum* sp. 2	
Solanaceae	*Solanum torvum*	Sikil much, x-tsay och
Solanaceae	*Solanum tuerckheimii*	Chilillo
Sapindaceae	*Thouinia canescens*	K'anchunuup
Arecaceae	*Thrinax radiata*	Chit, palma
Boraginaceae	*Tournefortia volubilis*	Beek ak', bejuco de mico,
Asteraceae	*Viguiera dentata*	Tah, tajonal

use (Roubik 1989, 1995), generalists may indeed specialize, even as presumed specialists may also broaden or alter their feeding niches, seasonally or according to resource availability (Roubik and Villanueva-Gutiérrez 2009) (Figs. 5.3 and 5.4).

The most abundant pollen species according to the number of pollen grains that appear in the pollen samples are as follows. For the Botanical garden of Puerto Morelos, the most abundant were *Senna* sp. 1, *Gymnanthes lucida*, *Myrcianthes fragrans*, *Solanum lanceifolium*, *Physalis pubescens*, *Thrinax radiata*, *Bursera simaruba*, and *Solanum torvum*. In Santa Teresa, the most abundant were *Bursera simaruba*, *Eugenia buxifolia*, *Eugenia axillaris*, *Gliricidia sepium*, and *Chrysophyllum cainito*. In Palmas, the most abundant pollen grains were *Bursera simaruba*, *Lonchocarpus rugosus*, *Lonchocarpus* sp., and *Solanum* sp. 1. For ECOSUR, the most abundant grains were *Gliricidia sepium*, *Cochlospermum vitifolium*, *Senna pallida*, *Cochlospermum ambrosioides*, and *Senna* sp. 1.

The specific recommendations for conservation of the preferred resources of *Melipona beecheii* in Quintana Roo would certainly include legumes (the Fabaceae, APG IV system of nomenclature, Missouri Botanical Garden) (Figs. 5.2, 5.3, and

5.4), most of which have nectar and pollen, although of the most important plants *Senna* ranked very highly (Fig. 5.5), but has no nectar (Roubik and Moreno 2013). Solanaceae and Bixaceae (formerly Cochlospermaceae), other key resources (Figs. 5.3 and 5.4), also lack nectar (Fig. 5.6).

Interestingly, Quintana Roo and the rest of the Yucatán Peninsula share none of the important bee plants in the Melastomataceae (see Brazilian studies, op cit.) because that family is absent, except in the southern part of Campeche state, where the conditions are more humid. The Burseraceae and Myrtaceae, the other resources of primary importance to the bee, have both nectar and pollen; each is also highly sought and utilized by the Africanized honey bee, with which *Melipona beecheii* competes for food (Villanueva-Gutiérrez et al. 2015; Villanueva-Gutiérrez and Roubik 2004) (see Figs. 5.3 and 5.4). The latter authors conclude the "Top 5" plant species resources of the Africanized honey bee in the Sian Ka'an Biosphere Reserve area, based upon quantitative techniques using grain counts and individual grain volume estimates, including almost half of their total diet consumption (Fig. 5.7). Remarkably, these are identical to those of *M. beecheii* insofar as the use of *Eugenia*

Fig. 5.3 Pollen importance semi-quantification. Families visited by *Melipona* ranked according to the number of pollen grains counted on slide samples

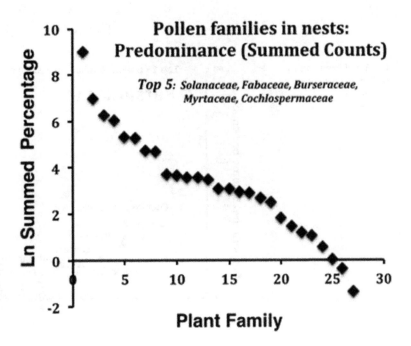

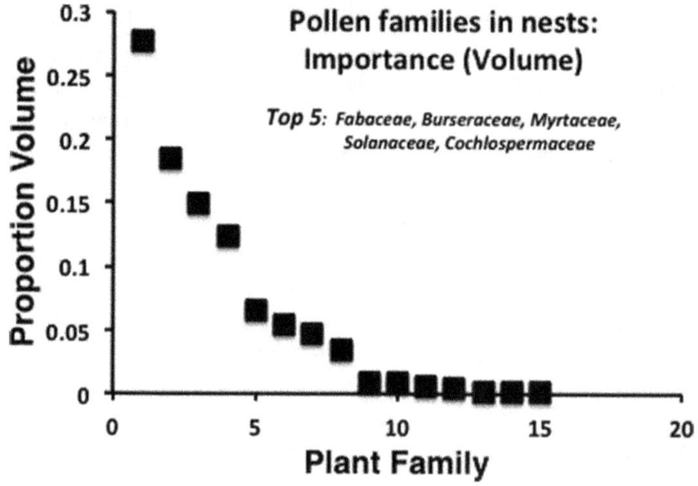

Fig. 5.4 Pollen importance by volume. Families visited by *Melipona* according to the volume of pollen taxa collected in pollen pots, calibrated with *Lycopodium* spore internal standards

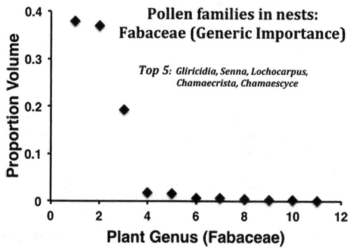

Fig. 5.5 Most important plant genera according to summed pollen volumes of all *Melipona* nest samples, collected from the Fabaceae family

Fig. 5.6 Plant families ranked by total pollen volume

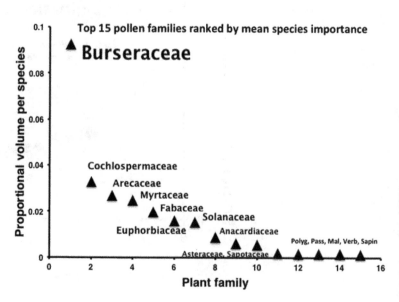

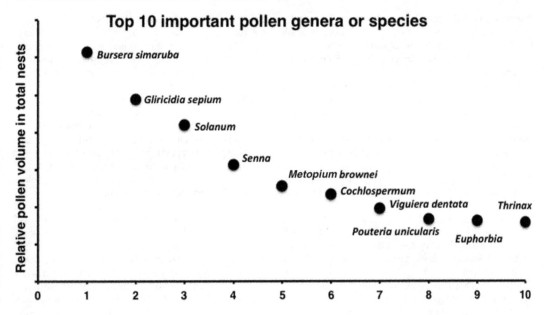

Fig. 5.7   Top 10 pollen taxa, genera, or species, by total volume

(Myrtaceae) and *Bursera* (Burseraceae) (Figs. 5.3 and 5.4, Table 5.1).

We suggest each bee is capable of pollinating these simple and open, bisexual flowers. Therefore these competitors, depending on which is the superior or more abundant pollinator, may actually feed one another (Roubik and Villanueva-Gutiérrez 2009). They are pollinators of their food plants, and when those seeds produce new plants with more flowers, competition may be largely converted to a mutualism. That is, competition at flowers is temporary, foraging adjustments lead to resource partitioning, and plant reproduction has the final effect in bee competitive interaction (Roubik 2009).

References

Erdtman G. 1943. An introduction to pollen analysis. Chronica Botánica Co; Waltham, USA. 239 pp.

Fonte L, Milera M, Demedio J, Blanco D. 2012. Foraging selectivity of the stingless bee *Melipona beecheii* Bennett at the EEPF "Indio Hatuey", Matanzas. Pastos y forrajes 35: 333–342.

Kleirnet-Giovannini A, Imperatriz-Fonseca VL. 1987. Aspects of the tropic niche of *Melipona marginata marginata* Lepeletier (Apidae, Meliponinae). Apidologie 18: 69–100.

Koeppen, W. 1936. Das gepographische System der Klimate In Handbuch der Klimatologie. Band I, Teil C; Berlin, Germany. 44 pp.

Labougle-Rentería JM, Zozaya- Rubio JA. 1986. La apicultura en México. Ciencia y Desarrollo 69: 17–36.

Leal-Ramos A, León-Sánchez LE. 2013. Antagonismo de *Apis mellifera* y *Melipona beecheii* por las fuentes de alimentación. Revista Cubana de Ciencias Forestales 1: 102–109.

Martins ACL, Rego MMC, Carreira LMM, Alburquerque PMC. 2011. Espectro polínico de mel de tiúba (*Melipona fasciculata* Smith, 1854, Hymenoptera, Apidae). Acta Amazonica 41: 183–190.

O'Rourke M K, Buchmann SL. 1991. Standardized analytical techniques for bee-collected pollen. Entomological Society of America 20: 507–513.

Palacios-Chavez R, Ludlow-Wiechers B, Villanueva-Gutiérrez R. 1991. Flora palinológica de la Reserva de la Biosfera de Sian ka'an, Quintana Roo, México. Centro de Investigaciones de Quintana Roo; Chetumal, Quintana Roo, México. 321 pp.

Rada JMD, Durán García R, García-Contrera G, Morín JA, Lugo EA, Méndez-García ME, Hernández MA. 2015. Conservationand use. pp. 169–196 in Islebe GA, Calmé S, Leon-Cortés JL, Schnook B, eds. Biodiversity and conservation of the Yucatán peninsula. Springer, New York. 401 pp.

Ramalho M, Kleirnet-Giovannini A, Imperatriz-Fonseca VL. 1990. Important bee plants for stingless bees (*Melipona* and Trigonini) and Africanized honey bees (*Apis mellifera*) in neotropical habitats: a review. Apidologie 21: 469–488.

Ramalho M. 1992. Food habits of eusocial bees: pollen analysis as a quantitative approach? Abstracts of the Eight International Palynological Congress. Aix- en – Provence, 1992. Aix- Marseille III University, France. pp. 119.

Roubik DW. Ed. 1995. Pollination of cultivated plants in the tropics. Agricultural Services Bulletin 118. Food and Agriculture Organization of the United Nations; Rome, Italy. 169 pp.

Roubik DW. 1989. Ecology and natural history of tropical bees. Cambridge University Press; New York, USA. 514 pp.

Roubik DW. 2009. Ecological impact on native bees by the invasive Africanized honey bee. Acta Biologica Colombiana.14: 115–124.

Roubik DW, Moreno JE. 2013. How to be a bee-botanist using pollen spectra. pp. 295–314. In: Vit P, Pedro SRM, Roubik DW, eds. Pot–honey: a legacy of stingless bees. Springer; New York, USA. 654 pp.

Roubik DW, Sakai S , Hamid Karim A , Eds. 2005. Pollination ecology and the rain forest: Sarawak studies. Springer-Verlag, Ecological Studies Series. No. 174. New York. 307 pp.

Roubik DW, Villanueva-Gutiérrez R. 2009. Invasive Africanized honey bee impact on native solitary bees: a pollen resource and trap nest analysis. Biological Journal of the Linnean Society 98: 152–160.

Sánchez-Sánchez O, Islebe GA, Ramírez Barajas PJ, Torrescano-Valle N. 2015. Natural and human induced disturbance in vegetation. pp. 153–167. In: Islebe GA, Calmé S, Leon-Cortés JL, Schnook B, eds. Biodiversity and conservation of the Yucatán peninsula. Springer, New York. 401 pp.

Stockmarr J. 1971. Tablets with spores used in absolute pollen analysis. *Pollen Spores* 13: 615–621.

Villanueva-Gutiérrez R, Roubik DW, Porter-Bolland L. 2015. Bee-plant interactions: Competition and phenology of flowers visited by bees. In: Biodiversity and conservation of the Yucatán Peninsula. pp. 131–152. In: Islebe, GA, Calmé S, León Cortéz JL, Schmook B, eds. Springer; New York. 401 pp.

Villanueva-Gutiérrez R, Roubik DW. 2004. Why are African honey bees and not European bees invasive? Pollen diet diversity in community experiments. Apidologie 35: 481–491.

Villanueva-Gutiérrez R. 2002. Polliniferous plants and foraging strategies of *Apis mellifera* in the Yucatán Peninsula, Mexico. Revista de Biología Tropical 50: 1035–1044.

Villanueva-Gutiérrez R. 1984. Plantas de importancia apícola en el ejido de Plan del Río, Veracruz, México. Biotica 9: 279–340.

Vossler FG, Tellería MC, Cunningham M. 2010. Floral resources foraged by *Geotrigona argentina* (Apidae, Meliponini) in the Argentine Dry Chaco forest. Grana 49: 142–153.

Melittopalynological Studies of Stingless Bees from the East Coast of Peninsular Malaysia

Roziah Ghazi, Nur Syuhadah Zulqurnain, and Wahizatul Afzan Azmi

6.1 Introduction

Stingless bees are known to be pollinators in Southeast Asian tropical rainforest (Eltz et al. 2002) and also good candidates for providing pollination services in agricultural ecosystems (Heard 1999; Slaa et al. 2006). Malaysia is home to a moderate number and diversity of stingless bee species that forage on various plants and vegetation zones including grasses, herbs, forest trees and cultivated plants. There are approximately 33 species that have been recorded so far in Peninsular Malaysia which varied in body sizes ranging from 2 to 14 mm (Mohd Norowi et al. 2008). In Thailand, over 30 species of Meliponini have been recorded (Schwarz 1939; Sakagami et al. 1985; Michener and Boongird 2004) and probably more than that will be found. A current report by Mohd Fahimee et al. (2016) revealed that a total of 35 species of stingless bees were collected from Peninsular Malaysia, and, thus, more may still be discovered. The size of Malaysian stingless bees can be divided into three categories, which are small (0.2–0.5 cm), medium (0.5–0.7 cm) and large (0.7–1.0 cm) (Mohd Fahimee et al. 2012). Their small sizes allow them to have access to many different kinds of flowers with openings too narrow to permit entrance by other flower visitors, such as most *Apis* (honey bee) species. Stingless bee workers usually visit and collect only one plant species in one trip but pollinate multiple plant species (Heard 1999; Jalil 2014).

Melissopalynology or pollen analysis of honey is generally important to trace the geographical origin of a particular type of honey (Ponnuchamy et al. 2014). Information on the pollen collected by stingless bees in Malaysia is important in meliponiculture, particularly in developing the premium marketable honey and bee products, and is called 'melittopalynology' (Roubik and Moreno, Chap. 1). It is also of considerable help for the potential farmers and entrepreneurs who wish to find ways to determine stingless bee resources to insure maximum quantity and quality of honey yield, as well as to boost the meliponiculture industry in Malaysia.

Pollen is the principal source of nitrogen and protein for most stingless bees and is collected in large quantities by workers for provisioning brood cells or for storage as colony 'pot-pollen'. Bee pollen contains a complex mixture of essential substances such as carbohydrates, proteins, amino acids, lipids, vitamins, mineral substances and trace elements which contribute to honey colour,

R. Ghazi
Agropolis Unisza, Universiti Sultan Zainal Abidin
Kampus Besut, 22200 Besut, Terengganu, Malaysia

N.S. Zulqurnain • W.A. Azmi (✉)
School of Marine and Environmental Sciences,
Universiti Malaysia Terengganu,
21030 Kuala Terengganu, Terengganu, Malaysia
e-mail: wahizatul@umt.edu.my

© Springer International Publishing AG, part of Springer Nature 2018
P. Vit et al. (eds.), *Pot-Pollen in Stingless Bee Melittology*, DOI 10.1007/978-3-319-61839-5_6

smell and flavour (Grout 1992; Küçük et al. 2007). In addition, it contains considerable amounts of polyphenolic substances, mainly flavonoids, which are regarded as potent antioxidants. Those flavonoids can be used to evaluate botanical origin and to set up quality standards for the assessment of nutritional and physiological properties of bee pollen (Ulusoy and Kolayli 2014).

However, the lack of reliable information about pollen types collected by stingless bees in Malaysia limits our knowledge about stingless bee food sources. Beekeeping in the east coast of Peninsular Malaysia, particularly Terengganu, has been taken up on a modest commercial scale in some districts. To date, little is known on the stingless bee products such as honey, propolis, bee bread (which is the pollen, processed by various microbes, and other ingredients stored in the brood cells) and bee wax in potential agricultural business for the people of Terengganu. Taman Tropika Kenyir (TTK) and Besut are regions with great meliponiculture potential. The main objective of this study was therefore to identify the pollen species collected by two commonly kept stingless bees in Terengganu, *Heterotrigona itama* and *Lepidotrigona terminata*, in Taman Tropika Kenyir and Besut, Terengganu. We hope that findings from this study will lead to new commercial strategies considering large-scale breeding of stingless bee colonies, mainly to secure bee pollination and to sustain biodiversity of tropical forest and agricultural ecosystems of Terengganu.

6.2 Pollen Collection by *Heterotrigona itama* in Tropical Island of Taman Tropika Kenyir, Terengganu

This study was conducted from April to September 2013 on an island called Taman Tropika Kenyir (TTK), which is owned by the Agrobiodiversity and Environment Research Centre, Malaysian Agricultural Research and Development Institute (MARDI). TTK is a garden for some species of indigenous plants of Malaysia. There are about 112 species of plants

that have been planted at TTK, which include a variety of herbs, exotic species, underutilised fruits, tropical fruits and landscape trees. In addition, TTK is known as a collecting place for "underutilised fruits" or fruit trees that are rarely found (M. Radzali Mispan, personal communication). Among the underutilised fruits that are planted in TTK are *Lepisanthes fruticosa* (Sapindaceae), *Rhodomyrtus tomentosa* (Myrtaceae), *Eugenia uniflora* (Myrtaceae) and *Passiflora edulis* (Passifloraceae), while the commercial fruit trees are *Citrus* spp. (Rutaceae), *Syzygium* spp. (Myrtaceae), *Averrhoa bilimbi* (Oxalidaceae) and *A. carambola* (Oxalidaceae) (see Table 6.1).

TTK was selected as the study site because there are active stingless beekeeping and honey production activities arising from programmes of the Management and Utilization of Biological Resources, or MARDI. There are nine species of stingless bees in TTK and most of them are native to the locality: *Tetragonilla collina*, *Tetragonilla atripes*, *Lepidotrigona terminata*, *Tetragonula fuscobalteata*, *Homotrigona fimbriata*, *Tetrigona apicalis*, *Heterotrigona itama*, *Geniotrigona thoracica* and *Tetragonula laeviceps*.

The reference samples of pollen grains were collected using a method adapted from Wahizatul et al. (2012). Flower buds from the dominant trees, shrubs and herbs found in TTK were plucked carefully and preserved in vials containing 70% ethanol to make sure that the flower buds were in good condition. The pollen grain samples were prepared on the slide by using a micropipette and slide cover. The slides were observed under a Moticam 1300 light microscope (400X magnification). The pollen image was captured, and the pollen length and width (in μm) of pollen were measured. The pollen grains from dominant species were used as a reference to compare with the pollen type collected from *Heterotrigona itama* foragers.

Pollen was collected from *Heterotrigona itama* using a method modified from Marques-Souza et al. (1996). Pollen was collected from the bees on hot sunny days between 0900 and 1100 h. This was done by briefly closing the hive entrance ($N = 6$) and capturing 10 arriving work-

Table 6.1 The pollen types visited and collected by *Heterotrigona itama* in Taman Tropika Kenyir (+) Present, (−) Absent

Scientific names	Common local names	Family	Observed	Acetolysis
Underutilised fruits				
Garcinia prainiana	Cerapu	Clusiaceae	+	−
Baccaurea lanceolata	Asam Pahong	Phyllanthaceae	−	+
Passiflora edulis	Markisa	Passifloraceae	+	−
Ardisia elliptica	Mempena	Primulaceae	+	−
Erioglossum rubiginosum	Mentajam	Sapindaceae	+	+
Lepisanthes fruticosa	Ceri Terengganu	Sapindaceae	+	+
Muntingia calabura	Ceri kampung	Muntingiaceae	+	+
Flacourtia jangomas	Kerkup	Salicaceae	+	+
Eugenia uniflora	Cermai Belanda	Myrtaceae	−	+
Rhodomyrtus tomentosa	Kemunting	Myrtaceae	+	−
Ornamental plants				
Ruellia simplex	Mexican petunia	Acanthaceae	+	−
Amischotolype griffithii	Purple ball	Commelinaceae	+	+
Costus woodsonii	Scarlet spiral flag	Costaceae	+	−
Bauhinia acuminata	Tapak kuda	Fabaceae, Faboideae	+	−
Mimusops elengi	Bunga Tanjung	Sapotaceae	+	+
Heliconia psittacorum	Sepit Udang	Heliconiaceae	+	−
Neomarica longifolia	Yellow walking iris	Iridaceae	+	−
Vegetables				
Benincasa hispida	Kundur	Cucurbitaceae	+	+
Cleome rutidosperma	Purple cleome	Cleomaceae	+	−
Capsicum sp.	Chili	Solanaceae	+	+
Shrubs				
Mikania sp.	Selapuk Tunggul	Asteraceae	+	+
Asystasia gangetica	Rumput Israel	Acanthaceae	+	+
Melastoma malabathricum	Sendudok	Melastomataceae	+	−
Cosmos caudatus	Ulam Raja	Asteraceae	+	+
Trees				
Polyalthia sp.		Annonaceae	−	+
Terminalia sp.	Terminalia	Combretaceae	−	+
Lithocarpus sp.	Mempening	Fagaceae	−	+
Litsea sp.		Lauraceae	−	+
Pentace sp.	Melunak Bukit	Malvaceae, Brownlowioideae	−	+
Agriculture fruits				
Coffea sp.	Kopi	Rubiaceae	+	+
Citrus sp.	Limau	Rutaceae	+	+
Tamarindus indica	Asam Jawa	Fabaceae, Caesalpinioideae	+	−
Syzygium sp.	Jambu	Myrtaceae	+	+
Averrhoa bilimbi	Starfruit	Oxalidaceae	+	+
Averrhoa carambola	Belimbing Buluh	Oxalidaceae	+	+
Others				
Unidentified 1		Dipterocarpaceae	−	+
Unidentified 2		Euphorbiaceae	−	+
Synsepalum dulcificum	Buah Ajaib	Sapotaceae	+	+
Phaleria macrocarpa	Mahkota Dewa	Thymelaeaceae	+	+

ers with an insect net. After the removal of their pollen loads with a blunt needle, the workers were released. The pollen clusters from the *H. itama* individuals were preserved in 70% ethanol, and the vials were labelled. Using a 10 μm micropipette, 1 μm of the preserved pollen in the vial was transferred to the haemocytometer and covered with a cover slip.

There were 59 plant species belonging to 30 families visited by *H. itama*, among the 360 pollen loads collected from returning foragers at TTK. However, only 27 plant species belonging to 24 families of flowering plants were successfully identified. They included ornamental trees, underutilised fruits, agricultural fruits and others (Table 6.1). Thus, there were 32 morphotypes of pollen that could not be identified.

Overall, the family Salicaceae (*Flacourtia jangomas*) had the most pollen grains collected by *H. itama*, followed by Fagaceae (*Lithocarpus* sp.) and Sapotaceae (*Mimusops elengi*). The scarcest pollen recorded was Annonaceae (*Polyalthia* sp.),

Asteraceae (*Mikania* sp.), Dipterocarpaceae (*Dipterocarpus* sp.), Fabaceae-Caesalpinioideae (*Tamarindus indica*) and Rubiaceae (*Coffea* sp.). However, there were 132 unidentified pollen grains, representing 26 pollen types which probably belong to several families (Fig. 6.1).

Underutilised fruit was the most abundant type of collected pollen for *H. itama*, followed by tree and ornamental plants, whereas agricultural fruit flowers provided the rarest pollen grains collected (Fig. 6.2). This is due to the flowering season of many underutilised fruit plants during April to September in TTK. During that period, *Flacourtia jangomas*, one such species, was a much utilised source of pollen.

Flower colour attracts flower visitors. Bright colours, for example, red and blue, are attractive (Brody and Mitchell 1997; Glover 2008), but not for bees (Willmer 2011). *H. itama* collected pollen grains mostly from the white and creamy white flowers (Fig. 6.3). For example, the flowers visited were *Capsicum* sp., *Citrus* sp., *Coffea*

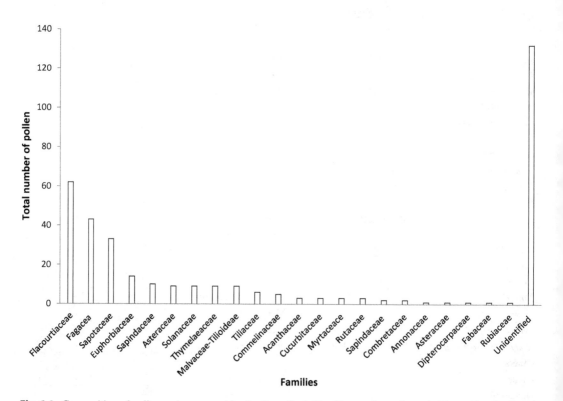

Fig. 6.1 Composition of pollen grains grouped by family collected by *Heterotrigona itama* in Taman Tropika Kenyir. Unidentified pollen consisted of 26 pollen types not classified to family

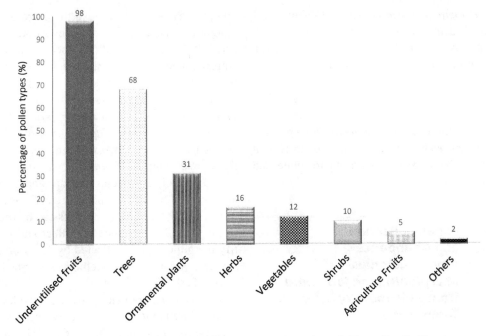

Fig. 6.2 Percentage of pollen types of plants collected by *Heterotrigona itama* in Taman Tropika Kenyir

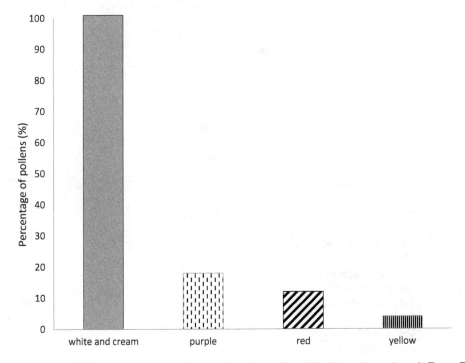

Fig. 6.3 Percentage of different colours among flowers with pollen collected by *Heterotrigona itama* in Taman Tropika Kenyir

sp., *Erioglossum rubiginosum, Eugenia uniflora, Flacourtia jangomas* and *Synsepalum dulcificum*. According to Kiew and Muid (1991), 66% of flowers that were visited by bees (honey bees) in Malaysia are white, creamy white or yellow in colour. In this study, *Flacourtia jangomas* was the most frequently collected by *H. itama,* followed by *Lithocarpus* sp. and *Mimusops elengi*. Our finding suggests that stingless bees are also attracted to white and yellow flowers.

6.3 Pollen Collection and Abundance Among Colonies of *Lepidotrigona terminata* from a Meliponary in Besut, Terengganu

The second study was conducted in a meliponary of Besut, in Terengganu, performed from November 2012 to February 2013. The meliponary was selected because there are active stingless beekeeping and honey production activities among the local people. Approximately 10 colonies of the stingless bee, *L. terminata,* were kept there. The area was an orchard that contains different species of trees such as mango (*Mangifera indica*), rambutan (*Nephelium lappaceum*), jackfruit (*Artocarpus heterophyllus*), lime (*Citrus hystrix),* and langsat (*Lansium parasiticum*). Approximately 27 species of crops and a large number of trees, climbers, lianas, shrubs, herbs, epiphytes and saprophytes can be found in the area. The methods to collect and identify the pollen types collected by *L. terminata* foragers are the same methods described in Sect. 6.2.

Although ten pollen types were collected from *L. terminata* workers, only eight were successfully identified (Table 6.2). The most frequently collected pollen grain was *Capsicum* sp., followed by *Citrus hystrix* and *Murraya paniculata* (Fig. 6.4). The other pollen types were

Table 6.2 Occurrences of pollen class percentages collected by the stingless bees, *Lepidotrigona terminata*, from an apiary in Besut, Terengganu

Habit	Pollen types (Order/Family/Species)	Common local name	Months		
			November 2012	January 2013	February 2013
Shrub	Sapindales/Rutaceae *Murraya paniculata*	Orange jasmine	I	–	–
Shrub	Sapindales/Rutaceae *Citrus hystrix*	Kaffir lime	D	–	I
Tree	Malpighiales/Calophyllaceae *Calophyllum inophyllum*	Balltree	I	A	–
Vegetable	Solanales/Solanaceae *Capsicum* sp.	Chilli	I	–	D
Shrub	Caryophyllales/Nyctaginaceae *Bougainvillea glabra*	Paper flower	–	A	–
Perennial	Fabales/Fabaceae, Mimosoideae *Mimosa pudica*	Sensitive plant	–	–	A
Herb	Lamiales/Acanthaceae *Asystasia gangetica*	Creeping foxglove	–	I	I
	Malpighiales/Euphorbiaceae *Suregada multiflora*	False lime	–	–	O
	Unidentified pollen 1		–	–	I
	Unidentified pollen 2		–	–	O

[Notes: *D* dominant pollen (>45%), *A* accessory pollen (15–45%), *I* isolated pollen (3–15%), *O* occasional pollen (<3%); – = Absent)]

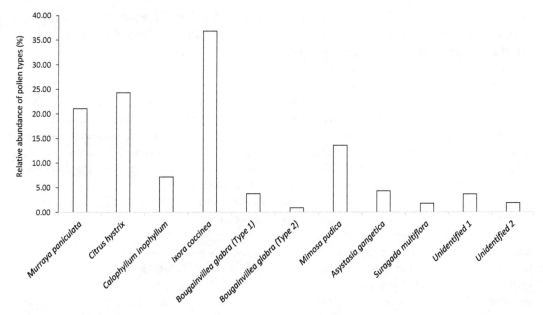

Fig. 6.4 Relative abundance (%) of pollen types collected by the stingless bee, *L. terminata* from an apiary in Besut, Terengganu

categorised as isolated pollen, which consisted of *Mimosa pudica*, *Callophyllum inophyllum*, *Asystasia gangetica*, *Bougainvillea glabra* and unidentified pollen 1. The results obtained indicate that in Besut the major source of pollen collected by *L. terminata* was provided by shrubs, as well as herbs, perennials and flowering trees. The study was published by Wahizatul et al. (2015).

As already mentioned, the most frequent pollen collected was *Capsicum* sp. which has several characteristics that might lead to a preference for its pollen by *L. terminata*. *Capsicum*, locally known as 'chilli', is a Neotropical vegetable used extensively in local food and traditional medicine. Chilli flowers are very attractive to a wide array of insects including honey bees (*Apis mellifera*) and stingless bees (Heard 1999; Klein et al. 2007). Chilli flowers, like those of most cultivated Solanaceae, have no nectar but produce pollen within poricidal anthers. Honey bees and stingless bee can collect pollen fallen on the petals and leaves. Flowers are pendant from leaf axils, show a white corolla, five to

seven stamens containing 1.0–1.5 mg of pollen, and one central style with a round sticky stigma on its tip. The anthers are tubular, and dehiscence occurs through lateral opening by vibratile or "buzz' collection by certain bees, not including *Apis* or most stingless bees. Both floral anthesis and anther dehiscence take place in the morning, between 0700 h and 0900 h (Dag and Kammer 2001). There are few studies on chilli pollination, and the role of stingless bees in producing quality fruits of this crop is still unknown.

Many studies indicate that the shape and odour of flowers play important roles as signals that facilitate bee recognition of rewarding resources. *Capsicum* sp., *Citrus hystrix* and *Murraya paniculata* have strong and agreeable odours. This might suggest why these three flowers were preferred by *L. terminata* in the current study. Another factor in pollen choice is probably the size of the flowers. This postulation was in accordance with Wille et al. (1983) who suggest that crops or flowers preferred by stingless bees are usually small, numerous flowers.

This is observed in *Citrus hystrix*, *Murraya paniculata* and *Mimosa pudica* which also possess the same characteristic of small size flowers (<1 inch in size). However, according to Roubik and Moreno (2013), the predominant pollen is not often an indication of the source of the nectar for colonial bees, and they often collect pollen from male flowers or species that have no nectar. Although the plant species listed in this study do not provide the nectar for honey, the pollen could possibly be important for provisioning brood cells or for storage (Küçük et al. 2007). However, to what extent the use of these pollen species for the brood cells or food storage and how the pollen influences the foraging behaviour of the stingless bee, such as the number of visits and their contributions to pollination, are still unknown. Thus, all of these research gaps need to be further studied in order to understand the preferred pollen species by native stingless bees in Malaysia.

6.4 Selected Flowers Producing Pollen Preferred by Stingless Bees in Terengganu

Flowers, pollen morphology and microscopic images of pollen grains often used by *H. itama* and *L. terminata* are presented in figures below: underutilised fruits (Fig. 6.5), ornamental plants (Fig. 6.6), vegetables (Fig. 6.7), shrubs (Fig. 6.8) and agriculture fruits (Fig. 6.9). The methods to collect and identify the pollen types collected by *H. itama* and *L. terminata* foragers are described in Sect. 6.2.

6.5 Conclusions

This study provides a brief list of frequent plant pollen types and species in the diets of *H. itama* and *L. terminata* in Terengganu. They may indicate valuable pollen and nectar sources

Fig. 6.5 Flowers visited by *Heterotrigona itama* and *Lepidotrigona terminata* for pollen collecting: underutilised fruits (Photos: S.A.R. Tuan Nek (flowers), R. Ghazi (pollen))

Garcinia praniana

Passiflora edulis

Flacourtia jangomas

Rhodomyrtus tomentosa

Fig. 6.6 Flowers visited by *Heterotrigona itama* and *Lepidotrigona terminata* for pollen collecting: ornamental plants (Photos: S.A.R. Tuan Nek (flowers), R. Ghazi (pollen))

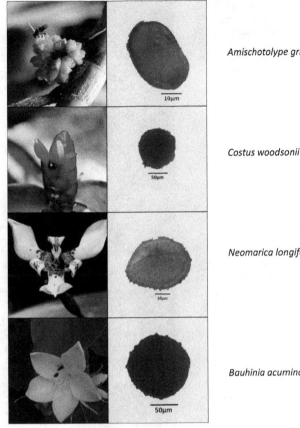

Amischotolype griffithii

Costus woodsonii

Neomarica longifolia

Bauhinia acuminata

for these stingless bees. The dominant pollen grains collected by the stingless bees were *Flacourtia jangomas*, *Lithocarpus* sp., *Mimusops elengi*, *Capsicum* sp., *Citrus hystrix* and *Murraya paniculata*. However, some plant species in this study are not honey sources because their flowers have no nectar, and most of the pollen types or species collected were not identified. It is hoped our experiences and results will help strengthen meliponiculture in Malaysia.

Acknowledgements We would like to thank the School of Marine and Environmental Sciences, Universiti Malaysia Terengganu, for the financial support of this study, Dr. Muhammad Radzali Mispan and Mohd Fahimee Jaapar from the Malaysian Agricultural Research and Development Institute (MARDI) for giving permission to conduct research in Taman Tropik Kenyir (TTK), Dr. Norul Badriah Hassan from Universiti Sains Malaysia Kubang Kerian for giving permission to conduct this study on her apiary, Tuan Haji Muhamad Embong for the assistance during the identification of pollens and Tuan Haji Muhammad Razali Salam, Mr. Johari Mohd Nor and Mr. Syed Ahmad Rizal for their guidance and assistance during sampling and laboratory session. Special thanks to Dr. D.W. Roubik for the critical review of the manuscript and Mr. J.E. Moreno-P for kindly checking the scientific names of all the plants. The study was supported by KETENGAH (Lembaga Kemajuan Terengganu Tengah), Terengganu State and Universiti Malaysia Terengganu (UMT).

Fig. 6.7 Flowers visited by *Heterotrigona itama* and *Lepidotrigona terminata* for pollen collecting: vegetables (Photos: S.A.R. Tuan Nek (flowers), R. Ghazi (pollen))

Capsicum sp.

Cleome rutidosperma

Benincasa hispida

Fig. 6.8 Flowers visited by *Heterotrigona itama* and *Lepidotrigona terminata* for pollen collecting: shrubs (Photos: S.A.R. Tuan Nek (flowers), R. Ghazi (pollen))

Cosmos caudatus

Melastoma malabathricum

Asystasia gangetica

Fig. 6.9 Flowers visited by *Heterotrigona itama* and *Lepidotrigona terminata* for pollen collecting: agricultural fruits (Photos: S.A.R. Tuan Nek (flowers), R. Ghazi (pollen))

References

Brody AK, Mitchell RJ. 1997. Effects of experimental manipulation of inflorescence size on pollination and pre- dispersal seed predation in the hummingbird-pollination plant *Ipomopsis agregata*. Oecologia 110: 86–93.

Dag A, Kammer Y. 2001. Comparison between the effectiveness of honey bee (*Apis mellifera*) and bumble bee (*Bombus terrestris*) as pollinators of greenhouse sweet pepper (*Capsicum annuum*). American Bee Journal 141: 447–448.

Eltz T, Brühl CA, Kaars SVD, Linsenmair KE. 2002. Determinants of stingless bee nest density in lowland dipterocarp forests of Sabah, Malaysia. Oecologia 131: 27–34.

Glover BJ. 2008. Understanding flowers and flowering: An integrated approach. Oxford University Press; Oxford, UK. 227 pp.

Grout RA. 1992. The hive and the honey bee. Dadant Sons; Hamilton, Illinois, USA. 52 pp.

Heard TA. 1999. The role of stingless bees in crop pollination. Annual Review Entomology 44: 183–206.

Jalil AH. 2014. Beescape for the meliponines: Conservation of Indo-Malayan stingless bees. Patridge Publishing; Singapore. 202 pp.

Kiew R, Muid M. 1991. Beekeping in Malaysia: Pollen atlas. United Selangor press Sdn Bhd; Kuala Lumpur, Malaysia. 186 pp.

Klein AM, Vaissiere BE, Cane JH, Steffan-Dewenter I, Cunningham SA, Kremen C, Tscharntke T. 2007. Importance of pollinators in changing landscapes for world crops. Proceedings of the Royal Society B 274: 303–313.

Küçük M, Kolayli S, Karaoğlu S, Ulusoy E, Baltaci C, Candan F. 2007. Biological activities and chemical composition of three honeys of different types from Anatolia. Food Chemistry 100: 526–534.

Marques-Souza AC, Moura CDO, Nelson BW. 1996. Pollen collected by *Trigona williana* (Hymenoptera: Apidae) in central Amazonia. Revista de Biologia Tropical 44: 567–573

Mohd Fahimee J, Hamdan S, Rosliza J, Suri R. 2012. Manual teknologi penternakan lebah kelulut. Institut Penyelidikan dan Kemajuan Asas Tani (MARDI). MARDI Publication; Kuala Lumpur, Malaysia. 20 pp.

Mohd Fahimee J, Madihah H, Muhamad Radzali M, Rosliza J, Mohd Masri S, Mohd Yusri Z, Roziah G, Idris AG. 2016. The diversity and abundance of stingless bee (Hymenoptera: Meliponini) in peninsular Malaysia. Advances in Environmental Biology 10: 1–7.

Mohd Norowi, H, Sajap AS, Rosliza J, Mohd Fahimie J, Suri R. 2008. Conservation and sustainable utilization of stingless bees for pollination services in agricultural ecosystems in Malaysia. Department of Agriculture; Kuala Lumpur, Malaysia. 5 pp.

Michener CD, Boongird S. 2004. A new species of *Trigona* from peninsular Thailand (Hymenoptera: Apidae: Meliponini). Journal of the Kansas Entomological Society 77: 143–146.

Ponnuchamy R, Bonhomme V, Prasad S, Das L, Patel P, Gaucherel C, Pragasam, A, Anupama K. 2014. Honey pollen: using melissopalynology to understand foraging preferences of bees in tropical south India. PLoS ONE 9(7): e101618. doi:10.1371/journal.pone.0101618

Roubik D, Moreno PJE. 2013. How to be a bee-botanist using pollen spectra. 295–314 pp. In: Vit P, Pedro SRM, Roubik D, eds. Pot-honey: a legacy of stingless bees. Springer, New York. 654 pp.

Sakagami SF, Inoue T, Salmah S. 1985. Key to the stingless bee species found or expected from Sumatra. In: RI Ohgushi, ed. Evolutionary ecology of insect in humid tropics, especially in central Sumatra. Kanazawa University, Japan. 37–43 pp.

Schwarz HF. 1939. The Indo-Malayan Species of *Trigona*. Bulletin of the American Museum of Natural History 76: 83–141.

Slaa EJ, Chaves LAS, Malagodi-Bragac KS, Hofsteded FE. 2006. Stingless bees in applied pollination: practice and perspectives. Apidologie 37: 293–315.

Ulusoy E, Kolayli S. 2014. Phenolic composition and antioxidant properties of Anzer bee pollen. Journal of Food Biochemistry 38: 73–82.

Wahizatul AA, Roziah G, Nor Zalipah M. 2012. Importance of carpenter bee, *Xylocopa varipuncta* (Hymenoptera: Apidae) as pollination agent for mangrove community of Setiu Wetlands, Terengganu, Malaysia. Sains Malaysiana 41: 1057–1062.

Wahizatul AA, Nur Syuhadah Z, Roziah G. 2015. Melissopalynology and foraging activity of stingless bees, *Lepidotrigona terminata* (Hymenoptera: Apidae) from an apiary in besut, terengganu. Journal of Sustainability Science and Management 10: 27–35.

Wille A, Orozco E, Raabe C. 1983. Polinización del chayote *Sechium edule* (Jacq.) Swartz en Costa Rica. Revista de Biologia Tropical 31: 145–154.

Willmer PG. 2011. Pollination and floral ecology. Princeton University Press, Princeton. 792 pp.

The Contribution of Palynological Surveys to Stingless Bee Conservation: A Case Study with *Melipona subnitida*

7

Camila Maia-Silva,
Amanda Aparecida Castro Limão, Michael Hrncir,
Jaciara da Silva Pereira,
and Vera Lucia Imperatriz-Fonseca

7.1 Introduction

Pollen is the principal protein source for larval development among the majority of bee species, whereas nectar is the primary energy source for individuals, required by both adults and larvae for growth and development. Social bees store both these food items inside the nest and use the hoarded reserves in times of reduced resource availability in the environment. Stingless bees (Apidae, Meliponini), in particular, store pollen and nectar in oval pots made of wax and resin (Ducke 1925; Michener 1974; Roubik 1989). Pollen grains may be found not only in the pollen storage pots but also in the honey pots. This is mainly due to the fact that, on landing on a flower

C. Maia-Silva (✉) • A.A.C. Limão • M. Hrncir
J. da Silva Pereira
Departamento de Ciências Animais, Universidade
Federal Rural do Semi-Árido,
Av. Francisco Mota 572, Mossoró, RN 59625-900,
Brazil
e-mail: maiasilvac@gmail.com

V.L. Imperatriz-Fonseca
Departamento de Ciências Animais, Universidade
Federal Rural do Semi-Árido,
Av. Francisco Mota 572, Mossoró, RN 59625-900,
Brazil

Instituto Tecnológico Vale,
Rua Boaventura da Silva 955, Belém, PA 66055090,
Brazil

to collect nectar, foragers dislodge some of the flower's pollen, which falls into the nectar. Consequently, pollen grains are sucked up by the bees together with the nectar stored in forager's stomach (Todd and Vansell 1942; Bryant 2001). Additionally, pollen grains adhere to the forager's body during nectar collection and are carried unintentionally to the colony, where they, occasionally, end up in the honey pots (Bryant 2001; Barth 2004). Pollen grains have often very specific morphology associated with the plant species from which they originate. Hence, through careful examination of their morphology, it is possible to determine the botanical origin of the resources collected by bees (Jones and Jones 2001; Silva et al. 2012; Silva et al. 2014). Pollen grains, therefore, may be considered reliable "fingerprints" of plants, leaving evidence in the colonies' food storage, on the foragers' bodies, and even in the bees' feces (Eltz et al. 2001b; Silva et al. 2012, 2014).

Melittopalynology, the analysis of pollen samples from bees, is an important tool to assess which plants are visited by the foragers. Moreover, it constitutes an accurate method to investigate important ecological questions by (1) characterizing the foraging preferences of bees, (2) providing information about the phenology and abundance of bee plants in the course of a year, (3) indicating food source overlap and competition among sympatric bee species, and (4) identifying important plant-pollinator interaction networks. (5) Such information forms a fundamental basis for developing management plans for preserving or restoring habitat quality for bees.

In order to assess the foraging preferences of colonies and to investigate temporal variation of diet composition of stingless bees, usually, pollen and honey samples are extracted from recently built pots inside the nests (Cortopassi-Laurino and Ramalho 1988; Wilms and Wiechers 1997; Malagodi-Braga and Kleinert 2009). This method, on the one hand, permits the identification of plant species harvested (qualitative analysis). On the other hand, it allows estimating the contribution of each plant to the bees' diet, particularly concerning the stored pollen, by calculating the relative abundance of each species in the pollen samples (quantitative analysis) – but see Roubik and Moreno (Chap. 4) and Villanueva-Gutiérrez et al. (Chap. 5) in the present volume. New pots are easily recognized by their lighter wax color compared to old pots. Moreover, owing to the fermentation process, pollen that has been stored for some time is darker in color than is freshly deposited pollen (Menezes et al. 2013).

Another way to investigate the floral resources harvested by bees is through collecting pollen samples from the bodies of foragers returning to the nest (Sommeijer et al. 1983; Roubik et al. 1986; Ramalho et al. 2007; Eltz et al. 2001a; Maia-Silva 2013; Maia-Silva et al. 2014, 2015a). In corbiculate bees, which include the Meliponini, the pollen collected on flowers is transported on bees, which move the pollen from other parts of their bodies in special structures on their hind tibiae, the corbiculae. In bees that collect only nectar, pollen grains incidentally adhere to the hairs on different parts of the foragers' bodies during nectar collection, and the analysis of these residuals helps to identify the nectar sources visited by the bees (Biesmeijer et al. 1999; Kajobe 2007). This analysis of pollen collected from foragers is a powerful method to assess bee-flower interactions in the course of the seasons (Biesmeijer et al. 1999; Eltz et al. 2001a; Kajobe 2007; Maia-Silva 2013; Maia-Silva et al. 2014, 2015a). However, the method is time-intensive and requires frequent sampling (Eltz et al. 2001b).

An interesting biome in which to study plant-bee interaction and resource preference of stingless bees is the Brazilian tropical dry forest, the Caatinga, of northeastern Brazil. The climate of this ecoregion is characterized by elevated annual temperatures year-round as well as extended periods of drought (Alvares et al. 2013). As an additional challenge for perennial social bee colonies, most flowering plants are in bloom only during a very short and unpredictable rainy season (Quirino and Machado 2014; Maia-Silva et al. 2015a). Thus, colonies need to respond quickly to any increase in floral resource abundance and collect as much food as possible within a short period of time, in order to store sufficient reserves for surviving later drought and potential resource scarcity. In the present chapter, we present a brief overview considering the floral origins of pollen and nectar harvested by colonies of the stingless bee "jandaíra," *Melipona subnitida* Ducke, 1911, one of the few social bee species naturally occurring in Brazilian tropical dry forest (Zanella 2000). These data are important to outline strategies for conserving native vegetation or for the restoration of degraded areas. These measures help to maintain suitable foraging habitats for *M. subnitida* and other native bee species. We will first discuss floral resource dynamics in the course of a year and then show the importance of melittopalynology as a tool in designing restoration strategies.

7.2 Floral Resources Dynamics: Pot-Pollen Versus Pollen from the Bees' Body

Temporal changes among resources visited by bees aids understanding the fluctuating bee dependencies on particular plant species in the course of a year, within a particular season, or even during a day. Moreover, it may help to identify the most important resources for a particular bee species, thereby forming a basis for conservation and management strategies.

Bees native to the dry forest have to be specially adapted to the unpredictable fluctuations in food availability in the course of the year. The analysis of pollen samples from the corbiculae of *M. subnitida* pollen foragers revealed that they collect pollen at only a few of the available pollen sources (between May 2011 and May 2012: 14 pollen types (usually species) collected, from a total of 63 pollen sources in bloom). The investigated colonies preferred the most profitable floral resources in the environment, such as mass-flowering trees[1] (*Pityrocarpa moniliformis*, *Mimosa arenosa*, *M. caesalpiniifolia*, *M. tenuiflora*, *Anadenanthera colubrina*). These plants provide large quantities of pollen for flower visitors and, consequently, are excellent resources, which allow stingless bees to hoard floral resources within their nests (Wilms et al. 1996; Ramalho 2004; Maia-Silva et al. 2014, 2015a). Among the trees visited by *M. subnitida*, two species are of particular importance for the maintenance of perennial colonies, *A. colubrina* and *M. tenuiflora* (Maia-Silva et al. 2015a). Whereas *P. moniliformis*, *Mimosa arenosa*, and *M. caesalpiniifolia* bloom in the rainy season, during which pollen sources are abundant, *M. tenuiflora* produces flowers mainly in the transition between rainy and dry seasons and *A. colubrina* during a very short period of time in the dry season (Fig. 7.1) (Maia-Silva et al. 2012; Quirino and Machado 2014). This particular flowering phenology allows the bee colonies to replenish part of their pollen reserves during a general dearth period (Maia-Silva et al. 2015a).

In addition to mass-flowering plants, important pollen sources for *M. subnitida* are shrubs with poricidal flowers (*Senna obtusifolia*, *Chamaecrista duckeana*, and *C. calycioides*) (Table 7.1, Fig. 7.1). This special flower type is characterized by anther dehiscence through apical slits or pores. Flowers release pollen only when vibrated by flower visitors (buzz-pollination; Buchmann 1983). Due to the fact that poricidal flowers contain large amounts of pollen, they are highly attractive for *Melipona* species and other bees

capable of buzz-pollination (Nunes-Silva et al. 2010).

Although *M. subnitida* is considered a generalist floral visitor (Pinto et al. 2014), the analysis of the pollen from corbicular loads of foragers demonstrated that the colonies maximize their food intake by focusing on the most profitable pollen sources available in the environment (mass-flowering trees and shrubs with poricidal anthers). Hence, the colonies are able to collect large amounts of resources within a short period of time.

When pollen samples from the corbicular loads of *M. subnitida* foragers were compared to samples taken from colony storage pots, most of those on the incoming bees were among those stored in the nest. We detected only five pollen types in the pot-pollen (between May 2011 and May 2012: 19 pollen types collected, while 63 potential pollen source species were in bloom) that had not been found on foragers. Those additional pollen types represented less than 15% of the total samples (data with volume correction coefficient of grains; Maia-Silva unpublished data). They originated from shrubs (*Spermacoce verticillata*, *Waltheria* sp., *Ipomoea asarifolia*, *Mimosa quadrivalvis*, *Sida cordifolia*) that bloom in the rainy season (Maia-Silva et al. 2012), a period characterized by high floral diversity, including mass-flowering trees and herbs (Reis et al. 2006; Santos et al. 2013; Quirino and Machado 2014). Owing to the elevated foraging activity of *M. subnitida* colonies during this time of the year (Maia-Silva et al. 2015a), some pollen types may have been easily missed when limited sampling was made of forager pollen loads. Thus, in order to determine the whole spectrum of pollen plants visited by bees, it is necessary to investigate both forager pollen loads and the stored pollen. Additionally, the colonies' waste can be used as an important source of information on the bees' pollen diet because the waste pellets contain pollen exine from larval feces (Eltz et al. 2001b). An alternative method could be the use of pollen traps that automatically sample pollen from all incoming foragers in the course of the day. However, this method is not very efficient for most stingless bee species, which rapidly deposit large quantities of resin on strange objects at their nest entrances, thus immobilizing the pollen traps (Menezes et al. 2012).

---

[1]Within a plant population, mass-flowering individuals of a certain species bloom synchronously for a short period of time, often less than a week. During this period, each individual produces a large number of new flowers each day (Augspurger 1980; Bawa 1983).

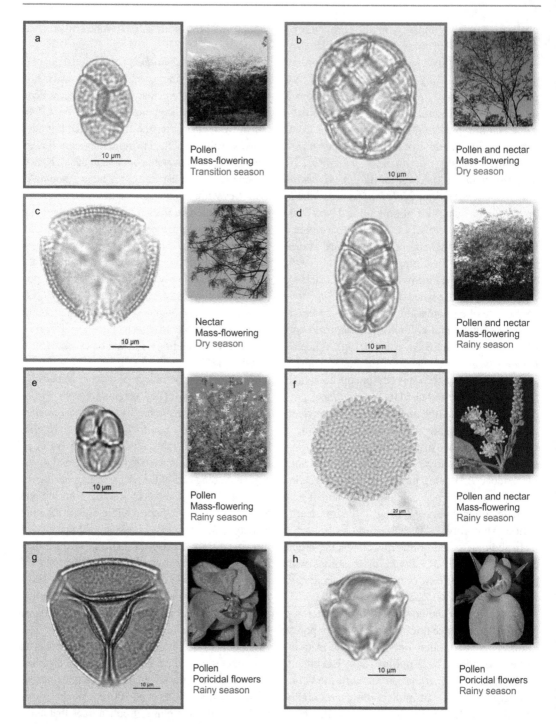

Fig. 7.1 Pollen types harvested by *Melipona subnitida* in the Brazilian tropical dry forest during the rainy and the dry season (pollen plants: Maia-Silva et al. 2015a and Limão et al., *unpublished data*; nectar plants: Limão 2015). (**a**) *Mimosa tenuiflora*. (**b**) *Anadenanthera colub-* *rina*. (**c**) *Myracrodruon urundeuva*. (**d**) *Pityrocarpa moniliformis*. (**e**) *Mimosa arenosa*. (**f**) *Croton sonderianus*. (**g**) *Senna obtusifolia*. (**h**) *Chamaecrista duckeana* (Photos M. Hrncir (plants) and C. Maia-Silva (pollen grains))

Table 7.1 Native plant species visited by *Melipona subnitida* in the Brazilian tropical dry forest

Families	Species	Stratum	R	FP	Pot-honey	Pot-pollen	Pollen Forager	Nectar Forager
Amaranthaceae	*Alternanthera ficoidea* Colla	Herb	N	R	(3)	No	No	No
Anacardiaceae	*Myracrodruon urundeuva* Allemão	Tree	N	D	(3)	(3)	No	(4)
Convolvulaceae	*Ipomoea asarifolia* (Desr.) Roem. & Schult.	Herb	P/N	R	No	(2)(3)	(5)	No
	Ipomoea bahiensis Willd. Ex Roem. & Schult.	Herb	P/N	R	No	No	No	(4)
Euphorbiaceae	*Croton sonderianus* Müll.Arg.	Tree	P/N	R	(3)	(3)	No	(4)
Fabaceae, Caesalpinioideae	*Chamaecrista calycioides* (DC. Ex Collad.) Greene	Herb	P	R	(3)	(2)(3)	(1)	No
	Chamaecrista duckeana (P.Bezerra & Afr.Fern.) H.S.Irwin & Barneby	Shrub	P	R	(3)	(2)(3)	(1)(5)	No
	Libidibia ferrea (Mart. Ex Tul.) L.P.Queiroz	Tree	N	R	No	No	No	(4)
	Senna macranthera (DC. Ex Collad.) H.S.Irwin & Barneby	Tree	P	R	No	No	(6)	No
	Senna obtusifolia (L.) H.S.Irwin & Barneby	Shrub	P	R	(3)	(2)(3)	(1)(6)	No
	Senna trachypus (Benth.) H.S.Irwin & Barneby	Shrub	P	R	No	No	(5)(6)	No
	Senna uniflora (Mill.) H.S.Irwin & Barneby	Shrub	P	R	No	(2)	(1)	No
Fabaceae, Mimosoideae	*Anadenanthera colubrina* (Vell.) Brenan	Tree	P/N	D	(3)	(2)(3)	(1)(5)	(4)
	Mimosa arenosa (Willd.) Poir.	Tree	P	R	(3)	(2)(3)	(1)(5)	(4)
	Mimosa caesalpiniifolia Benth.	Tree	P	R	(3)	(2)(3)	(1)(5)	(4)
	Mimosa quadrivalvis L.	Herb	P	R	(3)	(2)(3)	(5)	No
	Mimosa tenuiflora (Willd.) Poir.	Tree	P	T	(3)	(2)(3)	(1)(5)(6)	(4)
	Neptunia plena (L.) Benth.	Herb	P	R	No	No	(5)(6)	No
	Pityrocarpa moniliformis (Benth.) Luckow & R.W.Jobson	Tree	P/N	R	(3)	(2)(3)	(1)(5)	(4)
	Senegalia polyphylla (DC.) Britton & rose	Tree	P/N	R	(3)	No	No	No
Malvaceae, Malvoideae	*Sida cordifolia* L.	Shrub	P/N	R	(3)	(2)	No	No
Malvaceae, Byttnerioideae	*Waltheria* sp.	Shrub	P/N	R	(3)	(2)(3)	No	(4)
Passifloraceae, Turneroideae	*Turnera subulata* Sm.	Herb	P/N	R	No	(2)	(1)	No
Rubiaceae	*Spermacoce verticillata* (L.) G.Mey.	Herb	P/N	R	(3)	(2)(3)	No	(4)

Main resources (R) of the visited plants: *P* pollen, *N* nectar. Flowering periods (FP): *R* rainy season, *D* dry season, *T* transition season. Pollen samples were taken from pot-honey, pot-pollen, pollen foragers, and nectar foragers

(1) Maia-Silva et al. (2015a, b), *(2)* Maia-Silva unpublished data (data collected from the same colonies and in the same periods studied by Maia-Silva et al. 2015a), *(3)* Limão et al., *unpublished data* (pollen from pot-honey and pot-pollen), *(4)* Limão (2015) (pollen collected from nectar foragers bodies), *(5)* Pereira et al. (2014), Pereira (2015). *(6)* Limão et al. (2012), *no*: not observed

An important point to consider when analyzing pot-pollen is that these samples, usually, are composed of a mixture of pollen grains collected by foragers at a wide range of plant sources. When performing a quantitative analysis, the importance of each plant species is usually determined by counting the pollen grains in the sample. Yet, due to the high variability concerning the size (volume) of pollen grains from different plant species, this simple count overestimates the importance of small and abundant pollen types. Therefore, more critical studies consider including the grain volume as an essential parameter for determining the relative importance of each pollen type (Roubik 1989; Buchmann and O'rourke 1991; Silveira 1991; Biesmeijer et al. 1992; Malagodi-Braga and Kleinert 2009).

When analyzing the pollen in pot-honey samples, not all grains originate necessarily from nectar sources (Table 7.1). In a study conducted between January and December of 2012, we registered 17 pollen types in honey pots of *M. subnitida* colonies, among these grains from poricidal flowers (*Senna obtusifolia*, *Chamaecrista* spp.). Given that these flowers do not produce nectar, and pollen is their only reward for visitors, the occurrence of pollen from poricidal plant species in honey samples was rather unexpected, even more so because these pollen types were abundant in the pot-honey during several months of the study. A correlation analysis demonstrated an increasing incidence of pollen from poricidal flowers in pot-honey with increasing occurrence in pot-pollen (Limão et al., unpublished data) (Fig. 7.2). This finding indicates that, during periods with intense pollen foraging activity at flowers with poricidal anthers, the pollen in pot-honey of *M. subnitida* contains a mixture of grains from both nectar and pollen plants. In line with these observations, Pinto et al. (2014) registered the plants that provide only pollen for bees (*Chamaecrista ramosa*, secondary pollen, 15%–45%; *Senna* sp., minor pollen, <3%) in the pot-honey of *M. subnitida* colonies in the Brazilian coastal sand plains (Restinga).

When trying to identify the plant species visited by bees for nectar, a possibility to rule out sampling of non-nectar pollen types is through analyzing the pollen on the body or in the honey stomach of nectar foragers. Both these methods are considerably more time-consuming than is the extraction of samples from honey pots. In the case of pollen samples from the honey stomach,

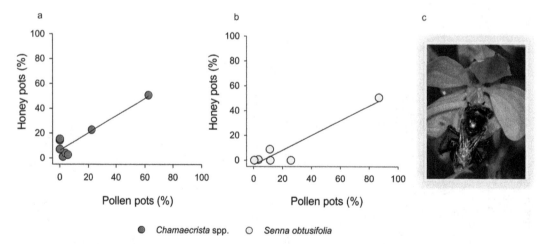

● *Chamaecrista* spp.   ○ *Senna obtusifolia*

Fig. 7.2 Correlation between the amount of pollen from flowers with poricidal anthers in pot-pollen and pot-honey of *Melipona subnitida* colonies. (**a**) Pollen from *Chamaecrista* spp.; *line* indicates linear regression (Pearson correlation, $R = 0.95$, $P < 0.05$). (**b**) Pollen from *Senna obtusifolia*; *line* indicates linear regression (Pearson correlation, $R = 0.93$, $P < 0.05$). Data were collected between January and December of 2012 (pot-pollen and pot-honey samples of five colonies), at the Experimental Field Station of the Brazilian Federal University at Mossoró (UFERSA). (**c**) *Melipona subnitida* collecting pollen at *S. obtusifolia* (Limão et al., *unpublished data*) (Photo M. Hrncir)

bees have to be dissected in order to extract the material (Todd and Vansell 1942). For the analysis of the pollen residues on the foragers' bodies, bees have to be captured individually, immobilized by chilling for better manipulation, and the pollen grains removed meticulously to pieces of glycerine jelly (Biesmeijer et al. 1999; Kajobe 2007). However, besides identifying the botanical origin of the nectar collected by bees, both these methods can provide detailed insights into the floral resource dynamics along the day and in the course of the year.

Limão (2015), analyzing pollen sampled from the bodies of *M. subnitida* nectar foragers, found grains predominantly from mass-flowering trees (rainy season, *Croton sonderianus*, *Libidibia ferrea*, *Pityrocarpa moniliformis*; dry season, *Myracrodruon urundeuva*) and from herbs that flower mainly in the rainy season (*Spermacoce verticillata*, *Ipomoea asarifolia*, *Waltheria* sp.). However, pollen grains from poricidal flowers, as those found in the pot-honey samples, were not detected on the bees' bodies. Hence, sampling the pollen from the nectar foragers' bodies proved to be more reliable than sampling pot-honey for identifying the main nectar sources in the course of the year (Table 7.1, Fig. 7.1).

7.3 Melittopalynology as Tool for Restoration Strategies: Suitable Foraging Habitats

The degradation and fragmentation of natural habitats, associated with agricultural intensification and urbanization, result in a reduced diversity and abundance of floral resources for bees (Kearns and Inouye 1997; Potts et al. 2010). Thus, conservation of areas with native vegetation together with efficient restoration practices are fundamental to preserve, as much as possible, the natural diversity of plant species.

For a successful conservation of bees in human-altered landscapes, habitat restoration strategies should focus on providing suitable foraging areas that offer floral resources in distinct periods of the year and suitable nesting sites within the flight range of species (Shepherd et al. 2003; Murray

et al. 2009; Jha et al. 2013). In this context, urban green spaces play an important role due the fact that they increase the availability of resources for many bee species in the environment (Aleixo et al. 2014; Baldock et al. 2015; Kaluza et al. 2016; Hicks et al. 2016). Since floral resource availability is indicated as a primary factor influencing bee population abundance (Roulston and Goodell 2011), pollinator-friendly practices in agricultural areas are an important contribution to bee conservation and are becoming more and more common in several parts of the planet. Here, an increasing plant biodiversity within and around agricultural crops improves the habitat for bees and other pollinators and, thus, helps to safeguard the pollination services necessary to enhance crop yield (Carvalheiro et al. 2011; Menz et al. 2011).

A knowledge of the foraging preferences of key pollinator species forms a solid background for the development of successful habitat restoration plans. The first step is to identify important food plants in order to create suitable resource patches in urban, suburban, and rural areas (Shepherd et al. 2003; Menz et al. 2011). Here, melittopalynological studies may provide an insight into the plant species exploited by bees as well as the plant flowering phenology at a population level (Table 7.1).

Native plants, usually, are the best option for habitat restoration, especially so in semiarid or arid regions like the Brazilian tropical dry forest, given that these plants are adapted to the abiotic conditions (e.g., soil type, relative humidity, air temperature, precipitation) and biotic factors of the environment (e.g., diseases, predators, pollinators, seed dispersers) and, therefore, grow with minimum attention (Shepherd et al. 2003). Moreover, native bees often prefer to forage on native rather than exotic plants (Morandin and Kremen 2013a; Hanley et al. 2014). For *M. subnitida*, this is underlined by the finding that colonies in urban areas preferentially collected pollen from native plants species despite an elevated abundance of exotic species compared to natural areas (Pereira et al. 2014, 2015), thus pointing to the importance of using native plants to compose urban green spaces.

In seminatural habitats of the Brazilian tropical dry forest, such as suburban and rural areas, the lack of native tree species that bloom during the dry season could be a breaking point for meliponiculture, the keeping of stingless bee colonies. It is common practice among beekeepers in this region to provide sugar syrup for the bees during the dry season, when floral resources are scarce (Koffler et al. 2015; Maia et al. 2015). However, management techniques that involve protein supplements for stingless bees are rarely employed. Thus, meliponine colonies must survive extended periods of dearth with only little or no protein supply, which leads to a strong decline in brood production and, eventually, results in colony death (Maia-Silva et al. 2014, 2016).

The results of our melittopalynological surveys indicate that suitable foraging habitats for *M. subnitida* should include plants that flower in the dry season as well as plants that flower in the rainy season, mass-flowering plants, and species with poricidal flowers (Maia-Silva et al. 2015a) (Table 7.1). In this context, plants that offer large amounts of nectar are very important resources, particularly concerning the commercialization of meliponine honey by beekeepers. Pollen plants, on the other hand, are crucial for the growth and development of the colonies, especially during periods of brood production, when the colonies need a large amount of protein as larval food (Roubik 1982; Maia-Silva et al. 2015a). When pollen storage in the nest declines, *M. subnitida* colonies decrease or, in extreme cases, even interrupt the production of new brood, which, in the long term, may lead to the collapse of the colonies (Maia-Silva et al. 2016). Thus, the availability of pollen sources in the environment is essential for maintaining or even increasing the productivity of meliponiculture, and, therefore, plants that provide this resource should be planted around meliponaries.

Appropriate management practices, such as bee-friendly gardens or hedgerows, may serve two purposes, the conservation of native bees and, given the significant contribution of wild pollinators to crop yield (Garibaldi et al. 2013, 2016), the local economy. In the case of *M. subnitida*, which is the main stingless bee species managed by beekeepers in the Brazilian tropical dry forest, local farmers are able to increase their income selling honey or colonies through adequate management techniques (Jaffé et al. 2015).

For the planning of bee-friendly environments, both herbaceous species and trees offer distinct advantages. Herbaceous species, on the one hand, grow quickly and produce flowers rapidly, in addition to offering large amounts of floral resources for bees (Hicks et al. 2016). Woody perennial plants like trees and hedgerows, on the other hand, may need a few years to initiate blooming and even longer to reach their peak productivity (Shepherd et al. 2003). However, once producing flowers, they constitute a major foraging opportunity for flower visitors. Moreover, trees provide suitable nesting sites for bees. *M. subnitida*, for instance, naturally nests in narrow cavities (diameter ≈ 10 cm; length ≈ 115 cm) preferentially in the native tree species *Commiphora leptophloeos*, *Poincianella bracteosa*, and *Myracrodruon urundeuva* (Cámara et al. 2004; Martins et al. 2004). The cavities of the trees provide thermal protection against the high ambient temperature. Among those trees, *M. urundeuva* is of particular importance because, in addition to offering nesting opportunities, it produces flowers that are an important nectar source during the dry season (Limão 2015). Thus, owing to their potential to provide both nesting sites and floral resources, native woody perennial plants are indispensable in bee-friendly habitats and constitute a crucial element in habitat restoration efforts (Menz et al. 2011; Morandin and Kremen 2013b).

7.4 Concluding Remarks and Future Steps

Information about the plants exploited by bees as food sources is an important cornerstone for planning conservation and restoration programs and in maintaining local bee biodiversity. Here, palynological surveys of the pollen collected by bees are considered a helpful tool. Pollen analyses require prior knowledge of the taxa and morphological characteristics of pollen available in a particular environment. Thus, an important first step is the establishment of a pollen grain reference library, with which pollen sampled from bees and/or nests can be compared (Silva et al. 2012, 2014).

The stingless bee species used in the present chapter as a case study, *M. subnitida*, lives under environmental conditions that can be considered extreme for social bees, such as high ambient temperatures, low and unpredictable precipitation, and very scarce food resources. Living under this environmental pattern has led to the evolution of a variety of behavioral responses that allow these bees to maintain perennial colonies. Colonies of *M. subnitida* react quickly to any increase in resource abundance by intensifying their foraging activity (Maia-Silva et al. 2015a). Focusing their collecting effort on the most profitable resources in the environment, they are able to hoard large amounts of nectar and pollen in their nests within a short period of time (Maia-Silva et al. 2015a). Variations in pollen storage are associated with variation in brood cell construction (Maia-Silva et al. 2016). Thus, during periods of high pollen availability within the nest, as a consequence of elevated resource abundance in the environment, *M. subnitida* colonies increase their brood production. However, when pollen is scarce, the colonies largely reduce their brood production, thus avoiding an accelerated consumption of this vital resource and preserving it longer during the periods of dearth of unpredictable duration (Maia-Silva et al. 2015a, 2016; for similar results on other meliponine species see: Roubik 1982).

Melipona subnitida is a key species with high ecological and economic potential. In addition to contributing to the maintenance of native plant biodiversity, stingless beekeeping with *M. subnitida* provides an important extra income for local farmers (Jaffé et al. 2015; Maia et al. 2015). Colonies are traditionally used for honey production (Koffler et al. 2015; Maia et al. 2015), but these stingless bees also have a noteworthy potential as pollinators of commercial crops (Cruz et al. 2004; Cruz et al. 2005; Bomfim et al. 2014).

Despite its economic potential and its ecological value as key pollinator, *M. subnitida* is threatened by environmental degradation and, to a minor extent, by predatory beekeepers and honey hunters (Rosso-Londoño and Imperatriz-Fonseca

2017). The loss of natural habitat in the Brazilian tropical dry forest (Leal et al. 2005) has contributed considerably to the decline of native bee populations, which demands urgent and efficient conservation/restoration measures. Here, patches of natural vegetation in combination with bee-friendly habitats that offer high plant diversity can maximize the availability of resources and contribute to the conservation of these bees.

Recent studies on stingless bees suggest important future directions for the use of melittopalynology associated with conservation management plans:

1. Potential shifts in the geographic distribution of bees are mainly related to shifts in the occurrence of their resources (Schweiger et al. 2008; Hegland et al. 2009; Giannini et al. 2013, 2015). Thus, knowledge about floral utilization by bees in a specific ecosystem allows modeling the future occurrence of bees in scenarios of climate warming predicted for coming decades (Giannini et al. 2012, 2015; Giannini, *unpublished data*).

2. Melittopalynology is an important tool to identify food niches of bees as well as bee-plant interaction networks (Biesmeijer et al. 1999; Bosch et al. 2009; Maia-Silva et al. 2014, 2015a, b). These interactions are of specific interest with regard to global warming, given that climate changes presumably result in shifts in ecological interactions, which, eventually, affect ecosystem functioning (Memmott et al. 2007; Tylianakis et al. 2008; Van der Putten et al. 2010; Zarnetske et al. 2012; Valiente-Banuet et al. 2015).

3. Concerning the economics of beekeeping, palynological studies may aid the assessment of floristic composition and landscape configuration as they contribute to the quality of pot-honey and pot-pollen (Felipe-Neto et al. 2017).

Such recent studies in stingless bees show that melittopalynological surveys do not necessarily have to remain on the level of a simple description of the floral resources collected by bees. These data, particularly if temporal resource dynamics are included, may form the basis for a vast range of studies aiming at a more profound understanding of bee-plant interactions, their shifts in response to ecological stressors, and the potential consequences of these changes in the ecosystem.

Acknowledgments We would like to thank David Roubik, Patricia Vit, and two anonymous reviewers for important suggestions and comments on the manuscript. We would like to thank the students of the Behavioural Ecology Laboratory (bee-LAB) for their help with the data collection. This study complies with current Brazilian laws and was financially supported by grants of the Brazilian Science Foundations CAPES (Coordenação de Aperfeiçoamento de Pessoal de Nível Superior) to CMS and CNPq (Conselho Nacional de Desenvolvimento Científico e Tecnológico) to VLIF (482218/2010-0, 406102/2013-9) and to MH (304722/2010-3, 309914/2013-2, 404156/2013-4).

References

Aleixo KP, Faria LB, Groppo M, Castro, MMN, Silva CI. 2014. Spatiotemporal distribution of floral resources in a Brazilian city: implications for the maintenance of pollinators, especially bees. Urban Forestry and Urban Greening 13: 689–696.

Alvares CA, Stape JL, Sentelhas PC, Gonçalves JLM, Sparovek G. 2013. Köppen's climate classification map for Brazil. Meteorologische Zeitschrift 22: 711–728.

Augspurger, C. K. 1980. Mass-flowering of a tropical shrub (*Hybanthus prunifolius*) - influence on pollinator attraction and movement. Evolution 34: 475-488.

Baldock KCR, Goddard MA, Hicks DM, Kunin WE, Mitschunas N, Osgathorpe LM, Potts SG, Robertson KM, Scott AV, Stone GN, Vaughan IP, Memmott J. 2015. Where is the UK's pollinator biodiversity? The importance of urban areas for flower-visiting insects. Proceedings of the Royal Society Biological Sciences 282: 20142849.

Barth OM. 2004. Melissopalynology in Brazil: a review of pollen analysis of honeys, propolis and pollen loads of bees. Scientia Agricola 61: 342–350.

Bawa KS. 1983. Patterns of fl owering in tropical plants. pp. 394–410. In Jones CE, Little RJ, eds. Handbook of Experimental Pollination Biology. Van Nostrand Reinhold Company Inc.; NewYork-NY, USA. 558 pp.

Biesmeijer JC, Vanmarwijk B, Vandeursen K, Punt W, Sommeijer MJ. 1992. Pollen sources for *Apis mellifera* L (Hym, Apidae) in Surinam, based on pollen grain volume estimates. Apidologie 23: 245–256.

Biesmeijer JC, Smeets MJAP, Richter JAP, Sommeijer MJ. 1999. Nectar foraging by stingless bees in Costa Rica: botanical and climatological influences sugar concentration of nectar collected by *Melipona*. Apidologie 30: 43–55.

Bomfim IGA, Bezerra ADM, Nunes AC, Aragão FAS, Freitas BM. 2014. Adaptive and foraging behavior of two stingless bee species in greenhouse mini watermelon pollination. Sociobiology 61: 502-509.

Bosch J, González AMM, Rodrigo A, Navarro D. 2009. Plant–pollinator networks: adding the pollinator's perspective. Ecology Letters 12: 409-419.

Bryant Jr VM 2001. Pollen contents of honey. CAP Newsletter 24: 10-24

Buchmann SL. 1983. Buzz pollination in angiosperms. pp. 73-113. In Jones CE, Little RJ eds. Handbook of Experimental Pollination Biology, Van Nostrand Reinhold; New York, USA. 558 pp.

Buchmann SL, O'rourke MK. 1991. Importance of pollen grain volumes for calculating bee diets. Grana 30: 591–595.

Cámara JQ, Sousa AH, Vasconcelos WE, Freitas RS, Maia PHS, Almeida JC, Maracajá PB. 2004. Estudos de meliponíneos, com ênfase a *Melipona subnitida* D. no município de Jandaíra, RN. Revista Biologia e Ciências da Terra 4: 20.

Carvalheiro LG, Veldtman R, Shenkute AG, Tesfay GB, Pirk CWW, Donaldson JS, Nicolson SW. 2011. Natural and within-farmland biodiversity enhances crop productivity. Ecology Letters 14: 251–259.

Cortopassi-Laurino M, Ramalho M. 1988. Pollen harvest by Africanized *Apis mellifera* and *Trigona spinipes* in São Paulo botanical and ecological views. Apidologie 19: 1–24.

Cruz DO, Freitas BM, Silva LA, Silva EMS, Bomfim IGA. 2004. Adaptação e comportamento de pastejo da abelha jandaíra (*Melipona subnitida* Ducke) em ambiente protegido. Acta Scientiarum. Animal Sciences 26: 293–298.

Cruz DO, Freitas BM, Silva LA, Silva EMS, Bomfim IGA. 2005. Pollination efficiency of the stingless bee *Melipona subnitida* on greenhouse sweet pepper. Pesquisa Agropecuária Brasileira 40: 1197-1201.

Ducke A. 1925. Die stachellosen Bienen Brasiliens. Zoologische Jahrbücher Abteilung für Systematik, Geographie und Biologie der Tiere 49: 335-448.

Eltz T, Brühl C, van der Kaars S, Chey VK, Linsenmair KE. 2001a. Pollen foraging and resource partitioning of stingless bees in relation to flowering dynamics in a Southeast Asian tropical rainforest. Insectes Sociaux 48: 273–279.

Eltz T, Brühl C, Kaars VD, Linsenmair K. 2001b. Assessing stingless bee pollen diet by analysis of garbage pellets : a new method. Apidologie 32: 341–353.

Felipe-Neto CAL, Pinheiro CGME, Tambosi LR, Imperatriz-Fonseca VL, Jaffé R. 2017. Como a estrutura da paisagem pode afetar a qualidade de mel da abelha jandaíra no semiárido brasileiro? pp 175-182. In Imperatriz-Fonseca VL, Koedam D, Hrncir M eds. A abelha jandaíra no passado, presente e futuro, EdUFERSA; Mossoró, Brasil. 269 pp.

Garibaldi LA, Steffan-Dewenter I, Winfree R, Aizen MA, Bommarco R, Cunningham SA, et al. 2013. Wild pollinators enhance fruit set of crops regardless of honey bee abundance. Science 339: 1608-1611.

Garibaldi LA, Carvalheiro LG, Vaissière BE, Gemmill-Herren B, Hipólito J, Freitas BM, et al. 2016. Mutually beneficial pollinator diversity and crop yield outcomes in small and large farms. Science 351: 388-391.

Giannini TC, Acosta AL, Garófalo CA, Saraiva AM, Alves-dos-Santos I, Imperatriz-Fonseca VL. 2012. Pollination services at risk: Bee habitats will decrease owing to climate change in Brazil. Ecological Modelling 244: 127-131.

Giannini TC, Acosta AL, Silva CI, Oliveira, PEAM, Imperatriz-Fonseca, VL, Saraiva, AM. 2013. Identifying the areas to preserve passion fruit pollination service in Brazilian Tropical Savannas under climate change. Agriculture, Ecosystems & Environment 171: 39–46.

Giannini TC, Tambosi LR, Acosta AL, Jaffé R, Saraiva AM, Imperatriz-Fonseca VL, Metzger JP. 2015. Safeguarding ecosystem services: A methodological framework to buffer the joint effect of habitat configuration and climate change. PLoS ONE 10: e0129225.

Hanley ME, Awbi AJ, Franco M. 2014. Going native? Flower use by bumblebees in English urban gardens. Annals of Botany 113: 1-8.

Hegland SJ, Nielsen A, Lázaro A, Bjerknes AL, Totland Ø. 2009. How does climate warming affect plant-pollinator interactions? Ecology Letters 12: 184-195.

Hicks DM, Ouvrard P, Baldock KCR, Baude M, Goddard MA, Kunin WE, Mitschunas N, Memmott J, Morse H, Nikolitsi M, Osgathorpe LM, Potts SG, Robertson KM, Scott AV, Sinclair F, Westbury DB, Stone GN. 2016. Food for pollinators: quantifying the nectar and pollen resources of urban flower meadows. PLoS ONE 11: e0158117.

Jaffé R, Pope N, Carvalho AT, Maia UM, Blochtein B, Carvalho CAL, Carvalho-Zilse GA, Freitas BM, Menezes C, Ribeiro MF, Venturieri GC, Imperatriz-Fonseca VL. 2015. Bees for development: Brazilian survey reveals how to optimize stingless beekeeping. PLoS ONE 10: e0121157.

Jha S, Burkle L, Kremen C. 2013. Vulnerability of pollination ecosystem services. Climate Vulnerability 4: 117-128.

Jones GD, Jones SD. 2001. The uses of pollen and its implication for entomology. Neotropical Entomology 30: 341–350.

Kajobe R. 2007. Botanical sources and sugar concentration of the nectar collected by two stingless bee species in a tropical African rain forest. Apidologie 38:110–121.

Kaluza BF, Wallace H, Heard TA, Klein, AM, Leonhardt, SD. 2016. Urban gardens promote bee foraging over natural habitats and plantations. Ecology and Evolution 6: 1304–1316.

Kearns CA, Inouye DW. 1997. Pollinators, flowering plants, and conservation biology. Bioscience 47: 297–307.

Koffler S, Menezes C, Menezes PR, Kleinert ADMP, Imperatriz-Fonseca VL, Pope N, Jaffe R. 2015. Temporal variation in honey production by the stingless bee Melipona subnitida (Hymenoptera: Apidae): long-term management reveals its potential as a commercial species in northeastern Brazil. Journal of Economic Entomology108: 858–867.

Leal IR, Silva JMC, Tabarelli M, Lacher TE. 2005. Changing the course of biodiversity conservation in the Caatinga of northeastern Brazil. Conservation Biology 19: 701–706.

Limão AAC. 2015. A influência dos fatores bióticos e abióticos no néctar coletado por Melipona subnitida (Apidae, Meliponini) na Caatinga. Master's thesis. Universidade Federal Rural do Semi-Árido. 60 pp.

Limão AAC, Maia-Silva C, Silva CI, Imperatriz-Fonseca VL. 2012. Pollen sources used by Melipona subnitida (Apidae, Meliponini) during the dry season in an urbanized landscape in the Brazilian semi-arid region. In Anais do X Encontro sobre Abelhas; Ribeirão Preto, Brasil.

Maia UM, Jaffé R, Carvalho AT, Imperatriz-Fonseca VL. 2015. Meliponicultura no Rio Grande do Norte. Revista Brasileira de Medicina Veterinaria 37: 327–333.

Maia-Silva C. 2013. Adaptações comportamentais de Melipona subnitida (Apidae, Meliponini) às condições ambientais do semiárido brasileiro. PhD thesis Universidade de São Paulo. 132 pp.

Maia-Silva C, Silva CI, Hrncir M, Queiroz RT, Imperatriz-Fonseca VL. 2012. Guia de plantas visitadas por abelhas na Caatinga. Editora Fundação Brasil Cidadão; Fortaleza, Brasil. 191 pp.

Maia-Silva C, Imperatriz-Fonseca VL, Silva CI, Hrncir M. 2014. Environmental windows for foraging activity in stingless bees, Melipona subnitida Ducke and Melipona quadrifasciata Lepeletier (Hymenoptera: Apidae: Meliponini). Sociobiology 61: 378–385.

Maia-Silva C, Hrncir M, Silva CI, Imperatriz-Fonseca VL. 2015a. Survival strategies of stingless bees (Melipona subnitida) in an unpredictable environment, the Brazilian tropical dry forest. Apidologie 46: 631-643.

Maia-Silva C, Silva C, Souza DA, Aleixo, K P, Imperatriz-Fonseca, V L, Hrncir M 2015b. Climate warming will increase direct competition for food among stingless bees. In XXXIII Encontro Anual de Etologia, Belém, Brasil.

Maia-Silva C, Hrncir M, Imperatriz-Fonseca VL, Schorkopf DLP. 2016. Stingless bees (Melipona subnitida) adjust brood production rather than foraging activity in response to changes in pollen stores. Journal of Comparative Physiology A. 202: 723-732.

Malagodi-Braga KS, Kleinert AMP. 2009. Comparative analysis of two sampling techniques for pollen gathered by Nannotrigona testaceicornis Lepeletier (Apidae, Meliponini). Genetics and Molecular Research 8: 596–606.

Martins CF, Cortopassi-Laurino M, Koedam D, Imperatriz–Fonseca VL. 2004. Tree species used for nidification by stingless bees in the Brazilian Caatinga (Seridó, PB; João Câmara, RN). Biota Neotropica 4: 1–8.

Memmott J, Craze PG, Waser NM, Price MV. 2007. Global warming and the disruption of plant–pollinator interactions. Ecology Letters 10: 710-717.

Menezes C, Vollet-Neto A, Imperatriz-Fonseca VL. 2012. A method for harvesting unfermented pollen from stingless bees (Hymenoptera, Apidae, Meliponini). Journal of Apicultural Research 51: 240–244.

Menezes C, Vollet-Neto A, Contrera FAFL, Venturieri GC, Imperatriz-Fonseca VL. 2013. The role of useful microorganisms to stingless bees and stingless beekeeping. pp. 153-171. In Vit P, Pedro SM, Roubik DW eds. Pot-Honey: a Legacy of Stingless Bees. Springer; New York, USA. 654 pp.

Menz MHM, Phillips RD, Winfree R, Kremen C, Aizen MA, Johnson SD, Dixon KW. 2011. Reconnecting plants and pollinators : challenges in the restoration of pollination mutualisms. Trends in Plant Science 16: 4–12.

Michener CD. 1974. The Social Behavior of the Bees. Cambridge, Harvard University Press, USA. 418 pp.

Morandin LA, Kremen C. 2013a. Bee preference for native versus exotic plants in restored agricultural hedgerows. Restoration Ecology 21: 26–32.

Morandin LA, Kremen C. 2013b. Hedgerow restoration promotes pollinator populations and exports native bees to adjacent fields. Ecological Applications 23: 829–839.

Murray TE, Kuhlmann M, Potts SG. 2009. Conservation ecology of bees: populations, species and communities. Apidologie 40: 211–236.

Nunes-Silva P, Hrncir M, Imperatriz-Fonseca VL. 2010. A polinização por vibração. Oecologia Australis 14: 140-151.

Pereira JS. 2015. Plantas importantes para a manutenção da abelha jandaíra (*Melipona subnitida*) em paisagem urbana do semiárido brasileiro. Bachelor's thesis. Universidade Federal Rural do Semi-Árido. 36 pp.

Pereira JS, Limão AAC, Silva AGM, Maia-Silva C, Hrncir M. 2014. Forrageamento de pólen no semiárido brasileiro: plantas visitadas pela abelha jandaíra (Meliponini, *Melipona subnitida*) em um ambiente urbano. In XXXII Encontro Anual de Etologia e V Simpósio Latino-americano de Etologia; Mossoró, Brasil.

Pereira JS, Silva AGM, Sá-Filho GF, Hrncir M, Maia-Silva C. 2015. Which environmental factors influence the pollen collection by *Melipona subnitida* (Apidae, Meliponini) in an urban landscape in the Brazilian semiarid region? In Anais do XI Encontro sobre Abelhas; Ribeirão Preto, Brasil.

Pinto RS, Albuquerque PMC, Rêgo MMC. 2014. Pollen analysis of food pots stored by *Melipona subnitida* Ducke (Hymenoptera: Apidae) in a restinga area. Sociobiology 61: 461–469.

Potts SG, Biesmeijer JC, Kremen C, Neumann P, Schweiger O, Kunin WE. 2010. Global pollinator declines: trends, impacts and drivers. Trends in Ecology and Evolution 25: 345–353.

Quirino Z, Machado I. 2014. Pollination syndromes in a Caatinga plant community in northeastern Brazil: seasonal availability of floral resources in different plant growth habits. Brazilian Journal of Biology 74: 62–71.

Ramalho M. 2004. Stingless bees and mass flowering trees in the canopy of Atlantic Forest: a tight relationship. Acta Botanica Brasilica 18: 37–47.

Ramalho M, Silva MD, Carvalho CALD. 2007. Dinâmica de uso de fontes de pólen por *Melipona scutellaris* Latreille (Hymenoptera: Apidae): uma análise comparativa com *Apis mellifera* L. (Hymenoptera: Apidae), no Domínio Tropical Atlântico. Neotropical Entomology 36: 38-45.

Reis AMS, Araújo EL, Ferraz EMN, Moura AN. 2006. Inter-annual variations in the floristic and population structure of an herbaceous community of "Caatinga" vegetation in Pernambuco, Brazil. Revista Brasileira de Botânica 29: 497–508.

Rosso-Londoño J, Imperatriz-Fonseca VL. 2017. "Abelha não serve só pra botar mel, não!": meleiros e conflito socioambiental na Caatinga potiguar. pp 101-108. In Imperatriz-Fonseca VL, Koedam D, Hrncir M eds. A abelha jandaíra no passado, presente e futuro, EdUFERSA; Mossoró, Brasil. 269 pp.

Roubik DW. 1982. Seasonality in colony food storage brood production and adult survivorship: studies of *Melipona* in tropical forest (Hymenoptera: Apidae). Journal of the Kansas Entomological Society 55: 789-800.

Roubik DW, Moreno JE, Vergara C, Wittmann D. 1986. Sporadic food competition with the African honey bee: projected impact on Neotropical social bees. Journal of Tropical Ecology 2: 97–111.

Roubik DW. 1989. Ecology and Natural History of Tropical Bees. Cambridge University Press, Cambridge, USA. 514 pp.

Roulston TAH, Goodell K. 2011. The role of resources and risks in regulating wild bee populations. Annual Review of Entomology 56: 293-312

Santos JMFF, Santos DM, Lopes CGR, Silva KA, Sampaio EVSB, Araújo EL. 2013. Natural regeneration of the herbaceous community in a semiarid region in Northeastern Brazil. Environmental Monitoring and Assessment 185: 8287–8302.

Schweiger O, Settele J, Kudrna O, Klotz S, Kühn I. 2008. Climate change can cause spatial mismatch of trophically interacting species. Ecology 89: 3472-3479.

Shepherd M, Buchmann S, Vaughan M, Black S. 2003. The Pollinator Conservation Handbook. The Xerces Society, Portland. 145 pp.

Silva CI, Maia-Silva C, Ribeiro FA, Bauermann SG. 2012. O uso da palinologia como ferramenta em estudos sobre ecologia e conservação de polinizadores no Brasil. pp. 369–383. In Imperatriz-Fonseca V, Canhos D, Alves D, Saraiva A eds. Polinizadores no Brasil - Contribuição e Perspectivas para a Biodiversidade, Uso Sustentável, Conservação e Serviços Ambientais. EDUSP; São Paulo, Brasil. 488 pp.

Silva CI, Imperatriz-Fonseca VL, Groppo M, Bauermann SG, Saraiva, AA, Queiroz EP, Evaldt ACP, Aleixo KP, Castro JP, Castro MMN, Faria LB, Caliman MJF, Wolff JL, Neto HFP, Garófalo CA. 2014. Catálogo Polínico das Plantas Usadas por Abelhas no Campus da USP de Ribeirão Preto. Holos; Ribeirão Preto, Brasil. 153 pp.

Silveira F. 1991. Influence of pollen grain volume the estimation of the relative importance of its source to bees. Apidologie 22: 495–502.

Sommeijer MJ, Rooy GA, Punt W, Bruijn LLM. 1983. A comparative study of foraging behavior and pollen resources of various stingless bees (Hym., Meliponinae) and honeybees (Hym., Apinae) in Trinidad, West-Indies. Apidologie 14: 205–224.

Todd FE, Vansell GH. 1942. Pollen grains in nectar and honey. Journal of Economic Entomology 35: 728-731.

Tylianakis JM, Didham RK, Bascompte J, Wardle DA. 2008. Global change and species interactions in terrestrial ecosystems. Ecology Letters 11: 1351-1363.

Valiente-Banuet A, Aizen MA, Alcántara JM, Arroyo J, Cocucci A, Galetti M, García MB, García D, Gómez JM, Jordano P, Medel R, Navarro L, Obeso JR, Oviedo R, Ramírez N, Rey PJ, Traveset A, Verdú M, Zamora R. 2015. Beyond species loss: the extinction of ecological interactions in a changing world. Functional Ecology 29: 299-307.

Van der Putten WH, Macel M, Visser ME. 2010. Predicting species distribution and abundance responses to climate change: why it is essential to include biotic interactions across trophic levels. Philosophical Transactions of the Royal Society B 365: 2025-2034.

Wilms W, Imperatriz-Fonseca VL, Engels W. 1996. Resource partitioning between highly eusocial bees and possible impact of the introduced Africanized honey bee on native stingless bees in the Brazilian Atlantic. Studies on Neotropical Fauna and Environment 31: 137–151.

Wilms W, Wiechers B. 1997. Floral resource partitioning between native *Melipona* bees and the introduced Africanized honey bee in the Brazilian Atlantic rain forest. Apidologie 28: 339–355.

Zanella FCV. 2000. The bees of the Caatinga (Hymenoptera, Apoidea, Apiformes): a species list and comparative notes regarding their distribution. Apidologie 31: 579–592.

Zarnetske PL, Skelly DK, Urban MC. 2012. Biotic multipliers of climate change. Science 336: 1516-1518.

Pollen Storage by *Melipona quadrifasciata anthidioides* in a Protected Urban Atlantic Forest Area of Rio de Janeiro, Brazil

8

Ortrud Monika Barth, Alex da Silva de Freitas, and Bart Vanderborgth

"If the bees disappear from the face of the earth, mankind will only have four more years of existence"

(Albert Einstein, 1901)

8.1 Introduction

The "Mata Atlântica Biome," including the Atlantic Forest, comprises part of the eastern and Atlantic side of Brazil. Several plant formations comprise a mosaic that consists of a tropical forest and other vegetation. Seasonal forests, dense, mixed, and open "ombrophilous" forests, are associated with mangroves and "restingas." An average of 4–12.5% remains of the original habitat, if confluent and/or isolated fragments, are considered. Human activities interfere strongly and contribute to the reduction of all these areas.

The municipality of Rio de Janeiro holds large fragments of Atlantic Forest (Fig. 8.1) composed of an ombrophilous forest, mainly on the mountains and hillsides, as well as mangroves and "restingas" on the lowlands. The city of Rio de Janeiro has one of the largest urban forests in the world, Tijuca National Park, with a continuous covering of trees, shrubs, and lowland vegetation typical of the Atlantic Forest. Nevertheless, there is a mix of endemic species, introduced exotic plant species, and deforested areas (Santana et al. 2015). Areas proximate to each other may be of low similarity (Peixoto et al. 2004).

The Tijuca National Park, part of the Serra da Carioca tropical forest, is surrounded by buildings from the coast to mountains at 1000 m.a.s.l. at the Christ Monument overlooking stunning scenery.

The preservation of most Neotropical plant species depends upon animal pollinators, including the native *Melipona* (Barth et al. 2013; Camargo, 1979 in Waldschmidt et al. 2000). *Melipona quadrifasciata* is a perpetually active Brazilian bee that reaches the tree canopy (Ramalho, 2004) and forages extensively when pollen is available. It comprises two subspecies, *M. quadrifasciata quadrifasciata*, which occurs in cooler regions, and *M. quadrifasciata anthidioides*, of warmer regions (Batalha-Filho et al. 2009; Waldschmidt et al. 2000)

In the present study, we aimed to investigate the pollen collected and the pollination activity of *M. quadrifasciata anthidioides* in an Atlantic Forest segment of Rio de Janeiro.

O.M. Barth (✉)
Instituto Oswaldo Cruz, Fiocruz, Brazil
e-mail: monikabarth@gmail.com

A. da Silva de Freitas
Universidade Federal Fluminense, Niterói, Brazil

B. Vanderborgth
Associação de Meliponicultores do Rio de Janeiro – AME-RIO, Rio de Janeiro, Brazil

© Springer International Publishing AG, part of Springer Nature 2018
P. Vit et al. (eds.), *Pot-Pollen in Stingless Bee Melittology*, DOI 10.1007/978-3-319-61839-5_8

Fig. 8.1 Sectors of the Tijuca National Park (*red color*) in Rio de Janeiro city and the four localities studied (Point 1 to Point 4)

8.2 Getting Pollen Loads and Pollen Grains by *M. quadrifasciata anthidioides*

Four localities were chosen for meliponary installation on the Atlantic side of Tijuca National Park. The meliponary of Point 1 (22°57′22.1″S and 43°16′46.7″W) was established inside the natural forest of the Tijuca National Park, next to the visitor center, Point 2 (22°59′51.1″S and 43°22′10.0″W) in the lowland area of this Park (Bosque da Barra) next to the Atlantic Ocean coast, Point 3 (22° 56′ 26″S and 43° 24′ 13″W) in the Atlantic lowland (Jacarepaguá) of a regenerating forest, and Point 4 (22°58′12.2″S and 43°17′43.4″W) at a higher altitude in a secondary and regenerating forest (Fig. 8.1). A total of 59 corbicular pollen samples were obtained.

Pollen loads of *M. quadrifasciata anthidioides*, the "mandaçaia" stingless bee, were obtained monthly from the corbiculae of five or more flying bees at entry to the hive in the morning. Pollen was thus collected around 10:00 h, from July 2014 to December 2015. The pollen pellets were kept frozen in Eppendorf tubes.

Samples were processed without acetolysis. The pellets of each vial were dispersed and cleaned with 70% EtOH once or twice, depending upon the amount of oil on the grains. After centrifugation, the pellet was washed in distilled water, centrifuged again, and left for a half hour using a mixture of water and glycerol, 1:1 by volume. After being shaken, a drop of the pollen grain suspension was spread between a microscope slide and cover glass, the volume of the drop depending upon the pollen thickness and suspension density. Sealing of the microscope slide sample was temporary (1–2 weeks) and employed nail polish. After pollen grain identification and counting the slides were discarded, but the reference batch was stored in glycerol at room temperature (Barth et al. 2010).

Counting and identification started by 400× magnification with 500 pollen grains. Besides these, attention was given to other structures detected, such as yeast, dark or colorless fungal spores, plant hairs, and organic and inorganic material.

The evaluation of the samples considered 90% or more of pollen grains (pollen sum) of one pollen type or plant species to be a monofloral batch or 60% if no accessory pollen (15–45% of the pollen sum) was present. Bifloral batches considered two accessory pollen types, while heterofloral samples had three or more pollen types (Barth et al. 2010).

8.3   Palynological Characteristics of Pollen Batches Collected from the Baskets of *M. quadrifasciata anthidioides*

A total of 56 samples were analyzed, 33 of which were monofloral, 19 bifloral, and 4 heterofloral (Fig. 8.2).

8.3.1   Monofloral Pollen Loads

Per month, Point 1 presented pollen-type batches of *Myrcia*, *Eucalyptus*, *Solanum*, and Melastomataceae. Point 2 showed *Myrcia* and *Eucalyptus* (Myrtaceae); Point 3 showed *Myrcia*, *Eucalyptus*, *Solanum*, *Anadenanthera colubrina*, and *Mimosa caesalpiniifolia* (Mimosaceae);

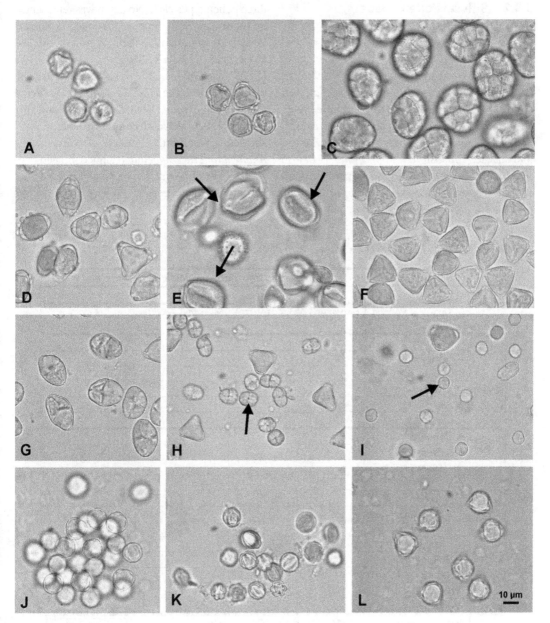

Fig. 8.2 Pollen types of monofloral and bifloral samples. (**a**, **b**) *Alchornea* sp. (**c**) *Anadenanthera colubrina*. (**d**) *Eucalyptus*. (**e**) *Geonoma* (*arrows*). (**f**) *Myrcia*. (**g**) *Mimosa invisa*. (**h**) *Mimosa caesalpiniifolia*. (**i**) *Mimosa scabrella*. (**j**, **k**) Melastomataceae. (**l**) *Solanum*

and Point 4 showed *Myrcia, Eucalyptus, Anadenanthera colubrina,* and Melastomataceae. During the very dry months of May and June in 2015, no monofloral pollen loads were collected by the bees, as well as in November when plant flowering was unexpectedly reduced (Table 8.1).

8.3.2 Bifloral Pollen Loads

Bifloral samples were composed mainly of pollen types among monofloral pollen load taxa. Point 1 displayed mixed batches of *Alchornea* (Euphorbiaceae), *Myrcia,* and *Mimosa caesalpiniifolia.* Point 2 had *Eucalyptus, Geonoma* (Arecaceae), Melastomataceae, *Mimosa caesalpiniifolia,* and *Myrcia.* Point 3 presented Melastomataceae, *Mimosa caesalpiniifolia, Mimosa scabrella, Myrcia,* and *Solanum* pollen types. Point 4 showed *Eucalyptus,* Melastomataceae, *Mimosa caesalpiniifolia* and *Myrcia* pollen types.

In August, September, and October 2014, as well as in July and October 2015, only monofloral samples occurred (Table 8.2).

8.3.3 Heterofloral Pollen Loads

Heterofloral pollen batches comprised more than three pollen types each. No taxon was represented by more than 60%. Samples obtained at Point 3 presented important pollen types of Melastomataceae (42.7%), *Mimosa caesalpiniifolia* (27.8%), and *Myrcia* (17.9%) in August 2014 and of Melastomataceae (26.8%), *Mimosa verrucosa* (51.5%) and *Myrcia* (14.4%) in September 2014.

Samples collected at Point 1 in January 2015 showed an important pollen contribution of *Anadenanthera colubrina* ("angico branco") of 42.4% together with *Solanum* and *Eucalyptus.* During the same time at Point 2, it comprises 11.9% only.

Table 8.1 Monofloral pollen loads of *Melipona quadrifasciata,* "mandaçaia" (Point 1 to Point 4)

Monofloral pollen loads of *Melipona quadrifasciata anthidioides,* "mandaçaia"						
Pollen types						
Year/Month Taxa	*Myrcia*	*Eucalyptus*	Melastomataceae	*Solanum*	*Anadenanthera*	*Mimosa caesalpiniifolia*
2014 July	X	X				
August	X	X				
September	X					
October	X			X		
November			X			
December	X				X	
2015 January					X	
February						X
March	X					X
April	X					X
May						
June						
July		X				
August		X				
September	X		X			
October	X					
November						
December			X			
Mimosa caesalpiniifolia = Mimosa caesalpiniaefolia (syn) Drought months						

Table 8.2 Bifloral pollen loads of *Melipona quadrifasciata anthidioides*, "mandaçaia (Point 1 to Point 4)

Bifloral pollen loads of *Melipona quadrifasciata anthidioides*, "mandaçaia"						
Pollen types						
Taxa Year/Month	Myrcia	Eucalyptus	Melastomataceae	Mimosa invisa	Mimosa caesalpiniifolia	Mimosa scabrella
2014 July	X				X	
August						
September						
October						
November	X		X			
December	X		X			
2015 January	X					X
February	X		X		X	
March	X				X	
April	X				X	
May	X		X		X	X
June		X	X			
July						
August			X		X	
September	X					
October						
November	X		X	X		
December	X			X		

Mimosa caesalpiniifolia = Mimosa caesalpiniaefolia (syn)   Drought months

Additional taxa: each one of *Solanum* (11/14), *Alchornea* (5/15), *Geonoma* (9/15)

8.3.4 Additional Pollen Types

Besides the plant taxa and pollen types presented in the Tables 1 and 2, there are additional 13 not well identified taxa of 3 to 45% frequency. Sometimes plant families could be recognized as Apocynaceae, Arecaceae, Asteraceae, and Fabaceae, and genera as *Acacia, Smilax, Triplaris,* and *Vernonia* and circa five unknown ones.

8.3.5 Additional Structured Elements

Eight samples only presented some structured elements besides the pollen grains being of low frequency. Colorless and dark-brown spores of fungi were detected in two samples. One monofloral sample of *Myrcia* collected in October included dark organic material, hyphae and yeast, soil fragments, and silica. One bee load (September/Point 2) was devoid of pollen grains, presenting soil and minerals only.

8.4 Plant Families, Genera, and Species Mostly Visited by *M. quadrifasciata anthidioides*

The palynological analysis of pollen loads revealed a trend in the behavior of *M. quadrifasciata anthidioides* remaining constant to a flowering species. Nearly 59% of the samples were monofloral and 34% bifloral. The strongest heterofloral samples were obtained from Point 3 during the winter months of August and September.

The most important monofloral pollen loads collected by *M. quadrifasciata anthidioides* in the Tijuca National Park came from the Myrtaceae family (nine samples), and pollen grains were indistinguishable among families thus grouped under the pollen-type *Myrcia* (Barth and Barbosa 1972), excluding *Eucalyptus*. *Myrcia* pollen covered the demand for protein of the bees during half of the study period and contributed some samples the rest of the time.

Species of *Eucalyptus* were introduced in Brazil as early as 1825 into the Botanical Garden, Rio de Janeiro, but a strong cultivation started only in the beginning of the twentieth century (Coelho et al. 2002). Being exotic species, they are unfamiliar to the native stingless bees and were visited by *M. quadrifasciata anthidioides* only when no known species were available, as in the winter months of July and August (four samples).

Melastomataceae pollen was present in the majority of the samples during a year. *Tibouchina* is the main genus of trees occurring in the forest, and vibratory pollen collection from anthers was used by foraging bees. Monofloral batches occurred uncommonly; in general, there were bifloral pollen loads.

In the early spring, only one monofloral sample of *Solanum* pollen was obtained. During the summertime, *Anadenanthera colubrina* ("angico branco") started to bloom, mainly in December and January (Lorenzi 1992). Its flowers were more attractive to the bees than those of any other taxon. It is a tall pioneer tree species, occurring at high elevations, usually >400 m. When *Anadenanthera* declines, *Mimosa caesalpiniifolia* starts blooming, and its monofloral batches were detected on bees in February through April.

Nine plant taxa composed the bifloral batches collected during one and a half year by *M. quadrifasciata anthidioides* in the study. The majority were detected in monofloral and bifloral samples. Besides these, *Alchornea* sp., *Geonoma*, *Mimosa invisa*, and *M. scabrella* pollen types occurred.

Four of the 56 samples (7%) studied showed only heterofloral pollen grain composition, with a large number of pollen types at low frequencies. Many of the taxa formerly cited were present, others new, and many unknown.

Some of the preferred floral sources of several species of *Melipona* in the southeastern Brazilian Atlantic Forest have been identified (Ramalho et al. 1989; Braga et al. 2012; Serra et al. 2012). Two investigations only considered *M. quadrifasciata anthidioides*. Antonini et al. (2006) record pollen and nectar resources on flowers in a semidecidual seasonal forest ("cerrado") of the Parque Estadual da Serra do Rola-Moça next to the city of Belo Horizonte, Minas Gerais State. Returning foragers captured at the nest entrance in Parque da Neblina, Mogi das Cruzes District, São Paulo State, Oliveira-Abreu et al. (2014) also provided information from pollen loads. In both areas, as well as in the present study of the Tijuca National Park of Rio de Janeiro, few plants were used intensely by the bees. Pollen grains of the *Myrcia* pollen type, comprising several genera of the Myrtaceae family, *Myrcia*, *Eugenia*, *Myrceugenia*, and more (Barth and Barbosa 1972), constituted the main resources, followed by plants of the Melastomataceae and Solanaceae. *Eucalyptus* pollen was used by this bee only when no other pollen was available. Ramalho et al. (1989) also observed such a tendency in foraging *Melipona*, mainly at the Juréia Ecological Station and Prudentópolis, São Paulo State.

Structured elements in pollen loads certainly can provide additional information about bee activity (Barth 1971), mainly considering environmental conditions. It is not advisable to apply acetolysis of pollen loads (also of honey samples) in this context because some spores, hyphae, organic material, soil constituents, and resin may be lost. The environmental conditions revealed by such material may help to inform where bees survive to pollinate our forests, fields, and crops.

8.5   Conclusion

The foraging activities of *M. quadrifasciata anthidioides* were similar at four localities studied here, three in forests of a hill and mountain, and one on a marine shoreline.

Nevertheless, the Atlantic Forest is very rich in tree and shrub species, and the native bee *M. quadrifasciata anthidioides* used one native flowering plant species while that plant species was producing pollen. This behavior allows efficiency in pollination and, finally, the perpetuation of our forests.

To obtain a more detailed behavior analysis for *M. anthidioides*, we suggest the following: (l) Start an investigation of the plant and pollen preferences of stingless bees, distinct on the Atlantic and continental side of the "Serra da Carioca" forest. (ll) Sample monthly for 3 to 5 years, while considering environmental conditions, mainly of temperature and rainfall. These two general approaches would certainly provide a more accurate picture of behavior and pollination potential of the bee in the tropical Atlantic Forest of Rio de Janeiro.

Acknowledgments Thanks to Dr. P. Vit for the invitation to write this chapter, carefully reviewed by Dr. D.W. Roubik, a thematic specialist. We are grateful to Mr. Carlos Ivan Siqueira and Mr. Luiz Alberto Medina from the "Associação de Meliponicultores do Rio de Janeiro – AME-RIO," Rio de Janeiro, Brazil, for a longtime dedication of collecting pollen load samples from the bees in several points of the project. AME-RIO was helpful in providing meetings with the beekeepers also, mainly by the initiative of its president, Mr. Gesimar C. dos Santos. Financial support was given by the Instituto Oswaldo Cruz, Fiocruz, Rio de Janeiro, and the National Counsel of Technological and Scientific Development – "Conselho Nacional de Desenvolvimento Científico e Tecnológico, CNPq."

References

Antonini Y, Soares SM, Martins RP 2006. Pollen and nectar harvesting by the stingless bee *Melipona quadrifasciata anthidioides* (Apidae: Meliponini) in an urban forest fragment in Southeastern Brazil. Studies on Neotropical Fauna and Environment 41: 209–215.

Barth OM 1971. Mikroskopische Bestandteile brasilianischer Honigtauhonige. Apidologie 2: 157–167.

Barth OM, Barbosa AF 1972. Catálogo sistemático dos pólens das plantas arbóreas do Brasil Meridional - XV. Myrtaceae. Mem. Inst. Oswaldo Cruz 70: 467–498.

Barth OM, Freitas AS, Oliveira, ES, Silva RA, Maester FM, Andrella RRS, Cardozo GMBQ 2010. Evaluation of the botanical origin of commercial dry bee pollen load batches using pollen analysis: a proposal for technical standardization. Anais da Academia Brasileira de Ciências 82: 893–902.

Barth OM, Freitas AS, Almeida-Muradian L, Vit P. 2013. Palynological analysis of Brazilian stingless bee pot-honey. In Vit P & Roubik DW (eds.). Stingless bees process honey and pollen in cerumen pots. Chapter 4. Facultad de Farmacia y Bioanálisis, Universidad de Los Andes; Mérida, Venezuela. http://www.saber.ula.ve/handle/123456789/35292

Batalha-Filho H, Gabriel AR, Melo GAR, Waldschmidt AW, Campos LAO, Fernandes-Salomão TM 2009. Geographic distribution and spatial differentiation in the color pattern of abdominal stripes of the Neotropical stingless bee *Melipona quadrifasciata* (Hymenoptera: Apidae). Zoologia (Curitiba, Impr.) 26 (2) Curitiba June 2009.

Braga JA, Sales EO, Neto JS, Conde MM, Barth OM, Lorenzon MC 2012. Floral sources to *Tetragonisca angustula* (Hymenoptera:Apidae) and their pollen morphology in a Southeastern Brazilian Atlantic Forest. International Journal of Tropical Biology 60: 1491–1501.

Coelho LG, Barth OM, Chaves HAF 2002. Palynological records of environmental changes in Guaratiba mangrove área, Southeast Brazil, in the last 6000 years B.P. Pesquisas em Geociências 29: 71–79.

Lorenzi H 1992. Árvores Brasileiras. Editora Plantarum, Nova Odessa, Brasil. 368 pp.

Oliveira-Abreu C, Hilário SD, Luz CFP 2014. Pollen and néctar foraging by *Melipona quadrifasciata anthidioides* Lepeletier (Hymenoptera: Apidae: Meliponini) in natual habitat. Sociobiology 61: 441–448.

Peixoto GL, Martins SV, Silva AF, Silva E 2004. Composição florística do componente arbóreo de um trecho de Floresta Atlântica na Área de Proteção Ambiental da Serra da Capoeira Grande, Rio de Janeiro, RJ, Brasil. Acta Botanica Brasilica 18: 151–160.

Ramalho M 2004. Stingless bees and mass flowering trees in the canopy of Atlantic Forest: A tight relationship. Acta Botanica Brasilica 18: 37–47.

Ramalho M, Kleinert-Giovannini A, Imperatriz-Fonseca, VL 1989. Utilization of floral resources by species of *Melipona* (Apidae, Meliponinae): floral preferences. Apidologie 20: 185–195.

Santana CAA, Freitas WK, Magalhães LMS 2015. Estrutura e similaridade em florestas urbanas na região metropolitana do Rio de Janeiro. Interciencia 40: 479–486.

Serra BDV, Luz CFP, Campos LAO 2012. The use of polliniferous resources by *Melipona capixaba*, an endangerd stingless bee species. Journal of Insect Science 12: 148: 1–14.

Waldschmidt AM, Barros EG, Campos LAO 2000. A molecular marker distinguishes the subspecies Melipona quadrifasciata quadrifasciata and Melipona quadrifasciata anthidioides(Hymenoptera: Apidae, Meliponinae). Genetics and Molecular Biology 23: 609–611.

Angiosperm Resources for Stingless Bees (Apidae, Meliponini): A Pot-Pollen Melittopalynological Study in the Gulf of Mexico

Elia Ramírez-Arriaga, Karina G. Pacheco-Palomo,
Yolanda B. Moguel-Ordoñez,
Raquel Zepeda García Moreno,
and Luis M. Godínez-García

9.1   Introduction

Pantropical and eusocial stingless bees belong to the monophyletic tribe Meliponini, included within the Apinae subfamily, in the Apidae family (Michener 2013). The relationship between Bombini and Meliponini is now strongly supported in the molecular phylogeny of bees (Cardinal and Danforth 2011). Historically, it is well known that bees are important pollinators of flowering plants (Roubik 1992). The oldest fossil bee is *Cretotrigona prisca*, tribe Meliponini, found in Late Cretaceous amber (approx. 70 Ma, Maastrichtian) in New Jersey, USA (Michener and Grimaldi 1988; Engel and Michener 2013). Specialised bee pollination must have played an important role in angiosperm diversification from the Middle to Late Cretaceous; this is supported by the age of crown group bees, which is consistent with tricolpate fossil pollen grains of eudicots (flowering plants) from approximately 123 to 125 Ma (Bell et al. 2010; Magallón 2010). Moreover, the majority of extant bee families have most likely been present since before the Cretaceous-Paleogene transition (Cardinal and Danforth 2013). Recent robust molecular studies have concluded that eusociality was developed from 78 to 95 Ma in corbiculate bees, with stingless bees and honey bees evolving independently

E. Ramírez-Arriaga (✉)
Laboratorio de Palinología: Paleopalinología y Actuopalinología, Departamento de Paleontología, Instituto de Geología, Universidad Nacional Autónoma de México, Ciudad Universitaria, Av. Universidad 3000, Coyoacán CP 04510, Distrito Federal, Mexico
e-mail: eliaramirezarriaga@gmail.com

K.G. Pacheco-Palomo
Universidad Autónoma de Campeche, Facultad de Ciencias Químico Biológicas,
Av. Agustín Melgar S/N entre Calle 20 y Juan de la Barrera, Col. Buenavista, C.P. 24039 San Francisco de Campeche, Campeche, Mexico

Y.B. Moguel-Ordoñez
Campo Experimental Mocochá, Instituto Nacional de Investigaciones Forestales, Agrícolas y Pecuarias, Km 25, antigua carretera Mérida-Motul, Mocochá CP 97454, Yucatán, Mexico

R. Zepeda García Moreno
Iniciativas para la naturaleza A.C. (INANA), Calle Tajín 33, Colonia Campo Viejo, Coatepec, CP 91540 Veracruz, Mexico

L.M. Godínez-García
Universidad Politécnica Mesoamericana, Departamento de Investigación y Desarrollo Sustentable, Tenosique, Tabasco, Mexico

(Cardinal and Danforth 2011). Although both corbiculate groups can exhibit generalist and specific floral preferences, Meliponini did not develop a communication system as complex as that of *Apis* (von Frisch 1967; Biesmeijer and Slaa 2004).

As a result of its geologic history, Mexico harbours high angiosperm richness (ca. 21,841 species) living under distinctive climatic regimes within its intricate physiography (Villaseñor and Ortiz 2014). Approximately 1840 bee species included in 144 genera have been reported in Mexico (Ayala et al. 1996; Ayala 2016); only 46 of these are stingless bee species (2.6% of Mexican bee fauna), which inhabit different ecosystems such as xerophytic vegetation, tropical deciduous forest, tropical evergreen forest and cloud forest (Ayala et al. 2013). Meliponini are of cultural, economic and ecologic interest (Yurrita et al. 2016); the Mayas created the god *Ah Mucen Kab* in honour of *Melipona beecheii* or *Xuna'an kab* (Vit et al. 2004). They also made a special beverage called *balché*, using stingless bee honey and the bark of the Fabaceae (Faboideae) *Lonchocarpus longistylus*. This beverage was consumed in religious ceremonies. Today, meliponiculture has become a common practice in many Mexican regions, and several stingless bee taxa are bred. A great diversity of rational beekeeping techniques has been developed (Quezada-Euán 2005; González-Acereto 2008; Guzmán et al. 2011).

In light of the importance of bees as pollinators of flowering plants in tropical and subtropical ecosystems, it is significant to contribute to the knowledge of the angiosperm resources foraged by Meliponini, in order to preserve and recover stingless bee populations in addition to various ecosystems that have been modified as a result of human activity and global climatic change.

This work aims to provide melittopalynological knowledge of *Melipona beecheii*, *Plebeia* sp. and *Scaptotrigona mexicana* pot-pollen samples in the states of Campeche and Veracruz located along the Gulf of Mexico and also to contribute to angiosperm data sets regarding the pot-pollen provisions of Mexican Meliponini. We also suggest sustainable activities with respect to the beekeeping of these native insects.

9.2   Background of Melittopalynological Studies in Mexico

Melittopalynology involves the pollen characterisation of honey, pollen loads, brood provisions, propolis and geopropolis, in order to determine botanic and geographic origin (Louveaux et al. 1978; Martínez-Hernández et al. 1993; Barth 2013). Mexican melittopalynological studies began in the 1980s, focused on *Apis mellifera* in tropical areas (Lobreau-Callen and Callen 1982; Villanueva-Gutiérrez 1984, 2002; Alvarado and Delgado 1985; Cárdenas-Chávez 1985). Nevertheless, it was not until the 1990s when there was an increase in systematic research related to the pollen grains contained in the honey of *A. mellifera* and in that of native stingless bees. Studies describing the pollen analysis of honey, pollen loads and brood provisions during an annual cycle in *Nannotrigona perilampoides*, *Plebeia* sp., *Scaptotrigona mexicana* and *Tetragonisca angustula* (Ramírez-Arriaga 1989; Martínez-Hernández et al. 1993; Ramírez-Arriaga et al. 1995) and the annual palynological characterisation of brood provisions in *Scaptotrigona hellwegeri* (Quiroz-García et al. 2011) contrast with other studies conducted only during important honey harvest periods of *Apis mellifera* (Martínez-Hernández and Ramírez-Arriaga 1998; Piedras and Quiroz-Gracía 2007; Quiroz-García and Arreguín-Sánchez 2008; Navarro 2008; Díaz 2008; Villanueva-Gutiérrez et al. 2009; Castellanos 2010; Ramírez-Arriaga et al. 2011, 2016) and *Scaptotrigona mexicana* in several Mexican states (Villamar 2004; Ramírez-Arriaga and Martínez-Hernández 2007).

Melittopalynology has been used as a tool to compare the foraging behaviours of European and Africanised bees (Medina 1992; Villanueva-Gutiérrez and Roubik 2004), in addition to those of *A. mellifera* and Meliponini (Martínez-Hernández et al. 1993; Ramírez-Arriaga and Martínez-Hernández 2007), and to describe the nourishment strategies of both solitary and communal bees such as *Centris inermis* (Quiroz-García and Palacios-Chávez 1999) and the orchid bee *Euglossa* (Ramírez-Arriaga and Martínez-Hernández 1998; Villanueva-Gutiérrez et al. 2013).

In summary, melittopalynological studies concerning only stingless bees have included Mexican regions such as south central Mexico, in the northern Puebla mountain range (Villamar 2004; Ramírez-Arriaga and Martínez-Hernández 2007); the Pacific coast, in Chamela, Jalisco (Quiroz-García et al. 2011); and south-eastern Mexico, in Chiapas (Martínez-Hernández et al. 1993), and new reports on the Gulf of Mexico from Campeche and Veracruz will be included in this work. Additionally, a list of angiosperm taxa identified in Meliponini pot-pollen samples analysed melittopalynologically is included at the end of this document. This compilation of pollen resources is aimed at exhibiting the diversity of angiosperms visited by stingless bees, in order to promote sustainable meliponiculture and reforestation using native plants for the recovery of altered plant communities.

9.3   Methods and Study Areas

A. Campeche state: Both colonies of *Melipona beecheii* were located near Chiná village at 19° 43′ 37.61″ N, 90° 24′ 57.10″ W (Fig. 9.1). In this region, the climate is Aw0(i')g (warm-humid with summer rains, winter rains between 5 and 10.2 mm, 5–7 °C yearly oscillation, hottest month earlier than June). The annual mean temperature is 26.9 °C and annual mean precipitation is 1127.2 mm, with rains from May to October (García 1981). The types of vegetation are high evergreen forest, secondary vegetation, grasses and savannah. A total of eight pot-pollen samples (hive 1 $n = 6$; hive 2 $n = 2$) were collected from June to September 2008. All pot-pollen samples were

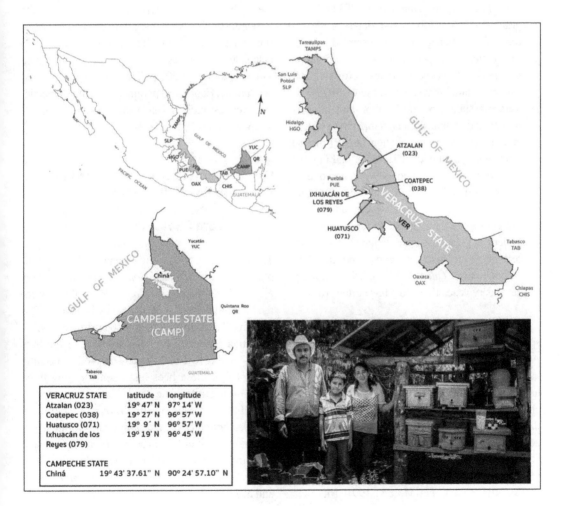

Fig. 9.1   Map showing the specific locations of the studied sites in Campeche and Veracruz

Table 9.1 Georeference and climate of each municipality in which meliponaries have been established in the state of Veracruz, Mexico (García 1981)

Municipality	Latitude	Longitude	Altitude	AMT	AMP	Climate
Atzalan (023)	19° 47′ N	97° 14′ W	1842 m	15.4	1943.4	Cb(fm)(i′) gw″
Coatepec (038)	19° 27′ N	96° 57′ W	1252 m	19.3	1800.4	(A)Cb(fm) (i′)gw″
Huatusco (071)	19° 9′N	96° 57′ W	1344 m	19.8	1951.6	(A)cam(f)(i′) gw″
Ixhuacán de los Reyes (079)	19° 19′N	96° 45′ W	1785 m	13.1	3026.4	Cb(fm)(i′) w″

A Warm-humid and subhumid
Cb Temperate-humid and subhumid, semi-cold
Cam Temperate, subtropical, monsoon summer rains
(fm) Irregular monsoon rains
(i′) Low temperature oscillation (between 5 and 7 °C)
g Hottest month in spring
w″ Two rainy seasons

processed using standard methods (Louveaux et al. 1978; Quiróz-García et al. 2011). Pollen grains were described and determined, and then 1200 pollen grains were counted in order to calculate the percentages of each taxon. Diversity and evenness indices were calculated (Shannon and Weaver 1949; Pielou 1977).

B. Veracruz state: Specific location and information related to annual mean temperature (AMT) and annual mean precipitation (AMP) of each municipality is shown in Table 9.1 and Fig. 9.1. Coatepec contains secondary vegetation of evergreen forest and coffee crops. Huatusco vegetation was originally cloud forest, although at present there are patches of this ecosystem, secondary vegetation of cloud forest undergoing restoration processes and crops. Ixhuacán de los Reyes houses *Quercus* forest, *Pinus* forestry, secondary vegetation and a few coffee crops.

A total of seven pot-pollen samples from *Plebeia* sp. and *S. mexicana* nests were collected in June and August 2016, directly from sealed pots at meliponaries located in Coatepec (*n* = 3), Huatusco (*n* = 2) and Ixhuacán de los Reyes (*n* = 2). Regarding chemical processes, 5 g of each sample were first diluted in distilled water, homogenised and then acetolysed (Louveaux et al. 1978; Quiroz-García et al. 2011). Pollen grains were described and determined using specialised literature (Palacios-Chávez et al. 1991; Roubik and Moreno 1991; Martínez-

Hernández et al. 1993; Alfaro-Bates et al. 2010) and the Palynological Reference Collection of the Laboratory of Palynology, although these descriptions are not included in this manuscript. In order to calculate percentages of taxa yielded in pot-pollen samples, a total of 500 pollen grains were counted per sample. Ecological parameters such as diversity and evenness indices were also calculated (Shannon and Weaver 1949; Pielou 1977).

9.4 Floral Resources Recorded in Pot-Pollen Samples of *Melipona beecheii* from Campeche

The eight pot-pollen samples analysed yielded 11 plant species, of which *Solanum verbascifolium*, *Physalis pubescens* and *Solanum lanceolatum* were the most important pollen sources in hive 1 (C1) and hive 2 (C2) from June to September (Fig. 9.2).

When *Melipona beecheii* resources from hive 1 (C1) and hive 2 (C2) were analysed, it became evident that these bees had been foraging three vegetation strata. The tree stratum was represented by *Bauhinia* sp., *Bursera simaruba*, *Mimosa bahamensis, Psidium guajava* and *Senna racemosa*. *Solanum verbascifolium* represented the shrub stratum, and from the herbaceous stratum, *Physalis pubescens, Solanum americanum* and *S. lanceolatum* were recorded (Fig. 9.2).

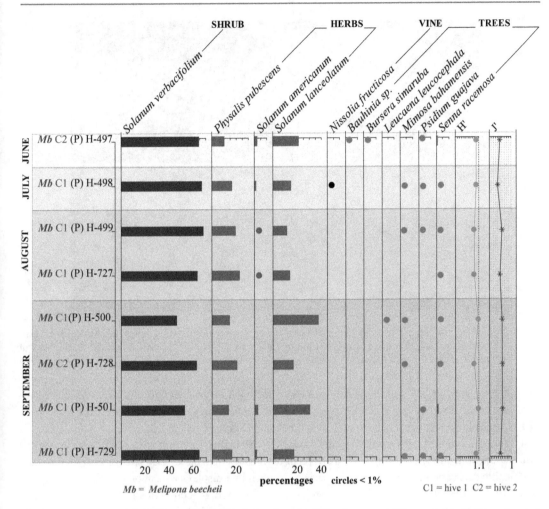

Fig. 9.2 Percentages of angiosperm pollen grains, index diversity (H′) and evenness (J′) recorded in *Melipona beecheii* pot-pollen samples from hive 1 (C1) and hive 2 (C2) in Campeche (June to September, 2008)

Mayan bees showed polylectic foraging behaviour, in addition to temporal or local monolectic and oligolectic specialisation. Temporal monolectic behaviour was demonstrated when the bees showed preference for *S. verbascifolium* (46.2–67.9%) from June to September. Local oligofloral foraging behaviour was also inferred when *Solanum* taxa such as *S. lanceolatum* and *S. verbascifolium* abounded in the pot-pollen melittopalynological assemblages; the sum of the percentages of both taxa ranged from 76.9% to 85.3%. The diversity index of Campeche pot-pollen samples varied from 0.9 to 1.1 with a mean of 1. Additionally, evenness values ranged from 0.4 to 0.6 with a mean of 0.4 (Fig. 9.2).

9.5 Meliponiculture and Melittopalynological Study of Pot-Pollen in Veracruz

9.5.1 INANA's Sustainable Meliponiculture

Several factors such as deforestation, agriculture, livestock farming, urbanism and the need for efficient techniques in sustainable meliponiculture have motivated programmes for the rescue of traditional Mayan meliponiculture, in place since 2004 in the Yucatán Peninsula (Villanueva-Gutiérrez et al. 2013), as well as recent meliponi-

Fig. 9.3 (**a**) *Melipona beecheii* guardian bee at the nest entrance. (**b**) Trophallaxis between two *M. beecheii* workers on the brood cells. (**c**) *Scaptotrigona mexicana* nest entrance; foragers arriving with pollen loads. (**d**) *S. mexicana* workers inside the nest (Photos **a–d**: Courtesy of Diana Caballero)

culture programmes led by Iniciativas para la naturaleza A.C. (INANA) in Veracruz. Since 2011, INANA's activity has been clearly and dynamically aimed at conserving our ecosystems there, while at the same time generating careful efforts in sustainable beekeeping development (Fig. 9.3).

In this sense, as part of INANA's project, meliponiculture produces various effects, such as the conservation of wild nests and the increase of the Meliponini population in Veracruz. At the same time, it supports pollination activities within tropical ecosystems. Knowledge of stingless bee nesting biology also leads to proper beekeeping methods and plays a key role in the educational programme known as '*school of meliponiculture*' in rural communities. This programme promotes conservation, and through it beekeepers share appropriate practices and utilise honey, cerumen and pollen to make several products that are then commercialised.

INANA's Meliponiculture Conservation Project has been launched with 12 meliponaries. Although *Scaptotrigona mexicana, Scaptotrigona pectoralis, Nannotrigona perilampoides, Melipona beecheii* and *Plebeia* sp. are considered to be among the species viable for beekeeping, the first phase, included in this work, exclusively corresponds to the pollen analysis of pollen loads from *Plebeia* sp. and *Scaptotrigona mexicana*.

Meliponaries comprising 10–20 nests are roofed with wooden and/or clay structures (Fig. 9.1), and each one displays educational signs regarding the importance of conservational meliponiculture in the sustainable management of evergreen forest. These meliponaries are located in the Antigua and Jamapa basins in the state of Veracruz, at altitudes from 800 to 1600 m.a.s.l. in order to document specific beekeeping needs according to climatic conditions and flowering periods.

9.5.2 Angiosperm Resources for Plebeia and Scaptotrigona Mexicana in Veracruz

Melittopalynological analysis, recently developed, is performed in order to characterise potpollen of native bees in Veracruz; its aim is to determine the floral resources being visited as a protein source (pollen loads). An additional objective of melittopalynological studies is to enable local inhabitants to protect natural areas.

As a result of their knowledge of bee-flora interaction, they will be able to understand the many benefits that bees provide not only to their natural habitats but also to family financial situations, with a constant focus on ecological awareness.

This section includes the melittopalynological analyses of seven samples collected from *Plebeia* sp. (pot-pollen *n* = 2) and *Scaptotrigona mexicana* (pot-pollen *n* = 5) nests in June and August 2016, from the municipalities of Coatepec, Huatusco and Ixhuacán de los Reyes in Veracruz (Fig. 9.3).

In general, Meliponini collected the following pollen resources from the tree stratum: *Alchornea latifolia*, Anacardiaceae, *Cecropia obtusifolia*, *Cocos nucifera*, *Cupania* sp., *Heliocarpus appendiculatus*, *Leucaena* sp., *Liquidambar* sp., *Lonchocarpus* sp., *Platanus mexicana*, *Quercus* sp. and *Thouinia paucidentata*. Shrubs of interest were *Acacia*, *Coffea arabica*, *Hedyosmum* sp. and *Ziziphus* sp. Finally, Araliaceae, Asteraceae and *Ocimum* sp. were important herbs (Figs. 9.4, 9.5 and 9.6).

The melittopalynological analysis of *Plebeia* sp. pot-pollen samples in Coatepec showed *Coffea arabica* (100%) as dominant. Additional analysis of pot-pollen indicated *Cocos nucifera* (17.8%), *Cecropia* (18.1%) and Asteraceae (33.3%) as relevant resources for this stingless bee. In this municipality, *Scaptotrigona mexicana* exploited *Heliocarpus appendiculatus* (96.1%) in August (Fig. 9.4).

On the other hand, in Huatusco, *Scaptotrigona mexicana* visited *H. appendiculatus* (39–100%), *Alchornea latifolia* (41.6%) and Araliaceae (16.7%) during June and August, all considered important resources. Predominant pollen grains detected in *S. mexicana* pot-pollen samples from Ixhuacán de los Reyes in June were *Quercus* (38.9%), *Liquidambar* (17.2%), *Ziziphus* (16.7%) and Asteraceae (13.9%). Additional floristic elements such as *Platanus mexicana* (12.2%), *Cupania* (39.7%) and Rubiaceae (24.8%) were foraged by this native bee in August (Fig. 9.4).

With regard to ecological parameters, *Plebeia* sp. samples recorded a diversity index (H') of 1.9 and an evenness (J') of 0.8 in Coatepec. The diversity indices estimated for *S. mexicana* pot-pollen samples ranged from 0.2 to 1.8, while J' values varied from 0.1 to 0.8. It is important to mention that the highest diversity values were found in the Ixhuacán de los Reyes samples (Fig. 9.4).

9.6 Analysis of the Polliniferous Plant Preferences of Stingless Bees in Campeche and Veracruz, Gulf of Mexico

Pollinators played a significant role within the geographically studied areas, and structural plant composition was important in establish-

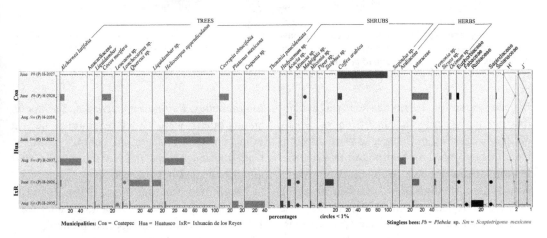

Fig. 9.4 Percentages of angiosperm pollen grains, index diversity (H') and evenness (J') recorded in pot-pollen samples from *Plebeia* sp. (*Pl*) and *Scaptotrigona mexi-cana* (*Sm*) at Coatepec (Coa), Huatusco (Hua) and Ixhuacán de los Reyes (IxR), Veracruz (June and August 2016)

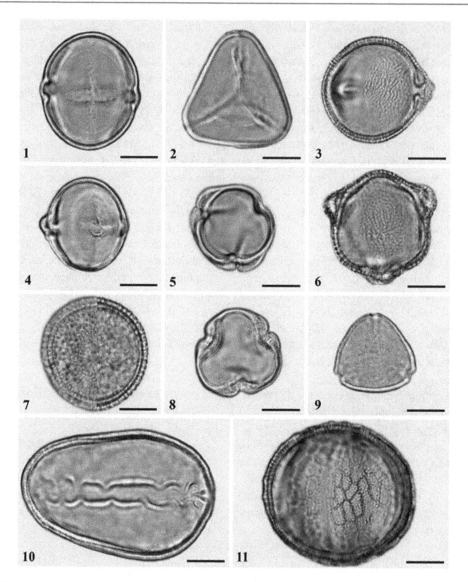

Fig. 9.5 Angiosperm pollen grains recovered from sting-less bee pot-pollen samples. 1. *Solanum verbascifolium*; 2. *Cupania* sp.; 3, 6. *Bursera simaruba*; 4–5. *Physalis* *pubescens*; 7. *Hedyosmum* sp.; 8. *Alchornea latifolia*; 9. *Psidium guajava*; 10. *Cocos nucifera*; 11. *Ocimum* sp.

ing bee interaction. In this sense, it was evident that the angiosperms that had been visited by *Melipona beecheii* in Campeche were different from those foraged by Meliponini in Veracruz. Dissimilarities related to plant composition surrounding meliponaries and bee foraging behaviour reflect original vegetation, the degree of alteration of an ecosystem, the existence of secondary vegetation and/or the presence of crops. For instance, several taxa from original cloud forest such as *Quercus*,

Liquidambar, *Platanus mexicana* and *Miconia*, among others, were visited by stingless bees at Ixhuacán de los Reyes, in contrast with the *Coffea arabica* crops and plantations of *Cocos nucifera* visited in Coatepec.

Faegri and van der Pijl (1979) described flower fidelity in bees. Ramalho et al. (1989, 1990, 1994) have also reported preferences in stingless bees exclusively for one floral source. Pot-pollen analyses suggested that the workers of *Scaptotrigona mexicana* transported unifloral

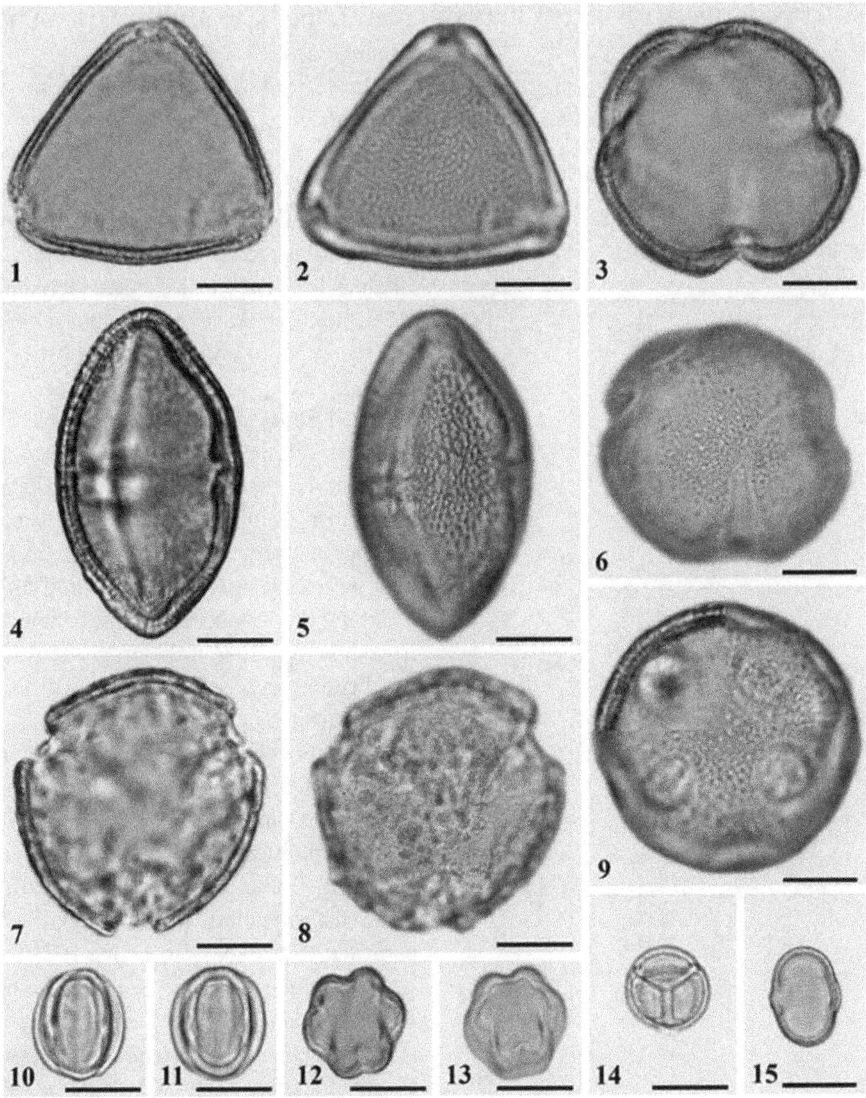

Fig. 9.6 Angiosperm pollen grains recovered from sting-less bee pot-pollen samples. 1–2. *Thouinia paucidentata*; 3, 6. *Coffea arabica*; 4–5. *Heliocarpus appendiculatus*; 7–8. *Quercus* sp.; 9. *Liquidambar* sp.; 10–13. *Miconia* sp.; 14. *Mimosa* sp.; 15. *Cecropia obtusifolia*

pollen loads of *Heliocarpus appendiculatus*, while *Plebeia* sp. workers preferred *Coffea arabica*. Oligolectic behaviour was observed in *Melipona beecheii*, especially when workers foraged on *Solanum* taxa in Campeche.

Based on field observations as well as melit-topalynological studies (Souza-Novelo et al. 1981; Villanueva-Gutiérrez 2002; Alfaro-Bates et al. 2010), *Bursera simaruba* is considered a nectaro-polliniferous tree and can be considered a constant pollen resource due to nearly perpetual

flower production (Villamar 2004; Villanueva-Gutiérrez et al. 2009).

Mimosa bahamensis has been cited as a nectaro-polliniferous plant by Villanueva-Gutiérrez et al. (2009) and Alfaro-Bates et al. (2008). According to Villanueva-Gutiérrez et al. (2009), *M. bahamensis* is more polliniferous than nectariferous.

The Myrtaceae *Psidium guajava* has also been reported as an important plant for *Melipona* in Guatemala and Mexico, and it is considered a sustainable resource for stingless bee colonies

Table 9.2 Ecological parameters of diversity (H′) and evenness (J′) indices reported in pot-pollen of Mexican stingless bees from Campeche (Camp, this work); Veracruz (Ver, this work) (Ixhuacán de los Reyes (IxR), Coatepec (Coa) and Huatusco (Hua)), Chiapas (Chis, Ramírez-Arriaga 1989) (Unión Juárez (UJ) and Santa Teresita (ST)), and Puebla (Pue, Ramírez-Arriaga and Martínez-Hernández 2007) (San Miguel Tzinacapan (Tzi) and Valle de Ateno, Ayotzinapan (Ayo))

	H′			J′		
	Mean	Min	Max	Mean	Min	Max
Melipona beecheii (Camp) pot-pollen	1.0	0.9	1.1	0.4	0.4	0.6
Plebeia sp. (Ver, Coa) pot-pollen	1.1	0.2	1.9	0.4	0.1	0.8
Plebeia sp. (Chis, UJ) pot-pollen		0.2	2.1		0.2	0.8
Plebeia sp. (Chis, ST) pot-pollen		0.5	1.9		0.02	0.7
S. mexicana (Ver, Hua) pot-pollen			1.2			0.7
S. mexicana (Ver, IxR) pot-pollen	1.75	1.7	1.8	0.7	0.7	0.7
S. mexicana (Pue, Tzi) pot-pollen		0.5	1.8		0.2	0.9
S. mexicana (Pue, Ayo) pot-pollen		0.2	1.6		0.1	0.7

(Enríquez and Dardón 2007; Catzín-Ventura et al. 2009).

Four species from the Solanaceae family were identified in *M. beecheii* pot-pollen samples from Campeche: *Physalis pubescens, Solanum americanum, S. lanceolatum* and *S. verbascifolium*. Stingless bees have been considered the most efficient bee pollinators (Landaverde et al. 2004; Villanueva-Gutiérrez 2002); in fact, *Melipona* is used as the specific pollinator of *Capsicum annum* crops (Slaa et al. 2006).

Many taxa found in Veracruz, such as Anacardiaceae, *Cocos nucifera, Lonchocarpus* sp., *Heliocarpus appendiculatus, Cecropia, Alchornea latifolia, Miconia, Coffea arabica* and *Sicyos* have been reported as nectariferous and/or polliniferous resources for stingless bees in Chiapas and Puebla (Martínez-Hernández et al. 1993; Ramírez-Arriaga and Martínez-Hernández 2007). Roubik and Moreno (2013) reported *Cecropia* as a polliniferous resource.

Upon comparison of the ecological parameters in different Mexican tropical regions, it can be observed that the highest diversity values in pot-pollen samples have been recorded in Chiapas, closely followed by Veracruz, in contrast with extremely low values in Campeche. In general, stingless bee foraging behaviour varied from homogeneous to heterogeneous, as shown by the evenness values for the different geographical sites in Mexico (Table 9.2).

Several of the plants of medicinal importance detected in Campeche and Veracruz stingless bee pollen load samples are *Mimosa bahamensis* for urinary problems (Biblioteca Digital de la Medicina Tradicional Mexicana 2016); *Psidium guajava* for healing scars and fighting intestinal parasites (Sousa-Novelo et al. 1981); *Solanum verbascifolium* for scrapes, hair loss, headaches and treatment of maladies of the stomach (Martínez et al. 1995); *Heliocarpus appendiculatus* for healing scars (Martínez et al. 1995); *Platanus* for the flu; and *Coffea arabica* important for domestic consumption in addition to stomach maladies and respiratory diseases.

9.7 Angiosperms Recorded in Systematic Mexican Pot-Pollen Melittopalynological Studies of Stingless Bees

Due to the fact that the information concerning the floral resources visited by stingless bees in Mexico is isolated, this chapter presents a synthesis compiling only systematic works covering the regions in which they have been studied, e.g. in cloud forest, tropical evergreen forest and secondary vegetation. These ecosystems are centres of biodiversity that have been preserved throughout space and time. Since these geographical areas are at risk and have been classified as hot spots (Myers et al. 2000), we are committed to their conservation in order to maintain natural plant-bee interaction.

A broad angiosperm data set for pot-pollen of stingless bees is presented in Table 9.3, including several systematic melittopalynological studies carried out in south central Mexico (Puebla: Villamar 2004; Ramírez-Arriaga and Martínez-Hernández 2007) and south-eastern Mexico (Chiapas: Martínez-Hernández et al. 1993), in addition to the new data from the Gulf of Mexico (Campeche and Veracruz, this work) analysed in the present chapter. In general, the list shows the flowering plants detected in pot-pollen (P) samples and the most important plant preferences represented by pollen percentages ≥10% (Table 9.3) for *Melipona beecheii* (*Mb*), *Nanotrigona perilampoides* (*Na*), *Scaptotrigona mexicana* (*Sc*), *Plebeia* sp. (*Pl*) and *Tetragonisca angustula* (*Te*) are marked with an asterisk.

The list comprises a total of 51 angiosperm families in which 101 genera and 82 species were recorded. Asteraceae (genus $n = 7$; species $n = 7$), Euphorbiaceae (genus $n = 6$; species $n = 4$) and Fabaceae-Mimosoideae (genus $n = 5$; species $n = 6$) were the most diverse plant families, followed by Urticaceae (genus $n = 4$; species $n = 3$), Sapindaceae (genus $n = 4$; species $n = 2$), Malvaceae-Grewioideae (genus $n = 4$; species $n = 3$) and Fabaceae-Faboideae (genus $n = 4$; species $n = 2$). In contrast, many flowering plant families registered ≤3 genera (Fig. 9.7, Table 9.3).

It is noteworthy that tree resources from original ecosystems, secondary vegetation and plant crops, which supply nectar and pollen to stingless bee populations, are also selected as nesting sites by *Tetragonisca angustula* and *Scaptotrigona mexicana*; these resources include *Mangifera indica*, *Cordia alliodora*, *Bursera simaruba*, *Cocos nucifera*, *Citrus*, *Cupania* and *Trichilia* (Fierro et al. 2012).

Finally, several angiosperm pollen grains found in the pollen loads of stingless bees are from plants with healing properties (Martínez et al. 1995). It is likely that the chemical properties of these plants are incorporated as part of the nectar and pollen evidenced by the fact that the honey and pollen of stingless bees is a peculiar product traditionally appreciated by many ethnic cultures (mainly Mayas, Nahuas and Totonacas) for its curative qualities.

9.8   General Considerations

Although *Melipona beecheii*, *Plebeia* sp. and *Scaptotrigona mexicana* are capable of visiting several angiosperm taxa, pot-pollen melittopalynological analyses show that they may also exhibit high flower fidelity for only a handful of resources. These temporal and local specific preferences make stingless bees quite vulnerable pollinators. A recent study demonstrated that places with local specialist interactions are less resilient to disturbance and that plants were in turn susceptible to the loss of bee species in Neotropical areas (Ramírez-Flores et al. 2015).

Mexico is home to abundant apifauna (Ayala et al. 1996; Ayala 2016). Although the culture of native corbiculate bees is of economic importance, beekeepers must protect natural site nesting. In this manner, sustainable meliponiculture will allow for the preservation of original habitats.

Only through comprehensive action, including the recovery of different types of vegetation and the protection of natural bee fauna, can Mexico's biodiversity be preserved. Such diversity is the expression of millions of years of evolution and complex interaction among bees and flowers, which began when angiosperm diversification first occurred in the middle Cretaceous and continued throughout the Paleogene and Neogene until today. This is a contribution in which researchers, regional society and not-for-profit organisations build comprehensive action in order to restore our biodiversity.

Table 9.3 Angiosperms reported in melittopalynological studies of pot-pollen (P) carried out in *Melipona beecheii* (Mb), *Scaptotrigona mexicana* (Sc), *Plebeia* sp. (Pl), *Nanotrigona perilampoides* (Na) and *Tetragona jaty* (Te) in Campeche (CAMP), Veracruz (VER), Puebla (PUE) and Chiapas (CHIS) states. 1–2. Ramírez-Arriaga et al. (this work); 3. Villamar 2004; 4. Ramírez-Arriaga and Martínez-Hernández 2007; 5. Martínez-Hernández et al. 1993. Important taxa are marked with an asterisk (*)

	CAMP	VER		PUE		CHIS			
	1	2		3	4	5			
	Mb	Sc	Pl	Sc	Sc	Na	Pl	Sc	Te
Magnoliids									
Lauraceae									
Nectandra salicifolia (Kunth) Nees*				P					
Piperaceae									
Piper hispidum Sw.								P	P
Piper spp.		P		P					P
Monocots									
Arecaceae									
Chamaedorea tepejilote Liebm.						P			
Cocos nucifera L. *			P						
Elaeis guineensis Jacq.						P			
Asparagaceae									
Cordyline fruticosa (L.) Kunth						P			
Commelinaceae									
Commelina sp.						P			
Tradescantia commelinoides Schult. & Schult.f.									P
Poaceae									
Muhlenbergia sp. *						P			
Poaceae types			P						
Eudicots									
Actinidiaceae									
Saurauia yasicae Loes.								P	
Altingiaceae									
Liquidambar sp.		P							
Amaranthaceae-Gomphrenoideae									
Iresine diffusa Humb. & Bonpl. Ex Willd.*				P					P
Anacardiaceae									
Anacardiaceae type *		P							
Mangifera indica L.				P				P	
Spondias mombin L.*				P					P
Spondias purpurea L.						P	P		
Apiaceae									
Arracacia sp.									P
Araliaceae									
Araliaceae type *		P							

(continued)

Table 9.3 (continued)

	CAMP	VER		PUE		CHIS			
	1	2		3	4	5			
	Mb	Sc	Pl	Sc	Sc	Na	Pl	Sc	Te
Oreopanax sp.		P							
Asteraceae									
Ageratum houstonianum Mill. *				P		P	P	P	P
Aldama dentata La Llave*				P				P	
Asteraceae types *		P							
Bidens pilosa L. *				P					
Tridax procumbens L.						P			
Vernonanthura deppeana (Less.) H.Rob.				P					
Vernonia arborescens (L.) Sw.*				P			P	P	
Vernonia sp. *		P				P		P	P
Wedelia acapulcensis Kunth *							P		
Balsaminaceae									
Impatiens walleriana Hook.							P		
Begoniaceae									
Begonia heracleifolia Schltdl. & Cham.				P					
Bignoniaceae									
Parmentiera aculeata (Kunth) Seem.							P		
Boraginaceae									
Cordia alliodora (Ruiz & Pav) Oken *				P			P		
Burseraceae									
Bursera simaruba (L.) Sarg*				P					
Cannabaceae									
Celtis iguanaea Sarg. *									P
Trema micrantha (L.) Blume *						P	P	P	P
Caricaceae									
Carica papaya L.							P		
Celastraceae									
Crossopetalum parvifolium L.O. Williams *				P				P	P
Chloranthaceae									
Hedyosmum sp.		P							
Cleomaceae									
Cleome guianensis Aubl. *				P					
Cleome parviflora Kunth				P					
Clethraceae									
Clethra occidentalis M. Martens & Galeotti *								P, B	H, P, B
Cucurbitaceae									

(continued)

Table 9.3 (continued)

	CAMP	VER		PUE		CHIS			
	1	2		3	4	5			
	Mb	Sc	Pl	Sc	Sc	Na	Pl	Sc	Te
Sicyos sp.							P		
Euphorbiaceae									
Acalypha sp.									P
Alchornea latifolia Sw.*		P	P						P
Cnidoscolus multilobus (Pax) I.M. Johnston *				P					
Croton sp.				P					
Euphorbia leucocephala Losty								P	P
Euphorbiaceae type		P	P						
Ricinus communis L.				P			P		
Fabaceae									
Fabaceae type		P							
Fabaceae-Caesalpinoideae									
Bauhinia sp.	P								
Parkinsonia praecox (Ruiz & Pav. ex Hook) Harms. *						P		P	
Senna racemosa (Mill.) H.S. Irwin & Barneby	P								
Fabaceae-Faboideae									
Dialium guianense (Aubl.) Sandwith *								P	P
Lonchocarpus spp. *		P					P	P	
Nissolia fructicosa Jacq.	P								
Phaseolus sp.				P					
Fabaceae-Mimosoideae									
Acacia angustissima (Mill.) Kuntze *				P				P	
Acacia cornigera (L.) Willd.						P		P	
Acacia sp.		P							
Inga sp.				P					
Leucaena leucocephala (Lam.) de Wit	P								
Leucaena sp. *		P		P					
Mimosa albida Humb. & Bonpl. Ex Willd. *				P					
Mimosa bahamensis Benth. *	P								
Mimosa orthocarpa Spruce ex Benth.						P			
Mimosa spp.		P							
Prosopis sp.								P	
Fagaceae									
Quercus sp. *		P		P					
Altingiaceae									
Liquidambar sp. *		P							

(continued)

Table 9.3   (continued)

	CAMP	VER		PUE		CHIS			
	1	2		3	4	5			
	Mb	Sc	Pl	Sc	Sc	Na	Pl	Sc	Te
Lamiaceae									
Hyptis mutabilis (Rich.) Briq.							P	P	
Hyptis suaveolens (L.) Poit.				P					
Ocimum sp.			P						
Salvia lamiifolia Vahl				P					
Loranthaceae									
Psittacanthus sp. *				P					
Struthanthus cassythoides Millsp. ex Standl. *				P		P			
Malpighiaceae									
Bunchosia nitida (Jaq-) A. Rich.							P		
Malpighia sp.			P						
Malvaceae-Byttnerioideae									
Guazuma ulmifolia Lam. *						P	P		P
Malvaceae-Grewioideae									
Corchorus siliquosus L. *				P					
Heliocarpus appendiculatus Turcz. *		P		P					
Heliocarpus donnellsmithii Rose *				P			P	P	
Trichospermum sp.							P		
Malvaceae-Malvoideae									
Sida sp.							P		
Melastomataceae									
Miconia sp. *		P							
Tibouchina tortuosa Cogn.				P					
Meliaceae									
Cedrela odorata L.	P						P		
Trichilia hirta L.						P		P	
Moraceae									
Brosimum alicastrum Sw. *				P					
Myrtaceae									
Pimenta dioica (L.) Merr. *				P					
Psidium guajava L. *	P			P					
Olacaceae									
Olacaceae type 1									P
Petiveriaceae									
Petiveria alliacea L.						P	P		
Platanaceae									
Platanus mexicana Moric. *		P							
Polygonaceae									
Coccoloba caracasana Meisn. *								P	

(continued)

Table 9.3 (continued)

	CAMP	VER		PUE		CHIS			
	1	2		3	4	5			
	Mb	Sc	Pl	Sc	Sc	Na	Pl	Sc	Te
Polygonum sp.				P					
Ranunculaceae									
Thalictrum dasycarpum Fisch., & Avé-Lall.						P		P	P
Rhamnaceae									
Gouania lupuloides (L.) Urb.						P			
Rhamnaceae type						P			
Ziziphus sp. *		P							
Rubiaceae									
Borreria sp.				P			P		
Coffea arabica L. *			P	P		P	P	P	P
Rondeletia spp.						P			
Rubiaceae type *		P							
Rutaceae									
Citrus limon (L.) Osbeck *						P			
Citrus sinensis (L.) Osbeck				P					
Sapindaceae									
Cupania sp.*		P							
Sapindaceae types		P							
Sapindus saponaria L. *				P			P	P	P
Sapindus sp.		P							
Talisia sp.		H							
Thouinia paucidentata Radlk. *		P							
Sapotaceae									
Manilkara zapota (L.) P. Royen				P					
Scrophulariaceae									
Capraria biflora L.									P
Solanaceae									
Datura metel L.						P			
Physalis pubescens L.	P								
Solanum americanum Mill.*	P								
Solanum lanceolatum Cav.*	P								
Solanum schlechtendalianum Walp.				P					
Solanum verbascifolium L.*	P								
Ulmaceae									
Ulmus mexicana (Liebm.) planch.				P		P	P		
Urticaceae									
Cecropia obtusifolia Bertol. *							P		
Cecropia sp. *			P						
Coussapoa purpusii Standl.							P		
Pilea sp.									P
Urtica mexicana Liebm.									P
Vitaceae									
Vitaceae type				P					

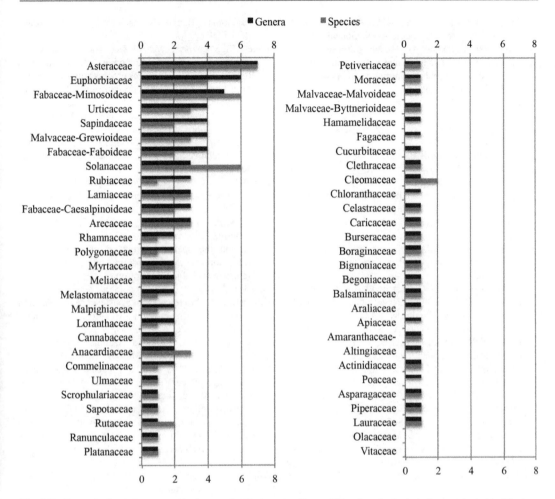

Fig. 9.7 Genera and species representation per family in pot-pollen melittopalynological studies conducted in Mexican tropical ecosystems

Acknowledgements The authors wish to acknowledge the valuable recommendations of David Roubik, Carmen L. Yurrita Obiols, Favio Vossler and Patricia Vit, which improved this manuscript. We express our deepest gratitude to Carlos Fuentes and Victoria Fernández, beekeepers from Tetlaxca, as well as to Faustino Flores Méndez, Adrían Cessa, Ingrid Estrada P., Alejandro Beltrán C. and Raquel Zepeda, and to all members of the INANA A.C. team. We would like to give special thanks to Diana Caballero for her beautiful photographs of stingless bees. We are grateful to S. Helen Ponce Wainer for revising and proofreading the English version of the manuscript. Finally, we also extend our thanks to Patrick Bennett Weill for revising the English version.

References

Alfaro-Bates R, González-Acereto JA, Ortíz-Díaz JJ. 2008. Recursos florales usados por *Melipona beecheii* Bennett (Apidae: Meliponinae) en Mérida, Yucatán, dentro del período de cosecha de miel. pp. 67-73. In: Memorias del V Congreso Mesoamericano sobre Abejas sin Aguijón. Universidad Autónoma de Yucatán; Yucatán, México. 198 pp.

Alfaro-Bates R, González-Acereto JA, Ortíz-Díaz JJ, Viera CA, Burgos PA, Martínez-Hernández E, Ramírez-Arriaga E. 2010. Caracterización palinológica de las mieles de la Península de Yucatán. Universidad Autónoma de Yucatán-Comisión Nacional pare el Conocimiento y Uso de la Biodiversidad; Mérida, Yucatán, México. 156 pp.

Alvarado JL, Delgado RM. 1985. Flora apícola de Uxpanapa, Veracruz, México. Biótica 10: 257–275.

Ayala R. 2016. Abejas (Apoidea). pp. 331–345. In: La biodiversidad en Colima. Estudio de Estado. Conabio; Colima, México. 766 pp.

Ayala R, Gonzalez VH, Engel MS. 2013. Mexican stingless bees (Hymenoptera: Apidae): Diversity, distribution and indigenous knowledge. pp. 135–152. In: Vit P, Pedro SRM, Roubik D, eds. Pot-Honey. A legacy of stingless bees. Springer; New York, USA. 654 pp.

Ayala R, Griswold TL, Yanega D. 1996. Apoidea. pp. 423–464. In: Llorente BJE, García AAN, González E, eds. Biodiversidad, taxonomía y biogeografía de artrópodos de México: hacia una síntesis de su conocimiento. Universidad Nacional Autónoma de México; Ciudad de México, México. 660 pp.

Barth OM. 2013. Palynology serving the stingless Bees. 285-294 pp. In: Vit P, Pedro SRM, Roubik D. eds. Pot-Honey. A legacy of stingless bees. Springer; New York, USA. 654 pp.

Bell CD, Soltis DE, Soltis PS. 2010. The age and diversification of the angiosperms re-revisited. American Journal of Botany 97: 1296 – 1303.

Biblioteca Digital de la Medicina Tradicional Mexicana. 2016. Universidad Nacional Autónoma de México. http://www.medicinatradicionalmexicana.unam.mx/index.php

Biesmeijer JC, Slaa EJ. 2004. Information flow and organization of stingless bee foraging. Apidologie 35:143–157.

Bogandov S, Lüllmann C, Martin P, von der Ohe W, Russmann H, Vorwohl G, Persano OL, Sabatini AG, Luigi MG, Pito R, Flamini MM, Lhéritier J, Botneck R, Marioleas P, Tsigouri A, Kerkvliet J, Ortiz A, Ivanov T, D'Arcy B, Mossel B, Vit P. 1999. Honey quality and international regulatory standards: review by the International Honey Commission. Bee World 80: 61–69.

Bogandov S, Ruoff K, Oddo LP. 2004. Physico-chemical methods for the characterisation of unifloral honeys: a review. Apidologie 35: S4–S17.

Cárdenas-Chávez S. 1985. Caracterización del ciclo apícola y flora nectarífera y polinífera, en la Chontalpa Tabasco, México. B.Sc.Thesis, Colegio Superior de Agricultura Tropical; Cárdenas, Tabasco, México. 120 pp.

Cardinal S, Danforth BN. 2011. The Antiquity and Evolutionary History of Social Behavior in Bees. PLoS ONE 6: e21086. doi:10.1371/journal.pone.0021086

Cardinal S, Danforth BN. 2013. Bees diversified in the age of eudicots. Proceedings of the Royal Society B- Biological Sciences 280: 20122686. http://dx.doi.org/10.1098/rspb.2012.2686

Castellanos PBP. 2010. Caracterización polínica estacional de miel inmadura de *Apis mellifera* L. en el estado de Tabasco. M.Sc. Thesis, Colegio de Postgraduados, Institución de Enseñanza e Investigación en Ciencias Agrícolas, Campus Tabasco; Tabasco, México. 114 pp.

Catzín-Ventura GR, Alfaro-Bates L, Medina M, Delgado HM. 2009. Actividad antimicrobiana y origen botánico en mieles de *Melipona beecheii*, *Scaptotrigona pectoralis* y *Apis mellifera* del Estado de Yucatán. pp. 86-92. In: Memorias del VI Congreso Mesoamericano sobre Abejas Nativas; Antigua Guatemala, Guatemala. pp.

Díaz CE. 2008. Estudio palinológico y fisicoquímico de la miel de *Apis mellifera* L. del municipio de San Pedro Tapanatepec, Oaxaca. B.Sc. Thesis, Escuela de Medicina Veterinaria y Zootecnia, Universidad Autónoma "Benito Juárez de Oaxaca"; Oaxaca, México. 91pp.

Engel MS, Michener CD. 2013. Geological history of the stingless bees (Apidae: Meliponini). pp. 1–7. In: Vit P, Roubik D eds. Stingless bees process honey and pollen in cerumen pots. Facultad de Farmacia y Bioanálisis, Universidad de Los Andes; Mérida, Venezuela.

Enríquez E, Dardón M. 2007. Caracterización de la miel de meliponinos de distintas regiones biogeográficas de guatemala. Dirección General de Investigación – Universidad de San Carlos de Guatemala; Guatemala, Guatemala. 44 pp.

Faegri K, van der Pijl L. 1979. The principles of pollination ecology. 3rd rev edn. Pergamon Press; Oxford, UK. 244 pp.

Fierro MM, Cruz-López L, Sánchez D, Villanueva-Gutiérrez R, Vandame R. 2012. Effect of biotic factors on the spatial distribution of stingless bees (Hymenoptera: Apidae, Meliponini) in fragmented Neotropical habitats. Neotropical Entomology 41: 95–104.

García E. 1981. Modificaciones al sistema de clasificación climática de Köppen (para adaptarlo a las condiciones climáticas de la República Mexicana). Instituto de Geografía, Universidad Nacional Autónoma de México, Distrito Federal, México. 252 pp.

González-Acereto JA. 2008. Cría y manejo de abejas nativas sin aguijón en México. Universidad Autónoma de Yucatán; Mérida, Yucatán, México. 177p.

Guzmán M, Balboa C, Vandame R, Albores ML, González-Acereto J. 2011. Manejo de las abejas nativas sin aguijón en México: *Melipona beecheii* y *Scaptotrigona Mexicana*. El Colegio de la Frontera Sur, Chiapas, México. 59 pp.

Landaverde V, Sánchez LA, Ruano C, Smeets M. 2004. Temporary dominance of pollen of nectariferous and polliniferous plants collected by *Melipona beecheii* in El Salvador and pollen of polliniferous plants collected by *Tetragonisca angustula* and *M. beecheii* in Costa Rica. pp. 44–52. In: Proceedings beekeeping in tropical countries, research and development for pollination and conservation. San José, Costa Rica. CD-ROM.

Lobreau-Callen D, Callen G. 1982. Quelle est la composition pollinique d´un miel exotique? I. Bulletin de la Societé Versaillaise de Sciences Naturelles, Série 4, 9: 70–85.

Louveaux J., Maurizio A., Vorwohl G. 1978. Methods of melisso- palynology. Bee World 59: 139–157.

Magallón S. 2010. Using fossils to break long branches in molecular dating: a comparison of relaxed clocks applied to the origin of angiosperms. Systematic Biology 59: 384–99. (doi:10.1093/sysbio/syq027)

Martínez AMA, Evangelista OV, Mendoza CM, Morales GG, Toledo OG, Wong LA. 1995. Catálogo de plantas útiles de la Sierra Norte de Puebla, México. Cuadernos del Instituto de Biología. Cuadernos 27, Universidad Nacional Autónoma de México, Distrito Federal, México. 303 pp.

Martínez-Hernández E, Ramírez-Arriaga E. 1998. La importancia comercial del origen botánico de las mieles por medio de su contenido de granos de polen (Melisopalinología). Apitec 10: 27–30.

Martínez-Hernández E, Cuadriello-Aguilar JI, Téllez-Valdez O, Ramírez-Arriaga E, Sosa-Nájera MS, Melchor-Sánchez JE, Mediana-Camacho M, Lozano-García MS. 1993. Atlas de las plantas y el polen utilizados por las cinco especies principales de abejas productoras de miel en la región del Tacaná, Chiapas, México. Instituto de Geología, Universidad Nacional Autónoma de México; México D.F., México. 105 pp.

Medina CM. 1992. Contribución al conocimiento de algunos aspectos ecológicos en relación a la flora apícola explotada por abejas europeas (*A. mellifera* L.), abejas africanizadas (*A. mellifera scutellata*) e híbridos en el Soconusco, Chiapas. M.Sc. Thesis. Universidad Nacional Autónoma de México. Ciudad de México, México. 155 pp.

Michener CD. 2013. The Meliponini. pp. 3–17. In Vit P, Pedro SRM, Roubik D, eds. Pot-Honey. A legacy of stingless bees. Springer; New York, USA. 654 pp.

Michener CD, Grimaldi DA. 1988 A *Trigona* from late Cretaceous amber of New Jersey (Hymenoptera: Apidae: Meliponinae). American Museum Novitates 2917: 1–10.

Myers M, Mittermeier RA, Mittermeier CG, Fonseca GAB, Kent J. 2000. Biodiversity hotspots for conservation priorities. Nature 403: 853–858.

Navarro CLA. 2008. Estudio palinológico y fisicoquímico de la miel de *Apis mellifera* L., en la Región costa de Oaxaca: Distritos Jamiltepec, Juquilla y Pochutla. B.SC. Thesis. Escuela de Medicina Veterinaria y Zootecnia, Universidad Autónoma "Benito Juárez" de Oaxaca, Oaxaca, México. 108 pp.

Palacios-Chávez R, Ludlow-Wiechers B, Villanueva-Gutiérrez R. 1991. Flora palinológica de la reserva de la biósfera de Sian Ka'an, Quintana Roo, México. Centro de Investigaciones de Quintana Roo, Chetumal; Quintana Roo, México. 321 pp.

Piedras GB, Quiroz-Gracía DL. 2007. Estudio melisopalinológico de dos mieles al sur del Valle de México. Polibotánica 23: 57–75.

Pielou E.C. 1977. Mathematical Ecology. John Wiley & Sons, Nueva York. 385 pp.

Quezada-Euán JJ. 2005. Biología y uso de las abejas sin aguijón de la Península de Yucatán, México (Hymenoptera: Meliponini). Universidad Autónoma de Yucatán. México. 112 pp.

Quiroz-García DL, Arreguín-Sánchez ML. 2008. Determinación palinológica de los recursos florales utilizados por *Apis mellifera* L. (Hymenoptera: Apidae) en el estado de Morelos, México. Polibotánica 26: 159–173.

Quiroz-García DL, Arreguín-Sánchez ML, Fernández-Nava R, Martínez-Hernández E. 2011. Patrones estacionales de utilización de recursos florales por *Scaptotrigona hellwegeri* en la Estación de Biología Chamela, Jalisco, México. Polibotánica 31: 89-119.

Quiroz-García DL, Palacios-Chávez R. 1999. Determinación palinológica de los recursos florales utilizados por *Centris inermis* Friese (Hymenoptera:

Apidae) en Chamela, Jalisco, México. Polibotánica 10: 59–72.

Ramalho M, Kleinert-Giovannini A, Imperatriz-Fonseca VL. 1989. Utilization of floral resources by species of *Melipona* (Apidae, Meliponinae): floral preferences. Apidologie 20: 185–195.

Ramalho M, Kleinert-Giovannini A, Imperatriz-Fonseca VL. 1990. Important bee plants for stingless bees (*Melipona* and Trigonini) and Africanized honeybees (*Apis mellifera*) in Neotropical habitats: a review. Apidologie 21: 469–488.

Ramalho M, Giannini TC, Malagodi-Braga KS, Imperatriz-Fonseca VL. 1994. Pollen harvest by stingless bee foragers (Hymenoptera, Apidae, Meliponinae). Grana 33: 239–244.

Ramírez-Arriaga E. 1989. Explotación de los recursos florales por *Plebeia* sp. (Apidae) en dos zonas con diferente altitud y vegetación en el Soconusco, Chiapas. B.Sc. Thesis. Facultad de Ciencias, Universidad Nacional Autónoma de México, México D.F. 159 pp.

Ramírez-Arriaga E, Martínez-Hernández E. 1998. Resources foraged by *Euglossa atroveneta* (Apidae: Euglossinae) at Unión Juárez, Chiapas, Mexico. A palynological study of larval feeding. Apidologie 29: 347–359.

Ramírez-Arriaga E, Martínez-Hernández E. 2007. Melissopalynological characterization of *Scaptotrigona mexicana* Gúerin (Apidae: Meliponini) and *Apis mellifera* L. (Apidae: Apini) honey samples in northern Puebla state, Mexico. Journal of the Kansas Entomological Society 80: 377–391.

Ramírez-Arriaga E, Martínez-Hernández E, Cuadriello-Aguilar E, Lozano-García S. 1995. Estrategias de pecoreo de *Plebeia* sp. (Apidae), basado en el análisis melisopalinológico y en parámetros ecológicos en Chiapas. Implicaciones Evolutivas. pp. 113–154. In Investigaciones Recientes en Paleobotánica y Palinología, Serie Arqueología, Instituto Nacional de Antropología e Historia; México D.F., México. 294 pp.

Ramírez-Arriaga E, Navarro-Calvo LA, Díaz-Carbajal E. 2011. Botanical characterization of Mexican honeys from a subtropical region (Oaxaca) based on pollen analysis. Grana 50: 40–45.

Ramírez-Arriaga E, Martínez-Bernal A, Ramírez-Maldonado N, Martínez-Hernández E. 2016. Análisis palinológico de mieles y cargas de polen de *Apis mellifera* (Apidae) de la región centro y norte del estado de Guerrero, México. Botanical Sciences 94: 141–156.

Ramírez-Flores VA, Villanueva-Gutiérrez R, Roubik DW, Vergara CH, Lara-Rodríguez N, Ferrer MEB, Rico-Gray V. 2015. Topological Structure of Plant-bee Networks in Four Mexican Environments. Sociobiology 62: 56–64.

Roubik DW. 1992. Ecology and natural history of tropical bees. Cambridge University Press; New York, USA. 514 pp.

Roubik DW, Moreno JE. 1991. Pollen and spores of Barro Colorado Island. Monographs in systematic botany No. 36. Missouri Botanical Garden; St. Louis, Missouri, USA. 268 pp.

Roubik DW, Moreno JE. 2013. How to be a bee-botanist using pollen spectra. pp. 295–314. In: Vit P, Pedro SRM, Roubik D, eds. Pot-Honey. A legacy of stingless bees. Springer; New York, USA. 654 pp.

Shannon CE, Weaver W. 1949. The mathematical theory of communication. University of Illinois Press; Urbana, USA. pp. 117.

Slaa EJ, Sánchez CLA, Malagodi-Braga KS, Hofstede FE. 2006. Singless bees in applied pollination: practice and perspectives. Apidologie 37: 293–315.

Souza-Novelo N, Suárez-Molina V, Barrera-Vázquez A. 1981. Plantas melíferas y poliníferas que viven en Yucatán. Fondo Editorial de Yucatán; Mérida, México. 61 pp.

Villamar EMI. 2004. Hábitos alimenticios de *Scaptotrigona mexicana* Guerin (Apidae, Meliponini) en el municipio de Cuetzalan del Progreso, Sierra Norte de Puebla. Tesis de Maestría, Facultad de Ciencias, Universidad Nacional Autónoma de México, México D.F., México. 77 pp.

Villanueva-Gutiérrez R. 1984. Plantas de importancia apícola en el ejido de Plan del Río, Veracruz, México. Biótica 9: 279–340.

Villanueva-Gutiérrez R. 2002. Polliniferous plants and foraging strategies of *Apis mellifera* (Hymenoptera: Apidae) in the Yucatán Peninsula, Mexico. Revista de Biología Tropical 50: 1035–1043.

Villanueva-Gutiérrez R, Moguel-Ordóñez YB, Echazarreta-González CM, Gabriela Arana-López G. 2009. Monofloral honeys in the Yucatán Peninsula, Mexico. Grana 48: 214–223, DOI: 10.1080/00173130902929203

Villanueva-Gutiérrez R, Quezada-Euan J, Eltz T. 2013. Pollen diets of two sibling orchid bee species, Euglossa, in Yucatán, southern Mexico. Apidologie 44: 440–446.

Villanueva-Gutiérrez R, Roubik DW. 2004. Why are African honey bees and not European bees invasive? Pollen diet diversity in community experiments. Apidologie 35: 481–491.

Villaseñor JL, Ortiz E. 2014. Biodiversidad de las plantas con flores (División Magnoliophyta) en México. Revista Mexicana de Biodiversidad 85: 134–142.

Vit P, Medina M, Enríquez ME. 2004. Quality standards for medicinal uses of Meliponinae honey in Guatemala, Mexico and Venezuela. Bee World 85: 2–5, doi:10.1080/0005772X.2004.11099603.

von Frisch K. 1967. The Dance Language and Orientation of Bees. Belknap Press of Harvard University Press; Cambridge, MA, USA. 566 pp.

Yurrita CL, Ortega-Huerta MA, Ayala R. 2016. Distributional analysis of *Melipona* stingless bees (Apidae: Meliponini) in Central America and Mexico: setting baseline information for their conservation. Apidologie doi:10.1007/s13592-016-0469-z.

Annual Foraging Patterns of the Maya Bee *Melipona beecheii* (Bennett, 1831) in Quintana Roo, Mexico

10

Juan Carlos Di Trani
and Rogel Villanueva-Gutiérrez

10.1 Introduction

In nature, food resources are typically distributed irregularly, both spatially and temporally (Pleasants and Zimmerman 1979; Zimmerman 1981; Possingham 1989; Hunter et al. 1992; Rathcke and Jules 1993; Williams and Kremen 2007). This poses a dilemma about where and when to forage a certain food resource, and how to ensure its continuous availability.

Food resources for social bees consist of honey and pollen, this second one, is used mostly for feeding the immature stages. Some studies with social bees, such as honey bees (*Apis mellifera*), stingless bees (*Melipona*), and bumblebees (*Bombus*), have determined the intensity of bees foraging a certain resource depends on both external and internal factors to the nest, as we will discuss ahead.

External conditions like the availability of food resources in the field, seasonality, and some climatic factors can strongly determine the foraging dynamics and the food choices bees make when foraging (Villanueva-Gutiérrez 2002; Villanueva-Gutiérrez and Roubik 2004; Villanueva-Gutiérrez et al. 2015).

First, bees tend to display most of the foraging activity when the external conditions are more favorable during the year (Roubik et al. 1986). In these periods, they forage large quantities of food, so they can use this surplus for surviving periods of during food scarcity, when there is not much food in the field. For temperate zone bees, these resource shortage occurs during the winter (Juliani 1967; Winston 1980, 1987) and for tropical bees, at the end of the rainy season (Roubik 1982; Woyke 1992; Echazarreta et al. 1997; Porter-Bolland 2001; Cortopassi-Laurino 2004; De Figueiredo-Mecca et al. 2013).

An analogous pattern can be observed in the daily foraging activity of the bees. Temperate zone bees generally avoid foraging early in the morning, when the temperature is still too low, and there are not many food resources available in the field. In contrast, for most of the bees in the tropics, the foraging activity is high in the morning, and decreases around midday (Roubik 1989; Pierrot and Schlindwein 2003; Borges and Blochtein 2005; Gouw and Gimenes 2013), probably because the ambient temperature is too warm, and the food resources are less abundant (pollen) or less energetically profitable (nectar).

J.C. Di Trani
Instituto de Investigaciones Científicas y Servicios de Alta Tecnología (INDICASAT), Condominio Don Oscar, C-6., Coco del Mar, San Francisco, Ciudad de Panamá, República de Panamá

R. Villanueva-Gutiérrez (✉)
El Colegio de la Frontera Sur. Unidad Chetumal. Av. Centenario km 5.5,
Chetumal, Quintana Roo, Mexico, C. P. 77014
e-mail: rvillanu@ecosur.mx

© Springer International Publishing AG, part of Springer Nature 2018
P. Vit et al. (eds.), *Pot-Pollen in Stingless Bee Melittology*, DOI 10.1007/978-3-319-61839-5_10

Other climatic factors such as humidity (Hilário et al. 2001; Kajobe and Echazarreta 2005; Peat and Goulson 2005; Souza et al. 2006), barometric pressure (Contreras et al. 2004; Hilário et al. 2007b), precipitation (Iwama 1977; Kleinert-Giovanini 1982; Hilário et al. 2007a), and wind (Iwama 1977; Kleinert-Giovanini 1982; Comba 1999; Hilário et al. 2007b) have also proven to correlate with the daily foraging activity and food choices of bees.

As we mentioned above, internal conditions in the nest also may play a fundamental role in bee foraging behavior. This could be due to the quantity of food reserves and the amount of larvae in the nest. Some experimental studies on honey bees have demonstrated when pollen reserves decreased, the pollen income by foraging trips intensifies, and when pollen reserves are overstocked, the pollen foraging trips decrease (Free 1967; Fewell and Winston 1992; Camazine 1993). A similar behavior has been observed in other social bees, such as *Bombus terrestris* (Molet et al. 2008) and *Melipona beecheii* (Biesmeijer et al. 1999). Curiously, Fewell and Winston (1996) find no intensification of nectar foraging activity when honey reserves are experimentally reduced.

Another important factor inside the nest is the offspring number. Free (1967) and Todd and Reed (1970) give evidence that honey bee foraging (especially for pollen) intensifies when the offspring number increases. This pattern has been confirmed later by many other studies (Hellmich and Rothenbuhler 1986; Eckert et al. 1994; Pankiw et al. 1998; Dreller et al. 1999; Pankiw 2004, 2007). It seems this behavior is triggered by pheromones, as Pankiw (2004, 2007) and Pankiw et al. (1998) demonstrated. By applying brood pheromones, the pollen foraging activity of honey bees was remarkably intensified. Other social bees show a similar behavioral pattern, as demonstrated by Van Veen et al. (2004) in *Melipona beecheii*, Hilário and Imperatriz-Fonseca (2009) in *Melipona bicolor*, and Nunes-Silva et al. (2010) in *Plebeia remota*.

Despite the many studies of honey bees and the influence of external and internal factors on foraging behavior, there are few studies on stingless bees.

Among the most remarkable of them we can find Biesmeijer et al. (1998) studies on the relation between external conditions and the foraging choices of *Melipona beecheii* and *M. fasciata*; Moo-Valle et al. (2001) studies about the influence of stored food and the production of reproductive individuals; Nunes-Silva et al. (2010), studies on the relation between the reproductive state and the foraging behavior of *Plebeia remota*; Roubik (1982) studies on annual brood production and foraging behavior of *M. favosa* and *M. fulva*; Van Veen et al. (2004) studies on the relation between brood and stored food in *M. beecheii*; Biesmeijer et al. (1999) studies on stored food and daily foraging activity of *M. beecheii*; and a recent study of Villanueva-Gutiérrez et al. (2015) comparing potential competition between honey bees and native bees in relation to their food resources.

Unfortunately none of the studies mentioned above combined offspring number and stored food to explore the foraging patterns of stingless bee colonies, and this was the aim of our study.

10.2   A Case Study

10.2.1   Field Observations: Registering Bee Activity

Eight nests of *Melipona beecheii* were used, at three sites in Quintana Roo, Yucatan Peninsula (Mexico): two nests in the city of Chetumal, three in the town of Carrillo Puerto, and three in the town of Tihosuco. Originally we selected three nests for each site, but one of the Chetumal colonies died during the observation year.

The area is dominated by lowland deciduous forest, medium deciduous forest, secondary vegetation and some fields (Villanueva-Gutiérrez 2002). A dry season occurs, usually between November and April, and the rainy season is between May and October. There is usually also a marked influence of cold fronts impacting the Yucatan Peninsula in the winter (November to March).

Foraging bee activity was recorded for 2 days at each nest and site, every 2 months between June 2008 and June 2009, for a total eight replicates. We recorded the bees entering each nest for

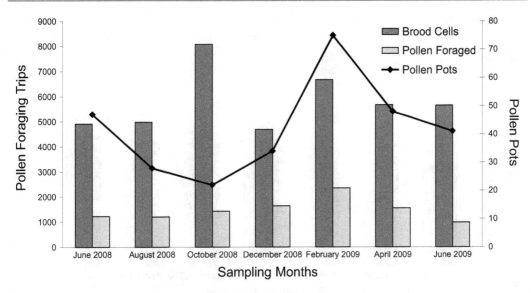

Fig. 10.1   Total pollen foraged per month in the eight nests, total pollen pots and total brood cells by date.

15 min every hour from 06:00 to 16:30, and the food resource (pollen or nectar) they were carrying. For analysis, the returning forager counts for each nest were totaled for each month.

After recording the foraging behavior of a colony, we opened the nest and counted the food pots containing pollen or nectar. When the pot content was not clear, we determined contents by lightly taping on the pot or by making a small hole. We also took pictures of the brood combs from different angles, and individually counted the brood cells. In a few cases, when some parts of the combs were not visible, we estimated the number of cells inside by measuring the area and calculating the cells it could include. Annual activity was registered for each resource foraged, comparing the number of cells and food pots (pollen or honey). Finally, we calculated the Pearson correlation index between foraging activity for each resource, the number of brood cells, and the number of food pots.

10.2.2   Foraging Activity to Collect Pollen and Nectar

Pollen foraging was highest during the first months of the year, during the dry season, from February to April (Fig. 10.1) when six of the eight nests presented their annual maximum. This peak coincides with the observed pattern in *Melipona colimana* (Macías-Macías and Quezada-Euán 2008) in México. In the remaining two nests the maximum occurred in August or October (Fig. 10.1).

After April, the pollen foraging flights decreased progressively until reaching a minimum in June and August. This decline could be the result of food scarcity in the field, the rain preventing or limiting the pollen foraging flights, or the increasing pollen "handling" time by the bees. During this period bees carried pollen loads that were very wet, and a pollen pellet would fall off the bee corbicula at the slightest touch. The bees may have been returning to the nest without full loads, or discarding the pollen loads upon arrival, because wet pollen tends to decompose rapidly, losing quality and also attracting the "Nenem" phorid fly (*Pseudohypocera kerteszi*, Enterlein 1912).

The nectar foraging was maximum in the long dry season and part of the wet season, February and June, where seven of the eight observation nests displayed maximum nectar foraging trips (Fig. 10.2). In contrast, the period of reduced nectar foraging was in the late wet to early dry season, December, when six of the eight studied nests had their minimum. In the other two nests,

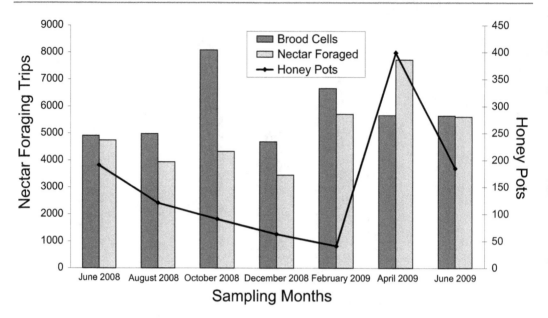

Fig. 10.2 Total nectar foraged per month in the eight nests, total honey pots and total brood cells by date

the minimum occurred in August or October. This decline could be attributed to the end of the rainy season, when there are few flowers in the environment (Villanueva-Gutiérrez et al. 2015) and when winter weather may restrict food foraging trips (Di Trani, unpublished data).

Our curves for pollen and nectar foraging trips (Fig. 10.2) show a very distinctive difference in annual foraging pattern for pollen and nectar, although both show high activity towards the early months of the year, between February and April, when the most important herbs for pollen in the area, *Viguiera dentata* and *Bidens pilosa* (Villanueva-Gutiérrez 2002; Villanueva-Gutiérrez and Roubik 2004), are flowering. During the year, nectar trips were always more numerous than pollen trips. Pollen trips were 20–30% of all foraging trips (excluding resin) for most of the year, even when the proportion was similar between December and February, when pollen trips counted for >40% of foraging trips. A proportion in favor of nectar trips could be attributed to bees foraging nectar regularly throughout the day, while pollen collection sharply declines after morning. This behavior occurs in most species of bees and has been confirmed for *Melipona* by

Pierrot and Schlindwein (2003), Borges and Blochtein (2006), and Macías-Macías and Quezada-Euán (2008), among others.

10.2.3 Stored Pot-Honey and Pot-Pollen Reserves

The number of pollen pots in the nests was highest in the month of February, found in seven of the eight nests observed. This is possibly due to the abundance of pollen resources in the field and subsequently a "boom" in pollen foraging trips. This pattern of food reserves on the nest contrasts with that observed for the same bee species in a Pacific Costa Rican rainforest by van Veen et al. (2004). In that work they found the maximum pollen reserves occurred in June and were lowest in December. They also found honey reserves had two peaks, one in March and one in August. This difference could be attributed to climatic differences between Costa Rica and the Yucatan Peninsula, especially considering the great impact cold fronts coming from North America have on Yucatan Peninsula during winter, and too many ecological differences between those places.

In contrast, the pot-pollen reserves were lowest in Quintana Roo during August–October, which coincides with the last months of the rainy season. In this period there are few species of plants flowering (Roubik 1982; Villanueva-Gutiérrez 2002; Villanueva-Gutiérrez et al. 2015), which decreases and colonies may often turn to their food supply.

The pot-honey reserves reached a peak during spring, especially in April, when all colonies had the most honey. This could be the result of the bees storing honey after most of the tree plants in the area have been blooming (Villanueva-Gutiérrez 2002; Villanueva-Gutiérrez et al. 2015). Later, the honey reserves gradually decreased, reaching a minimum at the end of the winter, in February. Apparently, bees were depleting their honey reserves during the first part of the winter, when there are very few plants flowering in the area.

It is remarkable that the peak of pollen collection occurs just before the peak of collecting nectar, although reserves of both resources are reduced by the year's end. Apparently, with the first flowers of the year, bees give priority to foraging for pollen, possibly by the urge to feed the larvae, prepare for potential colony swarming, and replenish the reserves. The intensification in pollen foraging activity, with depleted reserves, has also been observed after experimental reduction of stored pollen (Biesmeijer et al. 1999).

10.2.4 Offspring

The number of brood cells was highest in October, which coincides with the minimum amount of pollen reserves and one of the lowest values of pollen foraging, so probably the larvae are fed mostly with stored pollen. The next peak in the brood number occurred in February. This coincided with both maximum pollen foraging and pollen reserves. During the rest of the year the number of offspring was more or less stable (Fig. 10.1).

10.3 Correlations Between the Studied Factors

We found a weak positive correlation between the number of pollen pots and pollen foraging (Table 10.1), a negligible correlation between the number of pots of honey and pollen foraging, and a strong correlation between the brood number and pollen foraging.

Between the number of pollen pots and nectar foraging we found a weak positive correlation; between the number of honey pots and nectar foraging there was a moderately positive correlation, and the number of brood cells and nectar foraging showed a strong positive correlation.

Finally, the relationship between the number of offspring and the number of pollen pots and pots of honey was negligible in both cases. Between all the analyzed factors, the number of brood was the most correlated with the foraging pattern of the bees—specifically pollen foraging, which showed a high correlation coefficient (Table10.1). These results coincide with the observations previously reported for *M. beecheii* (Van Veen et al. 2004), *M. bicolor* (Hilário and Imperatriz-Fonseca 2009), *Plebeia remota* (Nunes-Silva et al. 2010), and the honey bee (Hellmich and Rothenbuhler 1986; Eckert et al. 1994; Pankiw et al. 1998; Dreller et al. 1999; Pankiw 2004, 2007).

A strong correlation between offspring and pollen foraging could suggest the highest priority in the nest is to maintain the immatures. Combined with the inverse correlation with

Table 10.1 Pearson Correlation Index (r) and p-value (p) for the parameters observed

	Pollen pots	Honey pots	Brood cells
Pollen foraging	0.2635	−0.1315	0.5021
	0.0384	0.1918	0.0002
Nectar foraging	0.2855	0.3573	0.4725
	0.0272	0.0074	0.0005

stored pollen in the nest, this suggests a priority for producing adult bees that later will become the working force of the colony, and will enhance colony survival. Another strong correlation was observed between the brood number and nectar foraging. Although nectar is not quantitatively as important to brood nutrition as pollen, a nest "in good condition" with an abundant food supply likely produces a greater number of offspring and has superior survival.

We observed a weak correlation between pot-pollen stocks and pollen foraging, because as already mentioned, the reduction in these reserves have been reported as one of the main stimuli for pollen foraging (Free 1967; Fewell and Winston 1992; Camazine 1993). All those studies, however, were in the temperate zone, where pollen availability is more or less the same as the active reproductive period of the honey bee. Conditions in the tropical forests are much different, with reproductive periods staged through multiple seasons and peak floral production by trees, shrubs, herbs or lianas (e.g. Roubik 1989). On the other hand, we found a moderate correlation between pot-honey stores and nectar foraging, while studies by Fewell and Winston (1996), in the temperate zone, found no correlation.

In our findings, when there was no significant positive relationship between pot-pollen reserves and the number of offspring, probably during some periods most of the pollen collected was immediately used for feeding the brood or the colony perhaps anticipated and provided for brood continuation during periods of scarcity.

References

Biesmeijer JC, Van Nieuwstadt MG, Lukács S, Sommeijer MJ. 1998. The role of internal and external information in foraging decisions of *Melipona* workers (Hymenoptera: Meliponinae). Behavioral Ecology and Sociobiology 42: 07–116.

Biesmeijer JC, Bom M, Lukács S, Sommeijer MJ. 1999. The response of *M. beecheii* experimental pollen stress, worker loss and different levels of information input. Journal of Apicultural Research 38: 29–37.

Borges FV, Blochtein B. 2005. Atividades externas *Melipona marginata obscurior* Moure (Hymenoptera, Apidae), em Distintas Épocas de ano, em Sao Francisco de Paula, Rio Grande do Sul, Brasil. Revista Brasileira de Zoololgia 22: 680–686.

Borges FV, Blochtein B. 2006. Variação sazonal das condições internas de colônias de *Melipona marginata obscurior* Moure, no Rio Grande do Sul, Brasil. Revista Brasileira de Zoololgia 23: 711–715.

Camazine S. 1993. The regulation of pollen foraging by honey bees: how foragers assess the colony's need for pollen. Behavioral Ecology and Sociobiology 32: 265–272.

Comba, L. 1999. Patch use by bumblebees (Hymenoptera Apidae): temperature, wind, flower density and traplining. Ethology Ecology and Evolution 11: 243–264.

Contreras F, Imperatriz-Fonseca VL, Nieh JC. 2004. Temporal and Climatological Influences on Flight Activity in the Stingless Bee *Trigona hyalinata* (Apidae, Meliponini). Revista de Tecnología Ambiental 10: 35–43.

Cortopassi-Laurino C. 2004. Seasonal strategies of harvesting by *Melipona* sp. in the Amazon region. In: Proceedings of the 8th IBRA International Conference on Tropical Bees and VI Encontro sobre Abelhas, Ribeirão Preto (SP), Brazil: 258-263.

De Figueiredo-Mecca G, Bego L, do Nascimento F S. 2013. Foraging behavior of *Scaptotrigona depilis* (Hymenoptera, Apidae, Meliponini) and its relationship with temporal and abiotic factors. Sociobiology 60: 267–282.

Dreller C, Page R, Fondrk M. 1999. Regulation of pollen foraging in honeybee colonies: effects of young brood, stored pollen, and empty space. Behavioral Ecology and Sociobiology 45: 227–233.

Echazarreta CM, Quezada-Euan JJ, Medina LM, Pasteur KL. 1997. Beekeeping in the Yucatan Peninsula: Development and current status. Bee World 78: 115–27.

Eckert C, Winston M, Ydenberg R. 1994. The relationship between population size, amount of brood, and individual foraging behaviour in the honey bee, *Apis mellifera* L. Oecologia 97: 248–255.

Fewell J H, Winston M L. 1992. Colony state and regulation of pollen foraging in the honey bee, *Apis mellifera* L. Behavioral Ecology and Sociobiology 30: 387–393.

Fewell JH, Winston ML. 1996. Regulation of nectar collection in relation to honey storage levels by honey bees, *Apis mellifera*. Behavioral Ecology 7: 286–291.

Free J B. 1967. Factors determining the collection of pollen by honey bee foragers. Animal Behavior: 15, 134–144.

Gouw MS, Gimenes M. 2013. Differences of the daily flight activity rhythm in two Neotropical stingless bees (Hymenoptera, Apidae). Sociobiology 60: 183–189.

Hellmich RL, Rothenbuhler WC. 1986. Relationship between different amounts of brood and the collection and use of pollen by the honey bee (*Apis mellifera*). Apidologie 17: 13–20.

Hilário SD, Imperatriz-Fonseca VL, Kleinert A. 2001. Responses to climatic factors by foragers of *Plebeia pugnax* Moure (in litt.) (Apidae, Meliponinae). Brazilian Journal of Biology 61: 191–196.

Hilário SD, Ribeiro M, Imperatriz-Fonseca VL. 2007a. Rain effect on flight activity of *Plebeia remota* (Holmberg, 1903) (Apidae, Meliponini). Biota Neotropica 7: 135–143.

Hilário SD, Ribeiro M, Imperatriz-Fonseca VL. 2007b. Wind effect on flight activity of *Plebeia remota* (Holmberg, 1903) (Apidae, Meliponini). Biota Neotropica 7: 225–232.

Hilário SD, Imperatriz-Fonseca VL. 2009. Pollen foraging in colonies of *Melipona bicolor* (Apidae, Meliponini): effects of season, colony size and queen number. Genetics and Molecular Research 8: 664–671.

Hunter MD, Ohgushi T, Price PW. 1992. Effects of Resource Distribution on Animal-Plant Interactions. Academic Press. 505 pp.

Iwama S. 1977. A influência dos fatores climáticos na atividade externa de *Tetragonisca angustula* (Apidae, Meliponinae). Bol. Zool. Univ. S. Paulo 2, 189–201.

Juliani, L. (1967) A descrição do ninho e alguns dados biológicos sobre a abelha *Plebeia julianii* Moure. 1962 (Hymenoptera, Apidae). Revista Brasileira de Entomologia 12: 1-58.

Kajobe R, Echazarreta C. 2005. Temporal resource partitioning and climatological influences on colony flight and foraging of stingless bees (Apidae, Meliponini) in Ugandan tropical forests. African Journal of Ecology 43(3): 267–275.

Kleinert-Giovanini, A. (1982). The influence of climatic factors on flight activity of *Plebeia emerina* Friese (Hymenoptera, Apidae, Meliponinae) in winter. Revista Brasileira de Entomologia 26: 1–13.

Macías-Macías JO, Quezada-Euán JJ. 2008. Comportamiento de pecoreo de *Melipona colimana* (Hymenoptera: Meliponini) en zonas de montaña del Sur de Jalisco, México. Cartel del V Congreso Mesoamericano de Meliponicultura, Mérida, Yucatán.

Molet M, Chittka L, Stelzer R J, Streit S, Raine NE. 2008. Colony nutritional status modulates worker responses to foraging recruitment pheromone in the bumblebee *Bombus terrestris*. Behavioral Ecology and Sociobiology 62: 1919–1926.

Moo-Valle H, Quezada-Euán J J, Wenseleers T. 2001. The effect of food reserves on the production of sexual offspring in the stingless bee *Melipona beecheii* (Apidae,Meliponini). Insectes Sociaux 48: 398–403.

Nunes-Silva P, Hilário SD, Imperatriz-Fonseca V L. 2010. Foraging Activity in *Plebeia remota*, a Stingless Bees Species, Is Influenced by the Reproductive State of a Colony. Psyche 16: 1–17.

Pankiw T, Page R, Fondrk M. 1998. Brood pheromone stimulates pollen foraging in honey bees (*Apis mellifera*). Behavioral Ecology and Sociobiology 44: 193–198.

Pankiw T. 2004. Brood pheromone regulates foraging activity of honey bees (Hymenoptera: Apidae). Journal of Economic Entomology 97: 748–751

Pankiw T. 2007. Brood pheromone modulation of pollen forager turnaround time in the honey bee (*Apis mellifera* L.). Journal of Insect Behavior 20: 173–180.

Peat J, Goulson D. 2005. Effects of experience and weather on foraging rate and pollen versus nectar collection in the bumblebee, *Bombus terrestris*. Behavioral Ecology and Sociobiology 58: 152–156.

Pierrot LM, Schlindwein C. 2003. Variation in daily flight activity and foraging patterns in colonies of uruçu – *Melipona scutellaris* Latreille (Apidae, Meliponini). Revista Brasileira de Zoologia 20: 565–571.

Pleasants JM, Zimmerman M. 1979. Patchiness in the dispersion of nectar resources: evidence for hot and cold spots. Oecologia 41: 283–288.

Porter-Bolland L. 2001. Landscape Ecology of Apiculture in the Maya Area of La Montaña, Campeche, México. PhD Thesis. University of Florida, Florida, USA. 310 pp.

Possingham HP. 1989. The Distribution and Abundance of Resources Encountered by a Forager. The American Naturalist 133: 42-60.

Rathcke BJ, Jules ES. 1993. Habitat fragmentation and plant-pollinator interactions. Current Science 65: 273–277.

Roubik D. 1982. Seasonality in colony food storage, brood production and adult survivorship, studies of *Melipona* in tropical forest (Hymenoptera: Apidae). Journal of Kansas Entomology Society 55: 789–800.

Roubik DW. 1989. Ecology and natural history of tropical bees. Cambridge University Press, New York, USA. 514 pp.

Roubik DW, Moreno JM, Vergara C, Wittmann D. 1986. Sporadic food competition with the African honey bee: projected impact on Neotropical social bees. Journal of Tropical Ecology 2 (2): 97–11.

Souza B, Carvalho C, Alves R. 2006. Flight activity of *Melipona asilvai* Moure (Hymenoptera: Apidae). Brazilian Journal of Biology 66: 731–737.

Todd FE, Reed CB. 1970. Brood measurement as a valid index to the value of honey bees as pollinators. Journal of Economic Entomology 63: 148–149.

Van Veen JW, Arce HG, Sommeijer MJ. 2004. Production of queens and drones in *Melipona beecheii* (Meliponini) in relation to colony development and resource availability. Proceedings of the Netherlands Entomological Society Meeting 15: 35-39.

Villanueva-Gutiérrez R. 2002. Polliniferous plants and foraging strategies of *Apis mellifera* (Hymenoptera: Apidae) in the Yucatán Peninsula, Mexico. Revista de Biología Tropical 50: 1035–1044.

Villanueva-Gutiérrez R, Roubik DW. 2004. Why are African honey bees and not European bees invasive? Pollen diet diversity in community experiments. Apidologie 35: 481–491.

Villanueva-Gutiérrez R, Roubik D W, Porter-Bolland L. 2015. Bee-plant interactions: competition and phenology of flowers visited by bees. 131-152. In: Islebe, G A, Calmé S, León-Cortéz J L, Schmook B, eds. Biodiversity and conservation of the Yucatán Peninsula. Springer International Publishing Switzerland. 401 pp.

Winston ML. 1980. Seasonal patterns of brood rearing and worker longevity in colonies of the Africanized honey bee (Hymenoptera: Apidae) in South America. Journal of Kansas Entomology Society 53: 157-165.

Winston, ML. 1987. The biology of the honey bee. Harvard University Press; Cambridge, Massachusetts. 281 pp.

Williams NL, Kremen C. 2007. Resource distributions among habitats determine solitary bee off-spring production in a mosaic landscape. Ecological Applications 17: 910–921.

Woyke J. 1992. Diurnal flight activity of African bees *Apis mellifera adansonii* in different seasons and zones of Ghana. Apidologie 23: 107–117.

Zimmerman M. 1981. Patchiness in the dispersion of nectar resources: Probable causes. Oecologia 49: 154–157.

Crop Pollination by Stingless Bees

Virginia Meléndez Ramírez, Ricardo Ayala,
and Hugo Delfín González

11.1 Introduction

Bees are important pollinators of flowering plants both wild and cultivated. They are essential to terrestrial ecosystems because they maintain the fundamental ecological processes involved in plant reproduction and are vital to food-producing agricultural systems (Ayala et al. 1996; Kearns et al. 1998; Michener 2007; Potts et al. 2010). Bees pollinate in most cultivated fruit- and seed-producing plants, many fiber crops such as cotton and flax, numerous medicinal plants, and principal forage species such as alfalfa and clover. By far their most valuable activity is the pollination of flowering plants. This becomes obvious when considering their functions in terrestrial ecosystems, particularly in the tropics where many flowering plants are bee pollinated (Ayala 2004; Michener 2007; Ollerton et al. 2011; FAO 2014).

Pollination is an environmental service required to maintain biodiversity. Of consequently great value to humanity, it has only recently been acknowledged (Kearns et al. 1998; FAO 2014, 2015; IPBES 2016). It may be in danger of disappearing in different regions as agricultural intensification has led to habitat loss and fragmentation, with significant impacts on biodiversity and pollinators in particular. A number of studies have linked loss of pollinator species to landscape changes such as the transition of native forests to other land uses, which causes the loss of resources important for bee species (e.g., pollen, nectar, oils, and nesting sites) (Ricketts et al. 2008; Brosi 2009; Meneses et al. 2010; Meléndez et al. 2013; Patricio and Campos 2014). Pollinators are also affected by common agricultural practices such as chemical (pesticide and herbicide) use, invasion of exotic plants and animals, and even the arrival of nonnative bees from other regions or continents, which can displace native bees and other pollinators (Kearns et al. 1998; Pinkus et al. 2005; Grajales et al. 2013).

The use of honey bees (*Apis mellifera*) for crop pollination is common in many countries even though they may not be native to a region. Recent and dramatic declines in the number of honey bee colonies due to disease and parasites, mostly in the United States and Europe, highlight the need for healthy pollination systems and for innovation in the management of bees and other crop-pollinating native animals (O'Toole 1993; Kearns et al. 1998; Meléndez et al. 2002).

V. Meléndez Ramírez (✉) • H. Delfín González
Departamento de Zoología, Campus de Ciencias
Biológicas y Agropecuarias, Universidad Autónoma
de Yucatán, AP 4-116, Col. Itzimná, 97100 Mérida,
Yucatan, Mexico
e-mail: virginia.melendez@correo.uady.mx

R. Ayala
Estación de Biología Chamela, Instituto de Biología,
Universidad Nacional Autónoma de México (UNAM),
AP 21, 48980 San Patricio, Jalisco, Mexico

© Springer International Publishing AG, part of Springer Nature 2018
P. Vit et al. (eds.), *Pot-Pollen in Stingless Bee Melittology*, DOI 10.1007/978-3-319-61839-5_11

In the tropics and subtropics, stingless bees are very diverse, consisting of nearly 600 species (Michener 2013). They are also very frequent visitors to plants in natural ecosystems and agro-ecosystems (Roubik 1992; Ayala 1999; Heard 1999; Meléndez et al. 2002), as well as effective crop pollinators (Heard 1999; Meléndez et al. 2000; Slaa et al. 2006; Eka et al. 2014). Their morphological, biological, and behavioral traits make them efficient crop pollinators. Stingless bee species differ widely in body size, number of individuals in a colony, and feeding strategies. Because their habitats are largely tropical, they can feed year-round and form perennial colonies (Roubik 2006; Slaa et al. 2006; Michener 2013).

Stingless bee morphology is adapted to the flowers on which they feed, with specialized structures for pollen and nectar collection. Like other social insects, they have special mechanisms for communication with individuals of the same species. In a colony's queen coordinates, functions and the older individual workers collect pollen and nectar. These bees are generalist foragers at a colony level, although at the individual level, they visit the same plant species, which is called floral constancy (Eickwort and Ginsberg 1980).

The generalist feeding habits and floral constancy suggest the profitable use of stingless species in field and greenhouse crops, a promising option for pollination efficiency (Slaa et al. 2006). A number of crop species can be pollinated by stingless bees; indeed, they are known to be effective pollinators of 18 crops and contribute to pollination of >60 agricultural plant species (Heard 1999; Meléndez et al. 2002; Slaa et al. 2006). The wide diversity of stingless bee species allows for selection of the most appropriate species for a crop and/or crop system, and they can be maintained and managed for trade (Cortopassi-Laurino et al. 2006).

A number of recent studies have addressed pollination with stingless bees. However, research is needed into the pollination efficiency of stingless bees in traditional and economically important crops. This chapter reviews biological traits of stingless bees that implicate them as the bee species with great potential for pollination in field and greenhouse crops, both in the tropics and other regions. We also emphasize the importance of habitat management to preserve local stingless bee populations.

11.2 Characteristics of Stingless Bees as Pollinators

As other bees in subfamily Apinae, stingless female bees exhibit a corbicula on the posterior tibia (the exceptions are queens, workers in cleptobiotic and obligate necrophage species, and in some genera the corbicula can be very small). This appears as a large smooth, often concave zone, with a fringe of long hairs that function to carry pollen and other substances to the colony (Michener 2013). This structure allows carrying to the bee colony great amount of pollen and at the same time leaves it between the flowers.

The stingless bees have no sting, or a rudimentary one, but by using other strategies, primarily biting or nest site selection, they can defend the nest (Roubik 2006; Michener 2007). The lack of stinging behavior allows for much easier handling of stingless bees than most honey bees. In addition, as part of their foraging behavior, some stingless bees, such as the genus *Melipona*, vibrate ("buzz") their bodies to extract pollen. This is vital for crops with poricidal anthers in their flowers (e.g., tomato or pepper) which usually require buzzing for pollination (Buchmann 1983; Michener 2007; De Luca and Vallejo-Marín 2013). Thus, research on vibration pollination, mainly of its mechanism, is important for the identification of commercially viable native pollinators for agricultural crops such as tomato (Nunes-Silva et al. 2010).

Plant communities benefit from pollinators with an ability to forage on various plant species. Stingless bees are true generalists and exhibit high coincidence in the floral resources they use (Heithaus 1979; Roubik 1992; Biesmeijer and Slaa 2006; Roubik and Moreno, in this volume). However, for some stingless bee species, the term generalist may be more qualitative than quantitative since a number of studies have that record quantitative pollen data demonstrate the high relative importance of different plant species for these bees (Roubik and Moreno 2013; Villanueva-Gutiérrez and Roubik 2004, and Roubik and Moreno, in this volume).

Stingless bees are social, implying that they consume large amounts of food, to constantly feed the brood and maintain their activity. Pollen is the main protein source for adults, for feeding larvae, and especially for the egg-producing female (the queen). Sugars from nectar are the main carbohydrate source for adults and contain amino acids, which contribute to nitrogen metabolism. Nectar is consumed as an energy source by adults and is mixed with pollen to make food for larvae (Michener 2007). To collect the food resources necessary for a colony, stingless bees forage at different distances depending on the size of individuals of the different species. However, many species can occupy a much smaller effective foraging range, <0.5 km, in response to other variables such as food demand and behavior linked to specialization in the search for specific floral resources, orientation ability, and utilization of odor trail on substrates, among other factors (Araújo et al. 2004). Their generally short foraging ranges facilitate the management of stingless bees in implementing pollination services in crops near natural habitats, or crops in zones requiring restoration, or if necessary placement of colonies near to the crop.

Pollinator occurrence and abundance are essential for many flowering plants (unless they are selfing, apomictic, or parthenocarpic). Stingless bee biology is particularly suited to pollination since they form colonies consisting of dozens, hundreds, and even thousands of workers. Being eusocial, they generally have only one reproductive queen per colony, which differs from the workers (all female) and is normally larger in size. In a stingless bee colony, social organization is controlled by the queen releasing pheromones that regulate the coordinated work of individuals to ensure nest survival. Males are similar to workers in size and appearance, but their main function is reproduction; very few males can be present in a colony at any given time (Michener 1974, 2013). Workers represent the largest number of individuals in a colony and are responsible for nest building,

resource collection, feeding (of the queen and larvae), and colony defense (Wille 1983). Food resources are stored in containers called pots, made from soft cerumen, and the pollen is stored separately from the honey (obtained from nectar). Most stingless bees have laying workers which commonly feed the queen by laying trophic worker eggs, and sometimes the workers can produce males by the laying of reproductive worker eggs (Sommeijer et al. 2003).

Stingless bees have complex communication systems that allow them to be used to study recruitment evolution. It is difficult to generalize about them because they are a diverse group that implements diverse strategies to attain the objective of recruiting nest members to visit food sources (Friedrich et al. 2008; Nieh 2004; Sánchez and Valdame 2013). *Scaptotrigona postica* recruits to a specific three-dimensional location by placing a trace of the odor, but some species of the genus *Melipona* have a recruiting system that communicates the distance through the sounds, within the nest and the direction with zigzag flights out of the nest. It has been observed in *M. panamica*, its forage can recruit to a specific direction, distance, and also canopy level (Nieh and Roubik 1995); they use optical flow information to measure not only distance traveled but also height above ground (Eckles et al. 2012). It is suggested that some stingless species have a more successful recruitment system than honey bees (Jarau et al. 2003). Nowadays, a number of studies have been done on the diversity of sensory and behavioral adaptations in stingless bees, especially on the sensory basis for foraging, communication chemistry, and the ecology of food search behavior (Hrncir et al. 2016).

11.3 Field Crop Pollination by Stingless Bees

Most flowering plants must be pollinated to produce fruits and seeds. Some can self-pollinate, but endogamic depression often results. Increasing fruit and seed production and quality is best achieved through pollinators, which is also vital to the genetic diversity needed to improve varieties of cultivated plants with fruits with better features (Roubik 1995; Michener 2007; Klein et al. 2007). Even plants mainly pollinated by wind produce more seeds when pollinated by bees (Sihag 1995).

Research on the use of honey bees in the pollination of cultivated plants in temperate zones was first documented and compiled in the 1970s (Free 1970; McGregor 1976). In contrast, data on cultivated plants in the tropics remained scattered until about 20 years ago when Roubik (1995) compiled data on the origin, use, reproductive system, and pollinators of 1330 tropical plant species. This mammoth effort included plants grown for medicinal ends, resins, tubers, fruit, cereals, forage, vegetables, spices, fibers, nuts, wood, pesticides, dyes, essences, oils, elastomers, waxes, sweeteners, drinks, tannins, erosion prevention, and genetic improvement. Nearly 70% of the documented plants were pollinated by animals and many by different bee species, including stingless bees.

A more recent detailed review addressing crop pollination by stingless bees found that they have been recorded as floral visitors in almost 90 of cultivated plants (Heard 1999). They were effective and important pollinators in nine crops but contributed to pollination in approximately 60 crops. A renovated review of the crops pollinated by stingless bees reported that 18 crops were effectively pollinated by them and that in 11 crops they have been used as pollinators under greenhouse conditions (Slaa et al. 2006).

At the present time, in a study done in Indonesia comparing the pollination efficiency of *Trigona laeviceps* and *Apis cerana* in fields of chili (*Capsicum annuum*), both species were found to improve fruit production per plant, fruit development, and quality (weight and size). Although both species would be effective pollination agents, the stingless *T. laeviceps* is more likely to be used because its small feeding zone can provide a high visitation rate, which is ideal in small-scale agricultural and enclosed conditions. In this study it is emphasized stingless bee has no functional sting and is less aggressive than honey bees, making it more appropriate for crop pollination near human settlements (Eka et al. 2014).

Nowadays, field research has documented stingless bees as field crop pollinators in 12 tropical countries worldwide. More than 25 species are known to contribute significantly to pollination in 14 economically important crops belonging to 12 different plant families (Table 11.1) (see Heard 1999; Slaa et al. 2006 and references therein for details on the crops and their pollinating species). The most common stingless bee species belong to the *Melipona* and *Trigona* genera. Genus *Trigona* contains the highest number of species (94) of Meliponini and for *Melipona* is less (72) but also with several species (Michener 2013). These are very conservative estimates, since verified species of Meliponini are now over 526, with dozens more known and being described (Rasmussen and Gonzalez 2013).

The high diversity of stingless bees in the tropics facilitates their use in field crop pollination. By planning habitat management in different locations, native bees can be used for this service, which is clearly beneficial in terms of biodiversity conservation. In areas where wild pollinators are absent, stingless bee colonies could be established and managed, and different stingless bee species combined to increase pollination efficiency.

A new study addressed stingless bee pollination of field crops in Bahia, Brazil, where Africanized honey bees (*A. mellifera*) are commonly used to increase pollination success in apple (*Malus domestica*) orchards. An evaluation was done of the effect on pollination and fruit and seed production of different hive densities of honey bee and stingless bee (*Melipona quadrifasciata anthidioides*). Using honey bees alone, no increase was observed in seed production at higher hive densities. It was concluded that the optimum strategy would be to use a combination of the stingless species at 12 hives ha^{-1} and the honey bees at seven hives ha^{-1}. This would improve apple blossom pollination since the presence of both bee species increases fruit and seed production (Viana et al. 2014). Application of this strategy would require evaluations of the number of stingless bee hives required in individual situations and whether additional honey bee hives would be necessary. When using native species from the area, it would also serve for ecological restoration programs.

Table 11.1 Stingless bees with important contribution in the pollination of the field crops

Crop	Plant species	Plant family	Bee species	Locality	Reference
Annatto, achiote	*Bixa orellana*	Bixaceae	*Melipona melanoventer*	Amazonas	Maues and Venturieri (1995)
			Melipona fuliginosa	Many regions	Wille (1976)
			Hypotrigona pothieri	Ivory Coast	Lobreau-Callen et al. (1990)
			Melipona seminigra merrillae	Amazonas	Absy and Kerr (1977)
			Melipona rufiventris	Amazonas	Absy et al. (1984)
Apple	*Malus domestica*	Rosaceae	*Melipona quadrifasciata anthidioides*	Brazil	Viana et al. (2014)
Avocado	*Persea americana*	Lauraceae	*Trigona nigra, Nannotrigona perilampoides*	Mexico	Can-Alonso et al. (2005)
			Geotrigona acapulconis, Trigona nigerrima, Partamona bilineata, Scaptotrigona pectoralis, Trigona nigra, Scaptotrigona mexicana, Trigona fulviventris, Plebeia frontalis	Mexico	Ish-Am et al. (1999)
Camu camu	*Myrciaria dubia*	Myrtaceae	*Melipona* sp., *Scaptotrigona postica*	Peru Amazonian	Peters and Vasquez (1986)
Carambola, starfruit	*Averrhoa carambola*	Oxalidaceae	*Melipona favosa*	Surinam	Engel and Dingemans-Bakels (1980)
Chayote, choko	*Sechium edule*	Cucurbitaceae	*Trigona corvina* and *Partamona cupira*	Costa Rica	Wille et al. (1983)
Chili	*Capsicum annuum*	Solanaceae	*Trigona laeviceps*	Indonesia	Eka et al. (2014)
Coconut	*Cocos nucifera*	Arecaceae	*Melipona* spp.	Costa Rica	Hedström (1986)
			Stingless bees	Surinam	Engel and Dingemans-Bakels (1980)
				Mexico	Meléndez et al. (2004)
Coffee	*Coffea arabica*	Rubiaceae	*Lepidotrigona terminata*	Indonesia	Klein et al. (2003a, b)
	Coffea canephora				
Cupuassu	*Theobroma grandiflorum*	Malvaceae	*Plebeia minima*	Brazil, Belem	Venturieri (1994)
			Trigona clavipes and *Trigona lurida*	Brazil Manaus	In Heard (1999)
Macadamia	*Macadamia integrifolia*	Proteaceae	*Tetragonula carbonaria*	Australia eastern	Heard and Exley (1994)
				Costa Rica	Masís and Lezama (1991)
Mango	*Mangifera indica*	Anacardiaceae	Meliponini and Diptera	Brazil	Cortopassi-Laurino et al. (1991), Iwama and Melhem (1979), Simão and Maranhão (1959)
			Meliponini	India	Singh (1989)
			Meliponini	Australia	Anderson et al. (1982)
			Tetragonisca angustula	Mexico	Sosa-Nájera et al. (1994)
Mapati, uvilla, Amazon tree grape	*Pourouma cecropiifolia*	Urticaceae	Meliponini	Brazil	Falcão and Lleras (1980)
Pumpkin	*Cucurbita moschata*	Cucurbitaceae	*Partamona bilineata*	Mexico	Meléndez et al. (2002)

11.4 Greenhouse Crop Pollination by Stingless Bees

Crop pollination by bees in greenhouses is common in many parts of the world. Some species of *Bombus* or *Apis mellifera* (honey bees) are used as pollinators. This represents an additional operating cost for producers and involves introducing exotic species and possible consequent disease contagion in native bee species (Torres-Ruiz et al. 2013). *A. mellifera* are not necessarily the most efficient pollinators of greenhouse crops (e.g., tomato; Banda and Paxton 1991), and their hives can sometimes be lost (Nicodemo et al. 2013). Evaluations are therefore needed of the pollination efficiency of native bees in greenhouses at a local level. Research has shown that stingless bees are an excellent option for pollinating a number of greenhouse crops.

Pollination by stingless bees is reported to be better than self-pollination in greenhouse crops such as habanero pepper (Cahuich et al. 2006) and bell pepper (Oliveira et al. 2005). In the case of habanero peppers (*Capsicum chinense*), some flower fecundation did occur without bee visits but was 71% higher with just one visit by *Nannotrigona perilampoides* and 112% higher with unrestricted visits. Bell peppers (*Capsicum annuum*) benefit from pollination by *Melipona subnitida*, which results in fruits that are significantly heavier, wider, and of better quality (fewer malformed fruits) and containing more seeds (Oliveira et al. 2005). Both *N. perilampoides* and *M. subnitida* can therefore be considered effective pollinators of pepper crops in greenhouses.

In Yucatan, Mexico, two stingless bee species have been proposed as pollinators in greenhouse conditions: *Partamona bilineata* for squash *Cucurbita moschata* and *N. perilampoides* for tomato *Solanum esculentum* and habanero pepper *Capsicum chinense*. These recommendations are based on research done in greenhouses and experimental fields. In the evaluation of *P. bilineata* and *A. mellifera* as pollinators of squash in greenhouses and open fields (control with various bee species; Meléndez et al. 2000), all the treatments involving bees (open fields, greenhouses with *P. bilineata*, and greenhouses with *A. mellifera*) resulted in greater fruit and seed production than self-pollination in greenhouses (control). Fruit quality (size and weight) was highest in the open parcels and greenhouses with *P. bilineata* treatments.

Considering the pollination effectiveness of the stingless bee *N. perilampoides* in greenhouse tomatoes (Cauich et al. 2004), the greatest effect was found with *N. perilampoides* and manual vibration compared to a control with no bees or vibration. This stingless bee species has been proposed as an alternative pollinator for greenhouse tomatoes in tropical climates instead of the highly defensive Africanized *A. mellifera* or nonnative bumblebees (i.e., *Bombus* spp). Further research is still needed to evaluate the cost/benefit ratios of large-scale use of *N. perilampoides* and/or other stingless bee species for pollination in greenhouses, versus mechanical pollination methods.

The effect of *N. perilampoides* as a pollinator of habanero pepper has also been studied (Cahuich et al. 2006). Fruit quality and productivity (kg m⁻²) did not differ between *N. perilampoides* worker visits and mechanical vibration treatments, although fruit quality and seed per fruit were lower in the latter treatment. Bee pollination increased fruit number and quality and produced more seeds, probably due to greater transfer of pollen grains by the bees. The use of *N. perilampoides* was deemed a better pollination alternative than mechanical vibration for habanero pepper grown in greenhouses in tropical climates.

In another study using *N. perilampoides*, tomato fruit production in greenhouses was evaluated using one *N. perilampoides* colony, one *Bombus impatiens* colony, or manual vibration (Palma et al. 2008a). Fruit production and quality and seed content had higher values in the *B. impatiens* treatment than the *N. perilampoides* treatment, while both these treatments were superior to the mechanical vibration treatment. Tomato plant density was high, and the *B. impatiens* colony provided more pollination than the *N. perilampoides* colony. Pollination by *N. perilampoides* could have had limitations, since, compared to *B. impatiens*, its colony had a smaller number of foragers for flower visits and shorter flight ranges. One suggested solution is to increase the *N. perilampoides* colony density to improve tomato yield in tropical greenhouses.

When the same type of study was done for habanero pepper grown in greenhouses, fruit quality was better with pollination by *N. perilampoides* and *B. impatiens* when compared to mechanical vibration (Palma et al. 2008b). However, fruit abortion rate was higher and fruit quality lower in the *B. impatiens* treatment compared to the *N. perilampoides* treatment, probably due to excessive flower visits by the bumblebee species. Pollination efficiency did not differ between the two species, despite the lower amount of pollen moved by *N. perilampoides*. It may be that the higher number of floral visits per forage trip together with adequate pollen transport loads and interfloral pollen transfer explain the superior pollination by *N. perilampoides* in greenhouse habanero pepper crops. This stingless bee species can therefore be seen as a viable alternative to *B. impatiens* in greenhouses in the tropics, where it naturally occurs.

Research has also been done in Brazil on the performance of stingless bees as greenhouse crop pollinators. In the state of Minas Gerais, a study evaluating *Melipona quadrifasciata* as a pollinator for greenhouse tomatoes reports that its use is feasible, but techniques for captive husbandry or colony multiplication would be needed to prevent serious depletion of wild populations for management schemes (Sarto et al. 2005). In Ribeirão Preto, Sao Paulo state, *M. quadrifasciata* in greenhouse tomato pollination is compared to *A. mellifera* by Santos et al. (2009). The use of *M. quadrifasciata* results in production of more fruit of better quality and with higher seed counts compared to *A. mellifera* and a control with no bees. The *A. mellifera* treatment produces fruit of the same size and weight as the control, highlighting the more efficient pollination service of *M. quadrifasciata* under the experimental conditions. In the same state, *M. fasciculata* was an efficient pollinator of eggplants, increased the fruit set by 29.5% and fruit quality (measured as fruit weight) compared to the control group, and may be a viable alternative to bumblebees in Brazil (Nunes-Silva et al. 2013).

In Australia, the contribution of stingless bees to pollination of commercially important field crops is known to be significant. To evaluate their potential under greenhouse conditions, a study was done of *Austroplebeia australis* and *Tetragonula carbonaria* in the pollination of Aries chili pepper (Greco et al. 2011). Compared to a control without bees, use of these two stingless species has the potential to increase fruit yield and quality, although the effects are inconsistent across three trials. Future studies will need to determine optimum colony density per flower, in order to maximize results.

11.5 Stingless Bee Management Under Greenhouse Conditions

Research into crop pollination with stingless bees has also generated useful data on improving management of colonies under greenhouse conditions. Adjustment of bees to the greenhouse environment is vital to their adequate performance. It can be measured in three ways: by the number of "lost bees" (those that do not return to the colony), by comparing colony population (larvae and adults) over time, and by foraging activity during the day (i.e., number of collectors per unit time, Cauich et al. 2004). Another illustrative and important variable in evaluation of stingless bees in greenhouse conditions is hive weight and larvae volume. These are indicative both of successful plant pollination and visitation, since they relate to pollination at sufficient rates (e.g., *A. australis* and *T. carbonaria*, Greco et al. 2011).

In a study of the stingless species *P. bilineata* in small greenhouse squash cultivation, the colonies foraged well for approximately 2 weeks and then had to be replaced occasionally (Meléndez 1997). With *N. perilampoides* in greenhouse tomato cultivation, the population decreased 25–33% throughout the experimental period but the colonies persisted. Both species began foraging from 5 to 9 days after introduction to the greenhouse. No correlation was identified between environmental variables and bee foraging activity, although water collection correlated positively with greenhouse temperature and negatively with humidity (Cauich et al. 2004).

Foraging activity results for *N. perilampoides* and *B. impatiens* in greenhouse tomato cultivation indicated that both species pollinated flowers (Palma et al. 2008a). However, individual *B. impatiens* collected pollen more rapidly, visited more flowers, and made more visits per flower than *N. perilampoides*, which spent more time on the flower. Correlations were found between environmental variables and the number of bees in a colony and the number of bees in flowers. The highest correlation for *N. perilampoides* was for light intensity, whereas in *B. impatiens* temperature had a negative effect. In another study comparing *N. perilampoides* and *B. impatiens* (Palma et al. 2008b), both species collected pollen at a similar number of flowers per time unit, but *N. perilampoides* visited more flowers per trip, spent more time on each flower, and used more time in feeding trips. Temperature and light intensity are correlated with foraging activity in *N. perilampoides*, but light intensity proves to be important for *B. impatiens*. Considering these results, use of stingless bees in greenhouses will require definition of the most adequate ranges of environmental conditions in each region and season and perhaps even for each crop.

Comparisons are also needed of each stingless bee species' pollination potential for any given crop since quite different foraging behavior exists between species. For example, in closed greenhouses in the state of Carabobo, Venezuela, the three stingless species *Tetragonisca angustula*, *Nannotrigona testaceicornis*, and *Plebeia* sp. did not pollinate tomato flowers but did pollinate cucumber (*Cucumis sativus*) and squash (*Cucurbita pepo*). All three species adapted to greenhouse conditions, with *T. angustula* being the least ill adapted. When using the same three stingless bee species under field conditions, none pollinated tomato flowers, but in cucumber and squash flowers, *N. testaceicornis* had a higher frequency as pollinator of flowers than *A. mellifera* and *T. angustula*. The two stingless bee species were promising pollinators under greenhouse and field conditions, even in the presence of *A. mellifera*. Variable flower visit frequency

has also been reported in a study of chili pepper in Australia in which *Austroplebeia australis* frequently foraged flowers, while *T. carbonaria* did so only occasionally (Greco et al. 2011).

The need to determine which stingless bee species are the most appropriate for a crop has been confirmed in two studies performed in Brazil. Using stingless bees and *A. mellifera* in greenhouse cultivation of parthenocarpic cucumber (*C. sativus*) in Ribeirão Preto, fruit production was 19.2% higher with *N. testaceicornis* than in the control without bees, the *A. mellifera* colonies were lost, and *T. angustula* did not visit crop flowers (Nicodemo et al. 2013). Another study evaluated the performance of the *Melipona subnitida* and *Scaptotrigona* sp. in diploid and triploid genotypes of miniature watermelon (*Citrullus lanatus*) under greenhouse conditions (Bomfim et al. 2014). *M. subnitida* clearly did not adapt to the greenhouse: it exhibited no interest in the crop, engaged in no foraging, and laid no eggs (it entered into a stage similar to diapause). In contrast, the *Scaptotrigona* adapted quite well, foraged, and collected nectar during visits to both flower types (stamens and pistils) and to flowers of both genotypes (seeded and seedless). This stingless bee species has potential applications in the pollination of greenhouse-cultivated miniature watermelons.

It is known that some stingless bee species can cause necrosis of the styles, which has been observed in *A. australis* (Greco et al. 2011) and *N. perilampoides* (Cauich et al. 2004; Palma et al. 2008b). This can be attributed to their foraging strategies in that they visit flowers in groups, but apparently it does not negatively affect pollen deposition on the stigmas or fruit and seed production. The same phenomenon also occurs with large bees such as *B. impatiens* (Palma et al. 2008b).

In summary, studies of stingless bee pollination in enclosed or greenhouse conditions have been conducted in eight countries and focused on ten crops from five families, including bell peppers and habanero chili peppers. Fourteen stingless bee species have been used most often in these studies, mostly from the genera *Melipona*, *Nannotrigona*, *Scaptotrigona*, *Tetragonula*, and *Tetrigona* (Table 11.2). In most of these studies,

Table 11.2 Studies reported on pollination of stingless bees under enclosed or greenhouse conditions

Crop	Plant species	Plant family	Bee species	Locality	Reference
Blue salvia	*Salvia farinacea*	Lamiaceae	*Nannotrigona perilampoides Tetragonisca angustula*	Costa Rica	Slaa et al. (2000a, b)
Cucumber	*Cucumis sativus (parthenocarpic)*[a]	Cucurbitaceae	*Scaptotrigona* aff. *depilis Nannotrigona testaceicornis*[a]	Brazil	Santos (2004b) Nicodemo et al. (2013)[a]
Eggplant	*Solanum melongena*	Solanaceae	*Melipona fasciculata*	Brazil	Nunes-Silva et al. (2013)
Green pepper, Habanero chili[b]	*Capsicum annuum*	Solanaceae	*Melipona subnitida* *Tetragonula carbonaria* *Nannotrigona perilampoides*[b]	Kenya Australia Mexico[b]	Kiatoko et al. (2014) Occhiuzzi (2000) Greco et al. (2011) Cahuich et al. (2006)[b] Palma et al. (2008b)[b]
Pumpkin	*Cucurbita moschata*	Cucurbitaceae	*Partamona bilineata*	Mexico	Meléndez et al. (2000)
Rambutan	*Nephelium lappaceum*	Sapindaceae	*Scaptotrigona mexicana + Tetragonisca angustula*	México	Rabanales (in Slaa et al. 2006)
Strawberry	*Fragaria x ananassa*	Rosaceae	*Plebeia tobagoensis*	Netherlands	Asiko (2004) Lalama (2001)
			Trigona minangkabau	Japan	Kakutani et al. (1993)
			Nannotrigona testaceicornis	Japan	Maeta et al. (1992)
			Tetragonisca angustula	Brazil	Malagodi-Braga and Kleinert (2004)
Sweet pepper	*Capsicum annuum*	Solanaceae	*Melipona favosa*	Netherlands	Meeuwsen (2000) Cruz et al. (2005)
			Melipona subnitida	Brazil	Cruz et al. (2005)
Tomato	*Lycopersicon esculentum*	Solanaceae	*Melipona quadrifasciata*	Brazil	Santos et al. (2004a, 2005)
			Nannotrigona perilampoides	Mexico	Cauich et al. (2004) Palma et al. (2008a)
Watermelon	*Citrullus lanatus*	Cucurbitaceae	*Scaptotrigona* sp. nov.	Brazil	Bomfim et al. (2014)

The superscripts indicate the type of plant (a) and species of bees or relationship with the common name of the crop (b), the species of bees, country, and reference

the evaluated stingless bee species were found to be efficient crop pollinators both in field crops and greenhouses. The use of local native stingless bees as pollinators is promising since in some instances they are more efficient than honey bees; it reduces the risk of the invasion of exotic species and the displacement of native ones.

11.6 Perspectives

Stingless bee diversity and abundance in tropical and subtropical regions provide the opportunity to sustainably utilize them in crop pollination, if their husbandry and conservation are implemented successfully. Their services as crop pollinators have clear economic benefits, but further research is needed to understand their possible contribution in crops not yet evaluated. Studies are also needed of the natural habitats and critical resources of stingless bees and other wild bees in their native ecosystems. These are critical components of conservation through habitat management planning, for example, conservation of areas with the required habitat, including floral resources and nesting sites. In this scenario, conservation of forest areas near orchards or fields could conserve wild populations while ensuring adequate populations of the stingless and other bee species that so effectively pollinate agricultural crops (Meléndez et al. 2002).

The absence of natural vegetation in agricultural fields is, not surprisingly, associated with declines in local stingless bee populations (Brown and Albrecht 2001; Meneses et al. 2010; Meléndez et al. 2013). Other human activities and natural phenomena such as hurricanes can transform or degrade ecosystems, with consequential pressures on pollinators (Meléndez et al. 2016). If multiple pressures interact, pollination can be interrupted, threatening both economically important crops and the ecological interactions vital to all terrestrial biotic communities (Potts et al. 2010; FAO 2015). Therefore, protecting stingless bees and other pollinators will require more effective short-term wildlife conservation programs and ecological restoration programs for habitat recovery.

Promotion of stingless bees as effective pollinators of field crops could help to revitalize the practice of meliponiculture, generating additional income for local human communities in tropical regions, as well as revive traditional cultural practices of indigenous cultures related to the use of Meliponini. In Mesoamerica, for example, stingless bees and ancient cultures in the region had a complex relationship in which humans exhibited a deep respect for certain bee species. Indeed, native stingless bees had a presence in their cosmologies and strongly influenced these cultures' spiritual life (Ayala et al. 2013).

Handling these bees will need to be improved to increase the availability of stingless bee hives and thus reduce dependence on wild populations. Further knowledge can be generated to identify and develop pollinators fitted to the needs of specific crops and habitats, because stingless bees have different feeding strategies and foraging behavior (Heard 1999). As already mentioned before a good example of a promising scenario for use of stingless bees as pollinators in greenhouses, because some of this exhibits a vibration (buzzing) behavior that has been shown effective in pollinating greenhouse tomatoes and peppers.

A pivotal measure for conserving stingless bee populations is to avoid introduction of exotic bees. Honey bees have been introduced in almost the entire world for honey production and in some areas for crop pollination, but this eusocial species can displace stingless bee species through floral resource competition (Meléndez 1997; Pinkus et al. 2005; Roubik 2009). Management of honey bees needs to be planned, and in areas with high native bee diversity and abundance, the use of *A. mellifera* and other introduced bees (e.g., *B. impatiens* and *B. terrestris*) should be avoided altogether. For imports of *Bombus* in several countries, an application of the "precautionary principle" when analyzing the feasibility of future introductions would be highly recommended, because of the impact they can have on native pollinators, and consider there are several species of native *Bombus* that could be used for the same purposes (Morales 2007).

Goulson (2003) provides five undesirable effects of the introduction of exotic pollinators into native ecosystems: (a) competition with native flower visitors for floral resources, (b) competition for nesting sites, (c) pathogen transmission to native organisms, (d) changes in native plant seed production, and (e) eventual pollination of nonnative weeds. For a good general discussion of the negative effects of introduced nonnative pollinators, see Morales (2007).

The current tendency toward decline in pollinator colonies (many of them introduced species) responds to multiple causes (e.g., habitat destruction, disease, parasites, pesticides, etc.) (Lever et al. 2014). This could increase interest in native stingless bee species as pollinators for field and greenhouse crops. To raise awareness of stingless bee potential, local and regional promotion is needed of the many positive and advantageous traits that make them such effective pollinators, including that they are native to the tropics, are generalist feeders, are adaptable, exhibit floral constancy, can be domesticated, live in perennial colonies that store food, can survive in conditions of low floral resource availability, have relatively short resource collection ranges, exhibit recruiting behavior, and carry out pollen exchange in the nest (Heard 1999, 2016). With these traits, appropriate habitat management can help to attract the pollination services of wild stingless bee colonies, or they can be managed in hives if necessary. In field crops, the best results are with the contribution of various bee species for flower pollination because each species contributes conspecific pollen grains (i.e., from the same plant species) (Meléndez et al. 2000).

Managing bee hives requires an evaluation of the different densities needed for each crop and per greenhouse. Wild colonies are restricted to subtropical and tropical regions but can be managed in colder climes; for example, stingless bees have successfully pollinated greenhouse crops in nontropical regions such as the Netherlands and Japan (Slaa et al. 2006). In these contexts, stingless bees are essentially exotics and need to be managed very carefully to prevent negative impacts on native bees.

References

Absy ML, Kerr WE. 1977. Algumas plantas visitadas para obtenção de pólen por operariãs de *Melipona seminigra merrillae* em Manaus. Acta Amazonica 7: 309-315.

Absy ML, Camargo JMF, Kerr WE, de A Miranda IP. 1984. Especies de plantas visitadas por Meliponinae (Hymenoptera; Apoidea), para coleta de pólen na regiao do médio Amazonas. Revista Brasileira de Biología 44: 227-237.

Anderson DL, Sedgley M, Short JRT, Allwood AJ. 1982. Insect pollination of mango in Northern Australia. Australian Journal of Agricultural Research 33: 541-548.

Araújo ED, Costa M, Chaud-Netto J, Fowler HG. 2004. Body size and flight distance in stingless bees (Hymenoptera: Meliponini): inference of flight range and possible ecological implications. Brazil Journal of Biology 64: 563-568.

Asiko AG. 2004. The effect of total visitation time and number of visits by pollinators (*Plebeia* sp and *Apis mellifera mellifera*) on the strawberry. MSc Thesis, Utrecht University; Utrecht, The Netherlands. 156 pp.

Ayala R, Griswold T y Yanega D. 1996. Apoidea (Hymenoptera). pp. 423-464. In: Llorente-Bousquets J, Garcia A and González E, eds. Biodiversidad, taxonomía y biogeografía de artrópodos de México: Hacia una síntesis de su conocimiento Universidad Nacional Autónoma de México (UNAM) - Comisión Nacional para el Conocimiento y uso de la Biodiversidad (CONABIO); Distrito Federal, México. 660 pp.

Ayala R. 1999. Revisión de las abejas sin aguijón de México (Hymenoptera: Apidae: Meliponini). Folia Entomológica Mexicana 106: 1-123.

Ayala R. 2004. Fauna de abejas silvestres (Hymenoptera: Apoidea). pp 193-219. In: Garcia-Aldrete AN, Ayala R, eds. Artrópodos de Chamela. Universidad Nacional Autonoma de México; Distrito Federal, México. 227 pp.

Ayala R, González VH y Engel MS. 2013. Mexican Stingless Bees (Hymenoptera: Apidae): Diversity, Distribution, and Indigenous Knowledge. pp. 135-152. In: Vit P, Silvia RMP, Roubik D, eds. Pot honey: A legacy of stingless bees. Springer; New York. 654 pp.

Banda HJ, Paxton RJ. 1991. Pollination of greenhouse tomatoes by bees. Acta Horticola 288: 194-198.

Biesmeijer JC, Slaa EJ. 2006. The structure of eusocial bee assemblages in Brazil. Apidologie 37: 240-258.

Bomfim I, Bezerra De MA, Nunes AC, De Aragão FAS, Freitas BM. 2014. Adaptive and foraging behavior of two stingless bee species (Apidae: Meliponini) in greenhouse mini watermelon pollination. Sociobiology 61: 502-509.

Brosi BJ. 2009. The complex responses of social stingless bees (Apidae: Meliponini) to tropical deforestation. Forest Ecology and Management 258: 1830-1837.

Brown JC, Albrecht C. 2001. The effect of tropical deforestation on stingless bees of the genus *Melipona* (Insecta: Hymenoptera: Apidae: Meliponini) in central Rondonia, Brazil. Journal of Biogeography 28: 623-634.

Buchmann SL. 1983. Buzz Pollination in Angiosperms. 73-113 pp. In: Jones CE, Little RJ, eds. Handbook of Experimental Pollination Biology, Van Nostrand Reinhold; New York. USA. 558 pp.

Can-Alonso C, Quezada-Euán JJG, Xiu-Ancona P, Moo-Valle H, Valdovinos-Nuñez GR, Medina-Peralta S. 2005. Pollination of "criollo' avocados (*Persea americana*) and the behaviour of associated bees in subtropical Mexico, Journal of Apicultural Research 44: 3-8.

Cauich O, Quezada-Euán JJG, Macias-Macias JO, Reyes Oregel V, Medina- Peralta S, Parra. 2004. Behavior and pollination efficiency of *Nannotrigona perilampoides* (Hymenoptera: Meliponini) on greenhouse tomatoes (*Lycopersicon esculentum*) in subtropical México. Journal of Economic Entomology 97: 475-481.

Cahuich O, Quezada-Euan JJG, Meléndez RV, Valdovinos-Nuñez O, Moo-Valle H. 2006. Pollination of habanero pepper (*Capsicum chinense*) and production in enclosures using stingless bee *Nannotrigona perilampoides*. Journal of Apicultural Research. 45: 125-130.

Cortopassi-Laurino M, Imperatriz-Fonseca VL, Roubik DW, Dollin A, Heard T, Aguilar, I, Venturieri GC, Eardley C, Nogueira NP. 2006. Global meliponiculture: challenges and opportunities. Apidologie 37: 275-292

Cortopassi-Laurino M, Knoll FRN, Ribeiro MF, van Heemert C, de Ruijter A. 1991. Food plant preferences of *Friesella schrottkyi*. Acta Horticola 288: 382–385.

Cruz DO, Freitas BM, Silva LA, Silva SEM, Bomfim IGA. 2005. Use of the stingless bee Melipona subnitida to pollinate sweet pepper (Capsicum annuum L.) flowers in greenhouse. Pesquisa Agropecuária Brasileira 40: 1197-1201

De Luca PA, Vallejo-Marín M. 2013. What's the 'buzz' about? The ecology and evolutionary significance of buzz-pollination. Current Opinion in Plant Biology 16: 1-7.

Eckles MA, Roubik DW, Nieh JC. 2012. A stingless bee can use visual odometry to estimate both height and distance. Journal of Experimental Biology 215: 3155-3160

Eka PR, Dana PA, and Kinasih I. 2014. Application of Asiatic Honey Bees (*Apis cerana*) and Stingless Bees (*Trigona laeviceps*) as Pollinator Agents of Hot Pepper (*Capsicum annuum* L.) at Local Indonesia Farm System. Psyche 2014: 1-5., http://dx.doi. org/10.1155/2014/687979

Engel MS, Dingemans-Bakels F. 1980. Nectar and pollen resources for stingless bees (Meliponinae, Hymenoptera) in Surinam (South America). Apidologie 11: 341-350.

Eickwort GC, Ginsberg HS. 1980. Foraging and mating behavior in Apoidea. Annual Review of Entomology 25: 421-446.

Falcão MA, Lleras E. 1980. Aspectos fenológicos, ecológicos e de produtividade do mapati (*Pourouma cecropiifolia* Mart.). Acta Amazonica 10: 711-724.

FAO. 2014. Principios y avances sobre polinización como servicio ambiental para la agricultura sostenible en países de Latinoamérica y el caribe. Organización de las Naciones Unidas para la Alimentación y la Agricultura; Santiago de Chile, Chile. 55 pp.

FAO. 2015. Crops, weeds and pollinators. Understanding ecological interaction for better management. Biodiversity and ecosystem services in agricultural production systems. Food and Agriculture Organization of the United Nations, Rome, Italy. 96 pp.

Free JB. 1970. Insect pollination of crops. Academic Press; London, UK. 544 pp.

Friedrich G, Barth MH, Stefan J. 2008. Signals and cues in the recruitment behavior of stingless bees (Meliponini). Journal of Comparative Physiology 194: 313-327.

Grajales CJ, Meléndez RV, Cruz LL, Sánchez GD. 2013. Native bees in blooming orange (*Citrus sinensis*) and lemon (*C. limon*) orchards in Yucatán, Mexico. Acta Zoológica Mexicana. Nueva serie. 29: 437-440.

Greco MK, Spooner HRN, Beattie AGAC, Barchia I, Holford P. 2011. Australian stingless bees improve greenhouse *Capsicum* production. Journal of Apicultural Research 50: 102-115.

Goulson D. 2003. Effects of introduced bees on native ecosystems. Annual Review of Ecology and Systematics 34: 1-26.

Heard TA, Exley EM. 1994. Diversity, abundance, and distribution of insect visitors to macadamia flowers. Environonmental Entomology 23: 91-100.

Heard TA. 1999. The role of stingless bees in crop pollination. Annual Review of Entomology 44: 183-206.

Heard TA. 2016. Alternative pollinators and the global pollination crisis. Available at https://www.griffith.edu.au/__data/assets/pdf_file/0004/447340/GU-alternative-pollinators.pdf

Heithaus ER. 1979. Flower visitation records and resource overlap of bees and wasps in Northwest Costa Rica. Brenesia 16: 9-52.

IPBES. 2016. The assessment report on Pollinators, Pollination and Food Production. Summary for policy makers. Available at http://www.ipbes.net/sites/default/files/downloads/pdf/SPM_Deliverable_3a_Pollination.pdf

Ish-Am G, Barrientos-Priego F, Castañeda-Vildozola A, Gazit S. 1999. Avocado (*Persea Americana* Mill.) pollinators in its region of origin, Revista Chapingo Serie Horticultura 5: 137-143.

Iwama S, Melhem TS. 1979. The pollen spectrum of the honey of *Tetragonisca angustula angustula* Latreille (Apidae, Meliponinae). Apidologie 10: 275-295.

Hedström I. 1986. Pollen carriers of *Cocos nucifera* L. (Palmae) in Costa Rica and Ecuador (Neotropical region). Revista de Biología Tropical. 34: 297-301.

Hrncir M, Jarau S, Barth FG. 2016. Stingless bees (Meliponini): senses and behavior. Journal of Comparative Physiology A 202: 597–601.

Jarau S, Hrncir M, Schmidt VM, Zucchi R, Barth FG. 2003. Effectiveness of recruitment behavior in stingless bees (Apidae, Meliponini). Insectes Sociaux 20: 365-374.

Kakutani T, Inoue T, Tezuka T, Maeta Y. 1993. Pollination of strawberry by the stingless bee, *Trigona minangkabau*, and the honey bee, *Apis mellifera*: an experimental study of fertilization efficiency. Researches on Population Ecology 35: 95-111.

Kearns CA, Inouye DW, Waser NM. 1998. Endangered mutualisms: the conservation of plant-pollinator interactions. Annual Review of Ecology and Systematics 29: 83-112.

Klein AM, Steffan-Dewenter I, Tscharntke T. 2003a. Fruit set of highland coffee increases with the diversity of pollinating bees. Proceedings of the Royal Society B 270: 955-961.

Klein AM, Steffan-Dewenter I, Tscharntke T. 2003b. Pollination of *Coffea canephora* in relation to local and regional agroforestry management. Journal of Applied Ecology 40: 837-845.

Klein AM, Vaissière BE, Cane JH, Steffan-Dewenter I, Cunningham SA, Kremen C, Tscharntke, T. 2007. Importance of pollinators in changing landscapes for world crops. Proceedings of the Royal Society B: Biological Sciences. http://dx.doi:10.1098/rspb.2006.3721

Kiatoko N, Raina SK, Muli E, Mueke J. 2014. Enhancement of fruit quality in *Capsicum annum* through pollination by *Hypotrigona gribodoi* in Kakamega, Western Kenya. Entomological Science 17: 106-110.

Lalama K. 2001. Pollination effectiveness and efficiency of the stingless bee *Plebeia* sp. and the honey bee *Apis mellifera* on strawberry *Fragaria × ananassa* in a greenhouse. MSc Thesis, Utrecht University; Utrecht, The Netherlands. 168 pp.

Lever JJ, Nes, EH, Scheffer M, Bascompte J. 2014. The sudden collapse of pollinator communities. Ecology Letters 17: 350-359.

Lobreau-Callen D, Thomas Ale, Darchen B, Darchen R. 1990. Quelques facteurs d´eterminant le comportement de butinage d' *Hypotrigona pothieri* (Trigonini) dans la végétation de Cóte-d'Ivoire. Apidologie 21: 69-83.

Maeta Y, Tezuka T, Nadano H, Suzuki K. 1992. Utilization of the Brazilian stingless bee, *Nannotrigona testaceicornis*, as a pollinator of strawberries. Honey bee Science 13: 71-78.

Malagodi-Braga KS, Kleinert AMP. 2004. Could *Tetragonisca angustula* Latreille (Apinae, Meliponini) be used as strawberry pollinator in greenhouses? Australian Journal Agriculchure Research 55: 771-773.

Meeuwsen FJAJ. 2000. Stingless bees for pollination purposes in greenhouses. pp. 143-147. In: Sommeijer MJ, Ruijter A, eds. Insect Pollination in Greenhouses: Proceedings Specialists' Meeting; Soesterberg, The Netherlands. 220 pp.

Maues MM, Venturieri GC. 1995. Pollination biology of anatto and its pollinators in Amazon area. Honey bee Science 16: 27-30.

McGregor SE. 1976. Insect Pollination of Cultivated Crop Plants. Agricultural Research Service; Washington. USA. 411 pp.

Masís CE, Lezama HJ. 1991. Estudio preliminar sobre insectos polinizadores de macadamia en Costa Rica. Turrialba 41: 520–523.

Meléndez RV. 1997. Polinización y biodiversidad de abejas nativas asociadas a cultivos hotícolas en el estado de Yucatán, México. Tesis de maestría. Universidad Autónoma de Yucatán, México. 90 pp.

Meléndez RV, Parra TV, Echazarreta CM, Magaña RS. 2000. Use of native bees and honey bees in hoticultural crops of *Cucurbita moschata* in Yucatan, Mexico. Management and diversity. pp. 65-70. In: Proceedings of the Sixth International Conference on Apiculture in Tropical Climates, Costa Rica. International Bee Research Association, Cardiff, UK. 226 pp.

Meléndez RV, Magaña RS, Parra TV, Ayala BR, Navarro AJ. 2002. Diversity of native bee visitor of cucurbit crops (Cucurbitaceae) in Yucatán, México. Journal of Insect Conservation. 6: 135-147.

Meléndez RV, Parra TV, Kevan PG, Ramirez MI, Harries H, Fernandez BM, Zizumbo VD. 2004. Mixed mating strategies and pollination by insects and wind in coconut palm (*Cocos nucifera* L. (Arecaceae)): importance in production and selection. Agricultural and Forest Entomology 6: 155-163.

Meléndez RV, Meneses CL, Kevan PG. 2013. Effects of human disturbance and habitat fragmentation on stingless bees. pp. 269-282. In: Vit P, Silvia RMP, Roubik D, eds. Pot honey: A legacy of stingless bees. Springer; New York, USA. 654 pp.

Meléndez RV, Ayala R y Delfín GH. 2016. Temporal variation in native bee diversity in the tropical subdeciduous forest of the Yucatan Peninsula, Mexico. Tropical Conservation Science 9: 718- 735.

Meneses CL, Meléndez RV, Parra TV, Navarro AJ. 2010. Bee diversity in fragmented landscape of the Mexican Neotropic. Journal of Insect Conservation 14: 323-334.

Michener CD. 1974. The social behavior of the bees. A comparative study. Harvard University Press; Cambridge, Massachusets, USA. 404 pp.

Michener CD. 2007. The bees of the world, 2nd edition. Johns Hopkins University Press; Baltimore, MD, USA. 953 pp.

Michener CD. 2013. The Meliponini. pp. 3-18. In: Vit P, Silvia RMP, Roubik D, eds. Pot honey: A legacy of stingless bees. Springer; New York, USA. 654 pp.

Morales C. 2007. Introducción de abejorros (*Bombus*) no nativos: causas, consecuencias ecológicas y perspectives. Ecología Austral 17: 51-65.

Nicodemo D, Braga ME, De Jong D, Nogueira CRH. 2013. Enhanced production of parthenocarpic cucumbers pollinated with stingless bees and Africanized honey bees in greenhouses. Semina: Ciências Agrárias, Londrina 34: 3625-3634.

Nieh JC, 2004. Recruitment communication in stingless bees (Hymenoptera, Apidae, Meliponini). Apidologie 35: 159-182.

Nieh JC, Roubik DW. 1995. A stingless bee (*Melipona panamica*) indicates food location without using a scent trail. Behavioral Ecology and Sociobiology 37: 63-70.

Nunes-Silva P, Hrncir M, Imperatriz-Fonseca VL. 2010. A polinização por vibração. Oecologia Australis 14: 140-151.

Nunes-Silva P., Hrncir M, Da Silva CI, Roldão YS, Imperatriz-Fonseca VL. 2013. Stingless bees, *Melipona fasciulata*, as efficient pollinators of eggplant (*Solanum melongena*) in greenhouses. Apidologie 44: 537-546.

Occhiuzzi P. 2000. Stingless bees pollinate greenhouse *Capsicum*, Aussie Bee 13: 15.

Oliveira CD de, Freitas BM, Da Silva LA, Sarmento da SEM, Abrahão BIG. 2005. Pollination efficiency of the stingless bee *Melipona subnitida* on greenhouse sweet pepper. Pesquisas Agropecuria Brasileira, Brasilia 40: 1197-1201.

O'Toole C. 1993. Diversity of native bees and agroecosystems. pp. 169-96. In: Hymenoptera and Biodiversity. LaSalle J, Gauld ID, eds. The Centre for Agriculture and Bioscience International, CABI; Wallingford, UK. 368 pp.

Ollerton J, Winfree R, Tarrant S. 2011. How many flowering plants are pollinated by animals? Oikos 120: 321–326.

Patricio GB, Campos MJO. 2014. Aspects of Landscape and Pollinators-What is Important to Bee Conservation? Diversity 6:158-175.

Palma G, Quezada Euán JJG, Meléndez RV, Irogoyen J, Valdovinos- Nuñez GR, Rejón M. 2008b. Comparative efficiency of *Nannotrigona perilampoides*, *Bombus impatiens* (Hymenoptera: Apoidea), and mechanical vibration on fruit production of enclosed habanero pepper. Journal of Economic Entomology 101: 132-138

Palma G, Quezada-Euan JJG, Reyes-Oregel V, Meléndez RV, Moo-Valle H. 2008a. Production of greenhouse tomatoes (*Lycopersicon esculentum*) using *Nannotrigona perilampoides*, *Bombus impatiens* and mechanical vibration (Hymenoptera: Apoidea). Journal of Applied Entomology 132: 9-85.

Peters C, Vasquez A. 1986. Estudios ecológicos de camu-camu (*Myrciariadubia*) I. Producción de frutas en poblaciones naturales. Acta Amazonica 16-17: 161-174.

Pinkus RM, Parra TV, Meléndez RV. 2005. Floral resources, use and interaction between *Apis mellifera* and native bees. The Canadian Entomologist 137: 441-449.

Potts SG, Biesmeijer JC, Kremen C, Neumann P, Schweiger O, Kunin WE. 2010. Global pollinator declines: Trends, impacts and drivers. Trends in Ecology and Evolution 25: 345-353.

Rasmussen C, Gonzalez VH. 2013. Stingless bees now and in the future. Prologue. In: Vit P, Roubik DW, eds. Stingless bees process honey and pollen in cerumen pots: vi–ix. Facultad de Farmacia y Bioanálisis, Universidad de Los Andes; Mérida, Venezuela; xii+170 pp.

Ricketts TH, Regetz J, Steffan DI, Cunningham SA, Kremen C, Bogdanski A, Gemmill Herren B, Greenleaf SS, Klein AM, Mayfield MM, Morandin LA, Ochieng A, Viana BF. 2008. Landscape effects on crop pollination services: are there general patterns? Ecology Letters 11: 499-515.

Roubik DW. 1992. Ecology and natural history of tropical bees. New York; Cambridge University Press. 514 pp.

Roubik DW. 1995. Pollination of cultivated plants in the tropics: Stingless bee colonies for pollination. Agricultural Services Bulletin 118. Food and Agriculture Organization; Rome, Italy. 198 pp.

Roubik DW. 2006. Stingless bee nesting biology. Apidologie 37: 124-143.

Roubik DW. 2009. Ecological impact on native bees by the invasive Africanized honey bee. Acta Biologica Colombiana 14: 115-124.

Roubik WD, Moreno P. 2013. How to be a bee-botanist using pollen spectra. pp. 295-314. In: Vit P, Silvia RMP, Roubik DW, eds. Pot honey: A legacy of stingless bees. Springer; New York. 654 pp.

Sánchez D, Valdame R. 2013. Stingless Bee Food Location Communication: From the flowers to the honey Pots. pp. 187-200. In: Vit P, Silvia RMP, Roubik D, eds. Pot honey: A legacy of stingless bees. Springer; New York, USA. 654 pp.

Santos SAB dos, Bego LR, Roselino AC. 2004a. Pollination in tomatoes, *Lycopersicon esculentum*, by *Melipona quadrifasciata* anthidioides and *Apis mellifera* (Hymenoptera, Apinae). pp. 688. Proceedings of the 8th IBRA International Conference on Tropical Bees and VI Encontro sobre Abelhas. 710 pp.

Santos SAB dos. 2004b. Pollination of cucumber - *Cucumis sativus* - by stingless bees (Hymenoptera, Meliponini). pp. 689. Proceedings of the 8th IBRA International Conference on Tropical Bees and VI Encontro sobre Abelhas. 710 pp.

Santos SAB dos, Roselino AC, Hrncir M, Bego LR. 2009. Pollination of tomatoes by the stingless bee *Melipona quadrifasciata* and the honey bee *Apis mellifera* (Hymenoptera, Apidae). Genetics and Molecular Research 8: 751-757.

Sarto MCL del, Peruquetti RC, Campos LAO. 2005. Evaluation of the neotropical stingless bee *Melipona quadrifasciata* (Hymenoptera: Apidae) as pollinator of greenhouse tomatoes. Apiculture and Social Insects 98: 260-266.

Sihag RC. 1995. Pollination, pollinators and pollination modes: ecological and economic importance. pp: 11-19. In: Roubik DW, ed. Pollination of Cultivated Plants in the Tropics. FAO Agricultural Services Bulletin 118. Food and Agricultural Organization; Rome, Italy. 195 pp.

Simão S, Maranhão ZC. 1959. Os insetos como agentes polinizadores da mangeira. Anais da Escola Superior de Agricultura Luiz de Queiroz 16: 299–304.

Singh G. 1989. Insect pollinators of mango and their role in fruit setting. Acta Horticola 231: 629-632.

Slaa EJ, Sanchez CLA, Malagodi BKS, Hofstede FE. 2006. Stingless bees in applied pollination: practice and perspectives. Apidologie 37: 293-315.

Slaa EJ, Sánchez LA, Sandí M, Salzar W. 2000a. A scientific note on the use of stingless bees for commercial pollinaton in enclosures, Apidologie 31: 141-142.

Slaa EJ, Sánchez LA, Sandî M, Salzar W. 2000b. Pollination of an ornamental plant (*Salvia farinacea*: Labiatae) by two species of stingless bees and Africanised honey bees (Hymenoptera: Apidae), pp. 209-215. In: Sommeijer M.J., Ruijter A. de, eds. Insect Pollination in Greenhouses: Proceedings specialists' meeting held in Soesterberg, The Netherlands. 220 pp.

Sommeijer MJ, de Bruijn LLM, Meeuwsen F. 2003. Reproductive behaviour of stingless bees: nest departures of non-accepted gynes and nuptial flights in *Melipona favosa* (Hymenoptera: Apidae, Meliponini). Entomologische Berichten 63: 7-13.

Sosa-Nájera MS, Martínez-Hernández E, Lozano-García MS, Cuadriello-Aguilar JI. 1994. Nectaropolliniferous sources used by *Trigona (Tetragonisca) angustula* in Chiapas, Southern Mexico. Grana 33: 225-30.

Torres-Ruiz A, Jones RW, Ayala-Barajas R. 2013. Present and Potential use of Bees as Managed Pollinators in Mexico. Southwestern Entomologist 38: 133-148.

Venturieri GA. 1994. Floral biology of cupuassu (*Theobroma grandiflorum* (Willdenow ex Sprengel) Schumann). PhD Thesis, University of Reading; Reading UK, 211 pp.

Viana BF, da Encarnação CJG, Garibaldi LA, Bragança GGL, Peres GK, Oliveira da SF. 2014. Stingless bees further improve apple pollination and production. Journal of Pollination Ecology 14: 261-269.

Villanueva-Gutiérrez R, Roubik DW. 2004. Why are African honey bees and not European bees invasive? Pollen diet determination in community experiments. Apidologie 35: 550-560.

Wille A. 1976. Las abejas jicotes del genero Melipona (Apidae: Meliponini) de Costa Rica. Revista de Biología Tropical 24:123–47.

Wille A. 1983. Biology of the stingless bees. Annual Review of Entomology, 28: 41-64.

Wille A, Orozco E, Raabe C. 1983. Polinizacion del chayote *Sechium edule* (Jacq.) Swartz en Costa Rica. Revista de Biología Tropical. 31: 145-54.

Stingless Bees as Potential Pollinators in Agroecosystems in Argentina: Inferences from Pot-Pollen Studies in Natural Environments

12

Favio Gerardo Vossler, Diego César Blettler,
Guillermina Andrea Fagúndez,
and Milagros Dalmazzo

12.1 Introduction

Wind and bees are the most important pollinating agents of the world (Michener 2007). Bees (group Apiformes or Anthophila sensu Michener (2013)) are the largest group of pollinators with >20,000 species (Michener 2007). They are either beneficial or actually essential for pollination, and therefore for the sexual reproduction, of much of the natural vegetation of the world, as well as for many agricultural crops (Michener 2007). In Brazil, for instance, one-third of crops depend on pollinators (Giannini et al. 2015a). Although the human diet is mostly based on anemophilous crops such as cereals, most crops are bee pollinated (Richards 2001; Ghazoul 2005).

Argentina has about 360,000 km^2 (36 M ha) of cultivated area out of 2.8 M km^2 (12.8%) of total continental surface. The most important extensive crops are soybean, corn, wheat, sunflower, sorghum, and rice. Their production was of more than 90 M tons during 2009/2010, while vegetable crops for the same period were only 10 M tons (from 500,000 ha). During 2002, fruit crop production was approximately 7 M tons (from 544,200 ha). In spite of the smaller cultivated area and lower volume of production, vegetables and fruit crops (1.4 and 1.2%, respectively, of total cultivated area in the country) are composed of a higher number of species with greater production per cultivated area and economic value per weight. The most important crops for the present survey are found in this smaller cultivated area, and most of them are moderately to highly dependent on biotic pollination.

The "honey bee" *Apis mellifera* L. has been considered the most important pollinator of monocultures in the world (McGregor 1976; Slaa et al. 2006; UNEP 2010), and this applies for Argentina as well (Torretta et al. 2010; Sáez et al. 2014). However, Garibaldi et al. (2013) showed that the honey bee supplements the pollination service of wild insects but cannot replace it. Moreover, the dependence on such a single pollinator throughout the world is an international concern as populations have been threatened by multiple factors and are in decline (Potts et al. 2010; UNEP 2010; Vandame and Palacio 2010). In addition, some specific crops such as many legumes and those having poricidal anthers

F.G. Vossler (✉) • D.C. Blettler • G.A. Fagúndez
Laboratorio de Actuopalinología, CICyTTP-
CONICET / FCyT-UADER,
Dr. Materi y España, E3105BWA Diamante, Entre
Ríos, Argentina
e-mail: favossler@yahoo.com.ar

M. Dalmazzo
Entomología, Facultad de Humanidades y Ciencias,
Universidad Nacional del Litoral, CONICET,
Paraje El Pozo s/n, 3000 Santa Fe, Argentina

© Springer International Publishing AG, part of Springer Nature 2018
P. Vit et al. (eds.), *Pot-Pollen in Stingless Bee Melittology*, DOI 10.1007/978-3-319-61839-5_12

(e.g., tomato) require pollination mechanisms different from that performed by *A. mellifera* (Matheson et al. 1996; Delaplane and Mayer 2000). This fact is important because crops dependent on pollinators are increasing in the world (Aizen et al. 2008). For these reasons, other bees began to be considered as alternative pollinators, the stingless bees (Meliponini) among them (Imperatriz-Fonseca et al. 2006; Giannini et al. 2015b).

There are >500 species of stingless bees in tropical and subtropical areas of the world (Michener 2013), and more than 33 species in 18 genera are present in Argentina (Roig-Alsina et al. 2013). The greater diversity of feeding habits and foraging behaviors in Meliponini can be explained by the higher number of genera compared to Apini (the genus *Apis*). The numerous flowers visited by stingless bee foragers from their populated colonies are represented in their pot-pollen and pot-honey stores (Roubik 1989). Thus, their pollen analysis allows us to infer plant preferences and other foraging behavioral aspects (Ramalho et al. 1985, 1989; Roubik 1989; Roubik and Moreno 2009; Vossler et al. 2010, 2014; Vossler 2015a). Pollen analysis is often, particularly considering legumes (Roubik and Moreno, Chap. 4), a good predictor of pollination. However, meliponines have been considered poor pollinators by some authors, as they moisten the transported pollen loads with nectar reducing the chances of losing individual grains to contact stigmas and perform pollination during floral visits (Thorp 2000). Nonetheless, this is also true for *Apis* and other corbiculate bees (*Bombus* and Euglossini), which are considered effective pollinators of many plants of economic importance (Dag and Kammer 2001; Maués 2002; Imperatriz-Fonseca et al. 2006; Briggs et al. 2013; Putra et al. 2014; Giannini et al. 2015b). A study performed on three stingless bee genera in a natural environment showed the predominance of small to medium grains in volume as well as in size in their stores; this fact presumably increased their chances of being transferred to stigmas as the number of grains transported per corbicula is higher when comparing it to larger grains (Vossler 2015a).

Stingless bees proved to be effective pollinators of strawberry, tomato, pepper, citrus fruits, and other crops and adapted well to greenhouse conditions in different regions around the world (Maeta et al. 1992; Sarto et al. 2005; Cruz et al. 2005; Antunes et al. 2007; Bispo dos Santos et al. 2009; Giannini et al. 2015b; Meléndez et al., Chap. 11).

In the present chapter, the suitability of Argentine crops for stingless bee pollination was assessed by identifying the mode of pollen transfer and plant breeding system or bee pollination dependence. Because warm climate conditions are needed for an optimal flight activity in stingless bees, only a small area of Argentina was found to be adequate for pollination in the open field. Greenhouses are an alternative option for temperate and cold regions of Argentina. Information on non-crop flowers present in agroecosystems that are beneficial for the maintenance of permanent colonies was also provided (see Sect. 12.7).

12.2 Potential Pollination by Stingless Bees in Argentina: Intrinsic and Extrinsic Factors

An individual bee makes its behavioral decisions based on information from a variety of sources (Biesmeijer and Slaa 2004). Among the intrinsic factors that influence bee individual behavior are the spontaneous preferences, memory, and innate preference, while the extrinsic ones are nest environment (colony state and nestmates), nestmates in the field, conspecific non-nestmates in the field, heterospecifics in the field, climate conditions, features of flowers, flower patches and other food sources, or the presence of predators. These extrinsic information sources provide a continuous flow of information to the forager who integrates it with its intrinsic information to make behavioral decisions (Biesmeijer and Slaa 2004).

Some of the characteristics that influence stingless bee ability as pollinators are polylecty and adaptability, floral constancy, domestication, perennial long-lived colonies, large food reserves stored in the nest, forager recruitment, etc. (Heard 1999; Slaa et al. 2006).

A likely cause of tropical and subtropical distribution of stingless bees is the sensibility of both individuals and colonies to low temperatures. The main abiotic factors that singly or in combination influence their flight activity are temperature, relative humidity, light intensity, and wind speed (Roubik 1989; Panizzi and Parra 2012). According to Fowler (1979), extreme values would affect the bees directly, while moderate values affect flight activity.

Temperature seems to be a determining factor to foragers, especially in small species, such as *Tetragonisca angustula* and *Plebeia*, which forage at 16 °C or above (Oliveira 1973; Imperatriz-Fonseca et al. 1985). Some species (*Plebeia*) reduce their flight activity at temperatures below 20 °C. Larger species of Meliponini, with a size between 8 and 12 mm such as *Melipona*, begin flight activity at lower temperatures, from 11 °C in *M. bicolor* (Hilário et al. 2000) to 13–14 °C in *M. quadrifasciata* and *M. marginata* (Guibu and Imperatriz-Fonseca 1984). Most species present optimal foraging activity between 20 and 30 °C (14–16 °C in *M. quadrifasciata* and 16–26 °C in *M. bicolor*) (Guibu and Imperatriz-Fonseca 1984).

Optimal values of relative humidity range between 30% and 70% for most species of Meliponini (Oliveira 1973; Hilário et al. 2001). The species *Plebeia remota*, *Schwarziana quadripunctata*, and *M. bicolor* present higher flight activity at higher levels, between 60% and 90% (Imperatriz-Fonseca et al. 1985).

Rain seems to decrease foraging activity, as shown in *Trigona fuscipennis* (Keppner and Jarau 2016). Black bees are more prone to overheat in sunlight than light-colored bees (Pereboom and Biesmeijer 2003).

In Argentina, studies on the extrinsic factors influencing the flight activity of stingless bees have not been carried out. A study of *Scaptotrigona depilis* in Southern Brazil, under fairly similar climate conditions to those of the subtropical area of North Argentina, showed that the foraging activity was ruled by the seasonal

pattern (Figueiredo-Mecca et al. 2013). For instance, foraging activity increased during the wet season in warm months (August/March) and decreased during the coldest and driest months (April/July). The general results showed that in lower temperature periods, bees began flight activity later, and the traffic was less intense, while at higher temperatures the bees began their flight activity earlier and more intensively. Foraging occurred only at temperatures above 15 °C. The end of the activity was correlated with decreasing light intensity, more than under the temperature effect. However, in warmer months, when the days were longer, the bees flew even in low light intensity all day long. Foraging occurred during the whole year but presented peaks in different months (Figueiredo-Mecca et al. 2013).

12.2.1 Advantages and Disadvantages of a Reduced to Moderate Flight Range in Stingless Bees

Flight range depends on bee size, especially, the wing size, and possibly also colony population size (Heard 1999). The flight range of *Melipona* species is greater than that of the less robust Meliponines, normally between 2 and 2.4 km for *Melipona*, 1.1–1.7 km for medium-sized species such as *Cephalotrigona* and *Scaptotrigona*, and 0.62–0.95 km for smaller species such as *Tetragonisca angustula* (Roubik and Aluja 1983; Araújo et al. 2004). The actual foraging distance also depends on the attractiveness of the resources in relation to distance from the nest, needs of the colony, and availability of alternative resources (Roubik 1982; Heard 1999). The level of distribution on floral resources during foraging activity (*polylecty, broad polylecty, degrees of polylecty*) could be of interest when selecting particular bee species for pollination (see Vossler, Chap. 2).

The reduced to moderate flight range of stingless bees (variable according to the genus) compared to *Apis* (>2 km, seldom up to 13 km) showed advantages and disadvantages as candidates for crop pollination and colony management. Among the advantages, it could avoid undesirable hybridization among crop varieties and contamination with agrochemicals from neighboring crops, while the disadvantages are that it requires colony management such as supplementary feeding in sites or during gaps and low flower availability.

12.2.2 Pollination Using Ground-Nesting Stingless Bees

Natural colonies of ground-nesting stingless bees (*Geotrigona*, *Paratrigona*, some *Plebeia*, etc.) could only be used for pollination in perennial crops (shrubs and trees), where soil remains undisturbed in a large area for many years. However, flowers should be available during long periods during the year when natural colonies are met. Alternatively, ground-nesting bees could be transferred to hives as it is done with trunk-nesting species; however, they are rather unsuccessful when kept in hives (Nogueira-Neto 1997; Slaa et al. 2006).

12.3 Pollen Spectra of Pot-Pollen in Colonies of Stingless Bees from Natural Environments

In Argentina, pot-pollen studies were carried out for *Tetragonisca*, *Scaptotrigona*, *Melipona*, and *Geotrigona* in an arid natural environment of north-central Argentina (Vossler et al. 2010, 2014; Vossler 2013, 2015a, b). Those palynological studies revealed that although a broad spectrum of pollen resources was present per nest, only a few of them were intensively foraged (Vossler, Chap. 2). Diversity and evenness indices of pollen and honey also helped to identify meliponine behavior (high diversity and heterogeneous use of resources) (Ramalho et al. 1985; Vossler et al. 2010; Vossler 2013). The most important resources were from the trees and woody shrubs most abundant in the forest. They provided generalized small flowers clustered in inflorescences offering both pollen and nectar (Ramalho 2004; Vossler 2015a).

When analyzing niche overlap among genera, different genera shared most of the resources, and only a few were exclusive (Kleinert-Giovannini and Imperatriz-Fonseca 1987;

Imperatriz-Fonseca et al. 1989; Wilms et al. 1996; Vossler 2013). The reason for such subtle differences was considered to be floral preference, or an innate preference, perhaps determined by flower or bee size, or other characteristics (Imperatriz-Fonseca et al. 1989; Ramalho 1990; Ramalho et al. 2007). In the Chaco forest of Argentina, *Tetragonisca* showed a more homogeneous resource use, in contrast to *Melipona* and *Geotrigona*. In *Geotrigona* it was markedly heterogeneous (Vossler 2013). In addition, pollen plants with poricidal anthers (such as *Solanum*) were abundantly found in *Melipona* stores due to the bees' exclusive capability to extract pollen by buzzing behavior (Vossler 2013).

Comparing pot-pollen samples from two types of dry Chaco forest, pollen composition varied according to floral resource availability (Vossler et al. 2014), indicating no particular floral preference, but nonetheless, local specialization occurred. The same result was found when studying samples from different periods in a same site (temporal specialization, i.e., temporal gathering of resources from only one or a few species in presence of other attractive melittophilous plants) (see Vossler, Chap. 2).

This foraging behavior inferred from palynological studies in natural ecosystems provides valuable information for pollination purposes (see Sect. 12.7.1).

12.4 Crops Potentially Pollinated by Stingless Bees in Argentina

For the present chapter, approximately 115 Argentine crops cultivated on large areas were selected to be analyzed (Fig. 12.1a, b). Out of the 115 crops, 85 are farmed for their seed or fruit (Table 12.1), and 65 of them require or benefit from maximum animal pollination and therefore could be addressed by stingless bee pollination. The remaining crops included those of anemophilous pollination, not dependent on vectors (parthenocarpic fruits), and those farmed for flowers or vegetative organs (chamomile, lettuce, pine, yerba mate, etc.). Seed production from vegetables farmed for foliage, stems, and roots amounts to <10% of the cultivated area in the

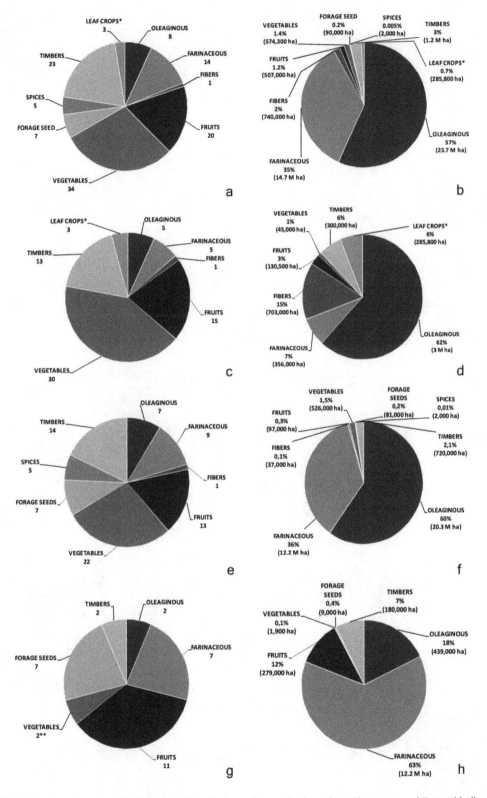

Fig. 12.1 Number of crop species (*left*) and cultivated area (*right*) for each category of market product in Argentina (**a**, **b**) and subtropical (**c**, **d**), temperate (**e**, **f**), and arid areas (**g**, **h**). *The category "Leaf crops" included only tobacco, tea, and yerba mate; "timbers" included Salicaceae, *Eucalyptus*, conifers, and other trees; "spices" included coriander, mint, chamomile, lavender, and oregano; and "vegetables" to 22 species mentioned in Table 12.1 and 12 whose market product were leaf, stem, or root. The other categories are detailed in Table 12.1. **As greenhouse microclimate conditions are similar in the three areas, only the number of vegetables and fruits cultivated on open field is represented in the graphs

Table 12.1 List of Argentina's crops whose marketed products are fruits and seeds

Type of crop	Mode of pollen transfer	Plant breeding system/bee pollination dependence	English name	Spanish name	Scientific name (Family)	Total cultivated area in Argentina (ha)	Cultivated area in subtropical area (%)	Cultivated area in temperate area (%)	Cultivated area in arid area (%)	Pollen (P) or nectar (H) foraging, flower visits (fv) and pollination (p)
OI	EN	HAU	Soybean	Soja or poroto de soja	Glycine max (Fabaceae)	20 M	13	85	2	fv(1)
OI	EN	HAL	Sunflower	Girasol	Helianthus annuus (Asteraceae)	1.5 M	10	90	–	p(2,3)
OI	EN	HAU	Peanut	Maní	Arachis hypogaea (Fabaceae)	388,500	–	90	10	?
OI	AN	HAU	Olive	Oliva	Olea europaea (Oleaceae)	110,000	92	8	–	–
OI	EN	HAL	Safflower	Cártamo or alazor	Carthamus tinctorius (Asteraceae)	100,000	95	5	–	?
OI	EN	HAL	Rape or rapeseed	Colza or canola	Brassica napus (Brassicaceae)	100,000	–	100	–	P(4), H(5), p(3)
OI	EN	HAL	Tung or tung tree	Tung	Vernicia fordii (Euphorbiaceae)	12,000	100	–	–	?
OI	EN	HAL	Flax or linseed	Lino or linaza	Linum usitatissimum (Linaceae)	10,000	–	100	–	?
Fa	AN	ND	Maize or corn	Maíz or choclo	Zea mays (Poaceae)	5.7 M	4	80	6	P(4)
Fa	AN	ND	Wheat	Trigo	Triticum aestivum (Poaceae)	4.6 M	–	85	15	–
Fa	AN	ND	Oat	Avena	Avena sativa (Poaceae)	1.5 M	–	85	15	–
Fa	AN	ND	Sorghum	Sorgo	Sorghum bicolor (Poaceae)	1.3 M	6	84	10	–
Fa	AN	ND	Barley	Cebada	Hordeum vulgare (Poaceae)	1 M	–	100	–	–
Fa	AN	ND	Rye	Centeno	Secale cereale (Poaceae)	0.3 M	–	50	50	–
Fa	AN	ND	Rice	Arroz	Oryza sativa (Poaceae)	233,600	20	80	–	–
Fa	AN	ND	Millet	Mijo, mijo perla, moha	Panicum miliaceum, Pennisetum glaucum, Setaria italica, Eleusine sp. (all Poaceae)	30,000	–	80	20	–

Fa	AN	ND	Canary seeds or canary grass	Alpiste	*Phalaris canariensis* (Poaceae)	19,300	–	100	–	–
Fa	AN	HAU	Amaranth	Amaranto	*Amaranthus* spp. (Amaranthaceae)	2,500	80	–	20	–
Fa	AN	ND	Quinoa or quinua	Quinoa or quinua	*Chenopodium quinoa* (Amaranthaceae)	1300	100	–	–	–
Fi	**EN**	**HAU**	**Cotton**	**Algodón**	***Gossypium hirsutum* (Malvaceae)**	**740,000**	**95**	**5**	**–**	p(3)
Fr	AN	ND	Grape or grape vine	Uva or vid	*Vitis vinifera* (Vitaceae)	218,500	7	3	90	H(6)
Fr	EN	HAU	**Citrus fruits (lemon, lime, orange, grapefruit, tangerine or mandarin)**	**Cítricos (limón, lima, naranja, pomelo, mandarina)**	***Citrus limon, C. aurantiifolia, C. sinensis, C. paradisi, C. reticulata* (Rutaceae)**	**161,800**	**54**	**46**	**–**	P(4,7,8,9), H (5,7,6), p(2,3)
Fr	**EN**	**HAL**	**Pome fruits (apple, pear, quince)**	**Frutas de pepita (manzana, pera, membrillo)**	***Malus domestica, Pyrus communis, Cydonia oblonga* (all Rosaceae)**	**50,000**	**–**	**5**	**95**	p(10)
Fr	**EN**	**HAL**	**Stone fruits (peach, nectarine, plum, apricot, cherry, sour cherry)**	**Frutas de carozo (durazno, pelón, ciruela, damasco, cereza, guinda)**	***Prunus persica, P. simonii, P. domestica* and *P. salicina, P. armeniaca, P. avium, P. cerasus* (Rosaceae)**	**29,700**	**5**	**10**	**85**	fv(11), p(3)
Fr	AN	HAU	Walnut	Nuez europea	*Juglans regia* (Juglandaceae)	16,446	65	5	30	?
Fr	AN	HAU	Pecan	Nuez pecán	*Carya illinoinensis* (Juglandaceae)	6,000	15	85	–	?
Fr	EN	PA	Banana	Banana	*Musa x paradisiaca* (Musaceae)	5,000	100	–	–	fv(2)
Fr	**EN**	**HAL**	**Raspberry**	**Frambuesa**	***Rubus idaeus* (Rosaceae)**	**3,500**	**–**	**20**	**80**	fv(11)

(continued)

Table 12.1 (continued)

Type of crop	Mode of pollen transfer	Plant breeding system/bee pollination dependence	English name	Spanish name	Scientific name (Family)	Total cultivated area in Argentina (ha)	Cultivated area in subtropical area (%)	Cultivated area in temperate area (%)	Cultivated area in arid area (%)	Pollen (P) or nectar (H) foraging, flower visits (fv) and pollination (p)
Fr	EN	HAL	Watermelon	Sandía	*Citrullus lanatus* (Cucurbitaceae)	3,000	85	10	5	p(3)
Fr	EN	HAL	Melon	Melón	*Cucumis melo* (Cucurbitaceae)	3,000	15	35	50	p(3)
Fr	EN	HAL	Highbush blueberry	Arándano	*Vaccinium corymbosum* (Ericaceae)	2,600	34	66	–	[a]p(12)
Fr	EN	HAU	Prickly pear	Tuna	*Opuntia ficus-indica* (Cactaceae)	2,000	97	2	1	?
Fr	EN	HAL	Strawberry	Frutilla, fresa	*Fragaria x ananassa* (Rosaceae)	1,300	41	52	7	p(2,3,13,14,15,16)
Fr	EN	HAL	Kiwifruit or kiwi	Kiwi	*Actinidia chinensis* (Actinidiaceae)	800	–	100	–	[a]?
Fr	EN	PA	Pineapple	Ananá	*Ananas comosus* (Bromeliaceae)	700	100	–	–	?
Fr	EN	HAL	Mango	Mango	*Mangifera indica* (Anacardiaceae)	500	100	–	–	P(7), p(2,3)
Fr	EN	HAL	Papaya	Mamón or papaya	*Carica papaya* (Caricaceae)	470	100	–	–	P(7), H(5,7)
Fr	EN	HAL	Black-seed squash	Cayote or alcayota	*Cucurbita ficifolia* (Cucurbitaceae)	350	50	–	50	?
Fr	EN	HAL	Fig	Higo	*Ficus carica* (Moraceae)	300	–	40	60	P(17)
Fr	EN	HAU	Red currant and blackcurrant	Corinto or grosella roja and cassis or grosella negra	*Ribes rubrum* and *R. nigrum* (Grossulariaceae)	300	–	–	100	?

Ve	EN	HAU	Winter legumes (chickpea, pea, lentil, lupin)	Legumbres invernales (garbanzo, arveja, lenteja, lupino)	*Cicer arietinum, Pisum sativum, Lens culinaris Lupinus albus* (Fabaceae)	500,000	–	100	–	?
Ve	EN	HAL	Pumpkin or squash	Zapallo or calabaza and zucchini, anco, calabaza rayada, zapallito de tronco	*Cucurbita pepo, C. moschata, C. argyrosperma, Cucurbita maxima* (Cucurbitaceae)	42,500	70	30	–	p(2,3)
Ve	EN	HAU	Tomato	Tomate	*Solanum lycopersicum* (Solanaceae)	14,000	40	50	10	[a]p(3,18,19)
Ve	EN	HAU	Pepper	Pimiento, morrón or ají	*Capsicum annuum* (Solanaceae)	13,000	60	40	–	fv(11), p(2,3,20,21)
Ve	EN	HAL	Avocado	Palta or aguacate	*Persea americana* (Lauraceae)	2,000	100	–	–	H(22), p(2,3)
Ve	EN	HAU	Eggplant	Berenjena	*Solanum melongena* (Solanaceae)	1,300	40	60	–	[a]P(7,8,9,23), H(7,23), fv(11,24), p(2)
Ve	EN	HAL	Cucumber	Pepino	*Cucumis sativus* (Cucurbitaceae)	1,000	65	30	5	p(2,3)
Ve	EN	HAU	Summer legumes (bean, cowpea)	Legumbres estivales (poroto)	*Phaseolus* spp. and *Vigna* spp. (Fabaceae)	500	100	–	–	?
Fos	EN	HAL	Grassland based on alfalfa, lucerne	Praderas base alfalfa	*Medicago sativa* and consocies (*Trifolium pratense, T. repens, Melilotus albus, M. officinalis, Lotus corniculatus, L. tenuis*) (all Fabaceae)	90,000	–	90	10	H(7,25), fv(24)

(continued)

Table 12.1 (continued)

Type of crop	Mode of pollen transfer	Plant breeding system/bee pollination dependence	English name	Spanish name	Scientific name (Family)	Total cultivated area in Argentina (ha)	Cultivated area in subtropical area (%)	Cultivated area in temperate area (%)	Cultivated area in arid area (%)	Pollen (P) or nectar (H) foraging, flower visits (fv) and pollination (p)
Sp	EN	HAL	**Coriander (seeds) or cilantro (foliage)**	**Coriandro or cilantro**	***Coriandrum sativum* (Apiaceae)**	**2,000**	–	**100**	–	p(2)

References of foraged resources at genus level (1–25): **1** Oliveira (2016), **2** Heard (1999), **3** Giannini et al. (2015b), **4** Carvalho et al. (1999), **5** Barth et al. (2013), **6** Obregón et al. (2013), **7** Ramalho et al. (1990), **8** Ramalho et al. (1989), **9** Imperatriz-Fonseca et al. (1989), **10** Viana et al. (2014), **11** Wilms et al. (1996), **12** Sezerino (2007), **13** Witter et al. (2012), **14** Antunes et al. (2007), **15** Maeta et al. (1992), **16** Kakutani et al. (1993), **17** Oliveira et al. (2009), **18** Sarto et al. (2005), **19** Bispo dos Santos et al. (2009), **20** Putra et al. (2014), **21** Cruz et al. (2005), **22** Ramalho (1990), **23** Vossler (2012), **24** Vossler (2013), **25** Basilio et al. (2013)

Some data on cultivated area were taken from: Beale and Ortiz (2013), Alcoba (2015), Doreste (2009, 2011), Cólica (2015), Gear (2006), Fernández Lozano (2012), Prataviera (2003), Caminiti (2013), Gómez Riera et al. (2014), Cogliatti (2014), Idígoras (2014), Dirección de Información Agropecuaria y Forestal (2016), Peralta and Liverotti (2012)

Type of crop (**Ol** = oleaginous, **Fa** = farinaceous, **Fi** = fiber, **Fr** = fruit, **Ve** = vegetable, **Fos** = forage seed, **Sp** = spice). They were primarily ordered according to their cultivated area. For each crop, the possibility of pollination by stingless bees is analyzed considering **pollen transfer mode** (**AN** = anemophilous, **EN** = entomophilous) and **plant breeding system** or **bee pollination dependence** (**HAU** = highly autogamous, **ND** = null dependence, **PA** = parthenocarpy, **HAL** = highly allogamous). In bold the selected crops suitable for stingless bee pollination. In addition, crops whose pollen or nectar is foraged by stingless bees were also recorded, as they are important in colony feeding (see Sect. 12.7)

[a]Crops requiring buzz pollination

country and is limited to arid areas; for this reason, they were not included in the analysis.

The highly autogamous crops such as soybean, tomato, and pepper are known to benefit from bee flower visits as they help to improve fruit or seed production and quality (Shipp et al. 1994; Cruz et al. 2005). The level of dependence on animal pollinators has been used to classify crops (Klein et al. 2007; Chacoff et al. 2010; Giannini et al. 2015a). However, as no data are available for many crops, or the values vary widely (from 0% to 50% for "soybean" (Chiari et al. 2005, 2008; Milfont et al. 2013; Santos et al. 2013)), the usage of the categories *highly autogamous* vs *highly allogamous* (Table 12.1) was chosen as an appropriate criterion.

12.5   Spatial Variation of Crops in Argentina

Three large areas in Argentina were analyzed and classified according to their main weather conditions favorable for crop growth and stingless bee survival and foraging activity, namely, subtropical, temperate, and arid (including irrigated valleys). For each area, a comparison was drawn between the number of crop species and cultivated area for each category of market produced (Figs. 12.1c–h and 12.2).

The subtropical area, northern Argentina, presents favorable climate conditions (long frost-free periods and high temperature) for stingless bee open-field pollination. Temperate and arid areas present a shorter period of favorable weather conditions; thus, use of meliponines in the greenhouse is an alternative option.

In Argentina, a total of 33 species in 18 genera of stingless bees have been recorded from the subtropical and part of the temperate area (some bees have favorable weather conditions in the temperate area). In the humid northeast of the subtropical area, 22 species in 16 genera are found (the richest stingless bee fauna); in the north-central dry region, 8 species in 7 genera; and in the northwestern 10 species in 8 genera (Roig-Alsina et al. 2013).

12.6   Temporal Variation of Flower Availability in Agroecosystems

In the following, some characteristics of temporal changes in floral availability, and crop life cycles in Argentine agroecosystems, are described. In extensive annual crops, one to two crop species are alternated throughout the year. Each crop life cycle is from 4 to 6 months (Díaz and Cabido 2001; Teubal 2006; Salado-Navarro and Sinclair 2009). Because weeds are effectively controlled by means of agrochemicals applied from just before sowing until the initial stages of development, flowers are not available for the bees when farming includes two species in a year (Satorre 2005; Salizzi 2014). Examples in the temperate area are soybean/sunflower/corn, sowed during spring, and wheat/oat/barley/rapeseed during autumn. When pasture crops are farmed, a greater flower diversity is available (Gaggiotti et al. 2014; Ciappini et al. 2009), either from the farmed crop (typically legumes of *Trifolium*, *Medicago*, *Melilotus*, and *Lotus* associated with grasses) and from weeds. Weeds are occasionally controlled using herbicides. Examples in the temperate area are associations of alfalfa/red clover/white clover/*Lotus* with grasses (*Bromus*, *Festuca*, *Lolium*).

Intensive vegetable agroecosystems can permit up to three crop cycles per year, and they can be associated with other crops, thus increasing the number of species in the area (Pérez 2009), some of them offering attractive flowers to bees throughout the year. In organic agroecosystems, weeds are often removed by hand (Stupino et al. 2007).

Perennial crops include woody fruits (citrus fruits, *Prunus*, etc.) and timber (*Eucalyptus*) that flower once per year for 20–30 days. As soil is barely managed and herbicide use is minimal, flowers from weeds are available during the year. Fruit crops are commonly accompanied by trees, at the margin of plantings, which provide flowers for bees (*Populus*).

Monocultures and polycultures farmed in greenhouses would offer flower diversity similar to that of extensive and intensive annual crops, respectively.

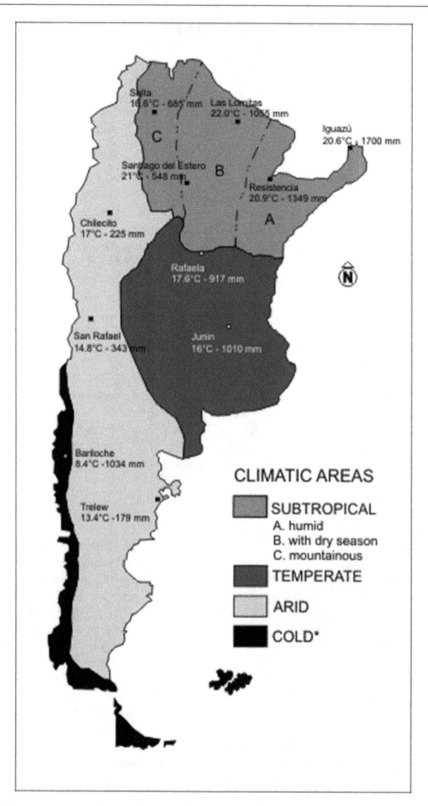

Fig. 12.2 Three main Argentine areas considered for the present study: subtropical (with three subareas), temperate, and arid. *Cold area was not analyzed. Data on mean annual temperatures and cumulative annual precipitations of some localities were provided

12.7 Crop and Non-crop Flowerings Present in Agroecosystems Beneficial for the Maintenance of Permanent Stingless Bee Colonies

12.7.1 Pollinating the Target Crop

For crop pollination purposes, bees should be concentrated on flowers of the target crop and not prefer other attractive plants. Methods to reduce floral competition have been described for *Apis* (Shivanna and Sawhney 1997). Palynological studies detected that although stingless bees forage on the available resources near the colonies, such as in the case of *Tetragonisca* (Vossler et al. 2014), plant family preferences have been recorded for different meliponine genera: *Scaptotrigona* with Myrtaceae (Ramalho 1990), *Melipona* with Solanaceae and Melastomataceae (Ramalho et al. 2007), *Plebeia* with Balsaminaceae and Euphorbiaceae (Imperatriz-Fonseca et al. 1989) and for stingless bees as a whole for Fabaceae (the three subfamilies), Myrtaceae, Asteraceae, Arecaceae, Anacardiaceae, Solanaceae, Rubiaceae, Euphorbiaceae, Sapindaceae, and Melastomataceae (Ramalho et al. 1990). The subtle floral preferences and slightly reduced foraging area in stingless bees would allow for an easier colony management than that for *Apis mellifera*. This is important for the placement of meliponine hives within a crop area for pollination purposes.

Information on foraging and floral visits (last column in Table 12.1) indicated that most exploited resources were from genera of native plants that lack cultivated species in Argentina, but a few of them are from the same genera or species as Argentinian crops (*Solanum*, *Citrus*, *Capsicum*, *Carica papaya*, *Persea americana*, *Mangifera indica*, and *Brassica*) (Table 12.1). Species of *Solanum* and *Citrus* were the most frequently recorded in these studies, being promising crops to be pollinated by stingless bees.

Feeding of permanent colonies during either before or after the flowering of the target crop to be pollinated can be achieved by crops in which pollination by stingless bees seems unnecessary (autogamous such as some legumes, *Solanum*, *Capsicum*, *Citrus*, *Gossypium*; anemophilous such as *Juglans*, *Carya*, and *Zea*; or parthenocarpic plants such as *Musa* and *Ananas*). Neighboring crops, edge vegetation, and weeds in the target crop also offer flowers for bees. If no alternative plants are available, artificial feeding or migratory beekeeping of hives is needed to avoid colony decline or mortality.

12.7.2 Weeds and Edge Vegetation as Complementary Flowerings for Permanent Stingless Bee Colonies

In regions of natural distribution of stingless bees (subtropical and part of temperate Argentina), woody species from edges of crops could provide food for wild nests of pollinators. However, most agroecosystems have destroyed such natural vegetation (Díaz and Cabido 2001; Cabido et al. 2005) which brought about the need for managed bees when the crop requires bee pollination. The maintenance of trees and shrubs with large trunks in crop edges is a good practice for both stingless bee nesting and crop pollination (Kearns et al. 1998; Brown and Albrecht 2001; Harris and Johnson 2004; Murray et al. 2009). Biological corridors within crops and at their edges should be considered when designing agroecosystems (Weyland and Zaccagnini 2008; Weyland et al. 2008). Although natural populations of meliponines are found in warmer areas in northern Argentina, these practices might best be performed at a large scale, to preserve other wild insects, as they can supplement the pollination activity of managed bees (Garibaldi et al. 2013).

12.7.3 Diversified Agroecosystems as Best Habitat for Stingless Bee Pollination and Colony Management

The diversity of floral resources is greater in intensive, perennial, and open-field crops (fruit and vegetable agroecosystems, smallholder, and subsistence farming) (Fernández and Marasas 2015), and it is beneficial for the maintenance of colonies without management.

On the other hand, in extensive crops, many different management practices are necessary for the survival of colonies, such as maintenance of natural vegetation on crop edges and along biological corridors (Sáez et al. 2014), the sowing of flowering plants, and temporal availability of flowering plants. Artificial feeding and movement of hives temporarily into more resource-rich areas are common practices in extensive crops.

Moreover, the richness and abundance of weeds have been drastically reduced in the last decades in central and northern Argentina (see Freitas et al. 2009), due to the technology associated with transgenic and non-transgenic crops (mainly due to the usage of herbicides and direct sowing) (Bilenca et al. 2009). Wild vegetation remnants of crop edges have also been farmed or removed. As a consequence, the possibilities of having permanent stingless bee colonies have become restricted to areas unsuitable for farming (river islands, river banks, rocky areas, arid areas, etc.).

12.8 Case Study: Pollination of Strawberries with *Plebeia catamarcensis* (Holmberg) in Santa Fe, Central Argentina

12.8.1 Strawberry Cultivation in Argentina

The strawberry currently farmed in Argentina has been identified as a hybrid between *Fragaria virginiana* Mill. and *F. chiloensis* (L.) Mill., both from the United States, but hybridized in Europe, and it is botanically known as *Fragaria* x *ananassa* (Weston) Duchesne. Due to its hybrid origin, it adapts to different climate conditions, from tropical and subtropical weather to the cold of the Nordic countries (Gariglio 1995).

The general agroecological requirements of commercial strawberry vary according to the cultivar. The flowering percentage of the cultivar depends on three factors: photoperiod, tempera-

ture, and photoperiod response of the cultivar. The cultivar photoperiod response may vary depending on the combination of the other two factors. By lengthening the photoperiod (from 8 to 24 h), the temperature at which flowering reaches 100% increases from 14 to 26 °C, and 20 °C with an 8 h photoperiod is the optimum condition for the "Rabunda" variety (Gariglio 1995). Temperatures between 10 and 12 °C are lethal for the plant, while temperatures between 20 and 25 °C are optimum for its growth. Flowers are totally or partially damaged by temperatures of 2–3 °C, with optimum conditions for pollination at 20 °C and 60% relative humidity. In cool temperatures, the pistils can remain receptive for up to 10 days (Gariglio 1995), but temperatures lower than 5 °C may result in embryo abortion and reduced pollen viability, thus causing fruit deformation (Ariza et al. 2011).

In Argentina, strawberries are grown on approximately 1,300 ha of land mainly distributed in three provinces: Tucumán, Buenos Aires, and Santa Fe, with an annual output of almost 50,000 tons. Annual production systems are predominant although there are regions with biannual crops in the Mar del Plata-Miramar horticultural corridor (Adlercreutz 2016).

The agroecological requirements vary according to whether plants are used in spring-summer-autumn production areas with cold winters and moderate summers (Buenos Aires, Mendoza, and south) or in autumn-winter-spring production areas with moderate winters (north-central areas such as Santa Fe, Tucumán, and Corrientes Provinces) (Adlercreutz 2016). In the autumn-winter production regions, cold or chilling protection equipment such as greenhouse tunnels (Gariglio 1995) is used.

12.8.2 The Strawberry in Santa Fe

An area of 414 ha is used in Santa Fe Province for strawberry farming, which represents 6% of the total land allocated to fruit and vegetable production, one of the most important fruit commercially cultivated in the central-eastern and southern region (36.7% of the fruit area). The production of strawberry is carried out at La

Costa (106 ha) and Coronda (308 ha) of the so-called Cadena Frutihortícola Santafesina (Report of the Ministry of Production, Government of Santa Fe 2010). On average, 30,000 kg of fruit per ha are produced per year, of which 60% is used for the domestic market as fresh fruit and the rest processed in regional factories (pulps, jams, and frozen strawberries) (Scaglia and Taborda 2004; Scaglia et al. 2004).

The season begins in autumn (March–April) with the planting of seedlings (Scaglia et al. 2004). The first strawberries are harvested (15% of the total) during winter (June–August), and harvesting continues up to the beginning of summer (December, 5% of the total). The strawberries harvested during September and October amount to 50% of the total (Scaglia and Taborda 2004; Scaglia et al. 2004). Micro-tunnel technology (a structure height of 70 cm) is used in most of the cultivated area, but in approximately 32 ha, greenhouse tunnel technology is used (Fig. 12.3). This technology is beneficial for the protection of flowers from the frost and also promotes ripeness, uniformity, and maturity at yield (Sordo 2014). Weed control is a common practice mainly achieved by polyethylene mulch on planting areas, manual extraction, and soil disinfection by chemical fumigants (Adlercreutz 2016).

12.8.3 Meliponini: Potential Pollinators in Santa Fe Strawberry Crops

Most of the cultivation systems are multi-varietal and of the cultivars "Camino Real," "Camarosa," and "Festival," the varieties with a greater product surface (Sordo 2014). Although the system of crop pollination, i.e., which insects supply pollination service, in the region is unknown, the producers have detected that lack of pollination causes poor fruit formation and lower yields (M. Dalmazzo, personal observation) (Fig. 12.3b, c). The problem of pollination is greater during the winter months as the crop is covered, preventing the entrance of floral visitors, which are scarce at this time of the year. Proper pollination during the winter is important for the producer as it determines the volume of "scoop" (first harvested fruit, therefore of greater value). Practices to mitigate the lack of pollination during this time of the year consist of generating wind currents using garden air sweepers and removing tunnel or micro-tunnel covers on sunny days to allow floral visitors and wind entry (M. Dalmazzo, personal observation).

The presence of flower-visiting insects becomes a key issue. In a survey carried out dur-

Fig. 12.3 Strawberry crop in Santa Fe Province, Argentina. (**a**) Strawberry crop in greenhouse tunnel in La Costa, Santa Fe. (**b**) Well-developed strawberry "fruit." (**c**) Defective strawberry "fruit." (**d**) Nest entrance of *Plebeia catamarcensis* (Holmberg) in an urban area of Susana, Santa Fe. (**e**) Strawberry flower and unripe "fruits" during the winter season in La Costa, Santa Fe, Argentina (Photos: M. Dalmazzo)

ing spring-summer (September–December) in an organic strawberry producer establishment, 27 species of insects visiting flowers have been found. Four were Hymenoptera of the superfamily Apoidea, including *Augochloropsis euterpe* (Holmberg) and *Apis mellifera* L. which presented the highest frequency of visits (Pacini et al. 2012). Among the 12 species of wild bees collected with pan traps in conventional strawberry cultivations and crop-growing borders with other vegetation, the meliponine *Plebeia catamarcensis* (Holmberg) (M. Dalmazzo, personal communication) was found. Although these preliminary results are part of a study on pollination network in horticultural crops currently underway, they have prompted management recommendations such as promoting natural vegetation and areas offering nesting substrate for wild bees (cavities, tillage-free soil, and decaying logs). Although such practices would encourage the presence of species recorded in spring and summer, they would not solve the problem of lack of visitors during the winter months; thus, another alternative should be found.

The management of meliponine nests for strawberry pollination has proven to be a successful alternative to improve fruit quality (Kakutani et al. 1993; Antunes et al. 2007; Witter et al. 2012). Another advantage of these colonies is that such species are harmless, unlike *Apis mellifera* or species of the genus *Bombus*.

In the case of Santa Fe, trials to evaluate the use of *Plebeia catamarcensis* for strawberry pollination in macro-tunnels during the winter months are starting to be implemented. This species has been selected for the trials because its natural distribution reaches the central region of Santa Fe Province (Dalmazzo 2010; Roig-Alsina et al. 2013). In this area, most of the cultivated areas are found, nests are abundant and relatively easy to find in both urban and suburban areas (Fig. 12.3d), and the meliponine has been recorded as a flower visitor in strawberry farming establishments (M. Dalmazzo, personal observation) as well as at ornamental plants (Table 12.2). The main challenge is the transfer of nests from their natural nests in building cavities (cracks in walls, window frames, holes in beams, and

Table 12.2 Flower-visiting records for *Plebeia catamarcensis* (Holmberg) in natural, urban, and peri-urban areas of the central region of the Santa Fe Province

Plant family	Plant species	Locality
Adoxaceae	*Sambucus* sp.	Susana, 31°21′S 61° 30′W
Asteraceae	*Aster squamatus* (Spreng.)[a] Hieron	Reserva Universitaria, Esperanza, 31°20′S 60°40′W
	Calendula sp.	Monte Vera, 31°31′S 60°40′W
	Cichorium intybus L.[a]	Reserva Universitaria, Esperanza, 31°20′S 60°40′W
	Sonchus oleraceus L.	Monte Vera, 31°31′S 60°40′W
	Taraxacum campyloides L.	Monte Vera, 31°31′S 60°40′W
Apiaceae	*Ammi visnaga* (L.) Lam.[a]	Reserva Universitaria, Esperanza, 31°20′S 60°40′W
	Apium sp.[a]	Susana, 31°21′S 61° 30′W
Bignoniaceae	*Campsis x tagliabuana* Rehder[a]	Susana, 31°21′S 61° 30′W
Xanthorrhoeaceae	*Hemerocallis* sp.[a]	Susana, 31°21′S 61° 30′W
Pittosporaceae	*Pittosporum tobira* (Thunb.) Aiton[a]	Susana, 31°21′S 61° 30′W
Euphorbiaceae	*Euphorbia* sp.	Santa Fe, 31°38′S 60°42′W
Geraniaceae	*Geranium* sp.	Rafaela 31°16′S 61° 29′W
Portulacaceae	*Portulaca oleracea* L.	Santa Fe, 31°38′S 60°42′W
Talinaceae	*Talinum* sp.	Rafaela, 31°16′S 61° 29′W

[a]Data from Dalmazzo (2010)

wood), from which extraction is difficult. As an alternative, trap nests made from plastic bottles have been used. Although these trap nests have been successfully used in obtaining nests from *Tetragonisca angustula* in southern Brazil and for *Tetragonisca fiebrigi* in the Chaco region of Argentina (F.G. Vossler, personal observation), the results with *P. catamarcensis* have not been encouraging as in no case was it possible to obtain the nest even when bee cerumen of the same species was used as bait. Different kinds of trap nests are being tested, and this is a key step to be solved to evaluate the pollinating potential of *P. catamarcensis* in strawberry crops under greenhouse tunnels.

Acknowledgments We are especially thankful to Patricia Vit for her kind invitation to participate in this book and David Roubik and Nora Brea for their help in English language, suggestions, and comments on the manuscript. This study was supported by CONICET (Consejo Nacional de Investigaciones Científicas y Técnicas).

References

Adlercreutz EGA. 2016. Cultivos anuales y bianuales de frutillas en el Sudeste de la provincia de Buenos Aires: Modificaciones en los parámetros de crecimiento. Master Thesis. Universidad del Litoral, Facultad de Ciencias Agrarias; Esperanza, Santa Fe, Argentina. 88 pp.

Aizen MA, Garibaldi LA, Cunningham SA, Klein AM. 2008. Long-term global trends in crop yield and production reveal no current pollination shortage but increasing pollinator dependency. Current Biology 18: 1572–1575.

Alcoba DL. 2015. Quinua. In Libro de Resúmenes V Congreso Mundial, II Simposio Internacional de Granos Andinos. Editorial de la Universidad Nacional de Jujuy-EDIUNJU, San Salvador de Jujuy, Jujuy, Argentina. 246 pp.

Antunes OT, Calvete EO, Rocha HC, Nienow AA, Cecchetti D, Riva E, Maran RE. 2007. Produção de cultivares de morangueiro polinizadas pela abelha jataí em ambiente protegido. Horticultura Brasileira 25: 94–99.

Araújo ED, Costa M, Chaud-Netto J, Fowler HG. 2004. Body size and flight distance in stingless bees (Hymenoptera: Meliponini): inference of flight range and possible ecological implications. Brazilian Journal of Biology 64: 563–568.

Ariza M, Soria C, Medina J, Martínez-Ferri E. 2011. Fruit misshapen in strawberry cultivars (*Fragaria* × *anan-*

assa) is related to achenes functionality. Annals of Applied Biology 158: 130–138.

Barth OM, Freitas AS, Sousa GL, Almeida-Muradian LB. 2013. Pollen and physicochemical analysis of *Apis* and *Tetragonisca* (Apidae) honey. Interciencia 38: 280–285.

Basilio AM, Spagarino C, Landi L, Achával B. 2013. Miel de *Scaptotrigona jujuyensis* en dos localidades de Formosa, Argentina. pp. 1–8. In Vit P, Roubik DW, eds. Stingless bees process honey and pollen in cerumen pots. Facultad de Farmacia y Bioanálisis, Universidad de Los Andes; Mérida, Venezuela. Available at http://www.saber.ula.ve/handle/123456789/3529

Beale I, Ortiz EC. 2013. El sector forestal argentino: eucaliptos. Revista de Divulgación Técnica Agrícola y Agroindustrial. Facultad de Ciencias Agrarias. Universidad Nacional de Catamarca 53: 1–10.

Biesmeijer JC, Slaa EJ. 2004. Information flow and organization of stingless bee foraging. Apidologie 35: 143–157.

Bilenca DN, Codesido M, González Fischer CM, Pérez Carusi LC. 2009. Impactos de la actividad agropecuaria sobre la biodiversidad en la ecorregión Pampeana. Ediciones INTA; Buenos Aires, Argentina. 42 pp.

Bispo dos Santos SA, Roselino AC, Hrncir M, Bego LR. 2009. Pollination of tomatoes by the stingless bee *Melipona quadrifasciata* and the honey bee *Apis mellifera* (Hymenoptera, Apidae). Genetics and Molecular Research 8: 751–757.

Briggs HM, Perfecto I, Brosi BJ. 2013. The role of the agricultural matrix: coffee management and euglossine bee (Hymenoptera: Apidae: Euglossini) communities in Southern Mexico. Environmental Entomology 42: 1210–1217.

Brown JC, Albrecht C. 2001. The effect of tropical deforestation on stingless bees of the genus *Melipona* (Insecta: Hymenoptera: Apidae: Meliponini) in central Rondonia, Brasil. Journal of Biogeography 28: 623–634.

Cabido M, Zak MR, Cingolani A, Cáceres D, Díaz S. 2005. Cambios en la cobertura de la vegetación del centro de Argentina. ¿Factores directos o causas subyacentes? pp. 271–300. In Oesterheld M, Aguiar MR, Ghersa CM, Paruelo JM, eds. La heterogeneidad de la vegetación de los agroecosistemas. Universidad Nacional de Buenos Aires; Buenos Aires, Argentina, 428 pp.

Caminiti A. 2013. Cultivo de grosellas. INTA Estación Experimental Agropecuaria Bariloche; San Carlos de Bariloche, Río Negro, Argentina. 37 pp.

Carvalho CAL, Marchini LC, Ros PB. 1999. Fontes de pólen utilizadas por *Apis mellifera* L. e algumas espécies de *Trigonini* (Apidae) em Piracicaba (SP). Bragantia 58: 49–56.

Chacoff NP, Morales CL, Garibaldi LA, Ashworth L, Aizen MA. 2010. Pollinator dependence of Argentinean agriculture: current status and temporal analysis. The Americas Journal of Plant Science and Biotechnology 3: 106–116.

Chiari WC, Toledo VAA, Ruvolo-Takasusuki MCC, Attencia VM, Costa FM, Kotaka CS, Sakaguti ES, Magalhães HR. 2005. Pollination of soybean [*Glycine max* (L.) Merril] by honeybees (*Apis mellifera* L.). Brazilian Archives of Biology and Technology 48: 31–36.

Chiari WC, Toledo VAA, Hoffmann-Campo CB, Ruvolo-Takasusuki MCC, Toledo TCSO, Lopes TS. 2008. Polinização por *Apis mellifera* em soja transgênica (*Glycine max* (L.) Merril) Roundup Ready cv. BRS 245 RR e convencional cv. BRS 133. Acta Scientiarum 30: 267–271.

Ciappini MC, Gattuso SJ, Gatti MB, Di Vito MV, Gómez G. 2009. Mieles de la provincia de Santa Fe (Argentina). Determinación palinológica, sensorial y fisicoquímica, según provincias fitogeográficas. Primera parte. Invenio 12: 109–120.

Cogliatti M. 2014. El cultivo de alpiste (*Phalaris canariensis* L.). Universidad Nacional del Centro de la Provincia de Buenos Aires UNICEN; Tandil, Buenos Aires, Argentina. 158 pp.

Cólica JJ. 2015. Producción de nueces en Argentina y Catamarca. pp. 1–5. In Resúmenes del III Simposio Internacional de Nogalicultura del Noroeste Argentino.

Cruz DO, Magalhães Freitas B, da Silva LA, Sarmento da Silva EM, Abrahão Bomfim IG. 2005. Pollination efficiency of the stingless bee *Melipona subnitida* on greenhouse sweet pepper. Pesquisa Agropecuária Brasileira 40: 1197–1201.

Dag A, Kammer Y. 2001. Comparison between the effectiveness of honey bee (*Apis mellifera*) and bumble bee (*Bombus terrestris*) as pollinators of greenhouse sweet pepper (*Capsicum annuum*). American Bee Journal 141: 447–448.

Dalmazzo M. 2010. Diversidad y aspectos biológicos de abejas silvestres de un ambiente urbano y otro natural de la región central de Santa Fe, Argentina. Revista de la Sociedad Entomólogica Argentina 69: 33–44.

Delaplane KS, Mayer DF, 2000. Crop pollination by bees. CABI Publishing; Wallingford, UK. 344 pp.

Díaz S, Cabido M. 2001. Vive la différence: plant functional diversity matters to ecosystem processes. Trends in Ecology and Evolution 16: 646–655.

Dirección de Información Agropecuaria y Forestal. 2016. Estimaciones Agrícolas, Informe Semanal, Ministerio de Agroindustria de la Nación. 24 pp.

Doreste P. 2009. El nogal y sus perspectivas. Alimentos Argentinos 45: 28–32.

Doreste P. 2011. Nuez de pecán. Alimentos Argentinos 51: 44–48.

Fernández V, Marasas ME. 2015. Análisis comparativo del componente vegetal de la biodiversidad en sistemas de producción hortícola familiar del Cordón Hortícola de La Plata (CHLP), provincia de Buenos Aires, Argentina. Revista de la Facultad de Agronomía 114: 15–29.

Fernández Lozano JF. 2012. La producción de hortalizas en Argentina. Informe técnico Secretaría de Comercio Interior, Corporación del Mercado Central de Buenos Aires. Buenos Aires, Argentina. 29 pp.

Figueiredo-Mecca G, Bego L. Nascimento F. 2013. Foraging behavior of *Scaptotrigona depilis* (Hymenoptera, Apidae, Meliponini) and its relationship with temporal and abiotic factors. Sociobiology 60: 277–282.

Fowler HG. 1979. Responses by a stingless bee to a subtropical environment. Revista de Biología Tropical 27: 111–118.

Freitas BM, Imperatriz-Fonseca VL, Medina LM, Kleinert MPA, Galetto L, Nates-Parra G, Quezada-Euán JJ. 2009. Diversity, threats and conservation of native bees in the Neotropics. Apidologie 40: 332–346.

Gaggiotti MC, Signorini M, Sabbag N, Wanzenried Zamora RA, Cuatrín A. 2014. Miel de abeja producida en un sistema lechero en base a pastura de alfalfa (composición fisicoquímica, palinológica y sensorial). Archivos Latinoamericanos de Producción Animal 22: 15–20.

Garibaldi LA, Steffan-Dewenter I, Winfree R, Aizen MA, Bommarco R, Cunningham SA, Kremen C, Carvalheiro LG, Harder LD, Afik O, et al. 2013. Wild pollinators enhance fruit set of crops regardless of honey-bee abundance. Science 339: 1608–1611.

Gariglio N. 1995. Frutilla. pp. 99–119. In Pilatti RA, ed. Cultivos Bajo Invernaderos. Centro de Publicaciones Universidad Nacional del Litoral y Editorial Hemisferio Sur S.A.; Buenos Aires, Argentina. 174 pp.

Gear JRE. 2006. El cultivo del maíz en la Argentina. pp. 4–8. In ILSI Argentina, ed. Maíz y nutrición: informe sobre los usos y las propiedades nutricionales del maíz para la alimentación humana y animal. Serie de Informes Especiales, Volumen II; Buenos Aires, Argentina, 80 pp.

Ghazoul J. 2005. Buzziness as usual? Questioning the global pollination crisis. Trends in Ecology and Evolution 20: 367–373.

Giannini TC, Cordeiro GD, Freitas BM, Saraiva AM, Imperatriz-Fonseca VL. 2015a. The dependence of crops for pollinators and the economic value of pollination in Brazil. Journal of Economic Entomology 108: 849–857.

Giannini TC, Boff S, Cordeiro GD, Cartolano Jr EA, Veiga AK, Imperatriz-Fonseca VL, Saraiva AM. 2015b. Crop pollinators in Brazil, a review of reported interactions. Apidologie 46: 209–223.

Gómez Riera P, Bruzone I, Kirschbaum DS. 2014. Visión prospectiva de la cadena de frutas finas al 2030. Ministerio de Ciencia, Tecnología e Innovación Productiva; Buenos Aires, Argentina. 78 pp.

Guibu LS, Imperatriz-Fonseca VL. 1984. Atividade externa de *Melipona quadrifasciata* Lepeletier (Hymenoptera, Apidae, Meliponinae). Ciência e Cultura 36: 623.

Harris LF, Johnson SD. 2004. The consequences of habitat fragmentation for plan-pollinator mutualisms. International Journal of Tropical Insect Science 24: 29–43.

Heard TA. 1999. The role of stingless bees in crop pollination. Annual Review of Entomology 44: 183–206.

Hilário SD, Imperatriz-Fonseca VL, Kleinert AMP. 2000. Flight activity and colony strength in the stingless bee *Melipona bicolor bicolor* (Apidae, Meliponinae). Revista Brasilera de Biología 60: 299–306.

Hilário SD, Imperatriz-Fonseca VL, Kleinert AMP. 2001. Responses to climatic factors by foragers of *Plebeia pugnax* Moure (in litt.) (Apidae, Meliponinae). Revista Brasileira de Biología 61: 191–196.

Idigoras G. 2014. Producción y procesamiento de productos frutihortícolas. Plan Nacional de Ciencia y Tecnología e Innovación Productiva, Argentina 2020. Ministerio de Ciencia, Tecnología e Innovación Productiva de Argentina. 81 pp.

Imperatriz-Fonseca VL, Kleinert-Giovannini A, Pires JT. 1985. Climate variations influence on the flight activity of *Plebeia remota* Holmberg (Hymenoptera, Apidae, Meliponinae). Revista Brasilera de Entomología 29: 427–434.

Imperatriz-Fonseca VL, Kleinert-Giovannini A, Ramalho M. 1989. Pollen harvest by eusocial bees in a non-natural community in Brazil. Journal of Tropical Ecology 5: 239–242.

Imperatriz-Fonseca VL, Saraiva AM, de Jong D. 2006. Bees as pollinators in Brazil: assessing the status and suggesting best practices. Holos, Editora; Ribeirão Preto, Brazil. 112 pp.

Kakutani T, Ioue T, Tezuca T, Maeta Y. 1993. Pollination of strawberry by the stingless bee, *Trigona minangkabau*, and the honey bee, *Apis mellifera*: an experimental study of fertilization efficiency. Researches on Population Ecology 35: 95–111.

Kearns C, Inouye DW, Waser N. 1998. Endangered mutualisms: the conservation of plant-pollinator interactions. Annual Review of Ecology and Systematics 29: 83–112.

Keppner EM, Jarau S. 2016. Influence of climatic factors on the flight activity of the stingless bee *Partamona orizabaensis* and its competition behavior at food sources. Journal of Comparative Physiology A 202: 691-699.

Klein AM, Vaissiere BE, Cane JH, Dewenter, IS, Cunningham SA, Kremen C, Tscharntke T. 2007. Importance of pollinators in changing landscapes for world crops. Proceedings of the Royal Society B: Biological Sciences 274: 303–313.

Kleinert-Giovannini A, Imperatriz-Fonseca VL. 1987. Aspects of the trophic niche of *Melipona marginata marginata* Lepeletier (Apidae, Meliponinae). Apidologie 18: 69–100.

Maeta Y, Tezuka T, Nadano H, Suzuki K. 1992. Utilization of the Brazilian stingless bee, *Nannotrigona testaceicornis* as a pollinator of strawberries. Honeybee Science 13: 71–78.

Matheson A, Buchmann SL, O' Toole C, Westrich P, Williams IH. 1996. The conservation of bees. Academic Press for the Linnean Society of London and the International Bee Research Association. 152 pp.

Maués MM. 2002. Reproductive phenology and pollination of the Brazil nut tree (*Bertholletia excelsa* Humb & Bonpl. Lecythidaceae) in Eastern Amazonia. pp. 245–254. In Kevan P, Imperatriz-Fonseca VL, eds. Pollinating bees: the conservation link between

agriculture and nature. Ministério Meio Ambiente; Brasília, Brazil. 313 pp.

McGregor SE. 1976. Insect pollination of cultivated crop plants, Agriculture Handbook N° 496. United States Department of Agriculture. Washington, DC, USA. 411 pp.

Michener CD. 2007. The bees of the world, 2 edn. The Johns Hopkins University Press; Baltimore, USA. 953 pp.

Michener CD. 2013. The Meliponini. pp 3–17. In Vit P, Pedro SRM, Roubik DW, eds. Pot honey: A legacy of stingless bees. Springer; New York, USA. 175 pp.

Milfont MO, Rocha EE, Lima AO, Freitas BM. 2013. Higher soybean production using honeybee and wild pollinators, a sustainable alternative to pesticides and autopollination. Environmental Chemistry Letters 11: 335–341.

Murray TE, Kuhlmann M, Potts SG. 2009. Conservation ecology of bees: populations, species and communities. Apidologie 40: 211–236.

Nogueira-Neto P. 1997. Vida e criação de abelhas indígenas sem ferrão. Edição Nogueirapis; São Paulo, Brasil. 445 pp.

Obregón D, Rodríguez-C Á, Chamorro FJ, Nates-Parra G. 2013. Botanical origin of pot-honey from *Tetragonisca angustula* Latreille in Colombia. pp. 337–346. In Vit P, Pedro SRM, Roubik DW, eds. Pot-Honey: A Legacy of Stingless Bees. Springer; New York, USA, 175 pp.

Oliveira MAC. 1973. Algumas observações sobre a atividade externa de *Plebeia saiqui* e *Plebeia droryana*. Master Thesis. Instituto de Biociências, Universidade de São Paulo; São Paulo, Brazil. 79 pp.

Oliveira FPM, Absy ML, Miranda IS. 2009. Recurso polínico coletado por abelhas sem ferrão (Apidae, Meliponinae) em um fragmento de floresta na região de Manaus - Amazonas. Acta Amazônica 39: 505–518.

Oliveira F. 2016. Efeitos da inter-relação entre a presença de visitantes florais e a produção na cultura da soja. Dissertação de Mestrado, Entomologia e Conservação da Biodiversidade, Universidade Federal da Grande Dourados; Dourados, Brasil. 40 pp.

Pacini AC, Rivero R, Dalmazzo M. 2012. Relevamiento de visitantes florales en cultivos hortícolas orgánicos de Santa Fe. pp. 1. In Libro de Resúmenes Primer Congreso Santafesino de Agroecología. Rosario, Santa Fe, Argentina.

Panizzi AR, Parra JRP. 2012. Insect bioecology and nutrition for integrated pest management. CRC Press; Boca Raton, FL, USA. 750 pp.

Pereboom, JJM, Biesmeijer JC. 2003. Thermal constraints for stingless bee foragers: the importance of body size and coloration. Oecologia 137: 42–50.

Peralta ME, Liverotti O. 2012. El consumo de ananá o piña y mango en Argentina. Gacetilla de Frutas y Hortalizas del Convenio INTA-CMCBA N° 19: 1–10.

Pérez M. 2009. Evaluación de la sustentabilidad en sistemas de producción hortícola alternativos en la región pampeana, Argentina. Cadernos de Agroecología 4: 1421–1424.

Potts SG, Biesmeijer JC, Kremen C, Neumann P, Schweiger O, Kunin WE. 2010. Global pollinator declines: trends, impacts and drivers. Trends in Ecology and Evolution 25: 345–353.

Prataviera AG. 2003. Una producción alternativa en marcha. El cultivo de la higuera. Revista de Información sobre Investigación y Desarrollo Agropecuario, serie IDIA XXI 3: 142–146.

Putra RE, Permana AD, Kinasih I. 2014. Application of Asiatic honey bees (*Apis cerana*) and stingless bees (*Trigona laeviceps*) as pollinator agents of hot pepper (*Capsicum annuum* L.) at local Indonesia farm system. Psyche. Article ID 687979, DOI: 10.1155/2014/687979.

Ramalho M. 1990. Foraging by stingless bees of the genus *Scaptotrigona* (Apidae, Meliponinae). Journal of Apicultural Research 29: 61–67.

Ramalho M. 2004. Stingless bees and mass flowering trees in the canopy of Atlantic forest: a tight relationship. Acta Botanica Brasilica 18: 37–47.

Ramalho M, Imperatriz-Fonseca VL, Kleinert-Giovannini A, Cortopassi-Laurino M. 1985. Exploitation of floral resources by *Plebeia remota* Holmberg (Apidae, Meliponinae). Apidologie 16: 307–330.

Ramalho M, Kleinert-Giovannini A, Imperatriz-Fonseca VL. 1989. Utilization of floral resources by species of *Melipona* (Apidae, Meliponinae): Floral preferences. Apidologie 20: 185–195.

Ramalho M, Kleinert-Giovannini A, Imperatriz-Fonseca VL. 1990. Important bee plants for stingless bees (*Melipona* and *Trigona*) and Africanized honeybees (*Apis mellifera*) in Neotropical habitats: a review. Apidologie 21: 469–488.

Ramalho M, Silva MD, Carvalho CAL. 2007. Dinâmica de uso de fontes de pólen por *Melipona scutellaris* Latreille (Hymenoptera: Apidae): uma análise comparativa com *Apis mellifera* L. (Hymenoptera: Apidae), no Domínio Tropical Atlântico. Neotropical Entomology 36: 38–45.

Report of the Ministry of Production, Government of Santa Fe. 2010. Cadena Frutihortícola Santafesina.. Available at http://www.santafe.gov.ar

Richards AJ. 2001. Does low biodiversity resulting from modern agricultural practice affect crop pollination and yield? Annals of Botany 88: 165–172.

Roig-Alsina A, Vossler FG, Gennari GP. 2013. Stingless bees in Argentina. pp. 125–134. In Vit P, Pedro SRM, Roubik DW, eds. Pot-Honey: A Legacy of Stingless Bees. Springer; New York, USA. 175 pp.

Roubik DW. 1982. Seasonality in colony food storage, brood production and adult survivorship: studies of *Melipona* in tropical forest (Hymenoptera: Apidae). Journal of the Kansas Entomological Society 55: 789–800.

Roubik DW. 1989. Ecology and natural history of tropical bees. Cambridge University Press; New York, USA. 514 pp.

Roubik DW, Aluja M. 1983. Flight ranges of *Melipona* and *Trigona* in tropical forest. Journal of the Kansas Entomological Society 56: 217–222.

Roubik DW, Moreno JE. 2009. *Trigona corvina*: An ecological study based on unusual nest structure and pollen analysis. Psyche. Article ID 268756, DOI:10.1155/2009/268756

Salado-Navarro LR, Sinclair TR. 2009. Crop rotations in Argentina: analysis of water balance and yield using crop models. Agricultural Systems 102: 11–16.

Salizzi E. 2014. Reestructuración económica y transformaciones en el agro pampeano: la expansión del cultivo de la soja y sus efectos sobre la apicultura bonaerense en los inicios del siglo XXI. Estudios Socioterritoriales 16: 13–46.

Santos E, Mendoza Y, Vera M, Carrasco-Letelier L, Díaz S, Invernizzi C. 2013. Aumento en la producción de semillas de soja (*Glycine max*) empleando abejas melíferas (*Apis mellifera*). Agrociencia Uruguay 17: 81–90.

Sarto MCL, Peruquetti RC, Campos LAO. 2005. Evaluation of the Neotropical stingless bee *Melipona quadrifasciata* (Hymenoptera: Apidae) as pollinator of greenhouse tomatoes. Journal of Economic Entomology 98: 260–266.

Satorre EH. 2005. Cambios tecnológicos en la agricultura argentina actual. Ciencia Hoy 15: 24–31.

Sáez A, Sabatino M, Aizen M. 2014. La diversidad floral del borde afecta la riqueza y abundancia de visitantes florales nativos en cultivos de girasol. Ecología Austral 24: 94–102.

Scaglia E, Taborda R. 2004. Frutilla: historia y evolución tecnológica de su cultivo en la zona de Coronda. Estación Experimental Agropecuaria INTA RAFAELA. Informe para extensión. Agencia Extensión Rural INTA SANTA FE; Rafaela, Santa Fe, Argentina. 18 pp.

Scaglia E, Sordo MH, Pernuzzi C. 2004. Cultivo de la Frutilla en la zona de Coronda, provincia de Santa Fe. Estación Experimental Agropecuaria INTA RAFAELA. Informe para extensión. Agencia Extensión Rural INTA SANTA FE. Rafaela, Santa Fe, Argentina. 7 pp.

Sezerino AA. 2007. Polinização do mirtilo (*Vaccinium corymbosum* L.) (Ericaceae) cultivares Misty e O'neal no município de Itá, Oeste de SC. Curso de Graduação em Agronomia, Centro de Ciências Agrárias, Universidade Federal de Santa Catarina; Florianópolis, SC, Brasil. 35 pp.

Shipp JL, Whitfield GH, Papadopoulos AP. 1994. Effectiveness of the bumble bee, *Bombus impatiens* Cr. (Hymenoptera: Apidae), as a pollinator of greenhouse sweet pepper. Scientia Horticulturae 57: 29–39.

Shivanna KR, Sawhney VK. 1997. Pollen biotechnology for crop production and improvement. Cambridge University Press; New York, USA. 448 pp.

Slaa EJ, Sánchez Chaves LA, Malagodi-Braga KS, Hofstede FE. 2006. Stingless bees in applied pollination: practice and perspectives. Apidologie 37: 293–315.

Sordo MH. 2014. Cultivo de frutilla en la provincia de Santa Fe año 2013. Informe para extensión. Agencia Extensión Rural INTA SANTA FE; Rafaela, Santa Fe, Argentina. 3 pp.

Stupino SA, Ferreira AC, Frangi J, Sarandón SJ. 2007. Agrobiodiversidad vegetal en sistemas hortícolas orgánicos y convencionales (La Plata, Argentina). Revista Brasileira de Agroecologia 2: 339–342.

Teubal M. 2006. Expansión del modelo sojero en la Argentina. Realidad Económica 220: 71–96.

Thorp RW. 2000. The collection of pollen by bees. Plant Systematics and Evolution 222: 211–223.

Torretta JP, Medan D, Roig-Alsina A, Montaldo NH. 2010. Visitantes florales diurnos del girasol (*Helianthus annuus*, Asterales: Asteraceae) en la Argentina. Revista de la Sociedad Entomológica Argentina 69: 17–32.

UNEP (United Nations Environment Programme). 2010. UNEP emerging issues: global honey bee colony disorder and other threats to insect pollinators.. Available at http://www.unep.org/dewa/Portals/67/pdf/Global_Bee_Colony_Disorder_and_Threats_insect_pollinators.pdf

Vandame R, Palacio MA. 2010. Preserved honey bee health in Latin America: a fragile equilibrium due to low-intensity agriculture and beekeeping? Apidologie 41: 243–255.

Viana BF, da Encarnação Coutinho JG, Garibaldi LA, Bragança Gastagnino GL, Gramacho KP, Oliveira da Silva F. 2014. Stingless bees further improve apple pollination and production. Journal of Pollination Ecology 14: 261–269.

Vossler FG. 2012. Flower visits, nesting and nest defence behaviour of stingless bees (Apidae: Meliponini): suitability of the bee species for Meliponiculture in the Argentinean Chaco region. Apidologie 43: 139–161.

Vossler FG. 2013. Estudio palinológico de las reservas alimentarias (miel y masas de polen) de "abejas nativas sin aguijón" (Hymenoptera, Apidae, Meliponini): un aporte al conocimiento de la interacción abeja-planta en el Chaco Seco de Argentina. Doctoral

Thesis. Universidad Nacional de La Plata; La Plata, Argentina. 152 pp.

Vossler FG. 2015a. Small pollen grain volumes and sizes dominate the diet composition of three South American subtropical stingless bees. Grana 54: 68–81.

Vossler FG. 2015b. Broad protein spectrum in stored pollen of three stingless bees from the Chaco dry forest in South America (Hymenoptera, Apidae, Meliponini) and its ecological implications. Psyche. Article ID 659538, DOI: 10.1155/2015/659538

Vossler FG, Tellería MC, Cunningham M. 2010. Floral resources foraged by *Geotrigona argentina* (Apidae, Meliponini) in the Argentine Dry Chaco forest. Grana 49: 142–153.

Vossler FG, Fagúndez GA, Blettler DG. 2014. Variability of food stores of *Tetragonisca fiebrigi* (Schwarz) (Hymenoptera: Apidae: Meliponini) from the Argentine Chaco based on pollen analysis. Sociobiology 61: 449–460.

Weyland F, Poggio SL, Ghersa CM. 2008. Agricultura y biodiversidad. Ciencia Hoy 106: 27–35.

Weyland F, Zaccagnini ME. 2008. Efecto de las terrazas sobre la diversidad de artrópodos caminadores en cultivos de soja. Ecología Austral 18: 357–366.

Wilms W, Imperatriz-Fonseca VL, Engels W. 1996. Resource partitioning between highly eusocial bees and possible impact of the introduced Africanized honey bee on native stingless bees in the Brazilan Atlantic rainforest. Studies on Neotropical Fauna and Environment 31: 137–151.

Witter S, Radin B, Brito Lisboa B, Galaschi Teixeira JG, Blochtein B, Imperatriz-Fonseca VL. 2012. Desempenho de cultivares de morango submetidas a diferentes tipos de polinização em cultivo protegido. Pesquisa Agropecuária Brasileira 47: 58–65.

Part II

Biodiversity, Behavior and Microorganisms of the Stingless Bees (Meliponini)

Stingless Bees (Hymenoptera, Apoidea, Meliponini) from Gabon

13

Edgard Cédric Fabre Anguilet, Taofic Alabi,
Bach Kim Nguyen, Toussaint Ndong Bengone,
Éric Haubruge, and Frédéric Francis

13.1 Introduction

Stingless bees are highly eusocial and are distributed in tropical and subtropical areas worldwide (Michener 2007). Africa is a region with a low richness of stingless bee species. For example, 22 stingless bee species have been recorded in Africa (Table 13.1) versus approximately 400 species in tropical America (Camargo and Pedro 2008). In Africa, stingless bees have been found in the sub-Sahara, with the greatest diversity occurring in Central Africa (Fabre Anguilet et al.

2015). Indeed, the Congo Basin forest constitutes a habitat where several African species are commonly encountered. Consequently, a number of African species have been identified in Gabon (see Fig. 13.1) (Pauly 1998; Eardley 2004; Eardley and Urban 2010; Fabre Anguilet et al. 2015).

Bees contribute to the pollination of native and crop plants (Michener 2007); thus, stingless bees are potentially efficient pollinators of crops like coffee (*Coffea*), avocado (*Persea*), and safou (*Dacryodes*) (Tchuenguem et al. 2001, 2002; Slaa et al. 2006; Munyuli 2014). Rural human populations also use honey and cerumen for traditional practices in Gabon. Consequently, stingless bees have important ecological, economic and cultural roles in these regions.

The abundance, ecology, nesting behaviour, nest structure and taxonomy of stingless bees have been studied in several African countries; however, clarification is required due to the presence of cryptic species, which are species that are morphologically similar and may only be distinguished by fine morphological characteristics (Michener 2007). The nest structure is also used to distinguish different species of stingless bees (Roubik 2006). Some species were reclassified by Eardley (2004) but were later re-established by Pauly and Fabre Anguilet (2013). The use of morphological identification associated with molecular tools for cryptic species could help resolve taxonomic and phylogenetic issues (Hoy 2013), allowing the taxonomy of African stingless bees

E.C. Fabre Anguilet (✉)
University of Liege – Gembloux Agro-Bio Tech.
Functional and Evolutionary Entomology,
Passage des Déportés 2, BE-5030 Gembloux,
Belgium

Centre National de la Recherche Scientifique et
Technologique (CENAREST), Institut de Recherches
Agronomiques et Forestières (IRAF),
Trois Quartier B, 3090 Libreville, Gabon
e-mail: efabre@doct.ulg.ac.be; efabreanguilet@gmail.com

T. Alabi • B.K. Nguyen • É. Haubruge • F. Francis
University of Liege – Gembloux Agro-Bio Tech.
Functional and Evolutionary Entomology,
Passage des Déportés 2, BE-5030 Gembloux,
Belgium

T. Ndong Bengone
Centre National de la Recherche Scientifique et
Technologique (CENAREST), Institut de Recherches
Agronomiques et Forestières (IRAF),
Trois Quartier B, 3090 Libreville, Gabon

Table 13.1 Stingless bee fauna recorded in Gabon (Darchen and Pain 1966; Brosset and Darchen 1967; Darchen 1969; Pauly 1998; Eardley 2004; Pauly and Fabre Anguilet 2013; Pauly and Hora 2013)

African species	Species recorded in Gabon
Cleptotrigona cubicep (Friese 1912)	×
Dactylurina staudingeri (Gribodo 1893)	×
Dactylurina schmidti (Stadelmann 1895)	
Hypotrigona araujoi (Michener, 1959)	×
Hypotrigona gribodoi (Magretti 1884)	×
Hypotrigona ruspolii (Magretti 1898)	×
Hypotrigona squamuligera (Benoist 1937)	
Meliponula (Meliplebeia) beccarii (Gribodo 1879)	×
Meliponula (Meliponula) bocandei (Spinola 1853)	×
Meliponula (Axestotrigona) cameroonensis (Friese 1900)	×
Meliponula (Axestotrigona) erythra (Schletterer 1891)	×
Meliponula (Axestotrigona) ferruginea (Lepeletier 1841)	×
Meliponula (Meliplebeia) griswoldorum Eardley 2004	
Meliponula (Meliplebeia) lendliana (Friese 1900)	×
Meliponula (Meliplebeia) nebulata (Smith 1854)	×
Meliponula (Meliplebeia) roubiki Eardley 2004	×
Meliponula (Axestotrigona) togoensis (Stadelman 1895)	×
Liotrigona bottegoi (Magretti 1895)	
Liotrigona bouyssoui (Vachal 1903)	×
Liotrigona gabonensis Pauly and Fabre Anguilet 2013	×
Liotrigona baleensis Pauly and Hora 2013	
Plebeina armata (Magretti 1895)	

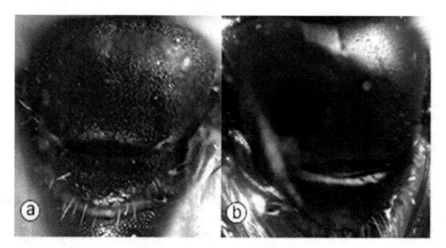

Fig. 13.1 *Scutum and scutellum*: (**a**) *Hypotrigona gribodoi and* (**b**) *Liotrigona gabonensis* (Photo: E.C. Fabre Anguilet © Fabre Anguilet E. C., 2016)

to be agreed upon. DNA barcoding is one of the methods used to distinguish species of bees. For instance, Koch (2010) demonstrated the existence of a new species of Western Malagasy *Liotrigona* in addition to the three already described species using morphometry and DNA barcoding.

Seventy-eight percent of the territory of Gabon is covered by rainforest, of which a great part is exploited by logging companies (Drouineau and Robert 1999; Fabre Anguilet et al. 2015). In addition, there are mining concessions, oil concessions and, essentially, rural agricultural practices. Stingless beekeeping is not practised and beekeeping using *Apis mellifera* is scarce. As a result, some beekeeping projects have been initiated by the Food and Agriculture Organization of the United Nations (FAO), National Higher Institute of Agronomy and Biotechnology (INSAB) and private companies, such as COLAS Gabon. To date, these projects have failed to achieve a wide dissemination or popularity in Gabon.

However, some initiatives in the sustainable management of biodiversity resources have been implemented. For example, the establishment of protected areas or sustainable management plans of forest resources by logging companies. These initiatives primarily take into account the sustainable management of flora and large mammals, due to the existence of significant scientific works. However, bees, particularly stingless bees, are rarely incorporated into these initiatives. Thus, research is needed to improve our understanding of the biology, ecology and diversity of stingless bees in Gabon, along with the impact of different human activities. To achieve this, the current status of stingless bees must be established. Here, we provide an overview of the diversity, distribution, biology, ecology and traditional knowledge and uses of stingless bees in Gabon.

13.2 Taxonomy and Morphological Diversity of Stingless Bees in Gabon

In Gabon, 16 stingless bee species have been listed (see Fabre Anguilet et al. 2015). These species are divided into five genera: *Cleptotrigona*, *Dactylurina*, *Meliponula*, *Hypotrigona* and *Liotrigona* (see Table 13.1).

These species were determined based on morphological characteristics and nest structure (see Sect. 13.4). Eardley (2004), Michener (2007), Pauly and Fabre Anguilet (2013) and Pauly and Hora (2013) provide full descriptions of stingless bee species encountered in Gabon. Thus, only a few of the main morphological characteristics are presented here.

There are two general size categories of stingless bees: species with small body size and relatively large stingless bees. Small bees generally have a body length less than 5 mm. This group includes *C. cubiceps*, *H. araujoi*, *H. gribodoi*, *H. ruspolii*, *L. bouyssoui* and *L. gabonensis*. These species have a dark-coloured integument, with a metasoma ranging from black to orange-yellow in colour (Eardley 2004; Michener 2007). *Hypotrigona* and *Liotrigona* have forewings with a few veins. Some cryptic species may be present in these and ratios to distinguish some species (Eardley 2004). *Liotrigona* resemble *Hypotrigona* but may be differentiated by a few morphological characteristics. For instance, the scutum of *Liotrigona* is smooth, shiny and sparsely punctate, while that of *Hypotrigona* is dull and densely punctate (Eardley 2004; Eardley et al. 2010; see Fig. 13.1). *Dactylurina* and *Meliponula* include species of intermediate to large body size. These species have a body length of 5–9 mm. *Dactylurina staudingeri* has a black integument, slender body and weakly plumose vestiture (Eardley 2004; Michener 2007). This species has forewings with a faintly black colour. The *Meliponula* are further subdivided into three subgenera: *Meliponula*, *Axestotrigona* and *Meliplebeia* (Eardley et al. 2010). *Meliponula bocandei* is the only species classified in the subgenus *Meliponula*. The integument of this species varies from black to red-black in colour, with orange and yellow-orange areas (Eardley 2004). The *M. (Axestotrigona)* subgenus includes species that lack yellow markings (*M. cameroonensis*, *M. erythra*, *M. ferruginea* and *M. togoensis*), while *Meliplebeia* have yellow markings on the face (*M. beccarii*, *M. lendliana*, *M. nebulata* and *M. roubiki*) (Eardley 2004; Eardley et al. 2010). In addition, *M. lendliana* has yellow markings on the proximal ends of the tibiae, whereas *M. nebulata* has black markings on the tip of the forewings (Pauly 1998).

13.3 Distribution of Stingless Bee Fauna in Gabon

Stingless bees are distributed throughout Gabon. Forests cover approximately 78% of the territory in Gabon (Drouineau and Robert 1999). The vegetation is essentially composed of secondary forests, dense primary forests, riparian forests, humid forests, gallery forests and savannahs (Drouineau and Robert 1999; Ambougou Atisso 1991). There are three climatic areas in Gabon: equatorial climate in the north, tropical climate in the south and a transition zone between the two (Drouineau and Robert 1999) (see Fig. 13.2). The equatorial climate is marked by two dry seasons per year, the tropical climate by 5 months of dry season and 7 months of rainy season and the transition zone between both by 3 months of dry season and 9 months of rainy season (Drouineau and Robert 1999). In general, the climate in Gabon is characterized by average annual rainfall greater than 1500 mm (range: 1200–3000 mm) and 25 °C average annual temperature (range: 22–30 °C) (Maloba Makakanga and Samba 1997; Drouineau and Robert 1999; Tsalefac et al. 2015). The soils are classified in the subclass of ferralitic soils highly desaturated (Chatelin 1968). The relief is characterized by a lack of high elevation (the highest peaks reach approximately 1000 m) and a low frequency of steep slopes (Drouineau and Robert 1999).

However, there is no published work considering distribution patterns of stingless bees based on vegetation cover or climate for equatorial Africa in general, including Gabon. Stingless bees have only been studied in a few locations in Gabon. For example, Brosset and Darchen (1967), Darchen (1966, 1969, 1973, 1977), Darchen and Pain (1966) and Ambougou Atisso (1990) studied the biology, ecology, nesting, nest structure and pollen composition of food reserves of stingless bees in Belinga and Makokou (eastern part of Gabon). Only Pauly (1998) studied the

distribution of stingless bees across this country between 1984 and 1987. Several species of stingless bee are widely distributed in Gabon (Pauly 1998; Eardley 2004; Pauly and Fabre Anguilet 2013) (see Fig. 13.2). Thus, *C. cubiceps, D. staudingeri, H. gribodoi, L. bouyssoui, L. gabonensi, M. bocandei, M. nebulata* and *M. togoensis* are distributed across numerous locations, whereas *M. cameroonensis* is only distributed in the eastern part of Gabon, while *M. lendliana* is most common in the western part of this country (Pauly 1998). When climatic zones are superimposed with the localities studied by Pauly (1998) (see Fig. 13.2), the climate transition zone has the highest species richness and the largest distribution of species. This phenomenon might be explained by two factors. First, forests constitute the preferred habitat of stingless bees (Brosi et al. 2008), and this climatic zone is largely covered by forests, unlike the tropical climatic zone, which is partly covered by gallery forests and savannah. Second, more locations were investigated in the transition climatic zone compared to the other two climate zones. This issue might explain the greater species richness and species distribution detected within the transition climatic zone.

Future studies should update knowledge about the distribution of stingless bees in Gabon to establish how human activities influence the habitats of stingless bees. For example, logging persists in about 51% of the territory of Gabon (Fabre Anguilet et al. 2015), and populations practise the clearing of the forest for the food crops. These activities are practised for many decades and might alter habitats, which, in turn, would impact stingless bee population dynamics. For instance, the species richness of stingless bees has declined substantially over the last three decades at Kougouleu (Fabre Anguilet, in preparation). Pauly (1998) identified eight stingless bee species in this location. Today, 60% of the *Meliponula* genus is no longer observed at Kougouleu, while the species that are still present occur at low frequency. This decline is probably linked to increasing habitat disturbance by human activities at this location. Indeed, the forests at Kougouleu have been subject to forest clearance,

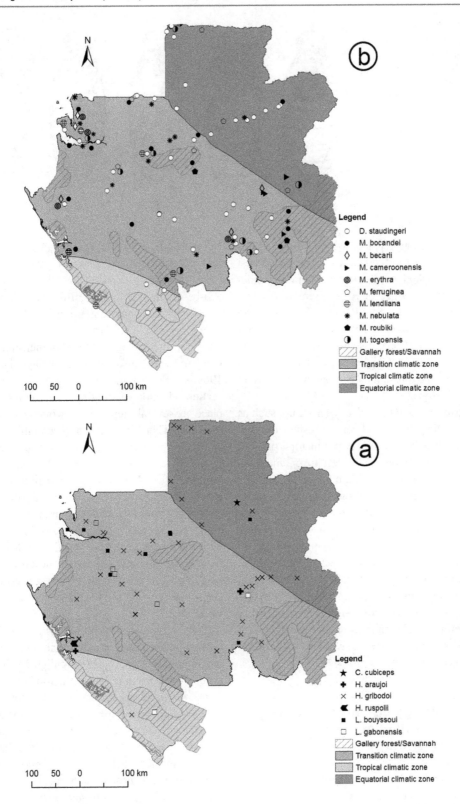

Fig. 13.2 Diversity of stingless bee fauna in Gabon: (**a**) Species with small body size, (**b**) species with larger body size (Modified from E.C. Fabre Anguilet et al. 2015)

Fig. 13.3 The change in metasomal coloration and length of *M. nebulata*: (**a**) at 1 day, (**b**) at 6 days, (**c**) at 13 days, (**d**) at 17 days and (**e**) at 28 days (Modified by E.C. Fabre Anguilet from Darchen 1969)

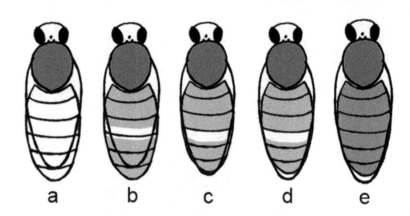

a   b   c   d   e

with forests being replaced with food crops. This huge pressure for food crops to replace forest habitats is linked to an increase in human populations in the province of Estuaire in Kougouleu. For instance, the population expanded from 102,577 inhabitants in 1970 to 895,689 inhabitants in 2013 (Lopez-Escartin 1991; Direction Générale de la Statistique 2015). These populations clear the forest, mostly along main roads, to establish crops.

Agribusinesses have also been set up, such as OLAM. No data have been published on how these agribusinesses impact pollinators, particularly bees in Gabon. The results obtained by Fabre Anguilet (Fabre Anguilet et al., unpublished data), along with those obtained 30 years earlier by Pauly (1998), show that logging has had no effect on the species richness of stingless bees in the concession of Precious Wood Gabon. This logging activity is conducted according to sustainable management rules, which include 25-year rotations. A better knowledge of the distribution and diversity of stingless bees and other pollinators would enhance their integration in sustainable management strategies of natural resources in Gabon.

13.4 Biology, Ecology and Nesting Behaviour of the Stingless Bees

The biology of stingless bee species in Africa has been poorly documented. The life cycle of these species is similar to that of other social species.

In brief, four steps exist: egg (fertilized egg to the new queens or workers), larva, pupa and adult stage (Kwapong et al. 2010). Workers exhibit age-based morphological differences. For instance, the metasoma colouring of *M. nebulata* varies from light to dark red-orange depending on age, with the length of the abdomen compared to the forewings also changing with age (Darchen 1969; see Fig. 13.3).

Reproduction in stingless bee colonies takes place in several steps. First, workers build the new nest. Then, an unmated queen (still capable of flight) and part of the colony occupy the new nest (Oliveira et al. 2013). Darchen (1977) described four steps in the occupation of a new nest by *H. gribodoi* in Gabon. First, the old workers begin to build a tubular entrance inside the new nest. Second, young workers join the old workers to continue nest building and to build food storage pots. Third, the new nest is occupied by a large number of young workers. Fourth, the virgin queen joins the new nest with the young workers.

Only Ambougou Atisso (1990) has published a study analysing the pollen in the food reserves of stingless bees. This study highlighted the presence of 14 types of pollen in the food reserves of *Hypotrigona* during the dry season in forest area of Makokou in the north-eastern part of Gabon.

Although forest and tree cavities constitute the preferred nesting habitat of stingless bees in Gabon, species such as *H. gribodoi* and *D. staudingeri* also nest in anthropogenic settings, such as buildings. *Hypotrigona gribodoi* nests in the cavities of the walls of houses (see Fig. 13.4),

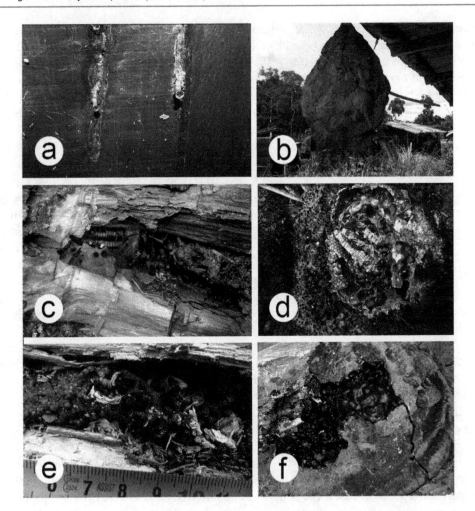

Fig. 13.4 Nests of stingless bees: (**a**) *Hypotrigona* in the wall of a house, (**b**) nest of *Dactylurina staudingeri*, (**c**) nest of *Meliponula nebulata* in a termite mound, (**d**) nest of *Meliponula nebulata* in a tree, (**e**) nest of *Liotrigona gabonensis* in a tree and (**f**) nest of *Meliponula ferruginea* in a tree (Photo: E.C. Fabre Anguilet © Fabre Anguilet E. C., 2016)

and *Dactylurina staudingeri* hangs its nests on the branches of *Theobroma cacao* (Darchen 1977; Darchen and Pain 1966). Two types of nest exist: exposed nests and nests in cavities (see Fig. 13.4). Only *D. staudingeri* builds exposed nests with wax, resin and plant material in Gabon (Smith 1954; Darchen and Pain 1966). The nests of this species were observed 2 m above the ground in secondary forests, primary forests and cocoa plantations in Belinga and Makokou (Darchen and Pain 1966).

Most species occupy tree cavities, including *Hypotrigona, Liotrigona, M. bocandei, M. nebulata, M. ferruginea, M. erythra* and *M. togoensis* (Smith 1954; Darchen 1969; Njoya 2009; Pauly and Fabre Anguilet 2013). *Meliponula bocandei* is the largest African stingless bee and occupies both the ground cavities and tree cavities of Uganda (Kajobe and Roubik 2006). Wille (1983) observed that relatively few stingless bee species build nests in underground cavities. Examples include *M. nebulata* in Gabon and *M. beccarii* and *M. lendliana* in Africa (Smith 1954; Portugal-Araujo 1963; Darchen 1969). Stingless bees also nest in unoccupied termite mounds, in the cavities left by colonies that have died (Darchen 1969; see Figure 4). Termite mounds provide a shelter that allows air circulation for the nests of stingless bees

(Darchen 1969). *Hypotrigona, Cleptotrigona, M. nebulata* and *M. beccarii* have been observed nesting in termite mounds (Darchen 1969). These bees also benefit from cavities excavated by birds and pangolins. The nests of *Crematogaster* ants are also occupied by *M. erythra* (Darchen 1969, 1971). Overall, the nesting biology of stingless bees varies in Gabon, as it does in other parts of the world (Roubik 2006).

13.5 Knowledge and Traditional Use of Stingless Bees in Gabon

Beekeeping with *Apis mellifera* for honey is not practised much in Gabon, and there is no meliponiculture. Rural populations mainly practice honey hunting. In general, only the honey is harvested, with all other products of the colony being thrown away. This activity has a cultural basis and is passed from one generation to the next. Indeed, Central African populations, especially those of Gabon, have a long history of hunter-gatherer-fishers (Oslisly 1998). The honey of stingless bees is primarily used for medicinal purposes. For instance, the honey produced by *M. bocandei* and *M. nebulata* is used to treat respiratory and stomach diseases. The honey collected through honey hunting appears to have been sufficient to satisfy the demand of traditional healers. As a result, rural populations have, perhaps, not seen the need to practice meliponiculture. Rural populations also retrieve the wild nests of *D. staudingeri* suspended from tree branches to suspend them on the walls of houses (see Fig. 4b; Darchen 1966). This practice is related to a local belief that the workers of *D. staudingeri* will protect the occupants of a house against people with malicious intentions.

Vernacular names are used in rural communities. For example, "Abè" is the name used for eusocial bees in the vernacular language of Fang. "Libundu" and "Lévéki" are the names used for stingless bees in the vernacular languages of Ndumu and Nzébi. "Divasou" and "Mvem" are the names used for *Hypotrigona* in the vernacular languages of Punu and Fang. *Meliponula bocandei* is

called "Dibouga" in the vernacular language Punu. However, two names are popular: "sweet honey" for *M. bocandei* and "sour honey" for honey produced by other stingless bees. Information remains limited on the knowledge of rural populations about stingless bees, because no work has been published on this subject in Gabon.

Darchen (1969) transferred *M. nebulata* to hives for stingless bees. Fabre Anguilet et al. (2017) also transferred *M. bocandei* to hives. These trials demonstrated the difficulties of setting up meliponiculture in Gabon. The transfers carried out by Darchen led to the bees deserting from the hives, due to the presence of fungi and parasites (phorids, other Diptera and Coleoptera). Similarly, the colonies of transferred *M. bocandei* were destroyed by the larvae of the small beetle hive *Aethina tumida*. These problems are also common in other African countries. For example, in Uganda, Nkoba (2012) identified *A. tumida* as a threat to stingless bees transferred to hives.

13.6 Conclusion

Gabon has one of the highest diversities of stingless bees in Africa. Species richness in this country must be further studied, because the taxonomy needs clarification, particularly for *Hypotrigona* and *Liotrigona*. The use of morphological methods coupled with molecular methods will facilitate this process. Stingless bees are common throughout Gabon, with the exception of a few species. Nesting behaviour varies within and across species. The effect of the exploitation of natural resources, such as logging, on the habitat and distribution of stingless bees in Gabon needs to be monitored, because most bee species nest in forest habitats and, therefore, depend on this resource.

To introduce and popularize stingless beekeeping in Gabon successfully, first, certain research is required. Information on thermoregulation, the splitting of colonies and control mechanisms against pests and enemies of stingless bees needs to be acquired. The study of the organoleptic and physicochemical characteristics of the honey of stingless bees in Gabon would also enhance its value in Gabon. At present, knowledge remains limited on stingless bees in Gabon, yet, they have the potential to provide economic benefits to local communities, provided appropriate knowledge is acquired. Targeted research on understanding the biology and meliponiculture is therefore required.

Acknowledgements The authors thank the Gabonese government and the Commission Economique du Bétail, de la Viande et des Ressources Halieutiques for their financial support and the Agence Nationale des Parcs Nationaux and Precious Woods Gabon for logistical support.

References

Ambougou Atisso V. 1990. Analyse pollinique des réserves alientaires *d'Apis mellifera adansonii* L. et *d'Hypotriogona sp* (Hym. Apidae sociaux de la région de Makokou (N-E) Gabon). Bulletin de la société Botanique de France 137, Actualité Botanique 2: 166–169.

Ambougou Atisso V. 1991. *Apis mellifera adansonii* Lat. et les plantes utiles mellifères gabonaises (Département de l'Ivindo), recherches palynologiques. Thèse de doctorat, Université Paris 6 ; Paris, France. 160 pp.

Brosi BJ, Daily GC, Shih, TM, Oviedo F, Durán G. 2008. The effects of forest fragmentation on bee communities in tropical countryside. Journal of Applied Ecology 45: 773–783.

Brosset A., Darchen R. 1967. Une curieuse succession d'hôtes parasites des nids de Nasutitermes. Biologia Gabonica 3: 153–168.

Camargo JMF, Pedro SRM. 2008. Meliponini Lepeletier, 1836. In Moure JS, Urban D, Melo GAR, eds. Catalogue of Bees (Hymenoptera, Apoidea) in the Neotropical Region - online version. Available at: http://www.moure.cria.org.br/catalogue .

Chatelin Y. 1968. Notes de pédologie gabonaise. V-géomorphologie et pédoloie dans le Sud Gabon, des Monts Birougou au Littoral. Cahiers ORSTOM Série pédologie 6: 3-20.

Darchen R. 1966. Sur l'éthologie de *Trigona (Dactylurina) staudingeri* Gribodoi (sic) (Hymenoptera Apidae). Biologia Gabonica 2: 37–45.

Darchen R. 1969. Sur la biologie de *Trigona (Apotrigona) nebulata* komiensis Cock. I. Biologia Gabonica 5: 151–183.

Darchen R. 1971. *Trigona (Axestotrigona) oyani* Darchen (Apidae, Trigoninae), une nouvelle espèce d'abeille

africaine – description du nid inclus dans une fourmilière. Biologia Gabonica 7: 407–421.

Darchen R. 1973. La thermorégulation et l'écologie de quelques espèces d'abeilles sociales d'Afrique (Apidae, Trigonini et *Apis mellifica* var. adansonii). Apidologie 4: 341–370.

Darchen R. 1977. L'essaimage Chez Les Hypotrigones au Gabon dynamique de quelques populations. Apidologie 8: 33–59.

Darchen R, Pain J. 1966. Le nid de *Trigona (Dactylurina) staudingeri* Gribodoi (sic) (Hymenoptera: Apidae). Biologia Gabonica 2: 25–35.

Direction Générale de la Statistique 2015. Résultats globaux du Recensement Général de la Population et des Logements de 2013 du Gabon (RGPL-2013). Libreville. 195 pp. Available at http://www.mays-mouissi.com/wp-content/uploads/2016/07/Recensement-general-de-la-population-et-des-logements-de-2013-RGPL.pdf

Drouineau S, Robert N. 1999. L'aménagement forestier au Gabon : historique, bilan, perspectives.. Available at http://forafri.cirad.fr/ressources/forafri/08.pdf

Eardley CD. 2004. Taxonomic revision of the African stingless bees (Apoidea: Apidae: Apinae: Meliponini). Plant Protection 10: 63–96.

Eardley CD, Urban R. 2010. Catalogue of Afrotropical bees (Hymenoptera: Apoidea: Apiformes). Zootaxa 2455: 1–548.

Eardley CD, Kuhlmann M, Pauly A. 2010. Les genres et sous-genres d'abeilles de l'Afrique subsaharienne. ABC Taxa 9: 1-144.

Fabre Anguilet EC, Nguyen BK, Bengone Ndong T, Haubruge E, Francis F. 2015. Meliponini and Apini in Africa (Apidae: Apinae): a review on the challenges and stakes bound to their diversity and their distribution. Biotechnologie, Agronomie, Société et Environnement, 19: 382–391.

Fabre Anguilet EC, Alabi T, Haubruge E, Nguyen BK, Bengone Ndong T, Francis F. 2017. Parasitisme d'*Apis mellifera adansonii* (Latreille 1804) et de *Meliponula bocandei* (Spinola 1853) par *Aethina tumida* (Murray 1867): premier recensement au Gabon et impact sur la domestication. Entomologie Faunistique 70: 3-11.

Hoy M. 2013. Insect molecular genetics, an introduction to principles and applications, 3rd Ed. Academic Press; San Diego, USA. 840 pp.

Kajobe R, Roubik, DW. 2006. Honey-making bee colony abundance and predation by apes and humans in a Uganda forest reserve. Biotropica 2: 210-218.

Koch H. 2010. Combining Morphology and DNA Barcoding Resolves the Taxonomy of Western Malagasy *Liotrigona* Moure, 1961 (Hymenoptera: Apidae: Meliponini). African Invertebrates 51: 413–421.

Kwapong P, Aidoo K, Combey R, Karikari A. 2010. Stingless Bees – importance, management and utilization: a training manual for stingless beekeeping. Unimax Macmillan Ltd; Accra, Ghana. 72 pp.

Lopez-Escartin N. 1991. Données de base sur la population : Gabon. Centre Français sur la Population et le Développement. Paris, France. 11 pp.

Maloba Makakanga JD, Samba G. 1997. Organisation pluviométrique de l'espace Congo-Gabon (1951-1990). Sécheresse 8 : 39-45.

Michener CD. 2007. The bees of the world, second edition. Johns Hopkins University Press; Baltimore, USA. 953 pp.

Munyuli T. 2014. Influence of functional traits on foraging behaviour and pollination efficiency of wild social and solitary bees visiting coffee (*Coffea canephora*) flowers in Uganda. Grana 53: 69–89.

Njoya MTM. 2009. Diversity of Stingless Bees in Bamenda Afromontane Forests – Cameroon: Nest architecture, Behaviour and Labour calendar. Dissertation, Hohen Landwirtschaftlichen Fakultät der Rheinischen Friedrich-Wilhelms-Universität zu Bonn, 138 pp.

Nkoba, K. 2012. Distribution, behavioural biology, rearing and pollination efficiency of five stingless bee species (Apidae: Meliponinae) in Kakamega forest, Kenya. Kenyatta University; Nairobi, Kenya. 237pp.

Oliveira RC, Menezes C, Soares AEE, Fonseca VLI. 2013. Trap-nests for stingless bees (Hymenoptera, Meliponini). Apidologie 44: 29–37.

Oslisly R, 1998. Hommes et milieux à l'Holocène dans la moyenne vallée de l'Ogooué (Gabon). Bulletin de la Société Préhistorique Française 95: 93–105.

Pauly A. 1998. Hymenoptera Apoidea du Gabon. Annales Sciences Zoologiques 282: 1–121.

Pauly A, Fabre Anguilet EC. 2013. Description de *Liotrigona gabonensis* sp. nov., et quelques corrections à la synonymie des espèces africaines de méli-pones (Hymenoptera: Apoidea: Apinae: Meliponini). Belgian Journal of Entomology 15: 1–13.

Pauly A, Hora ZA. 2013. Apini and Meliponini from Ethiopia (Hymenoptera: Apoidea: Apidae: Apinae). Belgian Journal of Entomology 16: 1–36.

Portugal-Araujo VD. 1963. Subterranean nests of two African stingless bees (Hymenoptera: Apidae). Journal of the New York Entomological Society 71: 130–141.

Roubik, DW. 2006. Stingless bee nesting biology. Apidologie 37: 124–143.

Smith FG. 1954. Notes on the biology and waxes of four species of African *Trigona* bees (Hymenoptera: Apidae). Proceedings of the Royal Entomological Society of London, Series A, General Entomology 29: 62-70.

Slaa EJ, Chaves LAS, Malagodi-Braga KS, Hofstede FE. 2006. Stingless bees in applied pollination: practice and perspectives. Apidologie 37: 293-315.

Tsalefac M, Hiol Hiol F, Mahé G., Laraque A., Sonwa D, Scholte P., Pokam W, Haensler A., Beyene T, Ludwig F, Mkankam FK, Djoufack VM, Ndjatsana M, Doumenge C. 2015. Climat de l'Afrique centrale : passé, présent et future. pp. 37-52. In Wasseige C, Tadoum M, Eba'a Atyi R, Doumenge C, eds. Les forêts du Bassin du Congo - Forêts et changements climatiques. Weyrich. Belgique. 128 pp.

Tchuenguem FN, Messi J, Pauly A. 2001. Activité de *Meliponula erythra* sur les fleurs de *Dacryodes edulis* et son impact sur la fructification. Fruits 56: 179–188.

Tchuenguem FN, Messi J, Pauly A. 2002. Recherches sur l'activité des Apoïdes sauvages sur le maïs à Yaoundé (Cameroun) et réflexions sur la pollinisation des graminées. Biotechnologie, Agronomie, Société et Environnement 6: 87–98.

Wille A. 1983. Stingless bee biology. Annual Review of Entomology 28: 41–64.

100 Species of Meliponines (Apidae: Meliponini) in a Parcel of Western Amazonian Forest at Yasuní Biosphere Reserve, Ecuador

14

David W. Roubik

14.1 Introduction

14.1.1 Yasuní Forest and Melittological Background

Yasuní National Park, recently designed a UNESCO Biosphere Reserve, is in the Francisco de Orellana Province of eastern Ecuador and lies south of the equator in western Amazonia, at 0°0.41′0.5″S, 76°0.23′58.9″W (Fig. 14.1). In its nearly million hectares a representative sample yielded 670 tree species in a single hectare (Pérez et al. 2014). Moreover, a permanent forest 50 ha study plot contains >1100 tree species (Davies 2014), and >3000 are expected in the entire forest preserve (Pérez et al. 2014). At least 500 liana species also grow there, with other flora and many fungi.

General arthropod abundance is spectacular in such Neotropical forest, with species in the tens of thousands, and those with no scientific name are common (Basset et al. 2012; Hanson and Nishida 2016). In Yasuní National Park, stingless honey bees (Meliponini) surpass expectations.

Collection efforts include the first that are systematic for the region and encompass roughly 130 field days and >180 collection sites during 20 years. I made an effort to visit and collect in the park throughout the year during 1998 to 2017, in part to detect the presence of the invasive Africanized honey bee (Roubik 1999, 2000).

Hyperdiversity has been encountered in all Yasuní life forms studied so far, as indicated by 610 bird species, 2274 trees and shrubs, 139 amphibians, 204 mammals, and 121 reptiles (Pérez et al. 2014; E. Baus, pers. comm.). Roubik (1999), from two trips in 1998, reported 67 stingless honey bee species at Yasuní, but provided no details. Those meliponines were useful in understanding the backbone of the bee community. They are permanent colonies, consisting of 100s to 1000s of workers, which forage throughout the year, but their biology and lifestyles vary considerably (Roubik 1989, 2006; Camargo 2013; Michener 2007, 2013).

One reason to study meliponine bees is that such arthropods, taken as a whole, are not only diverse in ecology and behavior, but also very numerous. The bees are readily seen and studied. Their collection at food and other forage, and nesting in trees or in the ground, allows them to be surveyed and sampled in a rigorous and efficient manner. As an indicator of biodiversity, their reliable encounter and identification should allow them to rank with the mammals, birds, trees, and other organisms that permit researchers

D.W. Roubik (✉)
Smithsonian Tropical Research Institute,
Calle Portobelo, Balboa, Ancón, Republic of Panama
e-mail: roubikd@si.edu

© Springer International Publishing AG, part of Springer Nature 2018
P. Vit et al. (eds.), *Pot-Pollen in Stingless Bee Melittology*, DOI 10.1007/978-3-319-61839-5_14

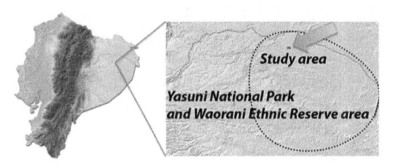

Fig. 14.1 Topographic map of Ecuador showing the Yasuní National Park and Waorani Ethnic Reserve, with the permanent forest plot at the Yasuní Scientific Research Center (PUCE-YSRS) shown in *red*, and the Napo River shown to the north (Graphic: D.W. Roubik)

to estimate species richness or local biodiversity as a whole (Basset et al. 2012). Moreover, Meliponini act relentlessly to favor forest reproduction, by moving pollen between flowers of the same plant species, and by competing with other potential pollinators, thereby increasing their dispersal between tree canopies or shrubs.

The stingless honey bees—meliponines—both exploit and move an enormous amount of pollen (Roubik 1993). Individual nests contain 1–10 billion pollen grains (Roubik and Moreno 2009), and the scutellum of an exposed nest (made by certain *Trigona* and *Partamona*) can comprise a thousand times more grains than that, owing to the pollen feces recycled as building material (Roubik and Moreno 2009, and chapter in this book).

Considering meliponine species richness, there has been a revolution in discovery, recognition, and documentation of species through careful comparative study, both morphological and molecular (e.g., Camargo and Pedro 2008; Rasmussen and Camargo 2008; Pedro 2014). In Brazil, 242 meliponine species have been discerned and another 86 collected new species await formal scientific description (see Pedro 2014). A similar large number of yet-to-be described species, and hundreds already known, are reported (Rasmussen and Gonzalez 2013). And in an extensive survey conducted for 1 year in the Brazilian frontier of Rondônia state, Brown and de Oliveira (2012) found 98 meliponines,

including 17 new to science. Of particular import for work discovering new species in western Amazonia, descriptions with keys have recently been given for Brazilian *Dolichotrigona* (Camargo and Pedro 2005), *Trigonisca* (Albuquerque and Camargo 2007), *Celetrigona* (Camargo and Pedro 2009), *Leurotrigona* (Pedro and Camargo 2009), *Camargoia* (Moure 1989) and in broader areas for *Oxytrigona* (Gonzalez and Roubik 2008), *Geotrigona, Ptilotrigona, Partamona, Paratrigona,* and *Aparatrigona* (Camargo and Moure 1994, 1996; Camargo and Pedro 2003a, b). Like other authors (Pedro and Camargo 2009; Brown and de Oliveira 2012), I categorize bees in genera, with no subgenera but instead, species groups. Groups may tend to erode with further information and research, as new species are discovered and described—which may lead to larger and more diverse groups. However, composites that name both a genus and one of its subgenera, like *Trigona (Geotrigona)*, *Plebeia (Schwarzula)*, or *Trigonisca (Leurotrigona)*[see Michener 2007], seem somewhat awkward or discordant with morphology. Regarding most supraspecific groups as genera is convenient where many species are un-named, and new genera are still being found or codified (Moure and Camargo 1982; Roubik et al. 1997; Camargo and Roubik 2005; Pedro and Cordeiro 2015; Melo 2016). Even without a specific epithet, aspects of biology of a species are often known from its morphological link to named species.

14.2 Discovering Meliponine Biodiversity

The present study was centered at the Pontífica Universidad Católica del Ecuador's scientific research station (PUCE-YSRS), where there is a trail system, and an access road, which allow surveys within a core forest area of approximately 3 × 3 km, close to 8 km², which include a flank of the Tiputini River. Some collections were made using the maintained gravel road system, 130 km in total extent, and also an area along the subterranean oil pipeline and road, with a 15 to 50-m strip kept clear manually—not with herbicide—at the border of woody vegetation. Almost all the collections evaluated here were made within 7 km of the field station, most <3 km.

Many secondary growth plants occur along forest edges and the road margin, including *Cecropia, Ochroma,* Melastomataceae, Piperaceae, *Cassia,* and *Heliconia.* In forest and near the Tiputini River, especially during sunny days, after 10,00 h, many "sweat bees", composed of worker meliponines and female halictid bees, land on the field researcher and imbibe salts from moist skin and clothing. The roadsides, similarly, often teem with butterflies, meliponines, an occasional *Apis,* and bees of many kinds, including *Centris,* megachilids, small *Xylocopa,* and halictids, which gather salts from moist soil along the pavement, and in areas where algae (which are rich in sodium) are found in puddles. Besides using such an "automatic collecting protocol", my standard collection methods included netting bees on all accessible flowers, resin sources (floral, on woody stems, or on fruit), and at baits made with honeywater. Combined with anise or peppermint flavoring, honeywater of approximately 40% sugar content was used. The honey was sometimes not of tropical origin, and results may differ from this factor, which seemed to negatively affect recruitment by *Melipona.* Honeywater was sprayed on vegetation, usually plants with large leaves, as the technique was originally applied in Costa Rica (Wille 1962; and Fig. 14.2; see also Brown and de Oliveira 2012) to attract a variety of honey bees and halictines.

In addition, an important "by-catch" from a butterfly food baiting study, using net cylinder traps placed in the canopy and understory, was included in 2016. Some collections were at natural nests (Fig. 14.3), and also made by taking bees collecting mineral salts, foraging or searching along the ground at various resources—presumably plant, fungal, or microbial, and at animal dung and remains. During October of 2016, a Waorani field helper provided bee names in the Waorani language. This was a young gentleman met through the field station staff. Since the Wao have occupied this part of western Amazonia for several thousand generations, the common forest bees, and their nests, have names. Whether these are universally applied is no different from taxonomy in its current, Western usage and application, but the two forms of nomenclature are driven by different goals, perhaps either practical or convenient. For example, while my "informant" applied different names to two different species of orange *Partamona,* his word for a large *Centris* or *Xylocopa* was the same. Future work will be required to verify the names reported here.

Yasuní Park contains an "aseasonal forest" in terms of monthly rainfall, but some seasonality occurs for certain non-meliponine bees (DWR, unpublished data). Year-long collections therefore remain important, as would be the case in more seasonal forest with predictable wet and dry seasons. Collections were made at PUCE–YSRS during April, 1998; November, 1998 (I. Tapia, F. Palomeque R., assistants); February, 2001; September, 2001 (E. Baus, assistant); December, 2002 (E. Baus); August, 2004; January, 2012 (Ascher, JS, Wyman ES, Webber D, AMNH); February, March and October, 2016, and during June and July, 2017.

Several species were found, in addition, in the collections at QMAZ (Quito-Católica, Zoología), many made by L. Coloma or G. Onore, pioneers of meliponine study in Ecuador. The monitoring of bees in the canopy and the understory, using food traps baited with mashed banana or rotten shrimp, for butterfly studies, was carried out by Sophia Nogales, PUCE, from March until August

Fig. 14.2 Example of group foraging by the large colony of *Trigona amazonensis*. (**a**) First arrival of "scout group" of ca. 150 workers, on a stem next to leaves sprayed with honeywater bait; (**b**) nest with typical nest entrance tube, at approximately 25 m height, on trunk of *Ceiba pentandra;* (**c**) part of large forager force (ca. 2000 workers) on honeywater bait. The original scouts had left the site; the larger group arrived approximately 10 min later, and another aggressive group forager, *T. amalthea*, was displaced immediately, after previous skirmishes with the scout group (Photos: D.W. Roubik)

of 2016. The work included 20 pairs of traps set within the permanent 50-ha CTFS/STRI (Center for Tropical Forest Studies, Smithsonian Tropical Research Institute) forest study plot (see Fig. 14.1), and captures from all traps of each height were pooled monthly. Herein is the taxonomic summary of stingless honey bees trapped and identified, using canopy to ground survey techniques (see Table 14.1), and brief discussion of some of the more salient aspects of their biology.

Readers who wonder about methods and approaches for study of bee species and their identification may wish to consider the following. One discovers "cryptic species" after becoming familiar with many related bees, after curating and examining hundreds of individual specimens, both alive and preserved. At the microscope, one often begins with size, measured with a micrometer, and then the wing color (which is not subject to much variation from resin or pollen on the integument, since it is not carried there, and it is mostly hairless). Certain wing veins, their color, size, and shape may also be diagnostic for species. Next, the scape and its hairs, and the color, size, and distribution of hair on the face in general, and the smoothness, sculpture, and reticulation on all body surfaces, along with general body and leg color, offer guidance in placing a specimen either with those already known and examined, or as another species. The mandibles may be opened, or removed, to be closely

Fig. 14.3 Examples of stingless honey bee nests in Yasuní Biosphere Reserve. (**a**) Nest entrance of *Partamona vicina* associated with live termite nest on live tree of *Cedrelinga*; (**b**) Nest of *Plebeia* sp. high in the slender trunk of *Cecropia* sp.; (**c**) hard resin nest entrance of *Ptilotrigona pereneae*, removed from live tree trunk; (**d**) exiting defensive workers at nest entrance of *Ptilotrigona lurida* in live tree; (**e**) exposed nest of *Trigona amalthea*, on *Cecropia* tree uprooted by wind (Photos: D.W. Roubik)

examined, since they often provide diagnostic characteristics, such as denticle size, position, form, and number. They are subject to wear and variation, however, and may differ greatly among bees from the same colony. The apical abdominal segments, both from ventral and dorsal views, often provide clear insight—considering hair, color, and form, as to the identity of species. The "hidden" seventh sternite is also very useful as part of the male genital capsule and associated structures. If a specimen is moldy, covered with resin or pollen, or broken, it can still be cleaned and made useful for taxonomic study.

14.3   Species Accounts and Frequency

The tiny Meliponini and *Trigona* are the two groups most speciose among Yasuní lowland forest meliponines; there are 23 or 24 genera at Yasuní; uncertainty stems from occurrence of *Nogueirapis* elsewhere in the Napo region, but with no confirmation in Yasuní forest. Eight of the genera consist of tiny bees (Table 14.1, Figs. 14.3, 14.4, and 14.5) with head widths of around 1 mm, and a metatibia length averaging

Table 14.1 Records of stingless honey bees at Yasuní forest, Ecuador

Aparatrigona impunctata (Ducke, 1916)[a]
Celetrigona euclydiana Camargo & Pedro, 2009
Cephalotrigona capitata (Smith, 1854)
Dolichotrigona browni Camargo & Pedro, 2005[a]
Dolichotrigona chachapoya Camargo & Pedro, 2005
Dolichotrigona moratoi Camargo & Pedro, 2005
Dolichotrigona 3 spp.
Duckeola ghilianii (Spinola, 1853)[a]
Frieseomelitta trichocerata Moure, 1990[a]
Geotrigona fulvohirta (Friese, 1900)
Lestrimelitta glabrata Camargo & Moure, 1989
Leurotrigona gracilis Pedro & Camargo, 2009
Leurotrigona 2 spp.
Melipona aff. *puncticollis*
Melipona captiosa Moure, 1962
Melipona crinita Moure & Kerr, 1950[a]
Melipona aff. *crinita*
Melipona eburnea Friese, 1900
Melipona aff. *fuscopilosa*
Melipona grandis Guérin-Méneville, 1844[a]
Melipona nebulosa Camargo, 1988
Melipona titania Gribodo, 1893
Nannotrigona melanocera (Schwarz, 1938)[a]
Nogueirapis mirandula (Cockerell, 1917) [Napo Prov.]
Oxytrigona huaoranii Gonzalez & Roubik, 2008
Oxytrigona mulfordi (Schwarz, 1948)
Oxytrigona obscura (Friese, 1900)[a]
Oxytrigona aff. *tataira*
Paratrigona aff. *eutaeniata*
Paratrigona aff. *lophocoryphe*
Paratrigona aff. *nuda*
Paratrigona prosopiformis (Gribodo, 1893)[a]
Paratrigona scapisetosa Gonzalez &Griswold, 2011
Partamona epiphytophila Camargo and Pedro, 2003a
Partamona testacea (Klug, 1807)[a]
Partamona vicina Camargo, 1980[a]
Partamona sp.
Plebeia aff. *alvarengai*
Plebeia flavocincta (Cockerell, 1912)
Plebeia minima (Gribodo, 1893)
Plebeia 4 spp.
Ptilotrigona lurida (Smith, 1854)[a]
Ptilotrigona pereneae (Schwarz, 1943)
Scaptotrigona tricolorata Camargo, 1988[a]
Scaptotrigona 2 spp.
Scaura latitarsis (Friese, 1900)[a]
Scaura longula (Lepeletier, 1836)[a]

(continued)

Scaura tenuis (Ducke, 1916)[a]
Scaura aff. *tenuis*
Schwarzula timida (Silvestri, 1902)[a]
Schwarzula sp.
Tetragona clavipes (Fabricius, 1804)[a]
Tetragona dorsalis (Smith, 1854)[a]
Tetragona goettei (Friese, 1900)[a]
Tetragona handlirschii (Friese, 1900)[a]
Tetragona truncata Moure, 1971[a]
Tetragonisca angustula (Latreille, 1811)[a]
Trigona albipennis Almeida, 1995[a]
Trigona amalthea (Olivier, 1789)[b]
Trigona aff. *amalthea*
Trigona amazonensis (Ducke, 1916)[a]
Trigona aff. *amazonensis*
Trigona aff. *branneri*
Trigona chanchamayoensis Schwarz, 1948[a]
Trigona cilipes (Fabricius, 1804)[a]
Trigona crassipes (Fabricius, 1793)[a]
Trigona dallatorreana Friese, 1900[a]
Trigona dimidiata Smith, 1854[a]
Trigona aff. *fuscipennis*
Trigona guianae Cockerell, 1910
Trigona aff. *fulviventris* Guérin-Méneville
Trigona hypogea Silvestri, 1902[a]
Trigona permodica Almeida, 1995[a]
Trigona recursa Smith, 1863[a]
Trigona truculenta Almeida, 1985[a]
Trigona williana Friese, 1900[a]
Trigonisca variegatifrons Albuquerque & Camargo, 2007[a]
Trigonisca 16 spp.

Scientific names for meliponines follow the checklist provided by Camargo and Pedro (2008), subgenera in Michener (2007) are given as genera, with the exception of *Melipona* (op. cit.). The term "aff." means similarity or affinity with another species or a group, at this working level. Voucher specimens are at STRI (Smithsonian Tropical Research Institute, Roubik collection), QCAZ, and AMNH

[a]Species known also from the state of Rondônia, Brazil (Brown and de Oliveira 2012)

[b]This species has not been collected near its reported type locality, Cayenne, French Guiana. Because the region is sparsely settled but naturally forested, the type specimen is probably mislabeled (Pauly et al. 2013; DWR, personal observation). Both Cockerell (1913) and Moure (1960) call the bee from French Guiana *T. fuscipennis* Friese, 1900, while Schwarz (1948) mentions the absence of *silvestriana* (Vachal, 1908) or *amalthea* in most of the Guiana Shield. Two photos of the actual type ZMUC00044527 (Courtesy C. Rasmussen) show it is neither *fuscipennis* nor *truculenta*

about 1.5 mm—the full range of species and genera suggest a 12× difference in tibia length and a 5× difference in headwidth, with *Leurotrigona* the smallest, and *Melipona*, *Duckeola*, certain *Trigona*, and *Cephalotrigona* the largest. The significance of some morphological features is discussed below, but the purpose for giving the measurements and photomicrographs is to help describe the singular bee biodiversity among Yasuní meliponines. The measurements of any particular series of a given species, using a fine-detailed photography system (Keyence, provided by the USDA-ARS bee biology and systematics lab at Logan, Utah), lead me to conclude that size variation, in excess of 10% the mean, is very unusual.

Both the largest and smallest meliponines coexist at Yasuní, along with the most varied biological repertoires of the tribe. There are approximately 50 named Meliponini in the forest by YSRS, and a total of >70 *Trigonisca, Dolichotrigona, Celetrigona, Leurotrigona* (Fig. 14.5), and other tiny species (see below). Recent phylogenetic studies, and also taxonomy (Camargo and Pedro 2003a, b, 2004, 2005, 2009; Michener 2007) in addition to nuclear gene studies (Rasmussen and Cameron 2010), suggest that *Trigonisca* and *Dolichotrigona* may be too similar to separate. Their normal morphology and behavior do not support this view, however (Figs. 14.3, 14.4, and 14.5). *Dolichotrigona* measure about 2–3 mm long and often run about on flowers or leaves, as though they were ants, with long, "elbowed" antennae, as do the even larger *Celetrigona* (Fig. 14.5a, e). *Trigonisca* (most ca. 2 mm in body length) have similar behavior, at least in rapid walks on vegetation sprayed with honeywater, while *Leurotrigona, Scaura, Schwarzula*, and most *Plebeia* are tiny but do not seem to mimic ants or move at such a rapid pace. The *Leurotrigona* (Fig. 14.5d) with two undescribed species and one also known from western Amazonia (Pedro and Camargo 2009) were previously unknown in Ecuador.

The largest single genus at Yasuní is *Trigonisca* (17 species, 16 undescribed) followed by *Trigona*

and *Plebeia*. *Melipona*, the largest meliponine and perhaps the largest single meliponine genus in the world, is endemic to the Neotropics. Eight *Melipona* have been found at Yasuní, about two-thirds the number known in lowland forest of French Guiana (Pauly et al. 2013) or in western Amazonian Brazil (Brown and de Oliveira 2012).

Although none of the tiny species display aggression toward other foragers, as a general rule, the other dominant species, genus *Trigona*, are organized group foragers and display both ritualized contests and attack (summary in Roubik 1989). Their group foraging was amply illustrated in arrival at honeywater bait (Fig. 14.2). During one morning, a typically large colony of *Trigona amazonensis* first arrived as a "scout group" of ca. 150 workers, on a stem next to leaves I sprayed with honeywater bait. A group of roughly 50–75 *T. amalthea* was already foraging at the bait, and several pairs of the arriving *T. amazonensis* and *T. amalthea* could be heard falling on leaves, locked venter to venter, as aerial skirmishes occurred. Abruptly, after ca. 10 min., most *T. amazonensis* left the mossy stem where they had landed, next to the bait on leaves. In approximately 10 min a much larger "expeditionary force" of ca. 2000 worker *T. amazonensis* arrived, and occupied the baited leaves to begin foraging. These were presumably bees led back to the foraging site by workers from the original scouting group, and the likely nest, located on a large *Ceiba pentandra* (evidently the tree of choice used by this species, from known nests $N = 5$, DWR, pers. obs.), was located <500 m from the site. *Trigona amazonensis* immediately abandoned the honeywater when the much larger group arrived. *Trigona* foraging groups fly above the canopy in small clouds of hundreds to thousands of workers, and may partition a tree canopy resource between different colonies (Roubik 2002). Only one colony of *T. amazonensis* appeared to find the bait I set out, but at one other site, two groups continued to battle for an entire morning, evidenced by pairs of hovering bees falling to the ground, locked in combat.

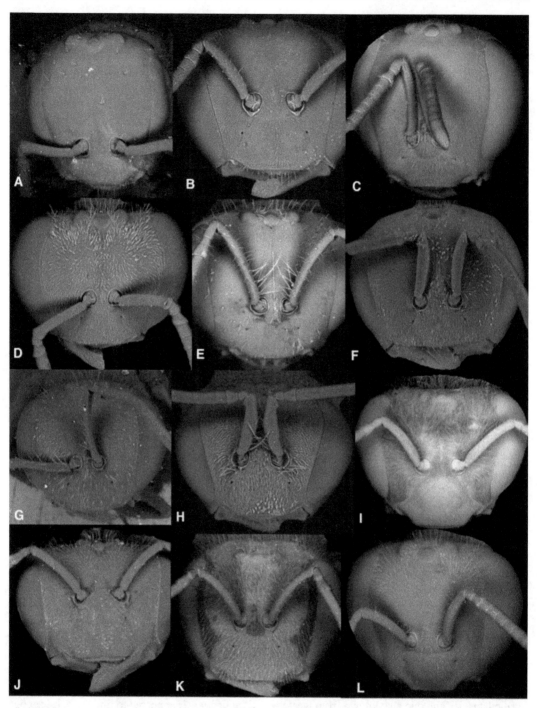

Fig. 14.4 Legs and heads of all genera of Meliponini already collected or likely to occur (*Nogueirapis*) in Yasuní. Not shown to a standard scale (see text). Scanning electron micrographs: J. Ceballos and D. W. Roubik. (**a**) *Leurotrigona*.;(**b**) *Nogueirapis mirandula*; (**c**) *Trigonisca variegatifrons* Albuquerque and Camargo 2007; (**d**) *Scaura longula*; (**e**) *Celetrigona euclydiana*; (**f**) *Schwarzula timida*; (**g**) *Dolichotrigona chachapoya*; (**h**) *Nannotrigona melanocera*; (**i**) *Melipona fuscopilosa*.; (**j**) *Cephalotrigona capitata*.; (**k**) *Duckeola ghilianii*; (**l**) *Partamona testacea*; (**m**) *Ptilotrigona lurida*; (**n**) *Frieseomelitta trichocerata*; (**o**) *Geotrigona fulvohirta*; (**p**) *Tetragona clavipes*; (**q**) *Trigona truculenta*; (**r**) *Tetragonisca angustula*.; (**s**) *Lestrimelitta* sp.; (**t**) *Paratrigona prosopiformis*. (**u**). *Aparatrigona impunctata*.; (**v**) *Scaptotrigona tricolorata*; (**w**) *Oxytrigona huaoranii*; (**x**) *Plebeia* sp.

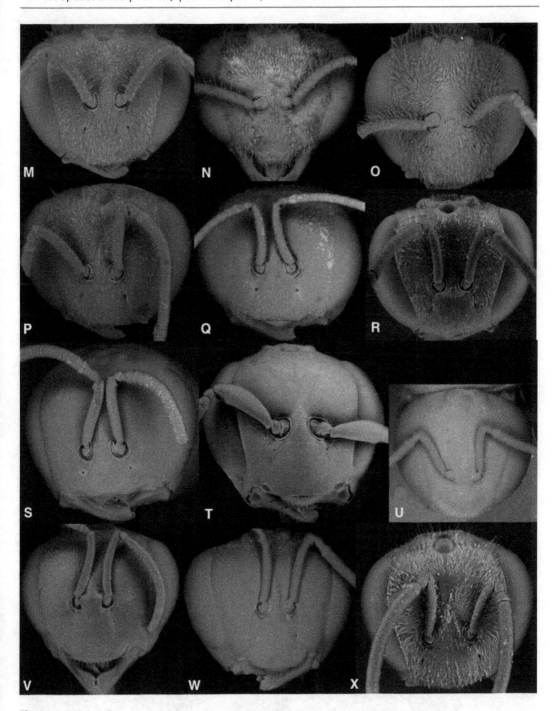

Fig. 14.4 (continued)

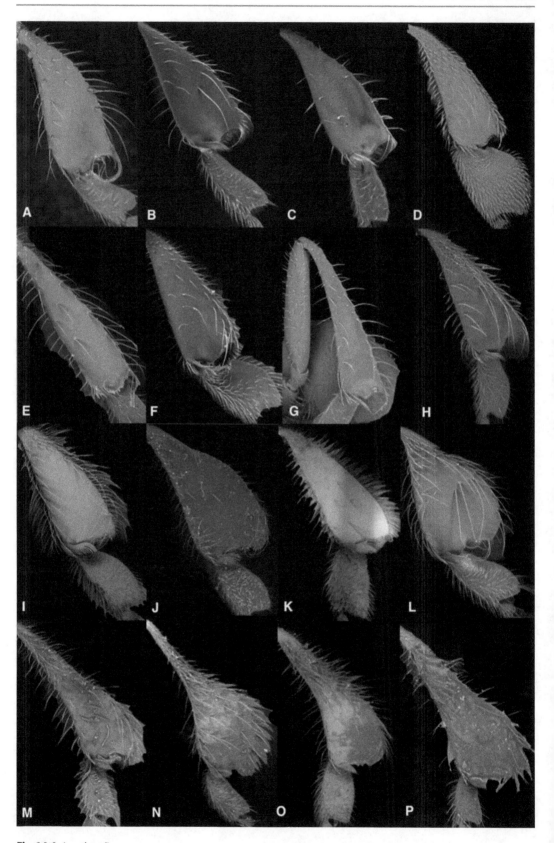

Fig. 14.4 (continued)

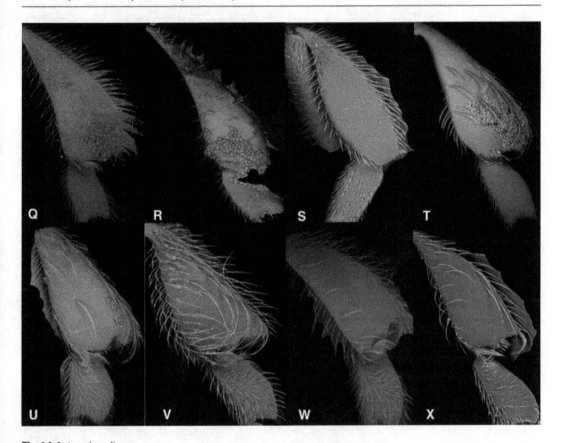

Fig. 14.4 (continued)

14.4 Insights from Comparative Morphology and Other Rich Amazonian Areas

Amazonian biogeography is complex, and three broad regions have been considered for meliponines, with the draining of the intercontinental seas 30 mya, and the rise of the Andes approximately 3 mya, playing major and yet scarcely explored roles (Camargo 2013). Here I include four lowland forest areas with extensive evergreen forest and considerable meliponine and botanical data, although only two are near major rivers and only one is near the Andes.

French Guiana, within its 8×10^4 km², and where annual rainfall is generally 2–3 m but reaches nearly 4 m in the northeast, and a 10,000 ha forest block north of Manaus, Brazil (average annual rainfall 2286 mm; climatemps.

com, accessed October, 2016) share a number of species with Yasuní (Oliveira 2001; Pauly et al. 2013). Moreover, the large systematically sampled state of Rondônia, Brazil, of 2.37×10^5 km² is perhaps the most similar. Nonetheless, no two sites share a majority of their species, and the reader may consult the original publications for further details on species or morphospecies. The Amazonian localities are separated from Yasuní by 1900–2700 km, and all are lowland; French Guiana is 3–5 °N, the Manaus site approximately 3 °S, and that of Ouro Preto, Rondônia 10 °S.

Almost all genera are shared among the 80 species listed for French Guiana, the 37 for the Manaus, Ducke Reserve, the 110 for the entire state of Rondônia, and the 100 found at Yasuní. Rondônia and Yasuní are identical in genera recorded (discounting *Nogueirapis*), and share approximately one-third of their meliponine species. French Guiana likely contains *Trichotrigona*

Fig. 14.5 Representative tiny Melipoini of the Yasuní forest.
(**a**) *Celetrigona*,
(**b**) *Trigonisca* sp. 1.,
(**c**) *Trigonisca* sp. 2,
(**d**) *Leurotrigona*,
(**e**) *Celetrigona*
(Photos: D.W. Roubik)

(Pauly et al. 2013), which occurs nearby in Brazil, as I also assume *Nogueirapis* may yet be found in Yasuní, judging from my collection of *N. mirandula,* at a higher elevation in western Napo (DWR and JMF Camargo, unpublished, from Rio Hollín). The genus *Camargoia* is absent in Yasuní but found in French Guiana and Manaus, but not in Rondônia, while *Schwarzula* and *Geotrigona* have not been found in French Guiana, but occur in Yasuní and the Brazilian sites. The Manaus area lacks *Nannotrigona, Paratrigona,* and *Dolichotrigona,* all relatively speciose at Yasuní and Rondônia. Further, 21 species (ca. 25%) are designated "sp. nov." or likely new to science, from French Guiana, as were 17 of 98 collected at Rondônia (17 %) using

year-long sampling and techniques much like those employed at Yasuní, while 43 of 100 species (43%) found in Yasuní are apparently new to science. Some perhaps already have been named, particularly *Trigona* and *Plebeia,* but require more study. There are only single collections or very small numbers of collected individuals for bees among some Yasuní resident genera, including *Leurotrigona, Partamona, Trigona, Tetragona, Duckeola, Lestrimelitta, Trigonisca,* and *Dolichotrigona*. In addition, *Duckeola* was found by a three-person collecting team during 1 month (Webber, Wyman and Ascher, courtesy American Museum of Natural History), while the present author, with about twice as much field time, over most of the year, has never seen this genus.

Its one western Amazonian species is known both in Rondônia and Colombia (Gonzalez and Nates-Parra 2004; Brown and de Oliveira 2012).

Far from being complete at any site, the species lists mentioned above show fewer species held in common than unique to the areas in question. As indicated in Table 14.1, there are at least 35 species held in common between the state of Rondônia and Yasuní—approximately one-sixth of the total of the two areas combined. Of interest, perhaps, is the large portion of *Trigona*, *Tetragona*, and *Scaura* that occur in each place. Other group comparisons show appreciably smaller numbers or proportions of species held in common.

At present, total bee collections made at Yasuní include approximately 220 sites, 180 by DWR, 1998–2017. Comparable collection at 187 localities in Ouro Preto, Rondônia, during large-scale deforestation for agrarian and urban development, was done in 1996–1997 (Brown and de Oliveira 2012) and did not encounter 12 species reported for the state. Those include 3 *Lestrimelitta*, 2 *Trigonisca*, 2 *Trigona*, 2 *Melipona*, and 1 *Scaptotrigona*. In Yasuní, an area only one two-hundred-thousandth the size of Rondônia, one might expect to find more species of *Lestrimelitta*, the robber Neotropical meliponine, which now has 19 known species (Camargo 2013). While they do not come to food baits, they collected at methyl salicylate, a euglossine scent bait (DWR, unpublished, see also Brown and de Oliveira 2012), but normally only take resources at the nests of other stingless bees, or Africanized honey bees (Roubik 1980). They never visit flowers. *Lestrimelitta* is an obligate eusocial bee colony robber, which primarily steals the brood provisions from within larval cells (Sakagami et al. 1993). Their lack of corbiculae attests to the absence of pollen gathering (Fig. 14.4).

Morphology and recorded behavior also confirm the non-pollinator status of other meliponines (Fig. 14.4). *Scaura* and *Schwarzula* and the *T. hypogea* group (Camargo and Roubik 1991) are not expected to be pollinators. The first two have a bladder-like hind basitarsus with long rows of bristles (Fig. 14.4d, f) which are applied to plant tissue surfaces to collect fallen pollen. The *T. hypogea* group uses dead animals, not pollen, for protein. Other *Trigona* that typically cut holes in the floral corolla or anthers to extract nectar and pollen, or use the tongue to extract pollen from poricidal anthers or those with pollen contained within a slit, are also unlikely to transfer pollen to the stigma of another flower. These, by and large, are genus *Trigona*, the only Neotropical meliponine bees besides *Paratrigona* with multiple, pointed apical denticles. Certain of the abovementioned groups do contain species that both rob and pollinate flowers, for example, *T. spinipes* (Fabricius, 1793) and *T. fulviventris* Guérin-Mélenville, 1844, and more field studies are certain to find additional variation. As already mentioned, 70% of the local meliponine species are tiny. The chances that they move between tree canopies, where most the local plants are trees that are dioecious, monoecious, or self-incompatible (see Bawa 1990, Roubik and Moreno, this book), seem remote.

The potential for carrying pollen, transporting it back to the nest as food, and then cleaning it from the body and grooming it from the hairs, all influence a bee's interaction with flowering plants, and particularly those with stamens. On the corbicula, that of *Scaptotrigona* with its dense hair (Fig. 14.4v), or the three long hairs in its center for *Partamona* (Fig. 14.4l) are noteworthy, while those of *Trigonisca* are very limited, and the corbicular pollen load is likely influenced by the tibial carrying capacity (see, e.g., Fig. 14.5b). The abundant long, bristle-like hairs on the antennal scape in several genera (Figs. 14.4e, g, h, o) may allow pollen to be carried or removed, particularly by tiny bees, from narrow passages in flowers. The *Dolichotrigona*, some *Trigonisca*, *Celetrigona*, *Nannotrigona*, and *Geotrigona* have several long hairs on one side of the scape, and when the two scapes are brought close together, pollen may be held, as if in a loose basket, by those and other hairs on the face.

The main carrying areas for pollen, on the outer hind tibia, are often gummed with pollen or concentrated nectar, or pollen mixed with resin (Fig. 14.4). This phenomenon includes most of the genera taxonomically affiliated (at the subgenus level, Michener 2007) with *Trigona*: *Trigona, Tetragona, Geotrigona. Tetragonisca, Pitlotrigona*, and *Frieseomelitta*. The possible adaptive reasons for such accumulations on the legs are obscure.

14.5 Bioprospecting for Pollination Knowledge and Sustainable Exploitation

In such hyperdiverse forest containing an estimated >3000 tree and 500 liana species, if each bee uses 30–40 species (an estimate from nest pollen identification, see Roubik and Moreno, Absy et al., in the present book) then almost 4000 plant species could be included in meliponine foraging. But this is unlikely. If 80% are potentially used by any meliponine species, other bees, or flower visitors, and 20% are regularly visited by more specialized meliponines, an ecological model can be constructed. The individual variation among members of a population is sometimes expressed as the 80/20 "rule." This rule embodies the Pareto principle (see Galvani and May 2005) and for epidemiology—the spread of an infectious organism among hosts—this principle appears suitable. It says that 80% of the transmission (e.g., of a disease or pollen) is done by 20% of the population. If we make an analogous model, whereby 80% of pollination is provided by 20% of the flower visiting species, and relative specialization is found only among 20% of bee species, the conclusion is that many hundreds of flowering species support the stingless bee community. The 20 relatively specialized meliponines, or 20% of the total, would use perhaps 32 pollen taxa (80% total pollen taxa for each bee species) × 20 bee species (=640), while the 80 more generalized (and presumably less consistently pollinating) meliponines might intensively use 32 species jointly. Further pollination insight requires field tests, particularly for unisexual plant species (Roubik and Moreno, present book), and the degree of dependency on

plants by bees is a subject of interest to bee conservation. Whether the stingless honey bees intentionally make their nests near suitable trees for pollen and nectar and resin, or other resources (e.g., Fierro et al. 2012), is a critical behavioral/ecological variable needing more study.

Why have so many more stingless honey bees been collected in <10 km^2 near the Yasuní scientific field station, compared to the rest of the world? There are more stingless bee species at Yasuní than in the Old World, and 25% more than at the best forest and the most collected portion of the eastern Amazon region, in French Guiana (80 species, Pauly et al. 2013) or in the Ouro Preto area of Rondônia, in western Amazonia (82 species; Brown and de Oliveira 2012). I hesitate to make a direct comparison with Ducke Reserve near Manaus, because its documented 37 species, corresponding with a low nest density, (1/km^2) make both estimates seem low (see Kajobe and Roubik 2006; Roubik 2006). There is more exploration to be done, as in the present study, which gradually added almost 50% to the original documented species richness (Roubik 1999, 2000). It is a general rule in tropical exploration that diversity is coupled with rareness, and the rare species in the hyperdiverse Yasuní biome may be quite rare. Also extraordinary is a claim (Bierregard et al. 2001, p. 172) that several species found only in "deforested" habitat in Manaus, but supposedly not in the 10,000 ha reserve plot are "recent arrivals." This suggestion is gratuitous, since it requires centuries for a stingless bee species to colonize 100 km of forest—their colonies do not freely swarm, but must make a nest close by, to permit colony dispersal to occur (Roubik 2006). They do not reproduce each year, at least by new colony formation, or by drone dispersal and mating (Slaa 2006; Roubik 2006). It is interesting, however, that some species reportedly inhabiting only the Manaus area lacking continuous forest are tiny meliponines, and those often seeking sweat on human skin. I believe this "degree of apparency" is linked both to exposed, sunny habitats and to diurnal fluctuations in such foraging activity—thus not necessarily a result of either biotope or forest cover. In contrast, the aseasonal forest of Yasuní, with no dry season, and annual rainfall ca. 3.4 m with mean temperature of 25 °C, also has a canopy of uneven height, due in part to low, rolling, hilly topography (Pérez et al. 2014). I suggest the light conditions, with improved sunlight in the understory (see Rincon et al. 1999), contribute to the flowering, and the salt-collecting possibilities (with large animals like deer, tapir, and jaguar contributing salt, and prey leavings) and thus make the habitat more conducive to proliferation of stingless bees. The accumulation of species in tropical forests, based on loose coevolution with particular host species of angiosperms—see the specializations noted in the present volume—without a comparable rate of extinction, seems the best available explanation for the mega-biodiversity of Meliponini at Yasuní. Moreover, the world's highest woody plant diversity, and possibilities for niche partitioning for floral resources, undoubtedly increase feeding, nesting, and foraging flexibility.

In addition, discovery of stingless bee nesting high on the thin trunk of *Cecropia* (DWR, unpublished) which had not been documented among Meliponini (Fig. 14.3), implies more bee species may be found, instead of solely ants, nesting in this and the other potential domiciles with hollow stems (Camargo and Pedro 2002), concentrated along forest edges, watercourses, roads, and in other patches where treefalls have occurred and pioneer trees are growing.

The subject of stingless bees inevitably leads to comparisons and contrasts, considering species that are aggressive and cut the hair, or attempt to—or produce a large amount of honey, and are not persistently aggressive after disturbance. For those of the second group, perhaps a dozen local species might be "domesticated" in hives, for honey or other hive products. The species are among the ~30 *Plebeia, Paratrigona, Nannotrigona, Cephalotrigona, Tetragona, Scaptotrigona,* and *Melipona*. This ratio (species

employed by humans, compared to total species present) is higher than that of Brazil, where the highest numbers of both meliponines and utilized species occur (Roubik, in press). The main requirement for successful meliponiculture is finding a species that responds well to being propagated in hives, such as *Tetragonula biroi* (Friese, 1898) or *T. carbonaria* (Smith, 1854) in Philippines or Australia, respectively, or *Melipona scutellaris* Latreille, 1811 and *M. beecheii* Bennett, 1831 in the Neotropics.

The *Melipona* at Yasuní are locally called "abejas reales" or "royal bees" in Spanish. The Waorani have species-specific names for many stingless honey bees, at least those that maintain large honey stores or are highly visible, and often aggressive at their nests. That knowledge is not encapsulated here, because the time spent in its pursuit has been brief, only a few days in October, 2016. Honey bees of the local species, with their Waorani names, are among those listed in Table 14.2. As managed portable pollination units, however, most of the native species might be useful, but only if maintained adequately and propagated.

The African or Africanized honey bee, *Apis mellifera* Linnaeus, 1758 exists in Yasuní (Roubik 2000). However, it is seldom foraging among the forest bees, and is best viewed as a transient, although abundant in highly diverse and extensive forest, elsewhere in Amazonia (Roubik 2009). Not a single worker was collected in the canopy and understory bait traps on the 50 ha plot, during one-half year, although over 500 bees of many meliponine and other species were thus collected. This contrasts with forest area surrounding the canopy crane and in Panama's Caribbean forest (Roubik et al. 2003, see also Basset et al. 2012). The invasive should continue to exist in Yasuní forest, but not thrive there, and be most readily observable at salts or other nonfloral resources, by the field station buildings, and at flowers of edge trees or those exposed to full sun, like *Cecropia, Ochroma,* and others.

Table 14.2 Waorani names given to some common meliponine bees in Yasuní Biosphere Reserve, Ecuador

Wao common name	Scientific name	Notes
auim	*Trigona amazonensis*	Exposed nest
auímo	*Melipona crinita*	
auñeta	*Partamona epiphytophila*	Exposed nest in epiphyte
awae	*Melipona captiosa*	
cowmuñi	*Trigona truculenta*	Exposed nest
eñamo	*Melipona titania*	
gihn	*Melipona eburnea*	
iñawe	*Ptilotrigona lurida*	
makawae	*Plebeia* sp.	*Cecropia* tree nesting
mingkaye	*Tetragona clavipes*	
ñabo	*Melipona nebulosa*	
oyo	*Melipona grandis*	
unatawe	*Trigona amalthea*	Exposed nest
uñepoi	*Scaptotrigona* spp.	

Information was provided by Mr. Nikimo Imo as nests and/or bees were examined in the field (see Figs. 14.1 and 14.2). All nests were in live trees, unless otherwise indicated.

Acknowledgments The SEM work of Jorge Ceballos, STRI, is gratefully acknowledged. Collection data and curation at QCAZ, PUCE, Ecuador, are appreciated, as is taxonomic help from S. R. M. Pedro and J. S. Ascher, and the field assistance of E. Baus, I. Tapia, and G. Onore. Any identification or other lapsus remains due to the author.

References

Albuquerque PMC, Camargo JMF. 2007. Espécies novas de *Trigonisca* Moure (Hymenoptera, Apidae, Apinae). Revista Brasileira de Entomologia 51: 160–175.

Basset Y, et al. 2012. Arthropod diversity in a tropical forest. Science 338: 1481– 1484. (2012); DOI: 10.1126/science.1226727.

Bawa KS. 1990. Plant-pollinator interactions in tropical rain forests. Annual review of Ecology and Systematics 1990: 399–422.

Bierregard Jr RO, Gascon C, Lovejoy TE, Mesquita R., eds. 2001. Lessons from Amazonia. The ecology and conservation of a fragmented forest. Yale University Press; New Haven, USA. 478 pp.

Brown JC, Oliveira ML de. 2012. Unpublished manuscript. The impact of agricultural colonization and deforestation on stingless bee (Apidae: Meliponini) composition and richness in Rondônia, Brazil. Accessed 10/11/2016 at: https://kuscholarworks.ku.edu/bitstream/handle/1808/12603/Brown.

Camargo JMF. 2013. Historical biogeography of the Meliponini (Hymenoptera, Apidae, Apinae) of the Neotropical region. pp. 19–34. In Vit P, Pedro SRM,

Roubik DW, eds. Pot-honey: a legacy of stingless bees. Springer; New York, USA. 654 pp.

Camargo JMF, Moure JS. 1994. Meliponinae Neotropicais: os gêneros *Paratrigona* Schwarz, 1938 e *Aparatrigona* Moure, 1951 (Hymenoptera, Apidae). Arquivos de Zoologia 32: 33–109.

Camargo JMF, Moure JS. 1996. Meliponinae Neotropicais: o gênero *Geotrigona* Moure, 1943 (Apinae, Apidae, Hymenoptera), com especial referência a filogenia e biogeografia. Arquivos de Zoologia 33: 95–161.

Camargo JMF, Pedro SRM. 2002. Mutualistic association between a tiny Amazonian stingless bee and a wax-producing scale insect. Biotropica 34: 446–451. DOI: 10.1111/j.1744-7429.2002.tb00559.x.

Camargo JMF, Pedro SRM. 2003a. Meliponini Neotropicais: o gênero *Partamona* Schwarz (Hymenoptera, Apidae, Apinae). Revista Brasileira de Entomologia 47: 311–372.

Camargo JMF, Pedro SRM. 2003b. Meliponini Neotropicais: o gênero *Ptilotrigona* Moure (Hymenoptera, Apidae, Apinae). Revista Brasileira de Entomologia 48: 353–377.

Camargo JMF, Pedro SRM. 2005. Meliponini Neotropicais: o gênero *Dolichotrigona* Moure (Hymenoptera, Apidae, Apinae). Revista Brasileira de Entomologia 49: 69–92.

Camargo JMF, Pedro SRM. 2008. Meliponini Lepeletier, 1836. In Moure JS, Urban D, Melo, GAR, eds. Catalogue of bees (Hymenoptera, Apoidea) in the Neotropical region – online version. Available at: http://www.moure.cria.org.br/catalogue.

Camargo JMF, Pedro SRM. 2009. Neotropical Meliponini: the genus *Celetrigona* Moure (Hymenoptera: Apidae, Apinae). Zootaxa 2155: 37–54.

Camargo JMF, Roubik DW. 1991. Systematics and bionomics of the apoid obligate necrophages: the *Trigona hypogea* group (Hymenoptera: Apidae; Meliponinae). Biological Journal of the Linnean Society 44: 13–39.

Camargo JMF, Roubik DW. 2005. Neotropical Meliponini: *Paratrigonoides mayri*, new genus and species from western Colombia (Hymenoptera, Apidae, Apinae) and phylogeny of related genera. Zootaxa 1081: 33–45.

Cockerell TDA. 1913. Meliponine bees from Central America. Psyche, February. pp. 10–14.

Davies SJ. 2014. Prólogo. pp. 9–10 In: Pérez AJ, Hernández C, Romero-Saltos H, Valencia R. 2014. Árboles emblemáticos de Yasuní, Ecuador. 337 especies. Publicaciones del herbario QCA. Escuela de Ciéncias Biológicas, Pontífica Universidad Católica del Ecuador, Quito, Ecuador. pp.

Fierro MM, Cruz-López L, Sánchez D, Villanueva-Gutiérrez R, Vandame R. 2012. Effect of biotic factors on the spatial distribution of stingless bees (Hymenoptera: Apidae, Meliponini) in fragmented neotropical habitats. Neotropical Entomology 41: 91–104. doi: 10.1007/s13744-011-0009-5.

Galvani AP, May RM. 2005. Epidemiology: Dimensions of superspreading. Nature 438: 293–295. PMID: 16292292.

Gonzalez VH, Nates-Parra G. 2004. *Trigona* subgenus *Duckeola* in Colombia (Hymenoptera: Apinae). Journal of the Kansas Entomological Society 77: 292.

Gonzalez VH, Roubik DW. 2008. Especies nuevas y filogenia de abejas de fuego *Oxytrigona* (Hymenoptera, Apidae, Meliponini). Acta Zoologica Mexicana 24: 43–71.

Hanson PE, Nishida K. 2016. Insects and other arthropods of tropical America. Comstock Publishing Associates, Cornell University Press; New York, USA. 374 pp.

Kajobe R, Roubik DW. 2006. Honey-making bee colony abundance and predation by apes and humans in a Uganda forest reserve. Biotropica 38: 210–218.

Melo GAR. 2016. *Plectoplebeia*, a new Neotropical genus of stingless bees (Hymenoptera: Apidae). *Zoologia (Curitiba)* [online]. 2016, vol.33, n.1, e20150153. Epub Mar 11, 2016. ISSN 1984-4670. http://dx.doi.org/10.1590/S1984-4689zool-20150153.

Michener CD. 2007. The bees of the world. 2nd Edn. Johns Hopkins University Press; Baltimore, USA. 953 pp.

Michener CD. 2013. The Meliponini. pp. 3–17 In Vit P, Pedro SRM, Roubik DW, eds. Pot-honey: a legacy of stingless bees. Springer; New York, USA. 654 pp.

Moure JS. 1960. Notes on the types of the Neotropical bees described by Fabricius (Hymenoptera: Apoidea). Studia Entomologica 3: 97–160.

Moure JS. 1989. *Camargoia*, um novo gênero neotropical de Meliponinae (Hymenoptera, Apoidea). Boletim do Mueeu Paraense Emilio Goeldi, Serie. Zoologia 5: 71–78.

Moure JS, Camargo JMF. 1982. *Partamona* (*Nogueirapis*) *minor*, nova espécie de Meliponinae (Hymenoptera: Apidae) do Amazonas e notas sobre *Plebeia variicolor* (Ducke). Boletim do Museu Paraense Emílio Coeldi, Série Zoología 120:1–10.

Nates-Parra G. 2004. Las abejas sin aguijón (Hymenoptera: Apidae: Meliponini) de Colombia. Biota Colombiana 2: 233–248.

Oliveira ML. 2001. Stingless bees (Meliponini) and orchid bees (Euglossini) in Terra Firme tropical forests and forest fragments. pp. 208–218 In Bierregard Jr RO, Gascon C, Lovejoy TE, Mesquita R., eds. Lessons from Amazonia. The ecology and conservation of a fragmented forest. Yale University Press; New Haven, USA. 478 pp.

Pauly A, Pedro SRM, Rasmussen C, Roubik DW. 2013. Stingless bees (Hymenoptera: Apoidea: Meliponini) of French Guiana. pp. 87–97 In Vit P, Pedro SRM, Roubik DW, eds. Pot-honey: a legacy of stingless bees. Springer; New York, USA. 654 pp.

Pedro SRM. 2014. The stingless bee fauna in Brazil (Hymenoptera: Apidae). Sociobiology 61: 348–354.

Pedro SRM, Camargo JMF. 2003. Meliponini neotropicais: o gênero *Partamona* Schwarz, 1939 (Hymenoptera, Apidae). Revista Brasileira de Entomologia 47 (Suppl.): 1–117.

Pedro SRM, Camargo JMF. 2009. Neotropical Meliponini: the genus *Leurotrigona* Moure— two new species (Hymenoptera: Apidae, Apinae). Zootaxa 1983: 23–44.

Pedro SRM, Cordeiro GD. 2015. A new species of the monospecific stingless bee genus *Trichotrigona* (Hymenoptera: Apidae, Meliponini). Zootaxa. 2015 May 11;3956(3):389–402. doi: 10.11646/zootaxa.3956.3.4.

Pérez AJ, Hernández C, Romero-Saltos H, Valencia R. 2014. Árboles emblemáticos de Yasuní, Ecuador. 337 especies. Publicaciones del herbario QCA. Escuela de Ciéncias Biológicas, Pontífica Universidad Católica del Ecuador, Quito. 394 pp.

Rasmussen C, Camargo JMF. 2008. A molecular phylogeny and the evolution of nest architecture and behavior in *Trigona* s. s. (Hymenoptera: Apidae: Meliponini). Apidologie 39: 102–118.

Rasmussen C, Cameron S. 2010. Global stingless bee phylogeny supports ancient divergence, vicariance, and long distance dispersal. Biological Journal of the Linnean Society 99: 206–232.

Rasmussen C, Gonzalez VH. 2013. Prologue. Stingless bees now and in the future. pp 1–7. In: Vit P, Roubik DW, eds. Stingless Bees Process Honey and Pollen in Cerumen Pots. Faculta de Farmacia y Bioanálisis, Universidad de los Andes, Mérida, Venezuela.

Rincon RM, Roubik DW, Finegan B, Delgado D, Zamora N. 1999. Regeneration in tropical rain forest managed for timber production: understory bees and their floral resources in a logged and silviculturally treated Costa Rican forest. Journal of the Kansas Entomological Society 72: 379–393.

Roubik DW. 1980. New species of *Trigona* and cleptobiotic *Lestrimelitta* from French Guiana. Revista de Biologia Tropical 28: 263–270.

Roubik DW. 1989. Ecology and natural history of tropical bees. Cambridge University Press, New York. 514 pp.

Roubik DW. 1993. Direct costs of forest reproduction, bee-cycling and the efficiency of pollination modes. J. Bioscience 18: 537–552.

Roubik DW. 1999. Grave-robbing by male *Eulaema*: implications for euglossine biology. Journal of the Kansas Entomological Society 71: 188–191.

Roubik DW. 2000. Pollination system stability in tropical America. Conservation Biology 14: 1235–1236.

Roubik DW. 2002. Tropical bee colonies, pollen dispersal and reproductive gene flow in forest trees. pp. 30–40 In Degen B, Loveless M, Kremer A, eds. Proceedings of the symposium: Modelling and experimental research on genetic processes in tropical and temperate forests. Documentos de Embrapa Amazonia, Belém. pp.

Roubik DW. 2006. Stingless bee nesting biology. Apidologie 37: 124–143.

Roubik DW. 2009. Ecological impact on native bees by the invasive Africanized honey bee. Acta Biologica Colombiana.14: 115–124.

Roubik DW (ed.). 2017 (In Press). The pollination of cultivated plants. A compendium for practitioners. Food and Agriculture Organization of the United Nations. Rome, Italy.

Roubik DW, Lobo JA, Camargo JMF. 1997. New endemic stingless bees genus from Central American cloudforests (Hymenoptera: Apidae; Meliponini). Systematic Entomology 22:67–80.

Roubik DW, Moreno JE. 2009. *Trigona corvina*: An ecological study based on unusual nest structure and pollen analysis. Psyche, vol. 2009, Article ID 268756, doi:10.1155/2009/268756.

Roubik DW, Sakai S, Gattesco, F. 2003. Canopy flowers and certainty: loose niches revisited. In Basset Y, Kitching R, Miller S, Novotny V, eds. pp. 360–368. Arthropods of tropical forests. Spatio-temporal dynamics and resource use in the canopy. Cambridge Univ. Press, London. pp.

Sakagami SF, Roubik DW, Zucchi R. 1993. Ethology of the robber bee, *Lestrimelitta limao*. Sociobiology 21: 237–277.

Schwarz HF. 1948. Stingless bees (Meliponidae) of the Western Hemisphere. Bulletin of the American Museum of Natural History 90: 1–546.

Slaa J. 2006. Population dynamics of a stingless bee community in the seasonal dry lowlands of Costa Rica. Insectes Sociaux 53: 70–79. doi:10.1007/s00040-005-0837-6.

Wille A. 1962. A technique for collecting stingless bees under jungle conditions. Insectes Sociaux 9: 291–293.

Diversity of Stingless Bees in Ecuador, Pot-Pollen Standards, and Meliponiculture Fostering a Living Museum Meliponini of the World

Patricia Vit, Silvia R.M. Pedro, Favian Maza, Virginia Meléndez Ramírez, and Viviana Frisone

15.1 Introduction

The oldest bee fossil, *Cretotrigona prisca*, is preserved in Cretaceous amber from New Jersey (Michener and Grimaldi 1988a, b). The chapter on Meliponini in the *Catalogue of Bees in the Neotropical Region* (Camargo and Pedro 2007) and recent inventories of meliponine faunas from Argentina (Roig-Alsina et al. 2013), Australia (Halcroft et al. 2013), Costa Rica (Aguilar et al. 2013a, b), French Guiana (Pauly et al. 2013), Guatemala (Yurrita Obiols and Vásquez 2013), Mexico (Ayala et al. 2013), Venezuela (Pedro and Camargo 2013), Brazil (Pedro 2014), El Salvador (Ruano Iraheta et al. 2015), Ecuador (here, Roubik Chap. 14), and Panama (Roubik and Moreno Chap. 4) are constantly updated for the scientific community.

This interest in the diversity of stingless bees is encompassed by the ancestral knowledge of meliponicultors who select and keep few species of these bees around the world. There is a need to document the existing Meliponini, to create standards for their products. In this chapter, we update entomological records of stingless bees in Ecuador, needed for normative assessment of the pot-honey and pot-pollen they produce. An important stingless bee window is revised to look at the impact of climate change on stingless bees. The role of a living eco-museum for Meliponini of the world and the stingless beekeepers is an option to care for intangible heritage and conservation of natural resources.

P. Vit (✉)
Apitherapy and Bioactivity, Food Science Department, Faculty of Pharmacy and Bioanalysis, Universidad de Los Andes, Mérida 5101, Venezuela

Cancer Research Group, Discipline of Biomedical Science, Cumberland Campus C42, The University of Sydney, 75 East Street, Lidcombe, NSW 1825, Australia
e-mail: vitolivier@gmail.com

S.R.M. Pedro
Departamento de Biologia, Faculdade de Filosofia, Ciências e Letras de Ribeirão Preto - FFCLRP, Universidade de São Paulo – USP, Av. Bandeirantes, 3900, CEP 14040-901 Ribeirão Preto, SP, Brasil

F. Maza
Academic Unit of Agro-Livestock Sciences, Universidad Técnica de Machala, Machala, El Oro Province, Ecuador

V.M. Ramírez
Departamento de Zoología, Campus de Ciencias Biológicas y Agropecuarias, Universidad Autónoma de Yucatán, AP 4-116, Col. Itzimná, 97100 Mérida, Yucatan, México

V. Frisone
Museum Curator, Museo di Archeologia e Scienze Naturali 'G. Zannato', 17 Piazza Marconi, 36075 Montecchio Maggiore, Vicenza, Italy

© Springer International Publishing AG, part of Springer Nature 2018
P. Vit et al. (eds.), *Pot-Pollen in Stingless Bee Melittology*, DOI 10.1007/978-3-319-61839-5_15

15.2 Megabiodiversity of Stingless Bees in Ecuador

Coloma (1986) made a thesis studying Ecuadorian stingless bees in the region around Quito, including the Pacific highlands and the Amazonian lowlands. His was the first contribution of stingless bee survey in Ecuador. His grouping criteria were based on substrate or location of colonies, separating stingless bees in three groups: (1) those making nests underground or "ground bees," (2) those making nests within tree hollows or "hollow bees," and (3) those making nests on tree branches. Within the stingless bees that make their colonies in tree hollows, he divided bees in two groups according to the presence or absence of entrance structure. Finally, within the group of bees without entrance structures, Coloma formed two groups defined by the behavior of bees at the entrance of the nest. He distinguished bees that "enter one at a time" with narrow entrance area and those that "enter all at once" with wide entrance area. The group that enters "one at a time" included bees with evasive behavior, which leave or enter the nest intermittently and stop entering the nest when they detect people and other potential predators. On the contrary, bees that enter "all at once" can be seen at a distance, as they keep a steady flow of individuals entering and leaving the nest. Finally, within the group of bees that do not sting, aggressive bees that attack their enemies without a sting were called bees that "bite but do not sting." Rasmussen (2004) also made a contribution on Ecuadorian stingless bees. In Table 15.1, we present the stingless bees collected by Coloma (1986) and Rasmussen (2004) and those reviewed by Camargo and Pedro (2007) and Vit (2015) – here, 54 species were collected and 26 were reported for Ecuador for the first time. For progress on the inventory of Ecuadorian stingless bees, see Chap. 14 by Roubik.

Ethnic names of Ecuadorian stingless bees retrieved by Vit (2015) are given in Table 15.2. Entomological collections, identification of the stingless bees, and their ethnic names are of great value for the continuation of this research. Ethnic names can be distinctive, but with such a great species richness, the same name is used to name several species – e.g., see "moroja" and "abeja real" in Table 15.2 – or sometimes there is no ethnic name.

15.3 A Revised Ecuadorian Honey Norm and Approach to Pot-Pollen Standards

Pot-honey and pot-pollen standards are needed in Ecuador. The recent revision of the NTE INEN 1572 Ecuadorian honey norm (INEN 2016) removed the annex for pot-honey chemical composition – wisely inserted in the Colombian honey norm NTC 1273 (ICONTEC 2007) – after the retirement of the Normalization Director. Nevertheless, a *lapsus calami* kept the references of the annex in this Ecuadorian norm. A new norm for pot-honey was suggested because *Apis mellifera* honey is a different product. Also, the word melissopalynology was removed from the reviewed Ecuadorian honey norm, but *Apis mellifera* unifloral and multifloral honeys remain unobserved in the market. Certainly, the diverse plants used by *Apis mellifera* to make honey confer particular chemical components and bioactive and sensory properties to characterize different products based on botanical origin, supported by scientific databases (Persano Oddo and Piro 2004). The unique norm for pot-honey in the world was created by the Brazilian State of Bahia and limited to the *Melipona* genus (ADAB 2014). Similarly, a pot-pollen norm should benefit from previous *Apis mellifera* bee pollen norms.

The compositional data in Table 15.3 on pot-pollen produced by 18 species of stingless bees from Brazil, Mexico, Thailand, and Venezuela, besides the review in Chap. 31 by Alves and Carvalho, contribute to the proposal of pot-pollen standards. It is worth noting that stingless bee products such as pot-honey and pot-pollen repeatedly "bounce" in normative processes. A social filter may prevent their insertion. Competitive *Apis mellifera* beekeepers – closer to capital cities where technical committees meet – do not represent interests of more isolated

Table 15.1 Stingless bees from Ecuador: scientific names and geographic records

No.	Stingless bee species	Geographic records			
		Coloma 1986 (BSc thesis)	Rasmussen 2004	Camargo and Pedro 2007 (catalogue)	Vit 2015 Prometeo-UTMACH Project[1]
1	*Aparatrigona impunctata* (Ducke, 1916)			NA, SU	PA
2	*Cephalotrigona* cf. *zexmeniae* (Cockerell, 1912)				EO
3	*Cephalotrigona* sp.	PI			
4	*Dolichotrigona browni* Camargo & Pedro, 2005			NA	
5	*Frieseomelitta silvestrii* (Friese, 1902)	PA			
6	*Geotrigona fulvohirta* (Friese, 1900)	NA	ZC	NA, PA	ZC
7	*Geotrigona fumipennis* Camargo & Moure, 1996[a2]	LO, MA		LO, MA	
8	*Geotrigona leucogastra* (Cockerell, 1914)[a]			GU, MA	SE
9	*Geotrigona tellurica* Camargo & Moure, 1996[a]			MS, SU	
10	*Lestrimelitta rufa* (Friese, 1903)[4]	NA, PA		NA, PA	
11	*Melipona (Melikerria) grandis* Guérin, 1834	NA, MS		NA	PA
12	*Melipona (Melipona)* sp. aff. *fuscata* Lepeletier, 1836		EO		
13	*Melipona (Melipona)* sp.				ES
14	*Melipona (Michmelia) eburnea* Friese, 1900			Ecuador (no province)	PA, ZC
15	*Melipona (Michmelia)* sp. aff. *eburnea* Friese, 1900[4]	MS, ZC			
16	*Melipona (Michmelia) fallax* Camargo & Pedro, 2008[5]	ES, IM		ES, GU	
17	*Melipona (Michmelia)* sp. aff. *fuscopilosa* Moure & Kerr, 1950	NA, MS, PA, TU			
18	*Melipona (Michmelia) illota* Cockerell, 1919			MS, NA	
19	*Melipona (Michmelia) indecisa* Cockerell, 1919	LO, IM			EO, ES, LO
20	*Melipona (Michmelia)* sp. aff. *indecisa* Cockerell, 1919				ES
21	*Melipona (Melikerria)* cf. *interrupta* Latreille, 1811				PA
22	*Melipona (Michmelia) mimetica* Cockerell, 1914[a]	MA	LO	GU, LO, MA	EO, ES, GU, LO

(continued)

Table 15.1 (continued)

No.	Stingless bee species	Geographic records			
		Coloma 1986 (BSc thesis)	Rasmussen 2004	Camargo and Pedro 2007 (catalogue)	Vit 2015 Prometeo-UTMACH Project[1]
23	*Melipona (Michmelia)* cf. *nebulosa* Camargo, 1988	NA, MS			
24	*Melipona (Michmelia) nitidifrons* Benoist, 1933[a]			ES, IM, NA, PI	
25	*Melipona (Michmelia) titania* (Gribodo, 1893)[6]	NA, PA		MS, PA, ZC	
26	*Melipona (Michmelia)* sp.1 (*fasciata* group)				MS
27	*Melipona (Michmelia)* sp.2 (*fasciata* group)	LO			
28	*Melipona* sp.		ZC		
29	*Nannotrigona melanocera* (Schwarz, 1938)	MS, NA			MS, NA, ZC
30	*Nannotrigona* cf. *mellaria* (Smith, 1862)				MA, EO
31	*Nannotrigona* sp. aff. *mellaria* (Smith, 1862)	LO	LO		
32	*Nannotrigona* cf. *perilampoides* (Cresson, 1878)				LO, EO, ES
33	*Nannotrigona tristella* Cockerell, 1922	MA			
34	*Nannotrigona* sp.	CO, ES, LO, MA			
35	*Nogueirapis mirandula* (Cockerell, 1917)			PI	
36	*Oxytrigona chocoana* Gonzalez & Roubik, 2008[a]			ES, IM	
37	*Oxytrigona huaoranii* Gonzalez & Roubik, 2008[a,7]	NA		NA	
38	*Oxytrigona mellicolor* (Packard, 1869)[a]	ES, LO		ES, LO, NA	ES, LO
39	*Parapartamona fumata* Moure, 1995[a,8]	NA		NA	
40	*Parapartamona tungurahuana* (Schwarz, 1948)[a]	TU	ZC	NA, SU, TU,	
41	*Parapartamona vittigera* Moure, 1995[a,9]	PI		CO, PI	
42	*Parapartamona* sp. [10]	MS			
43	*Paratrigona* sp. aff. *eutaeniata* Camargo & Moure, 1994				EO
44	*Paratrigona onorei* Camargo & Moure, 1994[a,11]	NA		NA	

(continued)

Table 15.1 (continued)

No.	Stingless bee species	Geographic records			
		Coloma 1986 (BSc thesis)	Rasmussen 2004	Camargo and Pedro 2007 (catalogue)	Vit 2015 Prometeo-UTMACH Project[1]
45	*Paratrigona opaca* (Cockerell, 1917)			NA	
46	*Paratrigona pacifica* (Schwarz, 1943)	MS, LO, NA	ZA	MS, NA	ZC
47	*Paratrigona prosopiformis* (Gribodo, 1893)	NA		NA	
48	*Paratrigona* cf. *rinconi* Camargo & Moure, 1994				EO, LO
49	*Paratrigona scapisetosa* Gonzalez & Griswold, 2011[a]			MS	
50	*Partamona aequatoriana* Camargo, 1980[a]	ES, PI		CO, ES, IM, PI	EO
51	*Partamona epiphytophila* Pedro & Camargo, 2003			MS, NA, PA, SU, ZC	MS, NA, OR, PA, ZC
52	*Partamona musarum* (Cockerell, 1917)				ES
53	*Partamona peckolti* (Friese, 1901)		EO	BO, CR, CH, CO, EO, ES, MA, PI	CH, GU, LO
54	*Partamona testacea* (Klug, 1807)	NA		NA, PA, TU	
55	*Partamona vicina* Camargo, 1980[12]	NA		NA	
56	*Partamona* sp. (probably = *Partamona peckolti*)[13]	MA, PI			
57	*Plebeia frontalis* (Friese, 1911)	MA			
58	*Plebeia* sp. 1				NA
59	*Plebeia* sp. 2				PA
60	*Plebeia* sp. 3				ES
61	*Plebeia* sp. 4				LO
62	*Plebeia* sp. 5 [14]	MS			
63	*Plebeia* sp. 6 [15]	NA			
64	*Plebeia* sp. 7 [16]	NA			
65	*Plebeia* sp. 8 [17]	CO			
66	*Plebeia* sp. 9	LO	LO		
67	*Plebeia* sp. 10		LO		
68	*Ptilotrigona lurida* (Smith, 1854)	NA		MS, NA, SU	
69	*Ptilotrigona occidentalis* (Schulz, 1904)[a]			ES	
70	*Ptilotrigona pereneae* (Schwarz, 1943)	NA		NA, PA, SU	

(continued)

Table 15.1 (continued)

No.	Stingless bee species	Coloma 1986 (BSc thesis)	Rasmussen 2004	Camargo and Pedro 2007 (catalogue)	Vit 2015 Prometeo-UTMACH Project[1]
		Geographic records			
71	*Scaptotrigona barrocoloradensis* (Schwarz, 1951)			ES, MA	
72	*Scaptotrigona* sp. aff. *barrocoloradensis* (Schwarz, 1951)	MA			
73	*Scaptotrigona tricolorata* Camargo, 1988			PA	MS
74	*Scaptotrigona* sp. (*limae* group) (Brèthes, 1920)	CO			
75	*Scaptotrigona* sp. 1				NA
76	*Scaptotrigona* sp. 2				EO, LO
77	*Scaptotrigona* sp. 3		EO		
78	*Scaptotrigona* sp. 4		ZC		
79	*Scaptotrigona* sp. 5 [18]	NA			
80	*Scaptotrigona* sp. 6 [19]	MS			
81	*Scaptotrigona* sp. 7 [20]	LO			
82	*Scaptotrigona* sp. 8	MA			
83	*Scaptotrigona* sp. 9	PI			
84	*Scaura tenuis* (Ducke, 1916)				NA, PA
85	*Scaura* sp. 1 aff. *latitarsis* (Friese, 1900)[21]	NA			
86	*Scaura* sp. 2 aff. *latitarsis* (Friese, 1900)				PA
87	*Scaura* sp. aff. *longula* (Lepeletier, 1836)				PA
88	*Schwarzula timida* (Silvestri, 1902)			SU	
89	*Tetragona dorsalis* (Smith, 1854)	PA			
90	*Tetragona* sp. (*clavipes* group) (Fabricius, 1804)	MS, NA			PA
91	*Tetragona* sp. aff. *essequiboensis* (Schwarz, 1940)				NA
92	*Tetragona* sp. aff. *ziegleri* (Friese, 1900)	LO			
93	*Tetragona* sp. 1	NA			
94	*Tetragona* sp. 2	ES			
95	*Tetragona* sp. 3		ZC		ZC
96	*Tetragonisca angustula* (Latreille, 1811)[22]	MS, NA, PA, ZC		Ecuador (no province)	MS, NA, PA, SU, ZC
97	*Tetragonisca* sp. 1 (*angustula* group) (Latreille, 1811)[23]	IM, MA			ES

(continued)

Table 15.1 (continued)

No.	Stingless bee species	Geographic records			
		Coloma 1986 (BSc thesis)	Rasmussen 2004	Camargo and Pedro 2007 (catalogue)	Vit 2015 Prometeo-UTMACH Project[1]
98	*Tetragonisca* sp. 2 (*angustula* group) (Latreille, 1811)				NA
99	*Tetragonisca buchwaldi* (Friese, 1925)[a]	CO, PI		GU, PI	
100	*Trigona* cf. *albipennis* Almeida, 1995				MS
101	*Trigona amalthea* (Olivier, 1789)	MS, NA, PA, TU		NA	MS
102	*Trigona amazonensis* (Ducke, 1916)	MS, PA		PA	ZC
103	*Trigona* sp. aff. *amazonensis* (Ducke, 1916)	NA, PA			
104	*Trigona branneri* Cockerell, 1912	PA		NA	NA, SU, ZC
105	*Trigona* cf. *branneri* Cockerell, 1912 [24]	PA			
106	*Trigona chanchamayoensis* Schwarz, 1948	NA, PA, TU		GU, NA, PA. SU	
107	*Trigona cilipes* (Fabricius, 1804)	MS, PA		MS	
108	*Trigona corvina* Cockerell, 1913	CO			
109	*Trigona crassipes* (Fabricius, 1793)[25]	PA		NA, PA	MS
110	*Trigona* sp. aff. *crassipes* (Fabricius, 1793)	NA			
111	*Trigona dallatorreana* Friese, 1900				NA
112	*Trigona dimidiata* Smith, 1854	MS		MS	
113	*Trigona ferricauda* Cockerell, 1917			CA, CH	EO
114	*Trigona fulviventris* Guérin, 1844	EL, LO	EO, LO		EO, LO
115	*Trigona* sp. (*fulviventris* group) Guérin, 1844				NA
116	*Trigona fuscipennis* Friese, 1900	CO, PI			EO, LO, MA, PA, SE
117	*Trigona* sp.1, aff. *fuscipennis* Friese, 1900[26]	CO, MA, PI			
118	*Trigona* sp.2, aff. *fuscipennis* Friese, 1900		LO		
119	*Trigona guianae* Cockerell, 1910	NA		NA, PA	NA, PA

(continued)

Table 15.1 (continued)

No.	Stingless bee species	Geographic records			
		Coloma 1986 (BSc thesis)	Rasmussen 2004	Camargo and Pedro 2007 (catalogue)	Vit 2015 Prometeo-UTMACH Project[1]
120	*Trigona* sp. (*guianae* group) Cockerell, 1910 [27]	PI			
121	*Trigona muzoensis* Schwarz, 1948	BO, CO			
122	*Trigona nigerrima* Cresson, 1878			Ecuador (no province)	
123	*Trigona recursa* Smith, 1863	NA		NA	
124	*Trigona* sp. aff. *recursa* Smith, 1863	MS			
125	*Trigona silvestriana* (Vachal, 1908)	BO, CA, CH, CO, ES, GU, LO, PI,	EO	AZ, CA, EO, ES, GU, LO, MA, PI	EO, ES, LO
126	*Trigona truculenta* Almeida, 1984				MS, NA, OR, PA, TU, ZC
127	*Trigona williana* Friese, 1900	NA		NA, SU	
128	*Trigona* sp. 1 [28]	LR			
129	*Trigona* sp. 2 [29]	GU			
130	*Trigonisca buyssoni* (Friese, 1902)			Ecuador (no province)	
131	*Trigonisca* sp.1				LO
132	*Trigonisca* sp. 2				ZC
	Total	**78**	**16**	**53**	**54**

The correspondence between species (e.g., sp.1, sp. 2) of different surveys was made tentatively (SRMP) wherever possible, as the specimens were not directly compared. Because of this, the total number of species listed here (132) is approximated and can be overestimated

Ecuadorian provinces: *AZ* Azuay, *BO* Bolívar, *CA* Cañar, *CR* Carchi, *CH* Chimborazo, *CO* Cotopaxi, *EO* El Oro, *ES* Esmeraldas, *GA* Galápagos, *GU* Guayas, *IM* Imbabura, *LO* Loja, *LR* Los Ríos, *MA* Manabí, *MS* Morona Santiago, *NA* Napo, *OR* Orellana, *PA* Pastaza, *PI* Pichincha, *SE* Santa Elena, *SD* Santo Domingo de los Tsáchilas, *SU* Sucumbíos, *TU* Tungurahua, *ZC* Zamora-Chinchipe

[a]Species described from Ecuador (type specimens from Ecuador)

[1]The stingless bees of the Prometeo-UTMACH Project (2014–2015) were kindly identified by Dr. SRM Pedro; *Geotrigona leucogastra* was kindly identified by Dr. C. Vergara

[2]Cited as *Geotrigona* sp. aff. *acapulconis* (Strand, 1919) by Coloma, 1986

[3]Cited as *Lestrimelitta limao* (Smith, 1863) by Coloma, 1986

[4]Cited as *Melipona* n. sp. aff. *eburnea* by Coloma, 1986

[5]Cited as *Melipona fuliginosa*, part, by Coloma, 1986

[6]Cited as *Melipona fuliginosa*, part, by Coloma, 1986

[7]Cited as *Oxytrigona mediorufa* (Cockerell, 1913) by Coloma, 1986

[8]Cited as *Parapartamona* sp. by Coloma, 1986

[9]Cited as *Parapartamona zonata zonata* (Smith, 1854) by Coloma, 1986

[10]Cited as as *Parapartamona zonata caliensis* (Schwarz, 1948) by Coloma, 1986

[11]Cited as *Paratrigona* n. sp. by Coloma, 1986

[12]Cited as *Partamona aequatoriana*, part, by Coloma, 1986

[13]Cited as *Partamona pearsoni* (Schwarz, 1938) and as *Partamona* sp. (gr. *cupira*) by Coloma, 1986

[14]Cited as *Plebeia minima* (Gribodo, 1893) by Coloma, 1986

[15]Cited as *Plebeia flavoscutellata* Moure, MS by Coloma, 1986

(continued)

Table 15.1 (continued)

[16]Cited as *Plebeia* n. sp. prope (*sic?*) *intermedia* by Coloma, 1986
[17]Cited as *Plebeia* n. sp. aff. *intermedia* (Wille, 1960) by Coloma, 1986
[18]Cited as *Scaptotrigona* aff. *affinis* Schwarz n.p. by Coloma, 1986
[19]Cited as *Scaptotrigona chorreroensis* Schwarz n.p. by Coloma, 1986
[20]Cited as *Scaptotrigona postica* (Latreille, 1807) by Coloma, 1986
[21]Cited as *Scaura latitarsis* (Friese, 1900) by Coloma, 1986
[22]Cited as *Tetragonisca angustula angustula*, part, by Coloma, 1986
[23]Cited as *Tetragonisca angustula angustula*, part, by Coloma, 1986
[24]Cited as *Trigona hyalinata* (Lepeletier, 1826) by Coloma, 1986
[25]Cited as *Trigona hypogea robustior* (Schwarz, 1940) by Coloma, 1986
[26]Cited as *Trigona setentrionalis* (Almeida, 1996) by Coloma, 1986
[27]Cited as *Trigona guianae*, part, by Coloma, 1986
[28]Cited as *Trigona pectoralis* [not *Scaptotrigona pectoralis* (Dalla Torre, 1896)] by Coloma, 1986
[29]Cited as *Trigona chanchamayoensis*, part by Coloma, 1986

meliponicultors; their voice was absent during the revision of the Ecuadorian honey norm (INEN 1988, 2016) (P. Vit, personal observation). Alternative strategies are needed to overcome administrative procedures and produce the required standards for the unique honey in the Americas before Columbus's discovery (Vit 2008) and standards for pot-pollen.

The reference values established in the Brazilian regulation for *Apis mellifera* fresh bee pollen (maximum 30% moisture) and dehydrated bee pollen (maximum 4% moisture) were fixed for dry basis m/m: minimum 8% proteins, minimum 1.8% lipids, minimum 2% crude fiber, maximum 4% ash, maximum free acidity 300 meq/kg, ranges of 14.5–55.0% total sugars, carbohydrates 14.5–55.0%, and pH 4.0–6.0 (Brasil 2001). Suggested reviewed regulatory values for dehydrated *Apis mellifera* pollen with percent content of ash < 6, proteins > 15, fat > 1.5, and sugars > 40. Corbicular pollen of *Apis mellifera* is the second product of the hive, after honey. However, for stingless bees, pot-pollen is harvested instead of corbicular pollen because pollen traps are not well adapted to the entrances of their nests.

In Table 15.3, values of moisture (16.1 ± 1.1–48.54 ± 0.41), ash (1.8 ± 0.2–4.9 ± 0.3), proteins (14.3 ± 0.6–24.72 ± 0.18), fat (0.89–7.4 ± 0.3), carbohydrates (24.48 ± 10.1–58.7 ± 3.5), pH (3.34 ± 0.02–3.75 ± 0.01), and fiber (0.87 ± 0.20–3.6 ± 1.4) are the ranges for fresh pot-pollen from Brazil, Mexico, Thailand, and Venezuela. Moisture is <30% for a group of seven species

(*Frieseomelitta* sp. aff. *varia*, *Lepidotrigona flavibasis*, *Lepidotrigona terminata*, *Melipona favosa*, *Scaptotrigona mexicana*, *Tetragonisca angustula*, and *Tetragonula laeviceps*). Ash content is <6% for all species; only *Melipona mandacaia* showed ash >4%. Proteins are >15% for all species except *Lepidotrigona terminata*. Fat is >1.5 for all species except *Scaptotrigona mexicana*. Pot-pollen produced by *Tetragonula laeviceps* has >55% carbohydrates; other species are between 15% and 55% as suggested in the Brazilian norm. Minor adjustments in the parameters used as pot-pollen quality indicators are needed for few stingless bee species to fulfill the suggested standards of *Apis mellifera* commercial bee pollen. We prefer to avoid reports on dry basis, as in the Brazilian regulation.

Functional properties of corbicular pollen and pot-pollen deserve further attention, considering the transformations of the floral pollen by the associated microbiota inside the nest. The bioactive indicators may be useful to describe the product and also to add value for their content of flavonoids, polyphenols, antioxidant, and antibacterial activities (see Chaps. 24, 26, 27, and 28).

Additional data on contents of elements in pot-pollen are given in Table 15.4. Micronutrients (Ca, Mg, P, K, Zn) and contaminants (As, Cd, Pb) from two previous studies and two chapters from this book comprise seven species from three countries Brazil, Mexico, and Thailand. Limits of detection (LOD), recommended daily intake (RDI), and maximum residue limits (MRL) are

Table 15.2 Ethnic names of Ecuadorian stingless bees

Table 15.2 (continued)

Stingless Bee Species	Ethnic names
Aparatrigona impunctata (Ducke, 1916)	"angelita negra"
Geotrigona fulvohirta (Friese,1900)	"moroja"
Melipona eburnea Friese, 1900	"abeja real"
Melipona eburnea Friese, 1900	"bunga amarilla"
Melipona eburnea Friese, 1900	"ergón"
Melipona eburnea Friese, 1900	"saramishki"
Melipona sp. (*fasciata* group)	"bunga amarilla"
Melipona grandis Guérin, 1834	"bunga negra"
Melipona grandis Guérin, 1834	"abeja real"
Melipona cf. *indecisa* Cockerell, 1919	"cananambo"
Melipona indecisa Cockerell, 1919	"cananambo"
Melipona indecisa Cockerell, 1919	"abeja real"
Melipona mimetica Cockerell, 1914	"bermejo"
Melipona mimetica Cockerell, 1914	"bermeja"
Nannotrigona melanocera (Schwarz, 1938)	"abejita suca"
Nannotrigona melanocera (Schwarz, 1938)	"abejita negra"
Nannotrigona melanocera (Schwarz, 1938)	"angelina"
Nannotrigona melanocera (Schwarz, 1938)	"llanaputan"
Nannotrigona melanocera (Schwarz, 1938)	"muruja suca"
Nannotrigona cf. *mellaria* (Smith, 1862)	"leticobe"
Nannotrigona cf. *mellaria* (Smith, 1862)	"suapillo"
Nannotrigona cf. *perilampoides* (Cresson, 1878)	"mosquitillo"
Nannotrigona cf. *perilampoides* (Cresson, 1878)	"pitón"
Oxitrigona mellicolor (Packard, 1869)	"mea fuego"

Stingless Bee Species	Ethnic names
Oxitrigona mellicolor (Packard, 1869)	"abeja amarilla"
Paratrigona aff. *eutaeniata* (Camargo & Moure, 1994)	"pirunga"
Paratrigona cf. *rinconi* Camargo & Moure, 1994	"abeja negra"
Paratrigona cf. *rinconi* Camargo & Moure, 1994	"abeja chiquita"
Paratrigona cf. *rinconi* Camargo & Moure, 1994	"pirunga"
Partamona aequatoriana Camargo, 1980	"abeja amarilla"
Partamona aequatoriana Camargo, 1980	"abeja negra"
Partamona aequatoriana Camargo, 1980	"barbacho"
Partamona aequatoriana Camargo, 1980	"chalaco"
Partamona epiphytophila Pedro & Camargo, 2003	"muruja hedionda"
Partamona epiphytophila Pedro & Camargo, 2003	"moroja"
Partamona epiphytophila Pedro & Camargo, 2003	"muruja"
Partamona peckolti (Friese, 1901)	"potolusho"
Partamona peckolti (Friese, 1901)	"moroja"
Plebeia sp.	"lambeojo"
Plebeia sp.	"mosco"
Plebeia sp.	"mosquito"
Scaptotrigona sp.	"catana"
Scaptotrigona sp.	"catana oreja de león"
Scaptotrigona sp.	"catiana"
Scaptotrigona sp.	"moroja"
Scaptotrigona sp.	"morojita"
Scaptotrigona tricolorata Camargo, 1988	"muruja grande"
Scaura sp. aff. *longula* (Lepeletier, 1836)	"abejita negra"
Tetragona sp. *gr. clavipes* (Fabricius, 1804)	"cuchiperro"
Tetragonisca angustula (Latreille, 1811)	"abeja ángel"
Tetragonisca angustula (Latreille, 1811)	"abeja finita"
Tetragonisca angustula (Latreille, 1811)	"angelina"

(continued)

(continued)

Table 15.2 (continued)

Stingless Bee Species	Ethnic names
Tetragonisca angustula (Latreille, 1811)	"chiñi"
Tetragonisca angustula (Latreille, 1811)	"chullumbo"
Tetragonisca angustula (Latreille, 1811)	"mosquito de miel"
Tetragonisca angustula (Latreille, 1811)	"trompetilla"
Trigona cf. *albipennis* Almeida, 1995	"moroja"
Trigona cf. *amazonensis* (Ducke, 1916)	"abeja amarilla"
Trigona cf. *branneri* Cockerell, 1912	"abejita negra"
Trigona cf. *crassipes* (Fabricius, 1793)	"pegón"
Trigona cf. *guianae* Cockerell, 1910	"angelita negra grande"
Trigona cf. *truculenta* Almeida, 1984	"moroja grande"
Trigona cf. *truculenta* Almeida, 1984	"pegón"
Trigona cf. *truculenta* Almeida, 1984	"pegón grande"
Trigona amalthea (Olivier, 1789)	"pegón-muruja"
Trigona dallatorreana (Friese, 1900)	"abeja amarilla"
Trigona ferricauda Cockerell, 1917	"abeja amarilla"
Trigona fulviventris Guérin, 1844	"abeja de tierra"
Trigona sp. gr. *fuscipennis* Friese, 1900	"abeja negra"
Trigona fuscipennis Friese, 1900	"chalaco"
Trigona fuscipennis Friese, 1900	"moruja"
Trigona silvestriana (Vachal, 1908)	"cortapelo"
Trigona silvestriana (Vachal, 1908)	"guarigane"
Trigona silvestriana (Vachal, 1908)	"moroja"
Trigona cf. *truculenta* Almeida, 1984	"abeja negra"
Trigonisca sp.	"lambeojitos"

From: Vit (2015)

needed for regulatory procedures, concerned with the methods, the nutritional reference, and the toxic harm, respectively. From the 21 elements measured in *Tetragonisca angustula* pot-honey and pot-pollen by Oliveira et al. (2017), the toxic elements As, Cd, and Pb are reviewed here. Surprisingly, lead is concentrated in pot-pollen but is below LOD in pot-honey.

15.4 Stingless Bee Keepers are Crucial for the Heritage and Conservation Mission

Stingless bees live in tropical and subtropical area around the world. These area include Africa, Asia, Australia, and Central and South America (Kwapong et al. 2010). Culturing stingless bees is often performed with native bees by indigenous people who may have all or part of the following criteria (Toledo 2001):

1. They are the descendants of the original inhabitants of a territory which has been overcome by conquest.
2. They are "ecosystem people," such as shifting or permanent cultivators, herders, hunters and gatherers, fishermen, and/or handicraft makers, who adopt a multi-use strategy of appropriation of nature.
3. They practice a small-scale, labor-intensive form of rural production which produces little surplus and has low energy needs.
4. They do not have centralized political institutions but organize their life at the community level and make decisions on a consensus basis.
5. They share a common language, religion, moral values, beliefs, clothing, and other identifying characteristics as well as a relationship to a particular territory.
6. They have a different world view, consisting of a custodial and non-materialistic attitude toward land and natural resources based on a symbolic interchange with the natural universe.
7. They are subjugated by a dominant culture and society.
8. They subjectively consider themselves to be indigenous.

Table 15.3 Review on average chemical composition of pot-pollen produced by 18 species of stingless bees from four countries (2009–2017)

Pot-pollen type Stingless bee species	N	Moisture	Ash	Proteins	Fat	Carbohydrates	pH	Fiber	Country	Reference
Frieseomelitta sp. aff. *varia*	2	29.96 (0.22)	3.13 (0.11)	**24.72** (0.18)	3.51 (0.09)	38.68 (0.20)	–	–	Venezuela	Vit et al. Chap. 26
Lepidotrigona flavibasis	3	22.8 (0.5)	2.2 (0.2)	16.7 (0.4)	4.9 (0.04)	53.3 (0.8)	–	–	Thailand	Chuttong et al. Chap. 22
Lepidotrigona terminata	3	25.3 (0.3)	**1.8** (0.2)	**14.3** (0.6)	5.3 (0.1)	53.4 (1.0)	–	–	Thailand	Chuttong et al. Chap. 22
Melipona eburnea	1	35.89 (0.31)	2.54 (0.15)	18.44 (0.29)	6.03 (0.20)	37.10 (0.25)	–	–	Venezuela	Vit et al. Chap. 26
Melipona sp. aff. *eburnea*	3	**48.54** (0.41)	2.33 (0.10)	18.32 (0.10)	3.19 (0.11)	27.62 (0.50)	–	–	Venezuela	Vit (2016)
Melipona favosa	1	29.01 (0.20)	2.92 (0.07)	22.31 (0.25)	4.38 (0.15)	41.38 (0.19)	–	–	Venezuela	Vit et al. Chap. 26
Melipona sp. *fulva* group	1	31.65 (0.32)	2.45 (0.12)	19.43 (0.10)	5.72 (0.19)	40.75 (0.30)	–	–	Venezuela	Vit et al. Chap. 26
Melipona interrupta	3	37.12[a] (0.60)	2.74[a] (0.02)	24.00[a] (0.01)	6.47[a] (0.15)	44.27[a] (1.55)	**3.34** (0.02)	13.65[a] (1.57)	Brazil	Rebelo et al. (2016)
Melipona lateralis kangarumensis	3	38.32 (0.45)	2.76 (0.23)	21.77 (0.17)	4.80 (0.24)	32.35 (0.48)	–	–	Venezuela	Vit et al. Chap. 26
Melipona mandacaia	21	36.0 (2.0)	**4.9** (0.3)	21 (2)	–	–	3.49 (0.04)	**3.6** (1.4)	Brazil	Bárbara et al. (2015)
Melipona paraensis	4	42.74 (0.51)	1.93 (0.19)	19.08 (0.23)	5.23 (0.31)	31.02 (0.48)	–	–	Venezuela	Vit et al. Chap. 26
Melipona scutellaris	6	44.71 (9.83)	1.84 (0.13)	23.88 (0.10)	4.25 (0.10)	**24.48** (10.1)	**3.75** (0.01)	**0.87** (0.20)	Brazil	Alves et al. Chap. 25
Melipona seminigra	3	53.39[a] (0.50)	4.03[a] (0.46)	37.63[a] (1.71)	10.81[a] (0.92)	25.66[a] (3.38)	3.70 (0.01)	9.30[a] (0.37)	Brazil	Rebelo et al. (2016)
Scaptotrigona mexicana	3	22.27	2.83	21.19	**0.89**	47.15	3.57	–	Mexico	Contreras-Oliva et al. Chap. 23
Scaptotrigona sp. cf. *ochrotricha*	3	43.49 (0.95)	1.94 (0.35)	16.80 (0.21)	6.72 (0.58)	31.03 (1.08)	–	–	Venezuela	Vit (2016)

Tetragonisca angustula	3	24.69 (0.78	2.06 (0.13)	22.97 (3.57)	4.58 (0.59)	45.98 (2.87)	–	–	Venezuela	Vit et al. Chap. 24
Tetragonisca angustula	3	23.34 (1.18)	2.13 (0.24)	22.43 (3.43)	4.42 (0.31)	46.68 (2.74)	–	–	Venezuela	Vit et al. Chap. 24
Tetragonula laeviceps	3	**16.1** (1.1)	2.3 (0.4)	15.5 (2.6)	**7.4** (0.3)	**58.7** (3.5)	–	–	Thailand	Chuttong et al. Chap. 22
Tetragonula testaceitarsis	3	31.7 (1.2)	2.2 (0.1)	17.9 (1.9)	5.4 (0.6)	43.1 (2.8)	–	–	Thailand	Chuttong et al. Chap. 22
Suggested standards for *Apis mellifera* dried pollen	–	<6	>15	>1.5	>40[b]					Campos et al.

Values are averages ± (SEM). Minimum and maximum values for each parameter are in boldface

Averages of groups for the same stingless bee in a publication were calculated for this table. Sum of percentages (Rebelo et al. 2016) >>100

[a]Values not considered in ranges because their addition is >>100%

[b]Sugars

Table 15.4 Review on average concentrations (µg/kg) of nutritional and toxic elements in pot-pollen produced by seven species of stingless bees in three countries (2009–2017)

Elements		Nutritional								Toxic			Country	References
Pot-pollen type Stingless Bee Species	N	Calcium Ca	Magnesium Mg	Potassium K	Phosphorus P	Selenium Se	Sodium Na	Zinc Zn		Arsenic As	Cadmium Cd	Lead Pb		
Lepidotrigona flavibasis	3	2719.7 (94.6)	1315.3 (2.5)	5125.7 (30.0)	–	–	81.7 (6.4)	–	–	–	–	–	Thailand	Chuttong et al.Chap. 7
Lepidotrigona terminata	3	2507.3 (4.2)	1176.0 (4.6)	4606.3 (75.6)	–	–	77.2 (5.9)	–	–	–	–	–	Thailand	Chuttong et al. Chap. 7
Melipona subnitida	1	1846.1	975.4	5918.5	–	–	–	36.4 (0.34)	–	–	–	–	Brazil	Silva et al. (2014)
	1	3424.9 (0.00)	2166.1 (170.34)	13,366.6 (0.00)	–	–	–	71.2 (7.06)	–	–	–	–	Brazil	Silva et al. (2014)
Scaptotrigona mexicana	3	–	–	249.85	320.87	–	–	–	–	–	–	–	Mexico	Contreras-Oliva et al. Chap. 5
Tetragonisca angustula	1	–	–	–	–	–	–	–	361.30 (18.88)	1.60 (0.01)	36.33 (2.31)	–	Brazil	Oliveira et al. (2017)
	1	–	–	–	–	–	–	–	330.16 (4.49)	1.64 (0.01)	99.79 (26.23)	–	Brazil	Oliveira et al. (2017)
	1	–	–	–	–	–	–	–	93.21 (25.16)	1.64 (0.01)	463.31 (35.16)	–	Brazil	Oliveira et al. (2017)
	1	–	–	–	–	–	–	–	51.10 (22.19)	1.60 (0.01)	137.13 (79.27)	–	Brazil	Oliveira et al. (2017)
	1	–	–	–	–	–	–	–	7.00 (0.45)	1.60 (0.01)	1.20 (0.01)	–	Brazil	Oliveira et al. (2017)
Tetragonula laeviceps	3	2566.0 (489.3)	1150.0 (222.6)	5656.0 (1274.9)	–	–	89.9 (20.0)	–	–	–	–	–	Thailand	Chuttong et al. Chap. 7
Tetragonula testaceitarsis	3	2904.0 (546.2)	1318.0 (95.3)	4594.7 (521.0)	–	–	133.5 (48.6)	–	–	–	–	–	Thailand	Chuttong et al. Chap. 7

Values are averages ± (SEM). Averages of groups for the same stingless bee in a publication were calculated

Indigenous and rural stingless beekeepers hold an ancestral knowledge which can be defined as intangible cultural heritage. In fact, it follows the domain of "knowledge and practices concerning nature..." defined by the Convention for the Safeguarding of the Intangible Cultural Heritage (UNESCO 2003). The Convention recognizes that communities, in particular indigenous communities, play an important role in the production, safeguarding, maintenance, and recreation of the intangible cultural heritage, thus helping to enrich cultural diversity and human creativity. Moreover, the Convention defines "safeguarding" the measures aimed at ensuring the viability of the intangible cultural heritage, including the identification; documentation; research; preservation; protection; promotion; enhancement; transmission, particularly through formal and informal education; as well as revitalization of the various aspects of such heritage. Last but not least, Article 15 of the Convention is about participation. Each State Party shall endeavor to ensure the widest possible participation of communities and involve them actively in its management. This is a recommendation that we should keep in mind also in the next draft of the Guide to the Management of the Route of Living Museums for Meliponi are bees in the World.

Moreover, the value of cultural heritage for the society is clearly expressed by the Faro Convention (2005) "the role of cultural heritage in the construction of a peaceful and democratic society, and in the processes of sustainable development and the promotion of cultural diversity."

It is in fact more and more accepted – at least in the scientific community – how it is necessary to adopt bio-cultural approaches to conservation (Gavin et al. 2015). The definition of bio-cultural heritage fits perfectly with our case: knowledge, innovations, and practices of indigenous and local communities that are collectively held, inextricably linked to, and shaped by the socio-ecological context of communities (Davidson-Hunt et al. 2012).

Wilder et al. (2016) clearly express how the indigenous knowledge in curbing the loss of biodiversity is crucial: "by not fully honouring the

real and potential value of indigenous science as it does professional academic and citizen science, our institutions risk ignoring the opportunity to consilience among the many sources of knowledge." The knowledge of indigenous and the rural population about stingless bees is evident (Camargo and Posey 1990; Ocampo-Rosales 2013; Reyes-González et al. 2014; Vit et al. 2015; Zamudio and Hilgert 2015; Wilder et al. 2016).

A recent Master's Thesis on meliponiculture in southern Ecuador carefully established technical basis as a rural initiative for sustainable development and conservation (Martínez-Fortún 2015; see Chap. 30). A group of 64 stingless beekeepers were interviewed to explore their perception on natural resources available for meliponiculture in Loja (Loja, Olmedo, Pindal, Zapotillo) and Zamora-Chinchipe (El Padmi) provinces of southern Ecuador, from July 2014 to March 2015. The ages of interviewees ranged from 22 to 72 years old, and they managed ten species – mostly the *Scaptotrigona* sp. "catana" and the *Melipona mimetica* "bermejo" and less frequent *Nannotrigona* sp. "alpargate"; *Tetragonisca angustula* "angelita"; *Melipona indecisa* "cananambo," "cojimbo," or "nichumbe"; *Paratrigona eutaeniata* "pirunga"; *Nannotrigona mellaria* "pitón"; and *Trigona fulviventris* "pulao." A rapid estimation of biodiversity of stingless bees from southern Ecuador – using no taxonomy or museum specimens with valid identification – was performed using geometric morphometry with 12 landmarks on anterior wing venation (García-Olivares et al. 2015). Pot-honey is harvested on a regular basis and consumed by all interviewees, compared to the lower yields of pot-pollen – consumed by 60% (Martínez-Fortún 2015). Another rural initiative on meliponiculture with *Scaptotrigona mexicana* Guérin-Meneville, 1845 provides technical and organizational support to increase the income of 125 families from two municipalities in Atzalán (mixed-race) and Zozocolco de Hidalgo (Totonacs) and conservation of natural resources in Eastern Mexico since 2012. In their 5-year report, stingless beekeepers participated in workshops of Good Manufacturing Practice (GMP) in meliponaries and received insights on people whose livelihoods depend on the rainforest, follow-up on quality control of pot-honey, marketing, brand development, the web page www.tiyatku.com, organic certification, and a proposal of a pot-honey norm needed for exports (Albalat Botana and Acosta Quijano 2016). Besides the nutritional and medicinal profit of pot-honey produced by *Scaptotrigona mexicana*, these communities use the cerumen for religious ceremonies (flowers and candles), but the pot-pollen that was mixed with corn dough in the past is now left for the bee colonies and not used for human consumption (N. Acosta Quijano, personal communication).

15.5 A Stingless Bee Window to Look at Climate Warming

Estimates of climate change predict disturbances worldwide, mainly changes in temperature and precipitation patterns, and generally more frequent extreme weather events. The tropics are so vulnerable, since the first climate changes without precedent will be there, due to tropical species that are more vulnerable to small changes in climate and because they have the greatest diversity of species. In these regions, the temperature will continue to rise, although relatively little in magnitude, and is likely to have more lethal consequences due to the limited physiological tolerance of tropical species compared to higher latitude species (FAO 2011; Mora et al. 2013; IPCC 2014).

It is expected that the changes in climatic conditions have impacts on the stingless bees, since they are insects that are distributed in the tropics. Most species occupy narrow geographical ranges and are strongly related to their environment, possibly due to physiological constraints and environmental specializations of each species (Michener 1974, 2013; Ayala et al. 2013, Maia-Silva et al. 2014). For example, in Ecuador although it is not known, all the diversity of bees is estimated to be high due to the presence of different ecosystems (i.e., the Coast, Sierra, and Amazonia), and some species of stingless bees are already known with restricted distribution (e.g., *Melipona mimetica* and *Geotrigona fumipennis* have only been recorded in southern Ecuador and northern Peru) (Rasmussen 2004).

For Mexico, 46 species of stingless bees are known: 13 (28%) are of wide distribution associated with tropical vegetation, both perennial and deciduous; 23 (50%) species are associated with evergreen tropical forest; and 10 (22%) species are endemic – they are associated with specific area of vegetation and altitude (Ayala 1999). Stingless bees depend on environmental conditions to carry out their activities, both outside and inside their nests. Therefore, they are an ideal group for the study of the environment windows. For example, the success of foraging by one species of bees is limited to a window of the environment, a combination of optimal environmental temperatures and the availability of resources (Maia-Silva et al. 2014; see Maia-Silva et al. Chap. 7 in this volume). Other abiotic conditions and biotic interactions (e.g., mutualism, competition, predation, parasitism, etc.) could also be added or become an environment window.

Temperature is probably the most important abiotic factor influencing on the biology of the insects (Menéndez 2007; Sable and Rana 2016) because they are ectothermal animals (i.e., the body temperature is variable and depends on the ambient temperature), and a forecast of negative effects is predicted if the high temperatures of the tropics increase beyond the limits that can be overcome by phenotypic plasticity or behavior (Bale and Hayward 2010). For the majority of the stingless bees, the temperature control depends on mechanisms related to the selection of the nesting site, the area where they breed, and the specialized nest structures such as the involucrum and the batumen area (Jones and Oldroyd 2006). Stingless bees have the ability to control the temperature of the nests within ranges that are optimal for the development of the immatures (Roubik 2006). The adult workers have physiological and behavioral responses to extreme variations in the environment to keep the internal temperature of the nest (Jones and Oldroyd 2006), although this adaptation has not been observed in certain species (Moo-Valle et al. 2000).

At present, there are few studies on the possible effects of climate change in the stingless bees. One of these reports considers that the changing climate could affect more the species that inhabit tropical area of low altitude in comparison to those that live in high altitude (Macías-Macías et al. 2011). When tolerance to thermal stress (cold and heat) is compared for stingless bee species thriving in low-lying tropical area (*Scaptotrigona pectoralis* Dalla Torre and *Melipona beecheii* Bennett) with a species from a sort of high tropical area (*Melipona colimana* Ayala), it is found that the species of low-lying area are more susceptible to abrupt changes in temperature. Therefore, climate change can affect in greater or less degree species of stingless bees depending on the distribution area and the altitude. The colonies have ranges of tolerance, as do the foragers. If they are too hot, they move up or use their behavior. If they really, really cannot tolerate a change of 1–2 deg. C, then they are not very well adapted, at all.

Another study reports the likely influence of climate change on the geographical distribution and the habitat reduction of the species, based on future scenarios of climate change in Brazil (Giannini et al. 2012). The results show that the total area of suitable habitat decreases for all species (ten species), except for one, under various future scenarios. The biggest reductions in the area of the habitat are found for *Melipona bicolor bicolor* Lepeletier and *Melipona scutellaris* Latreille, predominantly in area related to the Atlantic wet forest. Apart from the changes of distribution, it is important to consider that the stingless bees live in perennial colonies, where the production of individuals is continuous for long periods of time. These mismatches between blooming and the optimal foraging temperature may result in a reduction of the food intake of a colony and finally the production of offspring (Maia-Silva et al. 2014). The migration, the changes in its life cycle, and the behavior could help them to survive in new habitats, but possibly these bees will also have to adapt to a variety of new predators, parasites, and pathogens (Giannini et al. 2012).

It is essential to understand the impacts of climate change on these bees, as key species of ecological and economic importance for their role as pollinators of many species of native, wild, and cultivated plants (FAO 2011; Giannini et al. 2012; Meléndez et al. Chap. 11, in this volume). More studies are suggested to integrate the loss and fragmentation of habitat and the impact of agriculture together with natural phenomena (e.g., hurricanes) and climate change, since the combination of these phenomena can potentially produce more drastic scenarios (Giannini et al. 2012; Meléndez et al. 2013, 2016). With it, it will be possible to form solid base for the development and success of strategies and plans of conservation (Giannini et al. 2012; Maia-Silva et al. 2014; Meléndez et al. 2016). The impacts of the changes projected on agriculture, especially crops that are pollinated by bees, require further evaluation (Memmott et al. 2007; Giannini et al. 2012; Sable and Rana 2016). Moreover, the ecosystemic service of bees – pollination – can be assessed with a new proposed methodology to identify key area of conservation, protection, or restoration that could improve the connectivity of the habitat and safeguard the pollination service. This methodology has already been applied to the tropical species of stingless bees *Melipona quadrifasciata* Lepeletier, a pollinator of the native flora and of important agricultural crops of the Brazilian Atlantic Mata (Giannini et al. 2015), and can be implemented in other regions of the world.

15.6 Why a Living Museum to Embrace Meliponini of the World?

Entomological collections of Meliponini are necessary for taxonomic studies and reference of the species. The idea of a "Living Museum" was proposed at the National Museum of Apiculture "Ignacio Herrera" in Mérida, Venezuela, in 1988, and further expanded to the world in 2013 after great interactions with stingless beekeepers and feral stingless bee nests in Costa Rica (Aguilar et al. 2013a, b).

An informal "anthropologic research" by P. Vit on stingless beekeepers – their lifestyle, their relationships with stingless bee colonies, and harvesting of pot-honey, pot-pollen, and cerumen – was accomplished along the years by her empathetic way of listening to stingless beekeepers and developed awareness on their deep knowledge on stingless bees, the environment, and the impor-

Fig. 15.1 Route of Living Museums for Meliponini Bees in the World (Artwork design: C. Ruíz)

tance of preserving and sharing this ancestral knowledge (V. Frisone, personal observations). Local and international perceptions on how stingless beekeepers live – often with social and economic limitations – encouraged her proposal of a Route of Living Museums of Meliponini Bees in the World as a way to provide networking visibility within local communities and appreciation for preserving the art of keeping stingless bees. In 2014, a further step was taken during the Prometeo-SENESCYT-Universidad Técnica de Machala project on "Valorization of pot-honey produced by Meliponini in Ecuador," starting to think on how to structure this idea by joint efforts between international authorities on stingless bees and a Natural Science Museum curator in Italy. The Museo di Archeologia e Scienze Naturali "G. Zannato" (www.museozannato.it) operates its museum network in strict contact with the local population (citizen science, education, youth programs). A flexible structure was desirable to combine academic science with traditional ecological knowledge. Our objective was to propose an interactive bridge for scientific scholars and native practitioners. It consists of a network of formal institutions (scientific institutes, universities, livestock and forestry ministries, museums) and stingless beekeepers. The latter would be in charge of a station, together with culturing, research, and educational activities. From this idea, a first draft of the "Guide to the Management of the Route of Living Museums for Meliponini Bees in the redundant" (Vit 2015) was proposed. The creation of the portal Guide to the Management for the Route of Living Museums of Meliponini Bees in the World at the Universidad de Los Andes institutional repository (Vit 2016) is backed by APIMONDIA to provide further ubiquitous and asynchronous information. The guide is intended to be tailored to different socio-ecological contexts in the world. The name "Living

Museum" was proposed to stress the importance of the living beings which are the *conditio sine qua non* of this project: stingless bees and stingless beekeepers-providing visibility of their art to keep bees available in the ecosystem

From a museological point of view, the Route of Living Museums of Meliponini in the World complies with the definition of an eco-museum. An eco-museum is a museum focused on the identity of a place, largely based on local participation and aiming to enhance the welfare and development of local communities. The term "eco" is a shortened form of "ecology," but it refers especially to a new idea of holistic interpretation of cultural heritage, in opposition to the focus on specific items and objects, often performed by traditional museums (Davis 1999). An eco-museum for Meliponini would be an appropriate platform to present and expand conservation programs. See the logo in Fig. 15.1. The *Tetragonisca angustula* silhouette was chosen as the most widespread stingless bee in the Neotropics.

15.7 Conclusions

The biodiversity of Ecuadorian stingless bees reached 132 species in this chapter and >200 with the 100 in the megabiodiverse Yasuní Biosphere Reserve in Chap. 14 by Roubik. Chemical composition is reviewed for a proposal of pot-pollen standards, and the Ecuadorian progress for pot-honey standards is discussed. The role of stingless beekeepers in the intangible heritage and conservation missions is framed by criteria of people and biodiversity. Climate change can disturb species of stingless bees differently according to the distribution area and the altitude, and conservational roles of an eco-museum may help the situation.

Acknowledgments To Ecuadorian stingless beekeepers. To the Prometeo-UTMACH scholarship (2014–2015) to P. Vit for the research project "Valorization of Pot-Honey produced by Meliponini in Ecuador." Professor Med. Vet. F Maza (Planning Director) and his secretary Mrs. E. Brito kindly supported the administrative needs for the research successfully hosted at Universidad Técnica de Machala, Ecuador. To the late Professor JMF Camargo, Biology Department, Universidade de São Paulo, Ribeirão Preto, Brazil, for his scientific rigor and artistic imprint in observing stingless bees and his legacy with the Camargo Collection – RPSP. To Professor Giovanni Onore, retired from Pontificia Universidad Católica de Ecuador, for his academic struggle and success in a society known for a peculiar reduction of heads in the forest and his seminal work on Ecuadorian Meliponini. V. Frisone's work was supported by the Municipality of Montecchio Maggiore, Museo di Archeologia e Scienze Naturali "G. Zannato," Italy. To appreciated comments of reviewers. Dr. DW Roubik kindly edited the English manuscript.

References

ADAB. Agência Estadual de Defesa Agropecuária da Bahia. 2014. Regulamento Técnico de Identidade e Qualidade do Mel de Abelha social sem ferrão, gênero *Melipona*. Portaria ADAB N° 207 DE 21/11/2014; Bahia, Brasil. pp. 1-9.

Albalat Botana A, Acosta Quijano N. 2016. Fortalecimiento de la cadena productiva de miel de monte para elevar ingresos familiares y conservar los recursos naturales. Final Project Report. Asociación Istakuspinini AC; Xalapa, Veracruz, México. 8 pp.

Aguilar I, Herrera E, Vit P. 2013a. Acciones para valorizar la miel de pote. pp. 1-13. In: Vit P and Roubik DW, eds. Stingless bees process honey and pollen in cerumen pots. Facultad de Farmacia y Bioanálisis, Universidad de Los Andes; Mérida, Venezuela. http://www.saber.ula.ve/handle/123456789/35292

Aguilar I, Herrera E, Zamora G. 2013b. Stingless bees of Costa Rica. pp. 113-124. In: Vit P, Pedro SRM, Roubik D, eds. Pot honey: A legacy of stingless bees. Springer; New York, USA. 654 pp.

Ayala R. 1999. Revisión de las abejas sin aguijón de México (Hymenoptera: Apidae: Meliponini). Folia Entomológica Mexicana 106: 1-123.

Ayala R, González VH, Engel MS. 2013. Mexican stingless bees (Hymenoptera: Apidae): diversity, distribution, and indigenous knowledge. pp. 135-152. In: Vit P, Pedro SRM, Roubik D, eds. Pot honey: A legacy of stingless bees. Springer; New York, USA. 654 pp.

Bale JS, Hayward SAL. 2010. Insect overwintering in a changing climate. The Journal of Experimental Biology 213: 980-994.

Bárbara MS, Machado CS, Sodré GS, Dias LG, Estevinho LM, Carvalho CAL. 2015. Microbiological assessment, nutritional characterization and phenolic compounds of bee pollen from *Mellipona mandacaia* Smith, 1983. Molecules 20: 12525-12544.

Brasil. 2001. Instrução Normativa n.3, de 19 de janeiro de 2001. Ministério da Agricultura, Pecuária e Abastecimento aprova os regulamentos técnicos de identidade e qualidade de apitoxina, cera de abelha, geléia real, geléia real liofilizada, pólen apícola, própolis e extrato de própolis. In: Ministério da Agricultura, Pecuária e Abastecimento. Secretaria de Defesa Agropecuaria. Diário Oficial da União, Brasília, DF, 23 jan. 2001, Seção 1, p.18. 6 pp.

Camargo JMF, Pedro SRM. 2007. Meliponini Lepeletier, 1836. pp. 272-578. In: Moure JS, Urban D, Melo GAR, eds. Catalogue of Bees (Hymenoptera, Apoidea) in the Neotropical Region. Sociedade Brasileira de Entomologia; Curitiba, Brasil. 1958 pp.

Camargo JMF, Posey DA. 1990. O conhecimento dos Kayapó sobre as abelhas sociais sem ferrão (Meliponidae, Apidae, Hymenoptera): Notas adicionais. Boletim do Museo Paraense Emílio Goeldi, Série Zoologia. 6: 17-42.

Coloma LA. 1986. Contribución para el conocimiento de las abejas sin aguijón (Meliponinae: Apidae: Hymenoptera) de Ecuador. Tesis para obtener el título de Licenciado en Ciencias Biológicas. Pontificia Universidad Católica de Ecuador; Quito, Ecuador. pp. 146.

Davidson-Hunt IJ, Turner KL, Te Pareake Mead A, Cabrera-Lopez J, Bolton R, Idrobo CJ, Miretski I, Morrison A, Robson JP. 2012. Biocultural design: a new conceptual framework for sustainable development in rural indigenous and local communities. Sapiens 5: 33–45.

Davis P. 1999. Ecomuseums: a sense of place. Leicester University Press; London-New York, 271 pp.

FAO. 2011. Potential effects of climate change on crop pollination. In: Kjøhl M, Nielsen A, Stenseth NC, eds. Centre for Ecological and Evolutionary Synthesis (CEES), Department of Biology, University of Oslo, Norway. 38 pp.

Faro Convention. 2005. Council of Europe Framework Convention on the Value of Cultural Heritage for Society. 9 pp. http://www.coe.int/it/web/conventions/full-list/-/conventions/rms/0900001680083746

García-Olivares V, Zaragoza-Trello C, Ramirez J, Guerrero-Peñaranda A, Ruiz C. 2015. Caracterización rápida de la biodiversidad usando morfometría geométrica: Caso de estudio con abejas sin aguijón (Apidae: Meliponini) del sur de Ecuador. Avances en Ciencias e Ingenierías 7: B32-B38.

Gavin MC, McCarter J, Mead A, Berkes F, Stepp JR, Peterson D, Tang R. 2015. Defining biocultural approaches to conservation. Trends in Ecology & Evolution 30: 140-145.

Giannini TC, Acosta AL, Garófalo CA, Saraiva AM, Alves-dos-Santosa I, Imperatriz-Fonseca VL. 2012. Pollination services at risk: Bee habitats will decrease owing to climate change in Brazil. Ecological Modelling 244: 127- 131.

Giannini TC, Tambosi LR, Acosta AL, Jaffé R, Saraiva AM, Imperatriz-Fonseca VL, Metzger JP. 2015. Safeguarding Ecosystem Services: A Methodological Framework to Buffer the Joint Effect of Habitat

Configuration and Climate Change. PLoS ONE 10: e0129225. doi:10.1371/journal.pone.0129225

Halcroft M, Spooner-Hart R, Dollin A. 2013. Australian stingless bees. pp. 35-72. In: Vit P, Pedro SRM, Roubik D, eds. Pot honey: A legacy of stingless bees. Springer; New York, USA. 654 pp.

ICONTEC. Instituto Colombiano de Normas Técnicas y Certificación. 2007. Norma Técnica Colombiana. Miel de Abejas. NTC 1273; Bogotá, Colombia. 6 pp. http://www.sinab.unal.edu.co/ntc/NTC1273.pdf

INEN. Instituto Ecuatoriano de Normalización. 1988. Miel de Abejas, Requisitos. NTE INEN 1572:1988. INEN; Quito, Ecuador. 4 pp.

INEN. Instituto Ecuatoriano de Normalización. 2016. Norma Técnica Ecuatoriana Obligatoria. Miel de Abejas, Requisitos. NTE INEN 1572:2016. Primera Revisión. INEN; Quito, Ecuador. 12 pp.

IPCC. 2014. Climate change, synthesis report. Available in: http://www.ipcc.ch/pdf/assessment-report/ar5/syr/SYR_AR5_FINAL_full_wcover.pdf

Jones JC, Oldroyd BP. 2006. Nest thermoregulation in social insects. Advances in Insect Physiology 33: 153-191.

Kwapong P, Aidoo K, Combey R, Karikari A. 2010. Stingless bees, importance, management and utilisation. A training manual for stingless beekeeping. Unimax Macmillan Ltd; Accra North, Ghana. 72 pp.

Macías-Macías JO, Quezada-Euán JJG, Contreras-Escareño F, Tapia-González JM, Moo-Valle H, Ayala R. 2011. Comparative temperature tolerance in stingless bee species from tropical highlands and lowlands of Mexico and implications for their conservation (Hymenoptera: Apidae: Meliponini). Apidologie 42: 679-689.

Maia-Silva C, Imperatriz-Fonsec VL, Silva CI, Hrncir M. 2014. Environmental windows for foraging activity in stingless bees, Melipona subnitida Ducke and Melipona quadrifasciata Lepeletier (Hymenoptera: Apidae: Meliponini). Sociobiology 61: 378-385.

Martínez-Fortún S. 2015. Desarrollo sostenible y conservación etnoecológica a través de la meliponicultura, en el sur de Ecuador. Master dissertation, Universidad Internacional de Andalucía; Córdoba, Spain. 110 pp.

Memmott J, Craze PG, Waser NM, Price MV. 2007. Global warming and the disruption of plant-pollinator interactions. Ecology Letters 10: 710-717.

Meléndez RV, Meneses CL, Kevan PG. 2013. Effects of human disturbance and habitat fragmentation on stingless bees. pp. 269-282. In: Vit P, Silvia RMP, Roubik D, eds. Pot honey: A legacy of stingless bees. Springer; New York, USA. 654 pp.

Meléndez RV, Ayala R y Delfín GH. 2016. Temporal variation in native bee diversity in the tropical sub-deciduous forest of the Yucatan Peninsula, Mexico. Tropical Conservation Science 9: 718- 735.

Menéndez R. 2007. How are insects responding to global warming? Tijdschrift voor Entomologie 150: 355-365.

Michener CD. 1974. The Social Behavior of the Bees. A Comparative Study. Harvard University Press; Cambridge, Massachusets. 404 pp.

Michener CD. 2013. The Meliponini. pp. 3-18. In: Vit P, Pedro SRM, Roubik D, eds. Pot honey: A legacy of stingless bees. Springer; New York, USA. 654 pp.

Michener CD, Grimaldi DA. 1988a. A Trigona from late Cretaceous amber of New Jersey (Hymenoptera: Apidae: Meliponinae). American Museum Novitates 2917: 1-10 pp.

Michener CD, Grimaldi DA. 1988b. The Oldest Fossil Bee: Apoid History, Evolutionary Stasis, and Antiquity of Social Behavior. Proceedings of the National Academy of Sciences of the United States of America 85: 6424-6426.

Moo-Valle H, Quezada-Euán JJG, Navarro J, Rodriguez-Carvajal LA. 2000. Patterns of intranidal temperature fluctuation for Melipona beecheii colonies in natural nesting cavities. Journal of Apicultural Research 39: 3-7.

Mora C, Frazier AG, Longman RJ, Dacks RS, Walton MM, Tong EJ, Giambelluca TW. 2013. The projected timing of climate departure from recent variability. Nature 502: 183-187.

Ocampo Rosales GR. 2013. Medicinal Uses of Melipona beecheii honey, by the ancient Maya. pp. 229-240. In: Vit P, Pedro SRM, Roubik D, eds. Pot-Honey: A Legacy of Stingless Bees. Springer; New York, USA. 654 pp.

Oliveira FA, Abreu AT, Oliveira Nascimento N, Froes-Silva RES, Antonini Y, Nalini HA, Lena JC. 2017. Evaluation of matrix effect on the determination of rare earth elements and As, Bi, Cd, Pb, Se and In in honey and pollen of native Brazilian bees (Tetragonisca angustula – Jataí) by Q-ICP-MS. Talanta 162: 488-494.

Pauly A, Pedro SRM, Rasmussen C, Roubik DW. 2013. Stingless bees (Hymenoptera: Apoidea: Meliponini) of French Guiana. pp. 87-97. In: Vit P, Pedro SRM, Roubik D, eds. Pot honey: A legacy of stingless bees. Springer; New York, USA. 654 pp.

Pedro SRM. 2014. The stingless bee fauna in Brazil (Hymenoptera: Apidae). Sociobiology 61: 348-354.

Pedro SRM, Camargo JMF. 2013. Stingless bees from Venezuela. pp. 73-86. In: Vit P, Pedro SRM, Roubik D, eds. Pot honey: A legacy of stingless bees. Springer; New York, USA. 654 pp.

Persano-Oddo L, Piro R. 2004. Main European unifloral honeys: Descriptive sheets. Apidologie 25: S38-S81.

Rasmussen C. 2004. Bees from Southern Ecuador. Lyonia 7: 29-35.

Rebelo KS, Ferreira AG, Carvalho-Zilse GA. 2016. Physicochemical characteristics of pollen collected by Amazonian stingless bees. Ciência Rural 46: 927-932.

Reyes-González A, Camou-Guerrero A, Reyes-Salas O, Argueta A, Casas A. 2014. Diversity, local knowledge and use of stingless bees (Apidae: Meliponini) in the municipality of Nocupétaro, Michoacan, Mexico. Journal of Ethnobiology and Ethnomedicine10: 1-12.

Roig-Alsina A, Vossler FG, Gennari GP. 2013. Stingless bees in Argentina. pp. 125-152. In: Vit P, Pedro SRM, Roubik D, eds. Pot honey: A legacy of stingless bees. Springer; New York, USA. 654 pp.

Roubik DW. 2006. Stingless bee nesting biology. Apidologie 37: 1-20.

Ruano Iraheta CE, Hernández Martínez MÁ, Alas Romero LA, Claros Álvarez ME, Arévalo DR, Rodríguez González VA. 2015. Stingless bee distribution and richness in El Salvador (Apidae, Meliponinae). Journal of Apicultural Research. 54.: http://dx.doi.org /10.1080/00218839.2015.1029783

Sable MG, Rana DK. 2016. Impact of global warming on insect behavior - A review. Agricultural Reviews 37: 81-84.

Silva GR, Natividade TB, Camara CA, Silva EMS, Santos FDAR, Silva TMS. 2014. Identification of sugar, amino acids and minerals from the pollen of Jandaíra stingless bees (*Melipona subnitida*). Food and Nutrition Sciences 5: 1015-1021.

Toledo VM. 2001. Indigenous peoples and biodiversity. Vol. 3, pp. 451-463. In: Levin SA, ed. Encyclopedia of Biodiversity. Academic Press; San Diego, California, USA. 4666 pp.

UNESCO. 2003. Convention for the Safeguarding of the Intangible Cultural Heritage. 32nd Session of the General Conference, Paris, 29 September–17 October 2003. http://unesdoc.unesco.org/ images/0013/001325/132540e.pdf

Vit P. 2008. La miel precolombina de abejas sin aguijón (Meliponini) aún no tiene normas de calidad. Revista Boletín Centro de Investigaciones Biológicas 42: 415-423.

Vit P. 2015. Valorización de Mieles de Pote Producidas por Meliponini en Ecuador. Informe Final. Proyecto Prometeo-SENESCYT-Universidad Técnica de Machala. Machala, Ecuador. 102 pp + Annex.

Vit P. 2016. Guía para la Gestión de la Ruta de Museos Vivientes de Abejas Meliponini en el Mundo.. http:// www.saber.ula.ve/handle/123456789/42207

Vit P, Vargas O, López TV, Maza F. 2015. Meliponini biodiversity and medicinal uses of pot-honey from El Oro province in Ecuador. Emirates Journal of Food and Agriculture 27: 502-506.

Wilder BT, O'Meara C, Monti L, Nabhan GP. 2016. The importance of indigenous knowledge in curbing the loss of language and biodiversity. BioScience 66: 499-509.

Yurrita Obiols, Vásquez M. 2013. Stingless bees of Guatemala. pp. 99-111. In: Vit P, Pedro SRM, Roubik D, eds. Pot honey: A legacy of stingless bees. Springer; New York, USA. 654 pp.

Zamudio F, Hilgert NI. 2015. Multi-dimensionality and variability in folk classification of stingless bees (Apidae: Meliponini). Journal of Ethnobiology and Ethnomedicine 23: 11-41.

Nesting Ecology of Stingless Bees in Africa

16

Robert Kajobe and David W. Roubik

16.1 Introduction

The biology of stingless bees, also called stingless honey bees, varies widely considering both individual and colony size (Michener 1974, 2007; Roubik 1989; Jalil and Roubik 2017). The African meliponines are comprised of relatively few species and genera (Eardley 2004; Moure 1961; Darchen 1972; Kajobe and Roubik 2006). Unlike honey bees (*Apis*), the Meliponini produce brood with an egg placed on top of a food mass in a sealed cell. In general, colonies make far less honey, compared to honey bees. And in contrast to *Apis*, meliponines have no functional sting, mate only once, do not use water to cool their nest or pure wax to build it, and cannot freely swarm to reproduce. Stingless bees are important pollinators within tropical ecosystems and visit many flowering plants, with the intention of collecting pollen and/or nectar, and move from flower to flower before taking such resources to their nests. Unfortunately, very little is known about stingless bee ecology or economic value in most parts of Africa (Gikungu 2006; Kajobe 2008; Mogho Njoya 2009).

Significant advances in our understanding of bee evolution have occurred recently, following molecular studies on bee gene fragments, and in light of fieldwork. Bee origin, and that of the corbiculate bees in particular, has been scrutinized (Michener 2007; Rasmussen and Cameron 2010; Martins et al. 2014; Cardinal et al. in press). It is fascinating to contemplate stingless bee evolution—the earliest known highly eusocial anthophiline and corbiculate group—with their division of labor among castes and queen, storage of pollen or honey in waxen-resinous containers, perennial colony survival, and correspondence to the world's biological regions. Meliponine evolution can be traced to the present-day Brazil and *Epicharis* and *Centris*, in a basal group of mostly large, solitary apid bees, many of which harvest the oil and pollen from flowers of the plant family Malpighiaceae (Martins et al. 2014). Because our main purpose in the chapter is to discuss Afrotropical stingless bees and their nesting habits, we will delve little into general bee evolution but will reiterate a few points from the literature and attempt to clarify others.

R. Kajobe (✉)
National Agricultural Research Organisation (NARO), Rwebitaba Zonal Agricultural Research and Development Institute (Rwebitaba ZARDI), P.O. Box 96, Fort Portal, Uganda
e-mail: robertkajobe@gmail.com

D.W. Roubik
Smithsonian Tropical Research Institute, Calle Portobelo, Balboa, Republic of Panama

National Museum of Natural History, Washington, DC, USA

16.2 Meliponine Origin, Dispersal, and Richness

A meliponine origin in the mid- to late Cretaceous (96–81 mya) was proposed from 5 gene fragments, including nuclear genes, and 3596 aligned nucleotides studied in nearly 200 extant Meliponini and calibrated with some dated fossils (Rasmussen and Cameron 2010). As shown in previous work by those authors (Rasmussen and Cameron 2007), three clades—Afrotropical, Neotropical, and Indo-Malaysian-Australasian—with somewhat staggered major divergence from the original population, were evidently Gondwanan. They were alive and prospering before the Southern Continent, Gondwana, was broadly fragmented via tectonic plate movement. Among anthophilines (bees are classified as the apoid group, Anthophila) a large part of that early diversification occurred 125–60 million years ago, while recent radiation of familiar contemporary bee families took place considerably later (Poinar and Danforth 2006; Michener 2007; Rasmussen and Cameron 2010; Cardinal et al. in press).

Given an original bee clade (clan or group) of Late Cretaceous age (ca. 80–100 million years), which was entirely restricted to the New World, the corbiculate bees (those equipped with a "pollen basket" on the female hind leg) dispersed among continents. There was a notable extinction among corbiculates in the Eocene and Miocene, probably driven by drastic global cooling, aridity, and the reduction of moist tropical habitats (Flower and Kennett 1994; Michener 2007). In Miocene times bees attained a more modern aspect (Michener 2007, p. 101).

A problem related to dispersal of meliponine bees "out of Brazil" is that tectonic plates which compose Brazil or Africa separated 100 million years ago. The oldest estimated date for the existence of meliponines, 96 mya, and certainly the latest estimate, 80 mya (Rasmussen and Cameron 2010), are at a time when a deep ocean separated those continents. Furthermore, North America had no direct connection to South America, and Africa was separated by the Tethys Sea from the remainder of the Old World.

Nonetheless, fossils in amber attest to a Miocene, Eocene, or Late Cretaceous presence of meliponines or meliponoids in the Caribbean region, Europe, and North America (Rasmussen and Cameron 2010). Those authors explain that certain areas between Brazil and Africa possessed islands that could have made a "filter bridge." We suggest that colonies moved north or east, rather than originate in the present-day Old World and later disperse back to the Neotropics. The American tropics seem likely to have had a substantial head start on the diversification of meliponines and their persistence and adaptive radiation there. Further, paleopalynologists (e.g., Morley 2000) find ample evidence of tropical plant dispersal across the widening Tertiary Atlantic Ocean and also for the movement from a tropical Northern Hemisphere and into Asia, during ancient tropical climate phases.

Darlington (1957, p. 608), a zoogeographer mainly concerned with distribution of vertebrates like fish and mammals (and not convinced that vicariance, or continental movement and plate tectonics, were of resounding importance), wrote: "animals move more than land." We agree with the statement because the dispersal of stingless bees, from their ancestral habitat, occurred across the ocean and by "island-hopping." This was likely accomplished by rafting. That term implies floating on accreted islands of vegetation containing trees or other woody stems, which in turn contained entire nests and colonies. The pattern is observed in modern island fauna of stingless bees in the oceans surrounding Panama, and of course, individual meliponine queens or nests that are made underground, or largely made of mud or fibers or are exposed, cannot disperse over water (Michener 1974; Roubik 1992, 2006; Roubik and Camargo 2012, DWR, pers. obs.).

Individual meliponine queens, unlike female *Centris* or *Epicharis*, or even *Bombus*, the closest living relative of meliponines (Rasmussen and Cameron 2010; Cardinal et al. in press), are unable to establish new colonies or nests by themselves. However, both of these methods, fertile female and full colony dispersal, likely determined the spread of the corbiculate bees from an ancestral population in Brazil. The current number and composition of corbiculate bee species worldwide may, as opined by Roubik (1979) and Michener (1979, 2007, p. 103), also reflect competition and resource partitioning, and also extinction events, at community and biogeographic levels. The advanced social colonies with large perennial populations, opportunistic reproduction and nesting, rapid nest mate recruitment, and unaggressive but highly effective mass foraging influenced the kind and abundance of other bee species in the tropical zone (Roubik 1989, 2006). If those bees much resemble bees since the Eocene, they were the genus *Apis*, the honey bees. Their analog, the stingless honey bees (Meliponini), is seen as positioned between honey bees and the remaining bees, in a competitive hierarchy which, among other things, has made the tropics less rich in bee species than other areas (Roubik 1989).

The Old World Meliponini differed notably from those of the New World by 73 mya (million years ago), while their extant bees of higher groups (tribe, family, some genera) formed 50 mya and until as recently as 5 mya (Rasmussen and Cameron 2010; Cardinal et al. in press). Similarly, the Neotropics had its first major radiation in meliponines 71 mya, followed by another at 35 mya; African meliponines diverged first at 61 mya and then at two periods, 50 and 20 mya, while the Indo-Malayan and Australasian area had initially diverged 50 mya and then replaced their fauna during the next 40 mya and again near 5 mya. Clearly, entire lineages that would not fit any living genus were lost during this evolution, but some areas may have accumulated or retained genera and species more readily than others. Equatorial Africa's lowlands have maintained their tropical flora since 80 mya (Linder 2014), and as suggested below, many of their bees may also have persisted.

Did the evolution of *Apis* species in the Old World cause the extinction of competing meliponines, especially those of similar size to *Apis*? The current tropical bee fauna suggests this pos-

sibility. If the honey bees evolved in Asia and then dispersed to Africa after Eocene times (Michener 2007, p. 831), then two points seem important. First, Africa has a shorter history with *Apis* than does Asia, and, second, the relative shortage of stingless bee species in Africa is likely less related to the ecology of *Apis*. However, it is very difficult to trace the relative significance of size differences, differences in floral preference or bee species abundance and dominance, or extinction of species that left no fossils we have found. There are four species of stingless bees in the Old World that resemble sympatric *Apis*. The Asian *A. florea* and *A. andreniformis* are about the same size as *Geniotrigona* or *Homotrigona*, whereas in Africa, none of the stingless bees overlap with *A. mellifera* and only *Meliponula* approaches that size. African forests maintain tremendous numbers of tiny and medium meliponines and many *Apis mellifera*; Southeast Asia has tremendous numbers of medium-sized to small *Tetragonula* and other genera, and four or five *Apis* in potentially large number, while the Neotropics has many or several meliponine genera, of bees both large and small, and now a persistent and at times very abundant Old World *Apis mellifera*, from Africa (Roubik 1989, 1996, and the present volume).

The estimated number of meliponines continually grows, in conjunction with field work and museum research, and soon will include over 600 valid, described species in total (see Rasmussen and Gonzalez 2013). As these and other authors (e.g., Michener 2007; Camargo 2013; Halcroft 2013) make clear, the Neotropics contains roughly 85%, Southern Asia and Australia about 10%, and Africa 5% of bee species in the Meliponini. In Africa, one notes that the smallest meliponines are just as tiny as those in Asia or the Neotropics, measuring ca. 2 mm in length, while the largest are bigger than any worker meliponine of Asia or Australia but no larger than small *Melipona* and large *Scaptotrigona*, *Trigona*, *Duckeola*, or *Cephalotrigona* of the Neotropics. The truly large species all are Neotropical, several *Melipona* and a few *Trigona* being comparable to *Apis mellifera*.

16.3 Stingless Bee Species in Africa

Africa seems likely to have maintained some old stingless bee clades. It is the largest tropical continent and shares a few meliponine recent genera with the Asian tropics but none, as is also true of Asia, with the Neotropics (Michener 2007). Further, Africa was isolated from the rest of the Old World at least until the closing of the Tethys Sea, 50 mya. And thus, its meliponines had roughly 30 million years to evolve, after the breaking up of Gondwanaland, without any competition from similar bees living in perennial colonies (as far as we know). Moreover, tropical forest of Africa seems to have a remarkably long history of stability, dating from the mid-Cretaceous (Linder 2014).

Africa's botanical diversity, at a forest level, is substantially less than found in the most species-rich equatorial forests of the Americas or Asia (Linder 2014), and the same is true for meliponine species richness, in large, protected forest areas (Roubik 1996). It would perhaps be surprising, therefore, if Africa held as many stingless bees as the other two regions. And yet, as detailed below, its genera include both large and tiny meliponines, cleptobiotic meliponines, some that nest both in the ground and in tree cavities (Kajobe 2007; Fabre Anguilet et al. 2015), in termite or ant nests, and as exposed nest builders. No Asian meliponine assemblages approach such diversity, although those in large forest areas in Southeast Asia, or on the island of Borneo, have more species, by over 100%, than the richest African forests (Roubik 1990, 1996; Michener 2007).

Eardley (2004), in the most comprehensive taxonomic treatment of the African stingless bees to date, suggests there are 6 genera and 19 species in tropical Africa. The exact number of species of stingless bees in Africa is not yet known because of the gaps in research in this field of study (Mogho Njoya 2009). In five genera, *Dactylurina* Cockerell, *Meliponula* Cockerell, *Plebeina* Moure, *Hypotrigona* Cockerell, and *Liotrigona* Moure, workers collect pollen and nectar from flowers, but in one genus (*Cleptotrigona* Moure), they rob food from the nests of other stingless bees (Table 16.1). Before the study by Eardley (2004), Kerr and Maule (1964) had recognized 42 stingless bee species in Africa. We include the 24 species now recognized (Pauly and Fabre Anguilet 2013; Pauly and Hora 2013; Fabre Anguilet et al. 2015). In addition, we suggest eight genera, rather than the five listed by Michener (2007), because that author's subgenera or synonyms seem valid at the genus level: *Axestotrigona*, *Dactylurina*, *Hypotrigona*, *Liotrigona*, *Meliplebeia*, *Meliponula*, *Plebeiella*, and *Plebeina*. The group *Apotrigona* may be examined further and better classified.

The number of stingless bee genera and species in Africa varies from country to country. For example, in Tanzania, six stingless honey bee species were reported (*Dactylurina schmidti*, *Plebeina armata* (= *hildebrandti*, see Table 1), *Axestotrigona erythra*, *Meliplebeia beccarii* (= *ogouensis*), *Plebeiella lendliana*, and *Axestotrigona ferruginea*) (Njau et al. 2009). In Ethiopia, Pauly and Hora (2013) report six species, *Meliponula beccarii*, *Liotrigona bottegoi*, *L. baleensis*, *Hypotrigona gribodoi*, *H. ruspolii*, and *Plebeina armata*. In the Bamenda Afromontane Forests of Cameroon, six species of stingless bees grouped into four genera were found. They include *Meliponula* (three species), *Dactylurina* (one species), *Hypotrigona* (one species), and *Liotrigona* (one species) (Mogho Njoya 2009). In the Kakamega Forest of Kenya, six species of stingless bees were reported—*Meliponula bocandei*, *Plebeiella lendliana*, *Axestotrigona ferruginea*, *Meliplebeia beccarii*, *Hypotrigona gribodoi*, and *Plebeina armata* (Macharia et al. 2007). Aidoo et al. (2011) discuss 5 genera of stingless bees, comprising 11 species, from Ghana, *Dactylurina* (1 species), *Plebeina* (1 species), *Meliponula s. lat.* (5 species), *Hypotrigona* (3 species), and *Cleptotrigona* (1 species). A total of five stingless bee species belonging to two genera were found to occur in Bwindi Impenetrable National Park in Uganda (Kajobe 2008). *Meliponula s. lat.* had four species, while *Hypotrigona* had one. Roubik (1996) found a total of 14 species of Meliponini in the Makandé "Forét des Abeilles" in

Table 16.1 Stingless bee species in Africa

Species	East Africa	Central Africa	Southern Africa	West Africa
Cleptotrigona cubiceps (Friese, 1912)	Tanzania	Democratic Republic of Congo, Gabon	South Africa	Ghana, Sierra Leone
Dactylurina staudingeri (Gribodo, 1893)	Uganda	Central African Republic, Democratic Republic of Congo, Gabon	–	Cameroon, Ivory Coast, Ghana, Nigeria
Dactylurina schmidti (Stadelmann, 1895)	Kenya, Tanzania	–	–	–
Plebeina armata (Magretti, 1895)	Kenya, Rwanda, Tanzania, Uganda	Chad, Democratic Republic of Congo	Botswana, South Africa, Zimbabwe	Senegal, Guinea, Nigeria, Cameroun
Meliplebeia beccarii (Gribodo, 1879)	Eritrea, Tanzania, Ethiopia, Rwanda, Uganda, Kenya, South Sudan	Democratic Republic of Congo	Angola, Malawi, Namibia, South Africa	Cameroon, Ghana, Guinea, Nigeria
Meliponula bocandei (Spinola, 1853)	Kenya, Uganda	Central African Republic, Democratic Republic of Congo, Gabon	Angola, South Africa	Cameroon, Ivory Coast, Ghana, Nigeria
Axestotrigona cameroonensis (Friese, 1900)	Uganda	Democratic Republic of Congo, Gabon	–	Benin, Cameroon, Ghana, Nigeria
Axestotrigona erythra (Schletterer, 1891)	–	Gabon	–	–
Axestotrigona ferruginea (Lepeletier, 1841)	Kenya, Tanzania, Uganda	Democratic Republic of Congo, Gabon	Malawi, South Africa, Zambia	Cameroon, Ghana, Nigeria, Senegal
Axestotrigona togoensis (Stadelmann, 1895)	Tanzania, Uganda, Mozambique	Democratic Republic of Congo, Gabon	–	Benin, Cameroon, Ghana, Guinea, Ivory Coast, Senegal, Togo
Plebeiella griswoldorum (Eardley, 2004)	–	–	Zambia	–
Plebeiella lendliana (Friese,)	Kenya, Tanzania, Uganda	Democratic Republic of Congo, Gabon	–	Nigeria, Togo
Meliplebeia nebulata (Smith, 1854)	Uganda	Central African Republic, Democratic Republic of Congo, Equatorial Guinea, Gabon	Angola	Cameroon, Ghana, Liberia, Nigeria, Sierra Leone
Meliplebeia roubiki (Eardley, 2004)	–	Gabon	–	–
Hypotrigona araujoi (Michener, 1959)	Kenya, Tanzania	Democratic Republic of Congo, Gabon	Angola, Namibia, South Africa, Zambia, Zimbabwe	Ghana, Nigeria, Togo

(continued)

Table 16.1 (continued)

Species	East Africa	Central Africa	Southern Africa	West Africa
Hypotrigona gribodoi (Magretti, 1884)	Eritrea, Kenya, South Sudan, Uganda, Tanzania	Chad, Gabon, Democratic Republic of Congo	Angola, Mozambique, South Africa	Cameroon, Ghana, Nigeria, Senegal, Togo
Hypotrigona penna (Eardley, 2004)	–	–	–	Burkina Faso, Mali, Senegal
Hypotrigona ruspolii (Magretti, 1898)	Kenya, South Sudan, Tanzania, Uganda	Democratic Republic of Congo	Angola, Mozambique, South Africa, Zambia	Gambia, Ghana, Niger, Nigeria, Senegal, Sierra Leone
Hypotrigona squamuligera (Benoist, 1937)	Sudan	–	–	Burkina Faso, Cameroon, Mali, Niger, Senegal, Togo
Liotrigona baleensis (Pauly and Hora, 2013)	Ethiopia	Central African Republic, Democratic Republic of Congo	–	–
Liotrigona bottegoi (Magretti, 1894)	Ethiopia, Kenya	Democratic Republic of Congo	Angola, Malawi, Mozambique, Namibia	Cameroon, Nigeria
Liotrigona bouyssoui (Vachal, 1903)	–	Democratic Republic of Congo, Gabon	Angola	Cameroon, Senegal, Liberia, Burkina Faso, Ivory Coast, Togo
Liotrigona gabonensis Pauly and Fabre Anguilet, 2013	–	Congo Brazzaville, Democratic Republic of Congo, Gabon	–	–
Liotrigona parvula (Dachen, 1971)	–	Cameroon, Democratic Republic of Congo, Gabon	Angola, Botswana, Namibia, South Africa, Zambia	Ghana, Ivory Coast, Liberia

From Eardley (2004), Eardley et al. (2010), Pauly (1998), Pauly and Fabre Anguilet (2013), Pauly and Hora (2013), Kajobe (2008), Mogho Njoya (2009), Munyuli (2010), Fabre Anguilet et al. (2015), Smith (1954)

central Gabon but provides incomplete taxonomic data on those bees (Roubik 1996, 1999).

The recognition of stingless bee species can be undertaken successfully with the help of local guides and aboriginal peoples, who to some extent depend upon the honey, brood, or resin contained in these nests for their own livelihood (Roubik 1989). In the Brazilian Amazon, factors determining the Kayapó indigenous taxonomic system for stingless bees included habitat; preferred nesting substrate or niche; defensive behavior; size, forms, and color of adults; and the smell of the bees (Roubik 1989). In Uganda, the Batwa pygmies, who are the local indigenous honey hunters residing around Bwindi

Impenetrable National Park, helped in identification of stingless bees using body features like size, color, and spots on the bee. In some instances, the dwarf honey guide *Indicator pumilio*, a tiny bird endemic to the Albertine Rift Mountains, helped researchers locate stingless bee nest sites (Kajobe 2007). In Kakamega Forest, the Luhya community use characteristic features of the insects and their behavior and nesting ecology to permit their systematic naming (Nkoba 2012). In Africa, *Meliponula bocandei* is the best known of the Meliponini and is essential in obtaining a greater understanding of the phylogeny of the stingless bees in the continent. *M. bocandei* is found throughout most of tropical Africa (Liberia, Nigeria, Cameroon, Guinea, Congo, Uganda, Kenya, Tanzania, Angola, and Mozambique) where it is highly regarded by the people because of its honey. It is said that one nest can yield from 5 to 18 L of good honey, and the bees are easy to handle (Eardley and Urban 2010; Byarugaba 2004; Kajobe 2007).

16.4 Stingless Bee Nest Architecture

In Africa meliponiculture in a strict sense (husbandry and colony propagation) is not practiced, and honey harvesting is destructive. Perhaps, because a few of the more abundant species build subterranean nests or use the nests of other social insects (and do not adjust to being kept in box hives), the development of meliponiculture is somewhat hindered. To facilitate stingless bee rearing, there is a need to describe the nest architecture of the stingless bees. One of the few common species nesting in the ground is *Plebeina armata*, and nine of its nests were excavated in inhabited termite mounds. The bees modified cavities and tunnels dug by the termites. Entrance tubes protruding above the termite mounds had heights of 0–25 cm and 1 cm diameter. The tubes varied in concealment, rigidity, and perforations. Sticky resin droplets around the tubes trapped intruders such as ants, beetles, or termites. Entrance tubes through the termite mound were 44–120 cm long. Inside, tubes were amorphous and elongate bars of resin that deterred and provided resin to immobilize intruders. The nest cavity had a mean height of 18.8 cm and 12.4 cm wide, lined with 0.1 cm resin. In some nests, short pillars emerged from the resin lining and connected to the storage pots, while in others a soft involucrum layer separated the resin lining and the storage pots. Groups of pollen pots, average height 1.2 cm; width 0.7 cm, and nectar pots, height 1.2 cm; width 1.3 cm, were separated from brood combs by an involucrum layer. There were 6–10 circular, trapezoidal, or both combs in a nest. Comb diameter was 8–16 cm and separated by pillars with lengths ≤0.5 cm. In a nest, there were 3300–3775 brood cells with average height of 0.5 cm and diameter 0.3 cm. Queen cells had a mean height of 0.8 cm and diameter of 0.5 cm, located in a central opening in the combs. A slanting drainage tube with a mean length of 38.3 cm and 1 cm width originated from the base of the nest cavity.

Nest architecture is a species-specific trait that can support identification of species and provide information on how (e.g., nest cavity dimensions or volume, entrance tube dimensions, colony insulation, drainage and trash fixtures, brood comb and food storage pot dimensions and arrangement, temperature tolerance, defense mechanisms, response to particular nest location) a species can be reared for honey production and pollination. The internal nest architecture can also be used, in some cases, to trace lineages or the relatedness among meliponines (Roubik 2006; Rasmussen and Camargo 2008). The architecture of six species of stingless bees in Mogho Njoya (2009) reveals considerable design modification in their brood cell arrangements, storage pot arrangements, nest entrance shapes, and nest construction. They nest in the ground, in tree trunk cavities, in mud walls of houses, and in abandoned honey bee hives. Brood cells are arranged either in vertical combs, horizontal combs, or in clusters. Cells vary in shape and size, and the brood area is either protected with involucrum or not. Some nests are protected with batumen sheets, while others do not have that protective inner cover. The nests of *Meliplebeia beccarii* were built in the soil and exhibit architectural characters typical of all other genera of obligatory ground-nesting bees in Africa, like *Plebeiella* and *Plebeina*. In *Meliplebeia beccarii*, however, the nest proper consists of a brood area, an area of involucrum layers, and a storage pot area. The combs are horizontal and the mode of building comb is concentric, while cell construction is synchronous (elaborated in a daily batch, rather than sequentially). The nests are connected to the exterior by a short outer entrance of 0.5–0.6 cm above the soil and <0.1 cm wall thickness. A drainage tube, for excess moisture, measured 16.5 cm long and 0.7–1.2 cm diameter, below the nest cavity.

To solve the problem of water accumulation, a long drainage tube at the bottom of the bee nest serves to get rid of excess water that would otherwise accumulate (Smith 1954; Portugal-Araujo 1963; Wille and Michener 1973; Camargo and Wittmann 1989). The lining of the drainage tube with resin ensures that water is channeled away from the bee nest before it infiltrates the lower parts of soil or a nest substrate termite mound. In addition, the bees ensure that the runoff water from their nest is conducted not only downward but also to the side, through slanting tubes. Otherwise this water may move up by capillary forces and pass back to the nest, especially when the termite mound is heated up during periods of high temperature. Secondly, nest location in the nursery part of inhabited termite mounds has an additional advantage to the bees in that they can easily dig and modify nest cavities. The nursery region of the termite mound is typically soft due to more organic material, which is hygroscopic, hence has higher moisture absorbing capacity than the inner and the outer walls of the mound. The outer wall is made up of an impervious layer of sand and clay particles cemented together with the termite salivary secretions (summary in Wilson 1971).

Colonies of *P. armata* reared in wooden observation boxes did not adapt well. This was probably due to problems of temperature and humidity regulation in the hives, as the workers spent much time building soft involucrum sheets around the nest. No brood cells were constructed during this period, and the queen did not lay eggs, although foraging and the construction of storage pots continued. As *P. armata* has a short foraging range, it can nonetheless be useful for pollination purposes in plantations (Namu and Wittmann 2014).

16.5 African Stingless Bee Nesting Behavior

Recent major analyses of African and stingless bee nesting include Michener (1974, 2007), Wille and Michener (1973), Sakagami (1982), Wille (1983), Roubik (1989, 2006), and Nogueira-Neto

(1997). In Africa, some studies on stingless bee nesting ecology have been carried out (Kajobe 2008; Mogho Njoya 2009; Fabre Anguilet et al. 2015; Namu and Wittmann 2014). Most of the syntheses consider Neotropical stingless bees, while Afrotropical bees have not been effectively represented. However, in Africa, some studies on stingless bee nesting ecology exist (Kajobe 2008; Mogho Njoya 2009; Fabre Anguilet et al. 2015; Nkoba 2012; Namu and Wittmann 2014; Boesch et al. 2009).

The nest is the central place from which stingless bees mate, forage, and pass through life stages (Roubik 2006). Nests are immobile fixtures and potentially long-lived, much like trees in forests where meliponines live (Roubik 1989, 2006; Michener 1974; Eltz et al. 2003). Many data have been published on stingless bee nesting. Stingless bee nest entrances vary from species to species, ranging from simple holes to dome- or trumpet-shaped entrances (e.g., Roubik 2006; Cortopassi-Laurino and Nogueira-Neto 2016). Small bee colonies have small entrances that are easily defendable. One trade-off is that traffic jams and collisions may become common at very high foraging activity (Biesmeijer et al. 2007; Couvillon et al. 2008). However, the African stingless bee colonies have quite small or inconspicuous nest entrances, despite colony or bee size, and the species generally do not display biting behavior against large vertebrates (Michener 1974).

In Africa, of major import is the tool using ability of chimps that extract pot-pollen, brood and honey from various honey bee nests, using flexible sticks derived from lianas and others of more rigid wood (Kajobe and Roubik 2006; Boesch et al. 2009). The potential colony predation or harassment by primates is in general high. This pervasive factor may have led more uniformly to inconspicuous nest entrances and activity, in contrast to the stingless bees in Asia or the Americas (see Roubik 2006). We suggest that it has also played a role in the nesting diversity of African meliponines, particularly in using large termite or ant mounds, made by colonies that also must overcome threats by large animals, ranging from aardvarks to mustelids, civets, apes, and humans.

Stingless bees sometimes build nests in fallen trees, in bush that has been burned or that has been trampled or cut by man or other animals, in the earthen banks of road cuts, paths, fields, and in banks made by rushing water. These somewhat atypical settings have frequently provided the only opportunities to study bee nesting biology (Wilson 1971; Michener 1974; Roubik 1989). One of the attributes of most stingless bee nest sites, or in the case of exposed nests, of the nest itself, is excellent insulation. Nests in large trunks or in the soil are particularly well insulated. Many species, particularly those of the moist tropics, are unable to withstand chilling (Michener 1974).

Meliponula sensu *latu* (larger African meliponines, which we see as *Meliplebeia*, *Axestotrigona*, and *Meliponula*) use various structures in nature for nesting, both in the soil and in tree trunks and branches (Michener 2007; Eardley 2004). In Uganda, when collected in agricultural landscapes, these bees are, in addition, commonly found nesting in termite mounds and sometimes the walls of old buildings. In forest habitats, they nest in hollows in dead and living, standing trees. In savannah ecosystems, the bees are frequently found nesting underground or in the stumps of shrubs. Overall, the bees prefer making their nests near or inside primary, secondary, or degraded forest habitats. In Uganda, they are also seen nesting in leaves of wetland plant species (e.g., *Cyperus*).

Not much has been recorded on the stingless bee natural enemies in African forests, and somewhat more is known about which pests arrive at colonies kept by man. Nkoba (2012) provides data on natural enemies that include *Aethina tumida* (hive beetle), *Rhizoplatys* (dynastine scarab beetle), *Tenebroides* (darkling beetle), *Megaselia* (phorid fly), *Philanthus* (predatory wasp), *Myrmicaria* (ant), and *Elminia* (flycatcher). The first four consume stored pot-pollen and presumably larval cell provisions or occasionally larvae and pupae.

Stingless bees are known to be generalists with regard to selection of nest sites (Hubbell and Johnson 1977; Roubik 2006). In Cameroon, stingless bees were found to have huge variation in habitat preferences. Nests were found in tree

trunks, mud walls, traditional hives, and soils or even just attached to tree branches. Kajobe and Roubik (2006) discuss five stingless bees in Bwindi Impenetrable National Park—*Hypotrigona gribodoi*, *Meliponula bocandei*, *Axestotrigona ferruginea*, *Plebeiella lendliana*, and *Meliplebeia nebulata*. *Meliponula bocandei* nested both in the ground and in tree hollows. Nests of *H. gribodoi* were found in areas like walls of houses protected by roofs, tree trunks protected by the canopy, or in dry wood, which insulates the nest. Namu and Wittmann (2014) recorded four species of stingless bees, *Hypotrigona gribodoi*, *Plebeina armata*, *Axestotrigona ferruginea*, and "*Meliponula*" sp. in banana farms in Uganda. Bees nested in termite mounds, cavities in trees, houses, and other man-made structures. *Plebeina armata* nests were found in occupied termite mounds at the core of the mound (Namu and Wittmann 2014). That choice of location may be advantageous to the bees in several ways. The mound shields the nest from rain, flooding, fire, heat or cold, and attack by large enemies. *Meliplebeia beccarii* prefer to nest in *Eucalyptus* plantations and open farmlands with numerous lateral roots to be used as a site for their nests. *Meliplebeia beccarii* always cohabits with small white ants and some small beetles, likely scavenging on molds or pollen debris (Roubik 2006). In the banana farmlands in Central Uganda, there were four species of stingless bees, *Hypotrigona gribodoi*, *Plebeina armata*, *Axestotrigona ferruginea*, and "*Meliponula*" sp. They nested in cavities in trees, man-made structures, and termite mounds (Namu and Wittmann (2014)). *H. gribodoi* preferred nesting on the wall of the mud houses while *A. ferruginea* and *M. bocandei* nested in trees in the forest and farmlands.

The difference between the few published studies on stingless bees and their nesting biology in general might relate to differences in the habitat studied and the sampling effort employed in the different African countries. These studies are representative of the sparse data regarding social bees within the 54 countries of Africa. The lack of data could reflect a low level of interest among researchers or the difficulties encountered in studying the diversity and abundance of social bees in the field. Indeed, the difficulties that are involved in assessing the overall diversity and abundance of bees in this continent are numerous, and several are common to many African countries. One major difficulty is the uncertain taxonomy of particular meliponine species, which may hinder study of the diversity, biology, ecology, and other aspects of these interesting and important insects (Eardley and Kwapong 2013).

Regarding the abundance of Meliponini at the colony level, Darchen (1972) reported a nest density of 2.5 nests per hectare, mostly the genera *Hypotrigona*, *Liotrigona*, and *Dactylurina*, in Lamto savannahs, Ivory Coast. In Ugandan forests, Kajobe and Roubik (2006) report a density of 0.39 nests per hectare of highly eusocial bees (including the honey bee, *Apis mellifera*), while 0.27 nests per hectare were of Meliponini. In addition, Tornyie and Kwapong (2015) found Meliponini natural species nest density at 1.7–2.4 nests per hectare in Ghana. That study focuses on *M. bocandei*, *A. ferruginea*, and *D. staudingeri*.

In Africa, nest predation and habitat loss are the main threats that could cause the depopulation of certain meliponines. Although there is currently no evidence that diseases and pests endanger these bees, there is still a need for further studies for a better assessment of the risks that are associated with such potential threats.

Ways through which the nest can be defended, by the Meliponini in general, include: covering the intruder with resin and biting using mandibles, ejection of a burning liquid from the mandibular glands, camouflage of nesting sites, restricting the nest entrances, forming lines of sticky resin around the entrance tube, and positioning of guard bees in front of nest entrance (Wilson 1971; Michener 1974; Namu and Wittmann 2014). Certain stingless bees may even have a separate worker caste of guard bees, which are larger and more readily attack an intruder (Grüter et al. 2012).

The manner of protecting the open entrance of the nests at night hours varied among five African stingless bee species. The presence of a protective barrier at the open entrance of the nest to guard against intrusion during the night was found in three bee species, namely, *M. bocandei*, *P. lendli-*

ana, and *A. erythra* (reddish worker). In *H. gribodoi* and *A. togoensis* (black worker), the open entrances of nests were never closed during the night. The *M. bocandei* colony protected the open entrance at night by reducing its size with sticky resin deposited around the opening. The *P. lendliana* and *A. erythra* were the only ones among the five studied that completely sealed open entrances of their nests at night (Nkoba 2012).

A study by Fletcher and Crewe (1981) showed that *Plebeina armata* (= *hildebrandti*, *denoiti*) nesting in the ground was capable of thermoregulation through the entrance tube. At the same time, variation in the size, shape, camouflage, and firmness of the entrance tube indicated a defense role. This was more pronounced on termite mounds without vegetation cover, where all the entrance tubes were short and brown, or had mud added so that the color of the tube merged with the color of the mound. The color of the added plant material also merged with the color of the termite mound, camouflaging the entrance tube. On termite mounds with vegetation cover, the tubes were dark at the base and only light colored at the apex, making it difficult to distinguish them from plant stems and branches. Additional defense strategies were the deposit of resin droplets on and around the tubes, as well as depositing large quantities of resin in different spots inside the entrance tubes. The resin acted as a barrier to intruders and was also used to immobilize them.

References

Aidoo K, Kwapong P, Combey R, Karikari A. 2011. Stingless bees in Ghana. Bees for Development 100: 10–11.

Biesmeijer JC, Slaa EJ, Koedam D. 2007. How stingless bees solve traffic problems. Entomologische Berichten-Nederlandsche Entomologische Vereeniging 67: 7–13.

Boesch C, Head J, Robbins MH. 2009. Complex tool sets for honey extraction among chimpanzees in Loango National Park, Gabon. Journal of Human Evolution 56: 560–569.

Byarugaba D. 2004. Stingless bees (Hymenoptera: Apidae) of Bwindi impenetrable forest, Uganda and Abayanda indigenous knowledge. International Journal of Tropical Insect Science 24: 117–121.

Cardinal S., Buchmann, SL, Russell AL. In press. The evolution of floral sonication, a foraging behavior used by bees. Evolution.

Camargo JMF. 2013. Historical biogeography of the Meliponini (Hymenoptera, Apidae, Apinae) of the Neotropical region. pp. 19–34. In Vit P, Pedro SRM, Roubik DW, eds. Pot-honey: a legacy of stingless bees. Springer, New York, USA. 654 pp.

Camargo JMF, Wittmann, D. 1989. Nest architecture and distribution of the primitive stingless bee, *Mourella caerulea* (Hymenoptera, Apidae, Meliponinae): Evidence for the origin of *Plebeia* (s. lat.) on the Gondwana continent. Studies on Neotropical Fauna and Environment 24: 213–229.

Cortopassi-Laurino M, Nogueira-Neto P. 2016. Abelhas sem ferrão do Brasil. Editora da Universidade de São Paulo. 123 pp.

Couvillon MJ, Wenseleers T, Imperatriz-Fonseca VL, Nogueira-Neto P, Ratnieks FLW. 2008. Comparative study in stingless bees (Meliponini) demonstrates that nest entrance size predicts traffic and defensivity. Journal of Evolutionary Biology 21: 194–201.

Darchen R. 1972. Ecologie de quelques trigones (*Trigona* sp.) de la savane de Lamto (Cote D'Ivoire). Apidologie 3: 341–367.

Darlington PJ Jr. 1957. Zoogeography: the geographic distribution of animals. John Wiley and Sons, New York. 675 pp.

Eardley CD. 2004. Taxonomic revision of the African stingless bees (Apoidea: Apidae: Apinae: Meliponini). African Plant Protection 10: 63–96.

Eardley CD, Urban, R. 2010. Catalogue of Afrotropical bees (Hymenoptera: Apoidea: Apiformes). Zootaxa 2455: 1–548.

Eardley CD, Kuhlmann M, Pauly A. 2010. The Bee Genera and Subgenera of sub-Saharan Africa. ABC Taxa vol 7: i-vi, 138 pp. http://www.abctaxa.be/volumes/vol-7-bees

Eardley CD, Kwapong P. 2013. Taxonomy as a tool for conservation of African stingless bees and their honey. pp 261–268. In: Vit P, Pedro SRM, Roubik DW, eds. Pot-honey: a legacy of stingless bees. Springer. New York. 654 pp.

Eltz T, Bruhl CA, Imiyabir Z, Linsenmair KE. 2003. Nesting and nest trees of stingless bees (Apidae: Meliponini) in lowland dipterocarp forests of Sabah, Malaysia, with implications for forest management. Forest Ecology and Management 172: 301–313.

Fabre Anguilet EC, Nguyen BK, Bengone Ndong T, Haubruge E, Francis F. 2015. Meliponini and Apini in Africa (Apidae: Apinae): a review on the challenges and stakes bound to their diversity and their distribution. Biotechnologie, Agronomie, Société et Environnement 19: 382–391.

Fletcher DJC, Crewe RM. 1981. Nest structure and thermoregulation in the stingless bee, *Trigona (Plebeina) denoiti* Vachal (Hymenoptera: Apidae). J. Ent. Soc. South Africa 44 , 183-196.

Flower BP, Kennett JP. 1994. The middle Miocene climatic transition: East Antarctic ice sheet development, deep ocean circulation and global carbon cycling. Palaeogeography, Palaeoclimatology, Palaeoecology 108: 537–555.

Gikungu MW. 2006. Bee diversity and some aspects of their ecological interactions with plants in a successional tropical community. Ph.D. Dissertation, University of Bonn, 201 pp.

Grüter C, Menezes C, Imperatriz-Fonseca VL, Ratnieks FL. 2012. A morphologically specialized soldier caste improves colony defense in a neotropical eusocial bee. Proceedings of the National Academy of Sciences 109: 1182–1186.

Halcroft M. 2013. Australian stingless bees. pp. 35–72 In: Vit P, Pedro SRM, Roubik DW, eds. Pot-honey: a legacy of stingless bees. New York, Springer. 654 pp.

Hubbell SP, Johnson LK. 1977. Competition and nest spacing in a tropical stingless bee community. Ecology 58: 949–963.

Jalil AH, Roubik DW, eds. 2017. Handbook of Meliponiuclture. The Indo-Malayan Stingless bee clade. Akademi Kelulut Malaysia Sdn Bhd, Selangor, Malaysia. 560 pp.

Kajobe R. 2007. Nesting biology of equatorial Afrotropical stingless bees (Apidae; Meliponini)in Bwindi Impenetrable National Park, Uganda. J. Apic. Res. 464: 245–255.

Kajobe R. 2008. Foraging behaviour of equatorial Afrotropical stingless bees: habitat selection and competition for resources (Doctoral dissertation). Utrecht University Repository.

Kajobe R, Roubik DW. 2006. Honey-making bee colony abundance and predation by apes and humans in a Uganda forest reserve. Biotropica 38: 1–9.

Kerr WE, Maule V. 1964. Geographic distribution of stingless bees and its implications. Journal of the New York Entomological Society 72: 2–17.

Linder HP. 2014. The evolution of African plant diversity. Frontiers in Ecology and Evolution. doi: 10.3389/fevo.2014.00038

Macharia J, Raina SK, Muli E. 2007. Stingless bees in Kenya, Bees for Development Journal, 83: 1–9.

Martins AC, Melo GAR, Renner SS. 2014. The corbiculate bees arose from New World oil-collecting bees: Implications for the origin of pollen baskets. Molecular Phylogenetics and Evolution 80: 88–94.

Michener CD. 1974. The social behaviour of the bees: a comparative study. Belknap Press of Harvard University Press, Cambridge, Massachusetts. 404 pp.

Michener CD. 1979. Biogeography of the bees. Ann. Mo. Bot. Gard. 66: 277–347.

Michener CD. 2007. The bees of the world, 2nd Ed. Baltimore, 953 pp.

Mogho Njoya MT. 2009. Diversity of stingless bees in Bamenda Afromontane Forests–Cameroon: nest architecture, behaviour and labour calendar. Diss. PhD thesis: Wilhelms Universität Bonn-Institut für Nutzpflanzenwissenschaften und Ressourcenschutz

Rheinische Friedrich (Deutschland). http://hss.ulb. uni-bonn.de/2010/1993/1993.pdf

Morley RJ. 2000. Origin and evolution of tropical rain forests. John Wiley and Sons, New York. 309 pp.

Moure JS. 1961. A preliminary supra-specific classification of the old world meliponine bees (Hymenoptera, Apoidea). Studia Entomologica 4: 181–242.

Munyuli T. 2010. Pollinator population in the farmlands in Central Uganda. Ph.D. Thesis. Makerere University; Kampala, Uganda. 431 pp.

Namu FN, Wittmann D. 2014. Are stingless bees the primary vector in spread of banana *Xanthomonas* wilt in Central Uganda? International Journal of Ecology and Ecosolution 1: 52–60.

Njau MA, Mpuya PM, Mturi FA. 2009. Apiculture potential in protected areas: the case of Udzungwa Mountains National Park, Tanzania. International Journal of Biodiversity Science & Management 5: 95–101.

Nkoba K. 2012. Distribution, behavioural biology, rearing and pollination efficiency of five stingless bee species (Apidae: meliponinae) in Kakamega forest, Kenya. D. Phil. Thesis. Kenyatta University; Nairobi, Kenya. 237 pp.

Nogueira-Neto P. 1997. Vida e criação de abelhas indígenas sem ferrão. Editora Nogueirapis; São Paulo, Brazil. 445 pp.

Pauly A. 1998. Hymenoptera Apoidea du Gabon. Musée Royal de l'Afrique Centrale Tervuren, Belgique. Annales Sciences Zoologique, 282: 1–121.

Pauly A, Fabre Anguilet EC. 2013. Description de *Liotrigona gabonensis* sp. nov., et quelques corrections à la synonymie des espèces africaines de mélipones (Hymenoptera Apoidea Apinae : Meliponini). Belgian Journal of Entomology 15: 1–13.

Pauly A, Hora ZA. 2013. Apini and Meliponini from Ethiopia (Hymenoptera: Apoidea: Apidae: Apinae). Belgian Journal of Entomology 16: 1–35.

Poinar GO, Danforth BN. 2006. A fossil bee from Early Cretaceous Burmese amber. Science 314(5799): 614.

Portugal-Araujo V. 1963. Subterranean nests of two African stingless bees (Hymenoptera: Apidae). Journal of the New York Entomological Society 1:130–141.

Rasmussen C, Cameron SA. 2007. A molecular phylogeny of the Old World stingless bees (Hymenoptera: Apidae: Meliponini) and the non-monophyly of the large genus *Trigona*. Systematic Entomology 32: 26–39.

Rasmussen C, Cameron, SA. 2010. Global stingless bee phylogeny supports ancient divergence, vicariance, and long distance dispersal. Biological Journal of the Linnean Society 99: 206–232.

Rasmussen C, Camargo JMF. 2008. A molecular phylogeny and the evolution of nest architecture and behavior in *Trigona s.s.* (Hymenoptera: Apidae: Meliponini). Apidologie 39: 102. doi:10.1051/apido:2007051

Rasmussen C, Gonzalez VH. 2013. Prologue. Stingless bees now and in the future. pp 1–7. In Vit P, Roubik

DW, eds. Stingless Bees Process Honey and Pollen in Cerumen Pots. Facultad de Farmacia y Bioanálisis; Universidad de los Andes, Mérida, Venezuela. http://www.saber.ula.ve/handle/123456789/35292

Roubik DW. 1979. Africanized honey bees, stingless bees and the structure of tropical plant-pollinator communities. pp. 403–417 In: Caron D, ed. Proceedings V th International Symposium on Pollination. Maryland Agricultural Experimental Station Miscellaneous Publication No. 1 College Park, Maryland. 274 pp.

Roubik DW. 1989. Ecology and natural history of tropical bees. Cambridge University Press, New York. 514 pp.

Roubik DW. 1990. Niche preemption in tropical bee communities: a comparison of Neotropical and Malesian faunas. pp. 245-257 In: Sakagami SF, Ohgushi R, Roubik DW (eds.). Natural history of social wasps and bees in equatorial Sumatra. Hokkaido University Press, Japan. 274 pp.

Roubik DW. 1992. Stingless bees (Apidae: Meliponinae): a guide to Panamanian and Mesoamerican species and their nests. pp. 495–524 In: Quintero D, Aiello A (eds.) Insects of Panama and Mesoamerica. Oxford University Press. Oxford, UK. 692 pp.

Roubik DW. 1996. Order and chaos in tropical bee communities. pp. 122–132 In: Garófalo CA et al. (eds.). Segundo Encontro sobre Abelhas de Ribeirão Preto; São Paulo, Brazil. 216 pp.

Roubik DW. 1999. The foraging and potential pollination outcrossing distances flown by African honey bees in Congo forest. Journal of the Kansas Entomological Society 72: 394–401.

Roubik DW. 2006. Stingless bee nesting biology. Apidologie 37: 124–143.

Roubik DW, Camargo JMF. 2012. The Panama microplate, island studies and relictual species of *Melipona (Melikerria)* (Hymenoptera: Apidae: Meliponini). Systematic Entomology 37: 189–199.

Sakagami SF. 1982. Stingless bees. pp. 361–423 In: Hermann HR (ed.). Social insects Volume 3. Academic Press; New York, USA. 459 pp.

Smith FG. 1954. Notes on the biology and waxes of four species of African *Trigona* bees (Hymenoptera, Apidae). Proceedings of the Entomological Society of London 29: 62-70.

Tornyie F, Kwapong PK. 2015. Nesting ecology of stingless bees and potential threats to their survival within selected landscapes in the northern Volta region of Ghana. African Journal of Ecology doi: 10.1111/aje.12208.

Wille A. 1983. Biology of the stingless bees. Annual Review of Entomology 28: 41–64.

Wille A, Michener CD. 1973. The nest architecture of stingless bees with special reference to those of Costa Rica. Revista de Biologia Tropical 21: 1–278.

Wilson EO. 1971. The Insect Societies. Belknap Press; Cambridge, Massachusetts, USA. 562 pp.

On the Trophic Niche of Bees in Cerrado Areas of Brazil and Yeasts in Their Stored Pollen

17

Paula Calaça, Cláudia Simeão, Esther Margarida Bastos, Carlos Augusto Rosa, and Yasmine Antonini

17.1 Introduction

The need for information regarding the plants used by stingless bees is directly related to the threats they face in the Brazilian "Cerrado." Characterized as the world's most biologically rich highland savanna, its flatlands and high plateaus compose approximately 25% of Brazilian national territory. Human activities, such as deforestation, unsustainable honey harvesting, and intensive use of herbicides and pesticides (Larsen et al. 2005; Freitas et al. 2009; Johnson et al. 2010), are among the most important factors that affect the Cerrado wildlife. Stingless bees have perennial colonies, and the removal of native vegetation may affect their populations through the loss and/or alteration of floral resources and nesting sites (Kremen et al. 2004).

Relative to the number of Meliponini in Brazil, more than 240, according to Pedro (2014), the information about pollen sources and the trophic niche is scarce, especially regarding areas of Cerrado where deforestation has had a great influence on bee populations (Viana and Melo 1987). Eusocial bees, such as stingless bees, are mostly generalists and collect pollen from a variety of different plant species; however, among these sources, bees exhibit different levels of plant species "preference" (Michener 2007). The diversity of resources used should depend both on the number of foragers in the colony and their flight range, which together can determine foraging area and patterns of resource use (Guibu et al. 1988).

Because dozens of stingless bee species occur sympatrically, the mechanisms through which they partition resources are not well understood. The trophic niche of stingless bees can be studied

P. Calaça (✉)
Departamento de Botânica, ICB, CP 486, Universidade Federal de Minas Gerais, Belo Horizonte, MG 31270-901, Brazil

Serviço de Recursos Vegetais e Opoterápicos, Diretoria de Pesquisa e Desenvolvimento, Fundação Ezequiel Dias (Funed), Rua Conde Pereira Carneiro, 80, Gameleira, Belo Horizonte, MG 30510-010, Brazil
e-mail: paula.thiago@funed.mg.gov.br

C. Simeão • E.M. Bastos
Serviço de Recursos Vegetais e Opoterápicos, Diretoria de Pesquisa e Desenvolvimento, Fundação

Ezequiel Dias (Funed), Rua Conde Pereira Carneiro, 80, Gameleira, Belo Horizonte, MG 30510-010, Brazil

C.A. Rosa
Departamento de Microbiologia, ICB, CP 486, Universidade Federal de Minas Gerais, Belo Horizonte, MG 31270-901, Brazil

Y. Antonini
Departamento de Biodiversidade Evolução e Meio Ambiente, ICEB, Universidade Federal de Ouro Preto, Ouro Preto, MG 35400-000, Brazil

© Springer International Publishing AG, part of Springer Nature 2018
P. Vit et al. (eds.), *Pot-Pollen in Stingless Bee Melittology*, DOI 10.1007/978-3-319-61839-5_17

by analyzing stored pollen. In the Brazilian biome that we examine, there are extensive environmental heterogeneity and the occurrence of a wide variety of ecosystems, covering everything from savannas and grasslands to forest formations (Ribeiro et al. 1998; Brandão 2000), and thus great diversity and endemism (Reatto et al. 1998; Myers et al. 2000; Silva and Bates 2002) including the stingless bees therein.

The last update of Moure's bee Catalogue (http://moure.cria.org.br/) listed 417 species of stingless bees for the Neotropical region, grouped in 33 genera, at least 1 of which is extinct (Camargo and Pedro 2007, 2013). In Brazil, there are 244 species in 29 genera (Pedro 2014), of which 20 occur in the Cerrado (Camargo and Pedro 2013). In this biome, bees play a key role in pollinating native plant species. Of the zoophilic species of plants in the Cerrado studied by

Silberbauer-Gottsberger and Gottsberger (1988), 87.5% are melittophilous, i.e., exhibit morphological characteristics favoring pollination by bees. Martins and Batalha (2006) calculate the frequencies of pollination systems among woody plant species in fragments of Cerrado in the Central-West Region of Brazil and test whether these pollination systems are indeed related to floral traits. They find that among 99 plant species, 55.6% are mainly pollinated by bees.

In this chapter, we present a compilation of our own studies, along with others available in the literature, about the trophic niche of stingless bees considering exclusively their pollen sources in areas of Brazilian Cerrado. We also provide information on the relationships among yeast species and pot-pollen of those stingless bees (Fig. 17.1) and discuss the role of yeast in transforming pot-pollen inside bee nests.

Fig. 17.1 Pot-pollen from three different stingless bee species (*Tetragonisca angustula*, *Melipona quinquefasciata*, and *Scaptotrigona* sp.) studied in Cerrado areas of Minas Gerais, Brazil. (**a**) Open view of *Tetragonisca angustula* nest, showing some pot-pollen being sampled. (**b**) Detail of *Tetragonisca angustula* pollen pot. (**c**) *Melipona quinquefasciata* workers on a pollen pot. (**d**) Sampling pot-pollen from inside a *Scaptotrigona* sp. nest (Photos: **a**, **b**, Clemens Schlindwein; **c**, **d**, Paula Calaça)

17.2 Pollen Harvested by Native Bees of the Cerrado

Pollen pots contain different types of pollen from different plant species; thus, microscopic analysis of pollen grains contained in the pots can reveal the floral sources from which bees seek pollen to feed their young. Studies that have reported the pollen sources for stingless bees among the food pots or pollen loads of workers in areas of Cerrado are few. We have selected only studies that contain data on plant species visited by bees to collect pollen in different vegetation types in the Cerrado biome. The vegetation types include fragments in urban areas, dry forest, and Cerrado *sensu stricto*. In addition to published work (Antonini and Martins 2003; Antonini et al. 2006; Calaça 2011), we review additional data on pollen sources for nine stingless bee species (Table 17.1). Among these studies, six analyzed pollen loads of bees and four analyzed pot-pollen inside nests (Table 17.1).

Measurements of trophic niches can be performed using different analytical methods. In the literature, niche width has been measured using diversity indices that consider the frequency of occurrence of pollen types in pollen samples or the frequency of visits to flowers. Here we evaluate the trophic niche using the richness of pollen types mentioned in these articles. A matrix of presence/absence data containing a list of plant species/pollen types for each bee species was assembled. Pollen types not identified to at least the level of botanical family were excluded from the matrix, as were those with relative frequencies lower than 1%. Since we use only presence/absence data, we anticipate that the removal of species with very low frequencies will better represent the plants that are most important to the bees. Similarities among the diets of bee species were assessed using cluster analysis, with Ward's method and Euclidean distance, of three separate matrices: a matrix of botanical families (8 bee species × 59 botanical families), a matrix of plant genera (8 bee species × 156 plant genera), and a matrix of pollen types (8 bee species × 247 pollen types). Analyses were performed using software "R" (R Core Team 2015). The botanical nomenclature was updated according to Tropicos, the botanical information system at the Missouri Botanical Garden (www.tropicos.org) and the site Flora do Brazil 2020 development (http://floradobrasil.jbrj.gov.br).

Most studies on bee niches focus on the diversity of floral resources used and how they are partitioned (e.g., Wilms et al. 1996; Aguiar 2003; Goulson and Darvill 2004; Andena et al. 2012; Aguiar et al. 2013; Figueiredo et al. 2013). The level of complementarity or redundancy in the niches of flower-visiting insects is an important mechanism underlying the relationship between biodiversity and ecosystem functioning (Blüthgen et al. 2006; Hoehn et al. 2008; Santos et al. 2013). However, little empirical evidence exists about the level of complementarity among pollinators (Hoehn et al. 2008). Different species may have complementary niches and possess different functional roles (Santos et al. 2010), thereby realizing different niches (Begon et al. 2006). Niche complementarity requires some degree of trophic or temporal specialization, whereas niche redundancy is associated with high niche overlap and functional similarity (Blüthgen and Klein 2011). Two species may, however, present similarity in one dimension of their trophic niches (e.g., diet) and complementarity in another (e.g., temporal patterns of resource use) (Santos et al. 2013).

Here we present data on the richness of pollen types collected by 8 of the 20 known stingless bees from Cerrado areas. Being generalist social insects, it is not surprising that a total of 247 different pollen types were found among their pollen. Our database showed that the botanical families with the greatest number of pollen types were, in descending order, Fabaceae (50), Myrtaceae (49), Malvaceae (11), Euphorbiaceae (10), Arecaceae (9), Sapindaceae (7), Bignoniaceae (7), Anacardiaceae (6), and Poaceae (6). The plant genera with the greatest number of pollen types were, in descending order, *Eucalyptus*, *Mimosa*, *Serjania*, *Psidium*, *Poincianella*, *Anadenanthera*, and *Senna*. Finally, and also in descending order, the most frequent plant species/pollen types were *Poincianella pluviosa*, *Eucalyptus* sp., *Triumfetta rhomboidea*, *Anadenanthera colubrina*, *Myrcia* sp., *Serjania*

Table 17.1 References compiled in this chapter and each stingless bee species studied

Stingless bee species	References									
	Cortopassi-Laurino and Ramalho (1988)[b]	Calaça (2011)[a]	Soares (2003)[a]	Antonini and Martins (2003)[a]	Miranda et al. (2015)[b]	Faria et al. (2012)[b]	Ferreira et al. (2010)[b]	Duarte (2012)[b]	Antonini et al. (2006)[b]	Aleixo et al. (2013)[b]
Frieseomelitta varia										X
Melipona quadrifasciata				X					X	
Melipona quinquefasciata		X								
Melipona rufiventris			X							
Partamona rustica					X					
Scaptotrigona aff. depilis						X				
Scaptotrigona depilis							X			
Tetragona clavipes								X		
Trigona spinipes	X									

[a]Analysis of pot-pollen
[b]Analysis of pollen load of adult workers

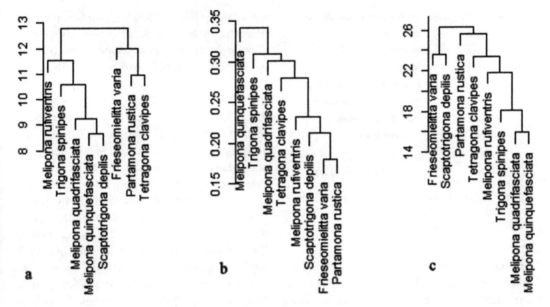

Fig. 17.2 Dendrograms of eight stingless bee species classified by cluster analysis using the Ward method and Euclidian distance considering three distinct matrixes: (**a**) a botanical family matrix (8 bee species × 59 botanical families), (**b**) a plant genera matrix (8 bee species × 156 plant genera), and (**c**) a pollen type matrix (8 bee species × 247 pollen types)

lethalis, Delonix regia, Guazuma ulmifolia, Leucaena leucocephala, and *Archontophoenix cunninghamiana*.

A statistical cluster analysis of the stingless bee species by botanical families (Fig. 17.2a) reveals one group composed of *M. rufiventris, T. spinipes, M. quadrifasciata, M. quinquefasciata*, and *S. depilis* and a second group composed of *F. varia, P. rustica*, and *T. clavipes*. The first group mostly visited plant species from the Fabaceae (40 pollen types) and Myrtaceae (35 pollen types), followed by Malvaceae, Asteraceae, and Sapindaceae, each one with only 5 pollen types. The second group used more equally a higher number of families: Fabaceae (26), Asteraceae (10), Malvaceae (10), Euphorbiaceae (10), Myrtaceae (7), and Arecaceae (7).

Cluster analysis by plant genera (Fig. 17.2b) separated *M. quinquefasciata* and *T. spinipes* from all other bee species, probably because the pollen diet of these two bees was composed of a low number of species (14 and 19, respectively) when compared to the other bees which used a total of 142 plant genera. Furthermore, *Trigona spinipes* was the only species that did not use

Mimosa. Similarly, *Eucalyptus* spp. were used by all the bees.

The cluster analysis by pollen types also revealed similarity in the pollen types identified in bee diets (Fig. 17.2c). Both *Frieseomielitta varia* and *Scaptotrigona depilis* used, among others, the same plant species as pollen sources: *Poincianella pluviosa, Machaerium aculeatum, Anadenanthera colubrina, Serjania lethalis, Delonix regia, Ligustrum lucidum, Duranta erecta, Muntingia calabura, Eucalyptus moluccana, Ricinus communis, Gallesia integrifolia, Triumfetta rhomboidea, Guazuma ulmifolia, Ixora chinensis*, and *Leucaena leucocephala*. The other bee species had lower overlap in plant species/pollen types, which reflected the composition of the local flora of each study site.

Despite the remarkable richness of plant species reportedly used by stingless bees, some species exhibited narrow niche breadth (Fig. 17.3). For example, we found that *M. quinquefasciata* collected pollen from 14 plant species but mainly from species that had mass blooming at the study site, such as *Mimosa* cf. *arenosa, Ouratea* sp., *Eugenia dysenterica*, and *Copaifera* sp., since

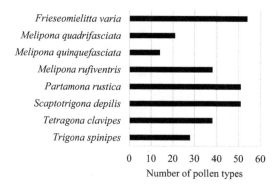

Fig. 17.3 Trophic niche breadth (number of pollen types) for the eight stingless bee species from Cerrado areas of Brazil

these species comprised more than 80% of the pollen stored inside nests (Calaça 2011). On the other hand, *M. rufiventris* collected pollen from more than 30 plant species. Soares (2003) found *M. rufiventris* had a broad trophic niche, but colonies had a low similarity in the use of plant species available in the study site, especially with regard to pollen sources, although this occurred mostly in months when there were a great number of flowering species. According to this author, the following plant species are a major pollen source for this bee: *Acacia* spp., *Copaifera langsdorffii*, *Desmodium barbatum*, *Senna silvestris*, *S. obtusifolia*, *Davilla elliptica*, *Lafoensia pacari*, *Protium* sp., and *Tabebuia* sp. Cortopassi-Laurino and Ramalho (1988) showed that *Trigona spinipes* had a broad pollen niche (H′) along with low evenness values (J′), thereby exhibiting great diversity of pollen collection with intense exploitation of just a few sources. These authors concluded that, in terms of floral resource and preference or specialization/generalization, this species, together with *M. marginata*, *T. angustula*, *Paratrigona subnuda*, *Plebeia remota*, and *M. quadrifasciata* (species cited by the authors), is differentiated from other bees, in plant species visited, by each using intensively just a few resources (Cortopassi-Laurino and Ramalho 1988).

As previously mentioned, although stingless bees are generalists, they may exhibit some degree of preference for certain plant families. This preference depends not only on the avail-

ability of the floral source but also how pollen is presented in the anthers. An example is the case of poricidal anthers, in which pollen is collected by vibrational pollen extraction by the bees. Ramalho et al. (1989) ask whether *Melipona* species have marked floral preferences and find that, in general, plants of Myrtaceae, Melastomataceae, Solanaceae, and Fabaceae are very important pollen sources. Melastomataceae and Solanaceae contain many species with poricidal anthers. However, pollen from plant species with such an anther is found in the pollen load of bee species that do not vibrate, such as the pollen of *Senna* in *Partamona rustica* (Miranda et al. 2015) and in *Scaptotrigona* aff. *depilis* (Ferreira et al. 2010; de Faria et al. 2012). According to Miranda et al. (2015), these "non-vibrating" bees obtain pollen from these plant species by "plundering material" which has fallen on the petals after the vibration of another bee or, in the case of species of *Trigona*, by cutting open the anthers with its mandibles. The same applies to species of other genera with poricidal anthers, as observed in the genera *Solanum*, *Tibouchina*, and *Miconia* (Marques-Souza et al. 2007; Ferreira et al. 2010; de Faria et al. 2012).

Many studies have shown that, in comparison with nectar sources, bees harvest pollen from relatively few plant species (Linsley 1958; Cortopassi-Laurino and Ramalho 1988; Soares 2003; Goulson and Darvill 2004; Antonini et al. 2006). While studying whether rarity in bumblebees was related to flower specialization, Goulson and Darvill (2004) reported that all species gathered nectar from a broader range of flowers than they did with pollen. According to Linsley (1958), constancy in the collection of pollen by bees is a general characteristic of this group of insects, which can be individual, as occurs in some eusocial bees but also in semisocial species (e.g., *Halictus*) and some solitary species (e.g., *Anthophora*, *Andrena*, *Megachile*). Thus, a worker bee performs one or more successive trips for collecting the pollen of one of few plant species, though the species is a generalist (Linsley 1958). Furthermore, communication between individuals indicating floral sources is believed to be efficient enough to lead to rapid concentration of foraging activity of colonies on a few productive sources (Ramalho et al. 2007).

17.3 Yeast in Stored Pollen: Diversity and Ecological Role

Yeast is combined with pollen that provides developing bee larvae with proteins, lipids, vitamins, and minerals (Penedo et al. 1976; Gilliam 1979; Hartfelder and Engels 1989; Zerbo et al. 2001; Michener 2007). Pollen taken directly from anthers is not consumed by bees (Herbert and Shimanuki 1978). Pollen loads stored in pots receive nectar or glandular secretions causing the pollen to perform lactic acid fermentation and transform into "bee bread," a term most often used for honey bee pollen stored in nests (Herbert and Shimanuki 1978; Gilliam 1979) and "saburá," "samora," or "samburá" in the case of the pot-pollen in stingless bee nests (Nogueira-Neto 1997).

The fermentation process makes stored pollen biochemically different from pollen in the flowers (Machado 1971; Gilliam 1979; Standifer et al. 1980; Gilliam et al. 1990; de Camargo et al. 1992; Nogueira-Neto 1997; Fernandes-da-Silva and Serrão 2000). This process is associated with microbial activity, mainly fermentation by bacteria and yeasts (Foote 1957; Haydak 1958; Herbert and Shimanuki 1978; Gilliam 1997). The microorganisms are associated with metabolic conversion of food, fermentation, and preservation of food stored within the nest (Gilliam and Prest 1987; de Camargo et al. 1992; Gilliam 1997). Fermentation can be responsible for enhancing the stabilization of this product or can produce chemical changes that enhance the digestibility and nutritional value of pollen to bees (Herbert and Shimanuki 1978; Gilliam 1979).

Gilliam (1979) observed that the number of yeast species in the bee bread of *Apis mellifera* was less than that of the pollen load, which was less than that of the pollen from flowers. Thus, Gilliam (1979) concluded that the biochemical changes that occur in pollen stored in nests throughout the fermentation process begin as soon as the bees collect the pollen from the flowers and carry it on their legs. Lactic acid bacteria also play an important role in the fermentation process when pollen is stored in cells or pots and are added by the worker bees when they mix small amounts of nectar with pollen they gather from flowers. Vásquez and Olofsson (2009) investigated the community of lactic acid bacteria both in the pollen load of honey bees and in bee bread, compared to bees stomachs, and found that bee bread is fermented by lactic acid bacteria from the honey bee stomach.

The microorganisms present in pollen and bee bread can produce compounds such as enzymes, vitamins, antimicrobial substances, organic acids, and lipids, all of which may contribute to the "ripening" of stored pollen and the stabilization of bee bread (Gilliam 1997). Among these microorganisms, 80% of the bee bread yeasts cause the fermentation of glucose and sucrose. One of the first reports on this subject was made by Casteel (1912), who observed that bee bread had more reducing sugars than freshly collected pollen and that these were, in turn, mixed by bees with honey or recently gathered nectar. Anderson et al. (2014) show that bee bread consists of 40–50% simple sugars. Bee bread contains vitamin K (Haydak and Vivino 1950) and a "milk-digesting enzyme" (Hitchcock 1956); corbicular pollen does not have this enzyme (Gilliam 1979). Chevtchik (1950) studied different stages of fermentation of pollen fresh from flowers and the bee bread of *Apis mellifera* and noted four phases of microbial development from the onset of lactic acid bacteria, yeasts, bacteria of the genus *Escherichia*, and sporulating aerobic bacteria. Chevtchik (1950) obtained 77 groups of yeasts from fresh and fermented pollen which may provide protein, lipids, and growth factors needed by bees. Pain and Maugnet (1966) observed that yeasts were the most abundant microorganisms in bee bread at the end of the fermentation process, which is initiated by lactic acid bacteria. Thus, these authors conclude that yeast has an important role in bee nutrition.

However, Anderson et al. (2014) studied bee bread in order to determine if it would undergo long-term nutrient conversion and become a more nutritious "bee bread" as microbes predigest it. They found that the bacterial communities of newly collected and hive-stored pollen did not differ. This finding indicates the lack of an emergent bacterial community which would digest stored pollen. They conclude that stored pollen has not evolved for microbial-mediated nutrient conversion, and yet it remains preserved in a favorable environment primarily due to added honey, nectar, bee secretions, and the properties of the pollen itself. It would be interesting to know if the same occurs with pot-pollen of stingless bees.

Yeasts are commonly found in the nectar of flowers, where they can reach high densities (Herrera et al. 2009). They also have the capacity to modify the physicochemical composition of nectar, especially by converting sucrose into the reducing sugars glucose and fructose (Herrera et al. 2009). Bees, among other flower visitors, are considered vectors that disperse yeasts among flowers (Brysch-Herzberg 2004), and moreover, recent studies have shown that pollinator behavior can be influenced by nectar yeast (Herrera et al. 2013).

Sandhu and Waraich (1985) studied the yeast community associated with the honey stomach of species of *Apis* and *Xylocopa* and with the nectary of plant species visited by the bees. The yeasts *Candida humicola* (*Vanrija humicola*), *C. incommunis*, *C. ishiwadae*, and *C. membranaefaciens* were prevalent, but no yeast species was held in common among all the bee species. The authors suggested that this distribution of yeast may be related to the diet of each individual species and their particular floral preferences (Sandhu and Waraich 1985).

Studies on yeast communities that live in association with particular species of stingless bees and solitary bees are few and in general have found new species of yeasts. Rosa and Lachance (2005) described *Zygosaccharomyces machadoi* isolated from "garbage pellets" of *Tetragonisca angustula*. However, the newly described species associated with bees frequently belong to the *Starmerella* clade (Rosa et al. 1999, 2003; Teixeira et al. 2003; Daniel et al. 2013). That clade currently contains about 30 species, most of which have been isolated from bees (Rosa

et al. 2003; Daniel et al. 2013). The specific phylogenetic relationships among the yeasts suggest that the group has adapted to bees and their nests or collected materials (Rosa et al. 2003; Daniel et al. 2013). Daniel et al. (2013) stated that the repeated isolation of *Starmerella neotropicalis* from honey and pollen of bees, mostly of *M. quinquefasciata*, confirms the known affinity of the yeasts of the *Starmerella* clade with bees, yet additional collections are likely to be necessary in order to establish the specificity to a bee species. *Starmerella meliponinorum* was described and isolated from various substrates associated with three species of stingless bee in Brazil and one in Costa Rica (Teixeira et al. 2003). The isolates of this species were found in garbage pellets, pollen provisions, honey, propolis, and on adult bees of *T. angustula*, in honey from *M. quadrifasciata*, and on adults of *M. rufiventris* and *T. fulviventris* (Teixeira et al. 2003).

Species of the *Starmerella* clade have also been isolated from solitary bees. *Candida riodocensis* (*Starmerella* clade) was isolated from larval provisions, larvae, and feces in nests of *Megachile* sp., and *Candida cellae* (*Starmerella* clade) was found in the nest provisions of *Centris tarsata* (Pimentel et al. 2005). Rosa et al. (1999), studying *Diadasina distincta* and *Ptilothrix plumata* in Minas Gerais, Brazil, report that the most prevalent yeast species is *Candida batistae* (*Starmerella* clade), found in larval provisions and on larvae and pupae. The authors suggest that this species may have a role in pollen maturation because it produces enzymes such as lipases and proteases. Table 17.2 summarizes work on the yeast species isolated from pot-pollen, larval provisions, and garbage pellets of native bees from Brazil. Due to the limited number of studies, we compiled all available data on the subject, including one study that took place in the Atlantic Rain Forest (Pimentel et al. 2005).

Thus far, there have been no reports of any species of the *Starmerella* clade found exclusively in pot-pollen, but they have been recovered from a great diversity of substrates within nests, such as honeypots, bodies of adult bees, garbage pellets, fecal material, discarded pollen or food, and propolis, as shown by Morais et al. (2013).

Table 17.2 A partial list of yeast belonging to the *Starmerella* clade and other yeast species, recovered from substrates of native eusocial and solitary bee species of Brazil: [a]Pot-pollen, [b]garbage pellets, [c]larval provisions

Yeast species	*Melipona quinquefasciata*[a] Calaça (2011)[a], Daniel et al. (2013)	*Tetragonisca angustula*[a,b] Rosa and Lachance (2005)[a], Rosa et al. (2003), Teixeira et al. (2003)[a]	*Melipona quadrifasciata*[b] Rosa et al. (2003)[a]	*Frieseomielitta varia*[b]	*Diadasina distincta*[c] Rosa et al. (1999)[a]	*Ptilothrix plumata*[c]	*Megachile* sp.[c] Pimentel et al. (2005)[b]	*Centris tarsata*[c]
Starmerella clade								
Candida batistae					X	X		
Candida cf. *etchellsii*	X							
Candida parapsilosis	X							
Starmerella meliponinorum		X						
Starmerella neotropicalis	X							
Candida apicola complex			X					
Candida riodocensis							X	
Candida cellae								X
Candida etchellsii		X						
Others								
Pseudozyma antarctica			X					
Aureobasidium pullulans	X							
Rhodosporidium toruloides	X							
Trichosporon sp.	X	X						
Zygosaccharomyces machadoi		X						
Zygosaccharomyces mellis	X							
Debaryomyces hansenii			X					
Kloeckera sp.				X				

[a]Studies made in Cerrado areas of Minas Gerais State, Brazil
[b]Study made in Atlantic Rain Forest, in Minas Gerais State, Brazil

As researchers seek to better understand the ecological relationship between yeast and *A. mellifera*, it has become clear that experimental studies are needed to elucidate the role that such microorganisms play in the lives of stingless bees. Studies that survey microbiota associated with meliponines are the first step toward understanding how this community functions. However, it is important to emphasize that understanding the keeping of stingless bees, which allows colonies to survive and continue to permit valid ecological and evolutionary studies, must advance in the naturally occurring locations of each species.

Acknowledgments We sincerely thank Dr. Patricia Vit for the opportunity of this chapter, for guidance, patience, and careful attention in every detail needed for the improvement of the manuscript. We thank Dr. David W. Roubik for his valuable and generous comments and corrections and for improving the English. We are grateful to Dr. Clemens Schlindwein for *Tetragonisca angustula* nest photos. We are also grateful to reviewers for their constructive suggestions in this chapter. The authors were financially supported by grants of the Brazilian science foundations, Conselho Nacional de Desenvolvimento Científico e Tecnológico (CNPq) and Fundação do Amparo a Pesquisa do Estado de Minas Gerais (FAPEMIG).

References

Aguiar CML, Santos GMM, Martins CF, Presley SJ. 2013. Trophic niche breadth and niche overlap in a guild of flower-visiting bees in a Brazilian dry forest. Apidologie 44: 153–162.

Aguiar CML. 2003. Utilização de recursos florais por abelhas (Hymenoptera, Apoidea) em uma área de Caatinga (Itatim, Bahia, Brasil). Revista Brasileira de Zoologia 20: 457–467.

Aleixo KP, Faria LB, Garófalo CA, Imperatriz-Fonseca VL, Silva CI. 2013. Pollen collected and foraging activities of Frieseomelitta varia (Lepeletier) (Hymenoptera: Apidae) in an urban landscape. Sociobiology, 60: 266–276.

Andena SR, Santos EF, Noll FB. 2012. Taxonomic diversity, niche width and similarity in the use of plant resources by bees (Hymenoptera: Anthophila) in a cerrado area. Journal of Natural History 46: 1663–1687.

Anderson KE, Carroll MJ, Sheehan TIM, Mott BM, Maes P, Corby-Harris V. 2014. Hive-stored pollen of honey bees: many lines of evidence are consistent with pollen preservation, not nutrient conversion. Molecular Ecology 23: 5904–5917.

Antonini Y, Martins RP. 2003. The value of a tree species (*Caryocar brasiliense*) for a stingless bee *Melipona quadrifasciata quadrifasciata*. Journal of Insect Conservation 7: 167–174.

Antonini Y, Soares SM, Martins RP. 2006. Pollen and nectar harvesting by the stingless bee *Melipona quadrifasciata anthidioides* (Apidae: Meliponini) in an urban forest fragment in Southeastern Brazil. Studies on Neotropical Fauna and Environment 41: 209–215.

Begon M, Townsend CR, Harper JL. 2006. Ecology: from individuals to ecosystems. Blackwell Malden, USA. 738 pp.

Blüthgen N, Klein AM. 2011. Functional complementarity and specialisation: the role of biodiversity in plant-pollinator interactions. Basic and Applied Ecology 12: 282–291.

Blüthgen N, Menzel F, Blüthgen N. 2006. Measuring specialization in species interaction networks. BMC Ecology 6:9.

Brandão M. 2000. Cerrado. pp. 55–63. In Mendonça MP, Lins LV (eds). Lista vermelha das espécies ameaçadas de extinção da flora de Minas Gerais. Fundação Biodiversitas e Fundação Zoo-botânica de Belo Horizonte; Belo Horizonte, Brasil. 157 pp.

Brysch-Herzberg M. 2004. Ecology of yeasts in plant-bumblebee mutualism in Central Europe. FEMS Microbiology Ecology 50: 87–100.

Calaça PSST. 2011. Aspectos da biologia de *Melipona quinquefasciata* Lepeletier (Mandaçaia do chão), características físico-químicas do mel, recursos alimentares e leveduras associadas., Dissertação de Mestrado. Instituto de Ciências Exatas e Biológicas. Departamento de Biodiversidade; Outro Preto, Brasil. 108 pp.

Camargo JMF, Garcia MVB, Júnior ERQ, Castrillón A. 1992. Notas prévias sobre a bionomia de *Ptilotrigona lurida* (Hymenoptera, Apidae, Meliponinae): associação de leveduras em pólen estocado. Boletim do Museu Paraense Emílio Goeldi 8: 391–395.

Camargo JMF (in memoriam), Pedro SRM. 2013. Catalogue of Bees (Hymenoptera, Apoidea) in the Neotropical Region, online version. Available at: URL http://moure.cria.org. br/catalogue?id=34135 . Last update: June 17, 2013.

Camargo JMF, Pedro SRM. 2007. Meliponini Lepeletier 1836. pp. 272-578. In Moure JS, Urban D, Melo GAR, eds. Catalogue of Bees (Hymenoptera, Apoidea) in the Neotropical Region. Sociedade Brasilera de Entomologia; Curitiba, Brasil. 1958 pp.

Casteel DB. 1912. The manipulation of the wax scales of the honeybee. US Dept. of Agriculture, Bureau of Entomology 161: 1–13.

Chevtchik V. 1950. Mikrobiologie pyloveho kvaseni. Publication from Faculty of Science, University of Masaryk, Czech Republic 323: 103–130.

Cortopassi-Laurino M, Ramalho M. 1988. Pollen harvest by Africanized *Apis mellifera* and *Trigona spinipes* in São Paulo. Botanical and ecological views. Apidologie 19: 1–24.

Daniel HM, Rosa CA, Thiago-Calaça PSS, Antonini Y, Bastos EM, Evrard P, Huret S, Fidalgo-Jiménez A, Lachance M-A. 2013. *Starmerella neotropicalis* f.a., sp. nov., a yeast species found in bees and pollen. International Journal of Systematic and Evolutionary Microbiology 63: 3896–3903.

Duarte RS. 2012. Aspectos da biologia destinados à criação Tetragona clavipes (Fabricius, 1804) (Apidae, Meliponini). Dissertação de Mestrado. Faculdade de Filosofia, Ciências e Letras. Ribeirão Preto, SP, Brasil. 82 pp.

Faria LB, Aleixo KP, Garófalo CA, Imperatriz-Fonseca VL, Silva CI. 2012. Foraging of *Scaptotrigona* aff. *depilis* (Hymenoptera, Apidae) in an urbanized area: Seasonality in resource availability and visited plants. Psyche: A Journal of Entomology 2012: 1–12.

Fernandes-da-Silva PG, Serrão J. 2000. Nutritive value and apparent digestibility of bee-collected and bee-stored pollen in the stingless bee, *Scaptotrigona postica* Latr. (Hymenoptera, Apidae, Meliponini). Apidologie 31: 39–45.

Ferreira MG, Manente-Balestieri FC, Balestieri JB. 2010. Pollen harvest by *Scaptotrigona depilis* (Moure) (Hymenoptera, Meliponini) in Dourados, Mato Grosso do Sul, Brazil. Revista Brasileira de Entomologia 54: 258–262.

Figueiredo N, Gimenes M, Miranda M, Oliveira-Rebouças P. 2013. *Xylocopa* bees in tropical coastal sand dunes: use of resources and their floral syndromes. Neotropical Entomology 42: 252–257.

Foote HL. 1957. Possible use of microrganisms in synthetic bee bread production. American Bee Journal 97: 476–478.

Freitas BM, Imperatriz-Fonseca VL, Medina LM, Kleinert AMP, Galetto L, Nates-Parra G, Quezada-Euán JJG. 2009. Diversity, threats and conservation of native bees in the Neotropics. Apidologie 40: 332–346.

Gilliam M. 1979. Microbiology of pollen and bee bread: the yeasts. Apidologie 10: 43–53.

Gilliam M. 1997. Identification and roles of non-pathogenic microflora associated with honey bees. FEMS Microbiology Letters 155: 1–10.

Gilliam M, Prest DB. 1987. Microbiology of feces of the larval honey bee, *Apis mellifera*. Journal of Invertebrate Pathology 49: 70–75.

Gilliam M, Roubik DW, Lorenz BJ. 1990. Microorganisms associated with pollen, honey, and brood provisions in the nest of a stingless bee, *Melipona fasciata*. Apidologie (France). 21: 89-97

Goulson D, Darvill B. 2004. Niche overlap and diet breadth in bumblebees; are rare species more specialized in their choice of flowers? Apidologie 35: 55–63.

Guibu L, Ramalho M, Kleinert-Giovannini A, Imperatriz-Fonseca V. 1988. Exploração dos recursos florais por colônias de *Melipona quadrifasciata* (Apidae, Meliponinae). Revista Brasileira de Biologia 48: 299–305.

Hartfelder K, Engels W. 1989. The composition of larval food in stingless bees: evaluating nutritional balance by chemosystematic methods. Insectes Sociaux 36: 1–14.

Haydak M. 1958. Pollen - pollen substitutes-beebread. American Bee Journal 98: 145–146.

Haydak MH, Vivino AE. 1950. The changes in the thiamine, riboflavin, niacin and pantothenic acid content in the food of female honeybees during growth with a note on the vitamin K activity of royal jelly and beebread. Annals of the Entomological Society of America 43: 361–367.

Herbert EW, Shimanuki H. 1978. Chemical composition and nutritive value of bee-collected and bee-stored pollen. Apidologie 9: 33–40.

Herrera CM, de Vega C, Canto A, Pozo MI. 2009. Yeasts in floral nectar: a quantitative survey. Annals of Botany 103: 1415–1423.

Herrera CM, Pozo MI, Medrano M. 2013. Yeasts in nectar of an early-blooming herb: sought by bumble bees, detrimental to plant fecundity. Ecology 94: 273–279.

Hitchcock J. 1956. A milk-digesting enzyme in pollen stored by honey bees. American Bee Journal 96: 487–489.

Hoehn P, Tscharntke T, Tylianakis JM, Steffan-Dewenter I. 2008. Functional group diversity of bee pollinators increases crop yield. Proceedings of the Royal Society of London B: Biological Sciences 275: 2283–2291.

Johnson RM, Ellis MD, Mullin CA, Frazier M. 2010. Pesticides and honey bee toxicity-USA. Apidologie 41: 312–331.

Kremen C, Williams NM, Bugg RL, Fay JP, Thorp RW. 2004. The area requirements of an ecosystem service: crop pollination by native bee communities in California. Ecology Letters 7: 1109–1119.

Larsen TH, Williams NM, Kremen C. 2005. Extinction order and altered community structure rapidly disrupt ecosystem functioning. Ecology Letters 8: 538–547.

Linsley EG. 1958. The ecology of solitary bees. Hilgardia 27: 543–599.

Machado J. 1971. Simbiose entre as abelhas sociais brasileiras (Meliponinae, Apidae) e uma espécie de bactéria. Ciência e Cultura 23: 625–633.

Marques-Souza AC, Absy ML, Kerr WE. 2007. Pollen harvest features of the central amazonian bee *Scaptotrigona fulvicutis* Moure 1964 (Apidae: Meliponinae), in Brazil. Acta Botanica Brasilica 21: 11–20.

Martins F, Batalha M. 2006. Pollination systems and floral traits in cerrado woody species of the Upper Taquari region (Central Brazil). Brazilian Journal of Biology 66: 543–552.

Michener CD. 2007. The Bees of the World. 2nd. Ed Johns Hopkins, Baltimore. 972 pp.

Miranda EA, Carvalho AF, Andrade-Silva ACR, Silva CI, Del Lama MA. 2015. Natural history and biogeography of *Partamona rustica*, an endemic bee in dry forests of Brazil. Insectes Sociaux 62: 255–263.

Morais PB, Calaça PSST, Rosa CA. 2013. Microorganisms associated with stingless bees. pp. 173-186. In Vit P, Pedro SRM, Roubik DW, eds. Pot-honey: a legacy of stingless bees. Springer; New York, USA. 654 pp.

Myers N, Mittermeier RA, Mittermeier CG, Fonseca GA, Kent J. 2000. Biodiversity hotspots for conservation priorities. Nature 403: 853–858.

Nogueira-Neto P. 1997. Vida e criação de abelhas indígenas sem ferrão. Editora Nogueirapis; São Paulo, Brasil. 446 pp.

Pain J, Maugnet J. 1966. Recherches biochimiques et physiologiques sur pollen emmagasiné par abeilles. Les Annales de l´Abeille 9(3): 209–236.

Pedro SRM. 2014. The stingless bee fauna in Brazil (Hymenoptera: Apidae). Sociobiology 61: 348–354.

Penedo MCT, Testa PR, Zucoloto FS. 1976. Valor nutritivo do gevral e do levedo de cerveja em diferentes misturas com pólen para *Scaptotrigona* (*Scaptotrigona*) *postica* (Hymenoptera, Apidae). Ciencia e cultura. 28(5): 536-538.

Pimentel MR, Antonini Y, Martins RP, Lachance MA, Rosa CA. 2005. *Candida riodocensis* and *Candida cellae*, two new yeast species from the *Starmerella* clade associated with solitary bees in the Atlantic rain forest of Brazil. FEMS Yeast Research 5: 875–879.

R Core Team. 2015. R: A language and environment for statistical computing [Internet]. Vienna, Austria: R Foundation for Statistical Computing; 2013. Available at: http://www r-project org.

Ramalho M, Kleinert-Giovannini A, Imperatriz-Fonseca VL. 1989. Utilization of floral resources by species of *Melipona* (Apidae, Meliponinae): Floral preferences. Apidologie 20: 185–195.

Ramalho M, Silva MD, Carvalho CA. 2007. Dinâmica de uso de fontes de pólen por *Melipona scutellaris* Latreille (Hymenoptera: Apidae): uma análise comparativa com *Apis mellifera* L. (Hymenoptera: Apidae), no Domínio Tropical Atlântico. 36(1): 38-45.

Reatto A, Correia JR, Spera ST. 1998. Solos do bioma cerrado: aspectos pedológicos. pp. 47–86. In Sano SM; Almeida SP, eds. Cerrado: ambiente e flora. EMBRAPA-CPAC, Planaltina, Brasil. 556 pp.

Ribeiro JF, Walter BMT, Sano S, Almeida S. 1998. Fitofisionomias do bioma Cerrado. pp. 89-166. In Sano SM; Almeida SP, eds. Cerrado: ambiente e flora. EMBRAPA-CPAC, Planaltina, Brasil. 556 pp.

Rosa CA, Lachance MA. 2005. *Zygosaccharomyces machadoi* sp. n., a yeast species isolated from a nest of the stingless bee *Tetragonisca angustula*. Lundiana 6: 27–29.

Rosa CA, Lachance MA, Silva JOC, Teixeira ACP, Marini MM, Antonini Y, Martins RP. 2003. Yeast communities associated with stingless bees. FEMS Yeast Research 4: 271–275.

Rosa CA, Viana EM, Martins RP, Antonini Y, Lachance MA. 1999. *Candida batistae*, a new yeast species associated with solitary digger nesting bees in Brazil. Mycologia 91: 428–433.

Sandhu DK, Waraich MK. 1985. Yeasts associated with pollinating bees and flower nectar. Microbial Ecology 11: 51–58.

Santos GMM, Aguiar CML, Mello MA. 2010. Flower-visiting guild associated with the Caatinga flora: trophic interaction networks formed by social bees and social wasps with plants. Apidologie 41: 466–475.

Santos RM, Aguiar CM, Dórea MC, Almeida GF, Santos FA, Augusto SC. 2013. The larval provisions of the crop pollinator *Centris analis*: pollen spectrum and trophic niche breadth in an agroecosystem. Apidologie 44: 630–641.

Silberbauer-Gottsberger I, Gottsberger G. 1988. A polinização de plantas do cerrado. Revista Brasileira de Biologia 48: 651–663.

Silva JMC, Bates JM. 2002. Biogeographic patterns and conservation in the south american cerrado: a tropical savanna Hotspot. BioScience 52: 225–234.

Soares SM. 2003. Utilização de recursos alimentares par *Melipona rufiventris* (Apidae, Meliponina) no cerrado de Brasilândia de Minas, MG. Dissertação de Mestrado. Instituto de Ciências Biológicas, Belo Horizonte, Brasil. 96 pp.

Standifer L, McCaughey W, Dixon S, Gilliam M, Loper G. 1980. Biochemistry and microbiology of pollen collected by honey bees (*Apis mellifera* L.) from almond, *Prunus dulcis*. II. Protein, amino acids and enzymes. Apidologie 11: 163–171.

Teixeira ACP, Marini MM, Nicoli JR, Antonini Y, Martins RP, Lachance MA, Rosa CA. 2003. *Starmerella meliponinorum* sp. nov., a novel ascomycetous yeast species associated with stingless bees. International Journal of Systematic and Evolutionary Microbiology 53: 339–343.

Vásquez A, Olofsson TC. 2009. The lactic acid bacteria involved in the production of bee pollen and bee bread. Journal of Apicultural Research 48: 189–195.

Viana LS, Melo GAR. 1987. Conservação de abelhas. Informe Agropecuário 13: 23–26.

Wilms W, Imperatriz-Fonseca VL, Engels W. 1996. Resource partitioning between highly eusocial bees and possible impact of the introduced Africanized honey bee on native stingless bees in the Brazilian Atlantic rainforest. Studies on Neotropical Fauna and Environment 31: 137–151.

Zerbo AC, Lúcia R, Moraes MS, Brochetto-Braga MR. 2001. Protein requirements in larvae and adults of *Scaptotrigona postica* (Hymenoptera, Apidae, Meliponinae): midgut proteolytic activity and pollen digestion. Comparative Biochemistry and Physiology Part B: Biochemistry and Molecular Biology 129: 139–147.

A Review of the Artificial Diets Used as Pot-Pollen Substitutes

18

Cristiano Menezes, Camila Raquel Paludo, and Mônica Tallarico Pupo

18.1 Introduction

18.1.1 Aim of the Chapter

The seasonality of food resources has forced bees to develop survival strategies to deal with periods of food scarcity (Roubik 1989). For example, solitary bees spend most of their lives as immature forms before emerging as adults prior to or during the peak bloom periods of specific plants. They generally reproduce quickly and die after provisioning brood cells, laying their eggs and closing the nests, with offspring emerging only in the next rewarding season (Michener 2000). In contrast, eusocial bees have more complex needs, because of their large population size, prolonged activity periods (several months to perennial), and inability to establish new colonies without workers (Michener 1974). During periods of low food availability, colonies adopt several strategies to save energy and food (Roubik 1982): colony population decreases, immature stages

receive less food, subsequent individual size is reduced, and workers reduce external activities in order to prolong their life spans (Roubik 1982; Ramalho et al. 1998, Veiga et al. 2013; Gomes et al. 2015). During periods of high resource availability, the colony population rapidly proliferates, and foraging activity intensifies in order to prepare bees for the next period of scarcity (Roubik 1982).

Of all the different strategies developed by eusocial bees to deal with a lack of food sources, storage was a key solution in their evolutionary process, as it allowed their perennial activity (Michener 1974). However, it also brought challenges, because pathogenic microbes can proliferate in stored food, decreasing its nutritional quality and causing health problems for the bees (Anderson et al. 2011). The current review aims to show what is known about artificial diet supplements and what needs to be done to improve their efficacy as a tool to maintain bee colony health in periods of food scarcity. Understanding how bees preserve stored food can yield significant benefits to human societies, as we also have to store food and deal with similar problems of spoilage and loss of nutritional quality. Stingless bees have spent more than 70 million years developing these strategies (Engel 2000). Knowledge about the microbiota, pollen composition, and exogenous contaminants in bee bread is also extremely useful for the development of artificial bee diets. With this information, new microorganisms, substances, and technologies

C. Menezes (✉)
Embrapa Amazônia Oriental, Belém, PA, Brazil
e-mail: menezes.cristiano@gmail.com

C.R. Paludo • M.T. Pupo
Faculdade de Ciências Farmacêuticas de Ribeirão Preto, Universidade de São Paulo, Ribeirão Preto, SP, Brazil

© Springer International Publishing AG, part of Springer Nature 2018
P. Vit et al. (eds.), *Pot-Pollen in Stingless Bee Melittology*, DOI 10.1007/978-3-319-61839-5_18

can be used to improve artificial diets and prevent or treat harmful bee pathogens and diseases.

18.1.2 How Do Stingless Bees Harvest and Store Their Food?

The typical food of stingless bees is nectar and pollen (Fig. 18.1). Nectar is the source of energy (carbohydrates) and mainly consumed by adult bees in the form of honey. Pollen is the source of protein, vitamins, and minerals used to feed their larvae and young adult bees (nurse bees). Newly emerged bees actively produce glandular secretions, which also play an important role as wax and larval food components (Michener 1974).

To harvest the nectar, the foragers collect it from flowers with their proboscis and store it inside their honey stomach. When this storage organ is full, bees return to the colony and give it to other workers through trophallaxis near the entrance tunnel of the colony. These workers stay aligned to each other inside the entrance tunnel fanning their wings continuously in the same direction to create airflow. They make a small drop of nectar under their mandibles and move the drop in and out of their stomach. In this process, the water content of the nectar, which is generally around 60–75% in the flowers, will decrease to 20–30% (Cortopassi-Laurino and Gelli 1991). When it reaches the desired concentration of sugars and water, workers deposit their crop contents inside honeypots made of cerumen. The pot is closed after being filled, and its con-

tents will ferment slightly and be opened again when needed (Menezes et al. 2013).

To harvest pollen, foragers usually grasp anthers, vibrate, or rub them to remove the pollen grains. The workers of stingless bees are hairy at the ventral surface of their body and legs, so most of the pollen remains attached to those hairs while the bees move between flowers (Fig. 18.2). During foraging, bees will comb the pollen from their hair with their legs and stick the pollen grains together with honey, carried from pot-honey in their stomach (Leonhardt et al. 2007). As they transfer the sticky pollen to the hind legs, it makes a small pellet on the corbicula, a concave structure on the hind legs, surrounded by hairs to hold the pollen pellet. When the worker bees arrive at the nest, they go straight to a pollen pot and dislodge the pollen pellet inside the pot, then return to the field (Fig. 18.3). When the pot is full, it is closed, and the pollen will ferment for a few weeks before being consumed (Menezes et al. 2013).

18.2 The Fermentation Process in Stingless Bee Storage Pots

18.2.1 General Characteristics of Pollen Fermentation

Microbial fermentation can improve the shelf life, nutritional value, flavor, consistency, and palatability of stored food (Tamang et al. 2016; Swain et al. 2014). Millions of years before

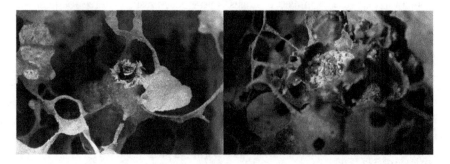

Fig. 18.1 Nectar (*left*) and pollen (*right*) storage inside cerumen pots by *Scaptotrigona depilis*, to form honey and bee bread after fermentation (Photos: C. Paludo)

Fig. 18.2 *Melipona scutellaris* foragers collecting pollen from *Eugenia uniflora* flowers. The bee holds the anthers and shakes them to harvest the pollen (*left*). The grains are caught by the abdominal hairs. Eventually, the bee remains stationary and brushes her hair and then moistens the pollen with honey in the corbicula (*right*) (Photos: C. Menezes)

Fig. 18.3 *Plebeia droryana* forager returning to the nest with pollen and depositing the pellet inside a pollen pot (Photos: C. Menezes)

human civilization emerged and began to use fermentation as a tool to help nutrition and food preservation, insects had already produced fermented food, an example being honey (DeWeerdt 2015). Nowadays, it is acknowledged that bees also use fermented pollen in their diet.

Pollen is a nutritious plant-derived material that contains large amounts of proteins, generally enzymes, which participate in the fertilization process of flowers (Roulston et al. 2000). Pollen is also the source of free amino acids, fatty acids and minerals (Yang et al. 2013), sterols and flavonoids (Silva et al. 2006), sugar (Herbert and Shimanuki 1978), and vitamins (Haydak and Palmer 1942; Loper et al. 1980). When bees collect pollen, glandular secretions containing enzymes are added to the pellet. Some enzymes, produced by bees, can help pollen predigestion (Costa and Cruz-Landim 2005). In addition, enzymes from plants, which are involved in pollen tube growth and fertilization (Roulston et al. 2000), may also participate in macronutrient degradation inside pot-pollen. Enzymatic activity, indicating the presence of proteases (Grogan and Hunt 1979), acid phosphatase, and esterase (Knox and Heslop-Harrison 1969), has been detected in pollen grains.

The presence of fermentation microbes in pot-pollen is remarkable (Gilliam 1979a, b; Gilliam et al. 1989, 1990; Vásquez and Olofsson 2009), and this specialized microbiota has been called responsible for the production of bee bread, which is obtained after fermentation of stored pollen. The fermentation process is beneficial to those insects. During microbial fermentation, enzymes may be secreted and amino acids become more available due to proteolysis. Pot-pollen microbiota can also produce vitamins and bioactive natural products.

18.2.2 Microbial Fermentation and Nutritional Enhancement of Pollen

The genus *Bacillus* is frequently found in association with pot-pollen (Gilliam 1979a). These bacteria may contribute to acetic and lactic acid fermentation (Woolford 1977) and produce enzymes that degrade macromolecules present in pollen, thus making it more nutritious for bees. Gilliam and collaborators (Gilliam et al. 1990) find that *Bacillus* spp., isolated from colonies of the stingless bee *Melipona fasciata* (Meliponini), produce a series of enzymes, including aminopeptidases and proteases, which can increase free amino acid concentration in bee bread, in agreement with the findings of Human and Nicolson (2006), who verified that the amino acid content in stored pollen is superior than that of fresh pollen from flowers. *Bacillus* strains also produce glucosidases and lipases that degrade complex carbohydrates and lipids (Gilliam et al. 1990).

Lactic acid bacteria (LAB) belonging to the family Pasteurellaceae, such as *Lactobacillus* spp. and *Bifidobacterium* spp., were described in *Apis mellifera* (Apini) as responsible for the acidification of pollen during bee bread formation. Lactic acidification of stored pollen is thought to be important for spoilage prevention (Vásquez and Olofsson 2009). These bacteria can also de novo biosynthesize B-group vitamins (LeBlanc et al. 2013), which are required for bee development and health (Anderson and Dietz 1976).

The presence of LAB in pot-pollen shows that the microbiota can contribute to the survival and maintenance of bee colonies by producing compounds to protect stored food and also for synthesis of important vitamins. However, Pain and Maugenet (1966) verified that *in vitro* lactic fermentation by LAB does not produce the same bee bread made by bees. Thus, the participation of other microorganisms or bee products is clearly essential.

Pot-pollen fermentation is a multispecies process in which each microorganism has its own contribution and importance. Besides bacteria, certain yeasts and filamentous fungi can also participate in bee bread fermentation. Fungi isolated from pollen are capable of producing a series of enzymes. *Penicillium* spp. and *Aspergillus* spp. produce aminopeptidases, esterases, phosphatases, and glucosidases (Gilliam et al. 1989), which aid in macronutrient degradation of pollen grains. Gilliam (1979b) relates how yeasts, isolated from bee bread and pollen, are capable of fermenting diverse carbohydrates and several isolates from pollen assimilate lactic acid, which are produced by LAB. An adequate diet is essential for larval development and bee health (Brodschneider and Crailsheim 2010), and the microbiota present in pot-pollen contribute to a balanced diet.

18.2.3 Impacts of Exogenous Compounds in Pollen

In a study of bee-associated microbiota, Gilliam and Morton (1978) described that the antibiotics oxytetracycline (TM-25) and fumagillin (Fumidil B), when consumed by caged bees, depressed the number of resident microorganisms. The herbicide 2,4-dichlorophenoxyacetic acid (2,4-D) decreased the amount of *Bacillus* strains isolated from the gut of treated bees and caused proliferation of intestinal yeasts. According to these results, bee-associated microbiota can shift due to the presence of chemicals used on cultivated crops and, consequently, present in pollen and nectar collected in the field. This means that the use of agrochemicals that target microorganisms could alter bee gut flora and colony microbiota and indirectly affect colony fitness.

Nowadays, there is great concern regarding Colony Collapse Disorder (CCD), an unexplained loss of adult worker bees in *A. mellifera* colonies. Several factors have been hypothesized to cause the phenomenon (van Engelsdorp et al. 2009), and the unspecific change in the symbiotic bee microbiota caused by exposure to xenobiotics present in the field could be important. Pollinators in decline signify an alarming problem that deserves attention and policies to control. Pesticides are important stressors that can cause insect health debilitation, and regulatory standards are necessary not just for *A. mellifera* but also for wild pollinators, such as Meliponini (Potts et al. 2016; Dicks et al. 2016).

Some experiments suggest that *Bacillus thuringiensis* (*Bt*)-modified corn can cause decline in monarch butterfly populations, *Danaus plexippus*, when larvae consume the transgenic pollen. The production of *Bt* toxins can be dangerous for nontarget insects in certain crops (Losey et al. 1999), including bees. In addition, some plants produce pollen containing toxic secondary metabolites that can affect pollinators (Irwin et al. 2014).

In summary, several xenobiotics, such as insecticides, herbicides, fungicides, antibiotics, and toxic secondary metabolites from plants, can be present in pollen and may cause serious problems for the entire colony (Johnson 2015). When an artificial diet is planned, it is therefore important to verify the chemical quality and origin of the pollen given to bees, in order to avoid contamination, mortality, and harm to bees and colonies alike.

18.3 Microorganisms Present in Pot-Pollen

18.3.1 Generalities of Host-Associated Microorganisms

Host-associated microbiota can provide significant insight into the life histories of their hosts and the biological systems present in nature. The remarkable capacity of coevolution between insects and microorganisms has been reported in studies involving ants, beetles, wasps, termites, and bees. These animals have distinct and particular food sources, humidity, acidity (pH), and osmolarity, associated with their nests and lifestyles. Such factors, allied to microbial competition for space and nutrients, contribute to the adaptive selection of microbiota and, consequently, the function of each microorganism in these "little jungles" that eusocial insects create and carry along in the evolutionary process, which is a never-ending phenomenon.

Microbiota studies are complex, and in many cases, the microorganisms present in a host can vary with time and be influenced by several factors. For example, the microbiota in pollen can vary, depending on where the pollen was collected, the season, and storage time inside the bee nest. According to Gilliam (1979a), 41 *Bacillus* strains are found in pollen collected from flowers, legs of foraging bees, and bee bread in different stages of fermentation. It was clear that the number of different strains increases with bee manipulation and, possibly, an inoculum of *Bacillus* spp. was added by foraging bees in pot-pollen. Gilliam (1979b) also describes 113 yeast strains from pollen collected in different periods of fermentation, but the number of strains decreases with time inside the open waxen comb of honey bees. Nonetheless, some yeast strains could be added to stored pollen by bees.

Occasionally, small changes in host microbiota initiate a pathological process or, in contrast, can be beneficial. A better understanding about the function of individual members in the pot-pollen microbiota is required to understand fermentation within bee nests. Such knowledge can readily be applied in the preparation of artificial bee diets.

18.3.2 Bacteria

The systematic isolation of *Bacillus* spp. from bee colonies indicates that the bacteria engage in symbiotic relationships with these insects (Wang et al. 2015; Sinacori et al. 2014; Gilliam 1979a, 1997). We have isolated different strains of *Bacillus* sp. from *Scaptotrigona depilis* (Meliponini) pot-pollen. Isolated strains showed antimicrobial activity against the entomopathogenic fungi *Beauveria bassiana* and *Metarhizium anisopliae*. These results suggested that *Bacillus* spp. could protect the colony by producing natural products with antimicrobial properties.

In order to verify that pot-pollen microorganisms were present in the larval gut, we performed a surface sterilization in *S. depilis* larvae, and using different culture media, we followed the isolation process, using macerated insects. The *Bacillus* sp. strain, SDLI1, was isolated using

ISP-2 medium, had its whole-genome sequenced, and was able to produce antifungal compounds with activity against entomopathogens and human pathogenic microorganisms. Its circular chromosome reveals eight biosynthetic gene clusters, which are responsible for the biosynthesis of various antimicrobial compounds, such as bacillomycin and surfactin lipopeptides (Paludo et al. 2016).

Apparently, the main source of *Bacillus* spp. for bees is pollen. However, bees such as *Trigona hypogea*, an obligate necrophage that uses dead animal tissue as a protein source, completely replacing pollen, also have *Bacillus* spp. that produce enzymes in brood provisions and help fermentation and protein degradation (Gilliam et al. 1985). In addition to enzymatic production and macronutrient degradation, the contribution of these bacteria can also be chemical defense, as *Bacillus* species are known to produce antibiotics, and may also prevent proliferation of competing, spoilage bacteria (Gilliam et al. 1990). Antimicrobial assays performed with *Bacillus* sp. SDLI1, isolated from the stingless bee *S. depilis*, and its whole-genome sequence certainly corroborate this proposition (Paludo et al. 2016).

Other bacteria have been described in pot-pollen, such as the LAB *Lactobacillus* spp. and *Bifidobacterium* spp., which participate in pollen fermentation (Vásquez and Olofsson 2009). However, additional research should be performed to investigate the role that each bacterial genus plays in this microbiota.

18.3.3 Yeasts

In agreement with observations on *Bacillus* spp., a number of different yeast strains can be found in pollen (Gilliam 1979b) and contribute to bee bread fermentation. Yeasts such as *Candida* spp., *Starmerella* spp., and *Torulopsis* spp., frequently isolated in association with bees (Rosa et al. 2003; Gilliam 1979b), perform alcoholic fermentation and also acidify the pollen. It is known that low pH and high ethanol levels are mechanisms to avoid food spoilage.

In addition, bee-associated yeasts, such as *Candida apicola* and *Candida bombicola*, are promising sources of sophorolipids, which are glycolipid biosurfactants of substantial commercial interest, due to their nonhazardous and environmentally friendly characteristics (Kurtzman et al. 2010; Van Bogaert et al. 2007). Some authors describe the pharmacological properties shown by sophorolipids, such as antimicrobial activity (Díaz De Rienzo et al. 2015). These compounds could participate in protecting bee bread against pathogens.

18.3.4 Filamentous Fungi

Gilliam and collaborators (Gilliam et al. 1989) isolated 148 filamentous fungi strains from pollen of *Prunus dulcis* in different stages of fermentation. They report that the majority of fungal isolates present in fermented pollen belong to *Penicillium* genus, followed by *Aspergillus*. *Mucorales* was the most frequent order of *Zygomycetes* fungi present in fresh material.

Penicillium spp. and *Aspergillus* spp. produce a very diverse and interesting group of secondary metabolites with biological activities. These filamentous fungi biosynthesize polyketides, nonribosomal peptides, terpenes, and natural products with complex structures that involve different biosynthetic gene clusters (Bladt et al. 2013). The natural products can also be advantageous for bees, in the prevention of pot-pollen spoilage.

18.4 Development of Artificial Diets

A key stage of any animal domestication program is finding a substitute for a natural food source or the means to supplement diets. In the case of stingless bees (Meliponini), the beekeeper can easily replace honey with artificial sugar solutions (Nogueira-Neto 1997). On the other hand, replacing pollen has not been an easy task because of the complexity of its nutritional composition and fermentation. An adequate artificial diet would drastically change the scenario of meliponiculture and improve the colony multiplication process, avoiding problems associated with the seasonality of floral resources. Efforts applied to develop pollen substitutes and some artificial diets are encouraging. Because fermentation is necessary for developing these diets, recent insights about the microbiota involved in pollen fermentation certainly help to improve artificial diet development.

Camargo (1976) was a pioneer who tried to define an artificial diet for stingless bees. Fresh pollen, mixed with honey and fermented pollen, was fermented for about 2 weeks and given to bees. The unfermented pollen was rejected by the workers. A more recent study again confirms that bees prefer fermented pollen, choosing bee bread instead of fresh material (Vollet-Neto et al. 2016). Applying fermented pollen as a natural inoculum for the unfermented pollen seems to be the key for bee acceptance. Using the fermented material, it was hypothesized that the established microbiota will be transferred to the unfermented pollen, and it will become more attractive and greater in nutritional value to the bees (Fig. 18.4).

Protein source substitutes have been tested and most recipes are simple. Generally, a mixture of different protein sources and honey or sugar is used, supplemented with natural fermented pollen from the species that will receive the artificial diet. Before using the mixture, it is desirable to ferment it for several days (Penedo et al. 1976; Fernandes-da-Silva and Zucoloto 1990; Costa and Venturieri 2009; Pires et al. 2009). Nevertheless, such pollen substitutes are frequently rejected by bees. Apparently the risk of causing problems to bees is not significant even if the diet is not accepted. However, beekeepers should carefully check the hived colonies while providing artificial diets, because in case of rejection, the nutrient-rich material can accumulate and may also attract enemies to the colonies, such as phorid flies.

Interestingly, there are exceptions, such as the stingless bee *Frieseomelitta varia* which shows a very peculiar behavior. Workers of *F. varia* were strongly attracted to artificial diets offered outside the nests, and they also accepted fermented pollen from another stingless bee species (Menezes et al. 2013). Negative effects in colonies that were fed with these diets were not seen. Because of this behavior, this species may have relatively great potential for being produced on a large scale.

A recent paper by Rebelo and collaborators (Rebelo et al. 2016) on the physicochemical characteristics of pollen collected by the stingless bees *Melipona seminigra* and *M. interrupta* has huge potential to contribute to the development of artificial diets for these insects in the future. The results reveal that the composition of collected pollen by each species is 53.39 and 37.12% moisture, 37.63 and 24.00% protein, 10.81 and 6.47% lipid, 4.03 and 2.74% ash, 9.30 and 13.65% crude fiber, 25.66 and 44.27% carbohydrates, 350.47 and 331.33 kcal energy, a pH of 3.70 and 3.34, total solid content of 46.60 and 62.87%, and water activity of 0.91 and 0.85, respectively.

Fig. 18.4 Fermented pollen from *Melipona scutellaris*, which is very humid and acidic (*left*), and fermented pollen from *Tetragonisca angustula*, which is dryer and sweeter than the first one (*right*). Each species has a different microbiome, and therefore, the stored pollen is different (Photos: C. Menezes)

Although we do have some alternatives for replacing pollen, as discussed, another nutritious and healthy way of improving the stingless bee diet is by propagating attractive plant species around the meliponary. This is a long-term investment but a necessary process, because even when an artificial diet is established, we may scarcely obtain a feasible substitute for natural pollen and which could replace its natural properties.

A recent study has shown that the stingless bee *S. depilis* uses 66 different plant species as food sources in a semi-urban area. Nevertheless, only nine plant species were responsible for 80% of all stored food during the study period (Aleixo et al. 2016). This indicates that we can increase the natural capacity of an area to support stingless bees by planting key plants around their hives. Moreover, we can choose species that bloom during periods of food scarcity and the ones that offer more pollen, such as plants of Myrtaceae family. In the case of *S. depilis* studied in one site in Brazil, for example, it would be highly recommended to propagate *Eugenia uniflora*, because it blooms at a time when few other resources are available and when environment stress is high and the flowers offer large amounts of pollen for bees.

Artificial diets allied to the propagation of plants that have different blooming periods are very important avenues of future research for the effective conservation, propagation, and management of stingless bees.

Acknowledgments The authors thank the Fogarty International Center, National Institutes of Health (NIH), USA (grant U19TW009872); the São Paulo Research Foundation (FAPESP), Brazil (grants 2013/50954-0, 2013/04092-7, 2013/07600-3, 2012/22487-6, 2012/51112-0, and 2014/23532-0); and the National Council for Scientific and Technological Development (CNPq), Brazil (grants 400435/2014-4 and 479710/2011-2).

References

Aleixo KP, Menezes C, Imperatriz-Fonseca VL, Silva CI. 2016. Seasonal availability of floral resources and ambient temperature shape stingless bee foraging behavior (*Scaptotrigona* aff. *depilis*). Apidologie *online first*: 1–11.

Anderson LM, Dietz A. 1976. Pyridoxine requirement of the honey bee (*Apis mellifera*) for brood rearing. Apidologie 7: 67–84.

Anderson KE, Sheehan TH, Eckholm BJ, Mott BM, DeGrandi-Hoffman G. 2011. An emerging paradigm of colony health: microbial balance of the honey bee and hive (*Apis mellifera*). Insectes Sociaux 58: 431–444.

Bladt TT, Frisvad JC, Knudsen PB, Larsen TO. 2013. Anticancer and antifungal compounds from *Aspergillus*, *Penicillium* and other filamentous fungi. Molecules 18: 11338–11376.

Brodschneider R, Crailsheim K. 2010. Nutrition and health in honey bees. Apidologie 41: 278–294.

Camargo CA. 1976. Dieta semi–artificial para abelhas da subfamilia Meliponinae (Hymenoptera, Apidae). Ciência e Cultura, 28: 430–431.

Cortopassi-Laurino M, Gelli DS. 1991. Analyse pollinique, proprietés physico–chimique et ac–tion antibactérienne des miels d'abeilles african–isées *Apis mellifera* et des Méliponines du Brésil. Apidologie 22: 61–73.

Costa AC, Cruz-Landim C. 2005. Hydrolases in the hypopharyngeal glands of workers of *Scaptotrigona postica* and *Apis mellifera* (Hymenoptera, Apinae). Genetics and Molecular Research 4: 616–623.

Costa L, Venturieri GC. 2009. Diet impacts on *Melipona flavolineata* workers (Apidae, Meliponini). Journal of Apicultural Research, 48: 38–45. DOI: 10.3896/IBRA.1.48.1.09.

DeWeerdt S. 2015. The beeline. Nature 521: S50–S51.

Díaz De Rienzo MA, Stevenson P, Marchant R, Banat IM. 2015. Antibacterial properties of biosurfactants against selected gram positive and negative bacteria. FEMS Microbiology Letters 363: fnv224.

Dicks LV, Viana B, Bommarco R, Brosi B, Arizmendi MC, Cunningham SA, Galetto L, Hill R, Lopes AV, Pires C, Taki H, Potts SG. 2016. Ten policies for pollinators. Science 354:975-976.

Engel, MS. 2000. A new interpretation of the oldest fossil bee (Hymenoptera: Apidae). American Museum Novitates 3296: 1–11.

van Engelsdorp D, Evans JD, Saegerman C, Mullin C, Haubruge E, Nguyen BK, Frazier M, Frazier J, Cox-Foster D, Chen Y, Underwood R, Tarpy DR, Pettis JS. 2009. Colony Collapse Disorder: A descriptive study. PLoS One 4: e6481.

Fernandes-da-Silva PG, Zucoloto FS. 1990. A semi–artificial diet for *Scaptotrigona depilis* Moure (Hymenoptera, Apidae). Journal of Apicultural Research, 29: 233–235.

Gilliam M, Buchmann SL, Lorenz BJ, Roubik DW. 1985. Microbiology of the larval provisions of the stingless bee, *Trigona hypogea*, an obligate necrophage. Biotropica 17: 28–31.

Gilliam M, Morton HL. 1978. Bacteria belonging to the genus *Bacillus* isolated from honey bees, *Apis mellifera*, FED 2, 4–D and antibiotics. Apidologie 9: 213–222.

Gilliam M, Prest DB, Lorenz BJ. 1989. Microbiology of pollen and bee bread: taxonomy and enzymology of molds. Apidologie 20: 53–68.

Gilliam M, Roubik DW, Lorenz BJ. 1990. Microorganisms associated with pollen, honey, and brood provisions in the nest of a stingless bee, *Melipona fasciata*. Apidologie 21: 89–97.

Gilliam M. 1979a. Microbiology of pollen and bee bread: The genus *Bacillus*. Apidologie 10: 269–274.

Gilliam M. 1979b. Microbiology of pollen and bee bread: The yeasts. Apidologie 10: 43–53.

Gilliam M. 1997. Identification and roles of non–pathogenic microflora associated with honey bees. FEMS Microbiology Letters 155: 1–10.

Gomes RLC, Menezes C, Contrera, FAL. 2015. Worker longevity in an Amazonia *Melipona* (Apidae, Meliponini) species: effects of season and age at foraging onset. Apidologie 46: 133–143.

Grogan DE, Hunt JH. 1979. Pollen proteases: their potential role in insect digestion. Insect Biochemistry 9: 309–313.

Haydak MH, Palmer LS. 1942. Royal jelly and bee bread as sources of vitamins B_1 B_2, B_6, C and nicotinic and pantothenic acids. Journal of Economic Entomology 35: 319–320.

Herbert Jr EW, Shimanuki H. 1978. Chemical composition and nutritive value of bee–collected and bee–stored pollen. Apidologie 9: 33–40.

Human H, Nicolson SW. 2006. Nutritional content of fresh, bee–collected and stored pollen of *Aloe greatheadii* var. *davyana* (Asphodelaceae). Phytochemistry 67: 1486–1492.

Irwin RE, Cook D, Richardson LL, Manson JS, Gardner DR. 2014. Secondary compounds in floral rewards of toxic rangeland plants: Impacts on pollinators. Journal of Agricultural and Food Chemistry 62: 7335–7344.

Johnson RM. 2015. Honey bee toxicology. Annual Review of Entomology 60: 415–434.

Knox RB, Heslop-Harrison J. 1969. Cytochemical localization of enzymes in the wall of the pollen grain. Nature 223: 92–94.

Kurtzman CP, Price NPJ, Ray KJ, Kuo TM. 2010. Production of sophorolipid biosurfactants by multiple species of the *Starmerella* (*Candida*) *bombicola* yeast clade. FEMS Microbiology Letters 311: 140–146.

LeBlanc JG, Milani C, de Giori GS, Sesma F, van Sinderen D, Ventura M. 2013. Bacteria as vitamin suppliers to their host: a gut microbiota perspective. Current Opinion in Biotechnology 24: 160–168.

Leonhardt SD, Dworschak K, Eltz T, Blüthgen N. 2007. Foraging loads of stingless bees and utilization of stored nectar for pollen harvesting. Apidologie 38: 125–135.

Loper GM, Standifer LN, Thompson MJ, Gilliam M. 1980. Biochemistry and microbiology of bee–collected almond (*Prunus dulcis*) pollen and bee bread. I– Fatty Acids, Sterols, Vitamins and Minerals. Apidologie 11: 63–73.

Losey JE, Rayor LS, Carter ME. 1999. Transgenic pollen harms monarch larvae. Nature 399: 214.

Menezes C, Vollet-Neto A, Contrera FAL, Venturieri GC, Imperatriz-Fonseca VL. 2013. The role of useful microrganisms to stingless bees and stingless beekeeping. 153–171 pp. In: Vit P, Pedro SRM, Roubik DW, eds. Pot–Honey, a legacy of stingless bees. Springer; New York, USA. 654 pp.

Michener CD. 1974. The social behavior of the bees. Harvard University Press, Cambridge, USA. 404 pp.

Michener CD. 2000. The Bees of the World. 2nd ed. The Johns Hopkins University Press, Baltimore, USA. 913 pp.

Nogueira-Neto P. 1997. Vida e criação de abelhas indígenas sem ferrão. Nogueirapis; São Paulo, Brasil. 445pp.

Pain J, Maugenet J. 1966. Recherches biochimiques et physiologiques sur le pollen emmagasiné par les abeilles. Les Annales de l'Abeille 9: 209–236.

Paludo CR, Ruzzini AC, Silva-Junior EA, Pishchany G, Currie CR, Nascimento FS, Kolter RG, Clardy J, Pupo MT. 2016. Whole–genome sequence of *Bacillus* sp. SDLI1, isolated from the social bee *Scaptotrigona depilis*. Genome Announcements 4: e00174–16.

Penedo MCT, Testa PR, Zucoloto FS. 1976. Valor nutritivo do gevral e do levedo de cerveja em diferentes misturas com pólen para *Scaptotrigona (Scaptotrigona) postica* (Hymenoptera, Apidae). Ciência e Cultura 28: 536–538.

Pires NVCR, Venturieri GC, Contrera FAL. 2009. Elaboração de uma dieta artificial protéica para *Melipona fasciculata*. Documentos/Embrapa Amazônia Oriental; Belém, PA, Brasil. 363 pp.

Potts SG, Imperatriz-Fonseca V, Ngo HT, Aizen MA, Biesmeijer JC, Breeze TD, Dicks LV, Garibaldi LA, Hill R, Settele J, Vanbergen AJ. 2016. Safeguarding pollinators and their values to human well-being. Nature 540:220-229.

Ramalho M, Imperatriz-Fonseca VL, Giannini TC. 1998. Within–colony size variation of foragers and pollen load capacity in the stingless bee *Melipona quadrifasciata anthidioides* Lepeletier (Apidae, Hymenoptera). Apidologie 29: 221–228.

Rebelo KS, Ferreira AG, Carvalho-Zilse GA. 2016. Physicochemical characteristics of pollen collected by Amazonian stingless bees. Ciência Rural 46: 927–932.

Rosa CA, Lachance MA, Silva JOC, Teixeira ACP, Marini MM, Antonini Y, Martins RP. 2003. Yeast communities associated with stingless bees. FEMS Yeast Research 4: 271–275.

Roubik, D.W. 1982. Seasonality in colony food storage, brood production and adult survivorship: studies of *Melipona* in tropical forest (Hymenoptera: Apidae). Journal of the Kansas Entomological Society 55: 789–800.

Roubik D. 1989. Ecology and natural history of tropical bees. Cambridge Univ. Press New York, USA. 514 pp.

Roulston TH, Cane JH, Buchmann SL. 2000. What governs protein content of pollen: Pollinator preferences, pollen–pistil interactions, or phylogeny? Ecological Monographs 70: 617–643.

Silva TMS, Camara CA, Lins ACS, Barbosa-Filho JM, Silva EMS, Freitas BM, Santos FAR. 2006. Chemical composition and free radical scavenging activity of pollen loads from stingless bee *Melipona subnitida* Ducke. Journal of Food Composition and Analysis 19: 507–511.

Sinacori M, Francesca N, Alfonzo A, Cruciata M, Sannino C, Settanni L, Moschetti G. 2014. Cultivable microorganisms associated with honeys of different geographical and botanical origin. Food Microbiology 38: 284–294.

Swain MR, Anandharaj M, Ray RC, Rani RP. 2014. Fermented fruits and vegetables of Asia: A potential source of probiotics. Biotechnology Research International 2014: 250424.

Tamang JP, Watanabe K, Holzapfel WH. 2016. Review: Diversity of microorganisms in global fermented foods and beverages. Frontiers in Microbiology 7: 377.

Van Bogaert INA, Saerens K, De Muynck C, Develter D, Soetaert W, Vandamme EJ. 2007. Microbial production and application of sophorolipids. Applied Microbiology and Biotechnology 76: 23–34.

Vásquez A, Olofsson TC. 2009. The lactic acid bacteria involved in the production of bee pollen and bee bread. Journal of Apicultural Research 48: 189–195.

Veiga JC, Menezes C, Venturieri GC, Contrera FAL. 2013. The bigger, the smaller: relationship between body size and food stores in the stingless bee *Melipona flavolineata*. Apidologie 44: 324–333.

Vollet-Neto A, Maia-Silva C, Menezes C, Imperatriz-Fonseca VL. 2016. Newly emerged workers of the stingless bee *Scaptotrigona* aff. *depilis* prefer stored pollen to fresh pollen. Apidologie *in press*.

Wang M, Zhao W-Z, Xu H, Wang Z-W, He S-Y. 2015. *Bacillus* in the guts of honey bees (*Apis mellifera*; Hymenoptera: Apidae) mediate changes in amylase values. European Journal of Entomology 112: 619–624.

Woolford MK. 1977. Studies on the significance of three *Bacillus* species to the ensiling process. Journal of Applied Bacteriology 43: 447–452.

Yang K, Wu D, Ye X, Liu D, Chen J, Sun P. 2013. Characterization of chemical composition of bee pollen in China. Journal of Agricultural and Food Chemistry 61: 708–718.

Yeast and Bacterial Composition in Pot-Pollen Recovered from Meliponini in Colombia: Prospects for a Promising Biological Resource

Marcela Villegas-Plazas, Judith Figueroa-Ramírez, Carla Portillo, Paola Monserrate, Víctor Tibatá, Oswaldo Andrés Sánchez, and Howard Junca

19.1 Introduction

Pollen is the key source of protein, amino acids, lipids, sterols, vitamins and minerals for bees (Brodschneider and Crailsheim 2010), while the nectar is the primary source of carbohydrates (Hydak 1970). In Colombia, bees collect these floral resources from hundreds if not thousands of plant species. While they appear to exhibit a wide preference, particular species of foraging bees exhibit preferences for pollen from certain plants (Rodríguez et al. 2011). The pollen product used by bees as a nutrient source, known as 'beebread' when it is derived from *Apis mellifera* or the so-called solitary bees or pot-pollen when it is made by bees of the Meliponini tribe

Marcela Villegas-Plazas and Judith Figueroa-Ramírez authors contributed equally to this work.

M. Villegas-Plazas • H. Junca (✉)
RG Microbial Ecology, Division of Ecogenomics & Holobionts, Microbiomas Foundation, Chia, Colombia
e-mail: info@howardjunca.com

J. Figueroa-Ramírez • C. Portillo • P. Monserrate
V. Tibatá • O.A. Sánchez
Research Group AYNI, Bee Science & Technology, Veterinary Microbiology, Faculty of Veterinary Medicine and Zootechnics, Universidad Nacional de Colombia, Bogotá, DC, Colombia

(Vit et al. 2016), experiences lactic acid fermentation and thus is a matured and stabilized pollen product (Gilliam 1979; Gilliam et al. 1989). This product is distinct from the pollen grains that are initially transported to the colony, whereby during fermentation, bees mix the floral pollen with nectar, glandular secretions, honey and bacteria from their own microbiome (Vásquez and Olofsson 2009; Martinson et al. 2011; Powell et al. 2014). This pollen is produced inside the nest in cerumen pots or carefully arranged in layers in the comb cells, a process occurring under controlled conditions by the bee to favour the substrate fermentation. The fermentation process not only provides stability to the product but also improves digestibility and adds nutritional value. It is known that in *Apis mellifera*, a decrease in pH and protein concentration is achieved by the fermentation process, together with an increase in bioavailability of all amino acids, except tryptophan, which demonstrates the potential of this product as a particularly rich source of essential amino acids (Donaldson-Matasci et al. 2013). Nevertheless, the transformation processes of pollen and honey within the nests are not fully understood, but conventional wisdom holds that they largely depend on metabolic transformations involving microbial communities, particularly those members of the intestinal microbiome of worker bees. Thus, microbial composition and associated functions have the potential to affect food production and health of the colonies. Given this, studying the composition of these microbial communities applying classical microbiological techniques and ecological concepts has been of great interest (Lee et al. 2015; Engel et al. 2016).

Studying the bee gut-associated microbiome and those microbial species involved in the natural fermentation of pollen generates valuable information that could be used for biotechnological applications. However, until recently, the dependence of cultivation on microbial species under laboratory conditions was a formidable limitation. Due to the advances in both microbiology and molecular biology methods, there is now a suite of sensitive and appropriate culture-dependent and culture-independent tools to better characterize microbial communities and thus address interactions between bees, pollen and microbiota. To this end, we can now extensively characterize the composition of microbial communities, particularly by culture-independent methods and with more sensitive culture-dependent methods. The aim of this study was to explore the inter-domain microbial communities of the so-called ripe pollen by cataloguing the yeast component using culture-dependent methods and the archaeal and eubacterial components using culture-independent methods, thus identifying those key members involved in the fermentation processes. From this work, along with that of the published literature, we propose some key insights into the production of pollen by Meliponini.

19.2   General Properties of Corbicular Bee-Derived Pollen

Foraging bees use the electrostatic potential of their hair's surface for the retrieval and collection of floral pollen. Similar to the structures and actions involved in grooming, the Apinae manipulate, agglomerate and compact the pollen by adding nectar, until they are able to accommodate all of the pollen load on their corbiculae for transportation to the hive (Thorp 1979). In this process, the native microbial communities of floral pollen are supplemented with bee-associated microorganisms of the surface of the body, tongue, saliva and nectar collected in honey from the hive that may also be regurgitated to moisten and facilitate in carrying the collected pollen. In the beebread of *A. mellifera*, the presence of *Nosema* spores, an intestinal pathogen of bees, has been found and suggests that intestinal microbial species are transferred to the pollen during the process of corbicular packing (Higes et al. 2008).

Bee pollen also has enzymatic activity. While the enzymes such as diastase (alpha- and beta-amylase) and invertase (alpha-glucosidase) vary between taxonomic groups of stingless bees (Vit and Pulcini 1996), the role of bee enzymes in pollen maturation has yet to be elucidated. Nonetheless, enzyme activity like that by fungal exoenzymes has been described to play an important role in pollen transformation by acting on lipids, proteins and carbohydrates (Gilliam et al. 1989).

Through apitherapy, health-associated applications of the corbicular and maturated pollen have been demonstrated in humans (Bogdanov 2004). Such therapeutic action has been attributed to various phenolic compounds that are present in pollen and act as antioxidants. The ethanol extracts of bee pollen contain >10 mg/g of phenolic compounds, with known inhibitory activity to *Bacillus subtilis*, *Pseudomonas aeruginosa* and *Klebsiella* sp. (Carpes 2008). The antimicrobial action of pollen extracts has also been studied on phytopathogenic bacteria such as *Agrobacterium tumefaciens*, *A. vitis*, *Erwinia amylovora*, *Pseudomonas corrugata* and *Xanthomonas campestris* pv. *campestris* (Basim et al. 2006), where its antifungal potential has also been reported against *Aspergillus fumigatus*, *A. niger*, *Candida albicans* and *C. glabrata* (Kacaniova et al. 2012). Furthermore, it has been shown that after pollen transformation into beebread, the antibacterial potential remains similar against bacteria such as *Escherichia coli*, *Salmonella enteritidis*, *Pseudomonas aeruginosa*, *Staphylococcus aureus*, *Streptococcus* and *Bacillus cereus* (Abouda et al. 2011).

To evaluate the possible therapeutic uses of pollen and its products, earlier work within our research group performed in vitro evaluation of antimicrobial potential of ethanolic extracts (EEPo) of *A. mellifera* and *Tetragonisca angustula* pollen (Monserrate 2015). The evaluation and comparison were performed using samples from corbicular pollen of *A. mellifera* from the Eastern 'Cundiboyacense' high Andean plateau region (Colombia) and with hive-stored pollen of *T. angustula* from the Antioquia Department (western Andes, Colombia). The antibacterial and antifungal potential was evaluated by the MIC method (AOAC 2007) with serial concentrations from 32 to 1 mg/mL of total soluble solids (TSS) against *Salmonella enterica* subsp. *enterica* (ATCC 14028), *Escherichia coli* (ATCC 31617), *Staphylococcus aureus* (ATCC 6538) and *Kocuria rhizophila* (ATCC 9341) and against the fungus *Candida albicans* (ATCC 10231). In addition, the antiparasitic and antiviral capacities of the EEPos were evaluated. The antiparasitic potential was determined by adjusting an in vitro model with concentrations of 4 and 8 mg/mL of EEPo TSS on the trophozoites of the intestinal parasite *Giardia* (ATCC-50803 WBC6), while the antiviral effect was performed using *Phi X174* (ATCC 13706-B1) on the host cell *Escherichia coli* (ATCC 700078). Finally, to complement the results, the total concentrations of phenols and the antioxidant activity of some EEPo samples were determined by TEAC-DPPH and correlated with their antimicrobial activity. The results of this study analysed the variability in biological activity of the EEPos evaluated from *A. mellifera* and *T. angustula* against pathogens and showed the effect in relation to the TSS tested and the geographic origin of the samples. The results demonstrated the greatest antibacterial activity with MICs between 32 and 16 mg/mL of TSS, varying according to the type of bacteria; 95% of EEPo from *T. angustula* inhibited all of the bacteria tested, while those from *A. mellifera* showed lower inhibition percentages, reaching 77% (from Cundinamarca samples) and 54% (from Boyacá samples) (Fig. 19.1). The effect on fungi was less significant and was present only in 26% of EEPo at 32 mg/mL, with only one sample

from hive-stored pollen of *T. angustula* displaying antifungal activity. Regarding the antiparasitic potential against *Giardia duodenalis*, the effect was similar at both concentrations evaluated, although a greater action of *A. mellifera* corbicular pollen from Boyacá was observed. Finally, with respect to antiviral activity, and although the results were not significantly different, in a comparison of the average inhibition values against the phage particles (*Phi* X174), a trend of higher inhibition was observed in the EEPo from *A. mellifera* from the Cundinamarca region. The average inhibition values were 32% ± 19.1% for 4 mg/mL and 39.6% ± 24.1% for 8 mg/mL, respectively, for *Apis* and *Tetragonisca*. These results underscore the potential of corbicular pollen and hive-stored pollen as products with value for controlling microbial pathogens (Monserrate 2015).

19.3 The Key Bacterial Assemblages Known to Be Associated with Bees and Pollen

Animals have an intimate association with diverse communities of microorganisms, commonly referred to as the microbiome, and bees are no exception. Bees are colonized by a diverse range of microorganisms including bacteria, archaea, fungi and protists, as well as viruses. The microbiome is known to be important for host health and functionality, despite comprising some opportunistic pathogen symbionts, yet the precise roles of most members of the bee microbiome are poorly understood. Overall, it has been reported that the specialized intestinal microbiome of bees is similar to the mammalian microbiota because it is composed of host-adapted, facultative anaerobic and microaerophilic bacteria (Mohr and Tebbe 2006; Kwong and Moran 2016), but it is much simpler. Nine bacterial species clusters are specific to bees (Engel et al. 2016) and are believed to be transmitted through social interactions between individuals (Powell et al. 2014). The specific bacterial clusters that are bee gut associated include *Bartonella apis*,

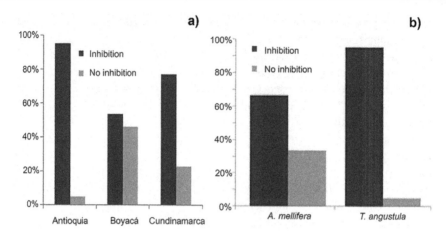

Fig. 19.1 Cumulative antimicrobial (antiviral, antibacterial and antiparasitic) percentages of ethanolic extracts of *A. mellifera* and *T. angustula* pollen collected from three different regions (**a**) and discriminated according to contributing bee source (**b**) (*Source*: Monserrate 2015)

Parasaccharibacter apium, Frischella perrara, Snodgrassella alvi, Gilliamella apicola, Bifidobacterium spp., *Lactobacillus firm-4* and *Lactobacillus firm-5* (Kwong and Moran 2016). To date, there have been multiple studies that explore the microbial communities of bees, mostly concentrating on the gut of honey bees (Ahn et al. 2012) and primarily focusing on *Apis mellifera* (Corby-Harris et al. 2014; Grubbs et al. 2015) and bumblebees (Kwong et al. 2014). While there are some reports characterizing the composition of products associated with bees, such as hive-stored pollen (Anderson et al. 2014) and beebread (Komosinska-Vassev et al. 2015), there are no reports comparing the microbial composition of the communities associated with bees, their pollen, the hive-stored pollen or the beebread.

The composition of microbial communities associated with pollen is assumed to change steadily during the process of transportation and maturation, reflecting a change in physical and chemical properties of the pollen products. The structure of the communities of the pollen products is a composite of the microbiome found in the pollen resources and the interacting bee's microbiome. Early studies using cultivation led to the first knowledge about microbial communities associated with bees and bee products, such as hive-stored pollen and beebread. Based on such methods, *Mucor* sp., which dominates

almond flowers (*Prunus dulcis*), were found not to be detected in the hive-stored pollen or beebread (Gilliam et al. 1989). Such findings may be the result of chemical changes in the pollen due to the process of regurgitation from bees, in which the honey and gland secretions are combined with antimicrobial peptides produced by the bees. During pollen maturation, bees preserve different taxonomic groups and abundances of microorganisms, as has been reported in *A. mellifera*. Many of the microorganisms involved in this process likely originate from the bee's intestinal microbiome. For example, some of the studies performed with worker *A. mellifera* showed that 4–6 days after adult eclosion, the gut is colonized by a core microbiome of Gram-negative bacteria: *Snodgrassella alvi* (Betaproteobacteria, Neisseriales), *Gilliamella apicola* and *Frischella perrara* (Gammaproteobacteria, Orbales). Transmission of these species depends on the presence of nurse or intestinal contents, while some Gram-positive bacteria are transferred mainly by exposure to the hive components as combs, honey and beebread (Powell et al. 2014). Three additional species are also now considered as part of the core microbiome, including phylogenetic lineages based on similarities of 31 marker proteins in metagenomic datasets, called *Alpha-1, Alpha-2.1* and *Alpha-2.2* belonging to class Alphaproteobacteria (Engel et al. 2012),

Bifidobacterium asteroides (Bottacini et al. 2012) and *Firme-4* and *Firme-5*, belonging to the *Firmicutes* phylum (Martinson et al. 2011; Powell et al. 2014). However, those microbial communities associated with beebread and hive-stored pollen have yet to be completely characterized, although it is known that the organoleptic characteristics of the final product are different, depending upon the species of bee (Camargo et al. 1992), which suggests the use of different fermentation pathways and initial substrates.

Considering the great interest in this field, a new consortium named BeeBiome was established in 2016, with an aim to develop an online database supporting applied and fundamental research on bees and their associated microbial communities (Engel et al. 2016). Since then, it has been suggested by culture-dependent techniques that the microbial composition of pollen changes quickly during the process of maturation, being also dependent upon the bees' own microbiome. In the sections below, we are reporting our original results, obtained from specimens collected from production farms in Colombia, about the taxonomical composition by culture-independent techniques, of the bacterial assemblages associated with two different stages of maturation in *A. mellifera*, the characterization of the total microbiome from the species *T. angustula* and two *Melipona* spp. (macerated specimens) and the microbial communities associated with their own hive-stored pollen. The hive-stored pollen of each one of these species was analysed in two different locations with varied organoleptic characteristics. For this study, we use the MiSeq Illumina platform and sequenced the V4 hypervariable region of 16S rRNA gene. All data were analysed using QIIME software (Kuczynski et al. 2011) and phyloseq (R complement; McMurdie and Holmes 2013).

A total of 66,250 sequences were clustered into 432 operational taxonomic units (OTUs) at 97% similarity, which were ascribed to nine described bacterial phyla (Proteobacteria, Firmicutes, Cyanobacteria, Actinobacteria, Bacteroidetes, Acidobacteria, Tenericutes, Fusobacteria, Verrucomicrobia) and one candidate phylum (GN02); no archaeal lineages were found in these communities. Taxonomic composition is shown at the order level, because at this level, the majority of sequences (>93%) could be classified in the hierarchy with >80% confidence in a RDP Bayesian classifier. Figure 19.2a shows the structure and composition of all the communities analysed at this taxonomic level and the clustering obtained through weighted UniFrac, which compares not only the presence of OTUs but also their relative abundance in each of the communities. Good's coverage values were >95% in all samples. Relative frequencies of all genera were inferred from 16S amplicon Illumina reads with an average of 7953 sequences per specimen (this study).

Comparing the similarity of the communities through a distance matrix, two main clusters are revealed, one containing the three bee specimens, together with the hive-stored pollen samples from *Melipona* spp., and the other cluster grouping the hive-stored pollen samples from *T. angustula* and the beebread of *A. mellifera*. The genera heat map in Fig. 19.2b clearly showed that the main differences are probably due to the distinct distribution of the three most prominent orders: Streptophyta and Rickettsiales, which are much more abundant in the pollen samples of *T. angustula* and *A. mellifera*, and the order *Lactobacillales*, which is present in a larger proportion in the samples from bee specimens and the *Melipona* pollen. Comparing the composition of all communities analysed and based on the number of OTUs present in each (Fig. 19.3), the *Melipona* species clearly comprised the greatest OTU richness, followed by *A. mellifera* and *T. angustula*. To this end, by now assessing the shared OTUs between samples, we can elucidate the influence of the bee's microbiome in the hive-stored pollen and the beebread.

To provide an overview of shared bacterial taxonomy, Table 19.1 lists the representative sequence and taxonomy from the OTUs that were present in all samples and then those that were exclusive to either bees or pollen. The two OTUs from the *Streptophyta* genus that are present in all samples are also some of the most abundant, but those that are exclusively associated with bees (all belonging to genus *Bacillus*) are in very low frequency. With respect to the exclusive OTUs from pollen samples, the most abundant is

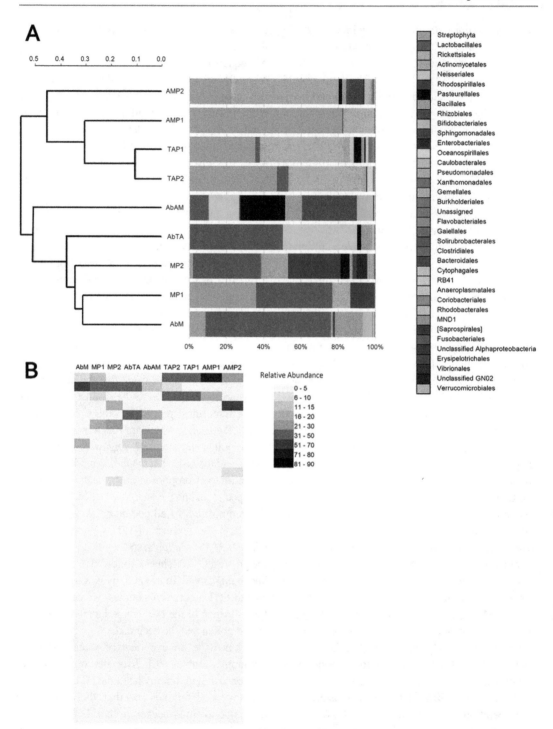

Fig. 19.2 Taxonomic composition of the microbiomes associated with bees and hive-stored pollen from *T. angustula* bee (*AbTA*) and pollen (*TAP1, TAP2*), *Melipona* spp. bee (*AbM*) and pollen (*MP1, MP2*), and *A.* *mellifera* bee (*AbAM*) and pollen (*AMP1, AMP2*). (**a**) At order level and UniFrac-weighted similarity dendrogram. (**b**) Corresponding OTU heat map

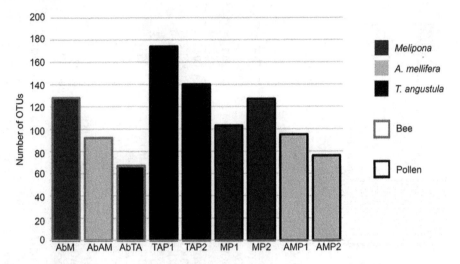

Fig. 19.3 Comparison of OTU number present in the microbial community associated with *Melipona* sp., *A. mellifera* and *T. angustula* pollen

an unclassified bacterium found in high frequency in samples of beebread from *T. angustula*.

A common notion that the bee microbiome affects microbial communities associated with matured pollen was then tested, as we expected to see a significant fraction of shared OTUs between the bees and their respective matured pollen. However, the distribution of shared OTUs in the communities associated with the bee and its derived pollen is clearly distinct (Fig. 19.4). The trend is similar in all three species, with a set of dominant OTUs that are bee associated and are different from the set of dominant OTUs in the pollen, while the set of shared OTUs between the three communities are represented by very low-frequency OTUs. In the case of *A. mellifera*, the set of predominant OTUs of the bee mostly comprised OTUs from the genera *Lactobacillus*, *Bifidobacterium*, *Bacillus* and other unclassified genera in orders Rhizobiales, Pasteurellales, Neisseriales and Rhodospirillales. The most abundant OTUs from the beebread of this species belonged mostly to the Streptophyta order, followed by Rickettsiales, Caulobacterales and the genera *Rhodococcus* and *Sphingomonas*. The set of eight shared OTUs (in low frequency) belonged to

the orders Caulobacterales, Sphingomonadales, Streptophyta, Enterobacteriales, Bacillales and Rickettsiales.

In the case of *Melipona* and hive-stored pollen, the highlighted genera from the set of representative OTUs in the bee's community were *Lactobacillus*, *Lactococcus*, *Bacillus* and some other unclassified species from the orders Lactobacillales, Streptophyta, Bifidobacteriales, Rhodospirillales and Rickettsiales. Those representatives of the hive-stored pollen belonged to the orders Rhodospirillales, Lactobacillales and Rickettsiales and the genera *Rhodococcus* and *Streptococcus*. In contrast to the case of *A. mellifera*, the set of shared OTUs between *Melipona* and hive-stored pollen comprised of 45 OTUs of intermediate relative frequency, mostly belonging to the *Lactobacillus* genus followed by the orders Streptophyta, Rickettsiales, Rhodospirillales and Enterobacteriales.

Lastly, in the case of *T. angustula*, the bee with a less rich community but having pollen with the largest number of associated OTUs, the distribution of shared OTUs between the bee and the hive-stored pollen was similar to that seen in

Table 19.1 Representative sequences and taxonomy from OTUs present in all samples and also exclusively in bees or pollen and beebread samples

OTUID	OTUs present in all samples	Taxonomy
1109768	GACAGAGGATGCAAGCGTTATCCGGAATGATTGGGCGTAAAGCGTCTGTAGGTGGCTTTTCAAGTCCG CCGTCAAATCCCAGGGCTCAACCCTGGACAGGCGGTGAAATCTACCAAGCTGGAGTACGGTAGGGGC AGAGGGAATTTCCGGTGGAGCGGTGAAATGCGTAGAGATCGGAAGAACACCAACGGCGAAAGCACT CTGCTGGGCCGCACTGACACTGACACTGAGAGACGAAAGCTAGGGGAGCGAAT	Bacteria [100%] *Cyanobacteria* [100%] Chloroplast [100%] Chloroplast [100%] *Streptophyta* [100%]
1106368	TACAGAGGATGCAAGCGTTATCCGGAATGATTGGGCGTAAAGCGTCTGTAGGTGGCTTTTCAAGTCCG CCGTCAAATCCCAGGGCTCAACCCTGGACAGGCGGTGAAATCTACCAAGCTGGAGTACGGTAGGGGC AGAGGGAATTTCCGGTGGAGCGGTGAAATGCGTAGAGATCGGAAGAACACCAACGGCGAAAGCACT CTGCTGGGCCGCACTGACACTGAGAGACGAAAGCTAGGGGAGCGAAT	Bacteria [100%] *Cyanobacteria* [100%] Chloroplast [100%] Chloroplast [100%] *Streptophyta* [100%]
1111419	TACGTAGGTGGCAAGCGTTATCCGGAATTATTGGGCGTAAAGCGCGCGTAGGCGGTTTCTTAAGTCTGAT GTGAAAGCCCACGGCTCAACCGTGGAGGGTCATTGGAAACTGGGAAACTTGAGTGCAGAAGAGGAAA GTGGAATTCCATGTGTAGCGGTGAAATGCGCAGAGATATGGAGGAACACCAGTGGCGAAGGCGACTTT CTGGTCTGTAACTGACGCTGAGTGTGCGAAAGCGTGGGGATCAAAC	Bacteria [100%] *Firmicutes* [100%] Bacilli [100%] *Bacillales* [100%] *Staphylococcaceae* [100%] *Staphylococcus* [100%]
	OTUs present only in bee samples	
1111451	TACGTAGGTGGCAAGCGTTGTCCGGAATTATTGGGCGTAAAGGGCTCGCAGGCGGTTTCTTAAGTCTG ATGTGAAAGCCCCCGGCTCAACCGGGGAGGGTCATTGGAAACTTGGGAACTTGAGTGCAGAAGAGG AGAGTGGAATTCCACGTGTAGCGGTGAAATGCGTAGAGATGTGGAGGAACACCAGTGGCGAAGGCG ACTCTCTGGTCTGTAACTGACGCTGAGGAGCGAAAGCGTGGG	Bacteria [100%] *Firmicutes* [100%] Bacilli [100%] *Bacillales* [98%] Bacillaceae 1 [96%] *Bacillus* [96%]
4349296	TACGTAGGTGGCAAGCGTTGTCCGGAATTATTGGGCGTAAAGGGCTCGCAGGCGGTTTCTTAAGTCTG ATGTGAAAGCCCCCGGCTCAACCGGGGAGGGTCATTGGAAACTTGGGAACTTGAGTGCAGAAGAGG AGAGTGGAATTCCACGTGTAGCGGTGAAATGCGTAGAGATGTGGAGGAACACCAGTGGCGAAGGCG ACTCTCTGGTCTGTTACTGACGCTGAGGAGCGAAAGCGTGGGAGCGAAC	Bacteria [100%] *Firmicutes* [100%] Bacilli [100%] *Bacillales* [100%] Bacillaceae 1 [89%] *Bacillus* [88%]
1098655	TACGTAGGTTGCAAGCGTTGTCCGGAATTATTGGGCGTAAAGCGCTCGCAGGCGGTTTCTTAAGTCT GATGTGAAAGCCCCCGGCTCAACCGGGGAGGGGCATTGGAAACTTGGGAAACTTGAGTGCAGAAG AGGAGAGTGGAATTCCACGTGTAGCGGTGAAATGCGTAGAGATGTGGAGGAACACCAGTGGCGAA GGCGGAC	Bacteria [100%] *Firmicutes* [99%] Bacilli [99%] *Bacillales* [90%] Bacillaceae 1 [67%] *Bacillus* [57%]
808534	TACGTAGGTGGCAAGCGTTGTCCGGAATTATTGGGCGTAAAGGGCTCGCAGGCGGTTTCTTAAGTCT GATGTGAAAGCCCCCGGCTCAACCGGGGAGGGTCATTGGAAACTGGGAAACTTGAGTGCAGAAGA GGAGAGGGGAATTCCACGTGTAGCGGTGAAATGCGTAGAGATGTGAGGAACACCAGTGGCGAAGG CGACTCTCTGGTCTGTAACTGACGCTGAGGAGCGAAAGCGTGGGGAGCGAAC	Bacteria [100%] *Firmicutes* [100%] Bacilli [100%] *Bacillales* [98%] Bacillaceae 1 [92%] *Bacillus* [91%]
4399441	TACGTAGGTGGCAAGCGTTGTCCGGAATTATTGGGCGTAAAGGGCTCGCAGGCGGTTTCTTAAGTCT GATGTGAAAGCCCCCGGCTCAACCGGGGAGGGTCATTGGAAACTTGGGAAACTTGAGTGCAGAAGA GGAGAGTGGAATTCCACGTGTAGCGGTGGTGAAATGCGTAGAGATGTGAGGAACACCAGTGGCGAAGG CGACTCTCTGGGCTGTAACTGACGCTGAGGAGCGAAAGCGTGGGGAGCGAAC	Bacteria [100%] *Firmicutes* [100%] Bacilli [100%] *Bacillales* [99%] Bacillaceae 1 [83%] *Bacillus* [76%]

	OTUs present only in pollen or beebread samples	
1627555	GACGGGGGGGGGCAAGTGTTCTTCGGAATGACTGGGCGTAAAGGGCACGTAGGCGGTGAATCGGGTT GAAAGTGAAAGTCGCCAAAAAGTGGCGAATGCTCTCGAAACCAATTCACTTGAGTGAGACAGAGG AGAGTGGAAATTTCGTGTGTAGGGGTGAAATCCGTAGATCTACGAAGGAACGCCAAAAGCGAAGGCAG CTCTCTGGGTCCCTACCGACGCTGGGGTGCGAAAGCATGGGAGCGAACAG	Bacteria [95%] (unclassified bacteria)
1110706	TACGTAGGGTGCAAGCGTTGTCCGGAATTACTGGGCGTAAAGAGTTCGTAGGCGGTTTGTCGCGTCG TTTGTGAAAACCAGCAGCTCAACTGCTGGCTTGCAGGCGATACGGGCAGACTTGAGTACTGCAGGGG AGACTGGAATTCCTGGTGTAGCGGTGAAATGCGCAGATATCAGGAGGAACACCGGTGGCGAAGGCGGG TCTCTGGGCAGTAACTGACGCTGAGGAACGAAAGCGTGGGTAGCGAAC	Bacteria [100%] 'Actinobacteria' [100%] Actinobacteria [100%] Actinobacteridae [100%] Actinomycetales [100%] Corynebacterineae [100%] Nocardiaceae [100%] Rhodococcus [100%]
1111646	TACGTAGGTGGCAAGCGTTATCCGGAATTATTGGGCGTAAAGCGCGCGCAGGTGGTTTCTTAAGTCT GATGTGAAAGCCCACGGCTCAACCGTGGAGGGTCATTGGAAACTGGGAGACTTGAGTGCAGAAGA GGAAAGTGGAATTCCATGTGTAGCGGTGAAATGCGTAGAGATATGGAGGAACACCAGTGGCGAAGG CGACTTTCTGGTCTGTAACTGACACTGAGGCGCGAAAGCGTGGGGAGCAAAC	Bacteria [100%] Firmicutes [100%] Bacilli [99%] Bacillales [99%] Bacillaceae 1 [51%] Bacillus [51%]

Fig. 19.4 Shared and unique OTUs of *A. mellifera*, *Melipona* and *T. angustula* between the bee gut microbiome (left bar pattern representing one composite sample analysed) and the microbial content in their corresponding derived pollen (bar patterns to the right from two independent composite samples)

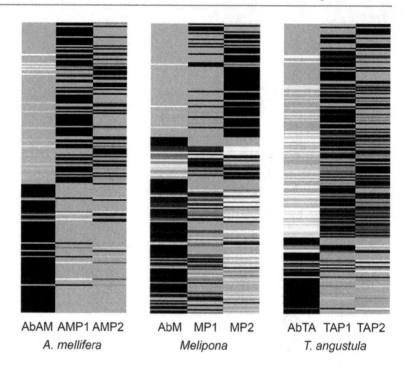

AbAM AMP1 AMP2
A. mellifera

AbM   MP1   MP2
Melipona

AbTA TAP1 TAP2
T. angustula

A. mellifera. The set of abundant OTUs comprising the bee's community were associated with the orders Lactobacillales, Bifidobacteriales, Neisseriales, Pasteurellales and Bacillales; those representatives of pollen were mostly classified as Streptophyta, followed by Lactobacillales and Rickettsiales. The set of shared OTUs comprised those of low frequency from the orders Streptophyta and Rickettsiales and the genera *Lactobacillus*, *Sphingomonas* and *Pediococcus*.

Some studies reveal that bacterial communities found in hive-stored pollen do not differ from those of the newly collected pollen (Anderson et al. 2014). Our results, however, suggest that this is not entirely true, first because the three types of hive-stored pollen analysed here have differing microbial composition and second because the number of OTUs present in each analysed stored pollen varies considerably between species (Fig. 19.3). Nonetheless, it is possible that if we were comparing the microbial communities of flower pollen with our data of hive-stored pollen samples, we would possibly find a more significant set of shared OTUs

than those shared between the bee and the maturated pollen, which would include taxonomic groups such as *Rosenbergiella*, *Pseudomonas*, *Lactococcus*, Flavobacteriaceae, Gammaproteobacteria and Enterobacteriaceae, among others, which have been reported in floral pollen (Ambika Manirajan et al. 2016) and were also found within the bacterial communities associated with hive-stored pollen from *A. mellifera*, *Melipona* and *T. angustula*.

A study following the microbial composition of pollen during the course of comb cell filling shows that the number of bacteria in hive-stored pollen decreases with storage time (Anderson et al. 2014). We found, as a complementary result, that for samples taken at different filling stages, the number of OTUs present in the beebread community decreased (Fig 19.3). The communities are becoming smaller and less rich, indicating that it is a very restrictive and selective medium that only allows a more specialized microbial growth, possibly mediated by competitive or antagonist traits on the microbes selected.

19.4 Yeast Communities Present in Pollen Collected by Colombian Bees

Yeasts are highly important in all processes carried out by bees. Yeast in the floral nectar not only alters its composition but also contributes to the emission of chemotactic volatiles (Herrera 2014), making flowers with yeasts more attractive to pollinators, as demonstrated with *Bombus terrestris* (Herrera et al. 2013). Within hives, yeast also plays important roles. In *T. angustula*, a strong association with *Starmerella meliponinorum* has been detected, and different strains from this species have been isolated from samples of different locations within the bee nest (Teixeira et al. 2003). Subsequent studies have detected other yeasts associated with *T. angustula* from adult bees, honey, pollen grains and propolis, which include *Aureobasidium pullulans*, *Candida etchellsii*, *C. versatilis*, *Cryptococcus albidus*, *Debaryomyces hansenii*, *Issatchenkia scutulata*, *Rhodotorula* spp., *S. meliponinorum* and *Zygosaccharomyces bisporus*. The same study found similar representatives for samples of *Melipona quadrifasciata*, including *A. pullulans*, *Candida apicola*, *C. catenulata*, *C. floricola*, *Cryptococcus laurentii*, *Cr. macerans*, *D. hansenii*, *Kodameae ohmeri*, *Pseudozyma antarctica*, *Rhodotorula* sp., *S. meliponinorum* and *Zygosaccharomyces* sp. From *Frieseomelitta varia*, strains of *A. pullulans*, *C. versatilis*, *Kloeckera* sp. and *S. meliponinorum* have also been isolated. From these species, only *S. meliponinorum* has been found in association with pollen (Teixeira et al. 2003).

Many yeast species associated with beebread are capable of fermenting glucose and sucrose, contributing to organoleptic characteristics of the product through factors of their own metabolism (Gilliam 1979). Studies on *A. mellifera* beebread after different storage periods have detected differences in the associated yeasts. In the beebread with 1-week colony storage, the presence of *Torulopsis magnoliae*, *Cryptococcus flavus*, *Cr.* *laurentii* and *Rhodotorula glutinis* was detected, but after 3–6 weeks in storage, only *Torulopsis magnoliae* and *Cr. albidus* were found. Considering that none of those species were isolated from floral pollen suggests that they were inoculated by the insects (Gilliam 1979).

The importance of yeast and its possible role in stingless bee colonies are associated with the ability to secrete enzymes that convert stored food and increase its nutritional value (Camargo et al. 1992; Menezes 2010). It is still unclear how yeast influences bee nutrition, but the observed changes within the stored pollen are remarkable (Menezes 2010). For example, a few groups of Meliponini, such as *T. angustula*, *Ptilotrigona* and *F. varia*, produce a relatively dry, fermented sweet pollen, while other stingless bees like *Melipona* and *Scaptotrigona* produce a somewhat wet and bitter hive-stored pollen. Such variation in texture and flavour can be associated with the pollen species, and the season, but can be also related to changes that occur in the corbiculae due to the action of different microorganisms that are provided by the bees during the foraging process, which include multiple yeasts. To this end, new species that are associated with the bees are still being described, such as *Zygosaccharomyces siamensis* (Saksinchai et al. 2012), *Z. machadoi* (Rosa et al. 2003) and *Candida* sp. *MUCL 4571*, found to be related to *C. apicola* and possibly having a mutualistic relationship with bees. The role and importance of yeast still require further understanding of these ecosystems.

A large number of yeasts have been isolated from different bee species, and some findings are referenced in Table 19.2. Some fungi also are essential for the survival of bees. In certain bee species, such as *Scaptotrigona* aff. *depilis*, *S. bipunctata* and *S. postica*, a filamentous fungus that grows in the provision of larval food has been detected. The larvae consume this microorganism. When it is eliminated, the survival of bees is drastically reduced (Menezes 2010).

To contribute to the knowledge of key features among the yeast assemblages present in bee

Table 19.2 Yeast isolations from different species of bees

Yeast	Isolate source	Reference
Aureobasidium pullulans	Intestinal microbiome of *Frieseomelitta varia, Tetragonisca angustula, Melipona quadrifasciata, M. rufiventris*	Rosa et al. (2003)
Pseudozyma antarctica		
Starmerella meliponinorum		
S. meliponinorum	Beebreads, honey, propolis and waste from *T. angustula, M. quadrifasciata, M. rufiventris*	Rosa et al. (2003)
Candida etchellsii	Intestinal samples from *T. angustula*	Rosa and Lachance (2005)
Cryptococcus albidus		
Issatchenkia scutulata		
Zygosaccharomyces bisporus	Intestinal and colony debris samples from *T. angustula*	
Candida apicola	Intestinal samples of *Melipona quadrifasciata* and beebread	Daniel et al. (2013)
Kodamaea ohmeri		
Candida catenulata	Intestinal samples of *Melipona quadrifasciata*	
Candida floricola		
Cryptococcus laurentii		
Cryptococcus macerans		
Kodamaea ohmeri		
Starmerella meliponinorum		
Starmerella neotropicalis	Bees and beebread of *Melipona quinquefasciata*	
Candida versatilis	*Frieseomelitta* gut samples	Rosa et al. (2003)
Candida cellae	*Centris tarsata*	Pimentel et al. (2005)
Candida batistae	*Diadasia distincta* and *Ptilothrix plumata*	Pimentel et al. (2005), Douglas and Sigler (2015)
Candida riodocensis	*Megachile* spp.	
Moniliella megachiliensis		
Metschnikowia reukaufii	Digestive tracts and food of *Bombus terrestris, B. pascuorum*	Brysch-Herzberg (2004)
Metschnikowia gruessii		
Metschnikowia pulcherrima		
Candida bombi		
Zygosaccharomyces rouxii		
Debaryomyces marasmus		
Candida kunwiensis	*Bombus terrestris, B. cryptarum, B. hortorum, B. lapidarius, B. pascuorum*	Hong et al. (2003)

pollen, we characterized yeasts involved in the transformation of corbicular pollen to beebread (Portillo Carrascal 2016). To carry out the study, samples of beebread from Africanized *A. mellifera* were obtained from the apiculture experimental greenhouse at Universidad Nacional de Colombia, Campus Bogotá D.C., Cundinamarca region, Colombia (4° 38′ 8″ N, 74° 4′ 58″ W). Samples from the Meliponini *Scaptotrigona*, *Plebeia*, *Paratrigona* and *T. angustula* were collected from meliponaries kindly provided by Asociación Apícola Comunera ASOAPICOM, located in Santander region, Colombia (6° 32′ 0″ N, 73° 12′ 0″ W). The samples from *A. mellifera* correspond to four different cell filling levels: ≤2 mm, 2.1–10 mm, 10.1–15 mm and ≥15.1 mm, the last one representing the filled comb cells. The study was conducted by conventional culture methods with biochemical characterization using Vitek2® system and by molecular techniques, sequencing the 5,8S–ITS region. The primers used were ITS1 (5'TCCGTAGGTGAACCTGCGG3') and ITS4 (5'TCCTCCGCTTATTGATATGC3') (White et al. 1990), and the sequences were classified using GenBank and the BLASTn tool and then validated with RDP classifier.

The Vitek ® system only allowed the identification of 60% of isolates, suggesting that for *A. mellifera*, there were differences in yeast assimilation of compounds, according to entomological origin and cell filling volume. Subsequent insights concerned the kinetics occurring during beebread maturation. The most frequently identified yeast by the Vitek ® system belonged to genus *Candida*. *Candida famata* was present in all stored stingless bee pollen, while *C. pulcherrima* was only detected in *T. angustula*, with *Rhodotorula glutinis* in *A. mellifera* and *C. parapsilosis* in Meliponini. Species such as *Kodamaea ohmeri* were present in *T. angustula*, *Scaptotrigona* sp. and *Paratrigona* sp. Through molecular methods, strains from *Starmerella bombicola*, *Kodamaea ohmeri*, *Metschnikowia pulcherrima*, *C. magnoliae*, *C. apicola* and *C. cellae* were identified from Meliponini samples. The last two were also predominantly isolated from *Plebeia* sp. and *Scaptotrigona* sp., while the other four were isolated only from *T. angustula* and *Paratrigona* sp. The specific results for *A. mellifera* beebread are shown in Table 19.3.

As shown in Fig. 19.5, there was a set of yeast species found only in the collection samples from beebread, although it is similar to *Zygosaccharomyces* spp., and their sequences suggest isolates possibly belonging to yet unidentified species. There is identical similarity to Saccharomycetaceae, and with such a genetic distance, it is possible to propose the isolates belong to a new genus—an observation that requires further detailed experimental and physiological tests on these isolates.

We built a sequence alignment from the ITS sequences obtained from yeasts isolated from *A. mellifera* and native meliponines (*Scaptotrigona*, *Plebeia*, *Paratrigona* and *T. angustula*) and the sequences from type of strains of yeast genera and species related to these sequences using the default parameters from MUSCLE at EBI service (http://www.ebi.ac.uk/Tools/msa/muscle/). The resulting tree from the second iteration, as shown in Fig. 19.5, has one identifier for each yeast isolate ITS beebread. The *A. mellifera* isolates are indicated with red dots, and 'APIS', and also the name of the beebread development, is categorized in four levels of percentage of pollen filling in the comb cells (25, 50, 75 and 100%); for the yeast isolates from beebread of meliponine bees, they have a distinctive green dot and the word 'Nativa' and P, PL, S and TA depending of the genera of the native, autochthonous bee originating from the beebread.

As can be observed, there are highly related yeast isolates in both beebreads from Africanized *Apis mellifera* and native bees, probably belonging to *C. apicola* (clade 1), and also related to Saccharomycetales, clustering near genus *Kodamaea* (*K. ohmeri*-type strain) (clade 4) concentrating yeasts mainly observed in mature (filled 75 and 100%) beebread. Clades 2, 3, 6 and 8 are composed exclusively by strains isolated from native bees; therefore, they could be a particular component of the yeast microbiome that is specifically related to those native bees, either absent or not detected in our current *A. mellifera* beebread survey and probably corresponding to strains closely related to yeast species such as

Table 19.3 Identification of isolations through VITEK and molecular tools (see text)

Alveolar filling volume	Identified yeast	
A. mellifera	Vitek®	Molecular tools
≤2 mm	*Rhodotorula glutinis*	*Kodamaea ohmeri, Candida apicola, Zygosaccharomyces siamensis*
2.1–10 mm	*Rhodotorula glutinis*	*Zygosaccharomyces siamensis*
10.1–15 mm	*Candida pulcherrima, C. famata*	*Metschnikowia pulcherrima, Kodamaea ohmeri, Zygosaccharomyces siamensis*
≥15.1 mm	*Kodamaea ohmeri Candida pulcherrima Rhodotorula glutinis Candida famata*	*Kodamaea ohmeri Metschnikowia pulcherrima Candida apicola*

Fig. 19.5 Phylogenetic relationships inferred from ITS sequences of yeast strains isolated from beebread of *A. mellifera*, Meliponini (*Scaptotrigona*, *Plebeia*, *Paratrigona* and *T. angustula*) and reference yeast species strains

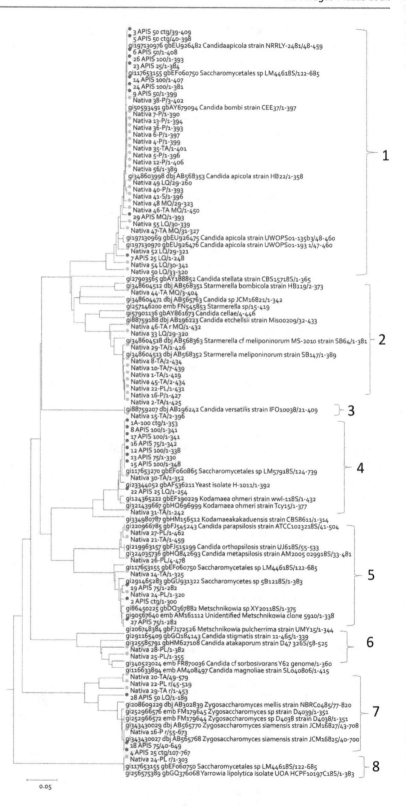

Starmerella meliponinorum (2), *Candida versatilis* (3), *C. atakaporum*, *C. sorbosivorans* or *C. magnoliae* (6) and some probably belonging to the *Yarrowia* genus or yet to be classified Saccharomycetales order (8). The results suggest that, for *A. mellifera*, there is a relationship between the presence of a particular genus of yeast and the beebread filling percentage of the comb cell, such as the presence of *Metschnikowia pulcherrima*, which can be due to the changes in the microenvironment of each filling stage, or the behaviour of the nurse bees, which promote the growth of a particular type of yeast. The data also suggest that there is an association between the genus of bee and yeast species involved in the process of transformation of pollen to beebread. These findings should be corroborated and verified for further efforts aiming to reproduce the process of developing beebread with microbial standardized microbial inoculum for a controlled fermentation of pollen for industrial applications.

19.5   Final Remarks

This chapter discusses the great potential for microbial biodiversity exploration and value of bee-processed pollen, highlighting some similarities and also distinct features resulting from literature search and our study of *A. mellifera* and Meliponini. Finer details, using culturing techniques and state-of-the-art molecular techniques, will come with further research. This work provides evidence that there are still many details to be studied, because it is indeed a rich and valuable natural resource. There may well be a potential for the bee industry, not to mention the importance of bee preservation and wider societal recognition considering pot-pollen, bee-collected pollen and its relevance today.

Acknowledgements We would like to thank Universidad Nacional de Colombia, Departamento Administrativo de Ciencia y Tecnología COLCIENCIAS for the financial support, Bee Research Lab (Laboratorio de abejas) Universidad Nacional LABUN for kindly providing the beebread samples from *Apis mellifera*, the Asociación Apícola Comunera for kindly providing samples of stingless bees and Compañía Campo Colombia SAS for kindly providing us the samples of stingless bees used in this report. We also thank the members of Research Group AYNI 'Microbiología Veterinaria, Facultad de Medicina Veterinaria y Zootecnia, Universidad Nacional de Colombia Sede Bogotá' for the continuous support and excellent technical help. We thank Erika García at Microbiomas Foundation, Colombia, and Nadim Ajami at Baylor College of Medicine, USA, for the excellent support on material preparation and technologies used for culture-independent analyses. We would like to thank the editors, Prof. Dr. Patricia Vit and Dr. David Roubik for their helpful and constructive remarks and the outstanding editorial work, and to Dr. Melissa L. Wos-Oxley for carefully proofreading the text and for her valuable suggestions and corrections which improved the manuscript.

References

Abouda Z, Zerdani I, Kalalou I, Faid MA. 2011. The antibacterial activity of Moroccan Beebread and bee-pollen (dresh and Dried) against Pathogenic Bacteria. Research Journal of Microbiology 64: 376-384.

Ahn JH, Hong IP, Bok JI, Kim BY, Song J, Weon HY. 2012. Pyrosequencing analysis of the bacterial communities in the guts of honey bees *Apis cerana* and *Apis mellifera* in Korea. Journal of Microbiology 50:735–745. doi: 10.1007/s12275-012-2188-0.

Ambika Manirajan B, Ratering S, Rusch V, Schwiertz A, Geissler-Plaum R, Cardinale M, Schnell S. 2016. Bacterial microbiota associated with flower pollen is influenced by pollination type, and shows a high degree of diversity and species-specificity. Environmental Microbiology Sep 9. Epub ahead of print. doi: 10.1111/1462-2920.13524.

Anderson KE, Carroll MJ, Sheehan TI, Mott BM, Maes P, Corby, Harris V. 2014. Hive-stored pollen of honey bees: Many lines of evidence are consistent with pollen preservation, not nutrient conversion. Molecular Ecology 23:5904–5917. doi: 10.1111/mec.12966.

Association of Official Analytical Chemists (AOAC). 2007. Official Methods of Analysis. 18th. Edition. AOAC; Arlington (TX), USA. 1383 pp.

Basim E, Basim H, Özcan M. 2006. Antibacterial activities of Turkish pollen and propolis extracts against plant bacterial pathogens. Journal of Food Engineering 77:992–996. doi: 10.1016/j.jfoodeng.2005.08.027.

Bogdanov S. 2004. Quality standards of bee pollen and beeswax. Apiacta 39:334–341.

Bottacini F, Milani C, Turroni F, Sánchez B, Foroni E, Duranti S, Serafini F, Viappiani A, Strati F, Ferrarini A, Delledonne M. 2012. *Bifidobacterium asteroides* PRL2011 Genome Analysis Reveals Clues for Colonization of the Insect Gut. PLoS One 7:1–14. doi: 10.1371/journal.pone.0044229.

Brodschneider R, Crailsheim K. 2010. Nutrition and health in honey bees - Review article. Apidologie 41: 278–294. doi: 10.1051/apido/2010012.

Brysch-Herzberg M. 2004. Ecology of yeasts in plant-bumblebee mutualism in Central Europe. FEMS Microbiology Ecology 50: 87–100. doi: 10.1016/j.femsec.2004.06.003.

Camargo J, García M, Júnior AC. 1992. Notas previas sobre a bionomia de *Ptilotrigona lurida* (Hymenoptera, Apidae, Meliponinae). Boletim do Museu Paraense Emilio Goeldi, 8: 391-394.

Carpes ST. 2008. Estudo Das Caracteristicas Fisico-Quimicas e Biologicas do Polen Apicola de *Apis mellifera* L. da regiao sul do Brasil. PhD Thesis, Universidade Federal Do Parana; Curitiba, Brasil. 255 pp.

Corby-Harris V, Maes P, Anderson KE. 2014. The bacterial communities associated with honey bee (*Apis mellifera*) foragers. PLoS One. doi: 10.1371/journal.pone.0095056

Daniel HM, Rosa CA, São Thiago-Calaça PS, Antonini Y, Bastos EM, Evrard P, Huret S, Fidalgo-Jiménez A, Lachance MA. 2013. *Starmerella neotropicalis* f. a., sp. nov., a yeast species found in bees and pollen. International Journal of Systematic and Evolutionary Microbiology 63:3896–3903. doi: 10.1099/ijs.0.055897-0

Donaldson-Matasci MC, DeGrandi-Hoffman G, Dornhaus A. 2013. Bigger is better: Honeybee colonies as distributed information-gathering systems. Animal Behavior 85:585–592. doi: 10.1016/j.anbehav.2012.12.020

Douglas G, Sigler L. 2015. *Trichosporonoides megachiliensis*, a new hyphomycete associated with alfalfa leafcutter bees, with notes on Trichosporonoides and Moniliellag. JSTOR 84: 555–570.

Engel P, Kwong WK, McFrederick Q, Anderson KE, Barribeau SM, Chandler JA, Cornman RS, Dainat J, de Miranda JR, Doublet V, Emery O. 2016. The bee microbiome: Impact on bee health and model for evolution and ecology of host-microbe interactions. MBio 7: e02164-15. doi: 10.1128/mBio.02164-15.

Engel P, Martinson VG, Moran N. 2012. Functional diversity within the simple gut microbiota of the honey bee. Proceedings of the National Academy of Sciences of the United States of America 109: 11002–11007. doi: 10.1073/pnas.1202970109.

Gilliam M. 1979. Microbiology of pollen and beebread: the genus *Bacillus*. Apidologie 10: 269–274. doi: 10.1051/apido:19790304

Gilliam M, Prest DB, Lorenz BJ. 1989. Microbiology of pollen and beebread: taxonomy and enzymology of molds. Apidology 20:53–68. doi: 10.1051/apido:19890106

Grubbs KJ, Scott JJ, Budsberg KJ, Read H, Balser TC, Currie CR. 2015. Unique honey bee (*Apis mellifera*) hive component-based communities as detected by a hybrid of phospholipid fatty-acid and fatty-acid methyl ester analyses. PLoS One 10:1–17. doi: 10.1371/journal.pone.0121697

Hydak M. 1970. Honey bee nutrition. Annual Reviews of Entomology. 15: 143-156. doi: 10.1146/annurev.en.15.010170.001043.

Herrera CM. 2014. Population growth of the floricolous yeast *Metschnikowia reukaufii*: effects of nectar host, yeast genotype, and host genotype interaction. FEMS Microbiology Ecology 88: 250–257. doi: 10.1111/1574-6941.12284

Herrera CM, Pozo MI, Medrano M. 2013. Yeasts in nectar of an early-blooming herb: sought by bumble bees, detrimental to plant fecundity. Ecology 94: 273–279. doi: 10.1890/12-0595.1

Higes M, Martín-Hernández R, Garrido-Bailon E, García-Palencia P, Meana A. 2008. Detection of infective *Nosema ceranae* (Microsporidia) spores in corbicular pollen of forager honeybees. Journal of Invertebrate Pathology 97: 76–78. doi: 10.1016/j.jip.2007.06.002.

Hong SG, Bae KS, Herzberg M, Titze A, Lachance MA. 2003. *Candida kunwiensis* sp. nov., a yeast associated with flowers and bumblebees. International Journal of Systematic and Evolutionary Microbiology 53: 367–372. doi: 10.1099/ijs.0.02200-0

Kacaniova M, Vuković N, Chlebo R, Haščík P, Rovna K, Cubon J, Dżugan M, Pasternakiewicz A. 2012. The antimicrobial activity of honey, bee pollen loads and beeswax from Slovakia. Archives of Biological Sciences 64: 927–934. doi: 10.2298/ABS1203927K

Komosinska-Vassev K, Olczyk P, Kaźmierczak J, Mencner L, Olczyk K. 2015. Bee pollen: chemical composition and therapeutic application. Evidence-Based Complementary and Alternative Medicine 2015:297425.

Kuczynski J, Stombaugh J, Walters WA, González A, Caporaso JG, Knight R. 2011. Using QIIME to analyze 16S rRNA gene sequences from microbial communities. Current Protocols in Microbiology. Bioinformatics Chapter 10: Unit 10.7. doi: 10.1002/0471250953.bi1007s36

Kwong WK, Engel P, Koch H, Moran NA. 2014. Genomics and host specialization of honey bee and bumble bee gut symbionts. Proceeding of the National Academy of Sciences of the United States of America 111:11509–11514. doi: 10.1073/pnas.1405838111

Kwong WK, Moran NA. 2016. Gut microbial communities of social bees. Nature Reviews Microbiology 14:374–384. doi: 10.1038/nrmicro.2016.43

Lee FJ, Rusch DB, Stewart FJ, Mattila HR, Newton IL. 2015. Saccharide breakdown and fermentation by the honey bee gut microbiome. Environmental Microbiology 17:796–815. doi: 10.1111/1462-2920.12526

Martinson VG, Danforth BN, Minckley RL, Rueppell O, Tingek S, Moran NA. 2011. A simple and distinctive microbiota associated with honey bees and bumble bees. Molecular Ecology 20:619–628. doi: 10.1111/j.1365-294X.2010.04959.x

McMurdie PJ, Holmes S. 2013. Phyloseq: an R package for reproducible interactive analysis and graphics of microbiome census data. PLoS One 8: e61217. doi: 10.1371/journal.pone.0061217

Menezes C. 2010. A produção de rainhas ea multiplicação de colônias em *Scaptotrigona* aff. *depilis* (Hymenoptera, Apidae, Meliponini). Ph.D. Thesis, University of São Paulo; Ribeirão Preto, Brazil. 97 pp.

Mohr KI, Tebbe CC. 2006. Diversity and phylotype consistency of bacteria in the guts of three bee species (Apoidea) at an oilseed rape field. Environmental Microbiology 8: 258-272.

Monserrate P. 2015. Valoración *in vitro* del potencial antimicrobiano de extractos etanólicos de polen de *Apis mellifera* y de *Tetragonisca angustula*, en busca de posibles usos terapéuticos. Universidad Nacional de Colombia. Master's Thesis, Universidad Nacional de Colombia; Bogotá, Colombia. 83 pp.

Pimentel MR, Antonini Y, Martins RP, Lachance MA, Rosa CA. 2005. *Candida riodocensis* and *Candida cellae*, two new yeast species from the Starmerella clade associated with solitary bees in the Atlantic rain forest of Brazil. FEMS Yeast Research 5: 875–879. doi: 10.1016/j.femsyr.2005.03.006

Portillo Carrascal C. 2016. Identificación de levaduras presentes en el proceso de transformación de polen corbicular a pan de abejas por métodos tradicionales y moleculares. Master's Thesis, Universidad Nacional de Colombia; Bogotá, Colombia. 109 pp.

Powell JE, Martinson VG, Urban-Mead K, Moran NA. 2014. Routes of acquisition of the gut microbiota of the honey bee *Apis mellifera*. Applied Environmental Microbiology 80: 7378–7387. doi: 10.1128/AEM.01861-14.

Rodríguez G, Chamorro A, Obregón F, Montoya D, Ramírez P, Solarte N. 2011. Guía ilustrada de polen y plantas nativas visitadas por abejas. Universidad Nacional de Colombia; Bogotá, Colombia. 230 pp.

Rosa CA, Lachance MA. 2005. *Zygosaccharomyces machadoi sp. n.,* a yeast species isolated from a nest of the stingless bee *Tetragonisca angustula*. Lundiana 6: 27–29.

Rosa CA, Lachance MA, Silva JO, Teixeira AC, Marini MM, Antonini Y, Martins RP. 2003. Yeast communities associated with stingless bees. FEMS Yeast Research 4: 271–275. doi: 10.1016/S1567-1356(03)00173-9

Saksinchai S, Suzuki M, Chantawannakul P, Ohkuma M, Lumyong S. 2012. A novel ascosporogenous yeast species, *Zygosaccharomyces siamensis*, and the sugar tolerant yeasts associated with raw honey collected in Thailand. Fungal Diversity 52: 123–139. doi: 10.1007/s13225-011-0115-z

Teixeira AC, Marini MM, Nicoli JR, Antonini Y, Martins RP, Lachance MA, Rosa CA. 2003. *Starmerella meliponinorum sp. nov.,* a novel ascomycetous yeast species associated with stingless bees. International Journal of Systematic and Evolutionary Microbiology 53: 339–343. doi: 10.1099/ijs.0.02262-0

Thorp RW. 1979. Structural, Behavioral and Physiological Adaptations of Bees (Apoidea) for collecting pollen. JSTOR 66: 788–812.

Vásquez A, Olofsson TC. 2009. The lactic acid bacteria involved in the production of bee pollen and beebread. Journal of Apicultural Research 48: 189–195. doi: 10.3896/IBRA.1.48.3.07

Vit P, Pulcini P. 1996. Diastase and invertase activities in Meliponini and Trigonini honeys from Venezuela. Journal of Apicultural Research 35: 57–62. doi: 10.1080/00218839.1996.11100913

Vit P, Santiago B, Pedro SRM, Ruíz J, Maza F, Peña-Vera M, Pérez-Pérez E. 2016. Chemical and bioactive characterization of pot-pollen produced by *Melipona* and *Scaptotrigona* stingless bees from Paria Grande, Amazonas State, Venezuela. Emirates Journal of Food and Agriculture. 28: 78-84.

White TJ, Bruns T, Lee ST. 1990. Amplification and direct sequencing of fungal ribosomal RNA genes for phylogenetics. pp. 315-322. In: Innis M, Gelfand D, Sninsky J, White T, eds. PCR Protocols: A guide to methods and applications, Academic Press, San Diego, California, USA, 482 pp.

Part III

Stingless Bees in Culture and Traditions

Cultural, Psychological, and Organoleptic Factors Related to the Use of Stingless Bees by Rural Residents of Northern Misiones, Argentina

20

Fernando Zamudio and Norma Ines Hilgert

20.1 Introduction

Ethnobiology can be understood as a discipline that studies the basis of human behavior and evolution in relation to the environment (Albuquerque and Muniz de Medeiros 2013). Hence, understanding why certain resources are used, while others are avoided or ignored, and which factors and variables explain human choices, are central to modern ethnobiology (Wolverton 2013). It is argued that understanding human drivers of exploitation of natural resources may improve our ability to make predictions regarding their more sustainable use and conservation (Stave et al. 2007; Albuquerque et al. 2009). This is important especially when human exploitation affects animals that provide ecosystem services such as pollination, nutrient cycling, or biological control, among others (e.g., bees, beetles, parasitic Hymenoptera). For example, pollination is a critically threatened ecosystem component due to global decreases and local extinctions of natural pollinators in the last decades (see Potts et al. 2010).

Ethnobiology provides scientists and resource managers with information on species biology, socio-ecological processes, and management practices that promote the understanding of the human dimension of natural resource management and collaboration between different forms of knowledge (Berkes 1999; Stave et al. 2007; Anadón et al. 2009; Zamudio and Hilgert 2012a). Another role of ethnoscience is to propose realistic and functional models for natural resource use, management, and policy planning (Albuquerque et al. 2009). In this regard, the evaluation of the "ecological apparency hypothesis" applied to ethnobiological questions allows analyzing the human drivers of natural resource exploitation. This theory proposes that some ecological variables related to a species (abundance, availability, biomass, dominance) or chemical variables (e.g., secondary compounds) may allow a resource to become more culturally important and therefore more widely used than others (Galeano 2000; Albuquerque and de Lucena 2005; Lucena et al. 2012). Interestingly, to date this hypothesis has not been tested on animals in the framework of ethnobiological studies.

Psychological variables have been scarcely considered as driving factors of human decisions on natural resource exploitation (Nolan and Robbins 2001). Animal behavior or morphology may determine how nature is perceived and conceptualized (Boster 1985; Berlin 1992) and

F. Zamudio (✉)
Instituto Multidisciplinario de Biología Vegetal (IMBIV), CONICET, UNC, Córdoba, Argentina
e-mail: zamufer@yahoo.com.ar

N.I. Hilgert
Instituto de Biología Subtropical (IBS), CONICET, Universidad Nacional de Misiones (UNaM). Facultad de Ciencias Forestales (UNaM),
Puerto Iguazú, Misiones, Argentina

© Springer International Publishing AG, part of Springer Nature 2018
P. Vit et al. (eds.), *Pot-Pollen in Stingless Bee Melittology*, DOI 10.1007/978-3-319-61839-5_20

therefore the way it will be used. In the same way as food or in medicinal resources, taste and other organoleptic characteristics can be pivotal in decisions about what to use or not (Almeida et al. 2005). The perception of natural components influences attitudes, tastes, and judgments (i.e., valuation processes), which affect our actions (Jodelet 1986). In that sense, the appraisal theories propose that human valuation processes always precede and elicit emotion; thus, valuations initiate the emotional process and encourage the physiological, expressive, and behavioral changes that define the emotional state (Roseman and Smith 2001). Another factor, scarcely considered in natural resource exploitation, is human ability to recognize the species in their environment. Sampling methods in ecological studies may fail in finding species with avoidance behavior or camouflage strategies (e.g., Karanth et al. 2003). Thus, the concept of "easily observed" coined by Bentley and Rodrıguez (2001) is useful for evaluating degree of apparency in human resource exploitation and to assess biases in detection. Ease of observation is a concept that takes into account the extent to which a given species stands out (perceptual salience) and how common it is (prevalence or commonness).

Stingless bees of the tribe Meliponini are social insects of pantropical distribution that produce honey (Michener 1969). They are important for human communities because of the utilitarian resources obtained from their colonies (e.g., waxy material cerumen), food and medicines (Posey 1983; Cortopassi-Laurino et al. 2006; Zamudio et al. 2010; Kamienkowski and Arenas 2012). In addition, stingless bees are commonly named in ethnobiological classification systems (Berlin 1992) and considered easy to see as they are abundant, have conspicuous behavior, and, on occasion, have very large nests and colonies (Bentley and Rodrıguez 2001). However, by analyzing the internal structure of a stingless bee's domain (i.e., an area of knowledge or activity), it can be noticed that some species might be culturally more important and/or easier to see than others. On one hand, some stingless bees produce honey that varies in organoleptic characteristics and potential uses (Zamudio et al. 2010; Deliza and

Vit 2013). On the other hand, the domain may include morphologically different bees (size, color, shape, etc.), particular colonies, and bees with very different behavior that can influence the ease of observation of each species (Wille 1983). We assumed that the species of stingless bees with the most valued honeys (e.g., in productivity or organoleptic characteristics) and therefore with the greatest cultural importance (see Reyes-García et al. 2006) are the best known, thus the most exploited. We expected the same to occur with the most psychologically prominent bees and/or easily observed bees (i.e., according to degree of apparency).

In this chapter, we propose to evaluate the relationship between cultural (e.g., utilitarian), psychological (e.g., prominence and ease of observation), and organoleptic factors (e.g., honey taste perceptions) related to the use of stingless bees by rural residents of northern Misiones, Argentina. For this, the relationships between eight variables are analyzed in order to explore linkages and identify factors correlated with the use and exploitation of these bees.

20.2 Southernmost Atlantic Forest Ecoregion

This study was conducted in the department of General Manuel Belgrano in the northwest of the province of Misiones (Argentina) in the border between Argentina and Brazil (Fig. 20.1). The area represents the southernmost distribution of the Atlantic Forest, called Upper Paraná Atlantic Forest (Galindo-Leal and Camara 2003). This is a subtropical semi-deciduous forest under a process of habitat loss due to the conversion of forest into agriculture and industrial forestry use (Izquierdo et al. 2008).

The current population of Misiones Province results from the conjunction and coexistence of indigenous population (Mby'a Guaraní), European and Asiatic immigrants who arrived between 1900 and 1940, together with Paraguayan and Brazilian families that fled to the province in the twentieth century (mixed cultural backgrounds). The region is characterized by precariousness in the legal pos-

Fig. 20.1 Map of the study region and study villages

session of the land and poor infrastructure development (Schiavoni 1995).

20.3   Ethnobiological Fieldwork

Fieldwork was carried out in rural villages inhabited by small farmers of mixed cultural backgrounds characterized by a strong Brazilian cultural influence (see details in Zamudio et al. 2010). All farmers interviewed are small producers (5–50 ha) that combine commercial cultivation of tobacco or Yerba mate (*Ilex paraguariensis*) with small agricultural plots for subsistence, supplemented by animal farming.

During the fieldwork, conducted between March 2007 and November 2009, we obtained information about the use, management, and ecological information of different stingless bee species. Open and semi-structured interviews and free listings were conducted (Bernard 2000) with a total of 68 inhabitants, 12 women (average age of 45) and 56 men (average age of 51), under oral informed consent. The selection of informants combined random sampling with the snowball technique, so as to include both experts and novices (defined as those who knew more and less than nine ethnospecies, respectively). Here, the term ethnospecies was used to refer to the organisms listed locally as distinguishable units with proper names regardless of formal academic taxonomic categories (see Zamudio and Hilgert 2015). Different topics were addressed: bees' vernacular names, criteria and descriptors, pot-honey and pot-pollen organoleptic characteristics, local ecological knowledge, and stingless bee use and management.

In addition, throughout the investigation period (5 years), colony location and collection were performed around the rural villages at different times and landscapes, either in the company of the interviewed inhabitants or on our own. Both procedures are equivalent to nonsystematic sampling commonly done by entomologists and taxonomists (e.g., Silveira et al. 2002).

The specimens were properly preserved and were deposited in the entomological collection of the research group in ethnobiology at the Instituto de Biología Subtropical and in the entomological collection of Museo de La Plata, Argentina (MLP). Most of the stingless bees collected were identified by specialists (Dr. Leopoldo Alvarez, Dr. Fernando A. Silveira, Dr. Claus Rasmussen). All the ethnotaxa mentioned by interviewees were collected except "mandacaia" (*Melipona quadrifasciata*). Collected species with no specific local names were not included in the analyses (e.g., *Nannotrigona testaceicornis*, among others).

20.4 Cultural, Psychological, and Organoleptic Factors

We performed a correlation between eight explanatory variables based mainly on information about bees provided by local people and partially published before (Zamudio and Hilgert 2012b, 2015). These variables can be arranged in cultural, psychological, and organoleptic factors. We did not use field ecological information about bees as generated with systematic sampling methods. We have detailed information about the species natural history through previously mentioned collections and observation in the field (Zamudio and Álvarez 2016). For conceptual description of cultural, psychological, and organoleptic factors, see Table 20.1. Following that, we explain the methodological foundations for the different explanatory variables and the quantitative analysis developed.

Table 20.1 Descriptions of cultural, psychological, and organoleptic factors

Analyzed factors	Descriptions
Cultural factors	Culture can be defined as socially shared beliefs, values, norms, expectations, and practices within a group, community, or society at large (Matsumoto 2000). Cultural factors can influence decision-making in relation to the use of human resources. Knowledge and use have been used as measures of the cultural significance of a particular resource (but not the only one; see Reyes-García et al. 2006)
Psychological factors	Refer to thoughts, feelings, and other cognitive characteristics that affect the attitude, behavior, and functions of the human mind. Some elements within a given knowledge domain are more psychologically prominent than others. These stand out often by implicit features not easily observable to the naked eye (see Rosch 1978). The "ease of observation" concept (Bentley and Rodrıguez 2001) takes into account the extent to which a given species stands out (perceptual salience), how common it is (prevalence or commonness), and it is proposed as a measure of the relevance of a resource (those which are named and used)
Organoleptic factors	Productivity and organoleptic properties of honey associated with stingless bees are molders of human behavior as they exert influence on which particular species or product to use and when (see Almeida et al. 2005). The process of evaluation precedes and elicits emotion so that valuations start the emotional process which encourage the physiological, expressive, and behavioral changes that define the emotional state (Roseman and Smith 2001). In this process, the underlying biological and psychological factors as well as the surrounding social and cultural context affect our preferences differently (Vabø and Hansen 2014)

20.4.1 Cultural Factors

Three explanatory variables were developed: the first is based on individual knowledge about ethnospecies, second is based on the perceived abundance (i.e., according to local expert opinions) of each ethnospecies, and third is based on the use of bee products by local people for a given ethnospecies. Local knowledge (LK) can be measured considering the number of elements or components of a domain of knowledge (e.g., bees, medicinal plants; see Reyes-García et al. 2006). In this article, we calculate a measure of global knowledge on stingless bees, according to the number of times a species was mentioned by all the interviewed participants.

Ethnobiological estimated abundance (EEA) was obtained through interviews in which each informant was requested ($N = 14$) to mention ethnospecies known to them in decreasing order, according to their own perception of abundance (1, most abundant). The value of EEA for each ethnospecies was obtained by averaging the number allocated by all respondents to each ethnospecies so that individual variations were contemplated.

Cultural value (CV) was estimated according to Reyes-García et al. (2006) using the following equation: $CVi = Uci x Ici \times \Sigma IUci$, where Uc is $_i$ $=$ number of uses/total uses, Ic_i is the number of useful mentions/number of participants, and ΣIUc_i is the number of uses 1+ use 2 + use 3/ number of participants. Uc_i expresses the versatility of the use of the species and is calculated as the total number of reported uses by ethnospecies i divided by the four possible uses considered in this study (medicinal, food, utility, and others); Ic_i expresses the consensus of the use of the species and is calculated as the number of participants who listed the ethnospecies i as useful divided by the total number of people participating in the study; IUc_i expresses the consensus of particular uses of the species and is calculated as the number of participants who mentioned each use of the ethnospecies i divided by the total number of participants.

20.4.2 Psychological Factors

Three quantitative explanatory variables were developed: the first is based on ethnospecies cognitive prominence, second is based on ecological characteristics of the ethnospecies reported by interviewees, the third is based on the sum of the mentioned characteristics together with the EEA value detailed before in cultural factors.

The cognitive salience index (CSI) was estimated according to Sutrop (2001). This index shows the psychological significance of a series of items (ethnospecies) in a free listing task combining the frequency and mean position of a term into one parameter. The principal assumptions of free listing are that the terms mentioned first tend to be locally more prominent (Bernard 2000). The CSI estimate has values between 0 and 1 (less to more prominent). It is calculated by the formula $S = F / (Nmp)$, where F is the frequency with which a term is named in the listing task, Nmp considers the weight of the mean position (mp) in which each ethnospecies is named, and N is the number of interviewees.

The ethnoecological salience index (ESI) measured a biological prominence of the ethnospecies and was calculated using the following formula: $ESI_i = [(Size_i \times Beha_i) + PC]$, where $Size$ is the tabulated size of ethnospecies (measuring average of five individuals), $Beha$ refers to the behavior of bees, and PC to characteristics of the colonies resulting from the combination of details about the nest entrance "piquera" and the location of the colonies (Fig. 20.2). Values were generated through Likert scales (Bernard 2000) taking into account the attributes that could cause some ethnospecies to be more easily observed than others. The following assumptions were considered:

(a) Larger bees are easier to see than smaller ones (1 = <3 mm, 2 = 3–5 mm, 3 = 5.1–10 mm).
(b) "Docile" bees (those that do not evade or attack humans) are more visible than "timid" ones (those that develop avoidance behavior toward humans), while "aggressive" bees (those that attack in large amounts, biting

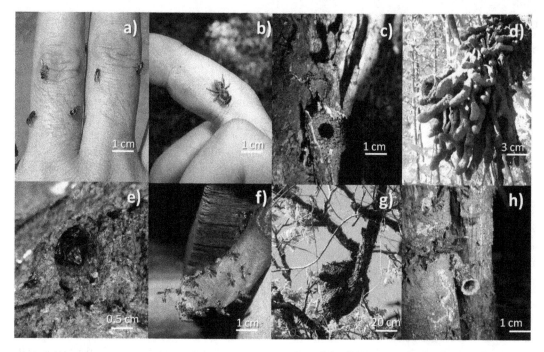

Fig. 20.2 Four most striking features of stingless bees used to develop salience and ease of observation explanatory variables (ESI and IEO). Size: small "mirĩ" bee (*Plebeia* spp.) (**a**) and big "mandurĩ" bee (*Melipona torrida*) (**b**). Entrance: colony of "mandurĩ" without tubular nest entrance (**c**) and colony of "iratĩn" (*Lestrimelitta rufipes*) with large entrance (**d**). Behavior: "mambuca" timid bee (*Cephalotrigona capitata*) (**e**) and "yateĩ" docile bee (*Tetragonisca fiebrigi*) (**f**). Colony location: external "carabozá" colony (*Trigona spinipes*) (**g**) and internal "mirĩ" colony (*Plebeia droryana*) within of palm trunk (**h**)

human hair and clothing) are more visible than docile and timid bees (1, timid; 2, docile; 3, aggressive).

(c) Ethnospecies with large nest entrances are easier to see than ethnospecies with small nest entrances and also easier to see than those without tubular nest entrances.

(d) External bee colonies (out of tree holes and of large size) are easier to see than internal colonies within cavities (tree holes or wall holes).

A new categorization (PC) was created based on assumptions (c) and (d). This categorization is composed of four levels that ranged from lower to higher detectability: 1, hollow tree stingless bee nests without tubular nest entrance; 2, hollow tree stingless bee nests with medium size entrance; 3, hollow tree stingless bee nests with large size entrance; and 4, external colony. Thus, the higher the value of ESI, the more striking or showy the ethnospecies will be and therefore the more readily detectable.

Finally the Index of Ease of Observation (IEO) (see Table 20.1) was calculated using the same formula used to calculate the ESI, complemented by the ethnobiological estimated abundance (EEA) value: IEO = ($Size_i$ × $Beha_i$) + PC + EEA. The higher the IEO value of an ethnospecies, the easier to see.

20.4.3 Organoleptic Factors

Honey productivity and per se organoleptic factors grouped as organoleptic factors were considered only for the purpose of analysis, although we know that productivity should not be included within organoleptic characteristics.

The range of honey production used in this study was estimated from literature sources (Holmberg 1887; Venturieri and Imperatriz-

Fonseca 2000; Rosso et al. 2001; Carvalho et al. 2003; Venturieri et al. 2003; Chagas and Carvalho 2005; Cortopassi-Laurino et al. 2006) and interviews conducted in northern Misiones. Since productivity data referenced in Holmberg (1887) were written based on old measurement units, the following equivalencies were used: 1 "demijohn" = 16 "fourths" and 1 "fourth" = 0.59 L. The productivity of each species was averaged according to the number of independent references in the literature, named productivity in bibliography (PB). In turn, this PB was averaged with the productivity by locals (PL, $N = 33$) to obtain a honey average productivity (HAP). The PL was obtained using two types of data, quantitative values expressed in metric units (kg., L, etc.) and qualitative information. The latter were classified into four categories to which a range of productivity based on the maximum and minimum productivity of the taxonomic genus of Meliponini (Table 20.2).

Finally, to characterize the organoleptic properties of the honey, we performed a qualitative analysis based on the perception on their taste (e.g., strong flavor) and ratings given to their honey. The information comes from sensory memory about the honey, so not all honey received a rating from all respondents. Conventional honey assays were not made.

Honey was classified as very good taste, good taste, intermediate fair taste and poor taste, based on the number of positive and negative ratings (e.g., "good tastes", "bad tastes", "is very rich", etc.). Very good taste and good taste are those not receiving any negative assessments. Very good taste are those receiving more than five positive ratings and good taste those which received less than five positive ratings. Intermediate fair taste are those with negative ratings not outweighing positive ratings, and poor taste are those with negative ratings outweighing positive ones. Next, honey was classified in a honey taste ranking (HTR) using a scale that ranged from 4 to 1 as follows: very good taste (<5 mentions), 4; good taste (>5 mentions), 3; intermediate fair taste, 2; and poor taste, 1.

20.5 Context of Exploitation of Stingless Bees

The rural inhabitants of northern Misiones obtain honey from feral colonies of European-derived honey bees (*Apis mellifera*) and from 17 species of stingless bees (Table 20.3). Harvesting of wild colonies is performed occasionally during rural activities, while intentional search for commercial purposes was not observed. Colony host trees are cut for honey extraction or perforated (machete, axe, or chainsaw) depending on nest accessibility.

In 63% of the visited households (N = 68), there was at least one colony of stingless bees, which had been transferred to housing. Colonies were found in different breeding substrates: "toras" (trunk section containing a colony), rustic wooden hive boxes without internal divisions, or dried gourds of *Lagenaria siceraria* (Molina Standl. syn. *L. vulgaris*). Breeding boxes with

Table 20.2 Qualitative and quantitative values used to transform productivity categories of stingless bees

Qualitative categories	Local expressions	Range of productivity (kg)	Quantitative productivity (kg)
A lot	"Large pots [honey]"/"is one that produces more honey"/"produce more than "yateí""	3.1–5.0	4.0
Medium	"Pots larger than "yateí"/"produce enough honey"	1.1–3.0	2.0
Little bit	"Makes little honey"/"doesn't produce much"/"has small honey pots"	0.31–1.0	0.65
Very little	"Produces almost nothing"/"produce very little honey"/"has small pots"	0.1–0.3	0.20

Table 20.3 Cultural (LK, CV, and EEA), psychological (CSI, ESI, and IEO), and productivity (HAP) explanatory variables related to stingless bees species of Misiones

Scientific name	Common name	LK	CV	CSI	HAP (kg)	EEA	ESI	IEO	Expressions about bees
Cephalotrigona capitata	"Mambuca"	13	0.02	0.04	6.8	3.5 (4)	6	14	"timid," "shy"
Lestrimelitta rufipes and *L.* aff. *limao*	"Iratín"	33	0.21	0.08	1.8	4 (1)	4	12	"wonderful," "beautiful," "accursed"
Melipona bicolor schencki	"Guaraipo"	23	0.06	0.09	4	4.6 (5)	4	9	"lively," "astute," "deceptive," "lazy," "unfriendly," "bashful"
Melipona quadrifasciata	"Mandasaia"	28	0.14	0.09	3	N/A	5	15	"astute," "funny"
Melipona torrida	"Mandurí"	23	0.03	0.07	2.7	4 (2)		16	"distrusful," "unfriendly," "gingerly,"
Oxytrigona tataira	"Cagafuego"	4	0.001	0.01	3.6	4 (1)	3	6	"shits a water that burns"
Plebeia droryana, P. emerinoides, P. remota, and *P. guazurary*	"Mirí"[a]	46	0.26	0.17	0.4	2.25 (6)	7	17	"dirty," "idle," "lazybones," "annoying"
Scaptotrigona depilis and *S. bipunctata*	"Tobuna"	24	0.03	0.08	3.3	3 (2)	7	21	"bad," "annoying"
Schwarziana quadripunctata	"Abeja del suelo"	36	0.07	0.09	2.7	3 (1)	4	4	"timid," "frisky"
Tetragona clavipes	"Borá"	47	0.28	0.22	2.8	2.8 (9)	3	5	"annoying," "shitty," "sticky"
Tetragonisca fiebrigi	"Yateí"	68	2.22	0.94	1.3	1.36 (11)	8	28	"interesting," "intelligent," "king of the forest"
Trigona spinipes	"Carabozá"	59	0.45	0.21	3	2.6 (9)	4	7	"despicable," "bad," "annoying," "disgusting"

LK local knowledge, *CV* cultural value, *CSI* cognitive salience index, *HAP* honey average productivity, *EEA* ethnobiological estimated abundance ($N = 14$), *ESI* ethnoecological salience index, *IEO* index of ease of observation
[a]The people recognized between 1 and 3 "types" of "mirí"; large, medium, and small (see Alvarez et al. 2016 for taxonomy of *Plebeia* of Argentina)

some degree of modernization (e.g., with internal division) were rarely observed. These colonies were usually located in the house veranda or near the house's main door (Fig. 20.3). "Yateí" (*Tetragonisca fiebrigi*) was the most commonly kept ethnospecies, while colonies of "mambuca" (*Cephalotrigona capitata*), "tobuna" (*Scaptotrigona* spp.), and "mirí" (*Plebeia* spp.) were also observed, although only once for each species. The reasons for the interviewees to move "yateí" wild colonies to their houses were varied. Some did so because they consider "yateí" aesthetically beautiful, others to prevent the colony from dying after the tree was cut (e.g., during clearing for agriculture) and others because they considered their honey as a good medicine

Fig. 20.3 Breeding of stingless bees: (**a**) bee colonies in "toras," a trunk section containing a colony, (**b**) rustic wooden breeding box located in a lemon tree in a garden, (**c**) bee colony dried gourds (*Lagenaria siceraria*) located in the house veranda, (**d**) breeding gourds located in the house veranda

(Zamudio et al. 2010; Zamudio and Hilgert 2011; Kujawska et al. 2012). Commercial breeding, like in other Latin-American locations, was not observed (see Cortopassi-Laurino et al. 2006). We do not document formal training activities related to meliponiculture. This is reflected in the low number of species managed in relation to the diversity of present bees, the low number of colonies observed per housing, the predominant use of rustic boxes, and the noncommercial purposes of the breeding of stingless bees.

Stories told by interviewees referred to a more intensive and frequent use of honey in the past, when the colonization of the region was beginning, ~1950. At that time, the greater use of stingless bee honey was due to two facts: families were in greater geographic isolation, with restricted access to markets and products, and there was a greater abundance of bees given the larger area of forest available. Villagers with longer residence in the area (50–60 years) and those identified by their peers as pioneers noted that at that time the forest was like a market. They obtained all they needed from it: bush meat, medicines, sweeteners, and other material resources. Such a past scenario contrasts with the present, with better access to local markets and health centers.

20.6 Cultural and Psychological Factors Related to the Use of Stingless Bees

The correlations between cultural and psychological explanatory variables of stingless bees indicate that bees perceived as more abundant (EEA) are the most frequently cited (LK) and therefore the best known by rural people (Table 20.4). The negative correlation is given because the EEA is an inverse ranking, and higher abundance values indicate lesser abundance. In turn, the significant correlation between CV and CSI (Table 20.4) shows that there is an overlap between the most used ethnospecies and the most salient ones, according to other cases in which this relationship was evaluated (Lozano et al. 2014).

The stingless bees perceived as more abundant (EEA) are the most culturally important (CV) and the most cognitively prominent (CSI) (Table 20.4). While these results would agree with the apparency hypothesis, in which some ecological variables related to species appearance may permit a resource to become more culturally important and more widely used than others, this case should not be interpreted as full evidence of the hypothesis. This is due to the fact that relationship between stingless bee-perceived abundance and both cultural importance and cognitive prominence is lost when "yateí," the most culturally important ethnospecies and the one perceived as most abundant, is removed from the analysis (an outlier value in relation to other bees). These results suggest two emerging themes: (a) if the abundance had been a molder of the cultural importance and knowledge of bees (as postulated in the apparency hypothesis), the relationship between the variables would not depend on a single species ("yateí") as was observed, and (b) ecological factors could not be unique molders of the relationship between cultural importance and species. That is, "yateí" abundance could be shaped by management

Table 20.4 Correlations between cultural and psychological factors related to stingless bees

	LK	CV	CSI	HAP	EEA	ESI	IEO	HTR
LK	1	1.40E−03	2.3×10^{-03}*	0.03*	0.01*	0.47	0.81	0.65
CV	0.96	1	2.1×10^{-03}*	0.03*	0.03*	0.52	0.76	0.53
CSI	0.92*	0.93*	1	0.11	0.03*	0.5	0.84	0.85
HAP	−0.67*	−0.64*	−0.48*	1	0.09	0.47	0.36	0.75
EEA	−0.8*	−0.7*	−0.71*	0.53	1	0.19	0.42	0.39
ESI	0.23	0.2	0.21	−0.23	−0.44	1	4.0×10^{-03}**	0.58
IEO	0.07	0.09	0.06	−0.28	−0.26	0.91**	1	0.51
HTR	−0.14	−0.19	−0.06	0.1	0.26	0.17	0.21	0.38

Probability values (p) are shown above the diagonal and correlation values (r) below. All variables are measured for each ethnospecies of bees

LK local knowledge, **CV** cultural value, **CSI** cognitive salience index, **HAP** honey average productivity, **EEA** ethnobiological estimated abundance ($N = 14$), **ESI** ethnoecological salience index, **IEO** index of ease of observation, **HTR** honey taste ranking

*Statistically significant correlations in shades of gray

**Statistically significant correlations between indexes associated without interpretative value (see Methods)

practices not considered in this analysis. The influence of human management and artificial selection on the morphology and structure of populations is widely known (e.g., Parra et al. 2010). The breeding of "yateí" promotes the generation of new colonies through the release of new queens or swarming bees (Michener 1969; Wille 1983; Roubik 2006); in this way, it could promote the establishment of new colonies and the increase in the number of nests in a given area. This is particularly true for ethnospecies as "yateí" that are very adaptable in their requirements for nesting and survival (Batista et al. 2003). Although it is not possible to establish whether its greater abundance promoted cultural importance or vice versa, "yateí" has an outstanding role in the local culture. It is the most used and known stingless bee, the most commonly kept in hives, and its honey has the widest medicinal uses and consequently is the most culturally valued (Zamudio et al. 2010; Zamudio and Hilgert 2011). Its breeding in domestic settings is closely associated with affective and symbolic factors while its colonies occupy important places in households (e.g., near the main door, galleries), and it is argued that it is raised because of its "beauty" or because it is "cute to look at" among other aesthetic appreciation, loaded with symbolism (Zamudio et al. 2010).

The results suggest no relationship between explanatory variables of psychological basis (ESI and IEO, respectively) and cultural importance and knowledge of the Meliponini. This shows that the easiest to find bees are not necessarily the most important, contrary to Bentley and Rodríguez (2001), who noted that cultural importance and ease of observation influence each other. Detectability varies markedly between animal species according to their behavior (Williams et al. 2002; Karanth et al. 2003) and may condition prey availability in consumptive activities such as wildlife hunting (e.g., Escamilla et al. 2008) and also in honey exploitation. Here, in an attempt to evaluate the ease of observation of each ethnospecies, we combined four of the most striking features of stingless bees (bee size, nest entrance, colony location, and behavior) in order to obtain salience and ease of observation gradi-

ents (Table 20.3). According to these explanatory variables, "carabozá," "iratín," and "tobuna" typified the showiest ethnospecies. They are aggressive ("tobuna" and "carabozá") and have striking nest entrances ("tobuna" and "iratín"), but only "carabozá" has a large external nest (~ 30 cm diameter) similar to a termite mound. Interestingly, the least showy bees are of the largest in size ("mambuca," "guaraipo," "mandurí," and "mandacaia"). These ethnospecies share a marked evasive behavior and have cryptic nests without tubular nest entrances. Meanwhile, the ethnospecies that reached intermediate values in both explanatory variables are those with medium-sized entrances and docile behavior as "yateí" and "mirí."

The lack of correlation between some explanatory variables suggests that the importance of ethnospecies is molded neither by perceived abundance nor by their striking character (ease of observation). However, the relative importance of the behavior of stingless bees has not been thoroughly assessed hereby. Indeed, the avoidance behavior of "timid bees" antagonistically influences their importance and use because it reduces detection and the possibility of using their honey, which is considered of very good taste (see below). Interestingly, this behavior is used as a mirror of local social conduct, to the point that the nickname "guaraipo" (*Melipona bicolor schencki*) is also used to refer to distrustful, suspicious, or lazy people.

20.7 Relationship Between Pot-Honey, Pot-Pollen, and Cultural, Psychological Factors

Counterintuitively, the least productive stingless bees (HAP) are the most known and culturally important (Table 20.3), which suggests that the productivity variable could not be a factor driving the use of these resources in the study area. In a similar way, honey taste ranking (HTR) did not correlate with the analyzed cultural and psychological variables (Table 20.4). That is, the bees with positively valued honey are not the culturally most important and therefore the most used by rural villagers of Misiones. The local perceptions of honey might better allow interpreting these seemingly counterintuitive results.

According to assessments of flavor, the honey of "abeja del suelo" (*Schwarziana quadripunctata*) and "mandacaia" (*Melipona quinquefasciata*) were classified as very good taste, having the highest number of positive ratings and appreciated flavor (Fig. 20.4). However, "yateí (*T. fiebrigi*) honey excels in this category as it received superlatives such as "It is the best honey found" or "It is special," among others. The remaining honey that received only positive reports were considered sweet and good to eat but classified as less valuable than the outstanding honey of the "abeja del suelo" or "yateí." The honey of "iratín" (*Lestrimelitta* spp.) and "borá" (*Tetragona clavipes*) was classified as good (e.g., "["iratín" honey] has a foul taste") as they received an equal number of positive and negative ratings. "Borá" honey is considered "strong" and is characterized by flavors such as sour, or sweet and sour, bitter, and spicy. In the descriptions of "iratín" honey, references to its mentholated flavor and "spicy" character predominated.

Among the poor honeys, those of "mirí" (not all species of genus *Plebeia*) and "carabozá" (*T. spinipes*) were identified. These bees have unhygienic habits; they settle on urine, feces, and on dead animals, and "mirí" also has the habit of "sucking" people's sweat. People usually suggest that these bees produce honey "out of crap" or "anything that is sujiera" [derived from "sujeira," which means dirt in Portuguese]. "The honey is characterized by a bitter, sour taste and, in the case of "carabozá," is recognized as honey of great intensity.

Although "carabozá" and "iratín" honey are in different groups according to our classification, they are not perceived as different by the local people. According to interviewees, both are thick because they "are mixed" with pot-pollen in the nest. Moreover, both ethnospecies have honey of different flavors and colors in the same nest. For example, people say that "carabozá" has "stocked honey" or "seven kinds of different honeys" including red honey "like blood" and white honey "like milk," among others.

Pot-pollen is a resource consumed among people of mixed cultural background. In Misiones, only at the moment of harvesting honey, it is as if it were candy. Unlike indigenous communities in Argentina like the Tobas (Arenas 2003) who appreciate and use pot-pollen in different ways, in Misiones the interviewees consume only fresh pot-pollen and never store or process it. Kamienkowski and Arenas (2012) review the activity of "meleo" (honey harvest) in Chaco indigenous communities, where they specify that pot-pollen is a widely exploited resource. The Toba-Pilagá and Wichi-lhuskutás consume pot-pollen when it is "nuevo" (fresh) and discard it when it is brownish or black. The Wichi-lhuskutás also prepare a drink called "aloja" with water and pot-pollen, leaving it to ferment for 3 days (Arenas 2003).

It is possible to state that the importance or salience of stingless bees was not necessarily associated with positive honey evaluations; conversely some ethnospecies were known and perceptually salient due to an underlying negative property. This is the case of "carabozá" and "mirí," two abundant, easy to see, and well-known ethnospecies that are considered dangerous and annoying, respectively, and whose honey is unpalatable. The negative characteristics are symbolically transferred to the ethnospecies and their honey from observing their unhygienic habits. Such facts are analogous to the transfer that occurs in the "circulation symptoms" proposed for the anthropology of disease (Laplantine 1999), i.e., a symbolic property that is transferred, in this case, to the honey.

20.8 Conclusions and Future Challenges

The results of this research reveal, contrary to expectations, that the use of stingless bees is not associated with their appearance or the characteristics of their honey (productivity and a positive evaluation of the honey). Other variables not considered in our analysis, such as historical or targeted breeding on "yateí" could have an impact on the bee's abundance and also generate positive feedback effects on its cultural value.

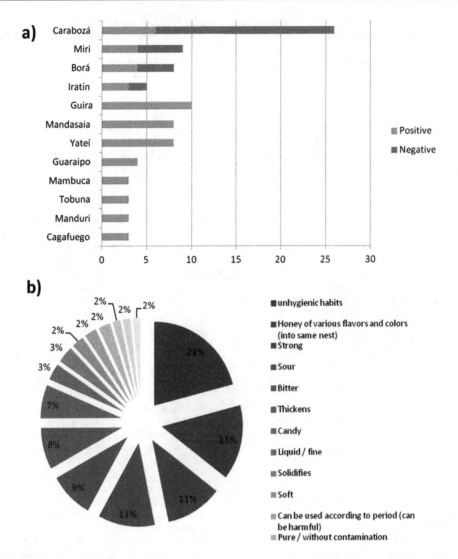

Fig. 20.4 Local perceptions on pot-honeys. (a) Number of positive and negative ratings about stingless bees' honey, (b) organoleptic properties of the honey, and other social components involved in their local definitions (e.g., hygienic or unhygienic habit of bees)

According to some emerging results, the process of evaluation of ethnospecies could, in turn, be influenced by bee aesthetic and behavioral variables, such as their beauty, unhygienic habits, or evasive behavior. These findings agree with those reported by Nolan and Robbins (2001) and Costa-Neto (2006) who show that the wide range of behavior and habits displayed by Hymenoptera has given rise to a wide variety of local representations, which are in turn embedded in the people's social and material life. The proposed analysis allowed identifying scarcely considered aspects in the assessment of natural resource exploitation and management—cultural and psychological factors. Characterizing the scale and intensity of human drivers of resource exploitation is relevant for conservation and management of stingless bees. The combined use of local and scientific knowledge, to solve complex socio-ecological problems, is promising but requires more effort. The study of which circumstances, under which methodology, and to what scale local ecological knowledge is useful for predicting exploitation of natural resources is a future necessity.

Acknowledgments The authors would like to thank the rural inhabitants of Misiones for sharing their knowledge and great cordiality; P. Tubaro, F. Silveira, C. Rasmussen, and L. Álvarez, for their contribution to the identification of the collected stingless bees; and CONICET and The Rufford Foundation for financial support. We thank Centro de Investigaciones del Bosque Atlántico (CeIBA) and IBS for institutional support and Conservación Argentina (CA) foundation and "yateí" reserve for logistic support. The authors are researchers from Consejo Nacional de Investigaciones Científicas y Técnicas (CONICET).

References

Albuquerque UP, de Lucena RFP. 2005. Can apparency affect the use of plants by local people in tropical forests? Interciencia 30: 506–511.

Albuquerque UP, de Sousa Araujo TA, Alves-Ramos M, Nascimento VT, de Lucena RFP, MonteiroJM, Alencar NL, Araujo EL. 2009. How ethnobotany can aid biodiversity conservation: reflections on investigations in the semi-arid region of NE Brazil. Biodiversity Conservation 18: 127–150.

Albuquerque UP, Muniz de Medeiros P. 2013. What is evolutionary ethnobiology? Ethnobiology and Conservation 2: 1–4.

Almeida CFCBR, Lima e Silva TC, Amorim ELC, Maia MBS, Albuquerque UP. 2005. Life strategy and chemical composition as predictors of the selection of medicinal plants from the caatinga (Northeast Brazil). Journal of Arid Environment 62: 127–142.

Alvarez LJ, Rasmussen C, Abrahamovich AH. 2016. Nueva especie de *Plebeia* Schwarz, clave para las especies argentinas de *Plebeia* y comentarios sobre *Plectoplebeia* en la Argentina (Hymenoptera: Meliponini). Revista del Museo Argentino de Ciencias Naturales 18: 65–74.

Anadón JD, Giménez A, Ballestar R, Pérez I. 2009. Evaluation of local ecological knowledge as a method for collecting extensive data on animal abundance. Conservation Biology 23: 617–625.

Arenas P. 2003. Etnografía y Alimentación entre los Toba ñachilamoleek y wichí-lhuku'tas del chaco central (Argentina). Arenas P; Buenos Aires, Argentina.562pp.

Batista MA, Ramalho M, Soares EA. 2003. Nesting sites and abundance of Meliponini (Hymenoptera: Apidae) in heterogeneous hábitat of the Atlantic Rain Forest, Bahia, Brazil. Lundiana 4: 19–23.

Bentley JW, Rodriguez G. 2001. Honduran folk entomology. Current Anthropology 42: 285–313.

Berkes F. 1999. Sacred Ecology: Traditional ecological knowledge and resource management. Taylor and Francis; Philadelphia, United States: 209 pp.

Berlin B. 1992. Ethnobiological classification: principles of categorization of plants and animals in traditional societies. Princeton University Press; Princeton, New Jersey, USA. 270 pp.

Bernard RH. 2000. Social research methods. Qualitative and quantitative approaches. Sage Publications; London, United Kingdom. 824 pp.

Boster JS. 1985. Requiem for the omniscient informant: there's life in the old girl yet. pp. 177–197. In Dougherty J, ed. Directions in cognitive anthropology. University of Illinois Press; Champaign, Illinois, United States. 451 pp.

Carvalho CAL, Alves RMO, Souza BA. 2003. Criação de abelhas sem ferrão: Aspectos práticos. DAS/DDP and Universidade Federal de Bahia; Salvador, BA, Brazil. 42 pp.

Chagas F., Carvalho S. 2005. Iniciação à criação de uruçu - Meliponário São Saruê. Chagas F, Carvalho S; Igarassu Pernambuco. 50 pp.

Cortopassi-Laurino M, Imperatriz-Fonseca VL, Roubik DW, Dollin A, Heard T, Aguilar I, Venturieri GC, Eardley C, Nogueira-Neto P. 2006. Global meliponiculture: challenges and opportunities. Apidologie 37: 275–292.

Costa-Neto EM. 2006. Cricket singing means rain: semiotic meaning of insects in the district of Pedra Branca, Bahia State, northeastern Brazil. Anais da Academia Brasilera de Ciência 78: 59–68.

Deliza R, Vit P. 2013. Sensory evaluation of stingless bee pot-honey. pp 349–361. In Vit P, Pedro SRM, Roubik D, eds. Pot-Honey a Legacy of Stingless Bees; Springer; New York, UE. 654 pp.

Escamilla A, Sanvicente M, Sosa M, Galindo-Leal C. 2008. Habitat mosaic, wildlife availability, and hunting in the tropical forest of Calakmul, Mexico. Conservation Biology 14: 1592–1601.

Galeano G. 2000. Forest use at the Pacific Coast of Chocó, Colombia: a quantitative approach. Economic Botany 54: 358–376.

Galindo-Leal C, Camara IG. 2003. Atlantic Forest hotspot status: An overview. pp. 3–11. In Galindo-Leal C, Camara IG, eds. The Atlantic Forest of South América: biodiversity status, threats, and outlook. Island Press; Washington DC, USA. 488 pp.

Holmberg EL. 1887. Viaje a Misiones. I Parte. Boletín de la Academia Nacional de Ciencias, Córdoba. 10:5–391.

Izquierdo AE, De Angelo CD, Aide TM. 2008. Thirty years of human demography and land-use change in the Atlantic Forest of Misiones, Argentina: an evaluation of the forest transition model. Ecology and Society 13: 1–18.

Jodelet D. 1986. La representación social: Fenómenos, conceptos y teoría. pp. 469–494. In Moscovici S, ed. Psicología Social II. Ediciones Paidós; Barcelona, España. 543 pp.

Kamienkowski N, Arenas P. 2012. La colecta de miel o "meleo" en el Gran Chaco: Su relevancia en etnobotánica. pp. 71–116. In Arenas P, ed. Etnobotánica en zonas áridas y semiáridas del cono sur de Sudamérica. CEFYBO-CONICET; Buenos Aires, Argentina. 270 pp.

Karanth KU, Nichols JD, Seidensticker J, Dinerstein E, Smith JLD, McDougal C, JohnsinghAJT, Chundawat RS, Thapar V. 2003. Science deficiency in conservation practice: the monitoring of tiger populations in India. Animal Conservation 6: 141–146.

Kujawska M, Zamudio F, Hilgert NI. 2012. Honey-based mixtures used in home medicine by non-indigenous population of Misiones, Argentina. Evidence-Based Complementary and Alternative Medicine. doi:10.1155/2012/579350.

Laplantine F. 1999. Antropología de la enfermedad. Ediciones del Sol; Buenos Aires, Argentina. 418 pp.

Lozano A, Lima-Araújo E, Medeiros TMF, Albuquerque UP. 2014. The apparency hypothesis applied to a local pharmacopoeia in the Brazilian northeast. Journal of Ethnobiology and Ethnomedicine 10: 2–17.

Lucena RFP, Medeiros PM, Lima-Araújo E, Alves AG, Albuquerque UP. 2012. The ecological apparency hypothesis and the importance of useful plants in rural communities from Northeastern Brazil: An assessment based on use value. Journal of Environment Management 96: 106–115.

Matsumoto, D. 2000. Culture and psychology. Brooks Cole Publishing Co.; Pacific Grove, California. 544 pp.

Michener CD. 1969. Comparative social behavior of bees. Annual Review of Entomology 14: 299–342.

Nolan JM, Robbins MCE. 2001. Emotional meaning and the cognitive organization of ethnozoological domains. Journal of Linguistic Anthropology 11: 240–249.

Parra F, Casas A, Peñaloza-Ramírez JM, Cortés-Palomec AC, Rocha-Ramírez V, González-Rodríguez A. 2010. Evolution under domestication: ongoing artificial selection and divergence of wild and managed Stenocereus pruinosus (Cactaceae) populations in the Tehuacan Valley, Mexico. Annals of Botany 106: 483–496.

Posey DA. 1983. Keeping of stingless bees by the kayapo' indians of Brazil. Journal of Ethnobiology 3: 63–73.

Potts SG, Biesmeijer JC, Kremen C, Neumann P, Schweiger O, Kunin WE. 2010. Global pollinator declines: trends, impacts and drivers. Trends in Ecology and Evolution 25: 345–353.

Reyes-García V, Huanca T, Vadez V, Leonard WR, Wilkie D. 2006. Cultural, practical and economic value of wild plants: a quantitative study in the bolivian amazon. Economic Botany 60: 62–74.

Roseman IJ, Smith CA. 2001. Appraisal theory: Overview, assumptions, varieties, controversies.pp. 3–19. In: Scherer KR, Schorr A, Johnstone T, eds. Appraisal processes in emotion: Theory, methods, research. Oxford University Press; New York, USA. 496 pp.

Rosch E. 1978. Principies of Categorization. pp. 28–49. In Rosch E, Lloyd B, eds. Cognition and Categorization. Laurence Erlbaum Associates, Inc; Hilldale, United States. 336 pp.

Rosso JM, Imperatriz-Fonseca VL, Cortopassi-Laurino M. 2001. Meliponicultura en Brasil I: Situación en 2001 y Perspectivas. pp. 28–35. In Memorias del II Seminario Mexicano sobre Abejas sin Aguijón. Mérida, México.

Roubik DW. 2006. Stingless bee nesting biology. Apidologie 37: 124–143

Schiavoni G. 1995. Organización doméstica y apropiación de tierras fiscales en la Provincia de Misiones (Argentina). Desarrollo Económico 34: 595–608.

Silveira FA, Melo GAR, Almeida EAB. 2002. Abelhas Brasileiras; Sistemática e Identificação. Min. Meio Ambiente/Fund. Araraucária; Bello Horizonte, Brasil. 253 pp.

Stave J, Oba G, Nordal I, Stenseth NC. 2007. Traditional ecological knowledge of a riverine forest in Turkana, Kenya: implications for research and management. Biodiversity Conservation 16: 1471–1489.

Sutrop U. 2001. List task and a cognitive salience index. Field Methods 13: 263–276.

Vabø M, Hansen H. 2014. The relationship between food preferences and food choice: a theoretical discussion. International Journal of Business and Social Science 5: 145–157.

Venturieri GC, Imperatriz-Fonseca VL. 2000. Scaptotrigona nigrohirta e Melipona melanoventer (Apidae, Meliponinae): Espécies amazónicas com potencialidade para a meliponicultura. pp. 356–356. In IV Encontro Sobre Abelhas. Faculdade de Ciências e Letras de Ribeirão Preto; Ribeirão Preto, Brasil. 356 pp.

Venturieri GC, Raiol VFO, Pereira CAB. 2003. Avaliação da introdução de da criação racional de Melipona fasciculata (Apidae: Meliponina) entre os agricultores familiares de Bragança, Belém. Biota Neotropica 3: 1–7.

Wille A. 1983. Biology of the stingless bees. Annual Review of Entomology 28: 41–64.

Williams BK, Nichols JD, Conroy MJ. 2002. Analysis of Management o Animal Populations. Academic Press; UE. 816 pp.

Wolverton S. 2013. Ethnobiology 5: Interdisciplinarity in an Era of Rapid Environmental Change. Ethnobiology letters 4: 21–25.

Zamudio F, Kujawska M, Hilgert NI. 2010. The honey as medicinal resource: Comparison between Polish and multiethnic settlements of the Atlantic Forest, Misiones, Argentina. The Open Complementary Medicine Journal 2: 1–16.

Zamudio F, Hilgert NI. 2011. Mieles y plantas en la medicina criolla del norte de Misiones, Argentina. Bonplandia 20: 59–78.

Zamudio F, Hilgert NI. 2012a. ¿Cómo los conocimientos locales aportan información sobre la riqueza de especies de abejas sin aguijón (Apidae: Meliponini) del norte de Misiones, Argentina? Interciencia 37: 36–43.

Zamudio F, Hilgert NI. 2012b. Descriptive attributes used in the characterization of stingless bees (Apidae: Meliponini) in rural populations of the Atlantic Forest (Misiones-Argentina). Journal of Ethnobiology and Ethnomedicine 8: 1–10.

Zamudio F, Hilgert NI. 2015. Multi-dimensionality and variability in folk classification of stingless bees (Apidae: Meliponini). Journal of Ethnobiology and Ethnomedicine 11:1–15.

Zamudio F, Álvarez LJ 2016. Abejas sin aguijón de Misiones: una guía etnotaxonómica para su identificación en el campo. Editorial de la UNC; Córdoba, Argentina. 218 pp.

The Maya Universe in a Pollen Pot: Native Stingless Bees in Pre-Columbian Maya Art

21

Laura Elena Sotelo Santos
and Carlos Alvarez Asomoza

On 7 Caban honey was first created, when we had none.
The Book of Chilam Balam of Chumayel

21.1 Introduction

In Mexico are 11 genera and 46 species of native stingless bees or *Meliponini* (Ayala 1999). Some of them have been intensively handled since pre-Columbian times to the present day (Mohar-Betancourt 1987). It is thus possible to focus on a meliponine cultural tradition, through which indigenous practices of the Nahua, Tarascs, or Chinantec for the catering of wax and honey are inscribed in the major Mesoamerican cultural tradition (Vásquez-Dávila and Solís-Trejo 1992). More than 32 species of the native stingless bees in Mexico (or 70%) inhabit the Maya region. However, Maya people don't use them all and only a few are handled or raised.

In order to trace the history of the honey culture among the Maya, a collective and interdisciplinary effort is needed. As it spans from the Preclassic period through Colonial era and even to the present day, archaeology, history, and ethnography converge to shed light on the meaning of wax and honey among the contemporary

L.E. Sotelo Santos (✉) • C. Alvarez Asomoza
Centro de Estudios Mayas, Instituto de
Investigaciones Filológicas, Universidad Nacional
Autónoma de México Ciudad Universitaria,
México 04510, D. F., Mexico
e-mail: biblos.2@att.net.mx

Maya. This simple act is part of a sociocultural background that inspires and guides a feeling of continuity and identity.

The most conservative Maya of Yucatan, Mexico, still farm *Melipona beecheii* with an ancestral technique that comprises an expert handling alongside a large corpus of beliefs. Today they refer to *Melipona beecheii* as *Xunan Kab* and *koolel kab*, or Lady Bee. *Chuu kab* in Yucatec Maya means a small pot of honey, literally *calabacita de miel* and the *Calepino de Motul* mentions "vejiguillas en que está la miel de las colmenas." Also the word *p'ool* makes reference to a honey receptacle or "pot of honey" (Álvarez 1980).

We know that honey and beeswax produced by *Melipona beecheii* and related species were traded in markets as far as Xicalango or the Gulf of Honduras and were considered, after salt, the main exchange products of Yucatan (Chapman 1959). Furthermore, during the Colonial period in the Maya area, these items were still held in high esteem from the Guatemalan highlands in the south to the lowlands in the north, mainly the Province of Campeche. Various chroniclers and historians, such as Gonzalo Fernández de Oviedo y Valdés, fray Diego de Landa, fray Diego López de Cogolludo, fray Francisco Ximénez, and Francisco Xavier Clavijero, among others, give accounts of the multiple kinds of honey and beeswax (Nárez 1988).

In this paper we analyze a cluster of images made during pre-Columbian times, which feature native stingless bees, from a double perspective:

P. Vit et al. (eds.), *Pot-Pollen in Stingless Bee Melittology*, DOI 10.1007/978-3-319-61839-5_21

first, data provided by modern biology (Michener 2013) and information from Maya beekeepers that use *Melipona beecheii*. Second, we present paintings and sculptures of small format, which provide a detailed knowledge of the insect's morphology and nests, as well as the singular handling of this particular species among the Maya, in their temporal and cultural context.

21.2 Maya Bee Myths

The only domestic animal (though not truly domesticated) the Maya had before the arrival of the Spaniards was the *Melipona beecheii*, called *Xunan Kab* or *Ko'olel Kab* as mentioned above. It was, and still is, considered a sacred bee or insect. On the eastern coast of the Yucatan Peninsula in Mexico, it is common for the Maya to refer to bees, in their domestic apiaries, as "people." They explain this is due to a myth in which both bees and humans share the same origin. As told by a traditional priest, *H-men*, from Tepich in Quintana Roo, Mexico: "When Adam and Eve had their first child, they brought him before God. But God punished them, for they had disobeyed Him. So they had to offer their first child, and cut off his head. Instead of blood, honey issued from the severed neck of the child. And from the spreading, different sorts of bees emerged, as "E'hol" (*Cephalotrigona zexmeniae*), "Bool" (*Nannotrigona perilampoides*), "Xiik" (*Frieseomelita nigra*), "Ko'olel Kab" (*Melipona beecheii*), "Xnuuk" (*Trigona fuscipennis*), "Kansas" (*Scaptotrigona pectoralis*), "P'uup" (*Trigonisca maya*), and "Muul" (*Trigona fulviventris*)" (De Jong 1999).

This myth refers to a primordial moment in which only the first couple existed: bees originate directly from the first sacrifice by decapitation and precede human kind. It points out as well honey's origin, a vital fluid from the first human being, full of fertility, which was offered in sacrifice so future progeny could live. Honey in this myth is a primordial liquid, which is natural and cultural at the same time; it is natural inasmuch as it has not been transformed by any direct activity from mankind, but has a cultural aspect due to

its care by and proximity to humans. It's a gift to humanity and, in addition, accomplishes the essential cosmic condition of being the sacred food for gods. Stingless bees directly descend from Adam and Eve, as the rest of human kind does. Following this line of thought, there exists a correlation between different kinds of bees and of people, as it seems that every single honey drop has the capacity to generate diversity.

According to a Yucatec Maya tradition, the *Xunan Kab, Melipona beecheii* produces honey of an incomparable quality because the *Xunan Kab* goes to *Xmaben* to collect honey. According to some, *Xmaben* is a honey-filled canoe at a place in the sky called "u Gloria" his Glory, which is guarded by the rain gods, the *Chako'ob*. There, among the gods, dwell the *Ba'baalo'ob*, giants of a sort, who love to bathe in the honey and tend to spoil it. For this reason, the dogs of the *Chako'ob* guard the canoe. According to other Maya, *Xmaben* is a celestial field of flowers which belongs to the *Chako'ob* and where the *Xunan Kab* collects honey. As this is the only species that visits *Xmaben*, their honey is said to be divine in both origin and quality (De Jong 1999). Honey from the *Xunan Kab* bees is deemed as sacred and has formed a substantial component in annual ceremonies, it is also said to have healing qualities. The bees' wax, known as *cera de Campeche*, has gained international recognition due to its attributes.

21.3 Small Format Modeled Sculptures

Here we include four examples modeled in clay. Two of them are on an exhibit in the Regional Museum of Yucatan, México, at the Palacio Canton in Merida, another in the National Museum of Anthropology at Mexico City, and a fourth was recently found at Nakum, in Guatemala. All of them bear testimony of the ancient Maya tradition of beekeeping using hollow tree trunks for the hives. These pieces were produced by the technique of modeling, as the potter gave shape to malleable clay, and later, when dried, fired them, thus obtaining stable pieces.

21.3.1 Small Clay Bee

This is a little rendering of a bee modeled in clay (2 inches long) which is now shown as part of a wider exhibit about Maya subsistence at Merida's museum. Its realism makes this piece an example of naturalist art. The anonymous Maya potter denotes a detailed knowledge of bee morphology, showing the three basic segments that modern taxonomy acknowledges: the head, thorax, and abdomen. Although no legs or antennae are featured, perhaps due to technical difficulties, it synthesizes in an accurate and balanced fashion anatomical elements the Maya deemed as significant: a round head, a pair of wings that comprise the thorax, and a segmented abdomen. Even though bees actually have two pairs of wings, they resemble only one in flight because both pairs, one in front and the other behind, are joined through a series of small hooks, the "hamuli." We infer that this piece portrays a bee during flight, for if it were standing still, the wings would be over the back of the thorax and part of the abdomen (Fig. 21.1). The piece would be viewed from a frontal and back view, as it is now shown in the museum. It seems it was painted, as there are traces of orange color and white pigments that could represent the wing veins and also the abdomen segments.

Fig. 21.1 Native stingless bees (*Melipona beecheii*) made from clay in natural scale Chen Mul type, from Structure 5 in Tipikal, Yucatán México. In exhibition at the Gran Museo del Mundo Maya de Mérida (Great Museum of the Maya World), Yucatán (Photo: F. Ávila)

21.3.2 Ceramic Beehive

The earliest evidence so far published stems from the recent find of a dedicatory cache dated to the Protoclassic period (ca. 50 BCE–ca. 300 CE) found at the city of Nakum in Peten, Guatemala. A small, fired clay cylinder in the shape of and with the features of a beehive, *hobon* in Yucatec Maya,[1] was excavated, among several other objects (Fig. 21.2). It has lateral covers to allow for the extraction of honey and cerumen wax, as well as a small central orifice that served as an entrance for the bees. It would be of great interest for our research to carry out chemical tests of this archaeological object. Such tests could be used to search for traces of wax, honey, and other by-products or remains of the insects—like wings, antennae, appendages, and other parts of the insect's body—in order to determine if this object functioned in fact as a beehive or if it was, instead, a votive replica in miniature, used exclusively as part of an elaborate ritual (Zrarlka and Wieslaw 2010).

This clay hollow cylinder was found, in 2007, deep inside Structure 99 in the North Group of Nakum, Peten, Guatemala. Right in the middle is a small round orifice (or circular perforation), and two clay disks seal the ends. Its shape and

Fig. 21.2 Scale model made from clay of an artificial hive (hobon) of *Melipona beecheii*. Nakum from Structure 99, Guatemala (From Zralka et al. 2014), used by permission of Estudios de Cultura Maya

---

[1] Modern Maya call *jobón* a man-made hollow piece of tree trunk that can shelter a *Melipona* hive.

archaeological context point to a ceremonial hive or *hobon*, which dates to the Protoclassic (50 BCE–AD 300). Photo: kindness of Zralka and Koszkul.

21.3.3 Censers

Among the ancient Maya ceremonial objects, incense burners and censers are highly relevant. Many examples have been found in archaeological contexts. Both were mainly used to burn aromatic resins and waxes. Some of them bear modeled effigies of deities to whom the ceremony was dedicated; thus, these are known as "effigy censers" (Cuevas 2007).

A Spanish document dating from the sixteenth century states that the Maya "praised some idols made of fired clay in the shape of small jars and pots for basil, made on them on the outer side disfigured faces; inside of these jars they burned a resin called copal of great odor. [...] There were idols of farming, idols of the sea and many other kinds for each thing, distinct the form of each idol from one another [...]" (de la Garza et al. 1983).

The following two examples are associated with deities that propitiate an abundance of honey. In the Maya showroom at the National Museum of Anthropology in Mexico City, within a glass cabinet that assembles works from the Late Postclassic period of Yucatan, a censer of an old god, from Cozumel Island, Mexico, is on display. This piece belongs to the censer tradition that spans 1250–1500 CE. The figure of the god is carrying a *hobon* as manner of necklace (Fig. 21.3). It is clear, even though the figure is quite realistic, that the *hobon* has a symbolic meaning given its size and the fact that the god is carrying it. In our point of view, the small hive on the chest is his emblem of power; hence, it is possible for it to be a mythological hive carrier or *ah cuch kab* in Yucatec.

Another incense burner, curiously modeled with the effigy of a "Diving God" (which carries on his hands *Melipona beecheii* brood cells) in the front, is also exhibited in the gods' case at the museum in Merida (Fig. 21.4). Inasmuch as the anthropomorphic figure on these censers depicts

Fig. 21.3 Censer with the figure of an elder god, with a *hobon* on his neck. From Cozumel island (near Yucatan western shore), elaborated between 1250 and 1500 C.E. In exhibition at the Museo Nacional de Antropología (National Anthropology Museum), in Mexico City (Photo: F. Ávila)

Fig. 21.4 Descending god that holds in his hands a *Melipona beecheii* brood comb. Chen Mul type censer with a pair of scale *hobon* at each side. In exhibition at the Museo Maya de Cancun, Quintana Roo, México (Photo: F. Ávila)

gods, it is clear that the breeding of honey-producing bees was considered a sacred labor. Some scholars had put forward the bees' god *Ah Mucen Kab* of the Maya as the deity portrayed on this unique piece (Thompson 1975).

There is evidence of some effigy censer fragments reported by Sydris (1983) in Crane, (Crane 1999) at the site of Chan Chen, in northern Belize, where some might also portray the god *Ah Mucen Kab*. Among them, at least one could be dated to the Late Preclassic (100 BCE – 250 CE). Although this is an isolated finding, if combined with the *hobon* found at Nakum, Guatemala, they give testimony of the ritual importance of beekeeping as known by the Maya thousands of years ago—perhaps as early as the domestication of maize. Furthermore, present-day native bees continue to receive great respect and reverence from beekeepers during breeding and honey harvest.

These archaeological Maya artifacts allow us to recognize at least three kinds of motifs related with the Maya meliponine cultural tradition: bees themselves, hives or *jobon'ob*, and brood cells. From a temporal perspective, we find the first evidence of beekeeping dated to the Late Preclassic, whereas the other examples come from Yucatan and date to the close of the pre-Columbian era.

21.4   Bees and Stingless Beekeeping in a Sacred Maya Book

Maya codices are folded books written in Maya hieroglyphic script on bark paper, made from the inner bark of a fig (*Ficus*) tree. The magnificent hieroglyphic book known as *Tro-Cortesianus Codex* or *Madrid Codex* (Fig. 21.5) was written before the Spanish Conquest during Late Postclassic and is now kept in the *Museo de America* at Madrid, Spain. It contains a specific section on the culture of Meliponini.

The religious specialist who achieved the realization of this codex knew the myths, and the ritual and symbolic traditions related to the sacred world of the bees. And, as such, he had every element needed for its unequivocal identification. The representations are inscribed within the symbolic and aesthetic codes of the ruling Maya elite. These figures rest on an extensive experience accumulated from the observation of the bees' world and the interior of hives, as well as the use of their products. Even when Maya priests had notions of wasps (Vespoidea) and other meliponine bees (*Trigona*), they only depicted *Melipona beecheii*, to such an extent that we can consider it the conventional way of

Fig. 21.5 Fragment of a Maya hieroglyphic book named *Tro-Cortesianus Codex*. Made of a strip of indigenous paper (6.82 m), folded forming sections similar to the pages of in Western book, since each has a size of approximately 22.6 cm length by 12.2 cm in width. The last pages show several aspects of Maya bee handling. Facsimile exhibited in the Gran Museo del Mundo Maya de Mérida (Great Museum of the Maya World) (Photo: L.E. Sotelo Santos)

rendering bees in Maya art. Even today, its accuracy in detail is surprising. These images are the first and most detailed forms representing these insects in global beekeeping history. The singularity of this section lies in its completeness and details. It comprises ten adjoining pages plus a small almanac, in which there are painted bees, hives, honeycombs, nests, meliponaries, and some aspects of bee handling such as the transfer from hollow trunks or hives, colony division, and the process of honey harvest, besides the rendering of plants this species visits or prefers for nesting. Deities, rituals, and offerings whose meaning still remains unclear are also represented. In other words, the so-called Bees Section contains several biological, technical, and cultural aspects of native stingless bees.

A visual inspection of the Bees Section of the *Tro-Cortesianus Codex* first results in the identification of bees, and with a further careful review of the almanacs, one can recognize brood combs or storage pots, beehives, apiaries, and procedures of wax and honey extraction or colony division. Reddish ochre was used as the main pigment, which reminds us of the color of the *achiote* (*Bixa orellana*). It confers to the figures a chromatic likeness with the insects, their hives, and products, giving it at the same time unity and also accentuating its importance in the text and visually emphasizing every component.

21.4.1 *Melipona beecheii* in the Tro-Cortesianus Codex

Renderings of bees in the codex show a front, dorsal, and ventral view (Tozzer and Allen 1910). The face and a pair of legs are seen from a frontal view and wings and abdomen from the back and a ventral view shows the egg-laying organ. This is a characteristic of two-dimensional Maya art: to show two or more views of what is depicted in order to synthesize a conventional form in this way (Fig. 21.6).

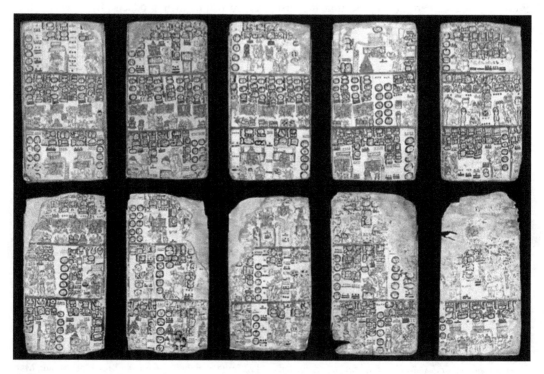

Fig. 21.6 Bee section from the *Tro-Cortesianus Codex* (p. 103–112). This pre-Hispanic manuscript contains the oldest written reference to *Melipona beecheii* beekeeping among the Yucatan Maya: their breeding in artificial hives, the hive interior, the queen bee, and the apiaries (Photo: I. Miceli)

21.4.2 Hobon

In the *Tro-Cortesianus Codex,* beehives inside hollow tree trunks were rendered as rectangles, most of the time with an inner frame. This was obviously a way to represent the trunk and the lateral lids (Fig. 21.7). There exist two variants: one is plain and shows an outside view of a beehive as an ochre rectangle (see Fig. 21.7a). The other is the glyphic variant that shows a *hobon* inside view (see Fig. 21.7b) represented by the glyph *Caban* (Fig. 21.8) that is also the sign for the 17th day in the Maya calendar (Calvin 2010).

The plain variant always shows a bee superimposed in the middle of the rectangle whose size surpasses that of the *hobon*, the glyphic variant sometimes also has a figure of a bee in the same position, so both forms are equivalent. The plain variant also shows an outside view of the hollow tree trunk, whereas the glyphic variant of the *hobon* is a hive's inner view.

To summarize, bee and hive representations in Maya art seem to share several elements through time. They are realistic images that sometimes synthesize features and in some others recreate them. The clearest example is the

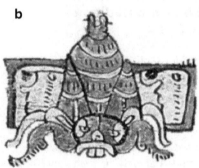

Fig. 21.7 Representation of Maya queen bees. (**a**) (*left*) A *Melipona beecheii* virgin queen bee on the foreground and a *hobon* seen from the outside, in the background, which represents a hollow log; this is a stingless bee colony. Detail from page 104 on the *Tro-Cortesianus Codex*. (**b**) (*right*) A queen bee, fertilized and about to lay an egg (white circle on top of its abdomen). In the background, a *hobon* interior, with the hive's differentiated areas, represented by a couple of lines: the *black circle* represents the honey and pollen storage pots, and the undulating ones, the waxen sheaths, and the overlapping horizontal lines, the different layers of the brood area. Detail from page 105 on the *Tro-Cortesianus Codex* (Photos: F. Ávila)

Fig. 21.8 Amphorae of honey and pollen. (**a**) (*left*) Glyph for the XVII day on the Mesoamerican 260-day calendar, *cabán*, "earth" in Yucatec Maya. It seemingly represents the storage area from a stingless beehive: the amphorae of honey and pollen. (**b**) (right) Detail from page 104 on the *Tro-Cortesianus Codex*, in which a deity holds in his hand a *hobon*. The artist placed special detail in the hive's interior and drew the honey and pollen storage pots along with the brood cells (Photos: L.E. Sotelo Santos)

hobon, while bees are treated more or less schematically, depending on the kind of figure rendered. In Maya plastic art, the size of the figures is not directly related to the actual size. This is evident in the *Tro-Cortesianus Codex,* where the figures of bees are larger than those of the *hobon*. A similar case is that of the effigy censers which have very little models of beehives, as they do not keep any anatomical proportion with either the anthropomorphic figures or the brood cells.

21.5 A World View of Humankind Through a Bee Model as Told by a *H-men*, a Traditional Maya Priest

At one of the last strongholds of *Meliponini* culture in Quintana Roo, Mexico, Maya farmers frequently refer to their bees as "people" (*gente* in Spanish). They explain this through a myth in which humans and bees share a common origin, as is told by *H-men* from Tepich, Quintana Roo.

A myth compiled already in the twenty-first century appears to synthesize both the role of bees in Maya world view, as well as different processes and epochs. It narrates a long tradition that goes back to pre-Columbian times and integrates two main meanings—opposing and complementary—of the honey- and wax-producing insects: culture and nature. And at the same time, it contrasts the autochthonous stingless bees with the imported honey bee, *Apis mellifera*.

The *H-men* states: "God placed the bees on earth: *Xunan Kab, E'hol, Kansak, Xiik, Bool, P'uup, Tsots* and others. They are like all the different races of humankind. The Maya are similar to *Ko'olel Kab* [i.e. *Xunam Kab*]. The Spaniards came and fought against the Maya. The same thing is happening between "*Americano Kab*" [the Western hive bee, *Apis mellifera*] and *Xunan Kab*. The Spaniards started it all, and that is why the bees are fighting now. But in the beginning, it was just between the Spaniards and the Maya. The forest bees do not fight with the foreigner. But whenever "*Americano Kab*" runs into

Xunan Kab, he bites into her wings and kills her; *Xunan Kab* can not exist" (De Jong 1999).

In other words, the myth seems to cover not only pre-Columbian times but also history after the arrival of the Spaniards: equating the Yucatec Maya to the *Melipona beecheii*, for they have shared a millenary history, and the Spaniards, and thus to *mestizos*, to the *Apis mellifera*.

Today, at the dawn of the twenty-first century, there are some people who express an opinion that native bees from the Maya area and their prodigious honey and wax are at the brink of extinction. Is the *Xunan Kab*—Lady Bee—sending a warning message about the necessity of protecting biodiversity, in order to guarantee the survival of our own species? A Maya from Tepich, Quintana Roo, says: "The Destiny of *Xunan Kab* is our own Destiny" (De Jong 1999).

21.6 The Maya Universe in a Pollen Pot

To talk about bees in the Maya area is to talk about the vegetation of the region, inasmuch as the bees live of the nectar and pollen collected from the flowers and the flowering plants depend on the bees for their pollination. In other words, there is a symbiotic relationship between plants and bees that has been in place for millions of years.

21.6.1 Once upon a Time, Long, Long Ago, There Were Bees…

On a geologic time scale, the fossil evidence for native stingless bees in the Maya area comes from the amber deposits in Simojovel, Chiapas, Mexico, that have been dated between 17 and 19 million years old. The amber fossil specimens are of *Problebeia silaceae* (Fig. 21.9), of the ancestral lineage to bees that now constitute the genus *Plebeia*. From those days, the stingless bees have collected nectar and pollen, thus contributing to pollination in the region.

Fig. 21.9 Specimens of *Proplebeia silaceae*, in amber from Simojovel, Chiapas. The age of Chiapas amber is 17–19 Ma, and once more radiometric dating has been undertaken that might prove to be around 18.5 Ma. Identification by Michael S. Engel (Photo: J. Mérida)

This fact seems to be synthesized in the Maya myth recorded on the *Libro de Chilam Balam de Chumayel*, a text written by some member of the indigenous nobility, which during the time of the viceroyalty learned to write the Yucatec Mayan language in Latin characters. The passage reads: The red wild bees are in the east. A large red blossom is their cup. The red *Plumeria* is their flower. The white wild bees are in the north. The white pacha is their flower. A large white blossom is their cup. The black wild bees are in the west. The black laurel flower is their flower. A large black blossom is their cup. The yellow wild bees are in the south. A large yellow blossom is their cup ...is their flower. (Roys 1933).

The story seems to make reference to the moment the world had just been arranged, taking the form of a square and flat space with four regions corresponding to the four directions. Symbolically, each of the parts is associated with a color and, in this way, with a different meaning related to the apparent movement of the sun around the Earth. On each cosmic sector is a flowering tree, in which we may assume there are colonies of wild stingless bees lodged in their trunks. This is a metaphor that tells us of the place that the bee has in pollination at a cosmic scale and of its fundamental place in the

conservation of the flora. The myth synthesizes in a couple of verses the tight relationship between plants and bees as a symbol of earthly fertility and of the bonds between plant and animal life. The four colored flowers represent the multiplicity of the plants in the Maya area, their blossoming, the plant diversity and abundance, the bees of varied colors, and the distinct species of indigenous bees. It also tells of a primeval time, in which humanity was still to be created. The myth points out that bees existed before men.

21.7   The Flower Dust

Even though the *Melipona beecheii* colonies lodge themselves in hollow tree trunks, inside they build spaces separated from each other by cerumen. This material is a characteristic of the stingless bees, a mixture of gums and the wax that worker bees produce in the dorsal part of their abdomen. With the cerumen they build the brood covering (involucrum) and the brood combs, and also the pots in which they store honey and pollen, alongside other structures, such as pillars and filaments that connect the different areas with the inner walls of the hive.

Melipona beecheii colonies store pollen and honey in wax "containers" that the workers build collectively during the time of nectar storage. The pollen pots are placed near the brood combs, for the protein in the pollen to serve as food for the future bees. These are the fundamental nutrients that allow the growth and development of the eggs into larvae, pupae, and, finally, bees.

According to colonial dictionaries, the term used for the pollen inside the cerumen pots is a word defined as "a kind of wax made by the bees with the powder from the flowers in order to raise their children." Each pollen pot contains the genetic information of the flowers of Yucatan, accumulated by the bees for millions of years.

Table 21.1 is a list of visited plant species in the Yucatan Peninsula by stingless bees, according to the Importance Value Index (IVI) in decreasing order.

Table 21.1 Importance Value Index of plants visited by stingless bees in Yucatan

No.	Species	Common name (Maya)	Importance Value Index (IVI)[a]
2	*Viguiera dentata*	Tajonal	92.64
19	*Lysiloma latisiliquum*	Tsalam	82.67
18	*Piscidia piscipula*	H' abin	65.69
7	*Jacquemontia pentantha*	Solen ak	59.30
1	*Gymnopodium floribundum*	Sakts' its 'ilche'	55.9
25	*Vitex gaumeri*	Ya' axnik	55.9
15	*Sabal yapa*	Huano	51.12
12	*Dalbergia glabra*	Ahmuk	50.45
20	*Haematoxylum campechianum*	Bu' ulch' ich	49.21
11	*Acacia gaumeri*	Box-catzim	39.19
13	*Caesalpinia gaumeri*	Kitam che'	37.0
10	*Bursera simaruba*	Chakah	34.8
21	*Hampea trilobata*	Hol	33.21
8	*Ipomoea crinicalyx*	Is' ak	32.17
17	*Leucaena leucocephala*	Guachim	31.60
9	*Ipomoea triloba*	Motul	31.02
3	*Turbina corymbosa*	Chukin siis	28.84
16	*Mimosa bahamensis*	Katzim	28.30
22	*Neomillspaughia emarginata*	Sakitsa'	26.78
5	*Merremia aegyptia*	Tsots ak'	19.90
23	*Coccoloba uvifera*	Uva de mar	19.75
24	*Tournefortia gnaphalodes*	Sikimay	18.50
4	*Bidens pilosa*	Kan mul	17.82
26	*Spondias mombin*	Jobo	15.75
14	*Caesalpinia yucatanensis*	K' anpok' olk' um	15.51
27	*Tridax procumbens*	Bakembox	13.14
28	*Herissantia crispa*	Sakle-sak-miis	12.64
29	*Waltheria indica*	Sak miisib	9.39
30	*Cordia gerascanthus*	Bohom	9.31
6	*Ipomoea nil*	Yaak' cal	6

[a]The IVI indicates the importance of a given species of plant within its environment. By seeing the importance of the plants that depend on the stingless bees for their pollination; we may have an appreciation of their ecology in the native world (Felfili and Silva 1993 in Porter-Bolland 2001)

Acknowledgments We acknowledge the financial support for our research by the Dirección General de Asuntos del Personal Académico, UNAM (Ixtli 400610) and PASPA.

References

Álvarez C. 1980. Diccionario etnolingüístico del idioma maya yucateco colonial, Vol. I: Mundo Físico, Centro de Estudios Mayas, Instituto de Investigaciones Filológicas, Universidad Nacional Autónoma de México, México. 386 pp.

Ayala BR. 1999, Revisión de las abejas sin aguijón de México (Hymenoptera: Apidae: Meliponini). *Folia Entomológica Mexicana*, Número 106: 1-20.

Calvin IE. 2010. Maya Hieroglyphics Study Guide, 3 Ed. Available at http://www.famsi.org/mayawriting/calvin/index.html

Chapman A. M. 1959 Puertos de Intercambio en Mesoamérica Prehispánica. INAH, Serie Historia III, México. 77 pp.

Crane E. 1999. The World History of Beekeeping and Honey Hunting, Routledge; New York, USA. 682 pp.

Cuevas García M. 2007. Los incensarios efigie de Palenque: deidades y rituales mayas. México, Centro de Estudios Mayas, Instituto de Investigaciones Filológicas, UNAM. 350 pp.

de la Garza M, Figueroa T, Izquierdo AL. 1983. Paleografía de María del Carmen León. Relaciones histórico-geográficas de la Gobernación de Yucatán, (Mérida, Valladolid y Tabasco), Tomo II, México, Centro de Estudios Mayas, Instituto de Investigaciones Filológicas, Universidad Nacional Autónoma de México; México. 189 pp.

De Jong H. 1999. The land of corn and honey: The keeping of stingless bees (meliponiculture) in the ethno-ecological environment of Yucatán (México) and El Salvador. Ph.D. Thesis, Utrecht University; Utrecht, The Netherlands. 342 pp.

Felfili JM, Silva MC. 1993. A comparative study of cerrado (*sensu stricto*) vegetation in Central Brazil. 9: 277-289.

Michener CD. 2013. The Meliponini. pp. 3-18. In: Vit P, Pedro SRM, Roubik D, eds. Pot honey: A legacy of stingless bees. Springer; New York, USA. 654 pp.

Mohar-Betancourt, LM. 1987. El tributo mexica en el siglo XVI: análisis de dos fuentes pictográfica, México, CIESAS (Cuadernos de la Casa Chata, 154). 388 pp.

Nárez J. 1988. Algunos datos sobre las abejas y la miel en la época Prehispánica, Revista Mexicana de Estudios Antropológicos, T. XXXIV: 123-140.

Porter-Bolland L. 2001. Landscape ecology of apiculture in the Maya área of La Montaña, Campeche, Mexico. PhD Thesis. University of La Florida; Florida, USA. 184 pp.

Roys LR. 1933. The Book of Chilam Balam of Chumayel, Carnegie Institution; Washington, D.C. 280 pp.

Tozzer A, Allen G. 1910. Animal Figures in the Maya Codices, Cambridge, Massachusetts, USA. 274 pp.

Thompson ES. 1975. Historia y Religión de los Mayas, Siglo XXI Editores; Ciudad de México. 485 pp.

Vásquez-Dávila MA, Solís-Trejo MB. 1992. La miel de los chontales. pp. 348-371. In Memorias del Primer Congreso Internacional de Mayistas, Vol. 1, México, Centro de Estudios Mayas, Instituto de Investigaciones Filológicas, UNAM.

Zralka J, Koszkul W, Radnicka K, Sotelo Santos LE, Hermes B. 2014. Excavations in Nakum structure 99: New data on Protoclassic rituals and Precolumbian Maya beekeeping. Estudios de Cultura Maya 44: 85-117.

Zrarlka J, Wieslaw K. 2010. New discoveries about the ancient maya. Excavations at Nakum, Guatemala. Expedition 52: 21-32.

Part IV

Chemical Composition, Bioactivity and Biodiversity of Pot-Pollen

Nutritional Composition of Pot-Pollen from Four Species of Stingless Bees (Meliponini) in Southeast Asia

22

Bajaree Chuttong, Rewat Phongphisutthinant,
Korawan Sringarm, Michael Burgett,
and Ortrud Monika Barth

22.1 Introduction

Stingless bees are distributed throughout the tropical regions of the world. There are more than 500 described species in 32 genera. The greatest species diversity is found in Central and South America (Michener 2013). For Thailand 32 species in 10 genera have been recorded (Rasmussen 2008). Not all species of stingless bees found in Thailand are adaptable for meliponiculture primarily due to the constraints of their mature tree cavity-nesting preferences. Several species are cultured in human-made hives; of these the most common and widely distributed species are in the *Tetragonula laeviceps* species complex. Additional managed species in Thailand include *Tetragonula fuscobalteata, T. testaceitarsis, Lepidotrigona*

doipaensis, L. flavibasis, L. terminata, and *Lisotrigona furva*. Chuttong et al. (2014) highlighted that meliponiculture in Thailand is in an incipient stage, but expanding. The utilization of stingless bees, as an economic activity, is largely confined to the southeastern region of Thailand, in the provinces of Chanthaburi and Trat. The estimated number of meliponiculturists in Thailand is more than 700 with approximately 5000 colonies under management (DOAE, Department of Agricultural Extension, Thai Ministry of Agriculture and Cooperatives, Region 3, Chanthaburi, 2014, personal communication).The keeping of stingless bees in this area was initiated by the use of stingless bees as supplemental pollinators for several tropical fruit species. A crucial step in the management of stingless bees in Thailand was the development of methodologies for the division of colonies allowing for colony increase. Presently the primary uses of managed meliponine colonies in Thailand are for pollination and the sales of hive products, principally honey and cerumen. The extraction and sale of stingless bee colony pollen stores is very limited in SE Asia. In Central and South America, honey, pollen, and cerumen of numerous stingless bee species are utilized for traditional purposes such as medicine and food resources (Ayala et al. 2013; Obiols and Vásquez 2013). In Mexico and Guatemala, *Melipona beecheii* pollen is used in traditional therapies (Vit et al. 2004). Menezes

B. Chuttong (✉) • R. Phongphisutthinant
Science and Technology Research Institute, Chiang Mai University, Chiang Mai 50200, Thailand
e-mail: bajaree@yahoo.com

K. Sringarm
Department of Animal and Aquatic Science, Faculty of Agriculture, Chiang Mai University,
Chiang Mai 50200, Thailand

M. Burgett
Department of Horticulture, Oregon State University, Corvallis, OR 97331, USA

O.M. Barth
Instituto Oswaldo Cruz, Fiocruz, Brazil

© Springer International Publishing AG, part of Springer Nature 2018
P. Vit et al. (eds.), *Pot-Pollen in Stingless Bee Melittology*, DOI 10.1007/978-3-319-61839-5_22

Fig. 22.1 Pot-pollen of *Tetragonula laeviceps* complex species (Photo: B. Chuttong)

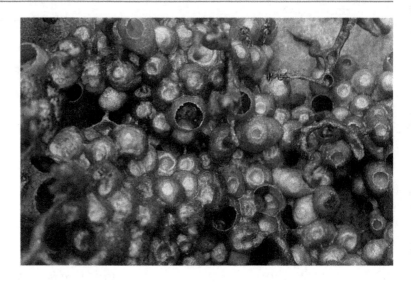

et al. (2013) discussed the use of pollen from *Scaptotrigona* spp. in human diets as an ingredient in beverages and the use of fresh, frozen, and dehydrated pollen. The amount of pollen produced from *Scaptotrigona* spp. colonies is much more than their honey. In Malaysia Omar et al. (2016) evaluated the antioxidant activity of *Lepidotrigona terminata* pollen and the use of its methanol extract to inhibit growth of breast cancer cells. The results showed that stingless bee pollen extract can be useful as a supplementary treatment for chemotherapy drugs. There are no regulations in Thailand regarding the medicinal use of pollen or extracts.

22.2 Shape and Volume of Stingless Bee Pollen Pots

Stingless bees store pollen within pots (made from cerumen) which differs from honey bees where pollen is stored in wax comb cells (Menezes et al. 2013). Stingless bees collect nectar and pollen from plants and place them into separate storage pots within the nest matrix. There is often a pronounced separation of honey pots from pollen pots within the food storage area, but such is not the case for all stingless bee species. When a pollen pot is full, the pot's narrowed entrance is sealed with a cerumen plug, which is unlike pollen storage cells of *A. mellifera.*

The characteristics of pollen pot shape and size vary between bee species, but food storage pots are always much larger than brood cells. The stored pollen varies in flavor, color, and texture dependent on floral species and microbiotic changes which occur during storage (Camargo et al. 1992). From ongoing research in Thailand with the *T. laeviceps* species complex, pollen pots and honey pots are very similar in shape and volume with an average pollen pot volume of 27.7 ± 6.2 mm^3 (Fig. 22.1) and an average honey pot volume of 27.9 ± 6.7 mm^3. Field observations of *T. testaceitarsis*, *L. terminata*, and *L. flavibasis* pollen pots have shown similarities in honey and pollen pot volumes albeit they are larger storage pots than those seen for the *T. laeviceps* species complex. These four stingless bee species have soft and sticky storage pots, but the storage pots of the *T. laeviceps* species complex and *T. testaceitarsis* are shown to be thicker than those for *L. terminata* and *L. flavibasis*. For the *T. laeviceps* species complex (Fig. 22.2) and *T. testaceitarsis*, their pollen pot and honey pot clusters are separated, while for *L. terminata* and *L. flavibasis*, their honey and pollen pots are intermixed (Fig. 22.3). Table 22.1 summarizes pollen pot volume of the four species we examined.

Fig. 22.2 Pot honey of
Tetragonula laeviceps
complex species (Photo:
B. Chuttong)

Fig. 22.3 Nest of
Lepidotrigona terminata
(*a*) honey pots, (*b*)
pollen pots, (*c*) brood
cells (Photo: C. Inkham)

Table 22.1 *T. laeviceps* complex species, *T. testaceitarsis*, *L. terminata*, and *L. flavibasis* pollen pot shape and volume

Stingless bee species	Location	Pollen pot shape	Pollen pot size width × height (mm)
Tetragonula laeviceps complex species	Chanthaburi and Chiang Mai	Ovoid-egg	6 × 7
Tetragonula testaceitarsis	Chanthaburi	Ovoid-egg	8 × 13
Lepidotrigona terminata	Chanthaburi	Ovoid-elongate	14 × 26
Lepidotrigona flavibasis	Chiang Mai	Ovoid-elongate	14 × 20

One closely examined colony of *T. laeviceps* species complex contained 1326 pollen pots and 1578 honey pots, with a brood nest containing 4929 capped cells. There were 2275 adult worker bees.

22.3 Nutritional Composition

Pollen is the male gametophyte formed in the anthers of flowering plants. Pollen is the major source of protein, lipids, vitamin, and minerals for bees (Almeida-Muradian et al. 2005; Campos et al. 2008; Krell 1996; Michener 1974). The chemical composition of pollen differs according to the plant species, nutrient condition of the plant when pollen is developed and storage within the nest (Herbert and Shimanuki 1978).

Research concerning the nutritional value of bee-collected pollen has been dominated by investigations with the Western honey bee (*Apis mellifera*) pollen (Bogdanov 2004; Campos et al. 2008; Herbert and Shimanuki 1978; Human and Nicolson 2006; Serra-Bonvehi and Escolà-Jordà 1997). The nutritional composition of bee-collected pollen shows a high concentration of reducing sugars, essential amino acids, unsaturated and saturated fatty acids, and the presence of minerals which would make bee pollen a valuable addition to the human diet (Almeida-Muradian et al. 2005; Campos et al. 2008; Serra-Bonvehi and Escolà-Jordà 1997). A recommended daily pollen intake as a dietary supplement is 5–10 g/day (M.T. Sancho, personal communication). Bee pollen has been marketed as a health food with an expansive range of putative nutritional and therapeutic properties (Campos et al. 1997, 2008; Wang et al. 1993). There are only a few investigations regarding the chemical makeup of stingless bee-collected pollen (Silva et al. 2006, 2009, 2014; Vit et al. 2016).

22.3.1 Macronutrients of Pot-Pollen

Stored pollen of four species of stingless bee, *T. laeviceps* species complex (*n* = 3), *T. testaceitarsis* (*n* = 3), *L. terminata* (*n* = 3), and *L. flavibasis* (*n* = 3), was taken from storage pots of living colonies located in the Chanthaburi and Chiang Mai provinces, Thailand. The macronutrient analyses, including moisture, ash, crude fat, and crude protein, followed standard methods of the Association of Official Analytical Chemists (AOAC 2005). The results are expressed as grams per 100 g dry weight.

Moisture content percentage was calculated by drying samples in an oven at 100 °C for 2 h. The dry samples were put into desiccators, allowed to cool, and then reweighed.

Ash percentage was calculated by placing crucibles in a 100 °C oven for 1 h. After cooling the crucible was weighed. Then, weighed samples were placed into a crucible and incinerated in a 500 °C muffle furnace for 2 h prior to weighing.

Crude protein was determined by the Kjeldahl method, and total protein content was calculated as the amount of total N multiplied by a nitrogen to protein conversion factor of 6.25.

Crude fat percentage was calculated by drying fats after extraction in a Soxhlet using diethyl ether.

Total carbohydrate and total energy calculations were performed according to Compendium of Methods for Food Analysis (2003). The total carbohydrate content was calculated by the following formula: Carbohydrate = 100 – percentage of (protein + fat + moisture + ash).

Total energy was calculated by the following equation: Total energy (kcal/100 g) = (% protein × 4) + (% carbohydrates × 4) + (% fat × 9).

The results of the macronutrient analysis (moisture, ash, protein, fat, and carbohydrate) of the four stingless bee species are shown in Fig. 22.4. These results are compared with two studies which looked at corbicular pollen as collected and stored by *A. mellifera* (Herbert and Shimanuki 1978; Human and Nicolson 2006).

The variation in macronutrients between pollen from the four Thai stingless bee species is small, with the exception of the moisture content of *T. laeviceps* complex species that we studied (16.1 ± 1.1 g/100 g), which may represent pollen that had undergone a longer storage period within the colony. Our results in moisture content of stored pollen of *T. testaceitarsis* (31.7 ± 1.2 g/100 g) were higher than the results of moisture content of *A. mellifera* stored pollen from Herbert and Shimanuki (1978) (23.8 g/100 g) and Human and Nicolson (2006) (21.0 g/100 g). But our results in moisture content of stored pollen of *L. terminata* (25.3 ± 0.3 g/100 g) and *L. flavibasis* (22.8 ± 0.5 g/100 g) are within the

Fig. 22.4 Macro-nutritional analysis of pot-pollen from four stingless bee species compared to corbicular pollen from *A. mellifera*. [1]Herbert and Shimanuki (1978); [2]Human and Nicolson (2006)

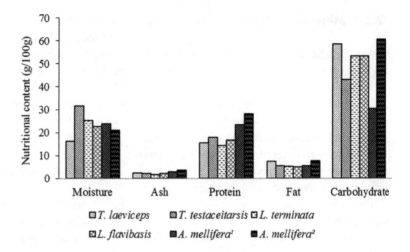

ranges as found in *A. mellifera* stored pollen. Our results of moisture content of stored pollen from four species were remarkably lower when compared to the report from Vit et al. (2016) who report the moisture content of *Melipona* sp. aff. *eburnea* (48.5 g/100 g) and *Scaptotrigona* sp. cf. *ochrotricha* (43.5 g/100 g) stored pollen.

The ash content in stingless bee stored pollen of the four species we analyzed was 2.3 ± 0.4 g/100 g (*T. laeviceps* species), 2.2 ± 0.1 g/100 g (*T. testaceitarsis*), 1.8 ± 0.2 g/100 g (*L. terminata*), and 2.2 ± 0.2 g/100 g (*L. flavibasis*); these results fall in the same range with the report of ash content in stored pollen of *Melipona* sp. aff. *eburnea* (2.3 g/100 g) and *Scaptotrigona* sp. cf. *ochrotricha* (1.9 g/100 g) (Vit et al. 2016), but are lower than the reports for *A. mellifera* stored pollen from Herbert and Shimanuki (1978) (2.8 g/100 g) and Human and Nicolson (2006) (3.6 g/100 g).

The stingless bee stored pollen protein content of the four species we examined was 15.5 ± 2.6 g/100 g (*T. laeviceps* species), 17.9 ± 1.9 g/100 g (*T. testaceitarsis*), 14.3 ± 0.6 g/100 g (*L. terminata*), and 16.7 ± 0.4 g/100 g (*L. flavibasis*). This compares to 23.3 g/100 g for *A. mellifera* stored pollen (Herbert and Shimanuki 1978) and 28.1 g/100 g from an *A. mellifera* monofloral stored pollen (Human and Nicolson 2006). The protein content from stingless bee stored pollen from Venezuela (Vit et al. 2016) gave protein contents of 18.3 g/100 g (*Melipona* sp. aff. *eburnea*) and

16.8 g/100 g (*Scaptotrigona* sp. cf. *ochrotricha*) which are within the boundaries of our results.

Our results of fat content in stored pollen showed small variation in four species we examined, 7.4 ± 0.3 g/100 g (*T. laeviceps*), 5.4 ± 0.6 g/100 g (*T. testaceitarsis*), 5.3 ± 0.1 g/100 g (*L. terminata*), and 4.9 ± 0.04 g/100 g (*L. flavibasis*), which are related to the reports of *A. mellifera* fat content in stored pollen from Herbert and Shimanuki (1978) (5.4 g/100 g) and Human and Nicolson (2006) (7.6 g/100 g). Vit et al. (2016) report the fat content of two species of stingless bee stored pollen: 3.2 g/100 g (*Melipona* sp. aff. *eburnea*) which is lower than our results and *Scaptotrigona* sp. cf. *ochrotricha* stored pollen which is in the same boundaries as our study.

Our results in carbohydrate content of stored pollen of *T. laeviceps* species complex (58.7 ± 3.5 g/100 g), *L. terminata* (53.4 ± 1.0 g/100 g), and *L. flavibasis* (53.3 ± 0.8 g/100 g) were related to Human and Nicolson (2006) who report *A. mellifera* stored pollen carbohydrate (60.7 g/100 g). Except for our result of carbohydrate in *T. testaceitarsis* (43.1 ± 2.8 g/100 g), stored pollen was lower. Our results in carbohydrate content of stored pollen were higher when compared to the report of Herbert and Shimanuki (1978) for *A. mellifera* stored pollen (30.4 g/100 g) and the report of Vit et al. (2016) for two species of stingless bee from Venezuela: *Melipona* sp. aff. *eburnea* (27.7 g/100 g) and *Scaptotrigona* sp. cf. *ochrotricha* (31.0 g/100 g).

22.3.2 Mineral Analysis

Mineral content was analyzed by an inductively coupled plasma optical emission spectrophotometer following the standard methods of AOAC (2005). The dried samples were digested with nitric acid and hydrochloric acid (1:3) at 200 °C for 30 min. Each sample was filtered using filter paper (0.45 micron) and stored in washed glass vials before analysis. The samples were diluted to provide concentrations in the proper absorption range.

The results of mineral analysis of the four stingless bee species are shown in Table 22.2. These results are compared to a recent Brazilian study analyzing pollen mineral content including a mineral analysis from the stored pollen of a Brazilian stingless bee species (Morgano et al. 2012). See Table 22.2.

The levels of each mineral examined (K, Ca, Mg, Na) were not significantly different in the stored pollen of the Thai stingless bee species studied here. Potassium is reported as the highest mineral concentration in pollen of *A. mellifera* (Stanley and Linskens 1974; Herbert and Shimanuki 1978; Serra-Bonvehi and Escolà-Jordà 1997) which is similar to that reported for the stingless bee *Melipona subnitida* (Silva et al. 2014) and our results. From our investigation the Na content appears in the same range as the

report of Na in Brazilian pollen, but it was not detected in pollen of *Melipona subnitida*. In comparing the Paleotropical pollens of SE Asia to the neotropical pollens examined by Morgano et al. (2012), Ca appears to be the only mineral which is significantly higher than those plant pollens examined from Brazil.

22.3.3 Fatty Acid Analysis

Pollen contains lipids and fatty acids, and their composition is variable depending on the plant species. Fatty acids are essential to honey bees as a source of energy, development, nutrition, and reproduction (Manning 2001). Fatty acids from numerous species of plants and pollen collect by *A. mellifera* have been reported by researchers (Serra-Bonvehi and Escolà-Jordà 1997; Herbert and Shimanuki 1978; Human and Nicolson 2006; Loper et al. 1980: Manning and Harvey 2002; Yang et al. 2013; Weiner et al. 2010). Fatty acids are also necessary for human nutrition. Although humans can synthesize saturated fatty acids and some monounsaturated fatty acids, some essential fatty acids including linoleic acid, alpha-linolenic acid, and omega-3 and omega-6 fatty acids cannot be synthesized and must be obtained from food (Mann and Truswell 2012; Robert and Maurice 1980).

Table 22.2 Comparison of mineral content of Thai pot-pollen and Brazilian references

| Bee species | Mineral content (mg/kg) | | | |
	Potassium (K)	Calcium (Ca)	Magnesium (Mg)	Sodium (Na)
Tetragonula laeviceps complex species	5656.0 ± 1274.9	2566.0 ± 489.3	1150.0 ± 222.6	89.9 ± 20.0
Tetragonula testaceitarsis	4594.7 ± 521.0	2904.0 ± 546.2	1318.0 ± 95.3	133.5 ± 48.6
Lepidotrigona terminata	4606.3 ± 75.6	2507.3 ± 4.2	1176.0 ± 4.6	77.2 ± 5.9
Lepidotrigona flavibasis	5125.7 ± 30.0	2719.7 ± 94.6	1315.3 ± 2.5	81.7 ± 6.4
Brazilian *Melipona subnitida*[a]	5918.5 ± 98.8	1846.1 ± 19.6	975.4 ± 25.4	ND
Brazilian *Apis mellifera* bee pollen[b]	5089.0 ± 1981.0	2215.0 ± 984.0	1179.0 ± 455.0	85.0 ± 177.0

ND not detected
[a]Silva et al. (2014)
[b]Morgano et al. (2012)

Our detailed fatty acid analysis was confined to the pollen stores from a single stingless bee species (*T. laeviceps* complex sp.). Our findings are shown in Table 22.3. Stingless bee pollen samples were collected from five individual colonies. The fatty acids were analyzed by gas chromatography-flame ionization detector following the standard methods of AOAC (2005).

Our analysis shows the amounts of unsaturated fatty acids to be higher than saturated fatty acids. The ratio of polyunsaturated to saturated fatty acids was 1.59. When the ratio of polyun-

Table 22.3 Fatty acid spectra of *Tetragonula laeviceps* species stored pollen

Fatty acids	Content of fatty acids (g/100 g)
Myristic acid (C14:0)	0.02 ± 0.01
Pentadecanoic acid (C15:0)	0.04 ± 0.04
Palmitic acid (C16:0)	1.65 ± 0.49
Stearic acid (C18:0)	0.20 ± 0.05
Arachidic acid (C20:0)	0.07 ± 0.06
Heneicosanoic acid (C21:0)	0.02 ± 0.01
Behenic acid (C22:0)	0.11 ± 0.03
Lignoceric acid (C24:0)	0.12 ± 0.06
Total saturated	*2.30 ± 0.59*
Palmitoleic acid (C16:1n7)	0.04 ± 0.03
cis-9-Oleic acid (C18:1n9c)	0.30 ± 0.07
cis-11-Eicosenoic acid (C20:1n11)	0.11 ± 0.10
Erucic acid (C22:1n9)	0.11 ± 0.12
Total monounsaturated	*0.56 ± 0.25*
trans-Linolelaidic acid (C18:2n6t)	0.13 ± 0.05
cis-9,12-Linoleic acid (C18:2n6)	1.52 ± 0.27
Gamma-linolenic acid (C18:3n6)	0.01 ± 0.01
Alpha-linolenic acid (C18:3n3)	1.40 ± 0.62
cis-11,14-Eicosadienoic acid (C20:2)	0.09 ± 0.12
Arachidonic acid (C20:4n6)	0.01 ± 0.01
Total polyunsaturated	*3.10 ± 0.82*
Total unsaturated fat	*3.66 ± 0.18*
Trans fat	*ND – 0.17*
Omega-3	*1.48 ± 0.73*
Omega-6	*1.54 ± 0.27*
Omega-9	*0.40 ± 0.18*

ND not detected

saturated and saturated fatty acid in human diet is greater than 1, HDL cholesterol can be slightly reduced by very high intakes of polyunsaturated fatty acids (Mann and Truswell 2012). Our results correspond to Serra-Bonvehi and Escolà-Jordà (1997) who indicated the ratio of unsaturated to saturated fatty acids was 1.96.

From our analysis of saturated fatty acids, palmitic acid is shown to be the most prevalent. This result corresponds to the report of Human and Nicolson (2006) who studied saturated fatty acids on *A. mellifera* stored pollen from *Aloe greatheadii* var. *davyana*. Serra-Bonvehi and Escola-Jorda (1997) examined *A. mellifera* collected pollen but did not specify the plant species. Their results showed palmitic acid was the prevalent fatty acid. Yang et al. (2013) investigated fatty acids of *A. mellifera* collected pollen from 12 different locations in China. Their results showed palmitic acid to be the predominant saturated fatty acid in all geographic locations except for one location where palmitic acid was slightly lower than stearic acid.

22.3.4 Amino Acid Analysis

Amino acids are simple organic compounds found in living organisms. About 20 amino acids are common in humans which are necessary for protein assembly (Hardy 1985). The dietary protein requirement for humans should contain sufficient and digestible amounts of nine essential amino acids (histidine, isoleucine, leucine, lysine, methionine, phenylalanine, threonine, tryptophan, and valine). Amino acids that can become essential under specific physiological or pathological conditions include cysteine, tyrosine, glycine, arginine, glutamic, and proline. Adequate total amino acid nitrogen can be derived from any of the above amino acids and from nonessential amino acids such as aspartic acid, asparagine, glutamic acid, alanine, and serine or other sources of nonessential nitrogen (WHO/FAO/UNU 2007).

The high protein content and variety of amino acids found in bee-collected pollen have been reported by many researchers (Herbert and Shimanuki 1978; Loper and Cohen 1987;

Szczesna 2006; Weiner et al. 2010), most of which examined pollen collected by *A. mellifera*. Silva et al. (2014) investigated the amino acid content of pollen from the stingless bee species *Melipona subnitida* from Brazil.

We report on the amino acid spectra for only one species of stingless bee (*T. laeviceps* species complex sp.), which is the most commonly encountered species in Thai meliponiculture. The results of our analysis are shown in Table 22.4. Five samples of stingless bee stored pollen were collected from five *T. laeviceps* species colonies. The amino acids were analyzed by mass spectrometry-gas chromatography following the standard methods of AOAC (2005).

Twenty amino acids are identified in our polyfloral pollen samples taken from five *T. laeviceps* colonies. For the essential amino acids, lysine, phenylalanine, and leucine were the dominant amino acids in our samples. Szczesna (2006) who examined amino acids of *A. mellifera* bee-collected pollen from Poland, Korea, and China found lysine and leucine to be the predominant amino acids. Human and Nicolson (2006) also reported leucine and lysine to be the predominant essential amino acids from *A. mellifera* stored pollen. Our results of nonessential amino acids show glutamic acid and tyrosine to be the dominant amino acids, while Human and Nicolson (2006) reported glutamic and aspartic acids at the highest concentration. Our results show lower overall amino acid levels compared to their report with the caveat that their research was based on monofloral pollen from *Aloe greatheadii* var. *davyana*. Silva et al. (2014) examined stingless bee pollen from *Melipona subnitida* in Brazil. This study found proline to be the predominant amino acids (0.11 mg/100 g) which is lower than our result of proline content. Overall Silva et al. (2014) found markedly lower levels of amino acids than our observations of our bee of the *T. laeviceps* complex in SE Asia.

The amino acid composition of plant pollen is recognized as being highly dependent on plant species; however, in a study of 142 plant-specific pollens, Weiner et al. (2010) comment that while amino acid composition varies between plant species, all examined pollen species contained a complete set of essential amino acids. Our examination of the stored pollen from *T. laeviceps* species illustrates that the entire essential as well as the nonessential amino acids are present in the colonial pollen stores of this species.

Table 22.4 Amino acids in *Tetragonula laeviceps* complex species pot-pollen

Amino acids	Amino acid content (g/100 g)
Essential amino acids	
Histidine	0.96 ± 0.07
Isoleucine	0.88 ± 0.06
Leucine	1.62 ± 0.12
Lysine	2.46 ± 0.14
Methionine	0.12 ± 0.03
Phenylalanine	1.63 ± 0.26
Threonine	0.17 ± 0.01
Tryptophan	0.13 ± 0.14
Valine	0.65 ± 0.29
Nonessential amino acids	
Alanine	0.40 ± 0.02
Arginine	<0.02
Aspartic acid	0.70 ± 0.06
Cysteine	0.15 ± 0.04
Glutamic acid	1.12 ± 0.01
Glycine	0.26 ± 0.02
Hydroxylysine	<0.02
Hydroxyproline	<0.02
Proline	0.60 ± 0.02
Serine	0.22 ± 0.02
Tyrosine	1.00 ± 0.08

22.4 Botanical Origin

Pollen samples were homogenized using ethanol, centrifuged and the sediment was resuspended in equal volumes of water and glycerin. One drop of this well-homogenized mixture was put on a microscope slide, covered with a cover glass, and sealed with nail polish. The durability of these preparations is about 10 days (Barth et al. 2010). More than 500 pollen grains per sample were counted at 400 × magnification. Evaluation follows the rule of monofloral patches when more than 90% of the pollen grains belong to a unique plant species or are of 60% when no accessory pollen type (15–45%) was present.

Table 22.5 Palynology analysis of stingless bee (Meliponini) pot-pollen

| Stingless bee species | Main pollen types | | | | Diagnosis |
| | Most frequent | Frequent | Less frequent | Rare | |
	++++	+++	++	+	
Tetragonula laeviceps complex species[a]	*Trema micrantha*	*Trema micrantha*, (d), (l)	(j), (k)	*Cocos nucifera*, (a), (d), (j), (k), (l), et al.	Heterofloral
Tetragonula testaceitarsis[a]	(−)	*Cocos nucifera*, (a), (b), (d), (e)	*Cocos nucifera*, (a), (b), (f), (g), (i)	*Cocos nucifera*, *Acacia*, (a), (h), (g), et al.	Heterofloral
Lepidotrigona terminata[b]	*Trema micrantha*	*Acacia*, (a)	(a), (b), (c)	*Cocos nucifera*, (b), (d), et al.	Heterofloral
Lepidotrigona flavibasis[b]	*Tapirira* sp. (a)	*Tapirira* sp.	(a)	Several	Bifloral

[a]Aggregate of a five-colony sample
[b]Aggregate of a three-colony sample
Unknown pollen types:
(a) = 1-colpate, medium size, reticulate (Monocotiledonea)
(b) = 1-colpate, small size, reticulate (Monocotiledonea)
(c) = 2-colpate, small size, psilate
(d) = 3-colporate, small size, psilate
(e) = 3-colporate, small size, microreticulate (Fabaceae-Faboideae type)
(f) = 1-colpate, large size, psilate (Arecaceae type)
(g) = 2-colpate, small size, spiculate
(h) = 2-porate, small size, psilate (Urticaceae type)
(i) = 4-colporate, small size, strongly reticulate
(j) = 3-colporate, small-medium size, strongly reticulate
(k) = 1-colpate, medium size, microreticulate (Monocotiledonea)
(l) = 3-colporate, very small size, psilate
(−) = absent

Table 22.5 shows the most frequent to the rarest pollen types. The right column gives the interpretation of the data obtained. Many pollen types remain taxonomically unknown, which is not unexpected as pollen keys for SE Asia are very limited, as exemplified by a study of pollen resources for the SE Asian night-flying carpenter bee *Xylocopa* (*Nyctomelitta*) *tranquebarica* (Fabricius, 1804) (Burgett et al. 2005). Pollen type assemblages can group some samples. Monofloral, bifloral and heterofloral pollen batches occurred. The monofloral samples proceed from a unique plant species and the bifloral are a mixture of pollen grains from two plant taxa. The heterofloral samples showed several pollen types that signified no predominant taxon.

Pollen composition was similar for individual bee colonies at specific geographic locations; samples from three colonies of *L. terminata*, as well as five samples from five *T. testaceitarsis* colonies, five colonial samples of *T. laeviceps* complex sp. and three colonial samples from *L. flavibasis* comprised mainly heterofloral pollen batches. Results expressed in Table 22.5 represent species conglomerates. The remaining samples are classified as heterofloral also presenting a variable number of pollen types, all of them at low frequencies.

Among the 16 pot-pollen samples, a total of four important plant taxa could be identified (*Acacia* sp., *Cocos nucifera*, *Tapirira* sp., and *Trema micrantha*) (Fig. 22.5), additionally twelve taxa were recognized and not identified, and there were a large number of sporadic species.

The most significant pollen types were *Trema micrantha* (Cannabaceae), a wind-pollinated

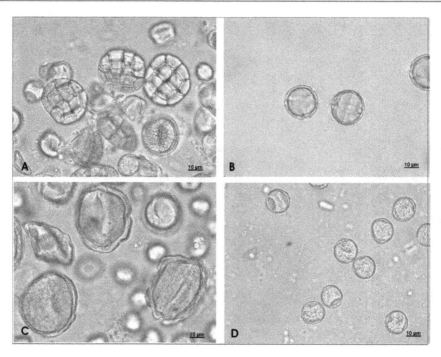

Fig. 22.5 Pollen types found in pot-pollen samples: (**a**) *Acacia* sp., (**b**) *Tapirira* sp., (**c**) *Cocos nucifera*, and (**d**) *Trema micrantha*

invasive, and *Tapirira* sp., a tree of the Anacardiaceae family. The third predominant pollen type belongs to an unknown plant species, possibly a palm tree (pollen type (a) of Table 22.5). Another plant that contributes greatly is *Cocos nucifera*, the coconut palm (Arecaceae), which was present in 11 of the 16 samples. Another important pollen type was *Acacia*, but less common.

As meliponines exhibit polylecty (Eltz et al. 2001; Ramalho et al. 1989, 1990; Roubik and Moreno 2013), the pollen stores found in stingless bee colonies will display great taxonomic variation dependent primarily on geographical location and seasonality. It should be stressed that the pollen identification from our samples represents a geographical and seasonal "snapshot."

22.5   Conclusions, Suggestion, and Future Research

Analyses of pot-pollen nutritional composition are consistent with the numerous researchers on honey beebread (*A. mellifera*) pollen. The macronutrients

and minerals are shown to be similar to those reported for beebread pollen. For mineral content, potassium is found to be the major component which correlates to the reports from several studies in honey bee and stingless bee pollen. The results of fatty acids and amino acids content from our stored pollen are compatible with previous findings. The composition of stored pollen from any bee species is primarily dependent on plant species, season, and the geographical location where bees collect pollen. The difference in plant species showed variability in their pot-pollen composition. Our study of the botanical origins shows a relatively restricted view of plants visited intensively by four species of stingless bee. Our study is the first report of the nutritional composition of pollen from four species of stingless bee in SE Asia: a species of the *T. laeviceps* complex, *T. testaceitarsis*, *L. terminata*, and *L. flavibasis*. There are many stingless bee species which have yet to be examined. It will be necessary to carry out further research on stingless bee pot-pollen from SE Asia to better understand a broader scope of nutritional qualities and host plant breadth.

Acknowledgments This research was supported by Chiang Mai University Short-Term Research in Overseas. We thank Dr. Hans Banziger, Chiang Mai University for his precise stingless bee identification. We thank all the stingless bee beekeepers who assisted us in the collection of pot-pollen samples. We are grateful to the technical assistance of MSc. Alex da Silva de Freitas in preparing the pollen slides with financial support of the Brazilian National Council of Research "*Conselho Nacional de Desenvolvimento Científico e Tecnológico.*"

References

Almeida-Muradian LB, Pamplona LC, Coimbra S, Barth OM. 2005. Chemical composition and botanical evaluation of dried bee pollen pellets. Journal of Food Composition and Analysis 18: 105-111.

AOAC International. 2005. Official methods of analysis of AOAC International. AOAC International; Virginia, USA. 771pp.

Ayala R, Gonzalez VH, Engel MS. 2013. Mexican stingless bees (Hymenoptera: Apidae): Diversity, distribution, and indigenous knowledge. pp. 135-152. In: Vit P, Pedro SRM, Roubik D, eds. Pot-Honey: A legacy of stingless bees. Springer. New York, USA. 654 pp.

Barth OM, Freitas AS, Oliveira S, Silva RA, Maester FM, Andrella RR, Cardozo G M. 2010. Evaluation of the botanical origin of commercial dry bee pollen load batches using pollen analysis: a proposal for technical standardization. Anais da Academia Brasileira de Ciências 82: 893-902.

Bogdanov S. 2004. Quality and standards of pollen and beeswax. Apiacta 38: 334-341.

Burgett DM, Sukumalanand P, Vorwohl G. 2005. Pollen species resources for *Xylocopa* (*Nyctomelitta*) *tranquebarica* (F.) – a night-flying carpenter bee (Hymenoptera: Apidae) of Southeast Asia. Science Asia 31: 61-64.

Camargo JMF, Garcia MVB, Junior ERQ, Castrillon A. 1992. Notas previas sobre a bionomia de *Ptilotrigona lurida* (Hymenoptera, Apidae, Meliponinae): associação de leveduras em pólenes tocado. Boletim do Museu Paraense Emílio Goeldi 8: 391–395.

Campos MG, Bogdanov S, de Almeida-Muradian LB, Szczesna T, Mancebo Y, Frigerio C, Ferreira F. 2008. Pollen composition and standardisation of analytical methods. Journal of Apicultural Research 47: 154-161.

Campos M, Markham KR, Mitchell KA, da Cunha AP. 1997. An approach to the characterization of bee pollens via their flavonoid/phenolic profiles. Phytochemical Analysis 8: 181–185.

Chuttong B, Chanbang Y, Burgett M. 2014. Meliponiculture: Stingless Bee Beekeeping in Thailand. Bee World 91(2): 41–45.

Compendium of methods for food analysis (in Thai). 2003. Department of Medical Sciences (DMSc), National Bureau of Agricultural Commodity and Food Standards (ACFS). Bangkok.

DOAE. 2014. Department of Agricultural Extension. Thai Ministry of Agriculture and Cooperatives. Region 3. Chanthaburi. Personal communication.

Eltz T, Brühl CA, Van der Kaars S, Chey VK, Linsenmair KE. 2001. Pollen foraging and resource partitioning of stingless bees in relation to flowering dynamics in a Southeast Asian tropical rainforest. Insectes Sociaux 48: 273-279.

Hardy PM. 1985. The protein amino acids. pp. 7-24. In: Barrett GC, ed. Chemistry and Biochemistry of the Amino Acids. Chapman and Hall Ltd, London, UK. 684 pp.

Herbert EW, Shimanuki H. 1978. Chemical composition and nutritive value of bee-collected and bee-stored pollen. Apidologie 9: 33-40.

Human H, Nicolson SW. 2006. Nutritional content of fresh, bee-collected and stored pollen of Aloe greatheadii var. davyana (Asphodelaceae). Phytochemistry 67: 1486-1492.

Krell R. 1996. Value-added products from beekeeping (No. 124). Food and Agriculture Organization of the United Nations. Rome, Italy, 409 pp.

Loper GM, Cohen AC. 1987. Amino acid content of dandelion pollen, a honey bee (Hymenoptera: Apidae) nutritional evaluation. Journal of Economic Entomology 80: 14-17.

Loper GM, Standifer LN, Thompson MJ, Gilliam M. 1980. Biochemistry and microbiology of bee-collected almond (*Prunus dulcis*) pollen and bee bread. I. Fatty acids, sterols, vitamins and minerals. Apidologie 11: 63–73.

Mann J, Truswell S. 2012. Essentials of human nutrition. Oxford University Press; Oxford, UK. 683 pp.

Manning R. 2001. Fatty acids in pollen: a review of their importance for honey bees. Bee World 82: 60-75.

Manning R, Harvey M. 2002. Fatty acids in honeybee-collected pollens from six endemic Western Australian eucalypts and the possible significance to the Western Australian beekeeping industry. Animal Production Science 42: 217-223.

Menezes C, Vollet-Neto A, Contrera FAFL, Venturieri GC, Imperatriz-Fonseca VL. 2013. The role of useful microorganisms to stingless bees and stingless beekeeping. pp. 153-171. In: Vit P, Pedro SRM, Roubik D, eds. Pot-Honey: A legacy of stingless bees. Springer. New York, USA. 654 pp.

Michener CD. 1974. The social behavior of the bees: a comparative study. Harvard University Press: Cambridge, Massachusetts, USA. 418 pp.

Michener CD. 2013. The Meliponini. pp. 3-17. In: Vit P, Pedro SRM, Roubik D, eds. Pot-Honey: A legacy of stingless bees. Springer. New York, USA. 654 pp.

Morgano MA, Martins MCT, Rabonato LC, Milani RF, Yotsuyanagi K, Rodriguez-Amaya DB. 2012. A comprehensive investigation of the mineral composition of Brazilian bee pollen: geographic and seasonal variations and contribution to human diet. Journal of the Brazilian Chemical Society 23: 727-736.

Obiols CLY, Vásquez M. 2013. Stingless Bees of Guatemala. pp. 99-111. In: Vit P, Pedro SRM, Roubik D, eds. Pot-Honey: A legacy of stingless bees. Springer. New York, USA. 654 pp.

Omar WAW, Azhar NA, Fadzilah NH, Kamal NNSNM. 2016. Bee pollen extract of Malaysian stingless bee enhances the effect of cisplatin on breast cancer cell lines. Asian Pacific Journal of Tropical Biomedicine 6: 265-269.

Ramalho M, Kleinert-Giovannini A, Imperatriz-Fonseca VL. 1989. Utilization of floral resources by species of Melipona (Apidae, Meliponinae): Floral preferences. Apidologie 20: 185-195.

Ramalho M, Kleinert-Giovannini A, Imperatriz-Fonseca VL. 1990. Important bee plants for stingless bees (Melipona and Trigonini) and Africanized honey-bees (*Apis mellifera*) in neotropical habitats: a review. Apidologie 21: 469-488.

Rasmussen C. 2008. Catalog of the Indo-Malayan/ Australasian stingless bees (Hymenoptera: Apidae: Meliponini). Zoo Taxa, New Zealand. 80 pp.

Robert SG, Maurice ES. 1980. Modern Nutrition in Health and Disease (6th ed.). pp. 134–138. Lea and Febinger; Philadelphia, USA. 80 pp.

Roubik DW, Moreno, PJE. 2013. How to be a bee-botanist using pollen spectra. pp. 295-314. In: Vit P, Pedro SRM, Roubik D, eds. Pot-Honey: A legacy of stingless bees. Springer. New York, USA. 654 pp.

Serra-Bonvehi SJ, Escolà Jordà R. 1997. Nutrient composition and microbiological quality of honeybee-collected pollen in Spain. Journal of Agricultural and Food Chemistry 45: 725-732.

Silva GR, da Natividade TB, Camara CA, da Silva EMS, dos Santos FDAR, Silva TMS. 2014. Identification of sugar, amino acids and minerals from the pollen of Jandaíra stingless bees (*Melipona subnitida*). Food and Nutrition Sciences 5: 1015.

Silva TMS, Camara CA, da Silva Lins AC, Barbosa-Filho JM, da Silva EMS, Freitas BM, dos Santos FDAR. 2006. Chemical composition and free radical scavenging activity of pollen loads from stingless bee *Melipona subnitida* Ducke. Journal of Food Composition and Analysis 19: 507-511.

Silva T, Camara CA, Lins A, Agra MDF, Silva E, Reis IT, Freitas BM. 2009. Chemical composition, botanical evaluation and screening of radical scavenging activity of collected pollen by the stingless bees *Melipona rufiventris* (Uruçu-amarela). Anais da Academia Brasileira de Ciências 81: 173-178.

Stanley, R. G., Linskens, H. F. 1974. Pollen: biology, chemistry, management. Springer-Verlag: Berlin. 310 pp.

Szczesna T. 2006. Protein content and amino acid composition of bee-collected pollen from selected botanical origins. Journal of Apicultural Science 50: 81-90.

Vit P, Medina M, Eunice Enríquez M. 2004. Quality standards for medicinal uses of Meliponinae honey in Guatemala, Mexico and Venezuela. Bee World 85: 2-5.

Vit P, Santiago B, Silvia P, Ruiz J, Maza F, Pena-Vera M, Perez-Perez E. 2016. Chemical and bioactive characterization of pot-pollen produced by *Melipona* and *Scaptotrigona* stingless bees from Paria Grande, Amazonas State, Venezuela. Emirates Journal of Food and Agriculture 28: 78-84.

Wang MS, Fan HF, Xu HJ. 1993. Effects of bee pollen on blood and hemopoietic system in mice and rats. Chinese Traditional Herbs and Drugs 588: 601.

Weiner CN, Hilpert A, Werner M, Linsenmair KE, Blüthgen N. 2010. Pollen amino acids and flower specialization in solitary bees. Apidologie 41: 476-487.

World Health Organization/Food and Agriculture Organization/United Nations University (2007) Protein and Amino Acid Requirements in Human Nutrition Report of a Joint WHO/FAO/UNU Expert Consultation. WHO Technical Report Series no. 935. World Health Organization; Geneva, Switzerland. 265 pp.

Yang K, Wu D, Ye X, Liu D, Chen J, Sun P. 2013. Characterization of chemical composition of bee pollen in China. Journal of Agricultural and Food Chemistry 61: 708-718.

Characterization of *Scaptotrigona mexicana* Pot-Pollen from Veracruz, Mexico

Adriana Contreras-Oliva, Juan Antonio Pérez-Sato,
Fernando Carlos Gómez-Merino,
Luz Anel López-Garay, Rogel Villanueva-Gutiérrez,
María Magdalena Crosby-Galván,
and Libia Iris Trejo-Téllez

23.1 Introduction

In Mexico, there are some 18 genera and 46 species of Meliponini (Ayala et al. 2013; Reyes-González et al. 2014), including *Scaptotrigona mexicana*. Although these species display a natural wide Neotropical distribution in eight biogeographic provinces within the country (1) Mexican Pacific Coast; (2) Trans-Mexican Volcanic Belt; (3) Balsas Basin; (4) Southern Sierra Madre; (5) Eastern Sierra Madre; (6) Gulf of Mexico, Chiapas; (7) Mexican Plateau; and (8) Yucatán Peninsula) (Yáñez-Ordóñez et al. 2008), they are meliponiculturally managed mainly in the states of Puebla, Veracruz, Guerrero, Tabasco, Tamaulipas, Chiapas, Campeche and Yucatán, and to a lesser extent in San Luis Potosí and Quintana Roo (Ayala et al. 2013). Importantly, most rural and indigenous communities worldwide harvest stingless bee honey from feral colonies but do not practice meliponiculture (Vit et al. 2004), which may endanger the preservation of such species.

In Mexico, genus *Scaptotrigona* is represented by three species, *Scaptotrigona hellwegeri*, *S. mexicana*, and *S. pectoralis*, out of a total 24 *Scaptotrigona* species known in Neotropical America, from Mexico to Argentina (Ayala et al. 2013; Hurtado-Burillo et al. 2013; Hurtado-Burillo 2015).

Meliponini are highly appreciated because the honey and pollen they produce not only have a high nutritional value but also, with the resin collected by bees, generally called "propolis," may be used in therapeutics to enhance several systems to control digestive, respiratory, skin, and visual disorders, as well as to treat female infertility (Vit et al. 2004).

A. Contreras-Oliva • J.A. Pérez-Sato
F.C. Gómez-Merino
Colegio de Postgraduados Campus Córdoba.
Carretera Córdoba-Veracruz km 348, Congregación Manuel León,
Amatlán de los Reyes, Veracruz, Mexico, C. P. 94946

L.A. López-Garay
Instituto Tecnológico Superior de Zongolica Campus Tequila. Carretera a la Compañía km 4,
Tepetlitlanapa, Zongolica, Veracruz,
Mexico, C. P. 95005

R. Villanueva-Gutiérrez
El Colegio de la Frontera Sur. Unidad Chetumal. Av. Centenario km 5.5,
Chetumal, Quintana Roo, Mexico, C. P. 77014

M.M. Crosby-Galván • L.I. Trejo-Téllez (✉)
Colegio de Postgraduados Campus Montecillo.
Carretera México-Texcoco km 36.5,
Montecillo, Texcoco, State of Mexico,
Mexico, C. P. 56230
e-mail: tlibia@colpos.mx

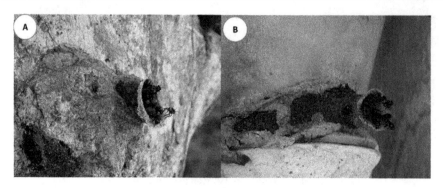

Fig. 23.1 Entrances of a *Scaptotrigona mexicana* natural nest and a managed hive in Amatlán de los Reyes, Veracruz, Mexico. (**a**) A natural nest. (**b**) A managed hive (Photo: J. A. Pérez-Sato)

Of the 46 stingless bees reported in Mexico, only two species, *Melipona beecheii* and *Scaptotrigona mexicana*, are extensively exploited for the production of honey, pollen, and cerumen (Albores-González et al. 2011; Guzmán et al. 2011). In the Yucatán Peninsula and southern Mexico, *Melipona beecheii* is by far the most representative and the most important cultivated species of native stingless bees, whose honey and pollen are used in traditional medicine and regionally marketed (González-Acereto 2008; Reyes-González et al. 2014). On the other hand, in the most northern tropical and mountainous regions in the country, *Scaptotrigona mexicana* is the most traditionally exploited stingless bee, which has an important regional market because of its high-quality honey, used as food and for therapeutic purposes (Ayala et al. 2013; Cano-Contreras et al. 2013; Reyes-González et al. 2014).

Scaptotrigona mexicana is native to Mexico, Belize, Guatemala, El Salvador, and Costa Rica (Arzaluz et al. 2002; Camargo and Pedro 2013; Hurtado-Burillo 2015). In Mexico, the species is distributed along the Atlantic Coast of the Gulf of Mexico and the Eastern Sierra Madre, from Tamaulipas to Veracruz. Furthermore, it is also found in the Trans-Mexican Volcanic Belt, the Southern Sierra Madre, and the Chiapas Sierra Madre, from sea level to an altitude of 1000 m (Hurtado-Burillo 2015; Yáñez-Ordóñez et al. 2008).

This species is black in color, of medium size (5.0–5.3 mm long), with 5.1–5.4-mm-long orange wings (Ayala 1999), and normally builds its nest in the hollows of tree trunks. The trumpet-shaped nest entrance is rather large, made of cerumen, and may contain worker bees guarding the nest from potential enemies (Fig. 23.1).

Scaptotrigona mexicana is the least aggressive stingless bee among the *Scaptotrigona* in Mexico (Hurtado-Burillo 2015). The queen probably mates with a single male and dominates reproduction (Palmer et al. 2002). Interestingly, Mueller et al. (2012) found diverse genetic composition among *S. mexicana* drones in a congregation, which avoid mating with close relatives. This species displays high precision during food recruitment of experienced (reactivated) foragers (Sánchez et al. 2004). Furthermore, both color and shape stimuli assist trained *S. mexicana* foragers in locating the target food (Sánchez and Vandame 2012). However, foragers choose significantly more often the correct target in the color experiments, rather than in shape experiments. Sánchez et al. (2016) investigated whether the most abundant colonies of *S. mexicana* are dominant, finding that colonies of this species do not monopolize resources; instead they seem to share food; however, some colonies had more foragers in a food patch or in a feeder, so some type of exclusion could be at work, though the final output of such interaction (i.e., if underrepresented colonies were eventually excluded, developed slower, or were overrepresented in other food patches) could not be determined. Resource partitioning within this species occurs peacefully, though further studies are needed to determine if threatening behavior or aggressions appear when resources are scarce and competition becomes harsher (Sánchez et al. 2016).

The main product of bees is honey, the natural sweet substance obtained from nectar of flowers and honeydew secretions from other insects (Codex Alimentations 2001). The properties and composition of bee honey depend on its geographical floral origin, season, environmental factors, and handling by beekeepers (Da Costa Leite et al. 2000; Kaškonienė et al. 2010; El Sohaimy et al. 2015).

Although an increasing number of studies on honey physicochemical characterization from *Apis* and stingless bees appear each year, an in-depth evaluation of the pollen stored and then fermented by microbes in stingless bee nests remains a daunting task. Moreover, the traditional values of meliponiculture with *Scaptotrigona mexicana*

have been little documented in Mexico. Therefore, in this chapter we first describe some salient experiences concerning the local use of honey and pollen produced by this stingless bee species and also report some of the chemical parameters, elemental composition, and palynological characterization of its pot-pollen from central Veracruz, one of the main regions where it is sustainably exploited in Mexico. We especially focused on characterization of *S. mexicana* pollen from hives located in three sites in the municipalities of Amatlán de los Reyes: Cañada Blanca (CB; 18° 57′ 10.5″ NL, 96° 51′ 40.4″ WL, 787 masl), Manuel León (ML; 18° 57′ 10.4″ NL, 96° 55′ 40.4″ WL, 650 masl), and Fortín de las Flores (FF; 18° 54′ 27.8″ NL, 96° 57′ 38.4″ WL, 884 masl) (Fig. 23.2). To our knowledge, this is the first detailed study gathering information on *Scaptotrigona mexicana* pollen.

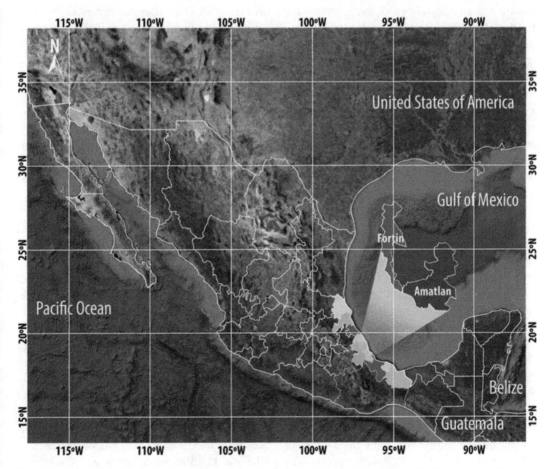

Fig. 23.2 Municipalities of Amatlán de los Reyes (Amatlán) and Fortín de las Flores (Fortín) in central Veracruz, Mexico, where *Scaptotrigona mexicana* pot-pollen samples were collected (Design: F. C. Gómez-Merino)

23.2 Traditional Values of *Scaptotrigona mexicana* Meliponiculture in Mexico

Stingless bee management and exploitation have been long associated with human history in the Americas. One of the most documented experiences has been the relationship between *Melipona beecheii* and the Mayan Culture in Mesoamerica. This species represented an important source of honey, wax, resin, larvae, and pollen and also formed part of Mayan cosmology and relationship with the world (Rosso et al. 2001; Souza et al. 2012). Interestingly, in spite of the presence of some 300 indigenous groups living in Mexico upon the arrival of the Europeans in the sixteenth century, the Maya were the only ones with a deeply rooted tradition in stingless beekeeping. Indeed, meliponiculture was an integral part of the economy and the social and religious life of the Mayan people. They traded honey for therapeutic and domestic purposes, as a sweetener and in the preparation of fermented beverages. Furthermore, they utilized pollen, bee nest cerumen, and resin for domestic and religious proposes (Cortopassi-Laurino et al. 2006; Souza et al. 2012).

The Totonac and Nahua indigenous groups, who currently inhabit mountainous areas in the states of Puebla and Veracruz, Mexico, have the longest tradition in *Scaptotrigona mexicana* meliponiculture. These groups use *S. mexicana* products as nutrient sources and in trade and also for medicinal purposes (Cortopassi-Laurino et al. 2006; Souza et al. 2012).

Within the Totonac culture in Papantla, Veracruz (the central northern region of the state), *S mexicana* is associated with the deity "Kiwikgolo," or Lord of the Forest. Hence, in the Totonac language, this species is named "Kiwitáxkat," or bee of the forest. According to Patlán-Martínez et al. (2015), in the Totonac world, these bees are highly appreciated and respected due to the benefits they bring to human beings and nature. They play a pivotal role as pollinators of important crop plants such as coffee (*Coffea arabica*), citrus (*Citrus* spp.),

and vanilla (*Vanilla planifolia*), the last of which is native to this region. Importantly, the "Kiwitáxkat" produces honey, pollen, and cerumen, among other products, which have nutritional or therapeutic uses. One of the most common beverages prepared using "Kiwitáxkat" honey includes cocoa, vanilla, and water, which has reputedly both stimulant and aphrodisiac effects (García-Flores et al. 2013).

While engaged in the activity of keeping "Kiwitáxkat" bees, the Totonac show great respect for life and nature, as a symbol of their cultural identity. Therefore, management of this stingless bee species is a tradition where Totonac cosmology and cosmovision converge. Accordingly, its management is carried out following certain natural processes, including the moon phases, and the honey harvest is performed basically when the moon is full. Nevertheless, hive products can also be harvested when a woman gives birth and applied in order to facilitate her labor. In the Totonac cosmovision, the association of "Kiwitáxkat" with the birth of a new human being means a connection between nature and life. Both in the Totonac and Nahua cultures, ceramic clay hives (Fig. 23.3a, b) are still used by indigenous people. Furthermore, the Totonac people sometimes use hollow logs as hives, and they harvest the honey and pollen in a more economic and less harmful (to the bees) manner.

Considering the Nahua culture in northern Puebla, in the vicinity of Veracruz, *Scaptotrigona mexicana* is also strongly rooted to ethnic traditions. According to Padilla-Vargas et al. (2014), the Nahua people of Cuetzalan, Puebla, respect the presence and rearing of bees, as shown by the ethical conduct that is so integral to their ethnic pride and identity. In Nahuatl (the language of the Nahua people), this stingless bee species is called "pitsilnekmej," or small bee. This meliponine represented a significant resource managed by Nahua people during pre-Columbian times and still remains important in certain areas today. The Nahua are aware of the plants that these stingless bees feed on. The honey and propolis are used to alleviate 11 different ailments and conditions, while ceremonial objects and plumes for ritual

Fig. 23.3 Hives made from clay for *Scaptotrigona mexicana* meliponiculture. (**a**) A ceramic clay hive employed in Papantla, Veracruz, Mexico, by Totonac people. (**b**) Clay hives usually employed in Cuetzalan, Puebla, Mexico, by Nahua people (Photo: J. A. Pérez-Sato)

dances are made from the cerumen. The most significant beekeeping practice consists of building hives by joining two pots at the mouth and their placement vertically on shelves beneath the roofs of houses (Fig. 23.3b).

Hence, both Nahua and Totonac cultures consider *Scaptotrigona mexicana* a part of their cosmology and relationship to the world. For these indigenous peoples, the bee represents an important source of food and income generated from the production of honey, pollen, propolis, cerumen, and other products derived from the waxes and resins that they collect and process.

23.3 Importance of Pot-Pollen in Meliponiculture

Pollen is the main food for young bees, and it may contain 10–40% protein, depending on environmental and biological factors (Menezes et al. 2012; Vossler 2015). Foragers harvest pollen from plant flowers by scraping or licking it up from the anther and then stick it together, on the corbicula, using nectar. The pollen is gradually formed into pellets on their corbiculae—baskets or polished cavities surrounded by a fringe of hairs and with hairs in the middle on the tibia of each hind leg. After harvesting pollen, bees return to the nest or hive, and pollen pellets are stored inside cerumen pots (Leonhardt et al. 2007). Inside the pots, pollen is basically processed by yeasts and bacteria. Such microorganisms in stored bee food promote biochemical changes that alter nutritional quality and enhance digestion and absorption of nutrients, but probably their main function is to prevent spoilage and diseases (Anderson et al. 2011; Menezes et al. 2013). This process is important for digesting the pollen grains and their nutrients and also for long-term storage (Rosa et al. 2003). After fermentation, organoleptic properties of pollen, including flavor, odor, color, and texture, may change considerably, while the fermented pollen characteristics vary according to bee species, since each species secretes different digestive enzymes from its salivary glands (Menezes et al. 2012; Rebelo et al. 2016). Most stingless bee species, including *Melipona* and *Scaptotrigona*, produce moist and extremely acidic pollen after fermentation, making it unpalatable unless mixed with honey or other sugar-rich foods or sweeteners (Menezes et al. 2012). As a result, this sort of pollen has not been widely exploited by meliponicultures, even though many stingless bee species may store large quantities of it (Cortopassi-Laurino et al. 2006).

Bee pollen is considered one of nature's most nutritionally valuable products. It contains most nutrients required by humans, and, importantly, nearly half of its protein is in the form of free amino acids that are ready to be used directly by the body (Feás et al. 2012). The amazing floristic and climatic diversity found in Mexico is largely responsible for the huge variety of flavors, aromas, and colors of pollen harvested and fer-

Fig. 23.4 Pollen pots of *Scaptotrigona mexicana* in managed hives in Amatlán de los Reyes, Veracruz, Mexico (Photo J. A. Pérez-Sato)

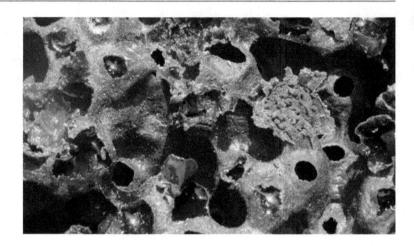

mented by these bees. Importantly, the botanic origin and the final chemical composition of metabolites influence all these pot-pollen organoleptic properties (Almeida-Muradian et al. 2005).

Apart from proteins and amino acids, bee pollen is rich in sugars and lipids, whereas vitamins, minerals, trace elements, and a variety of antioxidant compounds, including flavonoids and polyphenols, are found to a lesser extent (Kroyer and Hegedus 2001; Vit et al. 2016). According to Pascoal et al. (2014), pot-pollen contains crude carbohydrates, fiber, proteins, and lipids in the following percentages: 13–55, 0.3–2.0, 10–40, and 1–10, respectively. The ranges are very wide due to the variation in the bee species themselves and in the plant species from which they collect the pollen grains (Cimpoiu et al. 2013).

An increasing number of studies are conducted in order to characterize honey and pollen from *Apis mellifera* (i.e., Ziska et al. 2016; Simeão et al. 2016; Cornman et al. 2015), whereas information concerning honey and pot-pollen produced by native stingless bees is scarce. Consequently, detailed analyses of the botanical origin, active principles associated with any medicinal properties, and whether secondary metabolites of botanical origin originate in nectar, resins, or residual pollen await further research. To obtain a deeper insight into the chemical composition and parameters of pot-pollen produced by *S. mexicana* in central Veracruz, we collected samples from managed hives located in three different sites and performed the corresponding

analyses. Pollen pots sampled in these localities are depicted in Fig. 23.4.

Importantly, Di Pasquale et al. (2013) reported that both the quality and diversity of pollen can shape bee physiology and might aid in understanding the influence of agriculture and land-use intensification on bee nutrition and health. Therefore, the maintenance and development of floral resources within agroecosystems is needed to alleviate the negative impact of human activity and to sustain bee populations (Decourtye et al. 2010), which in turn may determine pollination efficiency and agricultural production.

23.4 Palynological Analyses of *Scaptotrigona mexicana* Pot-Pollen from Central Veracruz, Mexico

Pot-pollen samples produced by *Scaptotrigona mexicana*, in managed hives located in three different sites in central Veracruz, Mexico, were subjected to morphological pollen analysis, following the protocol described by Erdtman (1969), and pollen identification guide provided by Palacios-Chávez et al. (1991) (Table 23.1).

In the palynological analysis, we found 11 floral types in Cañada Blanca and 13 each in Manuel León and Fortín de las Flores. In Cañada Blanca, with 33.4% of all floral types, *Heliocarpus* (Malvaceae, Grewioideae) was the most abundant, followed by *Bursera simaruba* (Burseraceae)

Table 23.1 Palynological analyses of *Scaptotrigona mexicana* pot-pollen from central Veracruz, Mexico (*n* = 3)

Site	Plant family	Plant species	No. of pollen grains	Percentage of pollen grains
Cañada Blanca	Malvaceae, subfamily Grewioideae	*Heliocarpus* sp1	215	33.38
	Burseraceae	*Bursera simaruba*	102	15.83
	Fabaceae, subfamily Caesalpinioideae	*Chamaecrista* sp.	35	5.43
	Fabaceae, subfamily Faboideae	*Desmodium adscendens*	95	14.75
	Fabaceae, subfamily Faboideae	*Desmodium tortuosum*	38	5.9
	Myrtaceae	*Eugenia capuli*	60	9.31
	Asteraceae	*Bidens pilosa*	5	0.77
	Asteraceae	*Vernonia* sp1	12	1.86
	Sapotaceae	*Pouteria* sp.	22	3.41
	Asteraceae	*Verbesina* sp.	5	0.77
	Polygonaceae	*Coccoloba* sp.	55	8.54
Manuel León	Burseraceae	*Bursera simaruba*	45	5.34
	Asteraceae	*Parthenium fruticosum*	165	19.59
	Asteraceae, tribe Heliantheae	*Helianthus* sp.	50	5.93
	Anacardiaceae	*Spondias mombin*	95	11.28
	Solanaceae	*Solanum* sp1	75	8.9
	Solanaceae	*Solanum* sp2	52	6.17
	Fabaceae, subfamily Caesalpinioideae	*Chamaecrista* sp.	170	20.19
	Sapindaceae	*Serjania* sp.	12	1.42
	Malvaceae, subfamily Grewioideae	*Heliocarpus* sp3	22	2.61
	Asteraceae	*Vernonia* sp2	68	8.07
	Fabaceae, subfamily Mimosoideae	*Pithecellobium* sp.	15	1.78
	Polygonaceae	*Coccoloba* sp.	35	4.15
	Asteraceae	*Bidens pilosa*	38	4.51
Fortín de las Flores	Anacardiaceae	*Spondias mombin*	50	7.56
	Araliaceae	*Dendropanax arboreus*	45	6.8
	Solanaceae	*Solanum* sp1	110	16.64
	Asteraceae	*Verbesina* sp.	150	22.69
	Malvaceae, subfamily Grewioideae	*Heliocarpus* sp2	70	10.59
	Sapotaceae	*Pouteria* sp.	25	3.78
	Boraginaceae	*Cordia* sp.	55	8.32
	Asteraceae	*Vernonia* sp1	40	6.05
	Myrtaceae	*Eugenia* sp.	3	0.45
	Solanaceae	*Solanum* sp3	23	3.47
	Amaranthaceae	*Dysphania ambrosioides*	15	2.26
	Solanaceae	*Solanum* sp2	45	6.8
	Malvaceae, subfamily Grewioideae	*Heliocarpus* sp1	30	4.53

with 15.8%. In Manuel León, with nearly 20% of all floral types, *Chamaecrista* (Fabaceae, Caesalpinioideae) and *Parthenium fruticosum* (Asteraceae) exhibited higher presence than the others. Finally, in Fortín de las Flores, *Verbesina* (Asteraceae), with 22.7%, and *Solanum* (Solanaceae), with 16.6%, were the most abundant. In general, it seemed that there were no large differences among plant species foraged by *S. mexicana* in central Veracruz.

Because our palynological analysis revealed the presence of a many plant species, the pollen can be considered multifloral. Pollen characteristics are strongly influenced by botanical origin due to bee foraging preference, and it is useful to apply palynology for an understanding bee-plant interactions. Our pollen samples revealed a total of 16 botanical families foraged by *S. mexicana* (Table 23.1). In pollen samples obtained from Guatemala, pollen composition analyses allowed identification of four plant families, Fabaceae, Fagaceae, Melastomataceae, and Myrsinaceae (Dardón et al. 2013), foraged by *S. mexicana*, much lower than those we observed.

23.5　Chemical Parameters of *Scaptotrigona mexicana* Pot-Pollen from Veracruz, Mexico

Pollen pots of managed *Scaptotrigona mexicana* from three localities, situated in the municipalities of Amatlán de los Reyes (Cañada Blanca and

Manuel León) and Fortín de las Flores in central Veracruz, Mexico, were harvested from three nests on the same day. The cerumen pot was opened to collect the pollen with a spatula. Pot-pollen was kept frozen until analysis. Physical and chemical parameters were analyzed in triplicate, following the protocol described by Vit et al. (2016). Parameters measured included water content (gravimetric method), ash (gravimetric method), proteins (semimicro Kjeldahl), lipids (AOAC 1996), total soluble sugars (Lane-Eynon titration method), electrical conductivity (conductivity meter), and pH (titrimetric method). For protein calculation, nitrogen was first determined by the semimicro Kjeldahl method (AOAC 1996) (correction factor of 6.25). The analytical results on chemical composition of the three samples of *S. mexicana* pot-pollen are shown in Table 23.2.

The composition of bee pot-pollen showed variation among samples, which could be due to different botanical and geographical origin, plant growth condition, drying process conditions, and storage method (Barajas et al. 2012; Domínguez-Valhondo et al. 2011; Nogueira et al. 2012; Kostić et al. 2015). According to Krell (1996), water content in air-dried pollen produced by *Apis mellifera* may vary from 7 to 11%, which also agrees with the findings of Kostić et al. (2015), who reported a minimum of 4.3% and a maximum of 14.3% water content in bee pollen. In comparison to those reports, in our study *S. mexicana* pot-pollen displayed a high percentage of pollen moisture, ranging from 15.30 g/100 g pollen in

Table 23.2 Chemical parameters of *Scaptotrigona mexicana* pot-pollen from three localities in central Veracruz, Mexico ($n = 3$)

Physical and chemical parameters	Localities where nests were harvested		
	Cañada Blanca	Manuel León	Fortín de las Flores
Water content (g/100 g pollen)	24.6 ± 0.2	15.5 ± 0.1	26.7 ± 0.4
Ash (g/100 g pollen)	3.1 ± 0.2	2.5 ± 0.7	2.9 ± 0.3
Total soluble sugars (g/100 g pollen)	31.99 ± 3.09	33.10 ± 0.77	35.02 ± 0.58
Protein (g/100 g pollen)	22.01 ± 5.5	20.49 ± 3.0	21.06 ± 0.9
Lipids (g/100 g pollen)	0.46 ± 0.2	1.1 ± 0.1	1.1 ± 0.3
Electrical conductivity (mS/cm)	2.39 ± 0.12	2.32 ± 0.01	2.00 ± 0.03
pH	3.46 ± 0.01	3.61 ± 0.02	3.64 ± 0.01

samples harvested in Manuel León to 26.7 g/100 g pollen found in those from Fortín de las Flores. Nevertheless, Rebelo et al. (2016) and Vit et al. (2016) found a water content between 43.5 and 53.4% in pollen collected from stingless bee species of the genera *Melipona* and *Scaptotrigona*. The average of ash content of the samples ranged from 2.51 g/100 g for pot-pollen harvested in Manuel León to 3.07 g/100 g for those collected in Cañada Blanca. These results are within the ranges reported by other studies (i.e., Kostić et al. 2015; Rebelo et al. 2016; Vit et al. 2016).

Carbohydrates constitute the main fraction of collected *S. mexicana* pollen and represent an important component in terms of nutrition and energy. They comprise both reducing sugars, such as fructose, glucose, and maltose, and also nonreducing ones, such as sucrose. In our study, the maximum value of total soluble sugars was recorded in Fortín de las Flores (35.02 g/100 g pollen), and the minimum was obtained in Cañada Blanca (31.99 g/100 g pollen). In pollen samples produced by *Apis mellifera*, carbohydrates constitute between 13 and 55 g/100 g of the total pollen in dry bases (Rzepecka-Stojko et al. 2015), whereas Vit et al. (2008) reported sugar values of 15.0–50.0 g/100 g pollen. In pollen samples collected from *Melipona seminigra* and *M. interrupta* hives, the carbohydrate contents were 44.25 and 25.66 g/100 g pollen, respectively (Rebelo et al. 2016). Similarly, in *Scaptotrigona* cf. *ochrotricha* pollen samples, Vit et al. (2016) recorded carbohydrate values of 31.03 g/100 g pollen. By contrast, Nogueira et al. (2012) reported higher values, varying from 69.68 to 84.25 g/100 g pollen, while Kostić et al. (2015) found total carbohydrate concentrations varying from 64.4 to 81.8 g/100 g pollen in honey of *Apis mellifera*.

Proteins were the second most abundant component in the *S. mexicana* pot-pollen analyzed, and concentrations varied little among sampling sites. They were 22.01 g/100 g pollen in Cañada Blanca, 20.49 g/100 g pollen in Manuel León, and 21.06 g/100 g pollen in Fortín de las Flores. According to Vit et al. (2008), protein contents in *Apis mellifera* pollen may vary between 6.54 and 26.30 g/100 g pollen, whereas Rzepecka-Stojko

et al. (2015) report that pollen is a primary source of nutritious protein, with an average value of nearly 24 g/100 g pollen in dry bases. Similarly, Rebelo et al. (2016) find that pollen samples collected by *Melipona seminigra* and *M. interrupta* have protein percentages of 37.63 and 24.00 g/100 g pollen, respectively. Nogueira et al. (2012) report that the protein content in pollen samples varies from 12.20 to 25.15 g/100 g pollen. Accordingly, Vit et al. (2016) report average protein values of 16.8 g/100 g for *Scaptotrigona* cf. *ochrotricha* pot-pollen.

According to Rzepecka-Stojko et al. (2015), lipids constitute between 0.3 and 20 g/100 g bee pollen and comprise both unsaturated fatty acids (i.e., palmitoleic acid, oleic acid, α-linolenic acid, and arachidonic acid) and saturated fatty acids (i.e., caproic, caprylic, lauric, myristic, palmitic, and stearic acids). The ratio of unsaturated acids to saturated ones is 2.67, and bee pollen contains on average 2.7% essential unsaturated fatty acids. Lipid content ranged from 0.46 g/100 g pollen in samples harvested in Cañada Blanca to 1.27 g/100 g pollen in those from Fortín de las Flores. The latter value is in accordance with other studies previously reported for *Apis mellifera* pollen (i.e., Estevinho et al. 2012; Yang et al. 2013), but the former is considered a normal value (Campos et al. 2008). Lipids are partly responsible for the physicochemical properties of food, and those that are of major nutritional interest are the fatty acid esters (Estevinho et al. 2012). It is important to note that the determination of the lipids, which may have their origin in bee pollen, is not very common, which hinders comparison of results (Almeida-Muradian et al. 2013).

All in all, pot-pollen harvested from Fortín de las Flores displayed higher values of water content (26.7 g/100 h pollen), total soluble sugars (35.02 g/100 g pollen), and pH (3.64). On the other hand, the greatest value for protein content (22.01 g/100 g pollen) was recorded in Cañada Blanca pollen samples, though all three values were very close (i.e., 20.49 g/100 g pollen in Manuel León and 21.06 g/100 g pollen in Fortín de las Flores). Ash, electrical conductivity, and pH were very similar among the pollen samples analyzed. In general, Fortín de las Flores presented a high diversity of plant species foraged by *S. mexicana* (Table 23.1).

23.6 Elemental Composition of *Scaptotrigona mexicana* Pot-Pollen from Central Veracruz, Mexico

After digestion of fresh pot-pollen samples with salicylic acid at 3% in sulfuric acid and using a catalyst mixture (sodium sulfate anhydrous/copper sulfate pentahydrate/metallic selenium; 96:3.5:5, w) at 360 °C, phosphorus (P) and potassium concentrations were determined. The digested fresh pot-pollen samples were warmed on hot plates by using nitric acid, perchloric acid, and hydrogen peroxide (1.33:0.66:1.0; v) at 180 °C and then analyzed using inductively coupled plasma-optical emission spectrophotometry (ICP-OES). Concentrations of these two elements are expressed in mg/g fresh matter weight (FMW) in Table 23.3.

The order of element concentration in pot-pollen samples analyzed was P > K in the three sampling sites. This is an expected result, since pot-pollen is the main source of N for most stingless bees (Eltz et al. 2001). Moreover, P and K play pivotal roles for normal growth of bee colonies (Kostić et al. 2015). Concerning P concentrations, Grembecka and Szefer (2013) report values between 611 and 659 mg/100 g in *A. mellifera* pollen samples from Poland and Italy. Also, in pollen of *A. mellifera* samples collected in China, P concentration is 594.6 mg/100 g (Yang et al. 2013). On average, *Scaptotrigona mexicana* pot-pollen had a P concentration of 319.9 mg

P/100 g pollen, which is about half that reported in the studies cited above. Regarding K, Kostić et al. (2015) find values ranging from 246.2 to 423.6 mg K/100 g pollen in Serbia, with a mean value of 339.1 mg K/100 g pollen. In our analysis, K concentrations were between 222.50 and 283.69 mg K/100 g pollen. In *A. mellifera* pollen samples from China, Yang et al. (2013) find a mean of 532.4 mg K/100 g pollen. Conversely, in pollen samples collected from *A. mellifera* hives in Poland and Italy, Grembecka and Szefer (2013) report lower values, ranging from 69.3 to 70 mg K/100 g pollen. Water content and pH values were higher in pot-pollen from Fortín de las Flores, while P and K were lower in this sampling site.

23.7 Conclusions and Perspectives on *Scaptotrigona mexicana* Pot-Pollen in Mexico

Scaptotrigona mexicana is considered the second most productive stingless bee in Mexico, after *Melipona beecheii*. Within the country, this species is especially exploited by the Nahuas and Totonac people in the states of Veracruz and Puebla. Its pot-pollen is of high quality and displays different chemical and elemental composition. While we did not find strong differences in pollen chemical parameters among sampled sites, pollen from Fortín de las Flores exhibited higher water content and pH. Total soluble sugars were greatest in Fortín de las Flores samples, while ash, protein, and electrical conductivity were greater in Cañada Blanca pollen samples. Interestingly, P and K had higher concentration in pollen samples

Table 23.3 Concentration of the two most abundant elements in *Scaptotrigona mexicana* pot-pollen from three localities in central Veracruz, Mexico ($N = 3$)

Elements (mg/100 g FMW)	Localities where pot-pollen samples were harvested		
	Cañada Blanca	Manuel León	Fortín de las Flores
Phosphorus (P)	323.28 ± 14.26	365.73 ± 12.93	273.61 ± 10.14
Potassium (K)	242.76 ± 14.61	283.69 ± 19.76	222.50 ± 17.55

FMW fresh matter weight (i.e., weight of the fresh pollen, without drying)

harvested from Manuel León managed hives. After the palynological analysis, we were able to demonstrate that the plant families providing the most pollen were Tiliaceae and Burseraceae in Cañada Blanca, Fabaceae and Asteraceae in Manuel León, and Solanaceae and Asteraceae in Fortín de las Flores. Knowing the crucial roles that this stingless bee species plays in terms of food security and agriculture, we encourage bee-keepers, academics, and decision-makers to take effective measures in order to preserve it and promote its sustainable use and management.

Acknowledgments We acknowledge the Mexico's National Council of Science and Technology (CONACYT) for the M.Sc. scholarship No. 372038 granted to LALG. ACO also thanks the financial support from Colegio de Postgraduados through the Management and Investment Trust No. 167304. We are especially thankful to Dr. David W. Roubik for his helpful comments and suggestions; his expert advice and constructive criticism have been invaluable in ensuring the most careful scrutiny, which certainly served to improve this chapter substantially. We also are very much grateful to Dr. Patricia Vit, not only for her work as an editor but also for her sustained encouragement, interest, support, and valuable advice throughout the editorial process.

References

Albores-González ML, García-Guerra TG, Durán-Olguín L, Aguliar-Ayón A. 2011. Experiencia de la Unión de Cooperativas Tosepan en el fomento a la cría de las abejas nativas Pitsilnejmej (*Scaptotrigona mexicana*). pp. 95-99. In: Memorias del VII Seminario Mesoamericano sobre Abejas Nativas. San Cristóbal de las Casas, Chiapas, México. 242 p.

Almeida-Muradian LB, Pamplona LC, Coimbra S, Barth OM. 2005. Chemical composition and botanical evaluation of dried bee pollen pellets. Journal of Food Composition and Analysis 18: 105-111. DOI:10.1016/j.jfca.2003.10.008

Almeida-Muradian LB, Stramm KM, Horita A, Barth OM, da Silva de Freitas A, Estevinho LM. 2013. Comparative study of the physicochemical and palynological characteristics of honey from *Melipona subnitida* and *Apis mellifera*. International Journal of Food Science and Technology 48: 1698-1706. DOI:10.1111/ijfs.12140

Anderson KE, Sheehan TH, Eckholm BJ, Mott BM, DeGrandi-Hoffman G. 2011. An emerging paradigm of colony health: microbial balance of the honey bee and hive (*Apis mellifera*). Insectes Sociaux 58: 431-444. DOI:10.1007/s00040-011-0194-6

AOAC. 1996. Official methods of analysis of the Association of Official Analytical Chemists. Association of Official Analytical Chemists. Arlington, VA, USA. 937 pp.

Arzaluz A, Obregón F, Jones R. 2002. Optimum brood size for artificial propagation of the stingless bee *Scaptotrigona mexicana*. Journal of Apicultural Research 41: 62-63. DOI:10.1080/00218839.2002.11101070

Ayala R. 1999. Revisión de las abejas sin aguijón de México (Hymenoptera: Apidae: Meliponini). Folia Entomológica Mexicana 106: 1-123.

Ayala R, González V, Engel M. 2013. Mexican stingless bees (Hymenoptera: Apidae): diversity, distribution, and indigenous knowledge. pp. 135-152. In Vit P, Pedro-Silvia RM, Roubik D, eds. Pot-Honey: A Legacy of Stingless Bees. Springer; New York, NY, USA. 654 pp.

Barajas J, Cortes-Rodriguez M, Rodríguez-Sandoval E. 2012. Effect of temperature on the drying process of bee pollen from two zones of Colombia. Journal of Food Process Engineering 35: 134-148. DOI:10.1111/j.1745-4530.2010.00577.x

Camargo JMF, Pedro SRM. 2013. Meliponini Lepeletier, 1836. In Moure JS, Urban D, Melo GAR, orgs. Catalogue of bees (Hymenoptera, Apoidea) in the Neotropical Region – online version. Available at: http://moure.cria.org.br/catalogue?id=34932

Campos MGR, Bogdanov S, De Almeida-Muradian LB, Szczesna T, Mancebo Y, Frigerio C, Ferreira F. 2008. Pollen composition and standardisation of analytical methods. Journal of Apicultural Research 47: 154-161. DOI:10.3896/IBRA.1.47.2.12

Cano-Contreras EJ, Martínez-Martínez C, Balboa-Aguilar CC. 2013. La "Abeja de Monte" (Insecta: Apidae, Meliponini) de los Choles de Tacotalpa, Tabasco: Conocimiento local, presente y futuro. Etnobiología 11: 47-57.

Cimpoiu C, Hosu A, Miclaus V, Puscas A. 2013. Determination of the floral origin of some Romanian honeys on the basis of physical and biochemical properties. Spectrochimica Acta Part A: Molecular and Biomolecular Spectroscopy 100: 149-154. DOI:10.1016/j.saa.2012.04.008

Codex Alimentations. 2001. Draft revised standard for standard for honey (at step 10 of the Codex procedure). Alinorm 01/25 (2001). pp. 19-26.

Cornman RS, Otto CR, Iwanowicz D, Pettis JS. 2015. Taxonomic characterization of honey bee (*Apis mellifera*) pollen foraging based on non-overlapping paired-end sequencing of nuclear ribosomal loci. PLoS One 10: e0145365. DOI:10.1371/journal.pone.0145365

Cortopassi-Laurino M, Imperatriz-Fonseca VL, Roubik DW, Dollin A, Heard T, Aguilar I, Venturieri GC, Eardley C, Nogueira-Neto P. 2006. Global meliponiculture: challenges and opportunities. Apidologie 37: 275-292. DOI:10.1051/apido:2006027

Da Costa Leite JM, Trugo LC, Costa LSM, Quinteiro LMC, Barth OM, Dutra VML. 2000. Determination of oligosaccharides in Brazilian honeys of different botanical origin. Food Chemistry 70:93-98. DOI:10.1016/S0956-7135(99)00115-2

Dardón MJ, Maldonado-Aguilera C, Enríquez E. 2013. The Pot-Honey of Guatemalan Bees. pp. 395-408. In: Vit P, Pedro-Silvia RM, Roubik D, eds. Pot-Honey: A Legacy of Stingless Bees. Springer; New York, NY, USA. 654 pp.

Decourtye A, Mader E, Desneux N. 2010. Landscape enhancement of floral resources for honey bees in agro-ecosystems. Apidologie 41:264-277. DOI:10.1051/apido/2010024

Di Pasquale G, Salignon M, Le Conte Y, Belzunces LP, Decourtye A, Kretzschmar A, Suchail S, Brunet JL, Alaux C. 2013. Influence of pollen nutrition on honey bee health: Do pollen quality and diversity matter? PLoS ONE 8: e72016. DOI:10.1371/journal.pone.0072016

Domínguez-Valhondo D, Bohoyo Gil D, Hernández MT, González-Gómez D. 2011. Influence of the commercial processing and floral origin on bioactive and nutritional properties of honeybee-collected pollen. International Journal of Food Science and Technology 46: 2204-2211. DOI:10.1111/j.1365-2621.2011.02738.x

El Sohaimy SA, Masry SHD, Shehata MG. 2015. Physicochemical characteristics of honey from different origins. Annals of Agricultural Sciences 60: 279-287. DOI:10.1016/j.aoas.2015.10.015

Eltz T, Brühl CA, Van der Kaars S, Linsenmair KE. 2001. Assessing stingless bee pollen diet by analysis of garbage pellets: a new method. Apidologie 32:341-353. DOI:10.1051/apido:2001134

Erdtman, G. 1969. Handbook of Palynology (Morphology-Taxonomy-Ecology). An Introduction to the Study of Pollen Grains and Spores. Munksgaard, Copenhagen. 486 pp.

Estevinho LM, Rodrigues S, Pereira AP, Feás X. 2012. Portuguese bee pollen: palynological study, nutritional and microbiological evaluation. International Journal of Food Science and Technology 47: 429-435. DOI:10.1111/j.1365-2621.2011.02859.x

Feás X, Vázquez-Tato MP, Estevinho L, Seijas JA, Iglesias A. 2012. Organic bee pollen: botanical origin, nutritional value, bioactive compounds, antioxidant activity and microbiological quality. Molecules 17:8359-8377. DOI:10.3390/molecules17078359

García-Flores A, del Amo Rodríguez S, Hernández-Colorado MR. 2013. Taxkat, la abeja nativa de Mesoamérica. La Ciencia y el Hombre 26: Enero-Abril 2013. Available at: https://www.uv.mx/cienciahombre/revistae/vol26num1/articulos/las-abejas.html

González-Acereto J. 2008. Cría y manejo de abejas nativas sin aguijón en México. Universidad Autónoma de Yucatán. Planeta Impresores. Mérida, Yucatán, México. 177 pp.

Grembecka M, Szefer P. 2013. Evaluation of honeys and bee products quality based on the mineral composition using multivariate techniques. Environmental

Monitoring and Assessment 185: 4033-4047. DOI:10.1007/s10661-012-2847-y

Guzmán M, Balboa C, Vandame R, Albores MA, González-Acereto J. 2011. Manejo de las abejas nativas sin aguijón en México *Melipona beecheii* y *Scaptotrigona mexicana*. El Colegio de la Frontera Sur. 64 p. Available at: file:///C:/Users/Usuario/Downloads/ECO%20Manual%20meliponicultura%202011ecosur%20(1).pdf

Hurtado-Burillo M, Ruiz C, May-Itzá WJ, Quezada-Eúan JJG, de la Rúa P. 2013. Barcoding stingless bees: genetic diversity of the economically important genus *Scaptotrigona* in Mesoamerica. Apidologie 44:1-10. DOI:10.1007/s13592-012-0146-9

Hurtado-Burillo M. 2015. Caracterización molecular y morfométrica del género *Scaptotrigona* (Apidae: Meliponini) en Mesoamérica. Tesis de Doctorado. Universidad de Murcia-Facultad de Biología. Murcia, España. 167 p. Available at: http://www.tesisenred.net/bitstream/handle/10803/349217/TMHB.pdf?sequence=1

Kaškonienė V, Venskutonis PR, Čeksterytė V. 2010. Carbohydrate composition and electrical conductivity of different origin honeys from Lithuania. LWT – Food Science and Technology 43: 801-807. DOI:10.1016/j.lwt.2010.01.007

Kostić AŽ, Barać MB, Stanojević SP, Milojković-Opsenica DM, Tešić ŽL, Šikoparija B, Radišić P, Prentović M, Pešić MB. 2015. Physicochemical composition and techno-functional properties of bee pollen collected in Serbia. LWT – Food Science and Technology 2015:1-9. DOI:10.1016/j.lwt.2015.01.031

Krell R. 1996. Value-added products from beekeeping. FAO Agricultural Services Bulletin No. 124. Food and Agriculture Organization of the United Nations. Rome, Italy. Available at: http://www.fao.org/docrep/w0076e/w0076e00.htm

Kroyer G, Hegedus N. 2001. Evaluation of bioactive properties of pollen extracts as functional dietary food supplement. Innovative Food Science and Emerging Technologies 2: 171-174. DOI:10.1016/S1466-8564(01)00039-X

Leonhardt SD, Dworschak K, Eltz T, Blüthgen N. 2007. Foraging loads of stingless bees and utilisation of stored nectar for pollen harvesting. Apidologie 38:125-135. DOI:10.1051/apido:2006059

Menezes C, Vollet-Neto A, Imperatriz-Fonseca VL. 2012. A method for harvesting unfermented pollen from stingless bees (Hymenoptera, Apidae, Meliponini). Journal of Apicultural Research 51:240-244. DOI:10.3896/IBRA.1.51.3.04

Menezes C, Vollet-Neto A, León-Contrera FA, Venturieri GC, Imperatriz-Fonseca VL. 2013. The role of useful microorganisms to stingless bees and stingless beekeeping. pp. 153-171. In: Vit P, Pedro SRM, Roubik D, eds. Pot-Honey: A legacy of stingless bees. Springer. New York, NY, USA. 654 p.

Mueller MY, Moritz RF, Kraus FB. 2012. Outbreeding and lack of temporal genetic structure in a drone congregation of the Neotropical stingless bee *Scaptotrigona mexicana*. Ecology and Evolution 2: 1304-1311. DOI: 10.1002/ece3.203

Nogueira C. Iglesias A, Feas X, Estevinho LM. 2012. Commercial bee pollen with different geographical origins: a comprehensive approach. International Journal of Molecular Sciences 13: 11173-11187. DOI:10.3390/ijms130911173

Padilla-Vargas PJ, Vásquez-Dávila MA, García-Guerra TG, Albores-González ML. 2014. Pisilnekmej: una mirada a la cosmovisión, conocimientos y prácticas nahuas sobre *Scaptotrigona mexicana* en Cuetzalan, Puebla, México. Etnoecológica 10:37-40.

Palacios-Chávez R, Ludlow-Wiechers B, Villanueva-Gutiérrez R. 1991. Flora palinológica de la Reserva de la Biosfera de Sian ka'an, Quintana Roo, México. Centro de Investigaciones de Quintana Roo. Chetumal, Quintana Roo, Mexico. 321 pp.

Palmer KA, Oldroyd BP, Quezada-Euán JJ, Paxton RJ, May-Itza Wde J. 2002. Paternity frequency and maternity of males in some stingless bee species. Molecular Ecology 11:2107-2113.

Pascoal A, Rodrigues S, Teixeira A, Feás X, Estevinho LM. 2014. Biological activities of commercial bee pollens: antimicrobial, antimutagenic, antioxidantand anti-inflammatory. Food and Chemical Toxicology 63:233-239. DOI:10.1016/j.fct.2013.11.010

Patlán-Martínez E, Kañetas OKJ, Guerrero FH, López MS. 2015. Las abejas nativas: tradición totonaca en el cuidado de la Naturaleza. pp. 1525-1530. In Memorias del V Congreso Latinoamericano de Agroecología. La Plata, Argentina.

Rebelo KS, Ferreira AG, Carvalho-Zilse GA. 2016. Physicochemical characteristics of pollen collected by Amazonian stingless bees. Ciencia Rural 46. DOI:10.1590/0103-8478cr20150999

Reyes-González A, Camou-Guerrero A, Reyes-Salas O, Argueta A, Casas A. 2014. Diversity, local knowledge and use of stingless bees (Apidae: Meliponini) in the municipality of Nocupétaro, Michoacán, Mexico. Journal of Ethnobiology and Ethnomedicine 10: 47. DOI:10.1186/1746-4269-10-47

Rosa CA, Lachance MA, Silva JOC, Teixeira ACP, Marini MM, Antonini Y, Martins RP. 2003 Yeast communities associated with stingless bees. FEMS Yeast Research 4:271-275. DOI:10.1016/S1567-1356(03)00173-9

Rosso JML, Imperatriz-Fonseca VL, Cortopassi-Laurino, M. 2001. Meliponicultura en Brasil I: situación en 2001 y perspectivas. pp. 28-35. In Memorias del II Seminario Mexicano sobre Abejas sin Aguijón. Mérida, Yucatán, Mexico. 136 p.

Rzepecka-Stojko A, Stojko J, Kurek-Górecka A, Górecki M, Kabała-Dzik A, Kubina R, Moździerz A, Buszman E. 2015. Polyphenols from bee pollen: structure, absorption, metabolism and biological activity. Molecules 20:21732-21749. DOI: 10.3390/molecules201219800.

Sánchez D, Nieh JC, Hénaut Y, Cruz L, Vandame R. 2004. High precision during food recruitment of experienced (reactivated) foragers in the stingless bee *Scaptotrigona mexicana* (Apidae, Meliponini). Naturwissenschaften 91:346-349. DOI:10.1007/s00114-004-0536-6

Sánchez D, Vandame R. 2012. Color and shape discrimination in the stingless bee *Scaptotrigona mexicana* Guérin (Hymenoptera, Apidae). Neotropical Entomology 41: 171-177. DOI:10.1007/s13744-012-0030-3

Sánchez D, Solórzano-Gordillo E, Vandame R. 2016. A Study on Intraspecific Resource Partitioning in the Stingless bee *Scaptotrigona mexicana* Guérin (Apidae, Meliponini) Using Behavioral and Molecular Techniques. Neotropical Entomology DOI: 10.1007/s13744-016-0404-z

Simeão CM, Silveira FA, Sampaio IB, Bastos EM. 2016. Pollen analysis of honey and pollen collected by *Apis mellifera* Linnaeus, 1758 (Hymenoptera, Apidae), in a mixed environment of *Eucalyptus* plantation and native cerrado in Southeastern Brazil. Brazilian Journal of Biology 75:821-829. DOI:10.1590/1519-6984.23513.

Souza BA, Lopes MTR, Pereira FM. 2012. Cultural aspects of meliponiculture. pp. 1-6. In Vit P, Roubik DW, eds. Stingless bees process honey and pollen in cerumen pots. SABER-ULA. Universidad de Los Andes. Mérida, Venezuela. Available at: http://www.saber.ula.ve/handle/123456789/35619

Vit P, Herrera P, Rodríguez D, Carmona J. 2008. Characterization of fresh bee pollen collected in Cacute, in Venezuelan Andes. Revista del Instituto Nacional de Higiene "Rafael Rangel" 39: 7-11.

Vit P, Medina M, Enríquez E. 2004. Quality standards for medicinal uses of Meliponinae honey in Guatemala, Mexico and Venezuela. Bee World 85:2-5. DOI: 10.1080/0005772X.2004.11099603

Vit P, Santiago B, Pedro SRM, Perez-Perez E, Pena-Vera M. 2016. Chemical and bioactive characterization of pot-pollen produced by *Melipona* and *Scaptotrigona* stingless bees from Paria Grande, Amazonas State, Venezuela. Emirates Journal of Food and Agriculture 28:78-84. DOI:10.9755/ejfa.2015-05-245

Vossler FG. 2015. Broad Protein Spectrum in Stored Pollen of Three Stingless Bees from the Chaco Dry Forest in South America (Hymenoptera, Apidae, Meliponini) and Its Ecological Implications. Psyche 2015: Article ID 659538. DOI:10.1155/2015/659538

Yang K, Wu D, Ye X, Liu D, Chen J, Sun P. 2013. Characterization of chemical composition of bee pollen in China. Journal of Agricultural and Food Chemistry 61: 708-718. DOI:10.1021/jf304056b

Yáñez-Ordóñez O, Trujano-Ortega M, Llorente-Bousquets J. 2008. Patrones de distribución de las especies de la tribu Meliponini (Hymenoptera: Apoidea: Apidae) en México. Interciencia 33:41–45.

Ziska LH, Pettis JS, Edwards J, Hancock JE, Tomecek MB, Clark A, Dukes JS, Loladze I, Polley HW. 2016. Rising atmospheric CO_2 is reducing the protein concentration of a floral pollen source essential for North American bees. Proceedings of the Royal Society B: Biological Sciences 283. pii: 20160414. DOI:10.1098/rspb.2016.0414

Chemical Characterization
and Bioactivity of *Tetragonisca
angustula* Pot-Pollen from Mérida,
Venezuela

24

Patricia Vit, Bertha Santiago, María Peña-Vera,
and Elizabeth Pérez-Pérez

24.1 Introduction

Stingless bees belong to the Meliponini tribe (Hymenoptera: Apidae). They are distributed in tropical and subtropical regions of the world, mostly between 25° N and S latitude, in Mexico, Central and South America, sub-Saharan Africa, Southern Asia, and Northern Australia (Souza et al. 2006). In Venezuela there are 83 species of Meliponini recorded (Pedro and Camargo 2013); they are known as Creole stingless bees (Rivero Oramas 1972). Some of the stingless bee genera found in Venezuela are *Frieseomelitta, Melipona, Plebeia, Scaptotrigona, Scaura,* and *Tetragonisca.* A detailed review of the native stingless bees refers to the traditional breeding of

"rubitas," "conguitas," "guanotas," and "ericas" – locally known by these ethnic names (Vit 1994). The small, thin, and docile eusocial bee *T. angustula* (Latreille, 1811) flies like a ballerina. It is widespread from Mexico to Brazil (Camargo and Pedro 2013) and called "angelita," "españolita," "jataí," "mariola," "rubita," or little bee "abejita." For the layman this is a friendly urban bee that tolerates "disturbed" nature. Daily foraging activity of *T. angustula* from Medellín, Colombia (ca. 1400 m elevation), is little affected by urban heat, light, or noise. This fact, in addition to inconspicuous nests and nonaggressive defensive behavior against large intruders increases camouflage and reduces nest destruction by people in the city (Velez-Ruiz et al. 2013). This natural flora pollinator is not harmful for human and domestic animals (Slaa et al. 2006).

Foraging behavior of *Melipona favosa, Nannotrigona mellaria,* and *Frieseomelitta nigra* was compared with *Apis mellifera* in Trinidad, in an urbanized area, looking at their corbicular pollen (Sommeijer et al. 1983). Types of pollen harvested by *Trigona spinipes* (Cortopassi-Laurino and Ramalho 1988), by *T. angustula* (Carvalho and Marchini 1999; Novais and Absy 2013) from Brazil, and by *T. fiebrigi* (Vossler et al. 2014) from Argentina were examined to describe the pollen niche there. The composition of *T. angustula* propolis from Brazil was studied by high-temperature high-resolution gas chromatography (Pereira et al. 2003) and by electrospray ionization mass spectrometry (Sawaya et al. 2006). In

P. Vit (✉)
Apitherapy and Bioactivity, Food Science Department, Faculty of Pharmacy and Bioanalysis, Universidad de Los Andes, Mérida 5101, Venezuela

Cancer Research Group, Discipline of Biomedical Science, Cumberland Campus C42, The University of Sydney, 75 East Street, Lidcombe, NSW 1825, Australia
e-mail: vitolivier@gmail.com

B. Santiago
Apitherapy and Bioactivity, Food Science Department, Faculty of Pharmacy and Bioanalysis, Universidad de Los Andes, Mérida 5101, Venezuela

M. Peña-Vera • E. Pérez-Pérez
Laboratory of Biotechnological and Molecular Analysis, Faculty of Pharmacy and Bioanalysis, Universidad de Los Andes, Mérida 5101, Venezuela

© Springer International Publishing AG, part of Springer Nature 2018
P. Vit et al. (eds.), *Pot-Pollen in Stingless Bee Melittology*, DOI 10.1007/978-3-319-61839-5_24

Venezuela the cerumen types, honey and propolis, from a hive of *T. angustula* were investigated for antioxidant activity, isolated bacteria, and contents of flavonoid and polyphenols (Pérez-Pérez et al. 2013). *T. angustula* pollen has been studied for pollen spectra in honey (Roubik and Moreno Patiño 2013) and food provisions (Novais and Absy 2013; Novais et al. 2015) but not for chemical composition of pot-pollen.

The increasing interest regarding physiological functionality of natural foods is related to human health. Among natural products, several apicultural products such as honey, pollen, and propolis have been used for centuries in traditional medicine as well as in diets and supplemental nutrition. Bee pollen chemical composition helps to explain nutritional and health benefits to humans. The daily ingestion of bee pollen can regulate intestinal functions, can effectively reduce capillary fragility, and has beneficial effects on the cardiovascular system, vision, and skin (Pietta 2000). In addition, it triggers beneficial effects in the prevention of prostate problems, arteriosclerosis, gastroenteritis, and respiratory diseases; promotes allergy desensitization; improves the cardiovascular and digestive systems and body immunity; and may also delay aging (Estevinho et al. 2012). Scientific documentation of antioxidant (LeBlanc et al. 2009; Sardar et al. 2014), antibacterial (Barrientos et al. 2013; Sharaf et al. 2013), and cytotoxic (Kustiawan et al. 2014) effects of bee pollen is increasing, which suggests its usefulness in the prevention or delay of diseases caused by free radicals (Pascoal et al. 2014).

Flower pollen contains lipids, sugars, carbohydrates, proteins, amino acids, vitamins, carotenoids, and polyphenolics such as flavonoids (Human and Nicolson 2006; Vit 2009), as well as pot-pollen (Vit et al. 2016). The phenolic fraction of pollen principally consists of flavonol glycosides and hydroxycinnamic acids; this composition tends to be plant species specific and has been related to the therapeutic properties of corbicular pollen (pollen load from the hind leg) and bee pollen or bee bread (corbicular pollen processed inside the hive). The presence of antioxidants in the pollen reduces the harmful effects of the free radicals in the cell and decreases oxidative reactions in food (Almaraz-

Abarca et al. 2004). However, the total antioxidant activity can be attributed to a synergistic effect of the activity of antioxidant enzymes (mainly superoxide dismutase, peroxidase, and catalase) as well as to the content of low-molecular antioxidants such as carotenoids, tocopherols, ascorbic acid, and phenolic substances (Aličić et al. 2014). This synergistic action can be used in the treatment of several diseases. For example, Saral et al. (2016) identified the antioxidant properties of honey bee products from Turkey, in chestnut honey, pollen, propolis, and royal jelly, and their hepato-protective activity against CCl4-induced hepatic damage in rats. They found the highest antioxidant activities in propolis, followed by pollen, honey, and royal jelly. However, a similar preventive action of liver damage by the three bee products was explained by their variable bioavailability.

In this work, pot-pollen was collected three times every 4 months from two nests of "angelita" in the city of Mérida, Venezuela. The two colonies were previously observed in pot-honey research in Venezuela, located in gardens of the Faculty of Pharmacy (290 m.a.s.l. N 08° 34.509′ W 71° 09.278′) and the backyard of a private home (287 m.a.s.l. N 80° 33.250′ W 71° 12.634). See a nest's storage area of pollen pots in Fig. 24.1. The stingless bee *T. angustula* (Latreille 1811) was kindly identified by the late Professor JMF Camargo[†], who transferred one of the colonies from a nest in a Büchner funnel to a wooden hive during his visit at the Universidad de Los Andes

Fig. 24.1 Pot-pollen stored in a *T. angustula* "angelita" nest (Photo: P. Vit)

in 2008. The chemical and bioactive characterization of fresh *T. angustula* pot-pollen was done with proximal analysis (ash, carbohydrates, fat, moisture, crude protein), biochemical parameters (flavonoids, polyphenols, proteins), and antioxidant activity (ABTS radical cation, Fenton-type reaction, hydroxyl radical) on ethanolic extracts.

24.2   Proximal Analysis of *T. angustula* Pot-Pollen from Mérida

The proximal analysis was performed in duplicate following the official analytical methods (AOAC 1999). Moisture, ash, and fat were done by gravimetric methods. Proteins were measured by the micro-Kjeldahl method with the application of sulfuric acid (Merck, Darmstadt, Germany) digestion, ammonia distillation by vapor flow with sodium hydroxide 50% (Sigma-Aldrich, USA) and sodium thiosulfate 25% (Merck, Darmstadt, Germany), collection of boric acid 4% (Sigma-Aldrich, USA), and titration with hydrochloric acid 0.05 N (Sigma-Aldrich, USA); nitrogen content was corrected with the factor 6.25 X to express protein content. Carbohydrates were calculated by difference, after the addition of moisture, ash, fat, and protein percentages is subtracted from 100 carbohydrate percentage = 100 − (% moisture + % ash + % fat + % protein).

In Table 24.1, the results of the proximal analysis of pot-pollen processed by two colonies of *T. angustula* in Mérida, Venezuela, are shown.

Compared to data of Chap. 26 considering moistures of *Frieseomelitta* and *Melipona* pot-pollen from southern Venezuela (29.01–42.74 g/100 g), *T. angustula* pot-pollen has lower moisture values (23.34–24.69 g/100 g) in a narrower range. In the present study, *T. angustula* has an intermediate ash content (2.06–2.13 g/100 g) in contrast to a range between 1.93 and 3.13 g/100 g for *Frieseomelitta* and *Melipona*. Similarly, intermediate values of protein contents (22.43–22.97 g/100 g) were observed for *T. angustula* pot-pollen compared to 18.44–24.72 g/100 g for *Frieseomelitta* and *Melipona*, as well as the fat content (4.42–4.58 g/100 g) compared to 3.51–6.03 g/100 g. However, the carbohydrate contents in *T. angustula* pot-pollen (45.98–46.68 g/100 g) are higher than in *Frieseomelitta* and *Melipona* (31.02 and 41.38 g/100 g).

Pot-pollen produced by Amazonian *Melipona seminigra* and *Melipona interrupta* in Brazil was characterized with high moistures of 53.39 and 37.12 g/100 g, protein 37.63 and 24.00 g/100 g, lipids 10.81 and 6.47 g/100 g, ash 4.03 and 2.7 g/100 g, and low carbohydrates 25.66 and 44.27 g/100 g (Rebelo et al. 2016). These values are more similar to those of the Amazonian *Melipona* sp. *aff. eburnea* and *Scaptotrigona* cf. *ochrotricha* from Venezuela (Vit et al. 2016), with higher moisture (43.49–48.54 g water/100 g) pot-pollen and fat contents (3.19–6.72 g/100 g) pot-pollen than *T. angustula* from the Venezuelan Andes. However, lower percentages of protein (16:80–18.32 g), ash (1.94–2.33 g), and carbohydrate (27.62–31.03 g) were observed per 100 g of Venezuelan Amazonian pot-pollen. Pot-pollen of *Melipona mandacaia* from Brazil (Bárbara et al. 2015) had moistures of 36.0 ± 2.0 g, 4.9 ± 0.3 g ash, 12 ± 2 g total sugars, and 11 to 39 g protein, all per 100 g pot-pollen.

Table 24.1 Proximal analysis of *T. angustula* pot-pollen from Mérida, Venezuela

Pot-pollen nest location	*n*	Moisture	Ash	Protein	Fat	Carbohydrate
Faculty of Pharmacy garden	3	24.69 (0.78) [23.56–25.45]	2.06 (0.13) [1.90–2.22]	22.97 (3.57) [18.10–26.31]	4.58 (0.59) [3.98–5.43]	45.98 (2.87) [42.25–49.41]
Private home backyard	3	23.34 (1.18) [22.34–25.45]	2.13 (0.24) [1.80–2.42]	22.43 (3.43) [17.95–25.54]	4.42 (0.31) [3.68–5.14]	46.68 (2.74) [43.70–50.18]

Values in the table are averages ± (SEM), [minimum-maximum] values

24.3 Methods to Quantify Flavonoids, Polyphenols, Proteins, and Antioxidant Activity in Ethanolic Extracts of *T. angustula* Pot-Pollen

Flavonoid, polyphenols, and proteins were the biochemical indicators measured here in pot-pollen ethanolic extracts. The antioxidant activity was measured by three methods: ABTS$^{+\bullet}$ cation, AOA Fenton-type radical, and hydroxyl radical.

24.3.1 Preparation of Pot-Pollen Ethanolic Extracts

A mass of 100 ± 10 mg of pot-pollen was placed on a glass homogenizer (Thomas No. A3528, USA), and 5 mL of ethanol 95% (v/v) (Riedel-de Haën, Europe) was added and homogenized in an ice bath. Homogenates were centrifuged in a BHG Optima II (USA) centrifuge at 3000 rpm for 10 min, and supernatants with the ethanolic pot-pollen extract were used for biochemical and antioxidant analysis.

24.3.2 Flavonoid Content

The flavonoid content was measured using a modification of the aluminum chloride method (Woisky and Salatino 1998). Flavonoid concentration was measured with a calibration curve using standard solutions of quercetin (Sigma, Steinheim, Germany) diluted up to 25, 50, and 100 μg/mL in ethanol 80% (v/v). Standard solutions (0.5 mL) were mixed with 1.5 mL 95% (v/v) ethanol, 0.1 mL of aluminum chloride (Fisher Scientific, New Jersey, USA) 10% (w/v), 0.1 mL potassium acetate (Sigma-Aldrich, USA) 1 M, and 2.8 mL distilled water. After incubation at ambient temperature for 30 min, absorbance was recorded at 415 nm. In a similar way, 0.5 mL of ethanolic extracts of bee products reacted with aluminum chloride for flavonoid content determination. For the blank, 0.5 mL of ethanol 80% (v/v) was used instead of ethanolic extracts or standard.

24.3.3 Polyphenol Content

The polyphenol content was analyzed by spectrometry at 765 nm using Folin-Ciocalteu (Sigma-Aldrich, St. Louis, USA) reagent (Singleton et al. 1999). One hundred μL of ethanolic extract was mixed with 500 μL of Folin-Ciocalteu reagent diluted to 1/10 concentration in water, to which 400 μL of sodium carbonate 7.5% (w/v) was added (Sigma, Steinheim, Germany). Absorbance at 765 nm was recorded after 10 min of reaction at 37 °C, against a blank with MQ water instead of ethanolic extract. The polyphenol concentration was estimated with a calibration curve using a solution of 0.1 g/L of gallic acid (Sigma, Steinheim, Germany) as standard (0.00, 0.05, 0.10, and 0.25 g/L).

24.3.4 Protein Content

The protein content of the pot-pollen ethanolic extracts was determined by spectrometry at 750 nm by the method of Lowry et al. (1951) with bovine serum albumin (Riedel-de Haën, Europe) as the standard.

24.3.5 Antioxidant Activity by the ABTS$^{+\bullet}$ Method

The assay method of ethanolic decolorization with ABTS (Sigma, Canada) was dissolved in water to a 7 mM concentration. ABTS radical cation (ABTS$^{+\bullet}$) was produced by reacting 7 mM stock solution with potassium persulfate (Merck, Darmstadt, Germany) to a final concentration of 2.45 mM (in water), in the dark at room temperature (RT) for 12–16 h before use (Re et al. 1999). For pot-pollen analysis, ABTS$^{+\bullet}$ solution was diluted with 20% ethanol (v/v) until 0.60–0.70 absorbance units at 735 nm and 100 μL of homogenate were diluted in ethanol and mixed with 7.5 ml of ABTS$^{+\bullet}$ solution diluted in ethanol 20% (v/v). Absorbance values were measured 6 min after mixing. A solution of 8 mM Trolox (Sigma, Steinheim, Germany) was used as the antioxidant standard. Trolox was diluted to obtain 1, 2, 4, and 8 µM in 5 mM PBS buffer (pH 7.4). Decolorization percentage at 734 nm after 6 min was calculated and plotted as a function of different Trolox concentrations, and TAA was reported accordingly. TAA value for a given sample would be equivalent to Trolox concentration that produces the same decolorization percentage.

24.3.6 Antioxidant Activity (AOA) by Fenton-Type Reaction

The AOA was determined using the method developed by Koracevic et al. (2001). In this method a standardized solution of Fe-EDTA complex reacts with hydrogen peroxide by a Fenton-type reaction, leading to the formation of hydroxyl radicals. These radicals degraded benzoate, resulting in the formation of TBA-reactive substances (TBARS). The reaction was monitored spectrophotometrically, and the inhibition of color development was defined as the AOA.

24.3.7 Hydroxyl Radical Assay

Hydroxyl radical was generated by Fenton's reaction following the method described by Halliwell et al. (1987). The following reagents were added to a mixture of 0.1 mL of 28 mM deoxyribose: 0.5 mL of 40 mM phosphate buffer (pH 7.4), 0.1 mL of 1 mM $FeCl_3$, 0.1 mL of 1.04 mM EDTA, 0.1 mL of 1 mM H_2O_2, 0.1 mL

of 1 mM ascorbic acid, and 0.2 mL of aqueous extract. The mixture was incubated for a 1-hour interval at 37 °C. Then, 0.5 mL of 1% (wt/vol) TBA in 0.05 M NaOH and 0.5 mL of 2.8% (vol/vol) trichloroacetic acid were added to the mixture and left to react for 10 minutes at 100 °C. Absorbance at 532 nm was recorded.

24.4 Biochemical Components and Antioxidant Activity of Ethanolic Extracts from *T. angustula* Pot-Pollen

Flavonoid, polyphenol, and protein contents of ethanolic extracts of *T. angustula* pot-pollen from Mérida, Venezuela, are given in Table 24.2. Flavonoid content was between 104.6 and 676.4 mg quercetin equivalents/100 g pot-pollen. In the last column, protein concentration (118.9 and 811.4 mg protein/100 g pot-pollen) was similar to flavonoids. However, the polyphenol concentration (1053.1 to 2627.4 mg gallic acid equivalents/100 g pot-pollen) was higher than flavonoids and proteins. Higher contents of flavonoids, polyphenols, and proteins were found in the nest at the private home backyard and lower contents at the Faculty of Pharmacy garden. The same type of bee but different pollen sources may explain the observed differences, although no palynological analyses were performed.

The polyphenol concentrations of *T. angustula* from Venezuela (1053.1 to 2627.4 mg gallic

Table 24.2 Flavonoid, polyphenol, and protein contents of ethanolic extracts of *T. angustula* pot-pollen

Pot-pollen nest location	Flavonoid content (mg quercetin equivalents/100 g pot-pollen)	Polyphenol content (mg gallic acid equivalents/100 g pot-pollen)	Protein content (mg proteins/100 g pot-pollen)
Faculty of Pharmacy garden	219.0 ± 1.4[b]	1827.7 ± 23.5[c]	450.3 ± 4.5[d]
	278.7 ± 2.4[c]	2036.0 ± 33.6[d]	360,0 ± 4.2[c]
	104.6 ± 2.3[a]	**1053.1 ± 33.6[a]**	**118.9 ± 6.1[a]**
Private home backyard	256.3 ± 3.4b	1698.6 ± 26.4[b]	**811.4 ± 6.5**
	588.2 ± 2.7[c]	**2627.4 ± 43.6[e]**	317.0 ± 12.4[c]
	676.4 ± 9.4[d]	1848.8 ± 27.5[c]	276.8 ± 7.7[b]

Data are mean ± SEM values (*n* = 3). Columns within a sample sharing the same letter are not significantly different by ANOVA post hoc Scheffé test (*P* < 0.05). Values in bold are minimum and maximum for each parameter

acid equivalents/100 mg pot-pollen) are similar to those reported by Feás et al. (2012) for bee pollen from Portugal (1290 to 1980 mg of gallic acid equivalents/100 mL methanolic extract of bee pollen). Our results are also similar to those reported by Pascoal et al. (2014) for corbicular pollen samples from Portugal and Spain, ranging 1855–3215 mg/100 g extract. However, methanolic extracts of *Melipona mandacaia* from Brazil have higher polyphenol content 4000 ± 13 mg gallic acid equivalents/100 g and one tenth of the flavonoid content 100.0 ± 0.2 mg catechin equivalents/100 g (Bárbara et al. 2015). The polyphenol content of *T. angustula* pot-pollen is higher than that reported by Ulusoy and Kolayli (2014), with total phenolics of 0.5–2.6 mg/100 g bee pollen from Turkey. Those authors identified several phenolic compounds including *p*-OH benzoic acid, vanillic acid, caffeic acid, syringic acid, *p*-coumaric acid, ferulic acid, rutin, *trans*-cinnamic acid, and *cis,trans*-abscisic acid in Turkish bee pollen from Anzer.

Crude protein was determined by micro-Kjeldahl (AOAC 1999) in fresh pot-pollen and by Lowry et al. (1951) in ethanolic extracts. The interest to measure protein contents by two methods is to have information of interest for the standards with Kjeldahl. However, although this reference method is appropriate to measure all nitrogen content in food, also non-proteic nitrogen is added to the result, and it needs a correction factor. The Lowry spectrophotometric method measures aromatic amino acids and their proportion in proteins. The portion of total proteins in pot-pollen extracted with ethanol 80% was measured with Lowry to be compared with flavonoid and polyphenol contents in ethanolic extracts.

The antioxidant activity of pot-pollen ethanolic extracts measured by three methods is presented in Table 24.3. Hydroxyl radical inhibition is 30.0–60.1% inhibition/100 g pot-pollen. Experimental values were similar or even superior to commercial antioxidants used as the control (lipoic acid, melatonin, quercetin). AOA values are between 0.74 and 1.12 mM uric acid equivalent/100 g pot-pollen, and TEAC values were between 401.8 and 500.4 μmoles Trolox equivalents/100 g pot-pollen. Utilizing the AOA and TEAC methods, pot-pollen samples presented higher antioxidant activity than commercial antioxidants used in this work. For the three methods, pot-pollen from the private home backyard had the greatest antioxidant activity.

Feás et al. (2012) analyzed methanolic extracts of organic bee pollen from Portugal, with scavenger activity and β-carotene bleaching assays. They found (EC_{50}) values of 3.0 ± 0.7 mg/mL

Table 24.3 Antioxidant activity of ethanolic extracts of *T. angustula* pot-pollen and synthetic controls (lipoic acid, melatonin, and quercetin)

Pot-pollen nest location	Hydroxyl radical inhibition percentage (% inhibition/100 g pot-pollen)	Antioxidant activity (AOA) (mM uric acid equivalent/100 g pot-pollen)	Total equivalent antioxidant activity (TEAC) (μmoles Trolox equivalents/100 g pot-pollen)
Faculty of Pharmacy garden	48.8 ± 1.4[e]	1.05 ± 0.07[e]	466.3 ± 3.3[f]
	56.7 ± 2.5[f]	1.08 ± 0.04[f]	487.7 ± 4.5[f]
	30.0 ± 1.6[b]	**0.74 ± 0.02[b]**	**401.8 ± 9.5[d]**
Private home backyard	32.9 ± 6.8[c]	1.01 ± 0.01[d]	416.7 ± 5.5[e]
	60.1 ± 7.4[g]	**1.12 ± 0.02[g]**	**500.4 ± 5.2[g]**
	42.2 ± 2.5[d]	1.05 ± 0.04[e]	475.1 ± 2.7[f]
Commercial antioxidants			
Lipoic acid	43.8 ± 1.7[d]	0.67 ± 0.02[a]	176.8 ± 3.1[c]
Melatonin	24.8 ± 1.4[a]	0.87 ± 0.03[c]	87.6 ± 5.1[a]
Quercetin	34.6 ± 2.5[c]	0.89 ± 0.05[c]	113.2 ± 1.9[b]

Data are mean ± SEM values ($n = 3$). Columns within a sample sharing the same letter are not significantly different by ANOVA post hoc Scheffé test ($P < 0.05$)

Table 24.4 Correlation between antioxidant activity and physicochemical parameters

Parameters	Proteins	Flavonoids	Polyphenols	AOA	HR	TEAC
Proteins	1	0.056	0.029	0.126	0.105	0.096
Flavonoids		1	0.034	0.170	0.065	0.043
Polyphenols			1	**0.843**	**0.802**	**0.921**
AOA				1	0.192	0.098
HR					1	0.154
TEAC						1

AOA Fenton-type radical, *HR* hydroxyl radical, *TEAC* Trolox equivalent antioxidant activity

extract and 4.6 mg/mL ± 0.9 mg/mL extract, respectively, demonstrating the high antioxidant activity of Portuguese bee pollen. The antioxidant activity of Greek bee pollen EC_{50} 181.4 ± 1.7 μg/mL extract (Graikou et al. 2011) is comparable to values in several types of bee pollen. These authors found that Greek bee pollen was about tenfold more active than Brazilian bee pollen (Carpes et al. 2009), but it varies between slightly less active and tenfold less active compared to bee pollen of several plants from Arizona (LeBlanc et al. 2009). Recently, Nurdianah et al. (2016) measured antioxidant activities of ethanolic extracts from collected pollen of three species of Malaysian stingless bees, *Geniotrigona thoracica*, *Heterotrigona itama*, and *Tetrigona apicalis*, using the DPPH-HPLC method and gallic acid as a standard reference. They found the following percentages of DPPH inhibition by *Tetrigona apicalis* (39%, gallic acid equivalents to 0.3 mg/mL) compared with *Heterotrigona itama* (14.3%, gallic acid equivalents to 0.1 mg/mL) and *Geniotrigona thoracica* (6.7%, gallic acid equivalents to 0.05 mg/mL).

High positive correlations (0.802–0.921) between polyphenol contents and antioxidant activity measured by the three methods (Table 24.4) address mechanisms of action based on this group of metabolites tenfold more concentrated than flavonoids and proteins in pot-pollen (see Table 24.3).

There is a close relationship between pollen antioxidant bioactivity and phenolic compound concentration (Campos et al. 2003; Leja et al. 2007; LeBlanc et al. 2009; Pascoal et al. 2014). However, the mechanism of free radical scavenging ability between these two parameters is unclear (Marghitas et al. 2009). Bee pollen antioxidant activity is species specific and independent of geographical origin. Almaraz-Abarca et al. (2004) determined the antioxidant activities in a mixture of honey bee-collected pollen and its six constituent pollen types from Mexico. The antioxidant activities as radical scavenger substances and as inhibitors of lipid peroxidation varied with botanical origin and were associated to their flavonol content. Other examples of antioxidant activity related to botanical origin are the works developed with bee pollen of the United States (LeBlanc et al. 2009), Romania (Marghitas et al. 2009), Portugal, and Spain (Nogueira et al. 2012). In all cases, the antioxidant activity was not correlated with geographical origin but with polyphenol content.

24.5 Conclusions

Bee pollen is a natural product with chemical components conferring relevant biological properties, as a functional food product. The chemical composition and antioxidant activity of *T. angustula* pot-pollen from Mérida, Venezuela, were studied here for the first time. Percentages of 23.34–24.69 g water, 2.06–2.13 g ash, 22.43–22.97 g protein, 4.42–4.58 g fat, and 45.98–46.68 g carbohydrates characterized 100 g pot-pollen of *T. angustula*. Polyphenol concentrations (1053.1–2627.4) mg gallic acid equivalents, flavonoid contents (104.6–676.4) mg quercetin equivalents, and protein concentrations (118.9–811.4) mg proteins per 100 g pot-pollen in ethanolic extracts were found. The antioxidant activity/100 g pot-pollen of the ethanolic extract was (1) hydroxyl radical inhibition percentage (30.0–60.1 %) inhibition, (2) antioxidant activity varied between 0.74 and 1.12 mM uric acid equivalent, and (3) total antioxidant activity from 401.8 to 500.4 μmoles Trolox equivalents. The nutritional, biochemical, and bioactive assessment of *T. angustula* pot-pollen is a contribution to set quality standards. There is a strong positive correlation between the antioxidant activity (measured by three methods) and total polyphenol content (0.802–0.843–0.921). See these values in Table 24.4, in bold. The high flavonoid and polyphenol concentrations and a great antioxidant activity (measured by three different methods) suggest that pot-pollen of *T. angustula* can be used – by itself or possibly combined – to treat several diseases arising from free radical imbalance.

Acknowledgments Project FA-592-16-08-B and CVI-FA-04-97 from Consejo de Desarrollo Científico, Humanístico, Tecnológico y de las Artes at Universidad de Los Andes, Mérida, Venezuela. The Secretariat of the Universidad de Los Andes and Schullo Products from Ecuador supported the participation in the Congress APICENS, Okinawa, Japan, in 2016 with some of the results in this chapter. To comments of reviewers and Dr. DW Roubik for valued English editing.

References

Aličić D, Šubarić D, Jašić M, Pašalic H, Ačkar D. 2014. Antioxidant properties of pollen. Food in Health and Disease: Scientific-Professional Journal of Nutrition and Dietetics 3: 6-12.

Almaraz-Abarca N, Campos MG, Avila-Reyes JA, Naranjo-Jimenez N, Herrera Corral J, Gonzalez-Valdez LS. 2004. Variability of antioxidant activity among honeybee-collected pollen of different botanical origin. Interciencia 29: 574-578.

AOAC. 1999. Official Methods of Analysis. 15th ed. Association of Official Analytical Chemists; Arlington, VA, USA. 1093 pp.

Bárbara MS, Machado CS, Sodré Gda S, Dias LG, Estevinho LM, de Carvalho CA. 2015. Microbiological assessment, nutritional characterization and phenolic compounds of bee pollen from *Melipona mandacaia* Smith, 1983. Molecules 20: 12525-12544.

Barrientos L, Herrera CL, Montenegro G, Ortega X, Veloz J, Alvear M, Cuevas A, Saavedra N, Salazar LA. 2013. Chemical and botanical characterization of Chilean propolis and biological activity on cariogenic bacteria *Streptococcus mutans* and *Streptococcus sobrinus*. Brazilian Journal of Microbiology 44: 577-585.

Camargo JMF, Pedro SRM. 2013. Meliponini Lepeletier, 1836. In Moure JS, Urban D, Melo GAR, orgs. Catalogue of bees (Hymenoptera, Apoidea) in the Neotropical Region. On-line version. Available at: http://moure.cria.org.br/catalogue?id=34932 Accessed the 25.01.2017.

Campos MG, Webby RF, Markham KR, Mitchell KA, Cunha AP. 2003. Age-Induced Diminution of free radical scavenging capacity in bee pollens and the contribution of Consistent flavonoids. Journal of Agricultural and Food Chemistry 51: 742-745.

Carpes ST, Mourao GB, Alencar SM, Masson ML. 2009. Chemical composition and free radical scavenging activity of *Apis mellifera* bee pollen from Southern Brazil. Brazilian Journal of Food Technology 12: 220-229.

Carvalho CAL and Marchini LC. 1999. Tipos polínicos coletados por *Nannotrigona testaceicornis* e *Tetragonisca angustula* (Hymenoptera, Apidae, Meliponinae). Sciencia Agricola 56: 717-722.

Cortopassi-Laurino M, Ramalho M. 1988. Pollen harvest by Africanized *Apis mellifera* and *Trigona spinipes* in São Paulo. Botanical and ecological views. Apidologie 19: 1-24.

Estevinho LM, Rodrigues S, Pereira AP, Feás X. 2012. Portuguese bee pollen: palynological study, nutritional and microbiological evaluation. International Journal of Food Science and Technology 47: 429-435.

Feás X, Vázquez-Tato MP, Estevinho L, Seijas JA, Iglesias A. 2012. Organic bee pollen: botanical origin, nutritional value, bioactive compounds, antioxidant activity and microbiological quality. Molecules 17: 8359-8377.

Graikou K, Kapeta S, Aligiannis N, Sotiroudis G, Chondrogianni N, Gonos E, Chinou I. 2011. Chemical analysis of Greek pollen - Antioxidant, antimicrobial and proteasome activation properties Chemistry Central Journal 5: 33-42.

Halliwell B, Gutteridge J, Aruoma O. 1987. The deoxyribose method:a simple test-tube assay for determination of rate constants for reactions of hydroxyl radicals. Analytical Biochemistry 165: 215–219.

Human H, Nicolson SW. 2006. Nutritional content of fresh, bee-collected and stored pollen of *Aloe greatheadii* var. *davyana* (Asphodelaceae). Phytochemistry 67: 1486-1492.

Koracevic D, Koracevic G, Djordjevic V, Andrejevic S, Cosic V. 2001. Method for measurement of antioxidant activity in human fluids. Journal of Clinical Pathology 54: 356–361.

Kustiawan PM, Puthong S, Arung ET, Chanchao C. 2014. *In vitro* cytotoxicity of Indonesian stingless bee products against human cancer cell lines. Asian Pacific Journal of Tropical Biomedicine 4: 549-556.

LeBlanc B, Davis O, Boue S, DeLucca A, Deeby T. 2009. Antioxidant activity of Sonoran Desert bee pollen. Food Chemistry 115: 1299-1305.

Leja M, Mareczek A, Wyzgolik G, Klepacz-Baniak J, Czenonska K. 2007. Antioxidative properties of bee pollen in selected plant species. Food Chemistry 100: 237-240.

Lowry OH, Rosebrough NJ, Farr AL, Randall RJ. 1951. Protein measurement with the Folin phenol reagent. Journal of Biological Chemistry 193: 265-275.

Marghitas L, Stanciu O, Dezmirean D, Bobis O, Popescu O, Bogdanov S, Campos M. 2009. *In vitro* antioxidant capacity of honeybee-collected pollen of selected floral origin harvested from Romania. Food Chemistry 115: 878-883.

Nogueira C, Iglesias A, Feás X, Estevinho LM. 2012. Commercial bee pollen with different geographical origins: a comprehensive approach. International Journal of Molecular Sciences 13: 11173–11187.

Novais JS, Absy ML. 2013. Palynological examination of the pollen pots of native stingless bees from the Lower Amazon region in Pará, Brazil. Palynology 37: 218–230.

Novais JS, Garcez ACA, Absy ML, Santos FAR. 2015. Comparative pollen spectra of *Tetragonisca angustula* (Apidae, Meliponini) from the Lower Amazon (N Brazil) and caatinga (NE Brazil). Apidologie 46: 417-431.

Nurdianah HF, Ahmad Firdaus AH, Eshaifol Azam O, Wan Adnan WO. 2016. Antioxidant activity of bee pollen ethanolic extracts from Malaysian stingless bee measured using DPPH-HPLC assay. International Food Research Journal 23: 403-405.

Pascoal A, Rodrigues, S, Teixeira A, Feás X, Estevinho LM. 2014. Biological activities of commercial bee pollens: antimicrobial, antimutagenic, antioxidant and anti-inflammatory. Food and Chemical Toxicology 63: 233-239.

Pedro SRM, Camargo JMF. 2013. Stingless bees from Venezuela. pp. 73-86. In: Vit P, Pedro SRM, Roubik D, eds. Pot-Honey: A legacy of stingless bees. Springer. New York, USA. 654 pp.

Pereira AS, Bicalho B, Aquino Neto FR. 2003. Comparison of propolis from *Apis mellifera* and *Tetragonisca angustula*. Apidologie 34: 291-298.

Pérez-Pérez EM, Suárez E, Peña-Vera MJ, González AC, Vit P. 2013. Antioxidant activity and microorganisms in nest products of *Tetragonisca angustula* Latreille, 1811 from Mérida, Venezuela. pp. 1-8. In: Vit P & Roubik DW, eds. Stingless bees process honey and pollen in cerumen pots. Facultad de Farmacia y Bioanálisis, Universidad de Los Andes; Mérida, Venezuela. Available at: http://www.saber.ula.ve/handle/123456789/35292 Accessed the 25.01.2017.

Pietta PG. 2000. Flavonoids as antioxidants. Journal of Natural Products 63: 1035-1042.

Re R, Pellegrini N, Proteggente A, Pannala A, Yang M, Rice-Evans C. 1999. Antioxidant activity in improved ABTS radical cation decolorization assay. Free Radical in Biology and Medicine 26: 1231-1237.

Rebelo KS, Ferreira AG, Carvalho-Zilse GA. 2016. Physicochemical characteristics of pollen collected by Amazonian stingless bees. Ciência Rural 46: 927-932.

Rivero Oramas R. 1972. Abejas criollas sin aguijón. Colección Científica. Monte Avila Editores; Caracas, Venezuela. 110 pp.

Roubik DW, Moreno Patiño JE. 2013. How to be a bee-botanist using pollen spectra pp. 73–86. In: Vit P, Pedro SRM, Roubik D, Pot-Honey: A legacy of stingless bees. Springer, New York, USA. 654.

Saral Ö, Yildiz O, Aliyazicioğlu R, Yuluğ E, Canpolat S, Öztürk F, Kolayli S. 2016. Apitherapy products enhance the recovery of CCL4-induced hepatic damages in rats. Turkey Journal of Medical Science 36: 192-202.

Sardar AA, Akhan ZU, Perveen A, Farid S, Khan I. 2014. In vitro antioxidant potential and free radical scavenging activity of various extracts of pollen of *Typha domigensis* Pers. Pakistan

Sawaya ACFH, Cunha IBS, Marcucci MC, de Oliveira Rodrigues RF, Eberlin MN. 2006. Brazilian propolis of *Tetragonisca angustula* and *Apis mellifera*. Apidologie 37: 398-407.

Sharaf S, Higazy A, Hebeish A. 2013. Propolis induced antibacterial activity and other technical properties of cotton textiles. International Journal of Biological Macromolecules 59: 408-416.

Singleton VL, Orthofer R, Lamuela-Raventos RM. 1999. Analysis of total phenols and other oxidation substrates and antioxidants by means of Folin-Ciocalteu reagent. Methods in Enzymology 299: 152–178.

Slaa EJ, Chaves LAS, Malagodi-Braga KS, Hofstde FE. 2006. Stingless bees in applied pollination: practice and perspectives. Apidologie 37: 293-315.

Sommeijer MJ, de Rooy GA, Punt W, de Bruijn LLM. 1983. A comparative study of foraging behavior and pollen resouces of various stingless bees (Hym., Meliponinae) and honeybees (Hym., Apinae) in Trinidad, West Indies. Apidologie 14: 205-224.

Souza B, Roubik D, Barth O, Heard T, Enríquez E, Carvalho C, Marchini L, Villas-Bôas J, Locatelli J, Persano Oddo L, Almeida-Muradian L, Bogdanov S, Vit P. 2006. Composition of stingless bee honey: Setting quality standards. Interciencia 31: 867-875.

Ulusoy E, Kolayli S. 2014. Phenolic composition and antioxidant properties of Anzer bee pollen. Journal of Food Biochemistry 38: 73–82.

Velez-Ruiz RI, Gonzalez VH, Engel MS. 2013. Observations on the urban ecology of the Neotropical stingless bee *Tetragonisca angustula* (Hymenoptera: Apidae: Meliponini). Journal of Melittology. 15:1-8.

Vit P. 1994. Las abejas criollas sin aguijón. Vida Apícola 63: 34-41.

Vit P. 2009. Origen botánico y propiedades medicinales del polen apícola. Revista Médica de la Extensión Portuguesa ULA 3: 27-34.

Vit P, Santiago B, Pedro SRM, Peña-Vera M, Pérez-Pérez E. 2016. Chemical and bioactive characterization of pot-pollen produced by *Melipona* and *Scaptotrigona* stingless bees from Paria Grande, Amazonas State, Venezuela. Emirates Journal of Food and Agriculture 28: 78-84.

Vossler FG, Fagúndez GA, Bettler BC. 2014. Variability of Food Stores of *Tetragonisca fiebrigi* (Schwarz) (Hymenoptera: Apidae: Meliponini) from the Argentine Chaco Based on Pollen Analysis. Sociobiology 61: 449-460.

Woisky RG, Salatino A. 1998. Analysis of propolis: some parameters and procedures for chemical quality control. Journal of Apicultural Research 37: 99-105.

Chemical, Microbiological, and Palynological Composition of the "Samburá" *Melipona scutellaris* Pot-Pollen

Rogério Marcos de Oliveira Alves, Geni da Silva Sodré, and Carlos Alfredo Lopes Carvalho

25.1 Introduction

Brazil harbors an abundant and varied flora that provides an immense quantity of resources for bees. Among these resources, pollen is undoubtedly one of the most important; as an excellent source of protein, carbohydrates, minerals, and vitamins, it is considered a high-value food in the diet of both humans and bees and an additional source of income that can increase the value of beekeeping activities (Salamanca Grosso et al. 2000).

In the process of collecting pollen from flowers, social bees pack pollen into their corbiculae, and upon arriving at the colony, they deposit the product into combs or pots, compressing it with their heads to obtain a compact mass that undergoes transformations in response to temperature, humidity, and salivary enzymes and mixing it with nectar to form the beebread of *Apis mellifera* or the "samburá" of stingless bees (Moreti et al. 2007; Nogueira-Neto 1997).

The study of the plants visited by these bees for pollen collection is of fundamental importance to the conservation of the plant species, which provide the protein base for the diet of these bees (Moreti et al. 2007). Qualitative pollen analysis can provide important data on geographic origin, botanic origin, and harvest season (Barth 1989) because plant origin is a factor that directly affects the composition, color, aroma, and flavor of this product.

In general, pollen consists of 15–30% protein, and much of the protein composition is in the form of free amino acids (10–13%). In addition to protein, pollen has 20–40% total sugar; 20–26% reducing sugar, a low lipid content (between 1% and 5%); 3–5% fiber; and 2.5–3.5% mineral salts (Espina and Ordetx 1984; Donadieu 1979).

Some studies (Muradian and Penteado 2007; Campos et al. 2008; Carpes et al. 2009) have reviewed the composition of bee pollen as well as the analytical methods and have proposed quality criteria for the product produced by *A. mellifera*. The "samburá" of some bees within the genus *Melipona* from the Brazilian Mata (*M. scutellaris*) and Caatinga (*M. mandacaia* and *M. quadrifasciata*) biomes has been characterized using physicochemical, microbiological, and pollen analyses.

However, for the stored pollen or "samburá" of stingless bees (Fig. 25.1), there is a need for more information on the production, productivity, or physicochemical composition.

Knowledge of the characteristics of "samburá" is fundamental to promote the wider use of this product; to provide economic, nutritional, and therapeutic benefits for humans and for poor communities; to contribute to the conservation of plant species in different biomes; and to ensure food safety to consumers.

R.M. de Oliveira Alves (✉)
Instituto Federal de Educação, Ciência e Tecnologia Baiano, 48.110-000, Catu, Bahia, Brazil
e-mail: eiratama@gmail.com

G. da Silva Sodré • C.A.L. Carvalho
Universidade Federal do Recôncavo da Bahia, 44.380-000, Cruz das Almas, Bahia, Brazil

© Springer International Publishing AG, part of Springer Nature 2018
P. Vit et al. (eds.), *Pot-Pollen in Stingless Bee Melittology*, DOI 10.1007/978-3-319-61839-5_25

Fig. 25.1 Dry bee pollen (**a**); beebread, "samburá" (**b**); "samburá" of *Melipona scutellaris* (**c**); "samburá" of *Frieseomelitta* sp. (**d**) (Photos: R.M.O. Alves)

25.2   The "Samburá" of the True "Uruçú" Bee

The bee species *Melipona scutellaris* Latreille 1811, known as "uruçú," is found in northeastern Brazil where it is farmed on a large scale and considered a major producer of honey and pot-pollen ("samburá") (Carvalho et al. 2003). However, the characteristics of the pollen stored by these bees are still poorly known.

Pot-pollen samples were collected from a meliponary located in the municipality of Camaçari, Bahia (12° 48'38.66"S and 38°15'25.73"W, altitude 14 m), in an area comprising rainforest, pastures, and coconut plantations. The analysis of the "samburá" was performed at the Laboratory of the Center for Insect Studies, at the Center for Agricultural, Environmental and Biological Sciences of the Federal University of Recôncavo of the Bahia (UFRB), in Cruz das Almas, Bahia, Brazil.

Thirty colonies were placed in an isolated site within the meliponary and maintained there throughout the study. The pot-pollen was harvested using disposable knives and spoons. The procedure consisted of making a lateral incision into the pot-pollen and removing all of the contents with a spoon. Pot-pollen samples were stored in clean glass jars.

The pot-pollen samples were collected over 3 months during the rainy season (May to July) and 3 months during the dry season (December to February) and were grouped by month and season of harvest (rainy or dry), resulting in one sample per month and three samples per season, for a total of six samples. The procedure described was used due to the low production of pot-pollen in the colonies.

25.3   Physicochemical Characteristics of "Samburá"

The following parameters were quantified following the methods described below: total protein, water activity, carbohydrate (by difference), moisture, mineral and ash, acidity, pH, fiber, lipid, and total energy value. All reagents we used were from Sigma-Aldrich, USA.

Total Protein The protein content was determined according to AOAC method 928.080 (2000). Approximately 0.7 g of ground pot-pollen was digested in a macro-Kjeldahl flask with 4 g of a catalytic mixture (1:3 $CuSO_4$ and K_2SO_4) and 20 mL of concentrated H_2SO_4. A total of 80 mL of 40% NaOH was added to the digested solution to liberate ammonia, which was condensed and added to a solution of concentrated H_2SO_4 and then titrated with a standard solution of 0.1 M NaOH. To determine the total protein content, the nitrogen values were multiplied by the conversion factor of 6.25 (Roulston et al. 2000).

Water Activity (Aw) Water activity was determined at 22 °C with a PawKit water activity meter (Decagon) according to the methodology of the Instituto Adolfo Lutz (1985).

Moisture Approximately 2 g of ground pot-pollen was dried in a vacuum oven at 60 °C to a constant weight according to the methodology of the Instituto Adolfo Lutz (1985).

Fiber Two grams of ground pot-pollen was defatted in petroleum ether and hot digested in a solution of 0.113 M H_2SO_4 followed by 0.313 M NaOH (30 minutes for each digestion). After neutralization of the residue with hot water, the sample was washed with 20 mL of ethanol and 10 mL of ethyl ether. The residue was then incinerated at 550 °C in a muffle furnace, and the fiber content was quantified by gravimetry (Instituto Adolfo Lutz 1985).

Lipid Total lipid was determined by gravimetry according to the methodology of the Instituto Adolfo Lutz (1985) using 2 g of ground pot-pollen extracted with hot petroleum ether in a Soxhlet apparatus for approximately 4 h. Petroleum ether was evaporated from the collected extract in the boiling flask, and the amount of lipid recovered was calculated as percentage.

Ash Two grams of ground pot-pollen was placed in a porcelain crucible and incinerated in a muffle furnace at 550 °C until white ash was obtained (approximately 4 h). After cooling in a desiccator, the ash content was determined by gravimetry (AOAC 1990), and the results were expressed as percent ash.

Carbohydrates Carbohydrate content was determined by difference according to the following equation: G = 100 − (P + L + U + R + F), where P = total protein, L = ether extract (lipid), U = moisture, R = fixed mineral residue (ash), F = crude fiber, and G = carbohydrate (Pinheiro et al. 2012).

pH The potentiometric method was used to determine pH. Prior to the measurement, the concentration of the buffer solution with a constant pH value was determined (Moraes and Teixeira 1998).

Free Acidity Acidity was determined by titration with a 0.05 M NaOH solution according to Lorenzo (2002).

Total Energy Value (TEV) The energy value was calculated using the following formula (ANVISA 2003): TEV = $(P \times 4.0 + U \times 4.0 + L \times 9.0)$ kcal/100 g, where P = protein, U = moisture, and L = lipid.

The results obtained from the physicochemical analyses of the "samburá" of *M. scutellaris* are presented in Table 25.1.

25.3.1 Moisture

The "samburá" of *M. scutellaris* had a mean moisture content of 34.58 g/100 g for the rainy season and 54.54 g/100 g for the dry season. The mean across the two seasons was 44.71 g/100 g. Souza et al. (2004) analyzed the "samburá" of five species of *Melipona* from the Amazon and found a mean moisture content of 36.9% ± 11.1%. Pinheiro et al. (2012) documented values of 54.58% and 58.75% moisture for the "samburá" of *M. flavolineata* and *M. fasciculata*, respectively.

When analyzing samples of beebread from *A. mellifera*, Rubio (1959) and Sampaio and De Freitas (1993) found moisture contents of 25% and 20 to 30%, respectively. The higher moisture content from the "samburá" of *M. scutellaris* is probably due to the processing methods used by

Table 25.1 Physicochemical characteristics of pot-pollen ("samburá") from *M. scutellaris* collected during two seasons in Camaçari, Bahia, Brazil (*N* = 6)

Characteristic	Collection season		Mean ± SD
	Rainy	Dry	
Moisture (g/100 g pot-pollen)	34.88	54.54	44.71 ± 9.83
Ash (g/100 g pot-pollen)	1.72	1.97	1.84 ± 0.125
Lipids (g/100 g pot-pollen)	3.97	4.46	4.25 ± 0.1
Protein (g/100 g pot-pollen)	23.78	23.98	23.88 ± 0.1
Fiber (g/100 g pot-pollen)	1.07	0.67	0.87 ± 0.2
Carbohydrates (g/100 g pot-pollen)	34.58	14.38	24.48 ± 10.1
pH	3.75	3.76	3.75 ± 0.005
Free acidity (meq/kg pot-pollen)	152.64	148.51	150.57 ± 0.4
Water activity (Aw)	0.93	0.92	0.92 ± 0.005
Total energy value (kcal/100 g pot-pollen)	269.17	193.5	231.33 ± 37.83

the bees and the storage methods of the product in the bee nest.

The "samburá" undergoes mechanical agglomeration of pollen grains through the addition of saliva and honey, acquiring a pasty consistency, and is stored in closed pots of greater volume than the comb cells, which can hinder the loss of water from the product, even at internal temperatures above 30 °C. The beebread produced by *A. mellifera* is stored in unoccupied comb cells with small volumes, allowing for a greater moisture loss in this product.

The difference in the moisture content of the samples between seasons may be due to the environmental conditions in each season. In wet forest regions, the dry season corresponds to high temperatures, lower rainfall, and higher humidity due to high rates of evapotranspiration (BRASIL 1985). High ambient temperature increases the internal temperature of the hive (Nogueira-Neto 1997), which induces the bees to cool the colonies through ventilation and in turn causes more evaporation that increases the relative humidity in the hive environment and can hinder the loss of water from the stored pollen.

25.3.2 Ash

The total ash content is directly linked to the mineral content, and therefore, low ash values may correspond to lower levels of minerals in plant pollen (Martinez 2005). In the "samburá" of *M. scutellaris*, ash contents of 1.72 g/100 g and 1.97 g/100 g were obtained for the rainy season and the dry season, respectively, with an overall mean of 1.84 g/100 g. Souza et al. (2004) reported a mean of 2.1% for Amazon meliponines. Sampaio and De Freitas (1993) analyzed the beebread of *A. mellifera* and recorded a value of 2.77%, which is slightly higher than that observed in the "samburá" of *M. scutellaris*.

25.3.3 Lipids

The "samburá" of *M. scutellaris* showed lipid contents varying from 3.97 g/100 g for the rainy season to 4.46 g/100 g for the dry season, with a mean of 4.25 g/100 g across both seasons. Pinheiro et al. (2012) reported values of 3.55% and 2.26% in the pollen of *Melipona* in Pará, and Souza et al. (2004) calculated a mean of 4.0% for five species of Amazon meliponines. Sampaio and De Freitas (1993) determined a lipid value of 2.77 to 3.85% for the beebread from *A. mellifera*.

Bees obtain lipid exclusively from pollen, and the amount of lipid can vary among pollen types (Roulston and Cane 2000). Ivanivki et al. (2012) reported that the higher lipid content seen in winter may be associated with a higher percentage of pollen obtained from particular plant families (Myrtaceae and Brassicaceae).

25.3.4 Protein

Protein content is a very important variable in food, but its value depends on the plant species from which it originated (Espina and Ordetx 1984). The "samburá" from *M. scutellaris* had a mean of 23.88 g/100 g of protein with little variation between the two seasons: 23.78 g/100 g in the wet season and 23.98 g/100 g in the dry season.

The analysis of "samburá" from Amazon meliponine species performed by Souza et al. (2004) and Pinheiro et al. (2012) reported values of 19.5% and 17.52–23.43%, respectively. For the beebread of *A. mellifera*, Sampaio and De Freitas (1993) determined a surprisingly low protein content of 2.96–4.0%, and Rubio (1959) reported a value of 10–22%. The change in protein content observed in different seasons may be influenced by differences in pollen diversity (Ivanivki et al. 2012).

25.3.5 Fiber

The presence of crude fiber is nutritionally important. The highest concentration of fiber is found in the pollen exine. The analysis of the "samburá" of *M. scutellaris* produced fiber values of 1.07 and 0.67 g/100 g in the dry and rainy seasons, respectively, with a mean of 0.87 g/100 g across seasons. Sampaio and De Freitas (1993) documented a range of 1.85–2.39% for fiber in the beebread of *A. mellifera*, which is well above than that of the "samburá" analyzed by Pinheiro et al. (2012) for the pollen of bees from the genus *Melipona* in Pará (1.60 and 1.20%).

The low fiber content of "samburá" compared to *A. mellifera* bee pollen may be related to the types of pollen found in the samples or a fiber digestion by the associated microflora (see Chaps. 17 and 18). A comparison of moisture and fiber showed that the season with "samburá" of higher moisture corresponded to a lower fiber content.

25.3.6 Carbohydrates

The carbohydrate content is important because it is one of the indicators of the nutritional value of pollen and its energy content (Kroyer and Hegedus 2001). Baldi-Coronel et al. (2004) report that sugars represent the major constituents of honey-bee-collected pollen, reaching up to 60% in the pollen collected by bees. The origin of this sugar is probably from the honey or nectar added to the pollen after collection (Krell 1996).

For the "samburá" of *M. scutellaris*, an overall mean of 24.08 g carbohydrates/100 g pot-pollen, with values of 34.58 g/100 g for the rainy season and 14.38 g/100 g for the dry season, was determined. Pinheiro et al. (2012), investigated the "samburá" of *M. flavolineata* and *M. fasciculata*, reported values of 14.50% and 18.09% carbohydrate. Souza et al. (2004) evaluated the pollen of five species of *Melipona* from the Amazon and recorded a mean of 37.5% glucose. Rubio (1959) analyzed samples of beebread from *A. mellifera* and found 23 to 35% total sugars, and Sampaio and De Freitas (1993) obtained a mean of 24.76% in Paraná.

25.3.7 pH

The pH regulates many chemical, biochemical, and microbiological reactions and can vary according to cultivar diversity, maturity, seasonal variation, geographic area, and management and conservation practices prior to processing (Rahman 2003).

The "samburá" of *M. scutellaris* had a pH of 3.7. The low pH value of "samburá" from these bees could be due to the transformations that occurred throughout the transport and storage processes. Pinheiro et al. (2012) obtained a pH of 3.80 and 3.42 for the "samburá" of *M. flavolineata* and *M. fasciculata*, respectively. Sampaio and De Freitas (1993) obtained pH values of 4.3–4.4 for the beebread of *A. mellifera* in Paraná.

Ellis and Hayes (2009) reported that after fermentation in the presence of enzymes and sugars, pollen had a lower pH and a lower quantity of starch. It should be noted that a pH of 4.2 allows for the control of almost all microorganisms that result in food poisoning except lactic acid bacteria and some species of yeasts and fungi (Rahman 2003).

25.3.8 Free Acidity

The free acidity of the "samburá" of *M. scutellaris* had a mean value of 150.57 meq/kg. The values for the dry and rainy seasons were 152.64 and 148.51, respectively, which are close to the overall mean. Pinheiro et al. (2012), evaluating "samburá" from two meliponaries in Pará, found acidity values of 1781.0 and 1380.0 meq/kg, which is well above the maximum permissible value of 300 meq/kg. These very high values may be related to chemical reactions occurring during the processing of this product by the bees, the origin of the pollen, or the methods of analysis used.

25.3.9 Water Activity (Aw)

The Aw values obtained for the "samburá" of *M. scutellaris* were 0.93 (rainy season) and 0.92 (dry season). The pasty consistency of the "samburá" resulting from the formation process and the environment where it is stored is likely responsible for the high Aw values.

Aw is used to assess the susceptibility of the product to degradation. Aw values greater than 0.7 facilitate the proliferation of pathogenic microorganisms. Water activities from 0.6 to 0.7 do not create a good medium for the development of most microorganisms because pathogenic bacteria require an Aw of 0.85 and yeasts and molds require an Aw of 0.80 for growth (Martinez 2005).

Despite the high Aw value of *M. scutellaris* "samburá," no microorganisms harmful to human health were found in the samples analyzed (Table 25.2). This finding is probably due to the low pH and high free acidity, or the relation among other physicochemical characteristics of the product.

25.3.10 Total Energy Value

The seasonal analysis showed marked variation in energy value between the two time periods: 269 kcal/100 g (rainy) and 193.5 kcal/100 g (dry). This finding is possibly due to the nature of the TEV calculation, which uses values from the physicochemical analysis.

The mean total energy value of the "samburá" from *M. scutellaris* was 231.33 kcal/100 g. Souza et al. (2004), analyzing pollen from Amazon meliponines, reported a similar mean energy concentration of 264.4 kcal/100 g.

25.4 Microbiological Characteristics of "Samburá"

The microbiological analyses were performed according to the methods described in the international standards (Downes and Ito 2001) for each microbial group. In these analyses, standard counts were performed for molds and yeasts and for mesophilic and psychrotrophic aerobes, and the presence of total and thermotolerant coliforms in the pollen samples was documented.

A 5.0 g aliquot of each homogenized sample was used to prepare the first dilution (10^{-1}) in 45.0 mL of 0.1% buffered peptone water; the subsequent decimal dilutions were prepared in tubes containing 9.0 mL of the same diluent to obtain the 10^{-2} and 10^{-3} pollen concentrations.

For the molds and yeasts counts, the direct surface plating method was used with a potato dextrose agar (PDA) medium acidified with 10% tartaric acid to pH 3.5. Aliquots of 0.1 mL were seeded on the surface of the PDA agar, and the plates were incubated in a biological oxygen demand (BOD) chamber at 25 °C for 5 days. After this period, counts were performed to determine the number of colony forming units per gram of material (cfu/g).

For the analyses of mesophilic and psychrotrophic aerobes, the same decimal dilutions were used, and a 1.0 mL volume was deep plated on plate count agar (PCA) medium. After the homogenization and solidification of the agar at room temperature, the plates were incubated at 35 °C for 48 h for the count of mesophilic aerobic microorganisms. For the count of psychrotrophic microorganisms, the same plating procedures were used, and the incubation was performed at 7 °C for 10 days.

To evaluate the presence of total and thermotolerant coliforms, the most probable number (MPN) method, also known as the multi-tube method, was used. First, a presumptive test with

Table 25.2 Microbiological analysis of samples of "samburá" from *M. scutellaris* collected during two seasons in Camaçari, Bahia, Brazil ($N = 6$)

Pot-pollen season	Concentration	Molds and yeasts (cfu/g)			Total count of mesophilic aerobes in pollen (cfu/g)			Overall count of psychrotrophic aerobes (cfu/g)
		R1	R2	R3	R1	R2	R3	
Rainy	10^{-1}	–	–	–	8	7	15	–
	10^{-2}	–	–	–	5	8	4	
Dry	10^{-1}	–	–	–	30	22	50	
	10^{-2}	–	–	–	23	15	9	

R1, R2, R3 represent replicated samples

lauryl sulfate tryptose (LST) broth was performed for the incubation of the dilutions using three inverted Durham tubes containing 10 mL of LST for each dilution; these were kept in a BOD chamber at 35 °C for 48 h. The presence of coliform bacteria was documented via growth with gas production inside the Durham tubes.

The microbiological characterization of *M. scutellaris* "samburá" showed that the product remained within the control standards recommended by Brazilian food legislation (Table 25.2). The quantification of total and thermotolerant coliforms showed no contamination in the analyzed samples (<3 MPN/g).

The specifications for the presence of microorganisms in food in different Latin American countries show some variation in the allowable mesophile concentration: 150,000 cfu/g for Argentina and 10,000 cfu/g for Brazil, Mexico, and El Salvador. For fungi and yeast, Brazil and Argentina tolerate 100 cfu/g, whereas Mexico and El Salvador allow 300 cfu/g. There is a complete lack of legislation regarding fecal coliform. Ordinance 248 of 12/30/1998, which relates to the survey of spores of *Paenibacillus larvae*, states that an absence of these spores in 25 g of pollen is acceptable and that microbial criteria will be established in more specific regulations (BRASIL 2001).

The comparative analysis of the presence of mesophiles between the dry and rainy seasons showed an increase in the colonies of these microbes in the dry season. The dry season in the study region is known for high external temperatures (30–25 °C) and high humidity (80%).

The fact that "samburá" does not show contamination by the major microorganisms addressed in food legislation, even at a high Aw

value, could be related to the low pH, interactions among factors, and the nature of the storage site in the bee nest (cerumen pots and propolis/resin), as well as the harvest method used and best manufacturing practices. Microorganisms require water, nutrients, adequate temperature, and specific pH levels for their growth; however, there are combinations of Aw and pH that may favor the activity of microorganisms (Rahman 2003).

Martinez (2005) analyzed undried pollen from *A. mellifera* in Honduras and found high levels of mesophiles, fungi, yeast, and total coliform at a moisture level of 19.35% and an Aw of 0.65. Pinheiro et al. (2012) evaluated the pollen from two species of *Melipona* and found a fecal coliform content under 0.03 cfu/g, an absence of *Salmonella*, and molds and yeasts at concentrations of less than 10 cfu/g. Fresh samples of *A. mellifera* pollen were found to be free of *Salmonella* and *Staphylococcus*, but the population of molds and yeasts was relatively high.

Molds and yeasts are the most significant microbiological parameters for bee pollen, followed by bacteria and total coliforms (Hervatin 2009). Santos et al. (2010) analyzed pollen from *A. mellifera* and also detected a considerable population of molds and yeasts.

Even when using a drying temperature of 40 °C, fungi and bacteria may be present in a latent state as spores, as pollen is a nutritionally rich substrate that favors the recovery of biological contaminants in a moist conservation environment. If the collection, storage, and commercialization conditions are deficient, fungi may develop, as is observed in cereals (Gonzalez et al. 2005).

Although it is recognized as beneficial for human health, pollen consumption requires caution, as pollen is susceptible to environmental contaminants and the growth of microorganisms. In this context, it is necessary to highlight the importance of natural contaminants such as mycotoxins, which are the toxic products of the secondary metabolism of fungi and can colonize food during collection and storage (Rodrigues et al. 2008).

Fungi are known to produce mycotoxins. Among the most common and toxic are the aflatoxins, which are produced by fungi of the genus *Penicillium* and *Aspergillus* which can cause chronic intoxication in humans and animals. The high hygroscopicity of pollen also favors the reproduction of *Fusarium* spp. (Gonzalez et al. 2005). Rodrigues et al. (2008) confirmed that the microbiota found in the pollen of the "jataí" bee (*Tetragonisca angustula*) in Rio de Janeiro includes the dominant toxigenic and/or pathogenic genera for animals and humans; the genera *Aspergillus*, *Penicillium*, *Cladosporium*, *Fusarium*, *Mucor*, and *Curvularia* were isolated from this pollen.

Salamanca Grosso et al. (2000) found fungi of the genus *Alternaria* in samples of *A. mellifera* corbicular pollen, which is one of the more common fungi in the environment. However, the inhibition of fungal development could be attributed to the effect of antimycotic substances and bee secretions as well as the antimicrobial action of honey that is attributed to its low pH, high osmolality, peroxidase enzymatic activity, and the presence of aromatic acids and phenolic compounds (see Chap. 28 by Sulbarán-Mora et al.).

25.5 Pollen Analysis

To assist in the pollen analysis, samples of the plants present within a 500 m radius of the experimental site (meliponary) were collected and sent to the UFRB herbarium for identification. The dried specimens were also compared to lists and catalogs of the regional flora. The flowering period was observed, and the pollen grains were mounted on microscope slides for the assembly of a reference pollen library for the experimental area.

All pollen samples were prepared using the Erdtman (1952) acetolysis method and then subjected to quantitative and qualitative analysis. The quantitative analysis was performed by consecutive counts of 500 pollen grains/replicate/ sample and determining the relative abundance of each class occurring in the sample (Louveaux et al. 1978). The qualitative analysis (pollen types present in the samples) was performed by comparison with the reference slides and descriptions obtained from the published literature including Barth (1970a, 1970b, 1970c, 1971, 1989, 1990, 2004, 2006), Bastos et al. (2003), Barth et al. (2005, 2006), and Moreti et al. (2007).

A total of 23 pollen types belonging to 10 families and 19 genera were found in the pollen spectrum of *M. scutellaris* (Table 25.3). The genera with the highest number of species were *Solanum* (3) and *Eugenia* (2).

The families recorded were Myrtaceae (8 types); Solanaceae (4 types); Anacardiaceae, Melastomataceae, and Fabaceae-Mimosoideae (2 types each); and Fabaceae-Faboideae, Sapindaceae, Malvaceae, Cactaceae, Arecaceae (1 type each) (Fig. 25.2). The greater richness of Myrtaceae and Solanaceae is due to the vegetation characteristics of a relatively anthropogenic rainforest and the contribution from cultivated species. The Myrtaceae stand out as important producers of edible fruits, and the Solanaceae are pioneer plants in anthropogenic areas.

Of the 23 pollen types found, there was only one dominant type (*Campomanesia*) (DP), three accessory types (AP), six types of important isolated pollen (IIP), and 13 types of occasional isolated pollen (OIP). *Solanum* sp. and *Mimosa caesalpiniifolia* are considered to be accessory pollen types that bloom all year-round; the majority of the Solanaceae are considered polliniferous.

The seasonal analysis of the number of pollen types and families showed that diversity was higher during the rainy season (14 types) than the dry season (11 types). However, variation in the number of families was low (dry season, eight types; rainy season, six types).

During the dry season, only the families Malvaceae and Arecaceae did not occur in the pollen spectrum, despite the large number of Arecaceae present in the environment. Myrtaceae (53.53%), Anacardiaceae (16.69%), Solanaceae (14.18%), and Fabaceae-Mimosoideae (11.44%) occurred during this period, with Solanaceae and Fabaceae-Mimosoideae represented by only one pollen type each.

Table 25.3 Family, pollen type, harvest season, and occurrence classes of pollen in the "samburá" of *Melipona scutellaris* from a rainforest on the northern coast of the state of Bahia

Family	Pollen type	Season (%) Rainy	Dry	Overall mean (%)	Occurrence classes
Anacardiaceae	*Tapirira guianensis*	–	4.11	8.38	PII
	Spondias mombin	–	12.58		PII
Arecaceae	*Syagrus* sp.	8.00	–	4.00	PII
Cactaceae	*Hylocereus undatus*	–	0.28	0.16	PIO
Fabaceae-Faboideae	*Pueraria phaseoloides*	–	0.68	0.37	PIO
Fabaceae-Mimosoideae	*Mimosa caesalpiniifolia*	9.00	11.44	11.6	PA
	Pithecellobium dulce	2.00	–		PIO
Malvaceae	NI	1.00	–	0.50	PIO
Melastomataceae	*Ossaea* sp.	8.00	–	4.70	PII
	Miconia sp.	–	1.37		PIO
Myrtaceae	*Psidium* sp.	–	2.28	30.60	PIO
	Campomanesia sp.	–	50.11		PD
	Corymbia torrelliana	3.00	–		PII
	Myrcia obovata	1.00	–		PIO
	Syzygium samarangense	2.00	–		PIO
	Eugenia stipitata	1.00	–		PIO
	Eugenia uniflora	1.00	–		PIO
	Plinia cauliflora	–	1.14		PIO
Solanaceae	*Solanum macrocarpon*	30.00	14.18	39.10	PA
	Solanum stipulaceum	3.00	–		PII
	Solanum sp.	30.00	–		PA
	Cestrum sp.	1.00	–		PIO
Sapindaceae	*Allophylus* sp.	–	1.83	0.93	PIO

Note: *PD* dominant pollen (FR > 45%), *PA* accessory pollen (15% < FR < 45%), *PII* important isolated pollen (3% < FR < 15%), *PIO* occasional isolated pollen (FR < 3%), *NI* not identified, *FR* relative frequency (%)

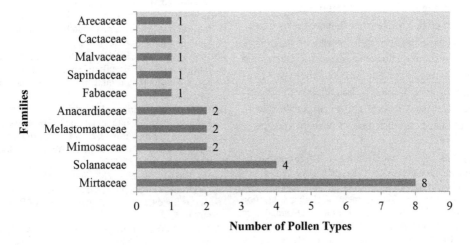

Fig. 25.2 Botanical families identified in samples of *M. scutellaris* "samburá" from a restricted area along the northern coast of the state of Bahia collected during two different seasons

During the rainy season, the family most frequently observed was Solanaceae (67.00%), with four pollen types, followed by Myrtaceae (53.49%) and Fabaceae-Mimosoideae (11.00%), with two types each. The families Anacardiaceae, Fabaceae-Faboideae, Sapindaceae, and Cactaceae did not occur during this period.

Comparing the two seasons, the dominant families were the Solanaceae, Myrtaceae, Anacardiaceae, Fabaceae-Mimosoideae, Melastomataceae, and Arecaceae. Oliveira et al. (2009) reported that the family Melastomataceae contributed the greatest quantity of pollen to the meliponine studied in Manaus, with the genus *Miconia* being more abundant in the dry season. In the Amazon, the Myrtaceae family concentrated its flowering in the periods that was less rainy. *Eugenia stipitata* and *Myrcia* sp. were the most important species of this family in the months of March to October. Modro et al. (2007) and Funari et al. (2003) reported that pollen types might vary depending on the region or season in which they are offered, influencing the composition of the stored pot-pollen.

Some species such as *Mimosa caesalpiniifolia* and *Solanum* sp. bloom year-round and appeared in both analyzed seasons. *Spondias mombin* is considered a polliniferous species that blooms during the same period as *Campomanesia* sp.; however, their contribution to pollen was lower, demonstrating that the composition of the pollen can reflect the bee's dietary preference.

Modro et al. (2007) cites Asteraceae as a family that has a low protein content in its pollen, unlike the family Fabaceae-Mimosoideae, which is a rich source of protein for bees. These two families are present in most pollen analyses and constitute the basis of beekeeping, primarily due to the abundance and distribution of these species.

Acknowledgements We thank the editors for the invitation, the meliponicultors for their knowledge and experience, and the Brazilian agencies of research development: Coordination for the Improvement of Higher Education Personnel (CAPES), the National Council of Technological and Scientific Development (CNPq), and State of Bahia Research Foundation (FAPESB) for the financial support to our studies.

References

ANVISA. 2003. Resolução RDC n. 360 de 23 de setembro de 2003. Regulamento Técnico sobre rotulagem nutricional de alimentos embalados. Diário oficial da União. Poder executivo. Brasília, Brasil. http://www.abic.com.br/publique/media/CONS_leg_resolucao360-03.pdf

Association of Analytical Chemistry (AOAC). 2000. Official Methods of Analysis. International 17th. Ed. AOAC; Gaithersburg, USA. 1375pp.

Association of Analytical Chemistry (AOAC). 1990. Official Methods of Analysis. International 15th. Ed. AOAC; Arlington, USA. 1018pp.

Baldi-Coronel B, Grasso D, Chaves Pereira S, Fernández, G. 2004. Caracterización bromatológica del polen apícola argentino. Ciencia, Docencia y Tecnologia 29: 145-181.

Barth OM. 1970a Análise microscópica de algumas amostras de mel. 1. Pólen dominante. Anais Academia Brasileira Ciências 42: 351-66.

Barth OM.1970b. Análise microscópica de algumas amostras de mel. 2. Pólen acessório. Anais Academia Brasileira Ciências 42: 571-90.

Barth OM. 1970c. Análise microscópica de algumas amostras de mel. 3. Pólen isolado. Anais Academia Brasileira Ciências 42: 747-72.

Barth OM. 1971. Análise microscópica de algumas amostras de mel. 6. Espectro polínico de algumas amostras de mel dos Estados da Bahia e do Ceará. Revista Brasileira de Biologia 31: 431-434.

Barth OM. 1989. O pólen no mel brasileiro. Luxor; Rio de Janeiro, Brasil. 152 pp.

Barth, OM. 1990. Pollen in monofloral honeys from Brazil. Journal of Apicultural Research 29: 89-94.

Barth OM. 2004. Melissopalynology in Brazil: a review of pollen analysis of honeys, propólis and pollen loads of bees. Scientia Agricola 61: 342-350.

Barth OM. 2006. Palynological analysis of geopropolis samples obtained from six species of Meliponinae in the Campus of the Universidade de Ribeirão Preto, USP, Brasil. Apiacta 1: 1-14.

Barth OM, da Luz CFP, Gomesklein VL. 2005. Pollen morphology of Brazilian species of *Cayaponia* Silva Manso (Cucurbitaceae, Cucurbiteae). Grana 44: 129 -136.

Barth OM, São Thiago LE, de Barros MA. 2006. Paleoenvironment interpretation of 1760 years B.P. old sediment in a mangrove area of the Bay of Guanabara, using pollen analysis. Anais da Academia Brasileira de Ciências 78: 227-229.

Bastos EMAF, Silveira VM, Soares AE. 2003. Pollen spectrum of honey produced in cerrado areas of Minas Gerais State (Brazil). Brazilian Journal of Biology 63: 599-615.

BRASIL. 2001. Instrução Normativa n.3, de 19 de janeiro de 2001. Ministério da Agricultura, Pecuária e Abastecimento aprova os regulamentos técnicos de identidade e qualidade de apitoxina, cera de abelha, geléia real, geléia real liofilizada, pólen apícola, própolis e extrato de própolis. In: MINISTÉRIO DA AGRICULTURA, PECUÁRIA E ABASTECIMENTO. Legislação. SisLegis – Sistema de Consulta à Legislação. Disponível em.: http://agricultura.gov.br

BRASIL. 1985. Ministério das Minas e Energia. Projeto Radam Brasil, folha SD24, Salvador: Geologia, morfologia, pedologia, vegetação e uso potencial da terra. Rio de Janeiro, Brasil. 624 pp.

Campos MGR, Bogdanov S.; Almeida-Muradian LB, Szczesna T,Mancebo Y, Frigerio C, Ferreira F. 2008. Pollen composition and standardisation of analytical methods. Journal of Apicultural Research 47: 156–163.

Carpes ST, Cabral ISR, Luz CFP, Capeletti JP, Alencar SM, Masson ML. 2009. Palinological and physicochemical characterizacion of *Apis mellifera* L. bee pollen in the southern region of Brazil. Journal of Food, Agriculture and Environment 7: 667-673.

Carvalho CAL, Alves RMO, Souza, BA. 2003. Criação de abelhas sem ferrão: Aspectos práticos. Série Meliponicultura 1. Ed. SEAGRI; Salvador, Brasil.42 pp.

Donadieu DY. 1979. El polen: Terapéutica natural, 4th Ed. Librairie Maloine; Paris, France. 32 pp.

Downes FP, Ito K. (eds.). 2001. Compendium of methods for the micobiologycal examination of foods. 4 ed. American Public Health Association (APHA); Washington, USA. 676 pp.

Ellis AM, Hayes GW jr. 2009. An evaluation of fresh versus fermented diet for Honey bees (*Apis mellifera*). Journal of Apicultural Research. 48: 216-216.

Erdtman G. 1952. Pollen morphology and plant taxonomy – Angiosperms. Almqvist & Wiksell; Stockholm. Sweden. 539 pp.

Espina D, Ordetx G. 1984. Apicultura Tropical. 4 Ed. Costa Rica. Editorial Tecnológica de Costa Rica, 506 pp.

Funari SRC, Rocha HC, Sforcin JM, Filho HG, Curi PR, Gomes Dierckx SMA, Funari, ARM, Oliveira Orsi R De. 2003. Composições bromatologica e mineral do pólen coletado por abelhas africanizadas (*Apis mellifera*) em Botucatu, Estado de São Paulo. Archivos Latino americanos de Produção Animal 11: 88-93.

Gonzalez G, Hinojo M, Mateo R, Medina R, Jimenez M. 2005. Ocurrence of mycotoxin producing fungi in bee pollen. International Journal of Food Microbiology 105: 1-9.

Hervatin HL. 2009. Avaliação microbiológica e físicoquímica do pólen apícola in natura e desidratado sob diferentes temperaturas. Master Thesis. Faculty of Food Engineering, Unicamp; Campinas, Brasil. 70 pp.

Instituto Adolfo Lutz. 1985. Normas analíticas do Instituto Adolfo Lutz, Instituto Adolfo Lutz, São Paulo 1: 21-43.

Ivanivki ASN, Silveira TA, Marchini LC, Moreti ACC. 2012. Relação entre parâmetros físico-químicos do pólen apícola e dados palinológicos. Anais do 19º

CBA. Gramado, 2012. Revista Mensagem Doce 58: 116.

Krell R. 1996. Vallue-added products from beekeeping. Rome: FAO Agricultural Services Bulletin N. 124. http://www.fao.org/docrep/w0076e/w0076e00.htm

Kroyer G, Hegedus N. 2001. Evaluation of bioactive properties of pollen extracts as functional dietary food supplement. Innovative Food Science and Emerging Technologies 2: 171-174.

Louveaux J, Maurizio A, Vorwohl G. 1978. Methods of melissopalynology. Bee World 59: 139-157.

Lorenzo C. 2002. La miel de Madrid. Instituto Madrileno de Investigacion Agraria, Madrid, España. 221 pp.

Martinez JVP. 2005. Caracterizacion físicoquimica y microbiológica del polen de abejas de cinco departamentos de Honduras. Thesis Agronomical Engineering. Universidad Zamorano; Tegucigalpa, Honduras. 36 pp.

Modro AFH, Message D, Da Luz CFP, Neto JAAM. 2007. Composição e qualidade de pólen apícola coletado em Minas Gerais Pesquisa Agropecuária Brasileira, Brasília 42: 1057-1065.

Moraes RMD, Teixeira EW. 1998. Análise de mel. Manual Técnico. Centro de Apicultura Tropical; Pindamonhangaba, Brasil. 42pp.

Moreti ACCC, Fonseca TC, Rodriguez APM, Monteiro-Hara ACBA, Barth OM. 2007. Fabaceae Forrageiras de Interesse Apícola. Aspectos Botânicos e Polínicos. Boletim Científico, 13, 1ª. ed. Instituto de Zootecnia; Nova Odessa, Brasil. 98 pp.

Muradian LB, Penteado MDVC. 2007. Vigilância sanitária: tópicos sobre legislação e análise de alimentos. Guanabara Koogan; Rio de Janeiro, Brasil. 203 pp.

Nogueira-Neto P. 1997. Vida e criação de abelhas indígenas sem ferrão. Editora Nogueirapis; São Paulo, Brasil. 446 pp.

Oliveira FPM, Absy ML, Miranda IS. 2009. Recurso polínico coletado por abelhas sem ferrão (Apidae-Meliponinae) em um fragmento de floresta na região de Manaus- Amazonas. Acta Amazônica 39: 505-518.

Pinheiro FDeM, Costa CVP, Das N, Baptista RDeC, Venturieri GC, Pontes MAN. 2012. Pólen de abelhas indígenas sem ferrão *Melipona fasciculata e Melipona flavolineata*: caracterização físico-química, microbiológica e sensorial. http://www.alice.cnptia.embrapa.br/bitstream/doc/408842/1/polendeabelhasindigenassemferraomelipona

Rahman M. 2003. Manual de conservación de alimentos. Editorial Acribia; Zaragoza, España. 845 pp.

Rodrigues MAA, Keller KM, Keller LAM, De Oliveira AA, Almeida TX, Marassi AC, Kruger CD, Barbosa TS, Lorezon MCA, Rocha Rosa CA. 2008. Avaliação micológica e micotoxicologica do pólen da abelha jataí (*Tetragonisca angustula*) proveniente de Ilha Grande, Angra dos Reis, RJ. Revista Brasileira de Medicina Veterinária 30: 249-253.

Roulston TH, Cane JH, Buchmann SL. 2000. What governs protein content of pollen: pollinator preferences, pollen pistil interactions, or phylogeny? Ecological Monographs 70: 617-643.

Roulston TH, Cane JH. 2000. Pollen nutricional content and digestibility for animals. Plant Systematics and Evolution 222: 187-209.

Rubio EM. 1959. Polinizacion. Instituto Biogenético Rubio, Tucuman, Argentina.174 pp.

Salamanca Grosso G, Hernandez Valero E, Fernando Vargas E. 2000. O pollen en el sistema de puntos criticos, cosecha, propriedades y condiciones de manejo.. http://www.apiservices.biz/es/articulos/ordenar-por-popularidad/738-el-polen-en-el-sistema-de-puntos-criticos

Sampaio EAB, De Freitas RJS. 1993. Caracterização do pólen apícola armazenado na colméia (pão de abelhas) de algumas localidades do Paraná. Boletim Ceppa, Curitiba 11: 81-87.

Santos LO, Silveira NFA, Leite RSF, Borghini RG. 2010. Avaliação microbiológica do pólen apicola comercializado no Estado de São Paulo.. www.iac.sp.gov/áreadoinstituto/pibic/anais/2010/artigos

Souza RC, Da Silva, Yuyama LKO, Aguiar JPL, Oliveira FPM. 2004. Valor nutricional do mel e pólen de abelhas sem ferrão da região amazônica. Acta Amazônica 34: 333-336.

Characterization of Pot-Pollen from Southern Venezuela

26

Patricia Vit, Giancarlo Ricciardelli D'Albore,
Ortrud Monika Barth, María Peña-Vera,
and Elizabeth Pérez-Pérez

26.1 Introduction

Pollen is the proteinaceous food for the bee brood, and its protein content varies (10–40%) according to botanical origin (Menezes et al. 2012; Vossler 2015). The botanical origin and medicinal properties of pollen have been reviewed (Vit 2009a). The pollen stored inside an *Apis mellifera* nest is known as beebread, and pot-pollen is that stored inside the meliponine nest, before being placed in a brood cell. Both beebread and pot-pollen origi-

nate from male gametophytes collected from flowers by the bees. When foraging, bees mix pollen with nectar to form a sticky mass deposited on the hind legs, up to the weight and volume possible to be carried back to the nest, which is stored and processed. This is the corbicular pollen, also known as pollen load or pollen pellet. Pollen collection by *Melipona quadrifasciata anthidioides* in Brazil occurred in peaks between 08:30 and 09:50 am, in contrast to nectar collection all day long (Oliveira-Abreu et al. 2014). Further transformation of this raw material continues inside the cerumen pots (Leonhardt et al. 2007).

Interactions with yeasts and bacteria extend the storage life of pot-pollen because spoilage and diseases are prevented in fermented pollen (Anderson et al. 2011; Menezes et al. 2013) and also facilitate digestion (Rosa et al. 2003). Fermented pollen can be very sour (Cortopassi-Laurino et al. 2006), and admixtures with pot-honey extracted by compression increase the free acidity (P. Vit, personal observation). The great diversity of Neotropical Meliponini (Camargo and Pedro 2013) deserves further studies on pot-pollen characterization.

During the 27th International Conference on Polyphenols held in Nagoya, Japan (ICP2014), the latest advances included polyphenol chemistry, physicochemistry, biosynthesis, genetics, metabolic engineering, and role in plant interactions with the environment, in nutrition and health, and in natural medicine (Cheynier et al.

P. Vit (✉)
Apitherapy and Bioactivity, Food Science Department, Faculty of Pharmacy and Bioanalysis, Universidad de Los Andes, Mérida 5101, Venezuela

Cancer Research Group, Discipline of Biomedical Science, Cumberland Campus C42, The University of Sydney,
75 East Street, Lidcombe, NSW 1825, Australia
e-mail: vitolivier@gmail.com

G.R. D'Albore
Universitá degli Studi, Perugia, Italy

O.M. Barth
Instituto Oswaldo Cruz, Fiocruz, Brazil

Laboratory of Palynology, Department of Geology, Institute of Geosciences, Federal University of Rio de Janeiro, Rio de Janeiro, Brazil

M. Peña-Vera • E. Pérez-Pérez
Laboratory of Biotechnological and Molecular Analysis, Faculty of Pharmacy and Bioanalysis, Universidad de Los Andes, Mérida 5101, Venezuela

2015). Polyphenols are diverse major secondary metabolites in plants. They confer sensory, nutritional, and medicinal properties to plant-derived foods, including nectar, pollen, and resins collected and processed by bees. Besides *Apis mellifera* honey, pollen, and propolis, commercial products of native stingless bees have their repertoire, including pot-honey, pot-pollen, and cerumen from the bee nest.

Only two species of stingless bees – *Melipona* sp. aff. *eburnea* and *Scaptotrigona* cf. *ochrotricha* – have been studied for Venezuelan pot-pollen chemical composition and antioxidant activity (Vit et al. 2016). The botanical origin of pot-honey and pot-pollen was compared (Vit and Ricciardelli 1994a, b). In this chapter we characterize chemical contents of pot-pollen and bioactive indicators such as flavonoids, polyphenols, and antioxidant activity in ethanolic homogenates. In addition to the proposal of pot-honey standards (Vit et al. 2004; Souza et al. 2006), pot-pollen standards also need to be set considering the entomological origin besides the indicators proposed for marketed *Apis mellifera* corbicular pollen (Campos et al. 2008).

Table 26.1 Genera and number of stingless bee species in Venezuela

No.	Genera of stingless bees	No. of species
1	*Aparatrigona*	1
2	*Cephalotrigona*	1
3	*Duckeola*	1
4	*Frieseomelitta*	2
5	*Geotrigona*	2
6	*Lestrimelitta*	2
7	*Melipona*	16
8	*Nannotrigona*	5
9	*Oxytrigona*	1
10	*Paratrigona*	3
11	*Partamona*	9
12	*Plebeia*	3
13	*Ptilotrigona*	1
14	*Scaptotrigona*	6
15	*Scaura*	1
16	*Tetragona*	2
17	*Tetragonisca*	2
18	*Trigona*	13
19	*Trigonisca*	2
Total	19	83

26.2 Venezuelan Stingless Bees

Venezuelan stingless bees were reviewed by Pedro and Camargo (2013) from the entomological identifications initiated in 1987, with all specimens deposited in the Camargo collection RPSP, located in the Biology Department, Faculty of Philosophy, Science and Literature at Universidade de São Paulo, Ribeirão Preto, Brazil. The 83 species representing 19 genera in Venezuela included material collected by RW Brooks and D Wittmann. However, the total number is probably underestimated because other collections were not considered. In Table 26.1 the number of species of each genus of stingless bees in Venezuela is presented. *Melipona*, *Trigona*, and *Partamona* are the most species-rich genera of Venezuelan stingless bees deposited in the Camargo collection.

Venezuelan pot-pollen has scarcely been studied. The pot-honey produced by a number of Venezuelan stingless bee taxa has been characterized for chemical, palynological, and bioactive parameters indicated in Table 26.2:

In this chapter, pot-pollen produced by seven species of stingless bees from southern Venezuela is analyzed. Major pollen types and chemical composition (ash, carbohydrates, fat, moisture, proteins) of pot-pollen were evaluated. Biochemical markers (flavonoids, polyphenols, proteins) and the antioxidant capacity by three methods (ABTS radical cation, Fenton-type reaction, hydroxyl radical) were measured in ethanolic homogenates of pot-pollen. In Table 26.3 the ethnic names and geographical origin of the bee species are summarized.

26.3 Botanical Origin of Venezuelan Pot-Pollen

Pollen is a unique resource that provides primary proteins and other nutrients required by bees to rear their larvae and to develop the ovaries of egg-laying females (Michener 2000). Pollen is collected from plants and prepared as larval provisions with minimum protein requirements to achieve adult bee body size (Vanderplanck et al. 2014). Therefore, bees are selective consumers in

Table 26.2 Studies on pot-honey from Venezuela

Stingless bee genera	Parameters analyzed	References
Frieseomelitta, Melipona, Plebeia, Scaptotrigona, Scaura, Tetragonisca	Melissopalynology	Vit and Ricciardelli (1994a)
Frieseomelitta, Melipona, Plebeia, Scaptotrigona, Scaura	Honey quality indicators[a]	Vit et al. (1994)
Frieseomelitta, Melipona, Plebeia, Scaptotrigona, Scaura	Sugars (erlose, fructose, glucose, maltose, glucose, turanose, trehalose) and electrical conductivity	Bogdanov et al. (1996)
Frieseomelitta, Melipona, Nannotrigona, Scaptotrigona, Tetragonisca	Diastase and invertase activity	Vit and Pulcini (1996)
Melipona	Flavonoids, phenolics	Vit et al. 1997
Frieseomelitta, Melipona, Nannotrigona, Scaptotrigona, Tetragonisca	Flavonoids	Vit and Tomás-Barberán (1998)
Frieseomelitta, Melipona, Nannotrigona, Scaptotrigona, Tetragonisca	Fructose, glucose, maltose, sucrose	Vit et al. (1998a)
Frieseomelitta, Melipona, Nannotrigona, Scaptotrigona, Tetragonisca	Honey quality indicators[a], multivariate analysis	Vit et al. (1998b)
Melipona, Scaptotrigona, Trigona	Standards	Vit et al. (2004)
Melipona	Honey quality indicators[a]	Vit et al. (2006a)
Melipona	Honey quality indicators[a]	Vit et al. (2006b)
	Honey quality indicators[a]	Souza et al. (2006)
Tetragonisca	Ethanol, moisture, sugars, antioxidant capacity (hydroxyl radical inhibition percentage and superoxide anion, benzoate degradation)	Pérez-Pérez et al. 2007
Melipona, Tetragonisca	Antioxidant capacity	Rodríguez-Malaver et al. (2007)
Melipona	Honey quality indicators[a]	Vit (2008a)
Melipona	Sensory descriptive	Vit (2008b)
Frieseomelitta, Plebeia, Scaptotrigona, Scaura, Tetragonisca, sp. 1	Honey quality indicators[a]	Vit (2009b)
Melipona, Tetragona	Sensory free-choice profile	Vit et al. (2011)
Melipona	Sensory free-choice profile	Vit et al. (2011)
Melipona	Flavonoids	Truchado et al. (2011)
Melipona	Honey quality indicators[a]	Vit et al. (2012)
Melipona	Honey quality indicators[a] Sensory descriptive	Vit (2013)
Melipona, Tetragona	Sensory free-choice profile	Deliza and Vit (2013)
Melipona	Anticancer activity in vitro	Vit et al. (2013)

[a]Honey quality indicators as in the honey norms: ash, diastase activity, free acidity hydroxymethylfurfural, moisture, reducing sugars, and sucrose. Invertase activity, nitrogen, pH, and water-soluble solids were included in some cases

a pollen-based economy, and forager decisions recognize cues associated with pollen rewards (Carr et al. 2015).

The proteins from pollen which have structural functions are necessary to produce and maintain adult bees; they control body fat and glands involved in wax and royal jelly production (Dabija 2010). Bee pollen selection based on protein content, and bee pollination, are currently debated, considering in the ecology of bees (Hanley et al. 2008). Bee foraging activity is also associated with the dynamics of collecting the

Table 26.3 Geographical origin and ethnic names of pot-pollen produced by stingless bees in four states of southern Venezuela

Stingless bee species	Ethnic names	Location	State
Frieseomelitta aff. *varia*	"angelita"	El Paují	Bolívar
Frieseomelitta aff. *varia*	"angelita"	Santa Elena de Uairén	Bolívar
Melipona compressipes (Fabricius, 1804)	"guanota"	Guasdualito	Apure
Melipona eburnea (Friese, 1900)	–	San Fernando de Atabapo	Amazonas
Melipona favosa (Fabricius, 1798)	"erica"	Barinas	Barinas
Melipona sp. *fulva* group	–	Guaramajé	Amazonas
Melipona lateralis kangarumensis (Cockerel, 1920)	–	Guaramajé	Amazonas
Melipona lateralis kangarumensis	–	San Juan de Manapiare	Amazonas
Melipona lateralis kangarumensis	–	San Juan de Manapiare	Amazonas
Melipona paraensis (Ducke, 1916)	–	Caño Tumo	Amazonas
Melipona paraensis	–	Maroa	Amazonas
Melipona paraensis	–	Carrizal	Amazonas
Melipona paraensis	–	Carrizal	Amazonas

pollen, which can depend on environmental temperature and humidity (Costa et al. 2015). Studies of botanical origin of *Apis mellifera* honey and pollen were done in Brazil (Barth et al. 2013).

Professor Camargo suggested pot-pollen was the first pollen stored – not merely placed in brood cells – by bees in our planet, dating back to the Late Cretaceous, before bee pollen was produced by *Apis mellifera* (JMF Camargo, personal communication). He mentioned the oldest-known bee fossil, *Cretotrigona prisca*, preserved in amber from New Jersey (Michener and Grimaldi 1988a, 1988b). Dinosaurs and stingless bees shared landscapes 97–74 million years ago.

Pot-pollen contains a story of stingless bee flights compacted into the cerumen pots. The transformation varies as in some pots powdery multilayers of varied pollen colors are visible, whereas in others a creamy brownish monotone resembles the color children get on their palettes after mixing all the paints together. Pollen forage returns to the nest for use in colony nutrition. Synchronous pollen presentation in the flowers, appropriate weather conditions, and an active bee in the correct place where pollen is offered are needed. Pollen grains from the plant are collected and moved by the bee to its specialized corbiculae, attached in such a way pollen is sel-

dom lost during the long collection and return flights. If the pollen is not collected, it will be recycled into the ecosystem. All the variables studied on the foraging bees returning to the nest (temperature, humidity, rain, sun radiation) with a corbicular colorful pollen mass end up in the pollen pot, which will be closed when full, and then another pot will be filled. Fast or slow filling occurs according to available pollen resources and also includes chemical and microbiological processing (see Chaps. 17 Calaca et al. and 18 Menezes et al.).

Melittopalynology is valuable for understanding the contribution of botanical resources in the composition of pot-pollen. Pollen resources from 13 stingless bee nests from southern Venezuela were studied, and the dominant botanical families are described. Fresh pot-pollen samples were mounted on a microscope slide and sealed with a cover slip after adding a drop of warm glycerin jelly (Louveaux et al. 1978). For identification under the microscope, a pollen reference collection and literature from tropical America with pollen illustrations were used: Barth (1969, 1989), Persano Oddo and Ricciardelli (1986), Palacios-Chávez et al. (1991), Roubik and Moreno (1991), Vit (2005), and also the bee flora from Venezuela by López-Palacios (1986). At least 300 pollen grains were counted for each pot-pollen sample to calculate frequency classes.

Pollen types belonging to nine predominant botanical families were recognized: Acanthaceae, Cunoniaceae, Fabaceae, Malvaceae, Myrtaceae, Polygonaceae, Rutaceae, Scrophulariaceae, and Solanaceae. The most representative taxa are illustrated in Fig. 26.1 and briefly described below (sizes refer to the polar axis x equatorial axis):

Antigonon (Polygonaceae): 3-colporate, reticulate pollen grains. A, polar view; B, equatorial view; size, 53.3 × 52.1 μm

Astronium (Anacardiaceae): 3-colporate, striate-reticulate pollen grains. C, polar view; D, equatorial view, colpus; E, idem, optical section; size, 25. × 23.3 μm

Avicennia (Acanthaceae): 3-colporate, reticulate pollen grains. F, polar view; G, equatorial view; size, 32.3 × 29.1 μm

Bidens (Asteraceae): 3-colporate, echinate pollen grains. H, polar view, surface; I, idem, optical section; J, equatorial view, colpus; K, idem, optical section; size, 27.5 × 30.7 μm

Eupatorium (Asteraceae): 3-colporate, echinate pollen grains. L, polar view, surface; M, idem, optical section; N, equatorial view, colpus; O, idem, optical section; size, 22.5 × 25.5 μm

Scrophularia nodosa (Scrophulariaceae): 3-colporate, finely reticulate pollen grains. P, polar view; Q, equatorial view, colpus; size, 33.1 × 34.6 μm

Solanum americanum (Solanaceae): 3-colporate, scabrate pollen grains. R, polar view, optical section; S, equatorial view, optical section; size, 27.4 × 24.8 μm

Pollen spectra were compared for honey produced by *Melipona* and non-*Melipona* species. Pollen diversity was higher in *Melipona* than in the non-*Melipona* stingless bee genera *Frieseomelitta*, *Nannotrigona*, *Scaptotrigona*, and *Tetragonisca* (Vit and Ricciardelli D'Albore 1994a, 1994b). Predominant pollen types from the northeastern Brazilian *Tetragonisca angustula* – Anacardiaceae, Dilleniaceae, Fabaceae, Hypericaceae, Lamiaceae, Malpighiaceae, Malvaceae,

Moraceae, Melastomataceae, Myrtaceae, Rubiaceae, and Urticaceae – were recognized by Novais et al. (2015). Three of these families are shared as dominant pollen found in the Venezuelan pot-pollen: Fabaceae, Malvaceae, and Myrtaceae. Pollen spectra of honeys from *Apis* and the *Tetragonisca angustula*, obtained on the same day in six apiaries of São Paulo state, Brazil, were compared by Barth et al. (2013). An overlap of food resources occurred, but each species displayed its particular choices.

26.4 Proximal Analysis of Venezuelan Pot-Pollen

Proximal analysis was done in duplicate following official analytical methods (AOAC 1999). Moisture, ash, and fat analysis is based on gravimetric methods; proteins were measured by the micro-Kjeldahl method following sulfuric acid (Merck, Darmstadt, Germany) digestion; ammonia distillation by vapor flow with sodium hydroxide (Sigma Aldrich, USA) and sodium thiosulfate (Merck, Darmstadt, Germany), collected in boric acid (Sigma Aldrich, USA); and titration with hydrochloric acid (Sigma Aldrich, USA); carbohydrates were calculated by difference. See Table 26.4.

The moisture contents in *Melipona* stored pollen varied between 29.01 and 42.74 g/100 g and in *Frieseomelitta* was 29.96 g/100 g, higher than *Tetragonisca angustula* moisture values (23.34–24.69 g/100 g) in Chap. 24. *Melipona* ash contents varied between 1.93 and 2.92 g/100 g and in *Frieseomelitta* 3.13 g/100 g, similar to *Tetragonisca angustula* ash values (2.06–2.13 g/100 g). The protein contents of *Melipona* pot-pollen varied between 18.44 and 22.31 g/100 g and were 24.72 g/100 g for *Frieseomelitta*. An intermediate protein content (22.43–22.97 g/100 g) was observed for *Tetragonisca angustula* pot-pollen, and a fat content (4.42–4.58 g/100 g), which varied between 4.12 and 6.03 g/100 g in *Melipona*, was lower in *Frieseomelitta* 3.51 g/100 g. Finally, carbohydrates in *Frieseomelitta* and *Melipona* spp. (31.02–41.38 g/100 g) were lower than in *Tetragonisca angustula* pot-pollen (45.98–

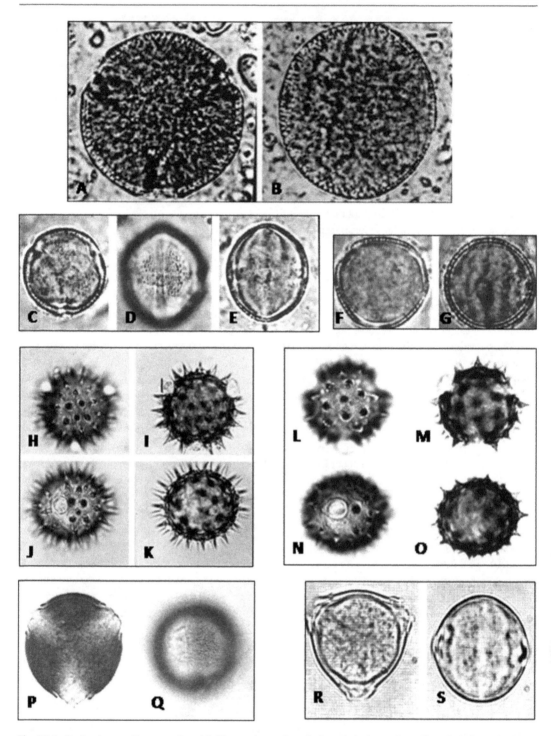

Fig. 26.1 Predominant pollen types found in Venezuelan pot-pollen. (**a, b**) *Antigonon* (Polygonaceae), (**c–e**) *Astronium* (Anacardiaceae), (**f, g**) *Avicennia* (Verbenaceae), (**h–k**) *Bidens* (Asteraceae), (**l–o**) *Eupatorium* (Asteraceae), (**p, q**) *Scrophularia nodosa* (Scrophulariaceae), (**r, s**) *Solanum americanum* (Solanaceae) (Photos: G. Ricciardelli D'Albore)

Table 26.4 Proximal analysis of pot-pollen from southern Venezuela

Pot-pollen type Stingless bee species	N	Moisture	Ash	Proteins	Fat	Carbohydrates
Frieseomelitta sp. aff. *varia*	2	29.96 (0.22)	**3.13** (0.11)	**24.72** (0.18)	**3.51** (0.09)	38.68 (0.20)
Melipona compressipes	1	32.75 (0.25)	2.85 (0.08)	21.01 (0.15)	4.12 (0.11)	39.27 (0.17)
Melipona eburnea	1	35.89 (0.31)	2.54 (0.15)	**18.44** (0.29)	**6.03** (0.20)	37.10 (0.25)
Melipona favosa	1	**29.01** (0.20)	2.92 (0.07)	22.31 (0.25)	4.38 (0.15)	**41.38** (0.19)
Melipona sp. *fulva* group	1	31.65 (0.32)	2.45 (0.12)	19.43 (0.10)	5.72 (0.19)	40.75 (0.30)
Melipona lateralis kangarumensis	3	38.32 (0.45)	2.76 (0.23)	21.77 (0.17)	4.80 (0.24)	32.35 (0.48)
Melipona paraensis	4	**42.74** (0.51)	**1.93** (0.19)	19.08 (0.23)	5.23 (0.31)	**31.02** (0.48)

Averages ± (SEM) values. Lowest and highest values of pot-pollen for each proximal analysis are in bold face

46.68 g/100 g) (Chap. 24). The carbohydrate content of pollen could be related to the fructose, glucose, and sucrose sugars found in nectar (Qian et al. 2008). A high content of mannitol (34.9 g/100 g dry pot-pollen) (Silva et al. 2006) and 20.8–31.0 g/100 g wet pot-pollen produced by *Melipona subnitida* in Brazil contrasts with the primary sugars glucose and fructose found in *Apis mellifera* corbicular pollen (Silva et al. 2014). Mannitol is a sweetener for low-calorie and diabetic food products (Ortiz et al. 2012). Silva et al. (2014) also find that proline and serine are the predominant amino acids in *Melipona subnitida* pot-pollen. In addition to the mineral content, mannitol concentration contributes to beneficial nutritional properties in the pot-pollen of *Melipona subnitida*.

Melipona seminigra and *Melipona interrupta* from the Brazilian Amazonia were characterized with high moistures of 53.39 and 37.12 g/100 g, very high proteins 37.63 and 24.00 g /100 g, high lipids 10.81 and 6.47 g/100 g, medium ash 4.03 and 2.7 g/100 g, and low carbohydrate contents 25.66 and 44.27 g/100 g (Rebelo et al. 2016). Similarly, pot-pollen from Venezuelan Amazonian *Melipona* sp. aff. *eburnea* and *Scaptotrigona* cf. *ochrotricha* (Vit et al. 2016) was characterized with higher moistures (43.49–48.54 g water/100 g pot-pollen) and fat contents (3.19–6.72 g/100 g pot-pollen) compared to *Tetragonisca angustula* from the Venezuelan Andes. However, lower per-

centages of protein (16.80–18.32 g), ash (1.94–2.33 g), and carbohydrate (27.62–31.03 g) were observed per 100 g of Venezuelan Amazonian pot-pollen. Pot-pollen of *Melipona mandacaia* from Brazil (Bárbara et al. 2015) had moisture contents of 36.0 ± 2.0, 4.9 ± 0.3 ash, 12 ± 2 total sugars, and 11–39 g protein/100 g pot-pollen. Campos et al. (2008) proposed standards for dehydrated *Apis mellifera* bee pollen: ash content lower than 6 g/100 g, fat content higher than 1.5 g/100 g, protein content higher than 15 g/100 g, and sugar content higher than 40 g/100 g. These are reference compositional standards for a bee pollen product obtained by the species most used in beekeeping, collected outside the nest in corbicular pollen traps, and dehydrated prior to packaging.

26.5 Bioactive Components and Antioxidant Activity of Pot-Pollen Ethanolic Extracts

26.5.1 Preparation of the Ethanolic Extract

For homogenate preparation, a weight of 100 ± 10 mg of the pot-pollen samples described in Table 25.2 was placed on a glass homogenizer (Thomas No. A3528, USA), 5 mL of ethanol 95% (v/v) (Riedel de Haën, Europe) was added, and the mixture was homogenized in an ice bath. Homogenates were centrifuged in a BHG Optima II (USA) centrifuge at 3000 rpm for 10 min, and supernatants were used for biochemical and antioxidant analysis.

26.5.2 Bioactive Components

Measurements of flavonoids (Woisky and Salatino 1998), polyphenols (Singleton et al. 1999), and proteins (Lowry et al. 1951) are thoroughly detailed in Chap. 24. The content of these bioactive components in pot-pollen is shown in Table 26.5.

A total of 13 samples of pot-pollen collected from southern Venezuela were analyzed in triplicate. For these pollen samples, flavonoid content varied between 75.7 and 656.3 mg quercetin/100 g pot-pollen, polyphenols between 1018.0 and 2085.0 mg gallic acid equivalents/100 g pot-pollen, and proteins between 75.3 and 426.5 mg proteins/100 g pot-pollen (Table 26.5).

Regarding the content of total flavonoids and polyphenols, the pot-pollen samples showed lower values than those found by Carpes (2008), LeBlanc et al. (2009), Mărghitas et al. (2009), Feás et al.

(2012), and Bárbara et al. (2015) and in pollen samples from Brazil, North America, and Romania, respectively. Higher values were reported for Indian mustard bee pollen by Ketkar et al. (2014) with flavonoid concentrations of 123.5 ± 3.1 mg quercetin equivalent/100 g pollen and polyphenol concentrations of 1828.1 ± 374.0 mg gallic acid equivalent/100 g pollen.

The protein values reported in ethanolic extracts (75.3–426.5 mg proteins/100 g) were lower than those reported by Morais et al. (2011) and Fatrcová-Šramková et al. (2013), from Portugal and Slovakia, respectively, but similar to the 182.2 ± 5.9 mg proteins/100 g reported by Ketkar et al. (2014).

Differences in the bee species, botanical origin, and geographical locations of the study contribute to variations on pollen antioxidant properties (Leja et al. 2007). Pollen composition is tightly associated with its botanical source and also to the geographical area, because soil and weather help to determine the flora visited by honey bees. Kostić et al. (2015) analyzed the mineral composition of bee pollen of different plant origins collected across Serbia using inductively coupled plasma optical emission spectrometry and concluded that the mineral composition

Table 26.5 Flavonoid, polyphenol, and protein contents of pot-pollen ethanolic extracts

Pot-pollen types Stingless bee species	N	Flavonoid content (mg quercetin equivalents/100 g pot-pollen)	Polyphenol content (mg gallic acid equivalents/100 g pot-pollen)	Protein content (mg proteins/100 g pot-pollen)
Frieseomelitta sp. aff. *varia*	2	690.8 ± 34.5[g]	1836.7 ± 76.4[c]	168.6 ± 4.4[c]
		149.3 ± 23.5[c]	**1018.0 ± 26.4**[a]	386.8 ± 6.3[e]
Melipona compressipes	1	210.6 ± 9.4[c]	1773.3 ± 98.5[b]	363.9 ± 5.2[e]
Melipona eburnea	1	606.0 ± 12.4[g]	1753.0 ± 25.4[b]	175.8 ± 7.5[c]
Melipona favosa	1	119.0 ± 1.4[b]	2055.6 ± 43.7[d]	279.7 ± 8.3[d]
Melipona sp. *fulva* group	1	284.1 ± 9.3[d]	1592.2 ± 45.3[b]	147.7 ± 5.5[b]
Melipona lateralis kangarumensis	3	**75.7 ± 1.4**[a]	1214.8 ± 23.5[a]	**75.3 ± 3.6**[a]
		499.7 ± 2.3[f]	**2085.0 ± 78.5**[d]	227.8 ± 3.3[d]
		288.0 ± 3.4[d]	1589.2 ± 28.7[b]	332.8 ± 5.5[e]
Melipona paraensis	4	105.0 ± 5.7[b]	1088.8 ± 56.3[a]	**426.5 ± 6.2**[f]
		656.3 ± 9.5[g]	2084.3 ± 87.4[d]	136.6 ± 6.8[b]
		363.1 ± 7.5[e]	1905.4 ± 65.3[c]	134.8 ± 95[b]
		420.4 ± 9.3[f]	1654.8 ± 235[b]	256.8 ± 9.1[d]

Data are mean ± SEM values ($n = 3$). Columns within a sample sharing the same letter are not significantly different by ANOVA post hoc Scheffé test ($P < 0.05$). Lowest and highest values of pot-pollen for each parameter are in bold face

of bee pollen depends more on the type of pollen-producing plant than on its geographical origin. In another example, Mărgăoan et al. (2014) determined total and individual carotenoids, fatty acid composition, and the main lipid classes of 16 fresh bee-collected pollen samples from Romania, finding the highest amount of total lipids in *Brassica* sp. pollen. Polyunsaturated fatty acids (PUFAs) were most abundant in the lipid profile, whereas saturated fatty acids were present in variable amounts. The variability of lipid and carotenoid contents was explained by the botanical origin.

26.5.3 Flavonoids by HPLC-UV

Collected bee pollen is rich in sugars, proteins, lipids, vitamins, and flavonoids, 3–5% of the dry weight (Tomas-Lorente et al. 1992). Flavonoids are chemotaxonomic markers for plants (Harborne and Turner 1984) and particularly useful to identify the botanical origin of bee pollen (Campos et al. 1997), which is also rich in flavones and flavonols (Campos et al. 1997, 2002). Flavonoids found in pollen loads of *Melipona subnitida* are aglycones (Silva et al. 2006). This flavonoid profile was characteristic in corbicular pollen collected by other stingless bee such as *Melipona rufiventris* (Silva et al. 2009) and *Scaptotrigona bipunctata* (Lins et al. 2003). Naringenin and isor-

hamnetin were isolated from yellow pollen loads, predominantly *Mimosa gemmulata*, and selagin and 8-methoxiherbacetin were found in brown pollen, predominantly Fabaceae (Silva et al. 2006). In a recent review, the polyphenolic spectra of bee pollen were characterized by two major classes: (1) phenolic acids, hydroxybenzoic acids (gallic acid, protocatechuic acid), and hydroxycinnamic acids (caffeic acid, ferulic acid, chlorogenic acid, *para*- and *ortho*-coumaric acid) and (2) flavonoids, flavones (apigenin, chrysin, luteolin), flavonols (quercetin, rutin, kaempferol, myricetin, galangin), flavanones (naringenin, pinocembrin), and isoflavones (genistein) (Rzepecka-Stojko et al. 2015).

Pot-pollen was prepared and analyzed by HPLC-UV following the method used by Tomás-Barberán et al. (1989) for bee pollen. Methanol pot-pollen extracted at room temperature overnight were filtered and then run on an HPLC coupled with a photodiode array detector, reversed-phase C-18 LiChrospher 100 RP (5 μm), and UV detection at 350 nm. The UV spectra were recorded with a photodiode array detector (Merck-Hitachi L-3000). Identifications of flavonoids were achieved by comparison with authentic reference compounds. In Table 26.6 the presence of major flavonoids is reported for pot-pollen of the stingless bee species studied here. Luteolin was present in 9 of the 13 pot-pollen samples, kaempferol in 6, quercetin in 5,

Table 26.6 Major flavonoids in Venezuelan pot-pollen

Pot-pollen types Stingless bee species	N	Major flavonoids				
		Genkwanin	Kaempferol	Luteolin	8 Methoxykaempferol	Quercetin
Frieseomelitta sp. aff. *varia*	2	–	X	X	–	X
Melipona compressipes	1	–	X	X	X	–
Melipona eburnea	1	X	–	–	X	–
Melipona favosa	1	–	–	–	–	X
Melipona sp. *fulva* group	1	–	X	–	X	X
Melipona lateralis kangarumensis	3	–	3X	3X	X	X
Melipona paraensis	4	2X	–	4X	–	–

X = found once in pot-pollen produced by each stingless bee species, 2X–4X = found from two to four times in the pot-pollen collected by the same species

8-methoxykaempferol in 4, and genkwanin in 3. All pot-pollen of *Melipona lateralis kangarumensis* and *Melipona paraensis* from the Amazon contained luteolin. Each pot-pollen type contained two to four identified flavonoids, but *Melipona favosa* had only quercetin. Silva et al. (2009) identified the aglycones isorhamnetin, luteolin, and quercetin with NMR spectra from Brazilian *Melipona rufiventris* pollen loads.

26.5.4 Antioxidant Activity of Venezuelan Pot-Pollen Ethanolic Extracts

The following methods were results derived by the ABTS radical cation (Re et al. 1999), Fenton-type reaction (Koracevic et al. 2001), and the hydroxyl radical (Halliwell et al. 1987), described in Chap. 24 by Vit et al. Antioxidant activity of pot-pollen ethanolic extracts is presented in Table 26.7. Hydroxyl radical inhibition percentage ranged between 54.08% and 97.32% inhibition/100 g pot-pollen, while AOA value ranged from 0.91 to 1.25 mM uric acid equivalents/100 g pot-pollen and TEAC value between 193.2 and 771.0 μmoles Trolox equivalents/100 g pot-pollen. The highest antioxidant activity was measured by the three methods in one of the two *Melipona lateralis kangarumensis* pot-pollen ethanolic extracts (Table 26.7). All pot-pollen samples presented antioxidant values higher than commercial antioxidants used as reference control: lipoic acid, melatonin, and quercetin. The TEAC values of pot-pollen produced by 13 species in southern Venezuela in this work were higher than those obtained by Pérez-Pérez et al. (2012) for *Apis mellifera* corbicular pollen of Mérida, Venezuela.

There is little information on antioxidant activity of pot-pollen. Solvents used for pollen extracts vary the activity in scavengers of active oxygen species in pot-pollen of *Melipona subnitida* from

Table 26.7 Antioxidant activity of Venezuelan pot-pollen ethanolic extracts and synthetic controls (lipoic acid, melatonin, quercetin) measured by three methods

Pot-pollen types Stingless bee species	N	Antioxidant activity (AOA) (mM uric acid equivalents/100 g pot-pollen)	Hydroxyl radical inhibition percentage (% inhibition/100 g pot-pollen)	Antioxidant activity (TEAC) (μmoles Trolox equivalents/100 g pot-pollen)
Frieseomelitta sp. *aff. varia*	2	1.09 ± 0.01[f]	82.06 ± 7.89[e]	498.2 ± 4.8[f]
		0.97 ± 0.04[d]	71.20 ± 2.78[d]	231.3 ± 2.7[d]
Melipona compressipes	1	1.02 ± 0.01[e]	86.14 ± 3.65[e]	347.5 ± 4.7[e]
Melipona eburnea	1	1.03 ± 0.02[e]	80.17 ± 5.87[e]	351.2 ± 10.3[e]
Melipona favosa	1	1.19 ± 0.02[g]	88.09 ± 0.45[g]	682.2 ± 3.6[g]
Melipona sp. *fulva* group	1	1.04 ± 0.03[e]	80.59 ± 4.65[e]	323.4 ± 12.3[e]
Melipona lateralis kangarumensis	3	0.92 ± 0.02[c]	60.94 ± 7.54[c]	**193.2 ± 5.7**[c]
		1.25 ± 0.03[h]	**97.32 ± 7.28**[h]	**771.0 ± 11.6**[h]
		1.00 ± 0.04[e]	79.99 ± 5.26[e]	359.5 ± 16.8[e]
Melipona paraensis	4	**0.91 ± 0.04**[c]	**54.08 ± 5.19**[c]	206.4 ± 7.3[c]
		1.21 ± 0.03[g]	90.74 ± 7.23[g]	667.6 ± 11.7[g]
		1.10 ± 0.01[f]	84.20 ± 4.44[e]	480.8 ± 12.8[f]
		1.02 ± 0.02[e]	83.13 ± 2.82[e]	357.1 ± 6.8[e]
Commercial antioxidants				
Lipoic acid		0.67 ± 0.02[a]	43.8 ± 1.7[c]	176.8 ± 3.1[c]
Melatonin		0.87 ± 0.03[b]	24.8 ± 1.4[a]	87.6 ± 5.1[a]
Quercetin		0.89 ± 0.05[b]	34.6 ± 2.5[b]	113.2 ± 1.9[b]

Data are mean ± SEM values ($n = 3$). Columns within a sample sharing the same letter are not significantly different by ANOVA post hoc Scheffé test ($P < 0.05$). Lowest and highest antioxidant values of pot-pollen for each method

Brazil (Silva et al. 2006) and bee pollen of *Apis mellifera* from Venezuela (Pérez-Pérez et al. 2012). According to the solvents, the magnitude order of free radical scavenging activity determined using a 1,1-diphenyl-2-picrylhydrazine (DPPH) assay was ethyl acetate > ethanol > hexane extract for pot-pollen of *Melipona subnitida* and ethanol > methanol > water for *Apis mellifera* bee pollen from Venezuela, using the ABTS radical cation assay.

Morais et al. (2011) measured antioxidant and antimicrobial properties of pollen from five Portuguese natural parks and found a free radical scavenging activity with EC_{50} 2.24 mg/mL, observing in the β-carotene bleaching assay the same behavior as in the DPPH method. Another recent example is the work of Ketkar et al. (2014) who investigated the nutraceutical potential of monofloral Indian mustard bee pollen, obtaining DPPH free radical scavenging activity with an inhibitory concentration of 54.79 μg/mL.

Correlations between antioxidant activity and phenolic contents are presented in Table 26.8 and suggest that phenolic compounds are active constituents of pot-pollen, as previously reported for *Apis mellifera* from pollen from Portugal (Morais et al. 2011; Feás et al. 2012), Greece (Graikou et al. 2011), India (Ketkar et al. 2014), and Turkey (Kim et al. 2015). Likewise for honey, despite the marked relationship demonstrated between bioactivity and polyphenols but not with flavonoid contents (Pérez-Pérez et al. 2013). The precise mechanisms of action for the positive correlation between antioxidant activity – measured by the tree methods – and polyphenol content are unknown. In a research with the Australian

Tetragonula carbonaria, the nonaromatic organic acids (citric, malic, and D-gluconic) present in pot-honey are known to chelate metals and may increase the antioxidant activity in contrast to a relatively low polyphenol content (Persano Oddo et al. 2008). Chemical composition of pollen loads collected by *Melipona subnitida* (Silva et al. 2006) and *Melipona rufiventris* (Silva et al. 2009) from Brazil would be wisely compared with the derived pot-pollen to understand further transformations inside the nest microenvironment within the "cerumen pot reactor." It is possible that pot-pollen undergoes an enrichment process inside the cerumen pot, and/or interaction with microbiota within the pollen pot produced active components added to pot-pollen loads (see Chaps. 17 and 18, this book).

Bee pollen has been referred to as the "only complete food from the hive" because it contains all the essential amino acids required by the human body. Of course, the bee pollen chemical composition varies with the plant and geographic origins, as well as climatic conditions, soil type, and beekeeper activities (Feás et al. 2012). Recently, Juszczak et al. (2016) evaluated the effect of supplementing multifloral honey with bee products that included phenolics and antioxidant activity. Royal jelly did not significantly affect these parameters, but supplementation of honey with other bee products (beebread, bee pollen, and propolis) resulted in a significant increase of total phenolics and flavonoid contents, antiradical activity, and reducing power, with the largest effect found in admixtures of honey and beebread. Recently, lyophilized bee pollen from Brazil

Table 26.8 Correlation between antioxidant activity and physicochemical parameters

Parameters	Proteins	Flavonoids	Polyphenols	AOA	HR	TEAC
Proteins	1	0.012	0.034	0.076	0.054	0,017
Flavonoids		1	0.123	0.120	0.076	0–054
Polyphenols			1	*0.977*	*0.901*	*0.934*
AOA				1	0.056	0.219
HR					1	0.195
TEAC						1

AOA Fenton-type radical, *HR* hydroxyl radical, *TEAC* Trolox equivalent antioxidant activity

was an effective natural antioxidant to retard lipid oxidation of refrigerated pork sausages, like current synthetic antioxidants (Florio-Almeida 2017). Beebread of *Apis mellifera* is analogous to pot-pollen in Meliponini, considered as transformed corbicular pollen stored in the nest. To the question "What is the future of bee pollen?" (Campos et al. 2010), we suggest inclusion of pot-pollen in the future of research on pollen collected and processed by stingless bees in their nests.

26.6   Conclusions

Chemical composition, major pollen types, and antioxidant activity were evaluated for pot-pollen from *Frieseomelitta* and six *Melipona* species collected in southern Venezuela. These results provide useful information about pot-pollen as a potential nutraceutical agent rich in carbohydrates (31.02–41.38 g/100 g fresh pot-pollen) and proteins (18.44–24.72 g/100 g fresh pot-pollen), with a role in the prevention of the free radical scavenging processes. A high antioxidant activity is explained by positive correlations with polyphenols (0.901–0.977) – principally due to their redox properties, important in neutralizing free radicals, quenching oxygen, or decomposing peroxides.

Acknowledgments To the memory of Professor João MF Camargo, Biology Department, Universidade de São Paulo, Ribeirão Preto, Brazil, for the identification of the Venezuelan stingless bees. Special thanks to stingless bee-keepers from southern Venezuela for their role as guardians of the tradition and facilitating the collection of pot-pollen. To project FA-127-93B from the Council for the Scientific, Humanistic, and Technological Development at Universidad de Los Andes, Mérida, Venezuela, for supporting field work needed to collect the pot-pollen in Venezuela; to the scholarship ULA-BID-CONICIT to PV, two stages at Universitá di Perugia, Italy, to study melissopalynology and two stages at CEBAS-CSIC, Murcia, Spain, to study flavonoids with Professor F.A. Tomás-Barberán and Professor F. Ferreres; to the support of ZG-AVA-FA-01-98-01 from the Council of Scientific, Humanistic, Technological, and Artistic Development at Universidad de Los Andes, to the Group Apitherapy and Bioactivity; to referees for their appreciated comments; and to Dr. D.W. Roubik for careful English editing.

References

Anderson KE, Sheehan TH, Eckholm BJ, Mott BM, DeGrandi-Hoffman G. 2011. An emerging paradigm of colony health: Microbial balance of the honey bee and hive (Apis mellifera). Insects Sociaux 58: 431–444.

AOAC. 1999. Official Methods of Analysis. 15th ed. Association of Official Analytical Chemists; Arlington, VA, USA. 1093 pp.

Bárbara MS, Machado CS, Sodré Gda S, Dias LG, Estevinho LM, de Carvalho CA. 2015. Microbiological assessment, nutritional characterization and phenolic compounds of bee pollen from *Mellipona mandacaia* Smith, 1983. Molecules 20: 12525-12544.

Barth OM. 1969. Pollenspektrum einiger brasilianischer Honige. Zeitschrift für Bienenforschung 9: 410-419.

Barth OM. 1989. O pólen no mel brasileiro. Editoria Luxor; Rio de Janeiro, Brasil. 151 pp.

Barth OM, Freitas AS, Sousa GL, Almeida-Muradian LB. 2013. Pollen, physicochemical and trophic analysis of paired honey samples of Apis and Tetragonisca bees. Interciência 38: 280-285.

Bogdanov S, Vit P, Kilchenmann V. 1996. Sugar profiles and conductivity of stingless bee honeys from Venezuela. Apidologie. 27: 445-450.

Camargo JMF, Pedro SRM. 2013. Meliponini Lepeletier, 1836. In Moure JS, Urban D, Melo GAR, orgs. Catalogue of bees (Hymenoptera, Apoidea) in the Neotropical Region. On-line version. Available at: http://moure.cria.org.br/catalogue?id=34932

Campos MGR, Bogdanov S, Bicudo de Almeida-Muradian L, Szczesna T, Mancebo Y, Frigerio C, Ferreira F. 2008. Pollen composition and standardisation of analytical methods. Journal of Apicultural Reserch 47: 154-161.

Campos MGR, Frigerio C, Lopes J, Bogdanov S. 2010. What is the future of bee pollen? Journal of ApiProduct and Apimedical Science 2: 131-144.

Campos M, Markham KR, Mitchell KA, Da Cunha AP. 1997. An approach to the characterization of bee pollens via their flavonoid/phenolic profiles. Phytochemical Analysis 8: 181–185.

Campos MG, Webby RE, Markham KR. 2002. The unique occurrence of the flavone aglycone tricetin in Myrtaceae pollen. Zeitschrift für Naturforschung C: A Journal of Biosciences 57: 944–946.

Carpes ST. 2008. Estudo das características físico-químicas e biológicas do pólen apícola de *Apis mellifera* L. da região Sul do Brasil. PhD Thesis, Tese apresentada ao Programa de Pós-Graduação em Tecnologia de Alimentos, Sector de Tecnologia da Universidade Federal do Paraná, Paraná, Brazil.

Carr DE, Haber AI, LeCroy KA, Lee DEA, Link RI. 2015. Variation in reward quality and pollinator attraction: the consumer does not always get it right. AoB Plants. 7: plv034; doi:10.1093/aobpla/plv034. http://www.ncbi.nlm.nih.gov/pmc/articles/PMC4417137/

Cortopassi-Laurino M, Imperatriz-Fonseca VL, Roubik DW, Dollin A, Heard T, Aguilar I, Venturieri GC,

Eardley C, Nogueira-Neto P. 2006. Global meliponiculture: challenges and opportunities. Apidologie 37: 275-292.

Costa SN, Alves RMO, Carvalho CAL, Conceição PJ. 2015. Fontes de pólen utilizadas por *Apis mellifera* Latreille na região semiárida, Ciencia Animal Brasileira, Goiânia 16: 491-497.

Cheynier V, Tomás-Barberán FA, Yoshida K. 2015. Polyphenols: from plants to a variety of food and non-food uses. Journal of Agricultural and Food Chemistry 63: 7589–7594

Dabija T. 2010. Study of amino acids in pollen's composition. Bulletin UASVM Animal Science and Biotechnologies 67: 1-5.

Deliza R, Vit P. 2013. Sensory evaluation of pot-honey. pp. 349-361. In: Vit P, Pedro SRM, Roubik D, eds. Pot-Honey: A legacy of stingless bees. Springer, New York, USA. 654 pp.

Fatrcová-Šramková K, Nôžková J, Kačániová M, Máriássyová M, Rovná K, Stričík M. 2013. Antioxidant and antimicrobial properties of monofloral bee pollen. Journal of Evironmental Science and Health B 48: 133-138.

Feás X, Vázquez-Tato MP, Estevinho L, Seijas JA, Iglesias A. 2012. Organic bee pollen: Bioactive compounds, antioxidante activity and microbiological quality. Molecules 17: 8359–8377.

Florio Almeida J, Soares dos Reis A, Serafini Heldt LF, Pereira D, Bianchin M, Moura C, Plata-Oviedo MV, Haminiuk CWI, Ribeiro IS, Luz CFP, Carpes ST. 2017. Lyophilized bee pollen extract: A natural antioxidant source to prevent lipid oxidation in refrigerated sausages. LWT - Food Science and Technology 76: 299-305.

Graikou K, Kapeta S, Aligiannis N, Sotiroudis G, Chondrogianni N, Gonos E, Chinou I. 2011. Chemical analysis of Greek pollen - Antioxidant, antimicrobial and proteasome activation properties. Chemical Center Journal 5: 33-37.

Halliwell B, Gutteridge J, Aruoma O. 1987. The deoxyribose method: a simple test-tube assay for determination of rate constants for reactions of hydroxyl radicals. Analytical Biochemistry 165: 215–219.

Hanley ME, Franco M, Pichon S, Darvill B, Goulson D. 2008. Breeding system, pollinator choice and variation in pollen quality in British herbaceous plants. Functional Ecology 22: 592–598.

Harborne JB, Turner BL. 1984. Plant Chemosystematics. Academic Press; London, UK 562 pp.

Juszczak L, Gałkowska D, Ostrowska M, Socha R. 2016. Antioxidant activity of honey supplemented with bee products. Natural Products Research 30: 1436-1439.

Ketkar SS, Rathore AS, Lohidasan S, Rao L, Paradkar AR, Mahadik KR. 2014. Investigation of the nutraceutical potential of monofloral Indian mustard bee pollen. Journal of Integral Medicine 12: 379-389.

Kim SB, Jo YH, Liu Q, Ahn JH, Hong IP, Han SM, Hwang BY, Lee MK. 2015. Optimization of extraction condition of bee pollen using response surface methodology: Correlation between anti-melanogenesis, antioxidant activity, and phenolic content. Molecules 20: 19764-19774.

Koracevic D, Koracevic G, Djordjevic V, Andrejevic S, Cosic V. 2001. Method for measurement of antioxidant activity in human fluids. Journal of Clinical Pathololology 54: 356–361.

Kostić AŽ, Pešić MB, Mosić MD, Dojčinović BP, Natić MM, Trifković JĐ. 2015. Mineral content of bee pollen from Serbia. Arhiv za Higijenu Rada i Toksikologiju 66: 251-258.

LeBlanc BW, Davis OK, Boue S, DeLucca A, Deeby T. 2009. Antioxidant activity of Sonoran Desert bee pollen. Food Chemistry 115: 1299–1305.

Leja M, Mareczek A, Wyżgolik G, Klepacz-Baniak J, Czekoń K. 2007. Antioxidative properties of bee pollen in selected plant species. Food Chemistry 100: 237–240.

Leonhardt SD, Dworschak K, Eltz T, Blüthgen N. 2007. Foraging loads of stingless bees and utilisation of stored nectar for pollen harvesting. Apidologie 38:125-135.

Lins ACS, Silva TMS, Camara CA, Silva EMS, Freitas BM. 2003. Flavonoides isolados do polen coletado pela abelha *Scaptotrigna bipunctata* (canudo). Revista Brasilera de Farmacognosia 13: 40–41.

López-Palacios S. 1986. Catálogo para una Flora Apícola Venezolana; Mérida, Venezuela, Consejo de Publicaciones - Universidad de Los Andes. 211 pp.

Louveaux J, Maurizio A, Vorwohl G. 1978. Methods of Melissopalynology. Bee World 51: 125-138.

Lowry OH, Rosebrough NJ, Farr AL, Randall RJ. 1951. Protein measurement with the Folin phenol reagent. Journal of Biological Chemistry 193: 265-275.

Mărgăoan R, Mărghitaş LA, Dezmirean DS, Dulf FV, Bunea A, Socaci SA, Bobiş O. (2014) Predominant and secondary pollen botanical origins influence the carotenoid and fatty acid profile in fresh honeybee-collected pollen. Journal of Agricultural and Food Chemistry 62: 6306-6316.

Mărghitas LA, Stanciu OG, Dezmirean DS, Bobiş O, Popescu O, Bogdanov S, Campos MG. 2009. *In vitro* antioxidant capacity of honeybee-collected pollen of selected floral origin harvested from Romania. Food Chemistry 115: 878–883.

Menezes C, Vollet-Neto A, Imperatriz-Fonseca VL. 2012. A method for harvesting unfermented pollen from stingless bees (Hymenoptera, Apidae, Meliponini). Journal of Apicultural Research 51: 240-244.

Menezes C, Vollet-Neto A, León-Contrera FA, Venturieri GC, Imperatriz-Fonseca VL. 2013. The role of useful microorganisms to stingless bees and stingless beekeeping. pp. 153-171. In: Vit P, Pedro SRM, Roubik D, eds. Pot-Honey: A legacy of stingless bees. Springer, New York, USA. 654 pp.

Michener CD. 2000. The bees of the world. Johns Hopkins Univ. Press; Baltimore, United States. 654 pp.

Michener CD, Grimaldi DA. 1988a. A *Trigona* from late Cretaceous amber of New Jersey (Hymenoptera: Apidae: Meliponinae). American Museum Novitates 2917: 10 pp.

Michener CD, Grimaldi DA. 1988b. The oldest fossil bee: Apoid history, evolutionary stasis, and antiquity of social behavior. Proceedings of the National Academy of Sciences of the United States of America 85: 6424-6426.

Morais M, Moreira L, Feás X, Estevinho LM. 2011. Honeybee-collected pollen from five Portuguese Natural Parks: palynological origin, phenolic content, antioxidant properties and antimicrobial activity. Food Chemistry and Toxicology 49: 1096-1101.

Novais JS, Garcêz ACA, Absy ML, Santos FAR. 2015. Comparative pollen spectra of Tetragonisca angustula (Apidae, Meliponini) from the Lower Amazon (N Brazil) and caatinga (NE Brazil). Apidologie 46: 417-431.

Oliveira-Abreu C, Hilario SD, Luz CFP, Alves-Dos-Santos I. 2014. Pollen and nectar foraging by *Melipona quadrifasciata anthidioides* Lepeletier (Hymenoptera: Apidae: Meliponini) in natural habitat. Sociobiology 61: 441-448.

Ortiz ME, Formaguera MJ, Raya RR, Mozzi F. 2012. *Lactobacillus reuteri* CRL 1101 highly produces mannitol from sugarcane molasses as carbon source. Applied Microbiology and Biotechnology 95: 991-999.

Palacios-Chávez B, Ludlow-Wiechers R, Villanueva R. 1991. Flora palinológica de la reserva de Sian Ka'an, Quintana Roo, México. Centro de Investigaciones de Quintana Roo; Quintana Roo, México. 321 pp.

Pedro SRM, Camargo JMF. 2013. Stingless bees from Venezuela. pp. 73-86. In: Vit P, Silvia RMP y Roubik D, eds. Pot honey: A legacy of stingless bees. Springer, New York. 654 pp.

Pérez-Pérez E, Rodríguez-Malaver J, Vit P. 2007. Efecto de la fermentación en la capacidad antioxidante de miel de *Tetragonisca angustula* Latreille, 1811. BioTecnología 10: 14-22.

Pérez-Pérez EM, Vit P, Huq F. 2013. Flavonoids and polyphenols in studies of honey antioxidant activity. International Journal of Medicinal Plants and Alternative Medicine 1: 63-72.

Pérez-Pérez EM, Vit P, Rivas E, Sciortino R, Sosa A, Tejada D, Rodríguez-Malaver AJ. 2012. Antioxidant activity of four color fractions of bee pollen from Mérida, Venezuela. Archivos Latinoamericanos de Nutrición 62: 375-380.

Persano Oddo L, Heard TA, Rodríguez-Malaver A, Pérez RA, Fernández-Muiño M, Sancho MT, Sesta G, Lusco L, Vit P. 2008. Composition and antioxidant activity of *Trigona carbonaria* honey from Australia. Journal of Medicinal Food 11: 789-794.

Persano Oddo L, Ricciardelli D'Arbore G. 1986. Spettro pollinico di alcuni mieli dell'America tropicale. Apicoltura 2: 25-66.

Qian WL, Khan Z, Watson DG, Fearnley J. 2008. Analysis of sugars in bee pollen and propolis by ligand exchange chromatography in combination with pulsed amperometric detection and mass spectrometry. Journal of Food Composition and Analysis 21: 78-83.

Re R, Pellegrini N, Proteggente A, Pannala A, Yang M, Rice-Evans C. 1999. Antioxidant activity in improved ABTS radical cation decolorization assay. Free Radical in Biology and Medicine 26: 1231-1237.

Rzepecka-Stojko A, Stojko J, Kurek-Górecka A, Górecki M, Kabała-Dzik A, Kubina R, Moździerz A, Buszman E. 2015. Polyphenols from bee pollen: Structure, absorption, metabolism and biological activity. Molecules 20: 21732–21749.

Rebelo KS, Ferreira AG, Carvalho-Zilse GA. 2016. Physicochemical characteristics of pollen collected by Amazonian stingless bees. Ciência Rural 46: 927-932.

Rodríguez-Malaver AJ, Pérez-Pérez EM, Vit P. 2007. Capacidad antioxidante de mieles venezolanas de los géneros *Apis*, *Melipona* y *Tetragonisca*, evaluada por tres métodos. Revista del Instituto Nacional de Higiene Rafael Rangel 38: 13-18.

Rosa CA, Lachance MA, Silva JOC, Teixeira ACP, Marino MM, Antonini Y, Martins RP. 2003. Yeast communities associated with stingless bees. FEMS Yeast Research 4: 271-275.

Roubik DW, Moreno PJE. 1991. Pollen and spores of Barro Colorado island. Monograph in Systematic Botany from the Missouri Botanical Garden. Number 36. Missouri Botanical Garden. Number 36. Missouri Botanical Garden; St. Louis, Missouri, USA. 270 pp.

Silva TMS, Camara CA, Lins ACS, Agra MF, Silva EMS, Reis IT, Freitas BM. 2009. Chemical composition, botanical evaluation and screening of radical scavenging activity of collected pollen by the stingless bees *Melipona rufiventris* (uruçu-amarela). Anais da Academia Brasileira de Ciências 81: 173-178.

Silva TMS, Camara CA, Lins ACS, Barbosa-Filho JM, Silva EMS, Freitas BM, Santos FAR. 2006. Chemical composition and free radical scavenging activity of pollen loads from stingless bee *Melipona subnitida* Ducke. Journal of Food Composition and Analysis 19: 507-511.

Silva GR, Natividade TB, Camara CA, Silva EMS, Assis Ribeiro dos Santos F, Silva TMS. 2014. Identification of sugar, amino acids and minerals from the pollen of jandaíra stingless bees (*Melipona subnitida*). Food and Nutrition Sciences 5: 1015-1021.

Singleton VL, Orthofer R, Lamuela-Raventos RM. 1999. Analysis of total phenols and other oxidation substrates and antioxidants by means of Folin-Ciocalteu reagent. Methods in Enzymology 299: 152–178.

Souza B, Roubik D, Barth O, Heard T, Enríquez E, Carvalho C, Marchini L, Villas-Bôas J, Locatelli J, Persano Oddo L, Almeida-Muradian L, Bogdanov S, Vit P. 2006. Composition of stingless bee honey: Setting quality standards. Interciencia 31: 867-875.

Tomás-Barberán TA, Tomás-Lorente F, Ferreres F, Garcia-Viguera C. 1989. Flavonoids as biochemical markers of the plant origin of bee pollen. Journal of the Science of Food and Agriculture 47: 337–340.

Tomas-Lorente F, Garcia-Grau MM, Niet JL, Tomas-Barberan FA. 1992. Flavonoids from *Cistus ladanifer* bee pollen. Phytochemistry 31: 2027-2029.

Truchado P, Vit P, Ferreres F, Tomás-Barberán F. 2011. Liquid chromatography-tandem mass spectrometry analysis allows the simultaneous characterization of C-glycosyl and O-glycosyl flavonoids in stingless bee honeys. Journal of Chromatography A 1218: 7601-7607.

Vanderplanck M, Moerman R, Rasmont P, Lognay G, Wathelet B, Wattiez R, Michez D. 2014. How does pollen chemistry impact development and feeding behaviour of polylectic bees? PLoS ONE 9:e86209 doi:10.1371/journal.pone.0086209

Vit P. 2005. Melissopalynology, Venezuela; Mérida, Venezuela, APIBA-CDCHT, Universidad de Los Andes. 205 pp.

Vit P. 2008a. La miel precolombina de abejas sin aguijón (Meliponini) aún no tiene normas de calidad. Revista Boletín Centro Investigaciones Biológicas 42: 415-423.

Vit P. 2008b. Valorización de la miel de abejas sin aguijón (Meliponini). Revista de la Facultad de Farmacia 50: 20-28.

Vit P. 2009a. Origen botánico y propiedades medicinales del polen apícola. Revista Médica de la Extensión Portuguesa ULA 3:27-34.

Vit P. 2009b. Caracterización fisicoquímica de mieles de abejas sin aguijón (Meliponini) de Venezuela. Revista del Instituto Nacional de Higiene Rafael Rangel 40: 7-12.

Vit P. 2013. *Melipona favosa* pot-honey from Venezuela. pp. 363-382. In: Vit P, Pedro SRM, Roubik D, eds. Pot-Honey: A legacy of stingless bees. Springer, New York, USA. 654 pp.

Vit P, Bogdanov S, Kilchenman V. 1994. Composition of Venezuelan honeys from stingless bees and *Apis mellifera* L. Apidologie 25: 278-288.

Vit P, Deliza R, Pérez A. 2011. How a Huottuja (Piaroa) community perceives genuine and false honey from the Venezuelan Amazon, by free-choice profile sensory method. Brazilian Journal of Pharmacognosy 21: 786-792.

Vit P, Enríquez E, Barth OM, Matsuda AH, Almeida-Muradian LB. 2006a. Necesidad del control de calidad de la miel de abejas sin aguijón. MedULA 15: 36-42.

Vit P, Fernández-Maeso MC, Ortiz-Valbuena A. 1998a. Potential use of the three frequently occurring sugars in honey to predict stingless bee entomological origin. Journal of Applied Entomology 122: 5-8.

Vit P, Medina M, Enríquez E. 2004. Quality standards for medicinal uses of Meliponinae honey in Guatemala, Mexico and Venezuela. Bee World 85: 2-5.

Vit P, Mejías A, Rial L, Ruíz J, Peña S, González AC, Rodríguez-Malaver A, Arráez M, Gutiérrez C, Zambrano A, Barth OM. 2012. Conociendo la miel de Melipona favosa en la Península de Paraguaná, Estado Falcón, Venezuela. Rev. Inst. Nac. Hig. Rafael Rangel 43: 15-19.

Vit P, Persano Oddo L, Marano ML, Salas de Mejías E. 1998b. Venezuelan stingless bee honeys characterised by multivariate analysis of compositional factors. Apidologie 29: 377-389.

Vit P, Pulcini P. 1996. Diastase and invertase activities in Meliponini and Trigonini honeys from Venezuela. Journal of Apicultural Research 35: 57-62.

Vit P, Ricciardelli D'Albore G. 1994a. Melissopalynology for stingless bees (Hymenoptera: Apidae: Meliponinae) in Venezuela. Journal of Apicultural Research 33: 145-154.

Vit P, Ricciardelli D'Albore G. 1994b. Palinología comparada en miel y polen de abejas sin aguijón (Hymenoptera: Apidae: Meliponinae) de Venezuela. pp. 121-132. X Simposio de Palinología. In: Mateu Andrés I, Dupré Ollivier M, Güemes Heras J, Burgaz Moreno ME, eds. Trabajos de Palinología Básica y Aplicada. Universitat de Valencia; Valencia, España. pp. 313.

Vit P, Rodríguez-Malaver A, Almeida D, Souza BA, Marchini LC, Fernández Díaz C, Tricio AE, Villas-Bôas JK, Heard TA. 2006b. A scientific event to promote knowledge regarding honey from stingless bees: 1. Physical-chemical composition. Magistra 18: 270-276.

Vit P, Santiago B, Pedro SRM, Perez-Perez E, Peña-Vera M. 2016. Chemical and bioactive characterization of pot-pollen produced by *Melipona* and *Scaptotrigona* stingless bees from Paria Grande, Amazonas State, Venezuela. Emirates Journal of Food and Agriculture 28: 78-84.

Vit P, Soler C, Tomás-Barberán FA. 1997. Profile of phenolic compounds of *Apis mellifera* and *Melipona* spp. honeys from Venezuela. Z. Lebensm. Unters. Forsch. 204: 43-47.

Vit P, Tomás-Barberán FA. 1998. Flavonoids in Meliponinae honey from Venezuela, related to their botanical, geographical and entomological origin to assess their putative anticataract properties. Zeitung Lebensmittel Unters. Forschung 206: 288-293.

Vit P, Yu JQ, Huq F. 2013. Use of honey in cancer prevention and therapy. pp. 481-493. In: Vit P, Pedro SRM, Roubik D, eds. Pot-Honey: A legacy of stingless bees. Springer. New York, USA. 654 pp.

Vossler FG. 2015. Broad Protein Spectrum in Stored Pollen of Three Stingless Bees from the Chaco Dry Forest in South America (Hymenoptera, Apidae, Meliponini) and Its Ecological Implications. Psyche 2015: Article ID 659538. doi:10.1155/2015/659538

Woisky RG, Salatino A. 1998. Analysis of propolis: some parameters and procedures for chemical quality control. Journal of Apiculture Research 37: 99-105.

Bioactivity and Botanical Origin of *Austroplebeia* and *Tetragonula* Australian Pot-Pollen

27

Elizabeth Pérez-Pérez, Miguel Sulbarán-Mora, Ortrud Monika Barth, Carmelina Flavia Massaro, and Patricia Vit

Ancestral knowledge of stingless bee keepers upholds scientific curiosity on pot-pollen.

27.1 Introduction

Pollen is the male gametophyte for sexual reproduction of plants (Edlund et al. 2004). Pot-pollen is the pollen stored by stingless bees inside cerumen pots, after collecting mature pollen from the flowers, mixing it with bee secretions and further microbial transformation, inducing fermentation. The botanical origin of pot-pollen has been studied in Brazil (Ramalho et al. 1990; De Novais

E. Pérez-Pérez • M. Sulbarán-Mora
Laboratory of Biotechnological and Molecular Analysis, Faculty of Pharmacy and Bioanalysis, Universidad de Los Andes, Mérida 5101, Venezuela

O.M. Barth
Instituto Oswaldo Cruz, Fiocruz, Brazil

C. Flavia Massaro
School of Earth, Environmental and Biological Sciences, Science and Engineering Faculty, Queensland University of Technology, Brisbane 4001, Australia

P. Vit (✉)
Apitherapy and Bioactivity, Food Science Department, Faculty of Pharmacy and Bioanalysis, Universidad de Los Andes, Mérida 5101, Venezuela

Cancer Research Group, Discipline of Biomedical Science, Cumberland Campus C42, The University of Sydney, 75 East Street, Lidcombe, NSW 1825, Australia
e-mail: vitolivier@gmail.com

et al. 2015), Malaysia (Azmi et al. 2015), and Venezuela (Vit and Ricciardelli 1994), as well as in stingless bee pollen loads (Barth 2004; Oliveira-Abreu et al. 2014). Bee pollen produced by *Apis mellifera* is used either as food or as a nutritional supplement (Cocan et al. 2005) that is sold as dry pellets obtained directly from foraging bees in pollen traps (Almeida-Muradian et al. 2005). Functional properties of bee pollen considered in food processing are water solubility, emulsification, foaming, water and oil absorption capacity, and antioxidant activity. These properties may also explain underlying mechanisms of medical value reviewed in bee pollen, such as antiproliferative, anti-allergenic, anti-inflammatory, anti-angiogenic, antibacterial, fungicidal, hepatoprotective, anti-atherosclerotic, immune-enhancing potential, and free radical scavenging activities (Denisow and Denisow-Pietrzyk 2016). Significant differences in the redox status, Nrf2 pathway, and endocannabinoid system between squamous cell lung carcinoma (SCC) and adenocarcinoma (AC) and understanding the relation between the various lipid mediators and antioxidants in different lung cancer subtypes may guide further research on effective anticancer therapy (Gegotek et al. 2016). Antioxidant status may play a part in anticancer activity, and phenolic compounds with antioxidant and genoprotective activities might be useful in the pharmaceutical industry (Singh et al. 2016).

Campos et al. (2008) proposed physicochemical parameters as standards in the quality control of corbicular dried pollen of *Apis mellifera*.

© Springer International Publishing AG, part of Springer Nature 2018
P. Vit et al. (eds.), *Pot-Pollen in Stingless Bee Melittology*, DOI 10.1007/978-3-319-61839-5_27

These include an ash content lower than 6 g/100 g, fat content higher than 1.5 g/100 g, protein content higher than 15 g/100 g, and sugar content higher than 40 g/100 g. In addition, lipids have been identified in bee pollen. They include fatty acids (linoleic, γ-linoleic, and archai), phospholipids, phytosterols like β-sitosterol (Szczsna 2006), and leukotrienes (Komosinska-Vassev et al. 2015). Nucleic acids found in bee pollen include ribonucleic acid (Lalhmangaihi et al. 2014). Phenolic compounds are another important chemical group found in bee pollen, including flavonoids (kaempferol, quercetin, and isorhamnetin), catechins, and phenolic acids (chlorogenic acid) (Asafova et al. 2001).

The chemical composition of bee-collected pollen varies with the plant and geographic origin, as well as climatic conditions, soil type, and bee keeper activities (Feás et al. 2012). In Australia, introduced honey bees *A. mellifera* and local native bees are pollinators of wild bush flora and crop vegetation. In particular, the social stingless bees (Meliponini) are used for commercial pollination of macadamia, avocado, lychee, blueberry, and strawberry crops. In the interaction with macadamia flowers, honey bees primarily collect nectar, while pollen is mainly collected by stingless bees (Heard 1994). Previous chemical research on Australian honey bee pollen has described high levels of lipids that originated from botanical species including *Brassica napus*, *Sisymbrium officinale*, *Rapistrum rugosum*, and *Hypochaeris radicata* as highly favored by foraging honey bees (Somerville 2005). Moreover, the protein and amino acid contents of honey bee pollen pellets have been analyzed from 33 floral species from southeast Australia, including *Eucalyptus*, *Corymbia*, and *Angophora* (Myrtaceae) (Somerville and Nicol 2006). Beyond the primary metabolites described in Australian bee products, several chemical studies have reported about secondary metabolites including flavonoids and phenolic derivatives in honey bee honeys (Yao et al. 2004) as well as in Meliponini pot-honeys (Tomás-Barberán et al. 2013). Interestingly, Australian stingless "sugar

bag" honeys showed antibacterial activities that did not originate from *Leptospermum* species (Myrtaceae) which has nectar foraged by honey bees producing a medicinal-grade honey variety (Massaro et al. 2014, 2015). Yet, the botanical origins as well as the chemical and biological properties of pot-pollens from Australian Meliponini bees remain unreported.

In this chapter we describe biochemical composition (flavonoids, polyphenols, proteins), botanical origin, and bioactive properties (antibacterial activity against Gram-positive and Gram-negative bacteria and antioxidant activity) using three analytic methods in ethanolic and methanolic extracts of pot-pollen produced by *Austroplebeia australis*, *Tetragonula carbonaria*, and *Tetragonula hockingsi* foraging in subtropical areas of eastern Australia.

Pot-pollen was collected in May 2016 (winter in the southern hemisphere) from nests within the apiary of Mr. Robert Luttrell in Highvale, Queensland, eastern Australia (see Fig. 27.1).

27.2 Nutraceutical Properties of Bee Pollen

Pollen is a paradoxical bee product, causing allergies but also used for its antiallergic properties. The immune response varies according to the mode of pollen penetration, causing allergies after contact with eye or nasal mucosae, or medicinal if ingested (fresh, dehydrated, combined with other nutraceuticals, etc.) or in injected extracts (A. Meléndez, personal communication). Human fatalities caused by anaphylactic shock after pollen ingestion have been reported (Callejo et al. 2002). Furthermore, air-transported anemophilous pollen is responsible for pollinosis via nasal inhalation, triggered by allergens of seasonal blooming plants (e.g., birch, hay, olive) acting on the immune system (Bartra et al. 2009). However, insect-transported entomophilous pollen – used for the pollination service of seed-producing plants and also harvested as a product from the hive of diverse species of

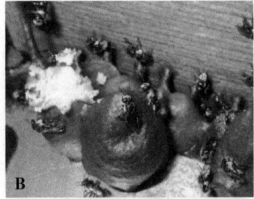

Fig. 27.1 Nests of southeast Australian stingless bees, with pollen stored in cerumen pots. (**a**) *Austroplebeia australis* (Photo: R. Luttrell). (**b**) *Tetragonula carbonaria* (Photo: C.F. Massaro). (**c**) *Tetragonula hockingsi* (Photo: C.F. Massaro)

bees – is additionally consumed as bee pollen or pollen loads, for human nutritional and medicinal purposes (Pascoal et al. 2014). The degranulation of mast cells releases histamine that causes allergic reactions, but bee pollen inhibited 62% of induced allergy by serum containing anti-IgE antibodies (Ishikawa et al. 2008). Empty pollen grains can be engineered for oral vaccination by filling their sporopollenin shell with bioactive molecules, which extend the antibody stimulation (Atwe et al. 2014).

The therapeutic application of bee pollen has been recently reviewed. It has analgesic, anti-athersclerotic, anticholesterolemic, anticlotting, anti-inflammatory, hepatoprotective, hypoglycemic, hypolipidemic, and hypotensive activities. Bee pollen also inhibits cyclooxygenase and lipoxygenase and improves fibrinolytic system performance, cerebral blood flow, detoxification, and visual acuity (Komosinska-Vassev et al. 2015; Denisow and Denisow-Pietrzyk 2016).

The average quality-of-life (QOL) score for bee pollen was 26.16 ± 2.78 in a study of complementary and alternative medicine (CAM) usage and QOL in patients undergoing cancer treatment. This value was intermediate compared to that of honey (25.00 ± 2.64) and grape seeds (26.50 ± 9.19), whereas the highest score (31.25 ± 5.96) corresponded to pollen combined with pomegranate juice and herbal tea (Korkmaz et al. 2016).

Potential applications of bee pollen relate to consistent biological activity in standardized pollen, considering aspects of pollen harvest, blends, and storage (Campos et al. 2010). Bee pollen is a multipurpose food ingredient with a high carbohydrate content, low protein water solubility, and better oil than water absorption capacity and exhibits emulsifying properties and foam-depressing activity (Kostic et al. 2015). Its surface active components interact with major pollen components. Previous bioassays determine the radical scavenging activity of Malaysian stingless bee pollen using 2,2-diphenyl-1-picrylhydrazyl combined the chromatographic technique –high performance liquid chromatography (DPPH-HPLC), with gallic acid as a standard reference (Nurdianah et al. 2016). Their findings were 39% inhibition percentages for *Tetrigona apicalis*, 14.3% for *T. itama*, and 6.7% *Geniotrigona thoracica* ethanolic pot-pollen extracts (1:10), corresponding to 0.3, 0.1, and 0.05 mg gallic acid equivalents / mL, respectively.

27.3 Botanical Origin of Australian Pot-Pollen

Stingless bees occur mainly in tropical countries around the world. Their pot-honey, pot-pollen, and geopropolis (Barth et al. 2009) are strongly appreciated by the Brazilian native people. Recently, research in pollination activities and preferences of several bee species has been performed in Australia (RIRDC Publication 2015). Melittopalynological investigation on Australian bees was relatively new. In general, studies refer to *Apis mellifera* – not Meliponini – and are focused on economic benefits.

Pollen morphology has been used as an instrument to recognize plant species used by the bees and, in addition, their pollination activity. However, it appears that stingless bees were considered wrongly to be relatively unselective. If floral resources are sufficiently available, they prefer single species for a period of time, as evidenced by monofloral honeys and pollen loads (Bazlen 2000). Pollen morphology was used as an instrument to recognize plant species used by the bees for nectar and pollen obtention and, in addition, for pollination activity.

As already mentioned, pot-pollen was collected in nests from *Austroplebeia australis*, *Tetragonula carbonaria*, and *Tetragonula hockingsi*. The samples were examined using natural pollen analysis without chemical treatment. Water washing was sufficient. Pollen grain content was preserved and was helpful for characterization. One sample was monofloral and two were heterofloral, proving that a wide spectrum of flowering plants was visited at the time. No crop could be detected. A difficulty was to recognize the plant species identity (Fig. 27.2).

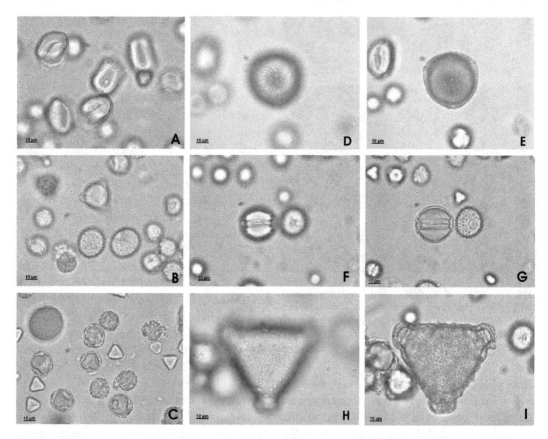

Fig. 27.2 Pollen grains of Australian pot-pollen. (**a, b**) *Austroplebeia australis*. (**a**) Arecaceae, (**b**) unknown. (**d–e**) *Tetragonula carbonaria*, (**d**) *Bauhinia* sp. (Caesalpinioideae) surface, (**e**) idem, optical section. (**c, f–i**) *Tetragonula hockingsi*, (**c, f, g**) unknown; (**h**) *Grevillea* sp. (Proteaceae) surface; (**i**) idem, optical section

The pollen sample of *Austroplebeia australis* presented Asteraceae (66.4%), a palm (Arecaceae) pollen (4.9%), and several others in low frequency. For *Tetragonula carbonaria*, Asteraceae was again the main pollen type (40.5%), followed by Myrtaceae (17.8) and *Eucalyptus* (6.8%), and several other pollen types also were in low frequency. For *Tetragonula hockingsi*, Asteraceae (40.4%) and Myrtaceae (26.8%) were the most common pollen types, followed by *Eucalyptus* (5.2%) and an unknown spheroidal, 1-porate, spiculate pollen type (5.8%). Several other pollen types were in low frequency. *Grevillea* pollen was present in all samples at low amount (single pollen). The pollen grains are large. It was found in propolis samples of *Tetragonula carbonaria* but not in honey samples (Lloyd-Prichard et al. 2016). It may be considered as an indicator, perhaps, of origin, but not a botanical source of pollen or nectar consumed by the bees.

27.4 Flavonoids, Polyphenols, and Antioxidant Activity

A weight of 100 ± 10 mg of pot-pollen was placed in a glass homogenizer (Thomas No. A3528, USA), and 5 mL of ethanol or methanol (Riedel de Haën, Europe) was added and homogenized on an ice bath. Homogenates were centrifuged in a BHG Optima II (USA) centrifuge at 3000 rpm for 10 min, and supernatants were used for biochemical and antioxidant analysis.

Measurements were performed in ethanolic and methanolic extracts. Flavonoids (Woisky and Salatino 1998) and polyphenols (Singleton et al. 1999) were the biochemical indicators measured here, besides protein content (Lowry et al. 1951). The antioxidant activity was measured by three methods: $ABTS^{+\cdot}$ radical cation (Re et al. 1999), AOA Fenton-type reaction (Koracevic et al. 2001), and hydroxyl radical (Halliwell et al. 1987). These methods are described in detail in Chap. 24.

Three chemical groups of bioactive components were investigated in two extracts of Australian pot-pollen (Table 27.1). For ethanolic extracts, flavonoid content varied between 282.9 and 698.0 mg quercetin equivalents/100 g pot-pollen, and the polyphenol content varied between 1281.2 and 2.683.2 mg gallic acid equivalents/100 g pot-pollen. Protein content varied from 571.2 to 832.3 mg proteins/100 g pollen. Similar variations were found in methanolic extracts, flavonoid content varied between 245.7 and 716.8 mg quercetin equivalents/100 g pot-pollen, and the polyphenol content varied between 1156.7 and 2.748.9 mg gallic acid equivalents/100 g pot-pollen. Moreover, protein

Table 27.1 Flavonoid, polyphenol, and protein contents of ethanolic and methanolic extracts of Australian pot-pollen

Extracts	Stingless bee species	Flavonoid contents (mg quercetin equivalents/100 g pollen)	Polyphenol contents (mg gallic acid equivalents /100 g pollen)	Protein contents (mg proteins/100 g pollen)
Ethanolic	*Austroplebeia australis*	448.3 ± 12.2^b	1281.2 ± 22.0^a	571.2 ± 1.9^a
	Tetragonula carbonaria	282.9 ± 3.6^a	1388.4 ± 43.6^a	657.3 ± 2.9^b
	Tetragonula hockingsi	698.0 ± 10.2^c	2683.2 ± 23.5^b	832.3 ± 3.3^c
Methanolic	*Austroplebeia australis*	417.8 ± 10.3^b	1156.7 ± 12.6^a	578.6 ± 2.6^a
	Tetragonula carbonaria	245.7 ± 2.7^a	1366.7 ± 25.6^a	648.2 ± 8.6^b
	Tetragonula hockingsi	716.8 ± 10.2^c	2748.9 ± 18.6^b	845.9 ± 10.5^c

Data are mean $\pm$ SE values ($n = 3$). Columns with values sharing the same superscript letter are not significantly different by ANOVA post hoc Scheffé test ($P < 0.05$)

content varied from 578.6 to 845.9 mg proteins/ 100 g pollen. No statistical differences were found between these values in ethanolic and methanolic extracts by ANOVA post hoc Scheffé test. However, chemical contents were slightly higher when ethanol was used as a solvent for extraction. *Tetragonula hockingsi* pot-pollen presented the highest flavonoid, polyphenol, and protein concentrations in both ethanolic and methanolic extracts, compared to pot-pollen from the other two bee species.

Our findings are similar in polyphenol and flavonoid contents to those found by Ketkar et al. (2014) who investigated the nutraceutical potential of monofloral Indian mustard bee pollen. A polyphenol content of 1828.61 mg gallic acid equivalents/100 g and flavonoids of 122.35 mg quercetin equivalents/100 g pot-pollen were reported in their study. However, lower polyphenol contents (17–66 mg gallic acid equivalents/g of pollen extract) were reported by Da Silva et al. (2013) in pollen samples produced by *Melipona* (*Michmelia*) *seminigra merrillae*, collected in the central and southern region from Amazonas state in Brazil. On the other hand, the results presented here are lower than those reported by Bárbara et al. (2008) who assessed the microbiological parameters and the chemical composition of 21 samples of *Melipona mandacaia* pot-pollen from two regions of Bahia, Brazil. The authors report total phenolic contents of 4000 mg gallic acid equivalents /100 g and flavonoid contents of 1000 mg catechin equivalents /100 g. Pollen was reportedly extracted from honeycombs of *Melipona mandacaia*, but this is not possible because *Melipona* bees store their pollen in cerumen pots, not in honeycombs. Also, in *Melipona* pot-pollen, the pH tends to be lower, while moisture is usually higher than their counterparts for *Apis mellifera* bee pollen (Vit et al. 2016) – in contrast to what was reported by Bárbara et al. abstract (2015).

The antioxidant activities were evaluated for pot-pollen samples of this study (Table 27.2).

Table 27.2 Antioxidant activity of ethanolic and methanolic extracts of Australian pot-pollen and synthetic controls (lipoic acid, melatonin, quercetin)

Extract	Stingless bee species	Hydroxyl radical inhibition percentage (% inhibition/ 100 g pot-pollen)	Antioxidant activity (AOA) (mM equivalent uric acid/100 g pot-pollen)	Total antioxidant activity (TEAC) (μmoles Trolox equivalents/100 g pot-pollen)
Ethanolic	*Austroplebeia australis*	58.0 ± 1.7^d	1.01 ± 0.03^c	248.8 ± 6.7^d
	Tetragonula carbonaria	49.3 ± 4.8^c	1.04 ± 0.04^d	203.0 ± 3.9^d
	Tetragonula hockingsi	74.8 ± 3.5^e	1.06 ± 0.08^d	$430.7 \pm 2.5e$
	Synthetic controls			
	Lipoic acid	43.8 ± 1.7^c	0.67 ± 0.02^a	176.8 ± 3.1^c
	Melatonin	24.8 ± 1.4^a	0.87 ± 0.03^b	87.6 ± 5.1^a
	Quercetin	34.6 ± 2.5^b	0.89 ± 0.05^b	113.2 ± 1.9^b
Methanolic	*Austroplebeia australis*	57.7 ± 2.4^d	1.00 ± 0.01^c	229.6 ± 1.5^d
	Tetragonula carbonaria	48.6 ± 1.6^c	1.05 ± 0.03^d	287.3 ± 9.2^d
	Tetragonula hockingsi	71.3 ± 4.8^e	1.05 ± 0.06^d	426.4 ± 1.8^e
	Synthetic controls			
	Lipoic acid	46.7 ± 3.9^c	0.71 ± 0.01^a	210.9 ± 6.0^c
	Melatonin	27.6 ± 2.8^a	0.92 ± 0.01^b	94.7 ± 3.9^a
	Quercetin	43.7 ± 5.9^c	0.87 ± 0.04^b	95.9 ± 6.9^a

Data are mean ± SE values ($n = 3$). Columns within a sample sharing the same letter are not significantly different by ANOVA post hoc Scheffé test ($P < 0.05$)

As in biochemical characterization (Table 27.1), no statistical differences between values of antioxidant activity were measured in methanolic and ethanolic extracts, being slightly superior for ethanolic extracts. In relation to hydroxyl radical, the inhibition percentage was 48.6–71.3%/100 g pollen; AOA values were between 1.00 and 1.05 mM uric acid/100 g pollen, and TEAC values were between 229.6 and 426.4 µmol TEAC/100 g of pollen. The antioxidant activities measured with the three methods in Australian pot-pollen of *Austroplebeia* and *Tetragonula* are higher than those induced by the pure compounds quercetin, melatonin, and lipoic acid. Finally, a positive correlation was found between the antioxidant activity measured with three methods and the polyphenol content of pot-pollen (Table 27.2).

The antioxidant activity of methanolic extracts of Australian pot-pollen measured by three methods is higher than that reported by Pérez-Pérez et al. (2012) who tested color fractions of *Apis mellifera* pollen loads (0.50–1.84 µmoles Trolox equivalents/100 g corbicular pollen). The content of total phenolic compounds in pollen extracts was solvent dependent in graded ethanol dilutions of 3.6–10.9 GAE/g dry pollen (Carpes et al.

2007). On the other hand, Silva et al. (2009) evaluated the chemical composition and the free radical scavenging activity of *Melipona rufiventris* pot-pollen ethanolic extracts from Brazil. These authors identified the following compounds: p-hydroxycinnamic acid, dihydroquercetin, isorhamnetin, isorhamnetin-3-O-(6′-O-E-p-coumaroyl)-beta-D-glucopyranoside, luteolin, and quercetin. They also measured an EC_{50} value for free radical scavenging activity assessed by DPPH radical scavenging method of 104.1 ± 1.2 µg/ml, with a strong positive correlation between antioxidant activity and polyphenol content ($R^2 = 0.990$). We also found a positive correlation between the antioxidant activity – measured by three methods – and the polyphenol contents within both the ethanolic (Table 27.3) and the methanolic extracts (Table 27.4).

Only one report in the literature compares the biochemical composition and antioxidant activity of ethanolic and methanolic extracts of bee pollen. Pérez-Pérez et al. (2012) measured polyphenol content and antioxidant activity of water, ethanol, and methanol soluble fractions of dried *Apis mellifera* bee pollen commercially available and produced by La Montaña farm (Merida, Venezuela). The authors report that ethanol

Table 27.3 Correlations between antioxidant activity and physicochemical parameters for ethanolic extracts

Parameters	Proteins	Flavonoids	Polyphenols	AOA	HR	TEAC
Proteins	1	0.298	0.049	0.023	0.076	0.254
Flavonoids		1	0.045	0.076	0.023	0.187
Polyphenols			1	**0.834**	**0.823**	**0.932**
AOA				1	0.054	0.140
HR					1	0.145
TEAC						1

AOA Fenton-type radical, *HR* hydroxyl radical, *TEAC* Trolox equivalent antioxidant activity

Table 27.4 Correlations between antioxidant activity and physicochemical parameters for methanolic extracts

Parameters	Proteins	Flavonoids	Polyphenols	AOA	HR	TEAC
Proteins	1	0.045	0.018	0.241	0.215	0.034
Flavonoids		1	0.100	0.165	0.067	0.089
Polyphenols			1	**0.912**	**0.934**	**0.865**
AOA				1	0.025	0.196
HR					1	0.254
TEAC						1

AOA Fenton-type radical, *HR* hydroxyl radical, *TEAC* Trolox equivalent antioxidant activity

extracts of bee pollen have the highest polyphenol content and antioxidant activity, comparable to other antioxidants found in human plasma. In contrast to the literature, polyphenol contents and antioxidant activities of Australian pot-pollen were not different in ethanolic and methanolic extracts.

The results presented here suggest that both the ethanolic the methanolic extracts of pot-pollen show a potent antioxidant activity, comparable to the three purified antioxidants tested in this work, probably due to flavonoid and total polyphenol content of pot-pollen. This is important because the pot-pollen would be beneficial not only as a dietary supplement but also as an antioxidant. In fact, the oxidative modifications of bioactive macromolecules have important roles in carcinogenesis, in particular lipid peroxidation products, which are involved in the activation of Nrf2 and endocannabinoids that affect cancer progression. As demonstrated by Gegotek et al. (2016), tumor tissues show lower antioxidant capacity than healthy tissues, accompanied by lower levels of fatty acids and higher levels of reactive aldehydes. Disturbances in antioxidant capacity and enhanced DNA oxidative modifications were observed in 88% of adenocarcinoma (AC) patients and 81% of squamous cell lung carcinoma (SCC) patients. These authors showed significant differences exist in the redox status, Nrf2 pathway, and endocannabinoid system between SCC and AC tissues. Understanding the relation between the various lipid mediators and antioxidants in different lung cancer subtypes may be a beginning for further research on effective anticancer therapy.

Altay et al. (2016) investigated the possible uses of glassworts as potential food ingredients, and their antiproliferative activity against colorectal adenocarcinoma cells together with their antioxidant and phytochemical profiles, as a promising source of therapeutics. Methanol extracts of five different taxa collected from different localities were screened for their antioxidant capacities by DPPH (IC_{50} 2.91–5.49 mg/ml) and ABTS (24.4–38.5 µmol Trolox equivalents/g extract) assays. *Salicornia freitagii* exhibited the highest DPPH radical scavenging activity.

LC-MS/MS analysis demonstrated that vanillic acid and p-coumaric acid were two main phenolic compounds in the extract. *Salicornia freitagii* methanol extracts also exhibited high antiproliferative activity against HT-29 (IC_{50}:1.67 mg/ml) and Caco-2 (IC_{50}: 3.03 mg/ml) cells for 72 h.

All these examples put in evidence the necessity to study the relationship between bioactive components such as flavonoids and polyphenols, with antioxidant activity of natural food such as pot-honey and pot-pollen, and to develop new "enhanced" foods that can provide this antioxidant activity as new therapeutic tool against cancer. In this scenario, pot-pollen is a promising functional food.

27.5　Antibacterial Activity of Australian Pot-Pollen

Within the multiple bioactivities attributed to bee pollen, antibacterial activity is generally explained by the disruption of bacterial metabolism, especially in *Staphylococcus aureus, S. epidermidis, Bacillus cereus, B. subtilis, Pseudomonas aeruginosa, Salmonella enterica, Listeria monocytogenes*, and *Escherichia coli* (Basim et al. 2006). It disrupts the structure of several macromolecules or inhibits synthesis reactions, especially in *Staphylococcus aureus, S. epidermidis, Bacillus cereus, B. subtilis, Pseudomonas aeruginosa, Salmonella enterica, Listeria monocytogenes*, and *Escherichia coli* (Basim et al. 2006), although the exact mechanism of antimicrobial action still remains to be elucidated. For honey samples, polyphenols with prooxidant activities were necessary intermediates that conferred oxidative action of H_2O_2, and phenolic/H_2O_2-induced oxidative stress constituted the mechanism of honey bacteriostatic and DNA-damaging activities (Brudzynski et al. 2012). One of the antimicrobial mechanisms well described for phenolic compounds is indiscriminant perturbation of the cell wall or membrane, leading to loss of cellular metabolites and/or ions, loss of membrane potential (respiration, electron transport chains), nutrient uptake, or membrane transport function (inactivation of outer membrane of membrane-associated proteins) (Tiwari et al. 2009; Nakamura et al. 2015).

The antibacterial activity of Turkish bee pollen was studied against 13 different bacterial species pathogenic to plants (*Agrobacterium tumefaciens, A. vitis, Clavibacter michiganensis* subsp. *michiganensis, Erwinia amylovora, E. carotovora pv. carotovora, Pseudomonas corrugata, P. savastanoi pv. savastanoi, P. syringae pv. phaseolicola, P. syringae pv. syringae, P. syringae pv. tomato, Ralstonia solanacearum, Xanthomonas campestris pv. campestris,* and *X. axonopodis pv. vesicatoria*), showing that Turkish bee pollen ethanolic extract had an inhibitory effect against all pathogens (Basim et al. 2006). The pollen and propolis extracts do not exhibit a toxic effect or impart unwanted taste/color to foods. Therefore, a potential use is suggested in the management of plant bacterial diseases caused by *A. tumefaciens, P. syringae pv. tomato, X. axonopodis pv. vesicatoria, P. syringae pv. phaseolicola, P. savastanoi pv. savastanoi, P. corrugata,* and *R. solanacearum.* These pathogens cause crown gall, bacterial speck, leaf spot, halo blight, olive knot, stem pith necrosis, and bacterial wilting diseases on monocots and dicots alike, tomato, tomato and pepper, bean, olive, tomato, and eggplant. *Pseudomonas syringae pv. tomato, X. axonopodis pv. vesicatoria,* and *P. syringae pv. phaseolicola* are known to be transmitted through seeds of the plants (Morkunas et al. 2013). The bee pollen extracts thus were potential seed protectants because some of the bacterial pathogens are known to be transmitted via seeds. (Basim et al. 2006). However, assays carried out with methanol extracts of Turkish bee pollen at concentrations from 0.02% to 2.5% had no activity against selected spoilage and pathogenic microorganisms (Erkmen and Ozcan 2008).

The antibacterial activity of Australian pot-pollen was measured with agar well diffusion and minimal inhibitory concentration (MIC) methods. A screening assay using well diffusion agar was carried out with some minor modifications (Bauer et al. 1966). Plates were incubated at 37 °C for 24–48 h. The presence of an inhibition

halo of the pot-pollen extract is indicative of antimicrobial activity in vitro. The following bacterial strains were used: Gram-positive *Staphylococcus aureus* (ATCC 25923) and *Bacillus subtilis* (ATCC 11778) and Gram-negative *Enterobacter cloacae* (ATCC 13047), *Escherichia coli* (ATCC 25922), and *Pseudomonas aeruginosa* (ATCC 27853), adjusted to 0.5 units McFarland (1 × 108 cfu/mL). Each assay was carried out in triplicate. Detailed methods are described in Chap. 28.

The pot-pollen samples with inhibition halos in the agar plates were analyzed to measure the minimum inhibitory concentration (MIC) adapting the procedure described by Patton et al. (2006) and Tan et al. (2009). The MIC was calculated using the formula:

$$MIC = \frac{1 - \left(\begin{array}{l} \text{Asorbance of sample well} \\ -\text{Absorbance control well} \end{array}\right)}{\left(\begin{array}{l} \text{Asorbance of growth control well} \\ -\text{Absorbance of sterility control well} \end{array}\right)} \times 100\%$$

The commercial antibiotic standards were prepared using the following dilutions: phenol (2 mg/L), ampicillin (10 mg/L), ciprofloxacin (5 mg/L), and tetracycline (30 mg/L) (Andrews 2001). Each assay was carried out in triplicate.

In Table 27.5 are results of agar diffusion method for ethanolic and methanolic extracts of Australian pot-pollen. All pot-pollen extracts had antibacterial activity against all microorganism tested in this work, except *Austroplebeia australis* methanolic extract against *Pseudomonas aeruginosa*, *Tetragonula hockingsi* methanolic extract against *Escherichia coli*, and *Tetragonula hockingsi* ethanolic extract against *Escherichia coli.*

Low MICs were found for the ethanolic and methanolic extracts of *Tetragonula hockingsi* pot-pollen (Table 27.6), indicating the highest antibacterial activity that was possibly explained by the high polyphenol and flavonoid concentrations (Table 27.1). In general, MIC values of ethanolic extracts were lower than methanolic extracts of pot-pollen. However, the antibacterial activity varied with the entomological origin of pot-pollen against the tested bacteria, e.g., the MIC of *Tetragonula hockingsi* pot-pollen metha-

Table 27.5 Inhibitory effects of Australian pot-pollen extracts by the agar well diffusion method in BHI medium

Extract	Stingless bee species	Gram-positive		Gram-negative		
		Staphylococcus aureus	Bacillus subtilis	Enterobacter cloacae	Escherichia coli	Pseudomonas aeruginosa
Ethanolic	Austroplebeia australis	11[c]	12[b]	6[a]	12[d]	5[a]
	Tetragonula carbonaria	16[d]	11[b]	5[a]	6[b]	8[b]
	Tetragonula hockingsi	11[c]	10[b]	8[b]	–	7[b]
Methanolic	Austroplebeia australis	8[b]	7[a]	8[b]	4[a]	–
	Tetragonula carbonaria	5[a]	12[b]	6[a]	8[c]	6[a]
	Tetragonula hockingsi	14[d]	12[b]	14[c]	–	12[c]

Table 27.6 Minimum inhibitory concentration (MIC) of Australian pot-pollen compared with MICs of commercial antibiotics

Extracts	Stingless bee species	Gram-positive		Gram-negative		
		Staphylococcus aureus	Bacillus subtilis	Enterobacter cloacae	Escherichia coli	Pseudomonas aeruginosa
Ethanolic	Austroplebeia australis	7.5 ± 0.5^d	7.5 ± 0.1^d	12.5 ± 0.1^d	7.5 ± 0.2^d	12.5 ± 0.2^d
	Tetragonula carbonaria	2.5 ± 0.2^b	7.5 ± 0.6^d	12.5 ± 0.2^d	12.5 ± 0.1^f	10.0 ± 0.3^c
	Tetragonula hockingsi	7.5 ± 0.3^d	7.5 ± 0.1^d	10.0 ± 0.3^c	NA	10.0 ± 0.2^c
Methanolic	Austroplebeia australis	10.0 ± 0.5^e	10.0 ± 0.4^e	10.0 ± 0.3^c	12.5 ± 0.2^f	NA
	Tetragonula carbonaria	12.5 ± 0.5^f	7.5 ± 0.3^d	12.5 ± 0.5^d	10.0 ± 0.5^e	12.5 ± 0.4^d
	Tetragonula hockingsi	5.0 ± 0.3^c	7.5 ± 0.2^d	5.0 ± 0.1^b	NA	7.5 ± 0.5^b
Commercial antibiotics	Phenol	5.0 ± 0.1^c	1.0 ± 0.1^a	1.0 ± 0.1^a	5.0 ± 0.1^c	$1.0 \pm 0.1a$
	Ampicillin	2.5 ± 0.1^b	4.0 ± 0.2^c	14.0 ± 2.6^e	26.0 ± 2.6^g	16.0 ± 0.4^e
	Ciprofloxacin	5.0 ± 0.1^c	1.0 ± 0.1^a	5.0 ± 0.2^b	2.0 ± 0.1^a	1.0 ± 0.1^a
	Tetracyclin	1.0 ± 0.1^a	2.0 ± 0.2^b	1.0 ± 0.1^a	4.0 ± 0.1^b	1.0 ± 0.1^a

NA not active. Data are mean ± SEM values ($n = 3$). Columns with values sharing the same superscript letter are not significantly different by ANOVA post hoc Scheffé test ($P < 0.05$)

nolic extract is lower than its ethanolic extract against *Staphylococcus aureus* and *Enterobacter cloacae* (Table 27.6). A further variation can be caused by the ethanol and methanol concentrations (70% and 96%) tested by Kacániová et al. (2012). In some cases, MIC values of pot-pollen are similar to MIC values of commercial antibiotics used like standards in this work. Ampicillin (10 mg/mL) was less antibacterial in Gram-negative than in Gram- positive bacteria. *Tetragonula carbonaria* ethanolic pot-pollen extract showed the lowest MIC against *Staphylococcus aureus* (2.5 mg/mL), whereas *Tetragonula hockingsi* methanolic pot-pollen extract was the most active with the lowest MICs against *Staphylococcus aureus* (5.0 mg/mL).

Our results are similar to those reported in literature for bee pollen, with evidence of a strong antibacterial activity of ethanolic extract against Gram-positive and Gram- negative bacteria, frequently related to polyphenol and flavonoid contents. Carpes et al. (2007) suggest the nature of phenolic compounds is more important than their concentration for the antibacterial activity of ethanolic pollen extracts against *Bacillus cereus*, *Bacillus subtilis*, *Klebsiella* sp., *Pseudomonas aeruginosa*, and *Staphylococcus aureus*. For example, Graikou et al. (2011) evaluate the antimicrobial activity of methanol and water extracts of Greek bee pollen and report an interesting antimicrobial profile, especially against Gram-positive strains and yeast. As another example, Kacániová et al. (2012) evaluated antimicrobial activity of propolis, pollen loads, and beeswax of *Apis mellifera* from two locations in Slovakia, against pathogenic bacteria, microscopic fungi and yeasts, using the agar well diffusion method, to compare preparations obtained with different solvent extractions using ethanol and methanol at 70% and 96% v/v, respectively. After 48 h, *Candida glabrata* was the most sensitive microorganism to the 70% methanol extract of pollen loads (the diameter of inhibition was 3.50 mm), while *Aspergillus fumigatus* was the most sensitive to the 70% ethanol extract (4.17 mm). This work was a clear example of differences between bioactivity of pollen loads based on the type and concentrations of the solvents used for extraction, probably due to differences in nature (polarity, hydrophilicity diffusing across the agar media and/or interacting with bacterial cell wall components) and quantity of bioactive molecules extracted with each solvent. Both methanol and ethanol are substances that increase the solubility of organic material with low polarity. The solubility of the organic active compounds in polar solvents is increased by decreasing the length of the hydrocarbon chain, the presence of polar functional groups, and the ability to form hydrogen bonds with the solvent. Therefore, the type of solvent and the concentration determine the kind of molecules extracted in the function of parameters previously described. Ethanolic and methanolic extracts share similarities and also cause differences in the quantity and quality of antioxidant and antibacterial molecules extracted. Fatrcová-Šramková et al. (2013) determined antibacterial activity of ethanolic and methanolic extracts of monofloral bee pollen samples to pathogenic bacteria, collected in different localities in Slovakia. The bacteria most sensitive to the ethanolic extracts of *Papaver rhoeas* (Papaveraceae) "poppy" and *Helianthus annuus* (Asteraceae) "sunflower" bee pollen were *Staphylococcus aureus* and *Salmonella enterica*, respectively. The methanolic extract of *Brassica napus* (Brassicaceae) "rapeseed" bee pollen inhibited the growth of *Salmonella enterica*. In our study, the Asteraceae botanical origins with other pollen types observed in Fig. 27.2 for *Austroplebeia australis* (Arecaceae), *Tetragonula carbonaria*

(*Bauhinia* sp., Fabaceae-Caesalpinioideae), and *Tetragonula hockingsi* (*Grevillea* sp., Proteaceae) were components of bioactivity of the three pot-pollen samples.

27.6 Conclusions

Our findings demonstrate that Australian pot-pollen ethanolic and methanolic extracts are rich in polyphenols and flavonoids. The polyphenol contents are positively correlated with the powerful free radical scavenging activity and antibacterial activity. There is potential for Australian pot-pollen to be used as a food supplement for health enhancement by decreasing the deleterious effects of free radicals, related to several degenerative diseases. Both the entomological origin and the botanical origin of the pot-pollen contribute to its biochemical composition and bioactivity. Additionally, the microbiological associations that process the corbicular pollen inside the cerumen pots also may be responsible of chemical transformations for the active substances into more active or less active forms. Possible products synthesized by the microbiota from botanical and entomological substrates that contribute to pot-pollen transformation need to be studied. Pollen grain analysis of stingless bee products in Australia must be intensified: firstly, to establish the regional and local vegetation capacity for stingless bee keeping; secondly, to obtain products of more constant properties; and, thirdly, to stimulate protective measures for vegetation – important to native pollinators, production of seeds, and perpetuation of plant species.

Acknowledgments The pot-pollen samples were freshly collected from local stingless beehives and kindly donated by Mr. Robert Luttrell, Highvale, Queensland, Australia. Special thanks to the support of ZG-AVA-FA-01-98-01 from the Council of Development of Scientific, Humanistic, Technological and Artistic, at Universidad de Los Andes; to Dr. F. Huq for sending pot-pollen from Australia to Brazil for palynological analysis; to Massimo Vit for his hospitality during two short stages of P. Vit at USYD in Sydney; and to the National Counsel of Technological and Scientific Development "*Conselho Nacional de Desenvolvimento Científico e Tecnológico, CNPq*" for a research fellowship. Dr. DW Roubik carefully commented and improved the manuscript.

References

Almeida-Muradian L, Pamplona LC, Coimbra S, Barth OM. 2005. Chemical composition and botanical evaluation of dried bee pollen pellets. Journal of Food Composition and Analysis 18: 105-111.

Altay A, Sagdicoglu Celep G, Yaprak AE, Basköse I, Bozoglu F. 2016. Glassworts as possible anticancer agents against human colorectal adenocarcinoma cells with their nutritive, antioxidant and phytochemical profiles. Chemical Biodiversity Oct 4. doi: 10.1002/cbdv.201600290.

Andrews JM. 2001. Determination of minimum inhibitory concentrations. Journal of Antimicrobial Chemotherapy 48: 5–16.

Asafova N, Orlov B, Kozin R, 2001. Physiologically Active Bee Products. YA Nikolaev; Nizhny Novgorod University, Russia. 221 pp.

Atwe SU, Ma Y, Gill HS. 2014. Pollen grains for oral vaccination. Journal of Control Release 194: 45–52.

Azmi WA; Zulqurnain NS; Ghazi R. 2015. Melissopalynology and foraging activity of stingless bees, *Lepidotrigona terminata* (Hymenoptera: Apidae) from an apiary in Besut, Terengganu. Journal of Sustainability Science and Management 10: 27-35.

Bárbara MS, Machado CS, Sodré GS, Dias LG, Estevinho LM, Carvalho CAL. 2008. Microbiological assessment, nutritional characterization and phenolic compounds of bee pollen from *Melipona mandacaia* Smith, 1983. Molecules. 20: 12525-12544.

Bárbara MS, Machado CS, Sodré GS, Dias LG, Estevinho LM, de Carvalho CA. 2015. Microbiological assessment, nutritional characterization and phenolic compounds of bee pollen from *Melipona mandacaia* Smith, 1983. Molecules 20: 12525-12544.

Barth OM. 2004. Melissopalynology in Brazil: a review of pollen analysis of honeys, propolis and pollen loads of bees. Sciencia Agricola 61: 342-350.

Barth OM, Barros MA, Freitas FO. 2009. Análise palinológica em amostras arqueológicas de geoprópolis do vale do rio Peruaçu, Januária, Minas Gerais, Brasil. Arquivos do Museu de História Natural e Jardim Botânico, Universidade Federal de Minas Gerais, Belo Horizonte, Minas Gerais 19: 277-290.

Bartra J, Sastre J, del Cuvillo A, Montoro J, Jáuregui I, Dávila I, Ferrer M, Mullol J, Valero A. 2009. From polinosis to digestive allergy. Journal of Investigative Allergology and Clinical Immunology 19: 3-10.

Basim E, Basim HS, Özcan M. 2006. Antibacterial activities of Turkish pollen and propolis extracts against plant bacterial pathogens. Journal of Food Engeniering 77: 992–996.

Bauer AW, Kirby WM, Sherris JC, Turck M. 1966. Antibiotic susceptibility testing by standardized single disk method. American Journal of Clinical Pathology 45: 493-496.

Bazlen K. 2000. Charakterisierung von Honigen stachelloser Bienen aus Brasilien. Thesis. Faculty of Biology, Eberhard-Karl University of Tübingen. 141 pp.

Brudzynski K, Abubaker K, Miotto D. 2012. Unraveling a mechanism of honey antibacterial action: polyphenol/H_2O_2-induced oxidative effect on bacterial cell growth and on DNA degradation. Food Chemistry 133: 329-336.

Callejo A, Sanchís ME, Armentia A, Moneoa I, Fernández A. 2002. A new pollen–fruit cross-reactivity. Allergy 57: 1088–1089.

Campos MGR, Bogdanov S, Almeida-Muradian LB, Szczesna T, Mancebo Y, Frigerio C, Ferreira F. 2008. Pollen composition and standardisation of analytical methods. Journal of Apicultural Research 47: 156-163.

Campos MGR, Frigerio C, Lopes J, Bogdanov S. 2010. What is the future of bee pollen? Journal of ApiProduct and Apimedical Science. 2: 131-144.

Carpes ST, Begnini R, De Alencar SM, Masson ML. 2007. Study of preparations of bee pollen extracts, antioxidant and antibacterial activity. Ciência e Agrotecnología 31: 1818-1825.

Cocan O, Marghitas LA, Dezmirean D, Laslo L. 2005. Composition and biological activities of bee pollen: review. Bulletin of the University of Agricultural Science and Veterinary Medicine 61: 221-226.

Da Silva IA, Silva TM, Camara CA, Queiroz N, Magnani M, Novais JS, Soledade LE, Lima Ede O, de Souza AL, de Souza AG. 2013. Phenolic profile, antioxidant activity and palynological analysis of stingless bee honey from Amazonas, Northern Brazil. Food Chemistry 141: 3552-3558.

De Novais JS, Garcez ACA, Absy ML, Santos FAR. 2015. Comparative pollen spectra of *Tetragonisca angustula* (Apidae, Meliponini) from the Lower Amazon (N Brazil) and caatinga (NE Brazil). Apidologie 46: 417-431.

Denisow B, Denisow-Pietrzyk M. 2016. Biological and therapeutic properties of bee pollen: a review. Journal of the Science of Food and Agriculture. 96: 4303-4309.

Edlund AF, Swanson R, Presuss D. 2004. Pollen and stigma structure and function: The role of diversity in pollination. The Plant Cell 16: 84-97.

Erkmen O, Ozcan MM. 2008. Antimicrobial effects of Turkish propolis, pollen, and laurel on spoilage and pathogenic foodrelated microorganisms. Journal of Medicinal Food 11: 587-592.

Fatrcová-Šramková K, Nôžková J, Kačániová M, Máriássyová M, Rovná K, Stričík M. 2013. Antioxidant and antimicrobial properties of monofloral bee pollen. Journal of Environmental Science and Health B. 48: 133-138.

Feás X, Vázquez-Tato MP, Estevinho L, Seija JA, Iglesias A. 2012. Organic bee pollen: botanical origin, nutritional value, bioactive compounds, antioxidant activity and microbiological quality. Molecules 17: 8359-8377.

Gegotek A, Nikliński J, Žarković N, Žarković K, Waeg G, Łuczaj W, Charkiewicz R, Skrzydlewska E. 2016. Lipid mediators involved in the oxidative stress and antioxidant defense of human lung cancer cells. Redox Biology 9: 210-219.

Graikou K, Kapeta S, Aligiannis N, Sotiroudis G, Chondrogianni N, Gonos E, Chinou I. 2011.Chemical analysis of Greek pollen: Antioxidant, antimicrobial and proteasome activation properties. Chemistry Central Journal 5: 5-13.

Halliwell B, Gutteridge J, Aruoma O. 1987. The deoxyribose method:a simple test-tube assay for determination of rate constants for reactions of hydroxyl radicals. Analytical Biochemistry 165: 215–219.

Heard TA. 1994. Behaviour and pollinator efficiency of stingless bees and honey bees on macadamia flowers. Journal of Apicultural Research 33: 191-198.

Ishikawa Y, Tokura T, Nakano N, Hara M, Niyonsaba F, Ushio H, Yamamoto Y, Tadokoro T, Okumura K, Ogawa H. 2008. Inhibitory effect of honeybee-collected pollen on mast cell degranulation in vivo and in vitro. Journal of Medicinal Food 11: 14–20.

Kacániová M, Vuković N, Chlebo, Haščík P, Rovná K, Cubon J, Dżugan M, Pasternakiewicz A. 2012. The antimicrobial activity of honey, bee pollen loads and beeswax from Slovakia. Archives of Biological Science 64: 927-934.

Ketkar SS, Rathore AS, Lohidasan S, Rao L, Paradkar AR, Mahadik KR. 2014. Investigation of the nutraceutical potential of monofloral Indian mustard bee pollen. Journal of Integrative Medicine 12: 379-389.

Komosinska-Vassev K, Olczyk P, Kazmierczak J, Mencner L, Olczyk K. 2015. Bee Pollen: Chemical Composition and Therapeutic Application. Evidence-Based Complementary and Alternative Medicine 1-6 pp. http://dx.doi.org/10.1155/2015/297425 (24.07.2016).

Koracevic D, Koracevic G, Djordjevic V, Andrejevic S, Cosic V. 2001. Method for measurement of antioxidant activity in human fluids. Journal of Clinical Pathology 54: 356–361.

Korkmaz M, Tavsanli NG, Ozcelik H. 2016. Use of complementary and alternative medicine and quality of life of cancer patients. Holistic Nursing Practice March/April: 88-95.

Kostic AZ, Barac MB, Stanojevic SP, Milojkovic-Opsenica DM, Tesic ZL, Sikoparija, B, Radisik P, Prentovic M, Pesic MB. 2015. Physicochemical composition and techno-functional properties of bee pollen collected in Serbia. LWT - Food Science and Technology 62: 301-309.

Kustiawan PM, Puthong S, Arung ET and Chanchao C. 2014. *In vitro* cytotoxicity of Indonesian stingless bee products against human cancer cell lines. Asian Pacific Journal of Tropical Biomedicine 4: 549-556.

Lalhmangaihi R, Ghatak S, Laha R, Gurusubramanian G, Kumar NS. 2014. Protocol for optimal quality and quantity pollen DNA isolation from honey samples. Journal of Biomolecular Techniques 25: 92-95.

Lloyd-Prichard D, Lucas S, Roberts T, Haberle S. 2016. Assessment of pollen assemblages from the hives of Tetragonula carbonaria for the presence of the threatened species Grevillea parviflora subsp. parviflora. Journal of Pollination Ecology 18: 23-30. Tetragonula carbonaria Grevillea parviflora … parviflora.

Lowry OH, Rosebrough NJ, Farr AL, Randall RJ. 1951. Protein measurement with the Folin phenol reagent. Journal of Biological Chemistry 193: 265-275.

LeBlanc BW, Davis OK, Boue S, DeLucca A, Deeby T. Antioxidant activity of Sonoran Desert bee pollen. 2009. Food Chemistry 115: 1299-305.

Luz CFP, Barth OM. 2012. Pollen analysis of honey and beebread derived from Brazilian mangroves. Revista Brasileira de Botânica 35: 79-85.

Massaro CF, Shelley D, Heard TA, Brooks P. 2014. *In vitro* antibacterial phenolic extracts from 'sugarbag' pot-honeys of Australian stingless bees (*Tetragonula carbonaria*). Journal of Agricultural and Food Chemistry 62: 12209-12217.

Massaro CF, Smyth TJ, Smyth WF, Heard TA, Leonhardt SD, Katouli M, Wallace HM, Brooks P. 2015. Phloroglucinols from anti-microbial deposit-resins of Australian stingless bees (*Tetragonula carbonaria*). Phytotherapy Research 29: 48-58.

Morkunas I, Formela M, Marczak L, Stobiecki M, Bednarski W. 2013. The mobilization of defence mechanisms in the early stages of pea seed germination against *Ascochyta pisi*. Protoplasma 250: 63-75.

Nakamura K, Ishiyama K, Sheng H, Ikai H, Kanno T, Niwano Y. 2015. Bactericidal Activity and Mechanism of Photoirradiated Polyphenols against Gram-Positive and -Negative Bacteria Journal of Agriculture and Food Chemistry 63: 7707-7713.

Nurdianah HF, Ahmad Firdaus AH, Eshaifol Azam O, Wan Adnan WO. 2016. Antioxidant activity of bee pollen ethanolic extracts from Malaysian stingless-bee measured using DPPH-HPLC assay. International Food Research Journal 23: 403-405.

Oliveira-Abreu C, Hilário SD, Luz CFP 2014. Pollen and néctar foraging by *Melipona quadrifasciata anthidioides* Lepeletier (Hymenoptera: Apidae: Meliponini) in natual habitat. Sociobiology 61: 441-448.

Pascoal A, Rodrigues S, Texeira A, Feás X, Estevinho LM. 2014. Biological activities of commercial bee pollen: review. Food Chemistry and Toxicology 63: 233-239.

Patton T, Barrett J, Brennan J, Moran N. 2006. Use of spectrophotometric bioassay for determination of microbial sensitivity to manuka honey. Journal of Microbiological Methods 64: 84–95.

Pérez-Pérez EM, Vit P, Rivas E, Sciortino R, Sosa A, Tejada D, Rodríguez-Malaver AJ. 2012. Antioxidant activity of four colour fractions of bee pollen from Mérida, Venezuela. Archivos Latinoamericanos de Nutrición 62: 375-380.

Ramalho M; Kleinert-Giovannini A; Imperatriz-Fonseca VL. 1990. Important bee plants for stingless bees (Melipona and Trigonini) and Africanised honeybees (*Apis mellifera*) in Neotropical habitats: a review. Apidologie 21: 469-488.

Re R, Pellegrini N, Proteggente A, Pannala A, Yang M, Rice-Evans C. 1999. Antioxidant activity in improved ABTS radical cation decolorization assay. Free Radical in Biology and Medicine 26: 1231-1237.

RIRDC Publication, 2015, 14/057. Honey bee and pollination program, five year research, developmentat and extension plan, 2014/15 – 2018/19.

Silva TM, Camara CA, Lins AC, Agra Mde F, Silva EM, Reis IT, Freitas BM. 2009. Chemical composition, botanical evaluation and screening of radical scavenging activity of collected pollen by the stingless bees *Melipona rufiventris* (Uruçu-amarela). Anais da Academia Brasileira de Ciências 81: 173-178.

Singh B, Sharma P, Kumar A, Chadha P, Kaur R, Kaur A. 2016. Antioxidant and in vivo genoprotective effects of phenolic compounds identified from an endophytic *Cladosporium velox* and their relationship with its host plant *Tinospora cordifolia*. Journal of Ethnopharmacology 194: 450-456.

Singleton VL, Orthofer R, Lamuela-Raventos RM. 1999. Analysis of total phenols and other oxidation substrates and antioxidants by means of Folin-Ciocalteu reagent. Methods in Enzymology 299: 152–178.

Somerville DC. 2005. Lipid content of honey bee-collected pollen from south-east Australia. Australian Journal of Experimental Agriculture 45: 1659.

Somerville DC, Nicol HI. 2006. Crude protein and amino acid composition of honey bee-collected pollen pellets from south-east Australia and a note on laboratory disparity. Animal Production Science 46: 141-149.

Szczsna T. 2006. Protein content amino acid composition of bee-collected pollen from selected botanical origins. Journal of Apicultural Research 50: 81-90.

Tan HT, Rahman RA, Gan SH, Halim AS, Hassan SA, Sulaiman SA, Kirnpal-Kaur B. 2009. The antibacterial properties of Malaysian tualang honey against wound and enteric microorganisms in comparison to manuka honey. Complementary and Alternative Medicine 9: 34-39.

Tiwari HK, Sapkota D, Das AK, Sen MR. 2009. Assessment of different tests to detect methicillin resistant Staphylococcus aureus Southeast Asian Journal of Tropical Medicine Public Health 40: 801-806.

Tomás-Barberán FA, Truchado P, Ferreres F. 2013. Flavonoids in stingless-bee and honey-bee honeys. pp. 461-474. In: Vit P, Pedro SRM, Roubik D, eds. Pot-Honey: A legacy of stingless bees. Springer; New York, USA. 654 pp.

Vit P, Ricciardelli D'Albore G. 1994. Palinología comparada en miel y polen de abejas sin aguijón (Hymenoptera: Apidae: Meliponinae) de Venezuela. pp. 121-132. In Mateu Andrés I, Dupré Ollivier M, Güemes Heras J, Burgaz Moreno ME, eds. ME. X Simposio de Palinología. Trabajos de Palinología Básica y Aplicada. Universitat de Valencia; Valencia, España., Septiembre. pp. 313.

Vit P, Santiago B, Pedro SRM, Peña-Vera M, Pérez-Pérez E. 2016. Chemical and bioactive characterization of pot-pollen produced by *Melipona* and *Scaptotrigona* stingless bees from Paria Grande, Amazonas State, Venezuela. Emirates Journal of Food and Agriculture 28: 78-84.

Woisky RG, Salatino A. 1998. Analysis of propolis: some parameters and procedures for chemical quality control. Journal of Apiculture Research 37: 99-105.

Yao L, Jiang Y, D'Arcy B, Singanusong R, Datta N, Caffin N, Raymont K. 2004. Quantitative high-performance liquid chromatography analyses of flavonoids in Australian Eucalyptus honeys. Journal of Agricultural and Food Chemistry 52: 210-214.

Antibacterial Activity of Ethanolic Extracts of Pot-Pollen Produced by Eight Meliponine Species from Venezuela

28

Miguel Sulbarán-Mora, Elizabeth Pérez-Pérez, and Patricia Vit

28.1 Introduction

Bees mix floral pollen with a small amount of saliva and nectar forming compacted loads on their rear legs and transported to the nest as corbicular pollen. The corbicula is a concave, smooth area, with fringes of long hairs in the borders that facilitate transportation of pollen or resins to the nest (Michener 2013).

In various climate zones, the chemical composition of pollen depends on the flora present (Nogueira et al. 2012). Pollen, obtained by bees from the flowers, constitutes a rich source of biologically active substances. Over 250 biologically active substances of plant origin have been isolated from bee pollen. It contains high concentrations of reducing sugars; essential amino acids and unsaturated/saturated fatty acids; minerals as Zn, Cu, and Fe; and high K/Na ratio and

M. Sulbarán-Mora • E. Pérez-Pérez
Laboratory of Biotechnological and Molecular Analysis, Faculty of Pharmacy and Bioanalysis, Universidad de Los Andes, Mérida 5101, Venezuela

P. Vit (✉)
Apitherapy and Bioactivity, Food Science Department, Faculty of Pharmacy and Bioanalysis, Universidad de Los Andes, Mérida 5101, Venezuela

Cancer Research Group, Discipline of Biomedical Science, Cumberland Campus C42, The University of Sydney, 75 East Street, Lidcombe, NSW 1825, Australia
e-mail: vitolivier@gmail.com

significant quantities of several vitamins: provitamin A, vitamin E (tocopherol), niacin, thiamine, folic acid, and biotin. The composition and the amount of these nutritious components are largely dependent on the botanical source of the pollen (Nogueira et al. 2012; Denisow and Denisow-Pietrzyk 2016), and the entomological origin has been scarcely studied (Vit et al. 2016).

Antimicrobial activity of stingless bee was found in propolis (Farnesi et al. 2009; Massaro et al. 2014), geopropolis (Liberio et al. 2011; da Cunha et al. 2013), and pot-honey (Boorn et al. 2010; Pimentel et al. 2013; Nishio et al. 2016). Therefore, it is expected that pot-pollen produced by stingless bees also has antimicrobial activity.

28.1.1 Biological Potential of Pollen Stored in Bee Nests

Bee pollen is characterized by a high antioxidant potential, which determines biological activity (Leja et al. 2007). Results of many tests conducted on animals have confirmed the hepatoprotective and detoxifying activities of bee pollen (Eraslan et al. 2009), its anti-inflammatory properties (Campos et al. 2010), and cytotoxic properties against many tumors (Wu and Lou 2007). Also, the polysaccharide fractions obtained from bee pollen stimulate immunological activity through an increase in macrophage phagocytic index, mainly the increase in the number of phagocytes, and they have beneficial effects on

© Springer International Publishing AG, part of Springer Nature 2018
P. Vit et al. (eds.), *Pot-Pollen in Stingless Bee Melittology*, DOI 10.1007/978-3-319-61839-5_28

splenocyte and NK lymphocyte proliferation (Li et al. 2009). Bee pollen may significantly decrease the negative effects of iron deficiency, thus demonstrating antianemic effects (Haro et al. 2000), and have a positive influence on osseous tissue (Hamamoto et al. 2006).

Basim et al. (2006) evaluated the antibacterial activity of pollen ethanolic extracts collected in the Hatay province in southeastern Mediterranean Turkey against 13 plant pathogenic bacteria that cause several kinds of diseases on many different plants, including vegetables and fruits. Among the tested bacteria, *Agrobacterium tumefaciens* was the most sensitive to 1/5 concentration of pollen extract and *Pseudomonas syringae* pv. *phaseolicola* was the most sensitive to 1/10 concentration of propolis extract. The authors correlated the antibacterial activity of pollen samples with their polyphenol content. This was the first report of antibacterial activities of pollen and propolis extracts against plant pathogenic bacteria. Later, Carpes et al. (2007) determined the antioxidant activity, phenolic content, and antibacterial activity of Brazilian pollen extracts obtained with different concentrations of ethanol. The authors concluded that extractive conditions (ethanol solutions 40–90%) were selective for the polyphenol profile and concentration, with >10 mg/g for 60–80% ethanol. Different solvents and concentrations used for extractions caused different antioxidant and antibacterial activities, because different types of phenolics may be extracted from pollen since they are solvent-dependent. Polyphenol and flavonoid effects involve the formation of complexes with bacterial cell walls by surface-exposed adhesin and polypeptides and/or cell membrane enzymes, which leads to the damage of cell wall integrity, blockage of ion channels, and inhibition of electron flow in the electron transport chain which determines adenosine triphosphate (ATP) synthesis, by scavenging electrons (Grajek 2007).

Ethanol extracts of bee pollen have quite powerful antibiotic properties against pathogenic Gram positive and Gram negative bacteria and pathogenic fungi. Such antibacterial activity has been assessed against many microorganisms, including *Pseudomonas aeruginosa, Listeria monocytogenes, Staphylococcus aureus, Salmonella enterica,* and *Escherichia,* with high sensitivity shown among all microorganisms (Knazovicka et al. 2009). In vitro antibacterial, antifungal, and antiparasitic capacities and the effect on the bacteriophage PhiX174 of ethanolic extracts of pot-pollen from *Tetragonisca angustula* in Antioquia, Colombia, showed MICs between 32 and 16 mg/m; there was a strong correlation between antimicrobial activity and polyphenol content (Monserrate 2015).

Bárbara et al. (2015) assessed the microbiological parameters and the chemical composition of *Melipona mandacaia* pot-pollen from Bahia, Brazil. The nutritional parameters (moisture, ash, water activity, pH, total acidity, protein, fiber, total phenolic, flavonoids, and reducing sugars) were within the bee pollen standards, except for pH and moisture content, which presented superior and inferior values, respectively, with an evident influence of the geographical origin on the assessed parameters, especially concerning the fatty acid profile. Another example is the work of Nogueira et al. (2012) who characterized eight commercial bee pollen types from Portugal and Spain. The moisture content, ash, water activity, pH, reducing sugars, carbohydrates, proteins, lipids, and energy were within the specifications in *Apis mellifera* pollen legislation and varied according to botanical and geographical origin. It is accepted that the antibacterial activity of bee pollen is due to its flavonoids and phenolic acids.

28.1.2 Aim of the Chapter

The antibacterial activity of ethanolic extracts of pot-pollen collected from eight meliponine species in four states of Venezuela was determined against two Gram positive and three Gram negative bacteria, using two standard methods (inhibition halo in agar well diffusion gel and minimal inhibitory concentration).

Table 28.1 Stingless bee species and locations of pot-pollen collection in Venezuela

Stingless bee species	N	Location (place/state)
Frieseomelitta aff. *varia*	2	El Paují/Bolívar
		Santa Elena de Uairén/Bolívar
Melipona compressipes	1	Guasdualito/Apure
Melipona eburnea	1	San Fernando de Atabapo/Amazonas
Melipona favosa	1	Barinas/Barinas
Melipona sp. group *fulva*	1	Guaramajé/Amazonas
Melipona lateralis kangarumensis	3	Guaramajé/Amazonas
		San Juan de Manapiare (2)/Amazonas
Melipona paraensis	4	Caño Tumo/Amazonas
		Maroa/Amazonas
		Carrizal(2)/Amazonas
Tetragonisca angustula	6	ULA(3)/Mérida
		La Mara (3)/Mérida

28.2 Pot-Pollen Samples and Ethanolic Extraction

The pot-pollen was collected from colonies of *Frieseomelitta, Melipona,* and *Tetragonisca* and kept frozen until analysis. The stingless bee species were kindly identified by late Professor JMF Camargo from Universidade de São Paulo, Ribeirão Preto, Brazil (Table 28.1).

A weight of 100 ± 10 mg of pot-pollen was placed in a glass homogenizer (Thomas No. A3528, USA), and 5 mL of ethanol (Riedel de Haën, Europe) was added and homogenized on an ice bath. Homogenates were centrifuged in a BHG Optima II (USA) centrifuge at 3000 rpm for 10 min, and supernatants were used for antibacterial analysis.

28.3 Well Diffusion Agar and Minimal Inhibitory Concentration Methods

A screening assay using well diffusion agar was carried out with some minor modifications (Bauer et al. 1966), in agreement with CLSI guidelines (Clinical and Laboratory Standards Institute 2009). Brain Heart Infusion Broth (BHI) (Sigma-Aldrich) agar plates were inoculated by rubbing sterile cotton swabs that were dipped into bacterial suspensions – overnight cultures grown at 37 °C on nutrient agar and adjusted to 0.5 McFarland units (a bacterial culture of 0.5 units of optic density with approximately 1×10^8 cfu/mL) in sterile saline, over the entire surface of the plate. After inoculation, 8.2 mm diameter wells were cut into agar surface using a sterile cork-borer. Eighty μL of test pollen extract at 25% (w/v) were added to each well. Plates were incubated at 37 °C for 24–48 h. The presence of an inhibition halo of the sample is indicative of antimicrobial activity. The bacterial strains used were Gram positive *Bacillus subtilis* (ATCC 11778) and *Staphylococcus aureus* (ATCC 25923) and Gram negative *Escherichia coli* (ATCC 25922), *Enterobacter cloacae* (ATCC 13047), and *Pseudomonas aeruginosa* (ATCC 27853). Each assay was carried out in triplicate.

The pot-pollen samples with inhibition halos in the agar plates were used to develop the minimum inhibitory concentration (MIC) method, adapting the procedure described by Patton et al. (2006) and Tan et al. (2009), conserving the standard methods developed by the Clinical and Laboratory Standards Institute (2009). Bacterial cultures were prepared during 24 h in BHI medium with continuous agitation and were adjusted to 0.5 units of McFarland (1×10^8 cfu/mL), then were diluted to

1/200 (1 mL of culture and 199 mL of BHI medium) at 5×10^5 cfu/mL. The first pot-pollen concentration was prepared at 50% (w/v) and sterilized with 2 μm sterile filters and was used to prepare serial concentrations of 5%, 10%, 15%, and 20% (w/v) of each pot-pollen sample. A volume of 190 μL of each dilution was transferred in sterile conditions to 96-well round-bottomed polystyrene microtiter plates, in 8 replicates by dilution – the first 2 wells of each dilution were the sterility controls, and the next six wells were inoculated with 10 μL of bacterial culture. Lines 11 and 12 were preserved for sterility tests and growth controls, adding 200 μL of BHI medium in line 11 and 10 μL of bacteria with 190 μL of BHI medium in line 12. The plates were incubated in orbital agitation at 120 rpm, 37 °C for 24 h. After incubation, absorbance was recorded at 590 nm using a microplate lector (Bio-Rad, US). The MIC was calculated using the formula:

$$MIC = \frac{1 - \left(\text{Asorbance of sample well} - \text{Absorbance of control well}\right)}{\left(\text{Absorbance of growth control well} - \text{Absorbance of sterility control well}\right)} \times 100\%$$

The antibacterial reference standards were prepared using the dilution method. Phenol (2 mg/L) (Sigma, USA), ciprofloxacin (5 mg/L), ampicillin (10 mg/L), and tetracycline (30 mg/L) were chosen as standards (Andrews 2001). Each assay was carried out in triplicate.

28.4 Antibacterial Activity of Venezuelan Pot-Pollen Ethanolic Extracts

28.4.1 Inhibition Zone Diameters

The antibacterial activity measured by the method of agar well diffusion shows that all pot-pollen ethanolic extracts were active against at least four of the bacterial strains tested. In general, *Melipona paraensis, Melipona lateralis kangarumensis* from San Juan de Manapiare (Amazonas), and two pot-pollen extracts of *Tetragonisca angustula* reached the widest inhibition halos (15–16 mm), indicating the highest antibacterial activities (Table 28.2). The antibacterial activity of pot-pollen produced by a stingless bee species is different for each bacterial strain tested in this work; for example, halos sizes 3–11 mm were caused by pot-pollen extracts of *Frieseomelitta* aff. *varia*, indicating different antibacterial potential according to the tested bacteria. Similarly, different actions are observed for pot-pollen extracts of *Melipona lateralis kangarumensis* (3–15 mm), *Melipona*

paraensis (3–15 mm), and *Tetragonisca angustula* (3–16 mm), including not detected halos (Table 28.2).

28.4.2 Minimal Inhibitory Concentrations

The MIC values of pot-pollen extracts, measured against different Gram positive and Gram negative pathogenic bacteria, are shown in Table 28.3. The lowest MIC values were reported for pot-pollen produced by *Melipona favosa, Melipona lateralis kangarumensis, Melipona paraensis,* and *Tetragonisca angustula* (2.5–5.0% w/v). Looking at the six pot-pollen extracts of *Tetragonisca angustula,* the two samples with the lowest MIC values between 2.5% and 5.0% (w/v) for all bacterial species used in this work were also the most active with inhibition halos of 10–16 and 7–15 mm diameter. The other four samples had lower antibacterial activity with MIC values between 7.5 and 12.5% (w/v) and also smaller inhibition diameters of 3–10 mm.

Escherichia coli was the bacterium most resistant to pot-pollen extracts, with lower MICs than other bacteria tested here. Ten pot-pollen extracts had no antibacterial activity against the Gram positive *Staphylococcus aureus* (1/1 *Melipona compressipes*) and Gram negative *Enterobacter cloacae* (1/2 *Frieseomelitta* aff. *varia,* 1/3 *Melipona lateralis kangarumensis*), *Escherichia coli* (1/1 *Melipona eburnea,* 1/4

Table 28.2 Inhibition zone diameter (mm)* in well diffusion agar of pot-pollen ethanolic extracts from eight meliponine species

Stingless bee specie	N	Gram positive		Gram negative		
		Bacillus subtilis	Staphylococcus aureus	Enterobacter cloacae	Escherichia coli	Pseudomonas aeruginosa
Frieseomelitta aff. varia	2	10.0 ± 0.9^d	8.0 ± 0.7^c	ND	7.0 ± 0.2^b	9.0 ± 0.4^c
		3.0 ± 0.4^a	3.0 ± 0.7^a	11.0 ± 0.5^d	4.0 ± 0.4^a	ND
		6.5 ± 1.3	**5.5 ± 0.9**	**11.0 ± 0.5**	**5.5 ± 1.0**	**9.0 ± 0.4**
Melipona compressipes	1	9.0 ± 0.7^c	ND	10.0 ± 0.6^d	9.0 ± 0.5^c	7.0 ± 0.7^b
Melipona eburnea	1	8.0 ± 0.4^c	9.0 ± 1.1^d	5.0 ± 0.4^b	ND	4.0 ± 0.2^a
Melipona favosa	1	13.0 ± 0.9^e	14.0 ± 0.9^f	11.0 ± 0.4^d	9.0 ± 0.6^c	ND
Melipona sp. fulva group	1	8.0 ± 1.1^c	11.0 ± 1.1^e	10.0 ± 1.1^d	12.0 ± 0.4^e	12.0 ± 0.8^d
Melipona lateralis kangarumensis	3	6.0 ± 0.6^b	3.0 ± 0.5^a	12.0 ± 0.5^e	6.0 ± 0.6^b	4.0 ± 0.2^a
		15.0 ± 1.3^f	14.0 ± 0.7^f	ND	13.0 ± 1.1^e	10.0 ± 1.1^c
		4.0 ± 0.1^a	5.0 ± 0.7^b	7.0 ± 0.5^c	8.0 ± 1.4^c	11.0 ± 0.9^d
		8.3 ± 1.0	**7.3 ± 0.9**	**9.5 ± 0.7**	**9.0 ± 1.4**	**8.3 ± 0.8**
Melipona paraensis	4	3.0 ± 0.2^a	5.0 ± 0.6^b	8.0 ± 0.5^c	11.0 ± 0.5^d	7.0 ± 0.4^b
		14.0 ± 0.8^f	11.0 ± 0.9^e	10.0 ± 1.2^d	ND	11.0 ± 0.7^d
		10.0 ± 0.2^d	11.0 ± 0.6^e	15.0 ± 1.4^f	10.0 ± 0.5^d	12.0 ± 0.6^d
		7.0 ± 0.7^b	6.0 ± 0.4^b	9.0 ± 0.4^d	10.0 ± 0.8^d	12.0 ± 0.5^d
		8.5 ± 0.6	**8.3 ± 1.1**	**10.5 ± 1.4**	**10.3 ± 0.9**	**10.5 ± 0.9**
Tetragonisca angustula	6	6.0 ± 0.2^b	4.0 ± 0.2^a	3.0 ± 0.3^a	ND	9.0 ± 0.5^c
		7.0 ± 0.3^b	3.0 ± 0.6^a	3.0 ± 0.2^a	ND	6.0 ± 0.7^b
		3.0 ± 0.2^a	10.0 ± 0.8^d	9.0 ± 0.9^d	ND	4.0 ± 0.5^a
		11.0 ± 0.8^d	16.0 ± 0.9^f	12.0 ± 1.2^e	13.0 ± 0.8^e	10.0 ± 0.9^c
		6.0 ± 0.2^b	7.0 ± 0.3^c	7.0 ± 1.1^c	ND	9.0 ± 0.7^c
		11.0 ± 0.7^d	15.0 ± 0.9^f	15.0 ± 2.5^f	13.0 ± 0.7^e	7.0 ± 0.4^b
		7.3 ± 1.0	**9.2 ± 1.6**	**8.2 ± 0.9**	**13.0 ± 0.1**	**7.5 ± 0.9**

Data are mean ± SE values ($n = 3$) for each pot-pollen sample, and mean ± SE values ($N = 2–6$) for stingless bee species, in boldface. *ND* not detected. Inhibition zone diameter in mm. Columns with values sharing the same superscript letter are not significantly different by ANOVA post hoc Scheffé test ($P < 0.05$)
*Including well

Melipona paraensis, and 4/6 *Tetragonisca angustula*), and *Pseudomonas aeruginosa* (1/2 *Frieseomelitta* aff. varia, 1/1 *Melipona favosa*) (Table 28.3).

28.4.3 Antibacterial Activity of Pollen and Polyphenols

The antibacterial activity of pot-pollen ethanolic extracts was positively correlated with the polyphenol content ($R = 0.876$ for *Bacillus subtilis*, $R = 0.768$ for *Staphylococcus aureus*, $R = 0.987$ for *Enterobacter cloacae*, and $R = 0.790$ for *Escherichia coli*, $R = 0.784$ for *Pseudomonas aeruginosa*) but not with the flavonoid content, $R < 0.3$ for all bacterial strains (See polyphenol and flavonoid contents in Table 24.1 and Table 26.6). Graikou et al. (2011) evaluated the chemical composition and the biological activity of Greek bee pollen rich in flavonoids and phenolic acids, which explains the observed free radical scavenging activity (see also Chaps. 24, 26, and 27) and the interesting antimicrobial profile. They demonstrated powerful antimicrobial activity against Gram positive bacteria (*Staphylococcus aureus* and *Staphylococcus epidermidis*), Gram negative bacteria (*Escherichia coli, Enterobacter*

Table 28.3 Minimum inhibitory concentration (MIC) of pot-pollen ethanolic extracts from eight meliponine species

Stingless bee specie	N	Gram positive		Gram negative		
		Bacillus subtilis	*Staphylococcus aureus*	*Enterobacter cloacae*	*Escherichia coli*	*Pseudomonas aeruginosa*
Frieseomelitta aff. *varia*	2	7.5 ± 0.4[d]	5.0 ± 0.3[c]	ND	5.0 ± 0.3[b]	5.0 ± 0.1[b]
		10.0 ± 0.4[f]	12.5 ± 0.3[f]	7.5 ± 0.3[d]	12.5 ± 0.1[d]	ND
		8.9 ± 0.5	**8.8 ± 1.0**	**7.5 ± 0.5**	**8.8 ± 1.0**	**5.0 ± 0.1**
Melipona compressipes	1	7.5 ± 0.2[d]	ND	7.0 ± 0.2[d]	7.5 ± 0.2[c]	5.0 ± 0.2[b]
Melipona eburnea	1	7.5 ± 0.2[d]	7.6 ± 0.2[d]	10.0 ± 0.4[e]	ND	10.0 ± 0.2[d]
Melipona favosa	1	2.5 ± 0.4[b]	2.5 ± 0.1[b]	2.5 ± 0.4[b]	5.0 ± 0.4[b]	ND
Melipona sp. *fulva* group	1	7.5 ± 0.3[d]	5.0 ± 0.4[c]	5.0 ± 0.2[c]	7.5 ± 0.2[c]	7.5 ± 0.3[c]
Melipona lateralis kangarumensis	3	10.0 ± 0.1[f]	12.5 ± 0.7[f]	5.0 ± 0.1[c]	12.5 ± 0.4[d]	12.5 ± 0.2[e]
		2.5 ± 0.5[b]	2.5 ± 0.1[b]	ND	2.5 ± 0.2[a]	5.0 ± 0.1[b]
		10.0 ± 0.6[f]	10.0 ± 0.2[e]	10.0 ± 0.3[e]	7.5 ± 0.3[c]	7.5 ± 0.3[c]
		7.5 ± 1.1	**8.3 ± 0.9**	**7.5 ± 1.1**	**7.5 ± 1.3**	**8.3 ± 1.8**
Melipona paraensis	4	12.5 ± 0.2[g]	12.5 ± 0.5[f]	10.0 ± 0.4[e]	5.0 ± 0.2[b]	10.0 ± 0.4d
		2.5 ± 0.3[b]	5.0 ± 0.3[c]	5.0 ± 0.5[c]	ND	5.0 ± 0.2[b]
		5.0 ± 0.1[c]	5.0 ± 0.3[c]	2.5 ± 0.1[b]	5.0 ± 0.5[b]	5.0 ± 0.4[b]
		10.0 ± 0.7[f]	10.0 ± 0.6[e]	7.5 ± 0.3[d]	7.5 ± 0.2[c]	10.0 ± 0.5[d]
		7.5 ± 1.2	**8.1 ± 1.0**	**6.3 ± 0.9**	**5.8 ± 0.9**	**7.5 ± 1.0**
Tetragonisca angustula	6	10.0 ± 0.5[f]	12.5 ± 0.3[f]	12.5 ± 0.9[f]	ND	5.0 ± 0.4[b]
		7.5 ± 0.6[d]	12.5 ± 0.5[f]	12.5 ± 0.6[f]	ND	10.0 ± 0.7[d]
		12.5 ± 0.7[g]	5.0 ± 0.3[c]	7.5 ± 0.4[d]	ND	12.5 ± 1.1[e]
		5.0 ± 0.2[c]	2.5 ± 0.6[b]	5.0 ± 0.4[c]	2.5 ± 0.2[a]	5.0 ± 0.6[b]
		10.0 ± 0.3[f]	10.0 ± 0.7[e]	10.0 ± 0.6[e]	ND	10.0 ± 0.9[d]
		5.0 ± 0.4[c]	2.5 ± 0.2[b]	2.5 ± 0.1[b]	2.5 ± 0.1[a]	5.0 ± 0.4[b]
		7.3 ± 1.3	**7.5 ± 1.0**	**8.3 ± 1.2**	**2.5 ± 0.1**	**7.9 ± 1.3**
Phenol		1.0 ± 0.1[a]	5.0 ± 0.1[c]	1.0 ± 0.1[a]	5.0 ± 0.1[b]	1.0 ± 0.1[a]
Ampicillin		4.0 ± 0.2[c]	2.5 ± 0.1[b]	14.0 ± 2.6[g]	126.0 ± 2.6[e]	16.0 ± 0.4[f]
Ciprofloxacin		1.0 ± 0.1[a]	5.0 ± 0.1[c]	5.0 ± 0.2[c]	2.0 ± 0.1[a]	1.0 ± 0.1[a]
Tetracycline		2.0 ± 0.2[b]	1.0 ± 0.1[a]	1.0 ± 0.1[a]	4.0 ± 0.1[b]	1.0 ± 0.1[a]

Data are mean ± SE values ($n = 3$) for each pot-pollen sample, and mean ± SE values ($N = 2$–6) for stingless bee species, in boldface. *ND* not detected. Columns within a sample sharing the same letter are not significantly different by ANOVA post hoc Scheffé test ($P < 0.05$)

cloacae, Klebsiella pneumoniae, and *Pseudomonas aeruginosa*), and pathogenic fungi (*Candida albicans, Candida tropicalis,* and *Candida glabrata*). Kacániová et al. (2012) tested the antimicrobial activity of ethanol (70%) and methanol (96%) extracts of Slovakian monofloral pollen loads from sunflower (*Helianthus annuus*), poppy (*Papaver somniferum*), and rape (*Brassica napus*) against pathogenic bacteria, microscopic fungi, and yeasts, using the agar well diffusion method. Among the tested bacteria, *Escherichia coli* was the most sensitive after 48 h of pollen methanolic and ethanolic extract, and the sensitivity of the bacteria was ranked as follows: *Staphylococcus aureus* > *Salmonella enterica* > *Pseudomonas aeruginosa* > *Listeria monocytogenes* –explained by phenolic spectra of diverse botanical and geographical origin of pollen. *Escherichia coli* sensitivity contrasts with the results reported in our work, where *Escherichia coli* was resistant to pot-pollen ethanolic extracts (Tables 28.2 and 28.3). Possibly the entomological origin adds variability to the botanical and geographical origin, in terms of microbiota associated with different species of stingless bees and transformations of pollen in the pots.

Graikou et al. (2011) established that antifungal and antibacterial activity of Greek bee pollen might be caused by high quercetin and kaempferol contents in the tested extracts of bee pollen. Indeed, quercetin and kaempferol were present in pot-pollen from *Frieseomelitta* sp. aff. *varia*, *Melipona* sp. *fulva* group, and *Melipona lateralis kangarumensis*, and kaempferol was found for *Melipona compressipes*, besides other combinations with genkwanin, luteolin, and 8-methoxikaempferol in these bees and the other four species (Vit et al., Chapt. 26). Moreover, Campos et al. (1998) found antibacterial activity of hydrophobic components of *Apis mellifera* bee pollen against *Streptococcus viridans*. Additionally, Tichy and Novak (2000) examined the presence of antimicrobials in ethanol extracts of propolis, pollen, and cappings of honeycomb cells, as well as unstrained honey containing honeycomb cappings and a mixture of antimicrobial compounds in various amounts. Thin-layer chromatography experiments with two different solvent systems differing in polarity suggested that major antimicrobials present in the bee products prepared from honeycomb cappings, honeycombs, pollen, and propolis have similar chemical properties. These active compounds were not extremely hydrophobic. Extracts from the tested samples of bee products exhibited antimicrobial properties at various levels depending on the nature of the sample and the microbes used for testing.

The content of flavonoids, polyphenols, and carotenoids of dried, frozen, and freeze-dried *Helianthus annuus* bee pollen varied according to the conservation method and was positively correlated with the in vitro antimicrobial activity against Gram positive and Gram negative bacteria and fungi. The best antibacterial effects were against *Escherichia coli, Paenibacillus larvae, Pseudomonas aeruginosa,* and *Enterococcus raffinosus*, and the best antifungal activity against *Aspergillus ochraceus*, and freeze-dried bee pollen extracts were effective against *Aspergillus niger* (Fatrcová-Šramková et al. 2016).

Due to the great biodiversity of pollen-producing plants and the scant literature about antimicrobial activity of pot-pollen extracts, more studies are necessary for a better understanding of the functional properties of pot-pollen. Moreover, due to its antibacterial activity, pollen-based products can be used in medicinal and veterinary fields. For example, Olczyk et al. (2016) evaluated the benefits and advantages derived from preparations based on extracts of bee pollen similarly to pharmaceuticals commonly used in the treatment of burns. The bee pollen ointment was applied for the first time in topical burn treatment. Clinical, histopathological, and microbiological effects were assessed in burn wounds inflicted on Polish domestic pigs treated with silver sulfadiazine and bee pollen ointment. The applied apitherapeutic agent reduced the healing time of burn wounds and improved general health with reduced number of microorganisms and an increased bactericidal activity of isolated strains. Based on the obtained bacteriological analysis, it was concluded that the applied bee pollen ointment facilitated the wound healing process for burns by preventing infection of newly formed tissue. Since several reports establish that bee pollen and pot-pollen antibacterial activity is related to polyphenol and flavonoid contents, the healing of wounds may conceivably involve these components.

28.5 Conclusions

The present research has shown that ethanolic extracts of pot-pollen produced by eight stingless bee species from Venezuela possess antibacterial effect against pathogenic Gram positive and Gram negative bacteria. *Escherichia coli* was the most resistant bacterium against pot-pollen ethanolic extracts. A variation of antibacterial activity (inhibition zone diameters and MIC values) in a stingless bee species with diverse geographical origin of pot-pollen was observed. Due to the great biodiversity of pollen sources, follow-up studies are necessary for a better understanding of the functional properties of pot-pollen. This is the first study on antimicrobial properties of pot-pollen ethanolic extracts from Venezuela. Pot-pollen can provide an alternative and effective antibacterial strategy against the continuous emergence of antibiotic resistant microorganisms.

Acknowledgments To the memory of Professor João MF Camargo, Biology Departement, Universidade de São Paulo, Ribeirão Preto, Brazil, for the identification of the Venezuelan stingless bees. To stingless bee keepers from southern Venezuela to provide their pot-pollen. To project FA-127-93B from Council for the Scientific, Humanistic and Technological Development at Universidad de Los Andes, Mérida, Venezuela, for supporting field work needed to collect the pot-pollen in Venezuela. To the support of ZG-AVA-FA-01-98-01 from the Council of Development of Scientific, Humanistic, Technological and Artistic, at Universidad de Los Andes, to the Group Apitherapy and Bioactivity. To referees for their timely comments. To Dr. D.W. Roubik for the careful English proofreading.

References

Andrews JM. 2001. Determination of minimum inhibitory concentrations. Journal of Antimicrobial Chemotherapy 48: 5–16.

Bárbara MS, Machado CS, Sodré Gda S, Dias LG, Estevinho LM, de Carvalho CA. 2015. Microbiological assessment, nutritional characterization and phenolic compounds of bee pollen from *Melipona mandacaia* Smith, 1983. Molecules 20: 12525-12544.

Basim E, Basim HS, Özcan M. 2006. Antibacterial activities of Turkish pollen and propolis extracts against plant bacterial pathogens. Journal of Food Engeniering 77: 992–996.

Bauer AW, Kirby WM, Sherris JC, Turck M. 1966. Antibiotic susceptibility testing by standardized single disk method. American Journal of Clinical Pathology 45: 493-496.

Boorn KL, Khor YY, Sweetman E, Tan F, Heard TA, Hammer KA. 2010. Antimicrobial activity of honey from the stingless bee *Trigona carbonaria* determined by agar diffusion, agar dilution, broth microdilution and time-kill methodology Journal of Applied Microbiology 108: 1534-1543.

Campos M, Cunha A, Markham K. 1998. Inhibition of virulence of *Pseudomonas aeruginosa* cultures, by flavonoids isolated from bee-pollen: Possible structure-activity relationships. pp. In: Polyphenol Communications 98. Proceedings of the XIX International Conference on Polyphenols, Lille, France, 1–4 September. Groupe Polyphenols; Bordeaux, France. pp.

Campos MG, Frigerio C, Lopes J, Bogdanov S. 2010. What is the future of Bee-Pollen? Journal of ApiProduct and ApiMedical Science 2: 131–144.

Carpes ST, Begnini R, Alencar SMD, Masson ML. 2007. Study of preparations of bee pollen extracts, antioxidant and antibacterial activity. Ciência e Agrotecnologia 31: 1818–1825.

Clinical and Laboratory Standards Institute. 2009. Methods for dilution antimicrobial susceptibility tests for bacteria that grow aerobically; approved standard. CLSI document M07-A8, Eighth edition, Wayne, Pennsylvania, USA. pp. 12-45.

da Cunha MG, Franchin M, de Carvalho Galvão LC, de Ruiz AL, de Carvalho JE, Ikegaki M, de Alencar SM, Koo H, Rosalen PL. 2013. Antimicrobial and antiproliferative activities of stingless bee *Melipona scutellaris* geopropolis BMC Complementary and Alternative Medicine 23: 1-9.

Denisow B, Denisow-Pietrzyk M. 2016. Biological and therapeutic properties of bee pollen: a review. Journal of the Science of Food and Agriculture 96: 4303-4309.

Eraslan G, Kanbur M, Silici S, Liman BC, Altinordulu S, Sarica ZS. 2009. Evaluation of protective effect of bee pollen against propoxur toxicity in rat. Ecotoxicology and Environmental Safety 72: 931–937.

Farnesi AP, Aquino-Ferreira R, De Jong D, Bastos JK, Soares AE. 2009. Effects of stingless bee and honey bee propolis on four species of bacteria. Genetics and Molecular Research 8: 635-640.

Fatrcová-Šramková K, Nôžková J, Máriássyová M, Kačániová M. 2016. Biologically active antimicrobial and antioxidant substances in the *Helianthus annuus* L. bee pollen. Journal of Environmental Science and Health B 51: 176-181.

Grajek W. 2007. Antioxidants in Food. WNT: Warsaw, Poland. pp. 258–259.

Graikou K, Kapeta S, Aligiannis N, Sotiroudis G, Chondrogianni N, Gonos E, Chinou I. 2011. Chemical analysis of Greek pollen - Antioxidant, antimicrobial and proteasome activation properties. Chemistry Central Journal 5: 33-41.

Hamamoto R, Ishiyama K, Yamaguchi M. 2006. Inhibitory effects of bee pollen *Cistus ladaniferus* extract on bone resorption in femoral tissues and osteoclast-like cell formation in bone marrow cells *in vitro*. Journal of Health Science 52: 268–275.

Haro A, López-Aliaga I, Lisbona F, Barrionuevo M, Alférez MJ, Campos MS. 2000. Beneficial effect of pollen and/or propolis on the metabolism of iron, calcium, phosphorus, and magnesium in rats with nutritional ferropenic anemia. Journal of Agricultural and Food Chemistry 48: 5715–5722.

Kacániová M, Vuković N, Chlebo R, Haščík P, Rovná K, Cubon J, Dżugan M, Pasternernakiewicz A. 2012. The antimicrobial activity of honey, bee pollen loads and beeswax from Slovakia. Archives of Biological Science 64: 927-934.

Knazovicka V, Melich M, Kacaniova M, Fikselova M, Hascik P, Chlebo R. 2009. Antimicrobial activity of selected bee products. Acta Fytotechnica et Zootechnica 12: 280–285.

Leja M, Mareczek A, Wyżgolik G, Klepacz-Baniak J, Czekońska K. 2007. Antioxidative properties of bee pollen in selected plant species. Food Chemistry 100: 237–240.

Li F, Yuan Q, Rashid F. 2009. Isolation, purification and immunobiological activity of a new water-soluble bee pollen polysaccharide from *Crataegus pinnatifida* Bge. Carbohydrate Polymers 78: 80–88.

Liberio SA, Pereira AL, Dutra RP, Reis AS, Araújo MJ, Mattar NS, Silva LA, Ribeiro MN, Nascimento FR, Guerra RN, Monteiro-Neto V. 2011. Antimicrobial activity against oral pathogens and immunomodulatory effects and toxicity of geopropolis produced by the stingless bee *Melipona fasciculata* Smith BMC Complementary and Alternative Medicine 108: 12-33.

Massaro CF, Katouli M, Grkovic T, Vu H, Quinn RJ, Heard TA, Carvalho C, Manley-Harris M, Wallace HM, Brooks P. 2014. Anti-staphylococcal activity of *C*-methyl flavanones from propolis of Australian stingless bees (*Tetragonula carbonaria*) and fruit resins of *Corymbia torelliana* (Myrtaceae) Fitoterapia 95: 247-257.

Michener CD. 2013. The Meliponini. pp. 3-17. In: Vit P, Pedro SRM, Roubik D. Pot-honey. A legacy of stingless bees. Springer, New York, USA. 654 pp.

Monserrate Y. 2015. Valoración *in vitro* del potencial antimicrobiano de extractos etanólicos de polen de *Apis mellifera* y de *Tetragonisca angustula*, en busca de posibles usos terapéuticos. Universidad Nacional de Colombia, Facultad de Medicina Veterinaria y de Zootecnia Bogotá, Colombia. pp 25-65.

Nishio EK, Ribeiro JM, Oliveira AG, Andrade CG, Proni EA, Kobayashi RK, Nakazato G. 2016. Antibacterial synergic effect of honey from two stingless bees: *Scaptotrigona bipunctata* Lepeletier, 1836, and *S. postica* Latreille, 1807. Scientific Reports doi: 10.1038/srep21641.

Nogueira C, Iglesias A, Feás X, Estevinho LM. 2012. Commercial bee pollen with different geographical origins: A comprehensive approach. International Journal of Molecular Science 13: 11173–11187.

Olczyk P, Koprowski R, Kaźmierczak J, Mencner L, Wojtyczka R, Stojko J, Olczyk K, Komosinska-Vassev K. 2016. Bee pollen as a promising agent in the burn wounds treatment. Evidence-Based Complementary and Alternative Medicine 2016. doi: 10.1155/2016/8473937.

Patton T, Barrett J, Brennan J, Moran N. 2006. Use of spectrophotometric bioassay for determination of microbial sensitivity to manuka honey. Journal of Microbiological Methods 64: 84–95.

Pimentel RB, da Costa CA, Albuquerque PM, Junior SD. 2013. Antimicrobial activity and rutin identification of honey produced by the stingless bee *Melipona compressipes* manaosensis and commercial honey BMC Complementary and Alternative Medicine 13: 1-13.

Tan HT, Rahman RA, Gan SH, Halim AS, Hassan SA, Sulaiman SA, Kirnpal-Kaur B. 2009. The antibacterial properties of Malaysian tualang honey against wound and enteric microorganisms in comparison to manuka honey. Complementary and Alternative Medicine 9: 34-39.

Tichy J, Novak J. 2000. Detection of antimicrobials in bee products with activity against viridans streptococci. Journal of Alternative and Complementary Medicine 6: 383–389.

Vit P, Santiago B, Pedro SRM, Perez-Perez E, Peña-Vera M. 2016. Chemical and bioactive characterization of pot-pollen produced by *Melipona* and *Scaptotrigona* stingless bees from Paria Grande, Amazonas State, Venezuela. Emirates Journal of Food and Agriculture 28: 78-84.

Wu YD, Lou YJ. 2007. A steroid fraction of chloroform extract from bee pollen of *Brassica campestris* induces apoptosis in human prostate cancer PC-3 cells. Phytotherapy Research 21: 1087–1091.

Metabolomics Analysis of Pot-Pollen from Three Species of Australian Stingless Bees (Meliponini)

Carmelina Flavia Massaro,
Tommaso Francesco Villa, and Caroline Hauxwell

We acknowledge the Aboriginal people as the Traditional Owners and Custodians of the land where Australian Meliponini have been described and pay respect to the Elders – past, present and future – for sharing their knowledge about native bees and their nest products.

29.1 Introduction

29.1.1 Historical Accounts of Australian Pot-Pollen

As described in Chap. 27, there are several species of highly social bees in Australia, including the introduced European honey bee (Hymenoptera: tribe Apini: *Apis mellifera* Linnaeus) and the native stingless bees of the Meliponini tribe. The Australian Meliponini include at least 14 species among which *Austroplebeia australis* (Friese), *Tetragonula carbonaria* (Smith) and *T. hockingsi* (Cockerell) are important pollinators of Australian wild vegetation and crops (Rasmussen 2008; Heard 1999; Heard 2016). Analysis of the pot-pollen of these three species are the focus of this chapter.

C. Flavia Massaro (✉) • T.F. Villa
C. Hauxwell
School of Earth, Environmental and Biological Sciences, Science and Engineering Faculty, Queensland University of Technology, Brisbane 4001, Australia
e-mail: cfmassaro@gmail.com

Pot-pollen is so-called because Meliponini store their pollen in oval, pot-shaped containers that differ in colour and size. Early historical accounts of pollen stored in Australian stingless bee nests date back to the late 1800s with Harold Hockings' reports of interacting with the local community of indigenous people inhabiting the areas of Mount Coot-tha, west of Brisbane, in southeastern Queensland, Australia (Hockings 1883). Interestingly, 'Coot-tha' derives from 'ku-ta', the Aboriginal name for 'wild stingless bee honey'. Hockings accurately described the features of native stingless bees and their nest morphology in *T. carbonaria* ('Karbi' or 'Keelar' in Aboriginal Turrbal language) and possibly *A. australis*/*A. cassiae* ('Kootchar'), resulting in the naming of *T. hockingsi* in his honour (Hockings 1883; Halcroft et al. 2013). Pot-pollen of *Tetragonula* nests was described as a 'damp, pasty and sour' bee product that bees stored in dark-coloured pots surrounding the brood and external to the involucrum, a resinous structure surrounding the reproductive core. A common feature of nest morphology in Meliponini is the organisation of pollen pots in clusters around the brood (Roubik 1979; Heard 2016). In contrast, *A. australis* reportedly stored pollen in light-coloured waxy pots positioned near the nest entrance (Hockings 1883).

The material used by the Meliponini to build pollen pots and other nest elements is bee cerumen, a mixture of beeswax and plant resin(s)

(Bankova and Popova 2007). In *Tetragonula* species, the cerumen is also known as 'stingless bee propolis' and is typically dark brown as a result of the incorporation of several plant resins into their wax (Massaro et al. 2014a, 2015). These resins can originate from a variety of botanical sources and are initially stored separately as creamy, red, orange and white deposits inside the nest. Stingless bees blend aliquots of the resin-deposits with their beeswax to make the building material used to construct the pots (Massaro et al. 2015). In contrast, *A. australis* bees appear to make limited or no use of plant resins in their cerumen, which is made primarily of beeswax and has a yellowish-dark colour attributed to the incorporation of *Eucalyptus* pollen grains (Milborrow et al. 1987).

29.1.2 Health Benefits of Bee Pollen

Pollen provides the bee colony with an important source of nitrogen in the form of vegetable protein, as well as lipids, minerals and vitamins that are essential for brood-rearing and multiple physiological processes, including promoting immunity to infection. Secondary plant metabolites, such as phenolic *p*-coumaric acid and several flavonoids, may function as nutraceuticals that regulate immune and detoxification processes (Berenbaum and Johnson 2015). Thus, the longevity, reproduction and survival of the colony, including thousands of individuals across multiple generations, may be affected by the quality and the quantity of pollen gathered from the environment.

Therapeutic bioactivity and multiple health benefits of pollen harvested from *A. mellifera,* both in human and in bee health, have been previously reported (Berenbaum and Johnson 2015; Komosinska-Vassev et al. 2015; Mărgăoan et al. 2014; Pérez-Pérez et al. Chapter 28; Rzepecka-Stojko et al. 2015). Products from the Australian Meliponini are not yet available for human consumption. Moreover, the secondary metabolites characteristic of Meliponini pot-pollen and their beneficial effects to bee health have not been studied. Chemical analyses are needed to provide

a reference for further botanical screenings as well as to test for possible bioactivity.

29.1.3 Botanical Sources within Flight Range

The composition of bee pollen varies with the botanical sources available to bees within their geographic location and flight range (Mărgăoan et al. 2014). The variability in composition also depends on the bees' preferences as well as on environmental factors such as the flowering season. A previous study analysed European pollen samples from honey bees *A. mellifera* by Liquid Chromatography Mass Spectrometry (LC-MS) and used Principal Component Analysis (PCA) to correlate botanical origins with the amounts of total lipids and fatty acid profiles (Mărgăoan et al. 2014). In chemometrics, PCA can be proficiently applied to reduce the dimensions of the space of the random variables with a measurable loss of the information about the data variance; the reduction of the dimensions allows for an efficient comparison of the analysed samples.

Bees' foraging preferences can affect the quality and quantity of pot-pollen. However, a recent study investigating the foraging flight range of *T. carbonaria* found that the typical homing range is 333 m with a maximum distance of 712 m and thus much shorter than that of honey bees (Smith et al. 2017). A comparison of foraging in *A. australis* and *T. carbonaria* found that *A. australis* prefer specific pollen resources of high quality if available, while *T. carbonaria* bees collect resources of both high and low qualities and *A. australis* collect pollen within a narrower radius than *T. carbonaria*, at least in periods of intense flowering (Leonhardt et al. 2014a).

29.1.4 Food Security

Toxic xenobiotics such as alkaloids can also be found in bee pollen. Toxic pyrrolizidine alkaloids (PAs) were reported in Australian *A. mellifera* honeys (Griffin et al. 2015) and pollen originating from *Echium plantagineum*, also known as

Patterson's curse (Somerville and Nicol 2006), and *E. vulgare* (Boraginaceae) (Lucchetti et al. 2016). However, plant-pollen contributed only marginally to PA in honey bee products, far less than was contributed by nectar to honey (Lucchetti et al. 2016).

Other toxic xenobiotics have been reported in pollen of *Ranunculus* (Ranunculaceae), which honey bees mix with nutritional pollen to mitigate toxic effects (Eckhardt et al. 2014). Toxic alkaloids and secondary plant metabolites can originate from Australian flora possibly including these botanical sources that might be visited by Meliponini bees. Chemical analysis can determine xenobiotics in Meliponini pot-pollen and indicate their botanical origins.

29.1.5 Research in Australian Meliponini Bee Products

Bee products are complex matrices of metabolites that have different properties of volatility, polarity and solubility in organic solvents, and these different features differ in the analytical technique required. For instance, essential oils and terpenoids of low molecular weight can be extracted by distillation or by Head Space Solid Phase Micro Extraction (HS-SPME) for separation by Gas Chromatography Mass Spectrometry (GC-MS). However, these compounds are of low polarity, non-ionizible and have few or no chromophores and are thus less amenable to separation by Liquid Chromatography (LC) and detection by Ultra Violet Electro Spray Mass Spectrometry (UV-ESI-MS). In contrast, phenolic compounds, including flavonoids with or without glycosylated moieties, are metabolites of medium molecular weight and features that make them more suitable for analysis by LC- UV-ESI-MS.

Former chemical analyses of bee products applied GC-MS to *A. australis* wax nest material to find hydrocarbons (C_{31}, C_{33} and C_{35}) and free acids (C_{10}–C_{18}), while *T. carbonaria* nest structures contained resin components mixed with beeswax (Milborrow et al. 1987). Stingless bee (*T. carbonaria*) propolis has been analysed by GC-MS, LC-MS and Nuclear Magnetic Resonance

spectroscopy (NMR) to identify flavonoids, phenolics, isoprenoids and phloroglucinols (Leonhardt et al. 2011; Massaro et al. 2011, 2014a, 2015; Nishimura et al. 2016). These techniques enabled the identification of chemical components in 'sugar bag' (stingless bee) honey to be glycosylated flavonoids with antimicrobial properties (Massaro et al. 2014b; Truchado et al. 2015). Chemical fingerprinting confirmed that *Corymbia torelliana* (Myrtaceae) fruit resin is stored in 'creamy' hive-deposits by *T. carbonaria* bees and subsequently incorporated into their propolis (Massaro et al. 2015). *C. torelliana* also provides *T. carbonaria* with a large quantity of pollen in a given season, but its contribution to pot-pollen is yet to be evaluated.

In the early 1930s, the naturalist Tarlton Rayment reported that the nitrogen and protein contents in pot-pollen of *T. cassiae* were 3.17% and 19.8%, respectively (Rayment 1935). In Sulbarán et al. Chap. 27, the total contents of polyphenols and proteins were reported for pot-pollen of the Australian Meliponini *T. carbonaria*, *T. hockingsi* and *A. australis*. In this chapter, chromatographic analyses were performed to determine levels of chemical similarity or difference between pot-pollen of those three bee species (*T. carbonaria*, *T. hockingsi* and *A. australis),* with access to the same botanical resources within overlapping flight ranges. Phenols and volatile organic compounds were identified and used to discriminate between the pot-pollen of the three species. The hypothesis tested was that Australian Meliponini bees would produce similar pot-pollen, based on the assumption that these stingless bees forage within a similar flight range (approx. 500 m radius) and would therefore have access to the same botanical sources located within an area abundant in plant-pollen sources.

29.1.6 Aim of the Chapter

In this chapter, two chemical profiles of pot-pollen from three Australian Meliponini bees were screened for the presence of volatile organic compounds and ethanol-extracted phenolics. Chemical fingerprints were compared, and characteristic markers were identified to create an online open source database for chemometric analysis of Australian pot-pollen.

29.2 Methods of Chromatographic Analysis of Pot-Pollen

29.2.1 Sampling Pot-Pollen from Bee Hives of Australian Meliponini

Nine beehives of *A. australis*, *T. carbonaria* and *T. hockingsi* were located within a 50 m radius in South East Queensland, Australia (27°21′59.5″S 152°49′38.2″E). Pot-pollen was harvested directly from the hives to obtain triplicates of pot-pollen from each bee species. Multiple pollen pots (up to 20) from inside each hive were pooled into one composite per nest to yield the nine samples (mean mass was 11.14 g). Collection was conducted at the end of December, 2016 (midsummer), and it was assumed that the pot-pollen was stored in the few weeks preceding sampling. Botanical species within 500 m from hives were identified from accounts and observations of a local beekeeper (R. Luttrell, personal communication). Pot walls were carefully removed during the sampling operations to prevent resin-derived compounds from interfering with the chemical analyses. All samples were immediately stored in glass containers at −20 °C until further use.

29.2.2 Extraction

Pot-pollen was homogenized before weighing for extraction. Raw samples (0.5 g) were subjected to HS-SPME with direct extraction followed by injection into the GC-MS for analysis. In the LC-UV-MS work, two sets of extractions were performed to target either the phenolic compounds or the alkaloids. Ethanol (EtOH), acetonitrile (ACN), methanol (MeOH) and formic acid were of analytical grade and available at the laboratories of the Centre for Analytical

Facility of Queensland University of Technology. Deionized demineralised water (dH$_2$O) was obtained from an in-house Ultrapure Water System.

Ethanolic extracts of the pollen samples were obtained from a modified protocol previously applied to *T. carbonaria* honey for the extraction and concentration of phenolics to remove the polar compounds (sugars and proteins) (Massaro et al. 2014b).

Briefly, the raw pollen was extracted in absolute ethanol (5%, w/v) at around 60 °C and allowed to cool at 24 °C for 2 h and then at 4 °C for 2 h, followed by storage at −20 °C overnight. The supernatant was decanted, and the precipitate (carbohydrates, proteins and debris) extracted again, following the same protocol. The combined supernatants were evaporated to dryness under reduced pressure and freeze-dried to yield three phenolic concentrates. Aliquots of the nine dry extracts were reconstituted in 100% ethanol for the LC analyses. A further extraction was performed on raw pollen samples to extract the alkaloids using a solvent system of 70% methanol, 29.5% ultrapure water and 0.5% formic acid, v/v (Lucchetti et al. 2016).

29.2.3 Volatiles by HS-SPME-GC-MS

Volatile organic compounds (VOCs) of pot-pollens were analysed using an SPME technique previously used in analyses of honey (Cuevas-Glory et al. 2007) and using the GC-MS technique with Electron Impact mode (EI) as formerly applied to VOC studies in bee products (Baroni et al. 2006; Nunes and Guerreiro 2012). SPME fibres with different coatings are available on the market, and the polydimethylsiloxane/divinylbenzene (PDMS/DVB) fibre (Sigma-Aldrich, Australia) was selected as the most suitable for the analyses. Pollen samples (0.5 g) were placed into 20 mL HS glass vials and then incubated at 70 °C for a time of 30 min. As the heated incubator also works as an agitator, an agitation interval of 10 sec (10 sec agitation, 10 sec resting) was selected during the incubation time.

In this study, the HS-SPME sampling procedure was automated using a TriPlus autosampler (Thermo Scientific, Australia) mounted onto a Thermo Scientific™ Single Quadrupole ISQ GC-MS. The SPME parameters included a fibre conditioning at 280 °C, an extraction time of 5 min and a desorption time of 2 min. The GC conditions follow herewith. Injection mode: splitless. Injection port temperature: 270 °C. Equilibration time: 0.5 min. Carrier gas: helium at a flow rate of 1.2 mL/min. Volatiles were separated on a DB-5 column ($L = 30$ m, ID 0.25 mm, 0.25 μm film thickness, PerkinElmer). Oven temperature: 40 °C held for 2 min. Programmed at 10 °C/min until 300 °C and held for 3 min. Total runtime: 30 min. Transfer line temperature: 280 °C. MS detector. Ionization mode: Electron Impact (EI) with 70 eV ionization. MS scan range: m/z 40–500.

GC-separated compounds were tentatively identified using spectral measurements of diagnostic ions and characteristic fragmentation patterns against the mass spectral database of the National Institute of Standards and Technology (NIST, v. 2.0, 2011). Moreover, non-isothermal-Linear Kováts Indices were calculated against a solution of standards C_6–C_{30} (Sigma-Aldrich, Australia) using the formula below:

$$KIx = 100n + 100(tx - tn)/(tn + 1 - tn)$$

where tn and tn + 1 are retention times of the reference n-alkane hydrocarbons eluting immediately before and after chemical compound 'X'; tx is the retention time of compound 'X'. Compound identifications were compared against reports of plant essential oils (Babushok et al. 2011) and honey (Baroni et al. 2006).

29.2.4 Chemical Constituents by HPLC-DAD-ESI(–)-MS/MS

Dry extracts were reconstituted in absolute EtOH (5 mg/mL) and aliquots (1 μL) were injected into a Kinetex XB-C18 column (Phenomenex, 4.6 × 100 mm, 2.6 μ particle size) mounted onto Ultra Performance Liquid Chromatographer with Diode Array Detector (HPLC-DAD) tandem Orbitrap for High Resolution Mass Spectrometry (HRMS) (Thermo Fisher™ Scientific, Australia).

Separation was conducted using LC-grade mobile phases, deionized water Mobile Phase A (MPA) and acetonitrile Mobile Phase B (MPB). The gradient method was 2.5% (MPB) at $T = 0.5$ min to 100% (MPB) at 13 min, held for 4.5 min and then re-equilibrated to the initial conditions for a total run time of 30 min. The flow rate was set at 0.4 mL/min. Orbitrap Fourier Transform Mass Spectrometry (FTMS) resolution was set at 120,000; in the MS/MS mode the activation was Collision Induced Dissociation (CID) type at normalized collision energy 30 eV for independent data acquisition from the most abundant parent ion in MS. Separate HPLC runs were carried out both in the negative and positive ionization modes at the scan ranges of m/z 50–700. DAD chromatograms were recorded at 205, 260, 290 and 340 nm.

LC-HRMS post-runs in the negative ESI(–)-mode were used for tentative compound identifications. Mono-isotopic masses, chemical formulae of isotopic ions and absolute errors in mDa were determined using the XCalibur™ software (Thermo Finnegan v. 3.0.63); MS/MS data were analysed to calculate elemental compositions and fragmentation patterns with high accuracy. DAD profiles were plotted using the same software. Targeted analyses of ESI(–)-MS ions was obtained from de-convoluted mass spectra showing moieties of phloroglucinols (m/z 181.0869, $C_{10}H_{13}O_3$), C-methyl-flavanones (m/z 269.0816, $C_{16}H_{13}O_4$) and previous reports of glycosylated flavonoids and other Australian Meliponini bee metabolites (de Rijke et al. 2006; Truchado et al. 2015; Massaro et al. 2014a, b, 2015; Nishimura et al. 2016). Accurate elemental compositions were generated using the Xcalibur in-built tool as well as the open access platform http://www.chemcalc.org/. Constituents were tentatively identified for known structures and bioactivities against the literature and the chemical database //scifinder.cas.org. The stereochemistry was not considered.

29.2.5 Targeted Analyses of Pyrrolizidine Alkaloids (PA) by ESI(+)-MS/MS

Separate analyses were performed by HPLC-DAD-HRMS in the positive ESI mode using the mobile phases 0.05% formic acid in water (MPA) and 0.05% formic acid in ACN (MPB), with a gradient method of 15–40% MPB. PA were targeted by extracting the Total Ion Current (TIC) at the mass ranges between m/z 398.2184 and 496.3402 as previously described (Lucchetti et al. 2016). DAD used one UV channel set at 254 nm.

29.2.6 Metabolomics of Pot-Pollen VOCs and Phenolics

Secondary metabolites present in pot-pollen of each bee species were tested as independent samples (each sample was from one nest, nine nests in total). Individual constituents were quantified as percentages (% w/w) of peak areas over total peak areas of compounds detected in the chemical analyses of VOC profiles and EtOH extracts. Values were presented as means of three samples ($n = 3$) with standard errors of the mean, and statistical differences between samples were assessed using two-way ANOVA and Tukey's multiple comparisons test (GraphPad Prism version 6.0, Graph Pad Software Inc., USA). The

post-run data-files were converted (http://prote-owizard.sourceforge.net/) and then plotted by PCA with centred unit of variance ($N = 9$ objects and $d = 2$ dimensions).

The analyses of VOCs and EtOH extracts have been made available as shared jobs # 1162211 and 1162222, respectively, via the open access platform for metabolomics https://xcmsonline.scripps.edu/.

29.3 Chemometrics of Australian Meliponini Pot-Pollen

The appearances of the pot-pollen were different: in *A. australis* samples formed a dry, compact material of a pale-yellow colour, whereas the *Tetragonula* specimens were of dark-orange to brown colours and showing presence of moisture, possibly indicative of having undergone fermentation (Fig. 29.1).

29.3.1 Volatile Organic Compounds (VOCs)

The individual volatile compounds were identified from the nine pot-pollen samples (Table 29.1), and the VOC chromatographic profiles of three representative samples were reported (Fig. 29.2). Quantities of the compounds were significantly different across the three bee

Fig. 29.1 Pot-pollen from (Aa) *Austroplebeia australis*, (Tc) *Tetragonula carbonaria* and (Th) *T. hockingsi* (Photo: C. F. Massaro)

Table 29.1 Volatile organic compounds from Australian Meliponini pot-pollen analysed by HS-SPME-GC-EI-MS. Relative abundances are percentages (% w/w) of individual compound peak area over total peak areas, mean values, $n = 3$. *KI were calculated as isothermal Kováts indices*

Rt (min)	KI	Compound name[b]	MW	Formula	*Austroplebeia australis*	*Tetragonula carbonaria*	*Tetragonula hockingsi*
2.36	632	Acetic acid	60	$C_2H_4O_2$	–	3.8 ± 3.8	24.9 ± 2.4
4.12	656	Hexanal	100	$C_6H_{12}O$	10.7 ± 6.5	0.7 ± 0.4	0.5 ± 0.5
6.38	687	α-Pinene	136	$C_{10}H_{16}$	1.6 ± 1.6	4.4 ± 2.0	1.5 ± 0.3
9.13	657	Nonanal	142	$C_9H_{18}O$	7.5 ± 1.2	0.1 ± 0.1	–
9.75	677	4-Ketoisophorone	152	$C_9H_{12}O_2$	0.7 ± 0.7	–	–
11.01	1220	*cis*-Geraniol	154	$C_{10}H_{18}O$	0.8 ± 0.8	–	–
11.36	1233	Geraniol	154	$C_{10}H_{18}O$	4.5 ± 0.7	0.2 ± 0.2	0.1 ± 0.1
11.4	1235	*p*-Anisaldehyde	136	$C_8H_8O_2$	6.7 ± 3.2	0.2 ± 0.2	0.2 ± 0.2
11.6	1242	α-Citral	152	$C_{10}H_{16}O$	0.7 ± 0.5	–	–
11.8	1249	Benzemethanol[a]	138	$C_8H_{10}O_2$	1.2 ± 1.0	0.1 ± 0.1	0.1 ± 0.1
13.12	1297	α-Copaene	204	$C_{15}H_{24}$	1.7 ± 1.2	1.8 ± 0.9	1.7 ± 0.5
13.58	1416	1H-Cycloprop[e]azulene	204	$C_{15}H_{24}$	1.4 ± 0.4	7.7 ± 0.3	6.4 ± 0.4
13.73	1422	Caryophyllene	204	$C_{15}H_{24}$	–	1.3 ± 0.9	1.0 ± 0.2
13.83	1426	Sesquiterpene	204	$C_{15}H_{24}$	1.3 ± 0.6	0.4 ± 0.2	0.2 ± 0.2
13.89	1429	Sesquiterpene	204	$C_{15}H_{24}$	1.0 ± 0.6	8.7 ± 0.2	7.4 ± 0.4
13.96	1432	Sesquiterpene	204	$C_{15}H_{24}$	1.1 ± 0.6	21.3 ± 0.6	17.1 ± 0.3
14.17	1440	Sesquiterpene	204	$C_{15}H_{24}$	0.1 ± 0.1	2.5 ± 0.2	2.1 ± 0.1
14.25	1443	Alloaromadendrene	204	$C_{15}H_{24}$	0.6 ± 0.6	15.1 ± 1.1	11.8 ± 0.4
14.56	1456	Hydrocarbon	–	–	5.0 ± 0.9	2.8 ± 1.0	4.0 ± 2.4
14.7	1462	(+)-Ledene	204	$C_{15}H_{24}$	–	6.9 ± 0.6	4.7 ± 0.3
14.89	1470	Sesquiterpene	204	$C_{15}H_{24}$	–	0.8 ± 0.1	0.6 ± 0.0
14.97	1473	Sesquiterpene	204	$C_{15}H_{24}$	0.8 ± 0.4	2.4 ± 0.3	1.3 ± 0.3
15.09	1478	Sesquiterpene	204	$C_{15}H_{24}$	1.7 ± 0.2	0.7 ± 0.2	–
15.15	1480	*cis*-α-Bisabolene	204	$C_{15}H_{24}$	0.6 ± 0.6	0.1 ± 0.1	–
15.46	1493	Epiglobulol	222	$C_{15}H_{26}O$	–	0.3 ± 0.0	0.1 ± 0.1
15.68	1602	(−)Spathulenol	220	$C_{15}H_{24}O$	0.5 ± 0.5	1.3 ± 0.4	1.0 ± 0.2
15.74	1605	Cayophyllene oxide	220	$C_{15}H_{24}O$	2.4 ± 1.8	3.8 ± 1.1	3.2 ± 0.4
15.91	1613	Hydrocarbon	–	–	0.1 ± 0.1	0.5 ± 0.4	0.1 ± 0.1
16.58	1643	Hydrocarbon	–	–	–	1.1 ± 0.9	2.3 ± 2.0
16.64	1646	Hydrocarbon	–	–	8.7 ± 6.2	0.5 ± 0.4	3.4 ± 2.0
16.87	1657	Hydrocarbon	–	–	1.5 ± 1.0	0.5 ± 0.2	0.4 ± 0.2
18.96	1857	Hydrocarbon	–	–	1.7 ± 1.0	0.1 ± 0.1	0.3 ± 0.2
20.16	1819	Methanone[c]	224	$C_{16}H_{16}O$	1.6 ± 1.6	3.2 ± 3.0	0.2 ± 0.1
20.31	1827	Labd-14-ene	290	$C_{20}H_{34}O$	–	0.8 ± 0.5	0.4 ± 0.1

[a]4-methoxy-
[b]Tentative identification against NIST library and literature (Babushok et al. 2011)
[c][3-(1-methylethyl)phenyl]phenyl-methanone against NIST library

species (Fig. 29.3, mean ± sem, $n = 3$) (Tukey's multiple comparisons test, number of families 34, **, $p < 0.05$).

Three characteristic compounds were found across the pollen samples: acetic acid (Rt = 2′36) was a predominant VOC detected in *T. hockingsi* pollen (24.9 ± 2.4% w/w) but found only in one sample from *T. carbonaria* (11.5, w/w). In contrast, *A. australis* pot-pollen did not contain acetic acid. The pot-pollen of *A. australis* was characterized by *p*-anisaldehyde (Rt = 11′40, 6.7 ± 3.2%), a floral marker previously identified by HS-GC-MS and electroantennography as mediating the plant-pollinator

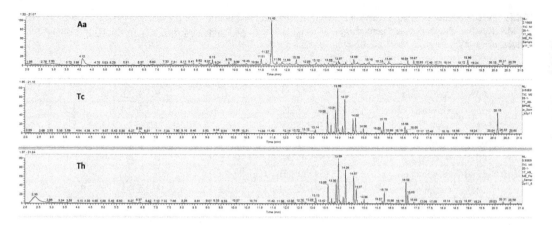

Fig. 29.2 GC-MS profiles of three representative samples of pot-pollen from *A. australis* (Aa), *T. carbonaria* (Tc) and *T. hockingsi* (Th) nests

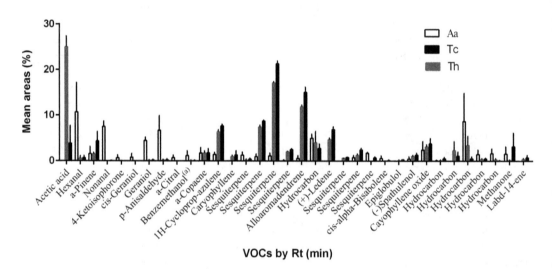

Fig. 29.3 Quantitative analyses of VOCs found in nine pot-pollen samples of *A. australis* (Aa), *T. carbonaria* (Tc) and *T. hockingsi* (Th) nests. The relative amounts (%) were plotted against the individual compounds sorted by their retention time (Rt) of elution by GC-MS using Prism GraphPad software

interactions of solitary wild bees foraging on strawberry crops (Klatt et al. 2013). Furthermore, one *T. carbonaria* sample contained a large quantity of a methanone derivative (Rt = 20.16, 9.12%) that was tentatively identified based on its fragmentation pattern and molecular fragment ion of *m/z* 224. The late elution of this VOC suggests that it has a large structure, and thus further work is warranted for a conclusive identification.

Some hydrocarbon compounds were detected from retention time (Rt 15.9 to 18.96 min) and fragmentation patterns in all samples in all three species but were not identified. The VOC hydrocarbons (nonanol, nonanal, nonanoic acid) have been identified as distinguishing markers of *Eucalyptus* honey by HS-SPME-GC (reviewed by Cuevas-Glory et al. 2007). Further investigation of the hydrocarbons in VOCs of pot-pollen might indicate specific plant associations.

29.3.2 Secondary Metabolites by LC-UV-HRMS/MS

Phenolics are less suited to direct GC-EI-MS analysis because they require a derivatization treatment to increase volatility, although multiple silylated derivatives can be detected to enable a tentative identification of the parent compound. In contrast, phenolic compounds including flavonoids with or without glycosylated moieties can be efficiently analysed by LC techniques (Ferretes et al. 2015). Pot-pollen constituents that were soluble in ethanol were extracted then separated by liquid chromatography (Fig. 29.4), tentatively identified (Table 29.2) and quantitated across the bee species (Table 29.3). There was no significant difference of the overall LC-ESI(−)-MS chemical profiles or individual constituents among bee species by two-way ANOVA.

Flavonoids have been studied previously as biochemical markers to determine the plant origin of honey bee pollen from almond and herbaceous species (Tomás-Barberán et al. 1989). Moreover, the occurrence of O-glycosyl-flavonoids and phenolics in Australian T. *car-*

bonaria pot-honey were formerly investigated by ESI(−)-MS/MS (Truchado et al. 2015; Massaro et al. 2014b). In this study, several pot-pollen constituents showed deprotonated molecular ions that were indicative of glycosylated phenolics or flavonoids with molecular fragments diagnostic of the loss of a glycosyl moiety as hexosyl (120 Da and 162 Da) or pentosyl (− 132 Da) unit. While a 120 Da loss can indicate an internal cleavage of the hexose, the number and the position of such sugar units remain a matter of further investigation, for conclusive identification of glycosides in Australian Meliponini pot-pollen. Flavonoid aglycones and phloroglucinol-derivatives were previously reported in T. *carbonaria* products (Massaro et al. 2014a, 2015; Nishimura et al. 2016). The mass spectra of trace constituents were found to contain these characteristic moieties obtained by deconvolution of the TIC chromatogram (Fig. 29.5). Further work is on-going to identify the unknown compounds.

Toxic pyrrolizidine alkaloids (PA) such as echimidine (m/z 398.2184, $C_{20}H_{31}NO_7$) were reported in honey bees' honey (Lucchetti et al. 2016)

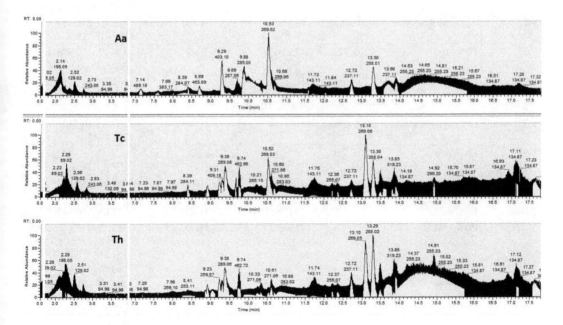

Fig. 29.4 HPLC-ESI(−)-MS profiles of three representative samples of pot-pollen from A. *australis* (Aa), T. *carbonaria* (Tc) and T. *hockingsi* (Th) nests

Table 29.2 Compounds of ethanol extracts from pot-pollen of Australian Meliponini analysed by HPLC-DAD-MS/MS

Rt (min)	Compound name	UV$_{max}$	Found [M-H]$^-$	Calculated formula	MS and MS/MS @ cid30.00 (m/z)
2.3	Glycosyl-phenolic[a]	205sh	431.1417	$C_{15}H_{27}O_{14}$	271.01040 ($C_9H_{19}O_9$); 269.0886 (Δ = 162 Da); 181.0722
7.14	Glycosyl-phenolic	–	533.0938	$C_{24}H_{21}O_{14}$	489.10: 285.0437
7.47	Phenolic	205, 260, 290	–	–	447.0935 ($C_{17}H_{21}O_{10}$ or $C_{22}H_{15}N_4O_7$)
7.59	Glycosyl-phenolic	260, 290	579.1426	$C_{26}H_{27}O_{15}$	349.12: 229.0070 (Δ = 120 Da)
7.71	Unknown	205, 260, 290	533.0936	–	–
7.93	Unknown	205, 260, 290	593.987	–	–
8.41	Glycosyl-flavonoid	260, 290	447.0932	$C_{21}H_{19}O_{11}$	447.09: 284.0683 (Δ = 162 + 1H Da, $C_{16}H_{12}O_5$); 151.0387 ($C_8H_7O_3$)
8.68	Glycosyl-flavonoid	205, 260, 290	463.0883	$C_{21}H_{19}O_{12}$	303.0512 (Δ = 162 + 2H Da, $C_{15}H_{11}O_7$); 463.09: 301.0083
8.91	Glycosyl-flavonoid	290	433.1139	$C_{21}H_{21}O_{10}$	433.11: 271.0499 (Δ = 162 Da); 151.0068 ($C_7H_3O_4$)
9.22	Glycosyl-phenolic	260	409.1706	$C_{17}H_{29}O_{11}$	379.17: 259.1056 (Δ = 120 Da, $C_8H_{19}O9$); 119.1869
9.27	Glycosyl-flavonoid	205, 260, 290	403.1036	$C_{20}H_{19}O_9$	403.10: 271.0197 (Δ = 132 Da)
9.37	Flavonoid	205, 260, 290	475.1651	$C_{17}H_{31}O_{15}$	439.19: 289.0782 ($C_8H_{17}O_{11}$); 135.0208
9.66	Glycosyl-phenolic	205, 260sh	–	–	582.2624 ($C_{36}H_{38}O_7$); 287 ($C_{15}H_{11}O_6$); 125.0242
9.72	Glycosyl-phenolic	205, 260, 290	–	–	287 ($C_{15}H_{11}O_6$); 125.0242
9.91	Unknown	205, 260, 290	285.0408	$C_{15}H_9O_6$	–
10.53	Flavonoid	205, 260–290[b]	269.0453	$C_{15}H_9O_5$	269.05: 148.9706
10.68	Flavonoid	–	271.0611	$C_{15}H_{11}O_5$	271.06: 150.9456
13.09	Flavonoid	–	459.1677	$C_{24}H_{27}O_9$	339.1998 ($C_{15}H_{31}O_8$); 269.0816 ($C_{16}H_{13}O_4$); 134.8661; 269.08: 164.9674
13.35	Flavonoid	–	339.2016	$C_{15}H_{31}O_8$	269.1078; 223.0974 ($C_{12}H_{15}O_4$); 134.8661; 223.10: 208.0392 ($C_{10}H_8O_5$)

(continued)

Table 29.2 (continued)

Rt (min)	Compound name	UV$_{max}$	Found [M-H]$^-$	Calculated formula	MS and MS/MS @ cid30.00 *(m/z)*
13.49	Unknown[b]	–	539.2085	C$_{33}$H$_{31}$O$_7$	473.1832; 319.2279 (C$_{20}$H$_{31}$O$_3$)
13.87	Unknown[b]	n.d.	552.2871	C$_{36}$H$_{40}$O$_5$	319.2280 (C$_{15}$H$_{27}$O$_7$); 237.1135; 134.8665
14.92	Unknown[a]	n.d.	539.2083	C$_{33}$H$_{31}$O$_7$	–
17.14	Phloro-flavanone[b, c]	n.d.	519.2399	C$_{31}$H$_{35}$O$_7$	277.2180; 134.8655; 519.24: 269.0824 (C$_{16}$H$_{13}$O$_4$); 181.0869 (C$_{10}$H$_{13}$O$_3$)
17.61	Ficifolidione[b, c]	n.d.	471.2763	C$_{28}$H$_{39}$O$_6$	385.2757 (C$_{25}$H$_{37}$O$_3$)

Legend:"–": not detected; "sh": shoulder; "Δ": loss of glycosyl moiety
[a]Broad peak did not allow area calculation
[b]Found in trace amounts within samples
[c](Massaro et al. 2014a, 2015)

Table 29.3 Ethanol extracts from Meliponini pot-pollen of *A. australis* (Aa), *T. carbonaria* (Tc) and *T. hockingi* (Th) analysed by HPLC-DAD-MS/MS. Relative abundances are percentages (% w/w) of individual compound peak area over total peak areas, mean values, *n* = 3. *Retention times and compound names as in* Table 29.2

Rt (min)	Compound name	Aa	Tc	Th
2.3	Glycosyl-phenolic[a]	–	–	–
7.14	Glycosyl-phenolic	–	–	–
7.47	Phenolic	–	–	–
7.59	Glycosyl-phenolic	–	–	–
7.71	Unknown	–	5.3 ± 3.3	1.3 ± 0.7
7.93	Unknown	1.5 ± 0.8	–	–
8.41	Glycosyl-flavonoid	2.9 ± 1.0	13.1 ± 3.2	7.0 ± 7.0
8.68	Glycosyl-flavonoid	4.6 ± 3.1	–	–
8.91	Glycosyl-flavonoid	5.0 ± 5.0	4.8 ± 2.5	1.6 ± 1.6
9.22	Glycosyl-phenolic	6.8 ± 4.0	11.1 ± 7.8	3.6 ± 3.6
9.27	Glycosyl-flavonoid	7.5 ± 6.0	1.6 ± 1.6	–
9.37	Flavonoid	–	4.2 ± 2.9	14.8 ± 7.3
9.66	Glycosyl-phenolic	–	17.0 ± 17.0	17.8 ± 8.9
9.72	Glycosyl-phenolic	5.1 ± 2.1	9.7 ± 5.1	16.9 ± 16.9
9.91	Unknown	37.4 ± 4.6	22.8 ± 9.5	3.7 ± 1.9
10.53	Flavonoid	28.1 ± 8.8	1.3 ± 1.3	–
10.68	Flavonoid	1.1 ± 1.1	5.2 ± 2.9	26.3 ± 11.0
13.09	Flavonoid	–	–	–
13.35	Flavonoid	–	–	–
13.49	Unknown[b]	–	–	–
13.87	Unknown[b]	–	–	–
14.92	Unknown[a]	–	–	–
17.14	Phloro-flavanone[b, c]	–	–	–
17.61	Ficifolidione[b, c]	–	–	–

[a]Broad peak did not allow area calculation
[b]Found in trace amounts within samples
[c](Massaro et al. 2014a, 2015)

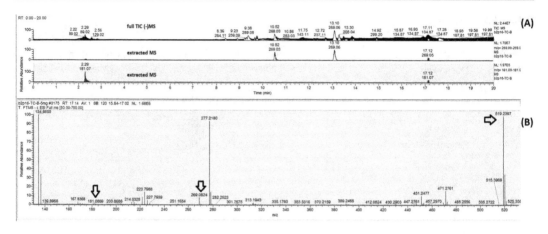

Fig. 29.5 ESI-(−)-HRMS of phenolic eluting at Rt 17.14 min. (**a**) Full and extracted chromatograms. (**b**) Mass spectrum of the deprotonated molecular ions at *m/z* 519 and the two diagnostic fragments (*m/z* 181 and *m/z* 269, *arrows*) matching syncarpic acid and a flavanone, respectively. TIC, total ion current

including Australian sources (Griffin et al. 2015). In separate analyses by ESI(+)-MS, we did not detect any protonated molecular ions that could suggest the presence of these compounds in Meliponini pot-pollens of this study (not shown). Further work using analytical grade standards is warranted. Moreover, whether *Echium* pollen or nectar are found within the Australian flora and available to stingless bees for foraging or could they be among their foraging choices, remain to be determined.

29.3.3 Chemometrics Using Open Source Data

An earlier study compared the pollen foraging activities of *A. australis* and *T. carbonaria* throughout 2012 and 2013, showing that these species allocated similar proportions of pollen foragers and yielded similar pollen intakes during summer (January in Australia) (Leonhardt et al. 2014a). The hypothesis tested in this chapter was therefore that Australian Meliponini bees would show no difference in pot-pollen composition, based on the assumption that these stingless bees forage within a similar flight range (approx. 500 m radius) and thus would have access to the

same botanical sources located within an area abundant in plant-pollen sources. Two-way ANOVA with Tukey's multiple comparisons confirmed that there was no difference in phenolic profiles across pot-pollen of the three Meliponine species ($P = 0.9945$, F (2, 66) = 0.005538). However, the VOC profiles differed significantly across pot-pollens of three bee species due to the predominance of characteristic chemical components in some samples by two-way ANOVA with Tukey's multiple comparisons ($P = 0.0052$, F (2, 204) = 5.404, **).

The PCA plots would indicate similar variances in the majority of samples from the three bee species in their VOC (Fig. 29.6) and phenolic profiles (Fig. 29.7). The first two principal components explained 40% and 46% of the cumulative variances of data, respectively.

Further studies are warranted to include other chemical profiles to better discriminate Australian pot-pollen by PCA. A similar approach was previously applied to honey bee pollen that was investigated for its lipid content, where the pollen samples were better discriminated after introducing data from the fatty acid and carotenoid fractions (Mărgăoan et al. 2014). Moreover, chemical analyses of pollen from floral sources and bees' corbiculae loads are recommended by the

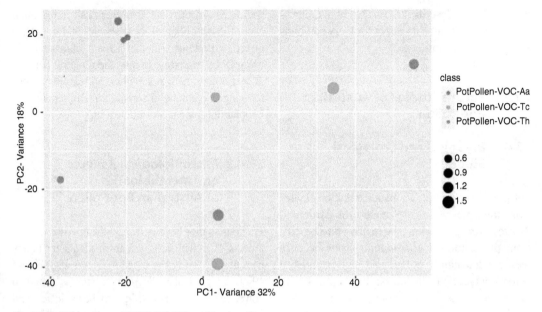

Fig. 29.6 PCA scores of VOC-GC-MS profiles ($n = 9$)

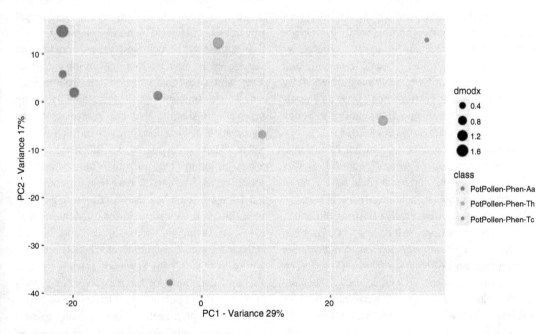

Fig. 29.7 PCA scores of LC-MS profiles ($n = 9$)

methodologies described in this study to contribute to the open source database for PCA comparison of chemical similarities.

29.4 Future Studies of Australian Pot-Pollen

29.4.1 Botanical and Ecological Studies

Pot-pollen is a complex substance that originates from multiple botanical sources. In particular, *Tetragonula* species have been shown to be efficient pollinators of *Macadamia* crops because they collect mainly pollen and can establish a better contact with the stigma in comparison to honey bees that are of larger size and collect mainly nectar from this plant (Heard 1999). Previous chemical research on Australian honey bee pollen has described high levels of lipids that originated from botanical species including *Hypochaeris radicata* (Asteraceae) and *Brassica napus*, *Sisymbrium officinale* and *Rapistrum rugosum* (Brassicaceae), as highly favoured by foraging Australian honey bees (Somerville and Nicol 2006). Similarly, polyunsaturated fatty acids and carotenoids have been used as chemical markers for botanical identifications in honey bees in Europe (Mărgăoan et al. 2014).

In this study, Meliponini foragers of the nine colonies had access to several botanical sources within their foraging range that could have contributed to their pot-pollen, including *Acacia* (Fabaceae, Mimosoideae), *Agapanthus* (Amaryllidaceae), *Backhousia citriodora* (Myrtaceae), *Callistemon* (Myrtaceae), *Crinum asiaticum* var. *pedunculatum* (Amaryllidaceae), *Corymbia torelliana* (Myrtaceae), *Davidsonia pruriens* (Cunoniaceae), *Dianella* (Asphodelaceae), *Eucalyptus* (Myrtaceae), *Grevillea robusta* (Proteaceae), *Hibbertia* (Dilleniaceae), *Jacaranda* (Bignoniaceae), *Leptospermum* (Myrtaceae), *Macadamia* (Proteaceae), *Salvia* (Lamiaceae), *Senna clavigera* and *S. acclinis* (Fabaceae, Caesalpinioideae) and *Syzygium smithii* (Myrtaceae). Findings of our work can provide an initial chemical tool for botanical and ecological studies in Australian Meliponini. More studies are warranted to increase the chemical information about the pot-pollen of these species and the plant-insect interactions. Another stream of research is using DNA-barcoding to map the foraging patterns in Australian native stingless and solitary bees foraging in natural and agricultural landscapes (SD Leonhardt personal communication).

29.4.2 Microbiological Aspects and Metabolomics of Australian Pot-Pollen

Previous studies have evaluated the microbiological aspects of pollen collected from plants by honey bees that store it as bee bread in their nests (Gilliam 1979). In particular, *Bacillus* organisms found in honey bee bread can contribute to the lactic acid and acetic acid contents in honey bee bread (Gilliam 1979). The authors demonstrated that honey bee pollen can inhibit the bacterial growth of *Bacillus* spp. by multiple factors that derive both from the bee and the plant; however, the authors reported that fungal growth was not affected by honey bee pollen.

In stingless bee colonies, yeasts, moulds and bacteria are the main microorganisms inhabiting the nests, and such non-pathogenic microorganisms can secrete antibiotic substances, directly or by enzymatic mediation, that can contribute to the preservation of their food provisions such as pot-pollen (Menezes et al. 2013). Biochemical processes including the microbial fermentation of pot-pollen are likely to increase both the nutritional and defensive values and thus contribute to the overall health of the colony (reviewed by Menezes et al. 2013). In Australian Meliponini, lactobacilli were found in the gut microbiota of adult bees (Leonhardt and Kaltenpoth 2014b). However, either the microbiome of their nests or the metabolomics aspects associated to pot-pollen are yet to be investigated.

In this chapter, acetic acid was found to be a characteristic metabolite of pot-pollen in nests of *Tetragonula* species and likely indicative of a fermentative process occurring during storage. Moreover, Australian Meliponini store pot-pollen within pot- walls made of stingless bee propolis that can preserve the pollen content possibly shape the microbial communities. Interestingly, the dry-stored pot-pollen of *A. australis* might result from a dehydration process that is yet to be defined as in other Meliponini species, i.e. *Frieseomelitta varia* and *Tetragonisca angustula*, or from the activity of *Candida* yeasts similarly to *Ptilotrigona lurida* pot-pollen (reviewed by Menezes et al. 2013). Future experimental work will elucidate the microflora associated to pot-pollen of the three Australian Meliponini of this chapter, their functional roles and the metabolomics characterizing their pollen-provisions in storage.

Beyond the presence of beneficial symbionts, bee pollen can harbour pathogenic microorganisms that can secrete potentially hazardous secondary metabolites. The presence of mycotoxins produced by strains of *Aspergillus* as well as *Fusarium* and *Penicillium* species releasing trichothecenes have been reported in honey bee pollen (Medina et al. 2004; Rodríguez-Carrasco et al. 2013). Any antifungal properties of the pot-pollen

or the potential levels of fungal toxins are yet to be investigated or described in Australian Meliponini.

29.4.3 Recent Trends in Australian Meliponiculture

Modern practices of Australian stingless beekeeping, or meliponiculture, have increased substantially in the last three decades (Heard 1988, 2016), with novel designs of boxes that have enabled colony splitting and propagation of domesticated native bees. Moreover, the opportunity to access the Australian Meliponini honey has attracted the attention of the public, demanding novel strategies that can ensure higher yields by sustainable means. In the case of pot-pollen collection, novel tools have been attempted.

Stingless beehive frames have been made available using 3D printing technology with

Fig. 29.8 Top view of the "Bob the Beeman" frame showing pot-pollen (*arrow*) inside a *T. hockingsi* beehive for meliponiculture in Australia (Photo: R. Luttrell)

FDA approved food-grade plastics (Fig. 29.8). They can be used for pollen collection from Australian stingless bees kept in hives, with the advantage of limiting the disturbance as well as the loss of bees during the harvest (R. Luttrell, personal communication). The pollen frames are positioned close to the entrance and the brood core in *Tetragonula* nests to enhance the storage of pot-pollen within the regular grid. Importantly, the frames encourage the bees to add layers of propolis within a framework to mimic the natural pots for pollen and honey storage for food preservation observed in a nesting layered structure. In terms of pollen production, anecdotal observations have indicated that *T. hockingsi* and *T. carbonaria* can have a high pollen production per hive unit and thus can afford a continuous harvest (R. Luttrell pers. comm.). Whether pollen production can be attempted from *A. australis* bees kept in hives, or the potential for harvest using the frames described, is yet to be determined.

29.5   Conclusions

A chemometrics approach was applied to the analysis of secondary metabolites found in pot-pollen from three species of Australian Meliponini that forage within the same geographical area during the Australian summer. The pot-pollen was compared by differential profiling and chemical contents of the volatile and phenolic extracts. No toxic pyrrolizidine alkaloids have been detected in pot-pollen samples harvested in areas of this study. The VOC profiles differed for the presence of characteristic compounds, including acetic acid as a byproduct of a likely fermentative process in *Tetragonula* pot-pollen that is preserved by the propolis-walls. Findings of this study are available via an open access platform to foster future data-sharing to implement the phytochemical information and metabolomics of Australian pot-pollen.

Acknowledgements The pot-pollen samples were freshly collected from local stingless bee hives and kindly donated by Mr. Robert Luttrell, Highvale, Queensland, Australia. The data reported in this paper were obtained at the Central Analytical Research Facility (CARF) operated by the Institute for Future Environments at Queensland University of Technology. We are thankful to David Marshall at CARF for technical support during the LC-MS analyses and to Fabio Borello for helping with XCMS-PCA analysis. Useful editorial annotations by P. Vit and D.W. Roubik are appreciated.

References

Babushok V, Linstrom P, Zenkevich I. 2011. Retention indices for frequently reported compounds of plant essential oils. Journal of Physical and Chemical Reference Data 40: 043101.

Bankova V, Popova M. 2007. Propolis of stingless bees: a promising source of biologically active compounds. Pharmacognosy Review 1: 88–92.

Baroni MV, Nores ML, Díaz MDP, Chiabrando GA, Fassano JP, Costa C, Wunderlin DA. 2006. Determination of volatile organic compound patterns characteristic of five unifloral honey by solid-phase microextraction– gas chromatography– mass spectrometry coupled to chemometrics. Journal of Agricultural and Food Chemistry 54: 7235–7241.

Berenbaum MR, Johnson RM. 2015. Xenobiotic detoxification pathways in honey bees. Current Opinion in Insect Science 10: 51–58.

Cuevas-Glory LF, Pino JA, Santiago LS, Sauri-Duch E. 2007. A review of volatile analytical methods for determining the botanical origin of honey. Food Chemistry 103: 1032–1043.

de Rijke E, Out P, Niessen WM, Ariese F, Gooijer C, Udo AT. 2006. Analytical separation and detection methods for flavonoids. Journal of Chromatography A 1112: 31–63.

Eckhardt M, Haider M, Dorn S, Müller A. 2014. Pollen mixing in pollen generalist solitary bees: a possible strategy to complement or mitigate unfavourable pollen properties? Journal of Animal Ecology 83: 588–597.

Gilliam M. 1979. Microbiology of pollen and bee bread: the genus Bacillus. Apidologie 10: 269–274.

Griffin CT, Mitrovic SM, Danaher M, Furey A. 2015. Development of a fast isocratic LC-MS/MS method for the high-throughput analysis of pyrrolizidine alkaloids in Australian honey. Food Additives & Contaminants: Part A 32: 214–228.

Halcroft M, Spooner-Hart R, Dollin LA. 2013. Australian stingless bees. pp. 35–72. In: Vit P, Pedro SRM, Roubik D, eds. Pot-Honey: A legacy of stingless bees. Springer. New York, USA. 654 pp.

Heard T. 1988. Propagation of hives of *Trigona carbonaria* Smith (Hymenoptera: Apidae). Austral Entomology 27: 303–304.

Heard TA. 1999. The role of stingless bees in crop pollination. Annual review of entomology 44: 183–206.

Heard TA. 2016. The Australian native bee book. Sugarbag Bees. Brisbane. 264 pp.

Hockings HJ. 1883. Notes on two Australian species of *Trigona*. Transactions of the Royal Entomological Society of London XI: 149–157.

Klatt BK, Burmeister C, Westphal C, Tscharntke T, von Fragstein M. 2013. Flower volatiles, crop varieties and bee responses. PLoS One 8: e72724.

Komosinska-Vassev K, Olczyk P, Kaźmierczak J, Mencner L, Olczyk K. 2015. Bee pollen: chemical composition and therapeutic application. Evidence-Based Complementary and Alternative Medicine. Article ID 297425: 1-6.

Leonhardt SD, Heard TA, Wallace H. 2014a. Differences in the resource intake of two sympatric Australian stingless bee species. Apidologie 45: 514–527.

Leonhardt SD, Kaltenpoth M. 2014b. Microbial communities of three sympatric Australian stingless bee species. PLoS One 9: e105718.

Leonhardt SD, Wallace HM, Schmitt T. 2011. The cuticular profiles of Australian stingless bees are shaped by resin of the eucalypt tree *Corymbia torelliana*. Austral Ecology 36: 537–543.

Lucchetti MA, Glauser G, Kilchenmann V, Dübecke A, Beckh G, Praz C, Kast C. 2016. Pyrrolizidine Alkaloids from *Echium vulgare* in honey originate primarily from floral nectar. Journal of Agricultural and Food Chemistry 64: 5267–5273.

Mărgăoan R, Mărghitaş LA, Dezmirean DS, Dulf FV, Bunea A, Socaci SA, Bobiş O. 2014. Predominant and secondary pollen botanical origins influence the carotenoid and fatty acid profile in fresh honey bee-collected pollen. Journal of Agricultural and Food Chemistry 62: 6306–6316.

Massaro C, Katouli M, Grkovic T, Vu H, Quinn RJ, Heard TA, Carvalho C, Manley-Harris M, Wallace H, Brooks P. 2014a. Anti-staphylococcal activity of *C*-methyl flavanones from propolis of Australian stingless bees (*Tetragonula carbonaria*) and fruit resins of *Corymbia torelliana* (Myrtaceae). Fitoterapia 95: 247–257.

Massaro CF, Shelley D, Heard TA, Brooks P. 2014b. In vitro antibacterial phenolic extracts from 'sugarbag' pot-honeys of Australian stingless bees (*Tetragonula carbonaria*). Journal of Agricultural and Food Chemistry 62: 12209–12217.

Massaro CF, Smyth TJ, Smyth WF, Heard T, Leonhardt SD, Katouli M, Wallace HM, Brooks P. 2015. Phloroglucinols from anti-microbial deposit-resins of Australian stingless bees (*Tetragonula carbonaria*). Phytotherapy Research 29: 48–58.

Massaro FC, Brooks PR, Wallace HM, Russell FD. 2011. Cerumen of Australian stingless bees (*Tetragonula carbonaria*): gas chromatography-mass spectrometry fingerprints and potential anti-inflammatory properties. Naturwissenschaften 98: 329–337.

Medina Á, González G, Sáez JM, Mateo R, Jiménez M. 2004. Bee pollen, a substrate that stimulates ochratoxin A production by *Aspergillus ochraceus* Wilh. Systematic and applied microbiology 27: 261–267.

Menezes C, Vollet-Neto A, Contrera FAFL, Venturieri GC, Imperatriz-Fonseca VL. 2013. 153-171 pp. In: Vit P,

Pedro SRM, Roubik D, eds. Pot-Honey: A legacy of stingless bees.Springer. New York, USA. 654 pp.

Milborrow B, Kennedy J, Dollin A. 1987. Composition of wax made by the Australian stingless bee *Trigona australis*. Australian Journal of Biological Sciences 40: 15–26.

Nishimura E, Murakami S, Suzuki K, Amano K, Tanaka R, Shinada T. 2016. Structure determination of monomeric phloroglucinol derivatives with a cinnamoyl group isolated from propolis of the stingless bee, *Tetragonula carbonaria*. Asian Journal of Organic Chemistry 5: 855–859.

NIST Standard Reference Database Mass Spectral Library with Search Program. Data Version 2.0, 2011.

Nunes CA, Guerreiro MC. 2012. Characterization of Brazilian green propolis throughout the seasons by headspace GC/MS and ESI-MS. Journal of the Science of Food and Agriculture 92: 433–438.

Rasmussen C. 2008. Catalog of the Indo-Malayan/ Australasian stingless bees (Hymenoptera: Apidae: Meliponini). Citeseer. Auckland.

Rayment T. 1935. A cluster of bees: sixty essays on the life-histories of Australian bees. Endeavour Press; Sydney, Australia. pp 539–551.

Rodríguez-Carrasco Y, Font G, Mañes J, Berrada H. 2013. Determination of mycotoxins in bee pollen by gas chromatography–tandem mass spectrometry. Journal of Agricultural and Food Chemistry 61: 1999–2005.

Roubik DW. 1979. Nest and colony characteristics of stingless bees from French Guiana (Hymenoptera: Apidae). Journal of the Kansas entomological Society 52: 443–470.

Rzepecka-Stojko A, Stojko J, Kurek-Górecka A, Górecki M, Kabała-Dzik A, Kubina R, Moździerz A, Buszman E. 2015. Polyphenols from bee pollen: structure, absorption, metabolism and biological activity. Molecules 20: 21732–21749.

Smith JP, Heard TA, Beekman, M., Gloag, R. 2017. Flight range of the Australian stingless bee *Tetragonula carbonaria* (Hymenoptera: Apidae). Austral Entomology 56: 50–53.

Somerville D, Nicol H. 2006. Crude protein and amino acid composition of honey bee-collected pollen pellets from south-east Australia and a note on laboratory disparity. Animal Production Science 46: 141–149.

Tomás-Barberán FA, Tomás-Lorente F, Ferreres F, Garcia-Viguera C. 1989 Flavonoids as biochemical markers of the plant origin of bee pollen. Journal of the Science of Food and Agriculture 47: 337–40.

Truchado P, Vit P, Heard TA, Tomás-Barberán FA, Ferreres F. 2015. Determination of interglycosidic linkages in *O*-glycosyl flavones by high-performance liquid chromatography/photodiode-array detection coupled to electrospray ionization ion trap mass spectrometry. Its application to *Tetragonula carbonaria* honey from Australia. Rapid Communications in Mass Spectrometry 29: 948–954.

Part V

Marketing and Standards of Pot-Pollen

Rural-Urban Meliponiculture and Ecosystems in Neotropical Areas. *Scaptotrigona*, a Resilient Stingless Bee?

30

Sol Martínez-Fortún, Carlos Ruiz,
Natalia Acosta Quijano, and Patricia Vit

30.1 Introduction

Ecuador is one of the countries with highest biodiversity and associated traditional knowledge according to the United Nations Environment Program (UNEP). In addition, it is considered one of the most diverse stingless bees hotspots, including 25% of the species catalogued worldwide, in a territory which occupies 1.5% of Latin America (Coloma 1986). In Chap. 14, Roubik made an extraordinary collection of 100 stingless bee species and 23 genera in a 50 ha parcel of the Yasuní Biosphere Reserve in Ecuador. The following Chap. 15 by Vit et al. reviews 132 species of Ecuadorian stingless bees in 23 genera. In southern Ecuador, more than 70 species in 17 genera occur (Ramírez Romero et al. 2013), some of them traditionally managed in rural-urban areas. However, in recent decades an awareness on decreased traditional stingless bee keeping points to inappropriate management (overharvest, unsuccessful nest transfer and division) and environmental conservation as major issues (Villanueva-Gutiérrez et al. 2005; Cortopassi-Laurino et al. 2006). Villanueva-Gutiérrez et al. (2013) surveyed 60 communities with 155 stingless bee keepers, owners of 1–100 colonies with experience of up to 50 years. They noted that 90% of the "xunan-kab" *Melipona beecheii* colonies of some 20% of the bee keepers in the zona Maya – southeastern Yucatán Peninsula in Mexico – collapsed in the last 25 years.

Meliponini pollen traps did not progress as those used to remove corbicular pollen from *Apis mellifera* hind legs at the nest entrance (Menezes et al. 2012). Therefore pot-pollen is extracted from the storage area inside the nest, not before.

S. Martínez-Fortún
Department of Natural Science, Universidad Técnica Particular de Loja,
San Cayetano Alto s/n C.P, 11 01 608 Loja, Ecuador

C. Ruiz
Department of Natural Science, Universidad Técnica Particular de Loja,
San Cayetano Alto s/n C.P, 11 01 608 Loja, Ecuador

Animal Biology Department, Veterinary Department, Campus de Espinardo, Universidad de Murcia, 30100 Murcia, Spain

N.A. Quijano
Posgraduate in Regional Rural Development, Universidad Autonoma Chapingo, Km 38.5 Carretera México-Texcoco, Texcoco, CP 56230, Mexico

Technical Coordination, Istaku Spinini A.C., Priv.Calle 1 No. 3 Colonia Buena Vista, Xalapa, Veracruz, CP 91080, Mexico

P. Vit (✉)
Apitherapy and Bioactivity, Food Science Department, Faculty of Pharmacy and Bioanalysis, Universidad de Los Andes, Mérida 5101, Venezuela

Cancer Research Group, Discipline of Biomedical Science, Cumberland Campus C42, The University of Sydney, 75 East Street, Lidcombe, NSW 1825, Australia
e-mail: vitolivier@gmail.com

In this chapter, we look at four aspects of meliponiculture in southern Ecuador, eastern Mexico, and southern Venezuela: species of stingless bees, knowledge in management, responsibility in the conservation of natural environments, and knowledge considering pot-pollen. Finally a multifactorial diagram links ecological, economic, and social key components needed to invigorate sustainable meliponiculture.

30.2 Initiatives to Revitalize Stingless Bee Keeping

During 2014–2015, a diagnosis of the state of the art in stingless bee keeping was carried out in the provinces of Loja and Zamora Chinchipe, Ecuador, with 64 stingless bee keepers within 13 districts who were surveyed using questions on (1) demographic and social information, (2) knowledge of stingless bees and uses of products, (3) knowledge on stingless bee management, and lastly (4) evaluation of conservation awareness (Martínez-Fortún 2015). Additionally, five focal groups were chosen, three of them with local stingless bee keepers (15 participants), one with technicians of institutions with projects on stingless bees (3 participants), and one with researchers involved in meliponiculture (5 participants). The main topics explored in these focal groups were (1) uses of bee products, (2) how meliponiculture can contribute to ameliorate the quality of life in rural areas and develop conservation actions of the natural environment, (3) involvement of public and private institutions, and (4) preference of meliponiculture compared to the management of *Apis mellifera* (S. Martínez-Fortún, unpublished data).

The revitalization of meliponiculture and a concomitant responsibility in forest conservation were assessed in three ecosystems of southern Ecuador: (1) Tropical dry forest (Loja Province), an area characterized by a long dry season during half the year, supporting high human pressures, and marked habitat loss and degradation (Portillo-Quintero and Sánchez-Azofeifa 2010), was declared a Biosphere Reserve in 2014 due to its high diversity and high levels of endemism (Parker and Carr, 1992); (2) the western slopes of tropical mountain forest (Loja Province), with a high degree of deforestation due to changes in land use (Tapia-Armijos et al. 2015) and potentially optimal conditions for stingless bee keeping in association with other productive activities such as coffee cultivation; and (3) the eastern slopes of tropical mountain forest (Zamora-Chinchipe Province) in the Amazonian region with high human pressure due to factors such as cattle expansion or extractive activities. In this area, habitat fragmentation and humidity are limiting factors for the stingless bee keeping.

In 2012–2016, a project in eastern Mexico to raise family income and conserve natural resources with *Scaptotrigona mexicana* honey focused on greater autonomy for 125 families from Atzalan (mixed-race) and Zozocolco de Hidalgo (Totonacs) backgrounds. The community promoters actively participated in the processes of harvest, gathering, packaging, and training for marketing. Pot-honey quality control and traceability to verify the origin were encouraged. It is expected that organized producers will continue with minimal external intervention in the next 3 years to achieve coordination from the field to the market.

From a different perspective without external interventions, a Wotuha (indigenous native people) initiative was created in southern Venezuela. Their mask-making-art is based on meliponine cerumen and other materials (clay, bark, palm fiber) from the Amazon. Mr. Alfonso Pérez is the president of the Warime Meliponiculture Cooperative (WMC) with 23 associated stingless bee keepers and 200 colonies in Praia Grande, Amazonas state, Venezuela. A further step was taken in 2013, with the donation of a *Melipona* sp. colony to the primary school Unidad Educativa Alberto Ravell (UEAR) in Praia Grande, directed by Mrs. Mary Luz Camico. A major objective is "Observing nature with a spirit of environmental protection" while also studying the plant origin of pot-honey and pot-pollen and sustainable technology applied to the legacy of stingless bee keeping (Aguilar et al. 2013).

30.3 Traditional Knowledge Involving *Scaptotrigona*

In Table 30.1, the ethnic names of the Ecuadorian species of stingless bees are compared in four studies (1987–2015). As generally observed in Venezuela (Vit 1994), species may have different ethnic names (e.g., *Melipona eburnea*, *Tetragonisca angustula*, *Scaptotrigona* sp., and *Trigona fulviventris*), or one ethnic name "abeja de tierra"

Table 30.1 Ethnic names of stingless bee species in Ecuador

Scientific names of stingless bees	Ethnic names	Coloma (1986)	Chieruzzi Löwenstein (1989)	Ramírez Romero et al. (2013)	Martínez-Fortún (2015)	Vit (2015)[a]
Geotrigona sp.	"abeja de tierra"	x	x	x	x	x
Lestrimellita limao[b] (Smith 1863)	"barbón"			x	x	
Melipona eburnea Friese 1900	"cojimbo"		x		x	
Melipona eburnea Friese 1900	"ergón"			x	x	x
Melipona indecisa Cockerell 1919	"cananambo"		x	x	x	x
Melipona mimetica Cockerell 1914	"bermejo"		x	x	x	x
Nannotrigona sp.	"pitón" "arepe"	x	x	x	x	
Oxitrigona mellicolor (Packard 1869)	"mea fuego"	x		x	x	x
Paratrigona sp.	"pirunga"	x	x	x	x	x
Plebeia sp.	"alpargate"	x		x	x	
Plebeia sp.	"lambeojo"				x	x
Scaptotrigona sp.	"catana"	x	x	x	x	x
Tetragonisca angustula (Latreille 1811)	"angelita"			x	x	
Tetragonisca angustula (Latreille 1811)	"nimbuche"				x	
Trigona sp.	"moroja"				x	x
Trigona fulviventris Guérin 1844	"pulao" "putulunga"			x	x	
Trigona septentrionalis Almeida (1995)	"pichilingue"			x	x	
Trigonisca sp.	"papito blanco"				x	
	"boca de sapo"				x	
–	"languacho"				x	
–	"guacoso"				x	
–	"serrano"				x	
–	"yapamal"				x	
–	"ollotongo"				x	
–	"suca"				x	

[a]See Vit et al. Chap. 15 for all stingless bee species and additional ethnic names
[b]Probably another species (D.W. Roubik, personal communication)

is given to more than one genus, e.g., *Geotrigona* and *Trigona* (Vit et al., Chap. 15). In Martínez-Fortún (2015), 11 ethnic names were new and previously unrecorded by Ecuadorian researchers. Either "new" names were elicited during the social work or they were created in the *rapport* to please overseas investigators. The word "suca" means "blond" in Ecuadorian folk Spanish and could be a descriptor of the bee instead of its ethnic name. However, "rubita" – meaning little blond – is an ethnic name for *Tetragonisca angustula* in Venezuela (Vit, Chap. 24). Similarly, "abeja amarilla" means "yellow bee" and is used to name several species – alone *Trigona* cf. *amazonensis* (Ducke 1916), *Trigona dallatorreana* (Friese 1900), and *Trigona ferricauda* (Cockerell 1917) or besides other ethnic names (*Oxitrigona mellicolor* (Packard 1869), *Partamona aequatoriana* Camargo 1980). Also, "abeja finita" is a descriptor for the thin bee. Knowledge of the richness and behavior of stingless bee species is useful for inventories, to empower traditional meliponiculture and to update management.

Despite the 29 ethnic names of stingless bees elicited in the Ecuadorian questionnaires, meliponiculture is mostly reduced to two species: "catana" *Scaptotrigona* sp. in 85% of cases and/or "bermejo" *Melipona mimetica* in 20% of cases.

Conventional production by *Apis mellifera* is largely standardized, compared to the management of stingless bees, which needs to be adjusted for each species (Heard 2016). For meliponines, particular techniques are based on certain species characteristics (body size, behavior, colony size, nest structures), the territory, and the communities that handle them. Although 70% of all respondents handle the same species as their ancestors, more than 65% have little or no knowledge on the management of the valued but declining *Melipona mimetica* "bermejo." It is noteworthy that 85% perceived harmful anthropic changes in forests where stingless bees thrive, while only 2% did not. All respondents agreed on the responsibility in the care of forests at family and community levels, a fact that must be considered when developing strategies for

recovery and conservation of natural areas and native bees. This implies that stingless bee species conservation requires an ecological and "ecosystem" approach which protects the environment (Reyes-García and Martí-Sanz 2007; Maia-Silva et al., Chap. 7). Traditional knowledge of meliponine species per se does not imply conservation (DW Roubik, personal communication), but sustainable stingless bee keeping may grow as a tool for forest conservation in southern Ecuador.

Guided learning by training or by trial and error (colony division, feeding and pest control) is needed after a colony is transferred into the beehive (Fig. 30.1). Although there is a ban on wild stingless bee nest collections in Ecuador, the unsustainable exploitation of wild colonies from nearby forest continues. This fact, together with the high rates of deforestation of the tropical dry forest, seems to be responsible for the decline and disappearance of *Melipona mimetica* as identified by the rural communities in Loja and Zamora Chinchipe (Ramírez Romero et al. 2013; Martínez-Fortún 2015).

The decrease of *Melipona eburnea* "ergón," *Melipona indecisa* "cananambo," *Melipona mimetica* "bermejo," and *Melipona grandis* "bunga negra" was also indicated by stingless bee keepers from the Ecuadorian provinces El Oro, Pastaza, and Zamora Chinchipe who lost colonies kept for more than 20 years. None or only one or two remain, out of one dozen, in their backyards (P. Vit, personal observation). However, in Santa Elena, a very dry province on the coast, stingless bee keepers surprisingly informed us that the number of *Melipona* colonies is increasing. How can this be true? One informant, a local development official Mr. Freire, provided a possible cause. There has been recent implementation of an irrigation project by the bi-national COOPI, Italy – GADPSE, Ecuador, coordinated by him. The increased water availability has a compulsory clause to keep bees and bee flora; otherwise, irrigation will be unavailable to low-income residents of the community. A Roubikian insight on conservation wisdom to serve the bees and their human "shepherds" arises from rigorous repli-

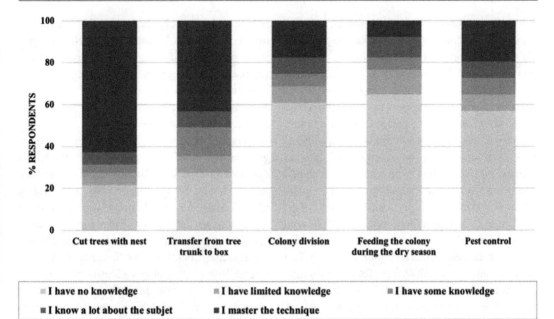

Fig. 30.1 Know-how in meliponiculture in southern Ecuador. (1) Protection of trees with wild nests, (2) nest transfer, (3) colony division, (4) colony feeding, (5) pest control

cated qualitative and quantitative data on pollen species used by bees (see Introduction of this book). Basic measurements on environmental abiotic factors need to be recorded before a "module" or "algorithm" to overcome climate warming is brought forth. The large *Melipona* bees are declining in the rest of rural Ecuador, but here the smaller *Scaptotrigona* bees – with more populated colonies – seem better adapted to environmental changes, because one can see that their colonies have not declined in the backyards of stingless bee keepers.

Food niche overlap between *Melipona mimetica* and *Scaptotrigona* sp. in Ecuadorian tropical dry forest was studied by using corbicular pollen loads stained with safranin. The pollen sample contained 35 genera of 28 plant families (Guerrero-Peñaranda 2016), with marked seasonal differences. In the dry season, a high dominance of a few key plant species was observed (e.g., pollen resources of *Scaptotrigona* sp.: Fabaceae 46%, Bombacaeae – now within Malvaceae – 40%) that contrasts with more diverse resources during rainy season (C. Ruiz, unpublished data).

30.4 One Stingless Bee, *Scaptotrigona*, Preferred over *Melipona*

Although stingless bee keepers of "catana" *Scaptotrigona* sp. may require a veil for protection of the face and head, this bee is relatively easy to manage (abundant feral nests, successful colony transfer, and nest division). Although stingless bee ability to thermoregulate nests is lower than *Apis mellifera*, the brood of *Scaptotrigona depilis* resisted to hot temperatures (Vollet-Neto et al. 2015). Compared to most stingless bee species, this genus is resilient to management and has very populous nests (up to 10,000 adult bees). Feral nests are abundant in disturbed environments and urban areas. *Scaptotrigona* is resistant to certain pests of stingless bees, such as phorid flies and ants. Additionally, *Scaptotrigona* stores large amounts of honey, pollen, and propolis and is useful for pollination of mass-flowering crops, such as coffee *Coffea arabica* (Rubiaceae) and açaí palm *Euterpe oleracea* (Arecaceae) (C. Menezes, personal communication).

"Catana" was chosen within the Ecuadorian project to promote stingless bee keeping in Loja Province, where the Loja Prefecture is working with more than 200 families (A. Rojas, personal communication). A pot-honey appreciated for its putative therapeutic properties is produced by *Melipona mimetica* – a species that is not reared because of its management requirements, a loss of traditional knowledge on rearing, and climate change or competition with exotic Africanized honey bees, now found throughout the drier parts of tropical America, in particular (Roubik 1989; Roubik and Villanueva-Gutiérrez 2009). *Melipona mimetica* is endemic to the dry forests of Ecuador and northern Peru (Coloma 1986). In Fig. 30.2, a feral nesting site is visible from the cerumen tube entrance on a tree trunk, a traditional hive and pot-pollen from storage area in a *Scaptotrigona* sp. nest. The Ecuadorian *Scaptotrigona* stingless bees were collected in isopropyl alcohol, dried, and sent to the Biology Department at the Universidade de São Paulo, Ribeirão Preto, Brazil. Specimens were housed in the Camargo Collection, RPSP, for entomological identification, kindly done by Dr. S.R.M. Pedro. This species is possibly an undescribed *Scaptotrigona*

(HF Schwarz, unpublished data) also widely distributed in Pacific Panama (D. W. Roubik, personal communication). The manuscript name is not used here, in keeping with the recommendations of the International Code of Zoological Nomenclature (ICZN 2016). See Table 30.1 with the ethnic name for these bees.

"Pisilnekmej" is the ethnic name of *Scaptotrigona mexicana* (Guérin-Meneville 1845) with a declared sanctuary in the municipality of Cuetzalan, Puebla, Mexico, since May 2011 during the VII Mesoamerican Seminar on Native Bees. According to the Nahuatl language, this small "pisil" bee "nektsin" is of ethnic pride because "pisilnekmej" helps to keep families united, and it is an icon of community identity, as it needs a peaceful environment. In contrast with the great biodiversity of stingless bees in Ecuador (Roubik, Chap. 14 Roubik, Vit et al., Chap. 15), the traditional selection of one bee – the small *Scaptotrigona* – exemplifies a different approach to meliponiculture. A project in eastern Mexico was initiated in 2012 with 125 families of two ethnic backgrounds: (1) mixed-race in Atzalan and (2) Totonacs in Zozocolco de Hidalgo. Stingless bee keepers participated in workshops of Good

Fig. 30.2 *Scaptotrigona* sp. nest from Ecuador. (1) Nesting site on tree trunk; (2, 3) Cerumen tube-entrances; (4) Pot-pollen; (5) Traditional hives (Photos: P. Vit)

Manufacturing Practice (GMP) in meliponaries (See Fig. 30.3). Pot-honey produced by *Scaptotrigona mexicana* is used as medicinal food (to treat respiratory, digestive, postpartum, and cutaneous diseases), and cerumen is modeled, like clay, for religious decorations (candles and floral motifs) used in traditional ceremonies. Pot-pollen is harvested to treat anemia and for bee colony material supplementation (to strengthen weak or newly divided colonies), just kept in drawers. However, in Puebla it is kept in airtight plastic containers to prevent the attack by the fly *Pseudohypocera kerteszi*, Diptera, and Phoridae (Padilla Vargas et al. 2014).

In southern Venezuela, the Wotuha Warime cooperative founded in 2005 operated with six types of stingless bees from the Amazonian forest: (1) "ajavitte" *Tetragona clavipes* (Fabricius 1804), (2) "isabitto" *Melipona* sp. *aff. fuscopilosa*, (3) "limoncita" *Lestrimelitta maracaia*

Marchi and Melo 2006, (4) "pico de loro" *Melipona* sp., (5) "sonquette" *Scaptotrigona* cf. *ochrotricha* (Buysson 1892), and (6) "tobillo morrocoy" *Melipona* sp. *aff. eburnea* Friese 1900. The species were kindly identified by late Professor J.M.F. Camargo at Universidade de São Paulo, Ribeirão Preto, Brazil. In 2013, more than 100 stingless beehives were from "sonquette" because *Scaptotrigona* was easier to keep. Although *Scaptotrigona* honey and pollen pots are smaller than those of *Melipona* spp., more colonies produce higher yields. The number of "ajavitte" colonies increased in 2017 because the management of *Tetragona clavipes* is easy. See pot-pollen and stingless beehive in Fig. 30.4.

In the Bolivian Amboró National Park, *Scaptotrigona* is more abundant than *Melipona* and produces 2–4 times more pot-pollen. *Scaptotrigona depilis* (Moure 1942) "obobosí,"

Fig. 30.3 *Scaptotrigona mexicana* in Zozocolco, eastern Mexico. (1) Open nest with exposed brood and pot-storage area for honey and pollen, (2) rational hives (Photos: A. Albalat Botana)

Fig. 30.4 *Tetragona clavipes* from Valle Opa, Playa Grande, Amazonas, Venezuela. (1) Pot-pollen and (2) characteristic entrance of a stingless beehive (Photos: P. Vit)

Scaptotrigona polysticta Moure 1950 "suro negro," and *Scaptotrigona* aff. *xanthotricha* (Moure 1950) "suro choco" are used for sustainable meliponiculture by 40 families of the Association of Native Honey Producers (APROMIN, "Asociación de Productores de Miel Nativa") (Ferrufino and Vit 2013). See meliponaries of *Scaptotrigona* and harvested pot-pollen (Fig. 30.5).

In Peru, *Scaptotrigona* is also resilient to anthropogenic changes and management, despite the biting defensive behavior that makes compulsory the use of veil or nighttime management. In a 15-year-old meliponary, *Scaptotrigona* is a robust genus, more resistant than others, and has abundant production of pot-honey, pot-pollen, and propolis (C. Rasmussen, personal communication). See Fig. 30.6 with the bee, her nest, and colonies kept in rectangular PNN (Paulo Nogueira Neto model) wooden boxes.

In Costa Rica, meliponaries are well organized with *Scaptotrigona pectoralis*, to harvest pot-honey and pot-pollen (Fig. 30.7).

30.5 Naming Pot-Honey and Pot-Pollen in Labels of Commercial Products

Attention to pot-pollen in Ecuador has focused on four themes: (1) the description of pots in the nests of *Paratrigona* sp., *Scaptotrigona* sp., and *Plebeia* sp. (Coloma 1986); (2) the traditional uses that are associated, in communities, with pot-pollen of *Melipona* cf. *fasciata* for stomach pain, *Geotrigona* sp. to treat rheumatism and leg pain, and *Nannotrigona* sp. as restorative (Chieruzzi Löwenstein 1989); (3) the differences in acidity of fresh and dried pot-pollen and its conservation (Ramírez Romero et al. 2013); and (4) chemical composition of *Tetragonisca angustula* presented in a congress (Vit 2016).

The volume of individual honey pots was measured in Costa Rica as 20.4 ± 5.6 mL for *Melipona beecheii* "jicote gato" and 2.8 ± 1.1 mL for *Scaptotrigona pectoralis* "soncuano" and in Venezuela 19.7 ± 5.9 mL for *Melipona* sp. "pico'e

Fig. 30.5 Meliponaries of *Scaptotrigona* kept in Carmen Surutú community, Amboró National Park, Santa Cruz de la Sierra Department, Bolivia. (1) *Scaptotrigona polysticta* in a hub, (2) rational hives of *Scaptotrigona depilis*. (3) Harvested pot-pollen is kept in airtight plastic containers (Photos: P. Vit)

loro" and 4.4 ± 0.7 mL for *Scaptotrigona* cf. *ochrotricha* "sonquette" (Aguilar et al. 2013). A similar difference among other *Melipona* and *Scaptotrigona* pollen pot volumes is expected. A larger size of storage pots facilitates pot-honey and pot-pollen harvest. However, Wotuha-Piaroa

Fig. 30.6 Meliponary of *Scaptotrigona* sp. kept near Tarapoto, Peru. (1) The bee; (2) Nest with brood (*center*), pot-pollen (*left*), and pot-honey (*right*) stores; and (3) Rational hives (Photos: C. Rasmussen)

Fig. 30.7 Meliponary of *Scaptotrigona pectoralis* in Santa Ana, San José, Costa Rica (Photo: I. Aguilar)

native people note strong advantages for colony division of *Scaptotrigona*; thus, their meliponaries in Praia Grande, Amazonas state, Venezuela, have become more populated by *Scaptotrigona* than *Melipona*. A Brazilian entrepreneur from Barra do Corda in Maranhão keeps more than

600 colonies of *Scaptotrigona postica* for pot-pollen production (Menezes et al. 2013a, b), and also the botanical sources of the propolis were studied (Souza et al. 2015).

Pot-honey – used as a sweetener, nutritional supplement and medicine, by 91% of Ecuadorian interviewees – and pot-pollen with 60% – are the most used products of meliponiculture in southern Ecuador. The consumption of propolis or cerumen (admixture of beeswax and resins) in these communities does not exceed 30% according to the surveys. Almost 92% of the stingless bee keepers use the bee products for their own consumption and direct sale to neighbors, while only 25% sell to stores in the community, while 18% sell outside the community. Such a situation, in which the main intention is not sale, delays the establishment of brands to explore commercialization channels. Therefore, students of the Faculty of Advertising at Universidad de Las Américas in Quito collaborated with stingless bee keepers. They conducted a market analysis of pot-honey, advertising strategies and brand proposals like "Stingfree," "Mel," and "Hanku." Concurrently, the Prefecture of Loja carried out an initial pot-honey gathering center in a strategic area of the Loja province, and efforts to achieve a pot-honey regulation began by modifying the Ecuadorian honey norm for products of *Apis mellifera* (see Chap. 15 Vit et al.).

In eastern Mexico, the brand "Tiyatkú" was developed for *Scaptotrigona mexicana* in the project "Strengthening the productive chain of bush honey to raise family income and conserve natural resources" (Albalat Botana and Acosta Quijano 2016). However the name for *Scaptotrigona mexicana* honey was "miel de melipona" and resulted in justified confusion. General preservation and commercialization of the entomological origin is complex, and the gap between science and the public needs special attention. There is no such term as "scaptotrigoniculture" for *Scaptotrigona* species because meliponiculture really derives from the Meliponini tribe to which all stingless bees are

assigned by taxonomic standard practice, not only for the species of *Melipona*.

In southern Venezuela, the *Scaptotrigona* pot-honey and pot-pollen have no brand, and after harvest they are kept refrigerated. Pot-honey is dark amber and is packed in clean recycled glass bottles and fresh pot-pollen in airtight plastic containers. Both honey and pollen undergo acid-lactic fermentation (Calaça et al., Chap. 17; Menezes et al., Chap.18). The distinctive sour taste developed along the fermentative process of raw nectar and pollen is appreciated in tropical countries. Indeed, in the lowlands, pot-pollen is mixed with cold water to produce a refreshing "lemonade" (Rivero 1972; Vit 1994). In contrast, the Wichi-lhuskutás ferment water and pot-pollen for 3 days to make "aloja" drink (see Zamudio and Helgert, Chap. 20). However, a pejorative perception is also given to pot-pollen in certain communities: "guateguán" means "guate" excrement and "guane" bee in Cumanagoto native language spoken on the eastern coast (Rivero 1972). Similarly, pot-pollen is so-called "shit of the bees" in central Amazonas state (P. Vit, personal observation). This different perception may refer to different stages of pot-pollen processing and botanical origin, with pollen pots filled either in layers of shiny cream-yellow-orange shades and citrus-like or, less frequently, with a thick shiny brownish material like excrement. Pollen that has been stored for some time is darker in color than is freshly deposited pollen (Menezes et al. 2013a, b).

The Venezuelan poet Guillermo De León Calles valued pre-Columbian honey and praised ancient bee keepers with the expression "honey-cultists" – extended now to "pollen-cultists." An underlying tribute for an ancient art comes from sociocultural anthropology as an interface considered "the most scientific of the humanities, and the most humanistic of the sciences" by Eric Wolf. A recurrent bond for indigenous knowledge and biodiversity in Posey's ethno-entomological legacy lives in a universe of relationships between human and insect societies (Zamudio 2016). This inherent aspect of stingless bee products may conduce the proposal of effective national standards still absent.

30.6 Future of Social Interventions in Meliponiculture

Stingless bee keeping within the Maya civilization (González-Acereto 2008; Jones 2013) has a high ethno-ecological value that integrates certain beliefs and worldviews of the community, traditional knowledge, and a set of productive practices (Toledo and Barrera-Bassols 2008; Vit et al., Chap. 15). Similarly to "xunan-kab" in Yucatan (Villanueva-G et al. 2005; Villanueva-Gutiérrez et al. 2013), "pisilnekmej" is a native bee with a cultural, economic, ecological, and conservational value. *Scaptotrigona* colonies kept by stingless bee keepers in Ecuador, Mexico, and Venezuela have a twofold meaning in rural-urban settlements and natural environments, as a bio-indicator for conservation and a source of nutraceuticals with economic value. Direct involvement of the rural-urban relationship between stingless bee and man is a factor we would like to see promoted in the activities of people residing in the rural-urban environment. Training courses held with stingless bee keepers and experts and free online video tutorials to spread by social networks would be very useful in this context. Furthermore, the practice of this activity in the area would support an increase in the presence of important pollinators for numerous plant species (Slaa et al. 2006, Ollerton et al. 2011), which in turn promotes the conservation of forests and their diversity (Vit 2000). Management of *Scaptotrigona* colonies would benefit with the knowledge on the biology of the bees, including queens (Menezes 2010, Menezes et al. 2013b). Proactive research on natural resources (nectar, pollen, resins), pot-honey (Vit et al. 1998), pot-pollen (Vit et al. 2016), and cerumen (Souza et al. 2015) supports chemical, bioactive, and botanical characterization. Most *Melipona* colonies are used for pot-honey extraction, but *Scaptotrigona* produces more pot-pollen than pot-honey (Menezes et al. 2013a; P. Vit, personal observation).

A diagram in Fig. 30.8 conveys multifactorial implications that include meliponiculture with *Scaptotrigona*. An icon involving *Scaptotrigona* was used for this chapter. Agricultural productivity and living standards should grow in parallel to ecological awareness, while environmental sustainability comprises the conservation of biodiversity and also pollination services. We believe that economic sustainability is fundamental for food sovereignty and contributes to income diversification. Social sustainability is a wide concept with two major rural-urban stingless bee components: health support with medicinal applications of bee products and gender empowerment with increasing participation of women in nonintensive meliponiculture – as well as children and the elderly – promoting intergenerational transfer. Any contribution for a better quality of life from this interactive triangle needs government attention. Otherwise, stingless bee keepers are not integrated into the productive chain and are left more vulnerable to lonely isolated efforts.

Bottom-up projects that involve existing institutions within local communities (Holl 2017) support correct management of stingless bee colonies and improve interest in the care and conservation of wildlands. A successful example of non-timber forest resource management by local communities in the dry Ecuadorian forest is the sustainable harvest of "palo santo" *Bursera graveolens* and distillation of its oil promoted by the Naturaleza y Cultura Internacional Foundation, awarded in 2014 by the UNDP (United Nations Development Program). *Bursera* is also a principal resource of *Melipona* in southeastern Mexico (Villanueva-Gutiérrez et al., Chap. 5).

Among the ideas generated by the surveys are several ecological-economic-social approaches: (1) promotion of the Meli Route project or Route of Living Museums of Meliponini Bees in the World (Vit et al., Chap. 15); (2) the use of stingless bees as pesticide bio-indicators, as in the protocol for neonicotinoid residue extraction and

Fig. 30.8 Meliponiculture
with *Scaptotrigona* has
multifactorial impacts
(Diagram: C. Ruiz)

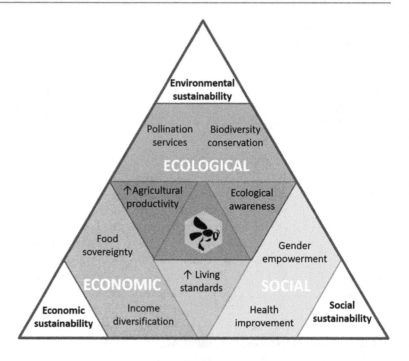

detection in the *Scaptotrigona* aff. *depilis* (Moure 1942) bodies (Souza Rosa et al. 2015); (3) the integration of meliponiculture in agroforestry systems that favor the diversification of resources in the economy (Wolff 2014); (4) the search for new marketing channels for bee products (Jaffé et al. 2015); (5) the update of an inventory of biodiversity of stingless bees in Ecuador using molecular and morphometric techniques (García-Olivares et al. 2015); (6) the valorization of medicinal uses of pot-honey (Ocampo 2013; Vit et al. 2015) and pot-pollen; (7) the chemical characterization of pot-pollen (Vit et al. 2016); and (8) the stingless bee keepers will benefit in economic, cultural, health, and food sovereignty.

An invigorated meliponiculture contributes to (1) rescue traditional knowledge damaged by acculturative processes, (2) provide alternatives for sustainable development, (3) empower rural-urban women and lessen the generational gap, (4) encourage a transition to organic farming, (5) promote care and conserve forests, taking advantage of the high degree of responsibility manifested by forest-residents, and (6) inter-institutional support is needed for larger-scale strategies. It is an opportunity to participate in the market with the stingless bee products (pot-honey and pot-pollen,

cerumen, pollination) and the social realities (traditional knowledge, food autonomy, conservation of the natural environment, sustainable development) and care as pets (Main 2012; Heard 2016). The interaction with the general public is a collective way to contribute toward consolidation of stingless bee keeping as a rural-urban practice with esteemed ethno-ecological roots.

Acknowledgments Authors thank Dr. D.W. Roubik for timely comments received and English editing. Dr. O. M. Barth's and Dr. T. Heard's proofreading was appreciated. Dr. C. Menezes improved our title. Ecuadorian academic institutions (Universidad Técnica Particular de Loja, Universidad Nacional de Loja, Universidad de Las Américas, Universidad Técnica de Machala), Universidad Autónoma de Yucatán from México, and the Council of Scientific, Humanistic, Technological and Art Development, from Universidad de Los Andes, Merida, Venezuela. Ecuadorian public institutions (Productive Management Department of the Prefecture of Loja, Parish governments of Pindal and Zapotillo), Venezuelan school Unidad Educativa Alberto Ravel, Praia Grande from Amazonas state, and non-profit organizations (Naturaleza y Cultura Internacional, Fundación Altrópico, Asociación la Restinga). Likewise, leading experts on stingless bee keeping participated from Ecuador and other countries (Mexico, Peru, Spain, and Venezuela). The local participants in these activities showed a great interest in contributing to the revitalization of the native stingless bee keeping, to increase farming income and to enhance their territory and life quality.

References

Aguilar I, Herrera E, Vit P. 2013. Acciones para valorizar la miel de pote. pp. 1-13. In: Vit P, Roubik DW, eds. Stingless bees process honey and pollen in cerumen pots. Facultad de Farmacia y Bioanálisis, Universidad de Los Andes; Mérida, Venezuela. http://www.saber.ula.ve/handle/123456789/35292

Albalat Botana A, Acosta Quijano N. 2016. Fortalecimiento de la cadena productiva de miel de monte para elevar ingresos familiares y conservar los recursos naturales. Final Project Report. Asociación Istaku Spinini AC; Xalapa, Veracruz, México. 8 pp.

Chieruzzi Löwenstein MC. 1989. Etnomeliponicultura y análisis químico de las mieles de cinco especies de abejas sin aguijón (Meliponinae). Biology Thesis. Biological Science Departament, Faculty of Exact and Natural Science, Pontificia Universidad Católica de Ecuador; Quito, Ecuador. 192 pp.

Coloma L. 1986. Contribución para el Conocimiento de las abejas sin aguijón (Meliponinae, Apidae, Hymenoptera) de Ecuador. Bachelor Thesis. Pontificia Universidad Católica del Ecuador; Quito, Ecuador. 146 pp.

Cortopassi-Laurino M, Imperatriz-Fonseca VL, Roubik DW, Dollin A, Heard T, Aguilar I, Nogueira-Neto P. 2006. Global meliponiculture: challenges and opportunities. Apidologie 37: 275-292.

Ferrufino U, Vit P (2013) Pot-honey of six species of Meliponini from Amboró National Park in Bolivia. pp. 409-416. In: Vit P, Pedro SRM, Roubik DWR, eds. Pot-honey, a legacy of stingless bees. Springer; New York, USA. 654 pp.

García-Olivares V, Zaragoza-Trello C, Ramirez J, Guerrero-Peñaranda A, Ruiz C. 2015. Caracterización rápida de la biodiversidad usando morfometría geométrica: Caso de estudio con abejas sin aguijón (Apidae: Meliponini) del sur de Ecuador. Avances en Ciencias e Ingenierías. 7: 40-46.

González-Acereto JA. 2008. Cría y manejo de abejas nativas sin aguijón en México. Impresiones Planeta; Mérida, México. 176 pp.

Guerrero-Peñaranda A. 2016. Competencia o partición de nicho por los recursos en abejas nativas *Melipona mimetica* y *Scaptotrigona* sp. en un bosque seco al sur de Ecuador. Bachelor Thesis.Universidad Técnica Particular de Loja; Loja, Ecuador. pp. i-ix, 1-50.

Heard TA. 2016. The Australian native bee book, keeping stingless bee hives for pets, pollination and sugarbag honey. Sugarbag Bees; Brisbane, Australia. 246 pp.

Holl KD. 2017. Restoring tropical forests from the bottom up. Science 355: 455-456.

International Code of Zoological Nomenclature (ICZN), Fourth Edition, online version. http://www.iczn.org/iczn/index.jsp, Accessed May 2016

Jaffé R, Pope N., Carvalho AT, Maia UM, Blochtein B, de Carvalho CAL, Imperatriz-Fonseca VL. 2015. Bees for development: Brazilian survey reveals how to optimize stingless beekeeping. PloS ONE 10: e0121157.

Jones R. 2013. Stingless bees: an historical perspective. pp. 219-227. In: Vit P, Silvia RMP, Roubik D ,

eds. Pot honey: A legacy of stingless bees. Springer; New York, USA. 654 pp.

Main D. 2012. A different kind of beekeeping takes flight. The New York Times. https://green.blogs.nytimes.com/2012/02/17/a-different-kind-of-beekeeping-takes-flight/

Martínez-Fortún S. 2015. Desarrollo sostenible y conservación etnoecológica a través de la meliponicultura, en el sur de Ecuador. Master's Thesis. Universidad Internacional de Andalucía; Córdoba, Spain. 110 pp.

Menezes C. 2010. A produção de rainhas e a multiplicação de colônias em *Scaptotrigona* aff. *depilis* (Hymenoptera, Apidae, Meliponini). Tese de doutorado, Departamento de Biologia, Faculdade de Filosofia, Ciências e Letras, Universidade de São Paulo, Ribeirão Preto, Brasil. 97 pp.

Menezes C, Vollet Neto A, Imperatriz Fonseca VL. 2012. A method for harvesting unfermented pollen from stingless bees (Hymenoptera, Apidae, Meliponini). Journal of Apiculturl Research 51: 240-244.

Menezes C, Vollet-Neto A, Contrera FAFL, Venturieri GC, Imperatriz-Fonseca VL. 2013a. The role of useful microorganisms to stingless bees and stingless beekeeping. pp. 153-171. In Vit P, Pedro SM, Roubik DW eds. Pot-honey: a legacy of stingless bees. Springer; New York, USA. 654 pp.

Menezes C, Vollet-Neto A, Imperatriz-Fonseca VL. 2013b. An advance in the in vitro rearing of stingless bee queens. Apidologie 44: 491-500.

Ocampo Rosales G. 2013. Medicinal uses of *Melipona beecheii* honey, by the ancient Maya. pp. 229-240. In: Vit P, Silvia RMP, Roubik D, eds. Pot honey: A legacy of stingless bees. Springer; New York, USA. 654 pp.

Ollerton J, Winfree R, Tarrant S. 2011. How many flowering plants are pollinated by animals? Oikos 120: 321-326.

Padilla Vargas PJ, Vásquez-Dávila MA, García Guerra TG, Albores González ML. 2014. Pisilnekmej: una mirada a la cosmovisión, conocimientos y prácticas nahuas sobre *Scaptotrigona mexicana* en Cuetzalan, Puebla, México. Etnoecológica 10: 37-40.

Parker T, Carr J. 1992. Status of forest remnants in the Cordillera de la Costa and adjacent areas of southwestern Ecuador; Washington, USA, Conservation International. 172 pp.

Portillo-Quintero CA, Sánchez-Azofeifa GA. 2010. Extent and conservation of tropical dry forests in the Americas. Biological Conservation 143: 144-155.

Ramírez Romero J, Ureña Alvarez J, Camacho A. 2013. Las abrejas sin aguijón en la región sur del Ecuador; Loja, Ecuador, Universidad Nacional de Loja. 119 pp.

Reyes-García V, Martí-Sanz N. 2007. Etnoecología: punto de encuentro entre naturaleza y cultura. Revista Ecosistemas 16: 46-55.

Rivero Oramas R. 1972. Abejas criollas sin aguijón. Colección Científica. Monte Avila Editores; Caracas, Venezuela. 110 pp.

Roubik DW. 1989. Ecology and natural history of tropical bees. Cambridge University Press; New York, USA. 514 pp.

Roubik DW, Villanueva-Gutiérrez R. 2009. Invasive Africanized honey bee impact on native solitary bees: a pollen resource and trap nest analysis. Biological Journal of the Linnean Society 98: 152–160.

Slaa EJ, Chaves LS, Malagodi-Braga KS, Hofstede FE. 2006. Stingless bees in applied pollination: practice and perspectives. Apidologie, 37: 293-315.

Souza HR, Correa AMS, Cruz-Barros MAV, Albuquerque PMC. 2015. Espectro polínico da própolis de *Scaptotrigona* aff. *postica* (Hymenoptera, Apidae, Meliponini) em Barra do Corda, MA, Brasil. Acta Amazonica 45: 307-316.

Souza Rosa A, I'Anson Price R, Ferreira Caliman MJ, Pereira Queiroz E, Blochtein B, Sílvia Soares Pires C, Imperatriz-Fonseca VL. 2015. The stingless bee species *Scaptotrigona aff. depilis* as a potential indicator of environmental pesticide contamination. Environmental Toxicology and Chemistry 34: 1851-1853.

Tapia-Armijos MF, Homeier J, Espinosa CI, Leuschner C, de la Cruz M. 2015. Deforestation and forest fragmentation in South Ecuador since the 1970s–losing a hotspot of biodiversity. PloS one, 10(9), p. e0133701.

Toledo V, Barrera-Bassols N. 2008. La memoria biocultural: la importancia ecológica de las sabidurías tradicionales. ; Barcelona, España, Icaria Editorial. 230 pp.

Villanueva-G R, Roubik DW, Colli-Ucan W. 2005. Extinction of *Melipona beecheii* and traditional beekeeping in the Yucatán Peninsula. Bee World 86:35-41.

Villanueva-Gutiérrez R, Roubik DW, Colli-Ucan W, Güemez-Ricalde FJ, Buchmann SL. 2013. A critical view of colony losses in managed Mayan honeymaking bees (Apidae: Meliponini) in the heart of Zona Maya. Journal of the Kansas Entomological Society 86:352-362.

Vit P. 1994. Las abejas criollas sin aguijón. Vida Apícola 63, 34-41.

Vit P. 2000. Una idea para valorizar la meliponicultura latinoamericana. Tacayá 10: 3-6.

Vit P. 2015. Valorización de mieles de pote producidas por meliponini en Ecuador. Informe Final. Proyecto Prometeo-SENESCYT-Universidad Técnica de Machala. Machala, Ecuador. 102 pp + Annex

Vit P. 2016. Pot-pollen is the bee pollen stored by Meliponini: *Tetragonisca angustula*, Mérida. Venezuela and Sucumbíos-Ecuador. The 5th Asia-Pacific International Congress on Engineering & Natural SciencesAPICENS 2016, Naha, Okinawa, Japan 2-5 August.

Vit P, Persano Oddo L, Marano ML, Salas de Mejías E (1998) Venezuelan stingless bee honeys characterised by multivariate analysis of compositional factors. Apidologie 29: 377-389.

Vit P, Santiago B, Pedro SRM, Peña-Vera M, Pérez-Pérez E. 2016. Chemical and bioactive characterization of pot-pollen produced by *Melipona* and *Scaptotrigona* stingless bees from Paria Grande, Amazonas State, Venezuela. Emirates Journal of Food and Agriculture 28: 78-84.

Vit P, Vargas O, López T, Maza F. 2015. Meliponini biodiversity and medicinal uses of the honey from El Oro province in Ecuador. Emirates Journal of Food and Agriculture 27: 502-506.

Vollet-Neto A, Menezes C, Imperatriz-Fonseca VL. 2015. Behavioural and developmental responses of a stingless bee (*Scaptotrigona depilis*) to nest overheating. Apidologie: 46: 455-464.

Wolff LF. 2014. Sistemas agroforestales apícolas: instrumento para la sustentabilidad de la agricultura familiar, asentados de la reforma agraria, afrodescendientes quilombolas e indígenas guaranies. Universidad de Córdoba; Córdoba, España. 426 pp.

Zamudio F. 2016. Tras los pasos de Darrel Posey; la etnoentomología y sus métodos. Boletín de la Sociedad Entomológica Argentina 27: 11-16.

Pot-Pollen 'Samburá' Marketing in Brazil and Suggested Legislation

31

Rogério Marcos de Oliveira Alves
and Carlos Alfredo Lopes Carvalho

31.1 Introduction

Beekeeping has developed substantially in Brazil in recent years. It is estimated that over 350,000 beekeepers and about 1,000,000 people are directly or indirectly involved in beekeeping activities throughout the country, generating 450,000 jobs in the field and 16,000 jobs in the industrial sector (Sampaio 2000; CBA 2012; IBGE 2012). These data are related to *Apis mellifera* bees mainly in the honey production chain.

The pollen collected and stored by stingless bees is a source of protein that can be exploited by meliponicultors, both for nutrition and income. In Brazil, although many stingless bee species have potential for pot-pollen exploitation, few have been studied. Important variables for the success of this activity include species, flowering, management, and type of pollen to be produced (fresh or fermented). Compared to production of other hive products, the amount produced and the economic profitability of pollen or pot-pollen are still insignificant, considering the diversity of the flora and the size of Brazil. Pollen plays very important roles in nature, and it is an excellent source of protein, carbohydrates, minerals, and vitamins. Also considered a food of high value in the human diet, it may provide a source of income that adds value to beekeeping.

Considering its territorial extension and diverse climatic conditions, Brazil harbors an abundant and varied flora that provides a great quantity of food resources for diverse species of bees. Honey, resin, and pollen are the main products collected by bees and exploited by humans. The consumer demand for pollen has increased greatly and requires studies aimed at knowledge of the physicochemical, microbiological, organoleptic, and palynological characteristics that contribute to human food security. Although pollen is considered a source of income and a food of high nutritional value, it is poorly produced, marketed, and consumed. According to Sampaio (2000), pollen commercialization in Brazil is performed by industry (50%), the pharmaceutical business (40%), and companies that work with the food chain (10%).

In the case of native bee pollen stored in pots and known by the name 'samburá' in Brazil, little has been done to characterize the product or review production techniques. That stimulus is needed for the development of research and technological projects. The study of the potential of

R.M. de Oliveira Alves (✉)
Instituto Federal de Educação, Ciência e Tecnologia Baiano, 48.110-000, Catu, BA, Brazil
e-mail: eiratama@gmail.com

C.A.L. Carvalho
Universidade Federal do Recôncavo da Bahia, 44.380-000, Cruz das Almas, BA, Brazil

© Springer International Publishing AG, part of Springer Nature 2018
P. Vit et al. (eds.), *Pot-Pollen in Stingless Bee Melittology*, DOI 10.1007/978-3-319-61839-5_31

'samburá' is important also for quality control, knowledge of production potential, or attempting standardization for industrial application (Barreto et al. 2006, Alves et al. Chap. 25).

31.2 Pot-Pollen Is Known as 'Samburá' in Brazil

The first inhabitants of Brazil called 'samburá' the pollen stored by stingless bees (Meliponini), which means basket in the Tupi language (Bordoni 1983), since this product is considered a delicacy (Posey 1983) and used in rituals. 'Samburá' consists of pollen that is collected by stingless bees and mixed with enzymes, and stored in pots built from a mixture of wax and resin (cerumen). In this container, the product is fermented and later used by the bees. It generally has a sour taste and a pasty texture. The pot-pollen of *Frieseomelitta doederlini* has a slightly sweet taste, distinct from other meliponine species (see Fig. 31.1).

'Samburá is widely used as a protein food to strengthen *Apis mellifera* colonies in the state of Piauí during times of food scarcity. In Pernambuco, the samburá of *Melipona scutellaris* is eaten mixed with *A. mellifera* honey. Currently, its use has been highly valued in gourmet cuisine as a substitute for mustard in salad dressings (R.M.O. Alves, personal observation).

31.3 Meliponine Species Used for the Production of 'Samburá'

Commercial opportunities have emerged in the broad spectrum of stingless beekeeping with the development of meliponiculture as a resource-generating activity for family farming communities. The pollen obtained through meliponiculture can be harvested as fresh pollen removed before the bee enters the nest or hive (fresh pollen loads similar to commercial honey bee pollen pellets) or as 'samburá' (stored pot-pollen). The production of 'samburá' propitiates the expansion of the market, since this product has its own characteristics, and thus somewhat avoids competition with *Apis* pollen pellets.

Many meliponine species are exploited for the production of pot-honey (*Melipona, Tetragonisca*, and *Scaptotrigona*) and resin or propolis (*Scaptotrigona*). Although they are well suited for honey production, bees of the genus *Melipona* present difficulties in management when developing production of samburá, due to low yields compared to other genera (*Scaptotrigona, Frieseomelitta*, and *Nannotrigona*).

Some stingless bee species are suitable for the harvest of fresh pollen, and others for the production of 'samburá.' Among the genera currently used for 'samburá,' *Scaptotrigona* has notable potential, with higher yields for both fresh pollen and pot-pollen. Fresh pollen harvest

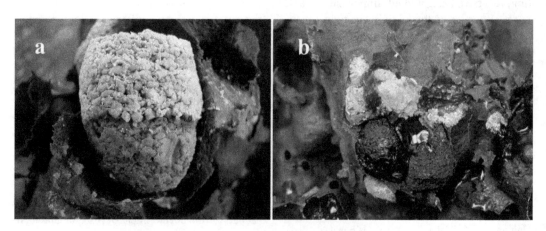

Fig. 31.1 Pot-pollen from two Brazilian species. (**a**) *Melipona scutellaris* 'Samburá' and (**b**) *Scaptotrigona* 'Samburá' (Photos: R.M.O. Alves)

is recommended for *Scaptotrigona* that are very populous and accumulate enough pollen. Furthermore, the harvest is accomplished more easily, and without unwanted contamination from other sources, like mud often collected by *Melipona*. In spite of their defensiveness (biting behavior), *Scaptotrigona* adjust to the use of a screen or trap in the nest entrance, which allows collection of fresh pollen. However, they also apply resin to the pollen traps—a problem that needs to be solved (see Fig. 31.2).

The choice of the largest pollen-producing species is critical to success in the enterprise. *Scaptotrigona xanthotricha* has good potential for accumulation of pot-honey and resin in the rainforest. Other species of *Scaptotrigona* from the dry forest have better potential for the production of pollen and resin. Studies are being carried out to evaluate the productive potential of several species of this genus. Recommendations to breed a species of this genus are often met with skepticism in Brazil due to preferences for *Melipona* species, which normally have a docile behavior, in contrast to *Scaptotrigona*, which has a more robust defensive behavior. However, for the production of resin and pollen, the management of *Scaptotrigona* constitutes a priority and goal for certain regions of Brazil, such as in the state of Maranhão. Fortún-Martínez et al., Chap. 30, further explain the idea of promoting *Scaptotrigona* in Neotropical meliponiculture.

Other species have shown promising potential for the production of fresh pollen and 'samburá', among them: *Frieseomelitta doederleini*, *F. varia*, *Tetragonisca angustula*, *Nannotrigona testaceicornis*, and *Tetragona clavipes*. *Trigona spinipes* is a species whose 'samburá' is exploited in an extractive way to be posteriorly commercialized and consumed. Some meliponicultors of the central region of the state of Bahia maintain nests in trees to extract pot-pollen throughout the year.

Species of the genus *Melipona* facilitate pot-pollen harvest due to the large size of their storage pots. However, in pollen production, it is necessary that the colony has a large population, which allows a greater number of worker bees to forage and store a great quantity of pollen. Usually, *Melipona* colonies have relatively low populations and therefore store limited 'samburá,' which is a disadvantage. Another negative aspect is the characteristic of collecting mud for the production of geopropolis, as already mentioned above, which can cause problems of contamination in 'samburá' harvest.

The monitoring of the pollen production of some species (*Melipona asilvai*, *M. mandacaia*, *M. quadrifasciata anthidioides*, *M. scutellaris*, and *M. mondury*) in the state of Bahia, Brazil, has shown that these species present low production of fresh pollen (3–5 kg/colony/year). However, preliminary evaluations show that some practices such as the management of 'samburá' production in 'supers' (hive compartments for storage), associated with intense pollen harvest by the

Fig. 31.2 Pollen trap for collecting fresh pollen from a nest of *Trigona spinipes* (Photos: P. Acioly)

colony, control of natural enemies, and artificial feeding with sugar syrup, can increase the production by colonies of these species, by up to 500% (R.M.O. Alves, personal observation).

31.4 Harvesting and Processing

The intensity of pollen collection by bees seems to adjust to three main variables: (1) Pollen chemistry due to protein content, (2) Accessibility, and (3) Predictability (Espina and Ordetx 1984; Martinez 2005). Bees collect pollen from the flowers, pack it in the corbiculae, and deposit the pellets in combs or pots in storage areas in the nest. Worker bees then compress the stored pollen with the head to obtain a compact mass that undergoes transformation under the action of microbes, temperature, humidity, and salivary enzymes, also mixed with nectar, to form bee bread by honey bees or 'samburá' by stingless bees (Carpes 2008).

In beekeeping with *A. mellifera*, bee pollen is collected by using a pollen trap consisting of a screen that allows workers to pass through but removes pollen pellets from the corbiculae, on the outer surface of the hind legs. Processing consists of drying corbicular pollen to eliminate excess moisture and reduce the growth of bacteria and fungi (Carpes 2008). Pot-pollen is collected directly from storage pots in meliponine colonies with a spatula, and readied for transportation, preferably under refrigerated conditions. Transfer of the product from harvesting to processing areas is accomplished using sealed containers, in the shortest possible time, to avoid any contamination. The harvest of *M. scutellaris* pot-pollen gave a net yield of approximately 60% to 70%. That is, 5 kg of 'samburá' in the storage pots provided 3.1 kg net weight of 'samburá' after removal of the cerumen (R.M.O. Alves, personal observation). In Fig. 31.3, pot-pollen processing is illustrated.

Fig. 31.3 Pot-pollen processing. (**a**) Hive super with 'samburá' from a *Melipona scutellaris* colony. (**b**) 'Samburá' in harvested cerumen pots. (**c**) Tray oven to reduce 'samburá' moisture. (**d**) Packed *Tetragonisca angustula* 'samburá' (Photos: R.M.O. Alves)

31.5   Marketing of Meliponine 'Samburá'

The harvesting and manufacture of fresh pollen and pot-pollen for marketing is a recent activity, used as an income supplement by some meliponicultors. In the state of Bahia, the 'samburá' is commercialized and slightly dehydrated by means of refrigeration, maintaining humidity above the recommended level for bee pollen (*A. mellifera*), in plastic bags or glass containers. The added economic value of pot-pollen to stingless beekeeping activity is recent. Mr. Chagas, a pioneer in pot-pollen production from Pernambuco state, Brazil, mixed 1 kg of *M. scutellaris* pot-pollen with 5 kg *A. mellifera* honey, marketing the final product at a 500% increased value.

The value of pot-pollen in the current market varies according to the bee species (Table 31.1) and can be composed of the following elements: (1) The gourmet market uses pot-pollen as an ingredient of salad dressings, for example, "Samburá Dijon" (olive oil + "uruçú" pot-honey + 'samburá' + black pepper + salt, created by R.M.O. Alves and the Bahian chefs C. Marinho and F. Lemos). If the pot-pollen is too sour, it is advisable to use a sweet honey with low free acidity, to balance the flavor. 'Samburá' is also part of the composition of dishes with other foods; (2) Supplemental feeding sugar syrup made with a spoon of pot-pollen in 1 liter of sugar-water (1:1); (3) Natural consumption of a daily teaspoon of the sweet *F. doederlini* sweet pot-pollen. Other stingless bee species produce sour pot-pollen consumed in juices, mixed with vitamins and yogurt.

31.6   Cultural Aspects of Pot-Pollen Consumption in Brazil

The inclusion of 'samburá' or pot-pollen in unconventional food is not a recent development in Brazil, since meliponicultors of the municipality of São Gabriel, Bahia, have used this fermented product as an ingredient of refreshment for many generations. The drink is considered a natural invigorant and aphrodisiac (R.M.O. Alves, personal observation). The preparation consists of mixing 1 l of cold water, one teaspoon of 'samburá,' and honey to taste. The container is shaken until the 'samburá' is completely diluted and served chilled before lunch or the evening meal.

31.7   Strategies to Increase the Production of 'Samburá'

Pollen foraged by stingless bees can result in two products: fresh corbicular pollen and 'samburá.' In order to obtain the fresh pollen, harvest must

Table 31.1 'Samburá' yields and value according to meliponine species

Pot-pollen type Stingless bee species "ethnic name"	Yield (kg/colony/year)	Value (R$)/kg	Value (USD)/kg
Melipona fasciculata "uruçú cinzenta" "tiúba"	4.00–5.00	140.00–200.00	45.00–64.00
Melipona mandacaia "mandaçaia"	3.00–4.00	120.00–180.00	39.00–58.00
Melipona scutellaris "uruçú"	4.00–5.00	200.00–800.00	64.00–257.00
Scaptotrigona spp. "tubi" "canudo"	5.00–6.00	120.00–250.00	39.00–80.00
Tetragonisca angustula "jataí" "mosquito"	3.00–4.00	100.00–500.00	32.00–161.00
Trigona spinipes "arapuá" "abelha cachorro"	0.70–0.90	120.00	39.00

be performed before the bee enters the hive, while the 'samburá' (stored pot-pollen) is harvested by opening the hive and withdrawing the contents directly from the interior of the cerumen pot using instruments, such as a spoon or a knife. The development of methods that allow the removal of pot-pollen from stingless bee species, either externally or internally, is still in its early stages, and it is thus associated with the availability of harvesting equipment. It is necessary to devise a management plan for production, and to perform selection of colonies with greater pollen production.

An important factor for the increase in production is the formation of gardens rich in stingless bee flora with plants known for their high pollen production. Some plant families are recognized in their prominence, like Myrtaceae, Melastomataceae, Fabaceae-Mimosoideae, Asteraceae, and Arecaceae. Other strategies may be based upon the bee species to be used for pot-pollen production, the management of the colonies, supplemental sugar feeding, and genetic improvement through the selection of highly productive colonies.

31.8 Seasonality of Pot-Pollen

The composition of bee pollen varies according to floral origin, air temperature, and soil chemical composition (Kroyer and Hegedus 2001; Carpes et al. 2007). A pollen harvest of a plant species, in different areas and seasons, will result in differences in the chemical composition of this food (Barreto et al. 2000).

The harvesting period during the beginning of the rains (tropical winter) corresponds to the highest peak of pollen storage by the bees. The presence of plant species may differ from one season to another. Both factors contribute to variable concentrations of nutritional constituents (Souza et al. 2004). Knowing the nutritional value of the different pollen types can contribute, for example, to choose the location of an apiary or a meliponary, since different floristic compositions influence the quality and the collection of different food sources by the bees and may contribute to a balanced diet (Modro et al. 2007). Certain botanical families predominate in the pollen spectrum of 'samburá,' and some of them contain higher concentrations of certain constituents. Modro et al. (2007) found Asteraceae pollen has a low protein content, unlike the Mimosoideae, which is rich in protein for bees. These two families are frequent in most of the pollen analyses.

31.9 Pot-Pollen Production Initiatives in Brazilian States

The production of fresh pollen and 'samburá' in Brazil is poorly developed. Initiatives have been promoted in some states (e.g., Bahia and Alagoas) to evaluate production, to obtain adequate equipment for harvesting and processing, to choose appropriate stingless bees species, and to market the product. Pot-pollen commercialization occurs normally when a mixture, with honey, is made for sale and generally is after harvest from wild colonies in the forest.

Peasants in the dry forest (state of Bahia) collect, process, and market pot-pollen from *T. spinipes* "arapuá" in the communities of the region. The extractive process consists of removing the 'samburá' while keeping the nests in place. Production is small, approximately 800 g/colony/year. In the rainforest (state of Bahia), pot-pollen of *M. scutellaris* is harvested after the pot-honey, obtaining higher productivity.

31.10 Suggested 'Samburá' Standards for Pot-Pollen Legislation

Although 'samburá' has been represented in traditional knowledge for a long time, interest in its physicochemical, microbiological, and sensorial characteristics has only recently been stimulated, necessitating a more detailed discussion by regulatory agencies regarding the development of technical standards for future regulation of the product.

The existing standards in the countries where meliponine species originated concern the pollen produced by Africanized bees (*A. mellifera*), which is produced in greater quantities and considered a source of income for the beekeeper. The problems faced by meliponiculture in Venezuela regarding the certification of 'samburá' are discussed in a recent article by Vit et al. (2016).

The parameters regulated by Brazilian legislation include sensory characteristics (aroma, color, appearance, and flavor) and physicochemical requirements (moisture, ash, lipid, protein, sugar, crude fiber, free acidity, and pH) (Brasil 2001). The requirements for pollen evaluation vary among Latin American countries. Only Argentina, Brazil, El Salvador, and Mexico have established quality parameters for bee pollen. In Brazil, the legislation includes both dried bee pollen and fresh bee pollen. Table 31.2 shows the main parameters used in pollen analyses for the four countries.

The technical regulations setting the identification and quality standards for bee pollen (Brasil 2001) were instituted in 2001 and have been revised to meet new requirements and the inclusion of new products including the 'samburá' from meliponine species. This inclusion is necessary to regulate the meliponiculture that produces 'samburá' and is based on the physicochemical and microbiological analyses carried out by several institutions and researchers (Table 31.3).

Bee pollen is defined as the product of the agglutination of floral pollen by worker bees using nectar and salivary substances, which is processed at the entrance of the hive and composed of protein, lipid, sugar, fiber, mineral salts, amino acids, vitamins, and flavonoids (Brasil 2001). Although the origin of the product is unique, the harvesting process often differentiates the products.

Both bees and microorganisms contribute to the process of fermentation of floral pollen, which is the basic material for the formation of the final product, to produce 'samburá.' Its sour taste and creamy texture result from fermentation, which makes the resulting product richer in lactic acid (Vit et al. 2011).

The inclusion of 'samburá' as a commercial product necessitates the standardization of this new product with its new characteristics and relatively novel use of bee pollen, as its production still lacks a consistent technology. However, the criteria that guide this regulation should be based primarily on the characteristics that distinguish the bee pollen product (moisture, pH, and ash content) from other pollen products.

The analysis of the physicochemical composition of 'samburá' by several authors in Brazil (Ribeiro and Silva 2007; Souza et al. 2004; Pinheiro et al. 2012; Santa Bárbara et al. 2015), in Honduras (Martinez 2005), and in Venezuela (Vit et al. 2016) demonstrates that the moisture content, fiber content, and pH of 'samburá' fall outside the standards required by the Brazilian legislation for nondried *Apis mellifera* bee pollen. Some variation in other criteria may have been due to the method of analysis. The requirements of the Brazilian standard for microbiological criteria (Brasil 2001) are based on the total product.

Acknowledgments We thank the editors for the invitation, the meliponicultors for their knowledge and experience, and the Brazilian agencies of research development: Coordination for the Improvement of Higher Education Personnel (CAPES), the National Council of Technological and Scientific Development (CNPq), and State of Bahia Research Foundation (FAPESB) for the

Table 31.2 Comparison of physicochemical parameters required for bee pollen by country

Country	Carbohydrates %	Protein %	Moisture %	Ash %	pH
Argentina	45–55	15–28	8	4	4–6
Mexico	NR	12–18	4–8	1.5–2.2	4–6
El Salvador	NR	NR	4	4	4–6
Brazil[a]	14.5–55.0	≥8	≤4	≤4	4–6
Brazil[b]	14.5–55.0	≥8	≤30	≤4	4–6

NR Not regulated
[a]Dried bee pollen
[b]Undried bee pollen

Table 31.3 Physicochemical parameters for bee pollen, fresh pollen, bee bread, and meliponine 'samburá' in Latin America

Physicochemial parameters	Apis mellifera			Meliponine pot-pollen						Requirements for nondried bee pollen
	Dried bee pollen	Fresh bee pollen	Bee bread	Amazon Melipona	Melipona flavolineata	Melipona fasciculata	Melipona aff. eburnea	Melipona mandacaia	Melipona scutellaris	Brasil (2001)
Authors	1, 2	3	4	5	6	6	7	8	9	
Moisture (g/100 g)	33–22	27.46	25	36.9	54.48	58.75	48.54	36.0	44.71	≤30.0
Ash (g/100 g)	1.59–4.98	1.87	2.77	2.1	2.26	2.12	2.33	4.9	1.84	≤4.0
Lipids (g/100 g)	3.2	5.28	2.77–3.85	4.0	3.55	2.26	3.19	–	4.25	≥1.8
Protein (g/100 g)	19.59–22.8	22.58	2.96–4.0	19.5	17.52	23.43	18.32	21.0	23.88	≥8.0
Fiber (g/100 g)	–	–	–	–	1.60	1.20	–	3.60	0.87	≥2.0
Carbohydrates (g/100 g)	49.24–31.0	–	24.76	37.5	14.50	18.09	27.62	12.0	24.48	14.5–55.0
pH	4.91–5.2	–	4.3–4.4	–	3.80	3.42	–	3.49	3.75	4.0–6.0
Free acidity (meq/kg)	–	–	–	–	1781	1380	–	146.0	150.57	≤300.00
Water activity (Aw)	0.65	–	–	–	–	–	–	0.86	0.92	Not required
Total energy value (kcal/100 g)	–	–	–	264.4	183.65	162.82	–	–	231.33	Not required

Legend: (1) Martinez (2005); (2) Marchini et al. (2006); (3) Ribeiro and Silva (2007); (4) Sampaio and De Freitas (1993); (5) Souza et al. (2004); (6) Pinheiro et al. (2012); (7) Vit et al. (2016); (8) Santa Bárbara et al. (2015); (9) See Table 25.1 by Alves et al.

financial support to our studies. To Prof. P. Vit for the translation of the Portuguese manuscript, and to Dr. D.W. Roubik for professional editing and English proofreading.

References

Barreto LMRC, Funari SRC, Orsi RO, Dib APS. 2006. Produção de pólen no Brasil; Cabral, Brazil, Taubté. 100pp.

Barreto LMRC, Rabelo P C, Belezia CO. 2000. Perfil proteico do pólen coletado por *Apis mellifera* (híbrida africanizada) no período outono- inverno no apiário do Centro de Estudos Apícolas da Universidade de Taubaté. In: Anais Congresso Brasileiro de Apicultura, 13. Confederação Brasileira de Apicultura; Florianópolis, Brazil (Cd-rom).

Bordoni O. 1983. A língua tupi na geografia do Brasil; Paraná, Brazil, Gráfica Muto. 803 pp.

BRASIL. 2001. Instrução Normativa n.3, de 19 de janeiro de 2001. Ministério da Agricultura, Pecuária e Abastecimento aprova os regulamentos técnicos de identidade e qualidade de apitoxina, cera de abelha, geléia real, geléia real liofilizada, pólen apícola, própolis e extrato de própolis. In: MINISTÉRIO DA AGRICULTURA, PECUÁRIA E ABASTECIMENTO. Legislação. SisLegis – Sistema de Consulta à Legislação. Disponível em: http://agricultura.gov.br

Carpes ST, Begnini R, Alencar SM, Masson ML. 2007. Study of preparations of bee pollen extracts, antioxidant and antibacterial activity. Ciencia Agrotécnica 31: 1818-1825.

Carpes ST. 2008. Estudo das características físico-químicas e biológicas do pólen apícola de *Apis mellifera* L. da região Sul do Brasil. PhD Thesis in Food Technology. Universidade Federal do Paraná-UFPR, Curitiba, Brazil. 248 pp.

CBA. 2012. Confederação Brasileira de Apicultura. Estratégias. Direcionamento Estratégico. 2007. Disponível em: http://www.brasilapicola.com.br. Acesso em: 23 de fevereiro de 2012.

Espina DY, Ordetx G. 1984. Apicultura Tropical. 4ª Ed. Costa Rica. Editorial Tecnológica de Costa Rica; Cartago, Costa Rica. 506 pp.

IBGE. 2012. Instituto Brasileiro De Geografia E Estatística. Economia, Agropecuária, Pecuária Municipal - PPM. Publicação Completa. Produção da pecuária municipal 2006 - 2007. Available at: http://www.ibge.gov.br

Kroyer G, Hegedus N. 2001. Evaluation of bioactive properties of pollen extracts as functional dietary food supplement. Innovative Food Science and Emerging Technologies 2: 171-174.

Marchini LC, Reis VDA, Moretti ACCC. 2006. Composição físico-química de amostras de pólen coletado por abelhas africanizadas (Hymenoptera, Apidae) em Piracicaba. Ciência Rural 36: 949-953.

Martinez JVP. 2005. Caracterizacion físicoquimica y microbiológica del polen de abejas de cinco departamentos de Honduras. Thesis Agronomical Engineering. Universidad Zamorano; Tegucigalpa, Honduras. 36 pp.

Modro AFH, Message D, Da Luz CFP, Neto JAAM. 2007. Composição e qualidade de pólen apícola coletado em Minas Gerais. Pesquisa Agropecuária Brasileira, Brasília 42: 1057-1065.

Pinheiro FDM, Costa CVP, Das N, Baptista RDC, Venturieri GC, Pontes MAN. 2012. Pólen de abelhas indígenas sem ferrão *Melipona fasciculata* e *Melipona flavolineata*: caracterização físico-química, microbiológica e sensorial. http://www.alice.cnptia.embrapa.br/bitstream/doc/408842/1/polendeabelhasindigenassemferraomelipona

Posey DA. 1983. Keeping of stingless bees by the Kayapó Indians of Brazil. Journal of Ethnobiololgy 3: 63-73.

Ribeiro JG, Silva RA. 2007. Estudo comparativo da qualidade de pólen apícola fresco, recém processado, não processado e armazenado em freezer e pólen de marca comercial através de análises físico-químicas. Tecnologia e Desenvolvimento Sustentável 2: 33-47.

Sampaio IM. 2000. Comércio nacional de produtos apícolas. In: Anais Congresso Brasileiro de Apicultura, 13. Confederação Brasileira de Apicultura; Florianópolis, Brazil (Cd-rom).

Sampaio EAB, De Freitas RJS. 1993. Caracterização do pólen apícola armazenado na colméia (pão de abelhas) de algumas localidades do Paraná. Boletim Ceppa, Curitiba 11: 81-87.

Santa Bárbara M, Machado CS, Sodré GS, Dias LG, Estevinho LM, Carvalho CAL. 2015. Microbiological assessment, nutritional characterization and phenolic compounds of bee pollen from *Melipona mandacaia* Smith, 1983. Molecules 20: 12525-12544.

Souza RCS, Yuyama LKO, Aguiar JPL, Oliveira FPM. 2004. Valor nutricional do mel e pólen de abelhas sem ferrão da região amazônica. Acta Amazônica 34: 333-336.

Vit P, Santiago B, Pedro SRM, Ruíz J, Peña-Vera MJ, Pérez-Pérez EM. 2016. Chemical and bioactive characterization of pot-pollen produced by *Melipona* and *Scaptotrigona* stingless bees from Paria Grande, Amazonas State, Venezuela. Emirates Journal of Food Agriculture 28: 78-84.

Vit P, Rojas PLB, Usubillaga AR, Aparicio AR, Meccia G, Fernando Muiño MA, Sancho MT. 2011. Presencia de ácido láctico y otros compuestos semivolátiles en mieles de Meliponini. Revista del Instituto Nacional de Higiene Rafael Rangel 42: 58-63.

Appendix A: Ethnic Names of Stingless Bees

"abeja amarilla" *Oxytrigona mellicolor*
Ecuador, 216,

"abeja amarilla" *Partamona aequatoriana*
Ecuador, 216

"abeja amarilla" *Trigona* cf. *amazonensis*
Ecuador, 217

"abeja amarilla" *Trigona dallatorreana*
Ecuador, 217

"abeja amarilla" *Trigona ferricauda*
Ecuador, 217

"abeja ángel" *Tetragonisca angustula*
Ecuador, 216

"abeja chiquita" *Paratrigona* cf. *rinconi*
Ecuador, 216

"abeja de tierra" *Geotrigona* sp. Ecuador, 423

"abeja de tierra" *Trigona fulviventris* Ecuador,
217, 423

"abeja del suelo" *Schwarziana quadripunctata*
Argentina, 290, 294

"abeja finita" *Tetragonisca angustula*
Ecuador, 216

"abeja negra" *Paratrigona* cf. *rinconi*
Ecuador, 216

"abeja negra" *Partamona aequatoriana*
Ecuador, 216

"abeja negra" *Trigona* sp. gr. *fuscipennis*
Ecuador, 217

"abeja negra" *Trigona* cf. *truculenta*
Ecuador, 217

"abeja real" *Melipona eburnea* Ecuador,
208, 216

"abeja real" *Melipona indecisa* Ecuador, 216

"abeja real" *Melipona grandis* Ecuador, 216

"abejas reales" *Melipona* spp. Ecuador, 203

"abelha cachorro" *Trigona spinipes* Brazil, 439

"abejita" *Tetragonisca angustula* Venezuela, 339

"abejita negra" *Scaura* sp. aff. *longula*
Ecuador, 216

"abejita negra" *Trigona* cf. *branneri*
Ecuador, 217

"abejita suca" *Nannotrigona melanocera*
Ecuador, 216

"ajavitte" *Tetragona clavipes* Venezuela, 427

"alpargate" *Plebeia* sp Ecuador, 423

"angelina" *Nannotrigona melanocera*
Ecuador, 216

"angelina" *Tetragonisca angustula* Ecuador, 217

"angelita" *Tetragonisca angustula* Ecuador,
Venezuela, 221, 340, 423

"angelita" *Frieseomelitta* aff. *varia*
Venezuela, 364

"angelita negra" *Aparatrigona impunctata*
Ecuador, 216

"angelita negra grande" *Trigona* cf. *guianae*
Ecuador, 217

"abejita negra" *Nannotrigona melanocera*
Ecuador, 216

"arapuá" *Trigona spinipes* Brazil, 439

"arepe" *Nannotrigona* sp Ecuador, 423

"auim" *Trigona amazonensis* Ecuador, 204

"auímo"*Melipona crinita* Ecuador, 204

"auñeta"*Partamona epiphytophila* Ecuador, 204

"awae" *Melipona captiosa* Ecuador, 204

"barbacho" *Partamona aequatoriana*
Ecuador, 216

"barbón" *Lestrimellita limao* Ecuador, 423

"bermeja" *Melipona mimetica* Ecuador, 216

"bermejo" *Melipona mimetica* Ecuador, 216,
221, 423, 424

"boca de sapo" undetermined Ecuador, 423

"bool" *Nannotrigona perilampoide*s
Mexico, 423

© Springer International Publishing AG, part of Springer Nature 2018
P. Vit et al. (eds.), *Pot-Pollen in Stingless Bee Melittology*, DOI 10.1007/978-3-319-61839-5

"borá" *Tetragona clavipes* Argentina, 290, 294

"bunga amarilla" *Melipona* sp. (gr. *fasciata*)
Ecuador

"bunga amarilla" *Melipona eburnea* Ecuador,

"bunga negra" *Melipona grandis* Ecuador,
216, 424

"cagafuego" *Oxytrigona tataira* Argentina, 290

"cananambo" *Melipona* cf. *indecisa*
Ecuador, 216

"cananambo" *Melipona indecisa* Ecuador, 216,
221, 423

"canudo" *Scaptotrigona* spp. Brazil, 439

"carabozá" *Trigona spinipes* Argentina, 288, 290

"catana" *Scaptotrigona* sp Ecuador, 216,
423–426

"catana oreja de león" *Scaptotrigona* sp.
Ecuador, 216

"catiana" *Scaptotrigona* sp Ecuador, 216

"chalaco" *Partamona aequatoriana* Ecuador, 216

"chalaco" *Trigona fuscipennis* Ecuador, 217

"chiñi" *Tetragonisca angustula* Ecuador, 217

"chullumbo" *Tetragonisca angustula* in Kichwa,
Ecuador, 217

"cojimbo" *Melipona eburnea* Ecuador, 423

"conguita" undetermined Venezuela, 339

"cortapelo" *Trigona silvestriana* Ecuador, 217

"cowmuñi" *Trigona truculenta* Ecuador, 204

"cuchiperro" *Tetragona sp. gr. clavipes*
Ecuador, 216

"dibouga" *Meliponula bocandei* Gabon, 186

"divasou" *Hypotrigona* Gabon, 186

"e'hol" *Cephalotrigona zexmeniae* Mexico, 300

"eñamo" *Melipona titania* Ecuador, 204

"ergón" *Melipona eburnea* Ecuador, 216,
423, 424

"erica" *Melipona favosa* Venezuela, 364

"españolita" *Tetragonisca angustula*
Venezuela, 339

"gihn" *Melipona eburnea* Ecuador, 204

"guacoso" undetermined Ecuador, 423

"guanota" *Melipona compressipes*
Venezuela, 364

"guaraipo" *Melipona bicolor schencki*
Argentina, 290, 293

"guarigane" *Trigona silvestriana* Ecuador, 217

"iñawe" *Ptilotrigona lurida* Ecuador, 204

"iratín" *Lestrimelitta rufipes* Argentina, 288, 290

"iratín" *Lestrimelitta* aff. *limao* Argentina, 290

"isabitto" *Melipona* aff. *fuscopilosa*
Venezuela, 427

"jandaíra" *Melipona subnitida* Brazil, 90

"jataí" *Tetragonisca angustula* Brazil, 356, 439

"kansas" *Scaptotrigona pectoralis* Mexico, 300

"karbi" *Tetragonula carbonaria* Australia, 401

"keelar" *Tetragonula carbonaria* Australia, 401

"kiwitáxkat" *Scaptotrigona mexicana*
Mexico, 328

"koolel kab", "ko'olel kab" *Melipona beecheii*
Mexico, 299

"kootchar" *Austroplebeia australis* /
Austroplebeia cassiae Australia, 401

"lady bee" *Melipona beecheii* Mexico, 299

"lambeojo" *Plebeia* sp. Ecuador, 216, 423

"lambeojitos" *Trigonisca* sp., Ecuador, 217

"languacho" undetermined Ecuador, 423

"leticobe" *Nannotrigona* cf. *mellaria*
Ecuador, 216

"lévéki" stingless bees in Nzébi language,
Gabon, 186

"lambeojo" *Plebeia* sp. Ecuador, 216, 423

"libundu" stingless bees in Ndumu language,
Gabon, 186

"limoncita" *Lestrimelitta maracaia*
Venezuela, 427

"llanaputan" *Nannotrigona melanocera* in
Kichwa, Ecuador, 216

"makawae" *Plebeia* sp. Ecuador, 204

"mambuca" *Cephalotrigona capitata*
Argentina, 290

"mandaçaia" *Melipona mandacaia* Brazil, 341,
344, 367, 382, 392, 439

"mandaçaia" *Melipona quadrifasciata anthidioides* Brazil, 103–109, 142, 143, 361

"mandasaia" *Melipona quadrifasciata*
Argentina, 290

"mandurí" *Melipona torrida* Argentina, 288, 290

"mariola" *Tetragonisca angustula* Costa
Rica, 339

"Maya bee" *Melipona beecheii* Mexico,
131–135, 300

"mea fuego" *Oxitrigona mellicolor* Ecuador, 423

"mingkaye" *Tetragona clavipes* Ecuador, 204

"mirí" *Plebeia droryana*, *P. emerinoides*, *P.
remota* and *P. guazurary* Argentina, 288,
290, 294

"mosco" *Plebeia* sp. Ecuador, 216

Appendix B: Microorganisms Associated with Stingless Bees or Used to Test Antimicrobial Activity (AM)

© Springer International Publishing AG, part of Springer Nature 2018
P. Vit et al. (eds.), *Pot-Pollen in Stingless Bee Melittology*, DOI 10.1007/978-3-319-61839-5

Index

---

[1]Note: List of Plant Taxa used by Bees—Plants used as source of nectar (N), pollen (P), trichomes (T) or resin (R), harvested from corbiculae (C), to make honey (H), build nests (B), pollinated (L) or on bee body (O) of stingless bees (S), and *Apis mellifera* (A), brood provisions (D).

© Springer International Publishing AG, part of Springer Nature 2018   451
P. Vit et al. (eds.), *Pot-Pollen in Stingless Bee Melittology*, DOI 10.1007/978-3-319-61839-5

Printed in the USA
CPSIA information can be obtained
at www.ICGtesting.com
LVHW081150221223
766874LV00043B/146